MATHEMATICS FOR TEACHERS

An Interactive Approach for Grades K–8

Thomas Sonnabend

Montgomery College

THOMSON

BROOKS/COLE

Australia • Canada • Mexico • Singapore • Spain
United Kingdom • United States

THOMSON

BROOKS/COLE

Editor: Jennifer Huber
Assistant Editor: Rebecca Subity
Editorial Assistant: Carrie Dodson
Project Manager, Editorial Production: Janet Hill
Marketing Manager: Tom Ziolkowski
Marketing Assistant: Jessica Perry
Advertising Project Manager: Bryan Vann
Print/Media Buyer: Rebecca Cross

Production Service: Hearthside Publishing Services
Text Designer: Kathleen Cunningham
Illustrator: Jade Myers/Hearthside
Cover Designer: Roy R. Neuhaus
Cover Image: Schultz/Premium Stock/PictureQuest
Compositor: UG / GGS Information Services, Inc.
Cover/Interior Printer: Quebecor World/Versailles

For more information about our products contact us at:
Thomson Learning Academic Resource Center 1-800-423-0563
For permission to use material from this text, contact us by:
Phone: 1-800-730-2214 Fax: 1-800-730-2215
Web: http://www.thomsonrights.com

Library of Congress Control Number 2003114645

Student Edition: ISBN 0-534-40374-3

Instructor's Edition: ISBN 0-534-40375-1

Brooks/Cole–Thomson Learning
10 Davis Drive
Belmont, CA 94002
USA

Asia
Thomson Learning
5 Shenton Way #01-01
UIC Building
Singapore 068808

Australia/New Zealand
Thomson Learning
102 Dodds Street
Southbank, Victoria 3006
Australia

Canada
Nelson
1120 Birchmount Road
Toronto, Ontario M1K 5G4
Canada

Europe/Middle East/Africa
Thomson Learning
High Holborn House
50/51 Bedford Row
London WC1R 4LR
United Kingdom

Latin America
Thomson Learning
Seneca, 53
Colonia Polanco
11560 Mexico D.F.
Mexico

Spain/Portugal
Paraninfo
Calle/Magallanes, 25
28015 Madrid, Spain

For my wife Celia

CONTENTS

PREFACE

The NCTM Standards state that "prospective teachers must be taught in a manner similar to how they are to teach—by exploring, conjecturing, communicating, reasoning, and so forth." This necessitates devoting more class time to discussing and discovering ideas and less time to lecturing.

Can a mathematics book actively involve students in developing and explaining mathematical concepts? Yes, by using a carefully organized, interactive lesson format that promotes student involvement and gradually leads the student to a deeper understanding of mathematical ideas. The interactive format also allows for more class discussion and small group work.

To implement the NCTM Standards, one needs a textbook that provides numerous opportunities for investigation and discourse. Rather than having a few special puzzle and investigation problems, this textbook presents a substantial collection of exercises in every lesson and homework set that involve reasoning, investigating, or communicating.

Most people do not enjoy reading mathematics textbooks. Can a mathematics textbook be interesting to read and study? Yes, if the mathematical presentation is straightforward and clear, and the lesson content is enriched with investigations, appropriate uses of technology, humor, history, and interesting applications. A text for this course should also make it clear how the topics in each chapter relate to both the NCTM Standards and the current school mathematics curriculum.

You are looking at the result of my effort to address all these concerns while covering new topics and examining underlying concepts and connections in elementary-school mathematics. This textbook is the culmination of 24 years of work with university, college, and community college students.

Teaching elementary school is one of the most important and challenging professions in our society. It is my hope that this textbook provides material for a more interesting course that produces more competent teachers with a better sense of what elementary-school mathematics is all about.

The third edition of *Mathematics for Teachers: An Interactive Approach for Grades K–8* retains the same goals as the first two editions. Changes in the third edition reflect the latest trends in elementary and middle school mathematics, feedback from students and professors, and a thorough review and refinement of the material on nearly every page of the second edition.

DISTINCTIVE FEATURES RETAINED IN THIS EDITION

This textbook places *greater emphasis* on

- discovery, discussion, and explanation of concepts that involve students and deepen their understanding;
- investigations and activities that require higher-level thinking;
- applications of mathematics that connect mathematical concepts to everyday life;
- explaining a concept or solving a problem in more than one way so that future teachers can see a variety of approaches;
- multiple representations of concepts so that students develop a broader understanding;

- how material relates to the current curriculum for grades K–8;
- difficult concepts in the mathematics curriculum for grades K–8 that future teachers have trouble explaining;
- connecting lessons, homework exercises, and chapter review exercises;
- diagnosing common student difficulties and learning how to help these students;
- inductive and deductive reasoning, showing how these two processes are used throughout mathematics;
- models for arithmetic operations that establish the connection between an operation and its everyday applications;
- common student error patterns in arithmetic and measurement that prepare future teachers for diagnosing student difficulties;
- spatial perception and perspective drawing so that students become more adept at three-dimensional geometry;
- algebra as a mathematical language and a generalization of arithmetic;
- important statistical applications including statistical deceptions, surveys, and standardized tests;

To make time for these topics, this textbook places less emphasis on (1) topics more students have studied or will study in other mathematics or education courses and (2) topics that are relatively more remote from the current elementary-school curriculum. Such topics include arithmetic and algebra skills, formal logic, geometry terminology, trigonometry, and isometries.

Other Features Retained in this Edition

- Charts at the end of each chapter show grade-level coverage of related elementary- and middle-school topics.
- Extensive coverage of geometry and measurement topics strengthens students' spatial and analytical abilities.
- Extensive coverage of statistics and probability incorporates many realistic applications.
- Mental computation and estimation with whole numbers, fractions, decimals, and percents make use of properties of operations and complement the use of calculators.
- Humor and historical vignettes humanize the mathematics.
- Selected lesson exercises that are especially suitable for discussion are labeled with the icon **D** **Discussion** .

FEATURES NEW TO THIS EDITION

The third edition has many new features.

- The new Classroom Connection feature includes challenging situations that occur in grades K–8 classrooms as well as more examples of textbook pages from children's textbooks.
- Revisions reflect current terminology and content in school mathematics more closely than other college texts.

- The book comes with eight activity cards that can be used with new lab activities in the text.
- Updated technology coverage includes more use of fraction calculators, graphing calculators, and dynamic geometry software.
- There is more emphasis on the mathematics of grades 7 and 8.
- There is more emphasis on inductive and deductive reasoning and problem solving throughout the text.
- Relevant objectives from 2000 NCTM Content Standards are listed at the beginning of each section.
- This edition contains new lesson opener and summary exercises that research suggests help students learn and retain the material.
- Lesson Exercises are now classified as opener, concept, skill, connection, communication, or summary.
- The many new and revised exercises include more exercises that require writing explanations of concepts.
- Basic homework exercises now have paired odds and evens, and the exercise sets are more carefully organized by topic.
- There are more basic homework exercises and fewer extension exercises. The level of the basic exercises is slightly higher.
- There are more skill exercises on integers, fractions, decimals, and measurement.
- The interactive video skillbuilder CD enables students to review basic arithmetic, algebra, and geometry on their own.
- Special exercises are now categorized by type.
- Alternate methods of assessment are covered at the end of each chapter.
- The InfoTrac® College Edition provides students with an online database of periodicals and academic journals.
- The four longest sections have been subdivided.
- Updated data, grades K–8 textbook pages, and reference lists.

Other major revisions listed by chapter are as follows.

- **Chapter 1,** Section 1.2 now covers affirming the hypothesis and denying the conclusion. Section 1.4 is extensively revised to reflect the current emphasis on arithmetic and geometric sequences in middle school. Section 1.5 has a new subsection on Japanese style "open-ended problems."
- **Chapter 2,** Section 2.2 has more challenging intersection/union exercises in the lesson. Functions are now covered in the new Section 2.3.
- **Chapter 3**, Section 3.1 has a new subsection on the "Mayan Numeration System." Sections 3.2 and 3.3 have revised pictorial models reflecting commonly used models in elementary school classrooms. The formerly long Section 3.4 has become three sections. Sections 3.5 and 3.6 include extensively revised algorithms for each operation with more emphasis on children's natural algorithms and one good alternate algorithm for each operation. The first three subsections of Section 3.7 (formerly 3.5) are substantially revised. Section 3.8 on other bases has a revised approach to algorithms similar to that of Sections 3.5 and 3.6.
- **Chapter 4,** Section 4.2 has new subsections on "Divisibility Test for 4" and "Divisibility Test for 6."

- **Chapter 5,** Section 5.1 has a revised presentation of both counters and number lines. In Section 5.2, the exercises are substantially revised. The textbook Web Site will have a new section on Modular Arithmetic.

- **Chapter 6** includes more material on mixed numbers and more use of geometric models. Section 6.4 combines the former Sections 6.4 and 6.5.

- **Chapter 7,** Section 7.3 has more emphasis on understanding ratio and proportion with models and comparisons. Section 7.4 now focuses on percent topics that are typically covered in grade 6, and Section 7.5 contains percent topics typically covered in grades 7 and 8.

- **Chapter 8,** Section 8.5 has a new subsection and an investigation on nets. Section 8.6 now focuses on different views of figures and drawing solids on isometric dot paper. Material on Logo will be placed on the textbook Web Site.

- **Chapter 9,** the former Section 9.1 has been split into two sections. Section 9.1 is substantially revised to reflect current coverage of rigid transformations in middle school. In Section 9.5, coverage of dilations reflects recent developments in the middle school curriculum.

- **Chapter 10,** Section 10.1 now includes coverage of the customary system and significant digits. The former Section 10.5 has been split into two sections. The two new sections feature new subsections on the surface area of pyramids and cones and the volume of a sphere.

- **Chapter 11** has been more extensively revised than any other chapter in response to the much greater attention being given to algebra topics in grades 5–8. The ordering and presentation reflect the new middle-school algebra curriculum. Section 11.3 covers exponential and quadratic functions as well as change patterns in linear and nonlinear functions. Section 11.4 includes new nonsymbolic problems for introducing systems of equations. Students also compare and contrast the use of graphs, tables, equations, and words to represent a problem situation. Section 11.5 has a new subsection on "Networks."

- **Chapter 12,** Section 12.1 has expanded coverage of line plots and circle graphs. The subsection on "Misleading Graphs" has been moved from 12.1 to 12.2. Section 12.3 has more discussion of models for the mode, median, and mean. Section 12.5 now includes systematic sampling instead of cluster sampling to reflect current coverage in middle school mathematics.

- **Chapter 13,** Section 13.1 "Experimental and Theoretical Probability" begins with a discussion of "certain," "likely," "unlikely," and "impossible," and also covers geometric probability. Section 13.4 uses area models for probabilities of independent and dependent events. Section 13.4 and 13.5 have been interchanged to reflect their relative importance in the current school curriculum for grades K–8.

TEACHING AND LEARNING AIDS

The following features make it easier for students and professors to make more effective use of this book.

- Chapter introductions set the stage for each new chapter.

- Subsection headings help students understand how each lesson is organized.

- Exercises within each lesson actively involve students in developing and reinforcing concepts.

- Lessons gradually lead students to a deeper understanding of difficult concepts in the curriculum for grades K–8.
- Answers to most lesson exercises provide students with the opportunity to check their work as they develop new ideas.
- Important definitions, theorems, and properties are boxed with boldfaced headings.
- All mathematical terms are set boldface.
- Color is used to highlight mathematical ideas and add clarity to figures,
- Sample textbook pages show newer topics in elementary-school mathematics, thus increasing student awareness of recent changes.
- A "friendly" writing style as well as cartoons, drawings, and photographs stimulate student interest.
- Homework exercises are categorized as basic, extension, and technology so that instructors can select the appropriate level and type of homework exercises.
- Answers to most odd-numbered homework exercises and all chapter review exercises provide students with adequate feedback.
- Calculator exercises are marked.
- Chapter study guides list important concepts and terms by section to help students recall major ideas.
- Chapter review exercises help students prepare for exams.
- Suggested reading lists for each chapter direct students to interesting resource material.

Lesson Organization

Each lesson presents a series of exercises that develop the basic ideas of the section. These exercises should be completed before attempting the related questions in the Homework Exercises. The Lesson Exercises may be done individually, in pairs, in small groups, in class discussion, or they can be completed at home. Answers to nearly all of the Lesson Exercises appear at the end of the lesson or in the lesson.

CHAPTER ORGANIZATION AND FEATURES

Chapter 1 introduces mathematical reasoning processes and problem solving techniques that are used throughout the course. Its broader view of mathematical reasoning includes inductive and deductive reasoning as well as patterns and problem solving.

Chapter 2 covers set and function concepts that are used later in the course to clarify other concepts. Venn diagrams are used to solve problems in Chapters 1 and 2.

Chapters 3–7 cover the number systems from whole numbers to real numbers. Categories for each whole number operation (e.g., compare and take away) are studied in depth in Chapter 3 and used with other number systems in Chapters 5–7. Mental computation and estimation are also used with each number system. In number theory (Chapter 4), students do proofs and learn two different methods for solving certain problem types. In Chapters 5–7, students learn how to explain difficult procedures in integer, fraction, and decimal arithmetic, using models and realistic applications. Students also compare different representations of real numbers.

Chapters 8–10 cover geometry and measurement. In Chapter 8, the van Hiele model is used to develop categories and definitions of quadrilaterals. The section on viewing and drawing solid figures strengthens students' spatial abilities. In Chapter 9, students connect transformation geometry to congruence, symmetry, and similarity, and students analyze a series of constructions using congruence properties. In Chapter 10, students perform a series of activities to learn about the metric system and to develop area formulas in a logical sequence.

Chapter 11, Algebra and Coordinate Geometry, follows and extends arithmetic from Chapter 3–7 and geometry and measurement from Chapters 8–10. Algebra is studied as a mathematical language. Students use models to solve equations and systems of equations. Multiple representations (words, tables, graphs, and formulas) are used in solving realistic application problems.

The final two chapters cover statistics and probability. Chapter 12 emphasizes choosing the most appropriate graph or statistic to summarize results. Choices include stem-and-leaf plots and box-and-whisker plots. The extensive applications of statistics include statistical deceptions, surveys, and standardized tests.

In Chapter 13, students learn the connection between theoretical and experimental probabilities and use simulations to generate experimental probabilities. Students see the importance of probability in insurance, drug testing, and gambling games.

COURSE OUTLINES

This textbook provides for some flexibility in organizing the course. The textbook contains more than enough material for two four-semester-hour mathematics courses for preservice teachers. The material in Chapters 1–7 should be studied in sequence; however, a number of sections and parts of sections can be omitted.

In Chapters 1–7 you should cover the following material.

1. Chapter 1.

2. In Section 2.1, review set notation and define whole numbers. In Section 2.2, cover "Intersection and Union."

3. In Section 3.1, cover all subsections beginning with "Models for Place Value." Cover Sections 3.2, 3.3, 3.4, and 3.7.

4. In Section 4.1, cover "factors." In Section 4.2, cover "Multiples," "Divisibility Tests for 2, 5, and 10," and "Divisibility Tests for 3 and 9." Cover Sections 4.3 and 4.4.

5. Cover Section 5.1 and "Inverses" in Section 5.3.

6. Cover Chapter 6.

7. In Section 7.1, cover "Place Value" and "Exponents." In Section 7.2, cover "Adding and Subtracting Decimals," "Multiplying Decimals," and "Dividing Decimals." Cover Section 7.3. In Section 7.4, cover "Percents, Fractions, and Decimals" and "Basic Percent Problems."

After studying the basic material in Chapters 1–7, the major ideas of Chapters 8–13 can be studied independently of one another with the following exceptions.

Material	Prerequisite
Chapter 10	Chapter 8
Section 10.7	Section 9.5
Section 11.5	Sections 8.3, 10.4
Section 13.5	Section 12.3

One-Semester Course

Design a one-semester course that suits your needs, with the following conditions.

1. Cover the required sections and parts of sections in Chapters 1–7 as already outlined.

2. You can skip optional Section 3.8, investigations, extension exercises, and special exercises.

3. Round out the course with additional material from Chapters 1–7, or select sections from Chapters 8–13. Depending upon the length of your course, the ability of your students, and the amount of time you spend on extension exercises, you might be able to cover about 5 to 15 more sections. Suggestions for additional material would include Sections 8.1–8.6, 9.1, 10.1–10.3, 12.1–12.3, 13.1 and 13.2

Two-Semester Course

In two four-semester one-hour courses, one can cover most of the sections in the book, depending upon the number of homework exercises covered. One could skip some of Sections 3.8, 9.3, 10.7, 11.5, 13.4, or 13.5.

SUPPLEMENTS FOR INSTRUCTORS
Instructor's Resource Manual

The instructor's resource manual contains teaching suggestions and complete solutions to textbook exercises.

Test Bank

The printed test bank includes sample tests for each chapter as well as final exams. The computerized test bank enables one to create different forms of chapter tests and final exams in IBM or Macintosh format.

BCA

Free access to the Brooks/Cole Assessment system for online testing, tutorials, and course management.

Companion Web Site

Contains transparency masters and additional lessons on Logo and modular arithmetic.

SUPPLEMENTS FOR STUDENTS
Student Activities Manual

It provides additional hands-on activities for each chapter.

Student Solutions Manual

This manual provides solutions to odd-numbered homework exercises.

ACKNOWLEDGMENTS

First, I would like to thank my wife Celia for her support, encouragement, and patience during the past 16 years of writing, not to mention her many suggestions based on her own extensive experience as an elementary-school teacher. I am also grateful for the encouragement and advice provided by my mother, my sister, my late stepfather, my friends, Judge Robert Mark, and author Dan Benice.

Next, I would like to thank the editorial staff at Brooks Cole Publishing, especially Senior Editor Jennifer Huber, Project Editor Anne Seitz of Hearthside Publishing Services, and Editorial Assistant Carrie Dodson, who put a great deal of time and effort into the project. Copyeditor Barbara Willette made a significant contribution to the improved exposition and accuracy in the third edition.

The attractive design of the book is due to the direction of Senior Art Director Vernon Boss, and the graphic artists who rendered the figures, with special thanks to Jade Myers. Thanks also to Rebecca Subity for putting together an excellent set of ancillary materials. After everyone worked so hard to put this textbook together, we were lucky to have Tom Ziolkowski and the sales force to get the message out about the new edition.

Next, I would like to thank my professors from graduate school, who provided many inspiring ideas that led to the development of this textbook. They included Neil Davidson, Jim Fey, Zal Usiskin, Stephen Willoughby, Jim Henkelman, and Mildred Cole—six outstanding mathematics educators.

Finally, I would like to thank my colleagues, my students, and the many reviewers whose helpful suggestions significantly improved this book. The names and affiliations of principal reviewers for the first three editions are as follows.

Principal Reviewers

James Arnold, *University of Wisconsin (Milwaukee)*
Rita Barger, *University of Missouri—Kansas City*
Susan Beal, *Saint Xavier University*
Judy Bergman, *University of Houston*
Peter Braunfeld, *University of Illinois (Urbana)*
Laura Cameron, *University of New Mexico*
Gail Cashman, *former Montgomery College student*
Don Collins, *Western Kentucky University*
Jane Crane, *Boise State University*
Terry Crites, *Northern Arizona University*
Richard E. Cowan, *Shorter College*
Cynthia Davis, *Truckee Meadows Community College*
Gregory Davis, *University of Wisconsin (Green Bay)*
Donald Dessart, *University of Tennessee (Knoxville)*
James Dudley, *University of New Mexico*
Billy Edwards, *University of Tennessee (Chatanooga)*
Ann Farrell, *Northern Arizona State University*
Joseph Ferrar, *Ohio State University*
Iris B. Fetta, *Clemson University*
Gail Gallitano, *West Chester University*
Cathy Gosler, *University of New Mexico*
Virginia Ellen Hanks, *Western Kentucky University*
Alan Hoffer, *Boston University*
John Koker, *University of Wisconsin (Oshkosh)*

Michael Krach, *Towson University*
Karen Marrongelle, *Portland State University*
Robert S. Matulis, *Millersville University of Pennsylvania*
Flora Metz, *Jackson State Community College*
William A. Miller, *Central Michigan University*
Barbara Moses, *Bowling Green State University*
John Neil, *Boise State University*
Leslie Paoletti, *Central Connecticut State University*
Pamela Pospisil, *Montgomery College*
Laura Ralston, *Floyd College*
Donald Ramirez, *University of Virginia*
Janice Rech, *University of Nebraska (Omaha)*
Alice I. Robold, *Ball State University*
Sharon Ross, *California State University (Chico)*
Carol Rychly, *Augusta State University*
Elizabeth Shearn, *University of Maryland*
Jane F. Sheilack, *Texas A & M University*
Keith Schuert, *Oakland Community College*
Peter Sin, *University of Florida*
Jim Trudnowski, *Carroll College*
C. Ralph Verno, *West Chester University*
Jeannine Vigerust, *New Mexico State University*
Richard Vinson, *University of South Alabama*
Stephen S. Willoughby, *University of Arizona*
Deon Woodward, *New Mexico State University*
John E. Young, *Southeastern Missouri State University*

INTRODUCTION

WHY LEARN MATHEMATICS?

As a teacher, you will introduce the next generation of students to mathematics, teaching them arithmetic, geometry, and statistics. This course will help you develop a deeper understanding of school mathematics. But it is only a beginning. The best teachers continue to learn more about mathematics all through their careers.

One objective of this course is to help you understand mathematics in a way that will make you confident about discussing mathematical ideas with your students. When you present solutions to problems in this course, be prepared to show the steps taken and to explain the thinking that was involved. This is exactly what an effective teacher does, and it is also a goal for your future students.

But why does anyone teach children mathematics as part of their education? For one thing, mathematics offers a unique way of looking at the world. Mathematics gives simple, abstract descriptions that illuminate general relationships among quantities or shapes while simplifying or ignoring qualities. Mathematics can show that, ounce for ounce, Brand A is a better buy than Brand B, focusing on the amount and the price while ignoring other differences in the products and the companies that produce them. Such mathematical information can help one decide which brand to buy.

Furthermore, the logical, objective approach of mathematics is applied to many situations. Mathematics has had an impact on nearly every area of life, from philosophy to economics to art. Philosophers apply the logic of mathematics to ultimate questions. Economists employ quantitative methods to describe financial trends. Artists use geometry when they represent our three-dimensional world on canvas.

Finally, mathematics is the language of technological societies. Businesses use mathematics in record-keeping and analysis. As a consumer and a citizen, each of us needs mathematics to make financial decisions and to interpret statistics in political and economic news.

What have your mathematical experiences been like? Is mathematics one of your favorite subjects? Or have you felt like Peppermint Patty in the following cartoon?

Your professor may ask you to complete the following survey.

Introductory Survey

Name_____

1. For your two most recent mathematics courses, list the course name, when and where you took it, and your final grade.

_____

_____

2. One of my favorite things about recent mathematics courses has been _____

_____

_____

3. One of my least favorite things about recent mathematics courses has been ____

_____

4. Do you feel adequately prepared for this class?
  (a) Yes   (b) No   (c) Not sure

5. (a) How many credit hours are you taking this term?   _____
  (b) How many hours will you work each week?   _____
  (c) Do you also have children to take care of?   _____

6. Rate your overall time schedule this term.
  (a) Very busy/may be too much
  (b) Busy
  (c) Comfortable

7. Would you be interested in meeting with other students in the class
  (a) to do homework?   Yes   No   Maybe
  (b) to study for exams?   Yes   No   Maybe

---

1 | MATHEMATICAL REASONING

How do we determine what is true in mathematics classes and in everyday life? You have been doing mathematics for many years, yet you may know little about the approaches you use over and over again in solving problems.

The two central processes of mathematics are inductive reasoning and deductive reasoning. Becoming aware of these reasoning processes will give you a better understanding of how you and your students learn mathematics. After studying these processes, you will use them throughout this course to make and prove generalizations.

Mathematics is sometimes called the study of patterns. Induction and deduction are used to generalize mathematical patterns, to extend them, and to prove generalizations about them.

In recent years, a greater emphasis has been placed on helping all students develop problem-solving abilities in a structured way. Elementary-school books now teach some version of Polya's four-step problem-solving procedure: Read, plan, solve, and check. When it comes to the second step, making a plan, the elementary-school curriculum now includes specific processes and strategies for problem solving. The most useful ones are induction, deduction, choosing the operation, making a table, and drawing a picture. More specialized strategies include guessing and checking and working backwards. These processes and strategies are introduced in Chapters 1 and 3.

What should children be learning about mathematics in grades pre-K through 12? The National Council of Teachers of Mathematics (NCTM) presents its vision for school mathematics in the *Principles and Standards for School Mathematics* (2000). To help familiarize you with this document, each section of this book begins with a list of objectives from the *Principles and Standards*. These objectives relate to the material in the section. The inside front cover lists the overall standards for grades pre-K through 12. For more information on the standards, visit the NCTM Web site.

1.1 INDUCTIVE REASONING

NCTM Standards
- describe, extend, and make generalizations about geometric and numerical patterns (3–5)
- make and investigate mathematical conjectures (pre-K–12)
- recognize and apply mathematics in contexts outside of mathematics (pre-K–12)

How do we form our beliefs? For example, do you believe that your friend Alan is trustworthy?

Alan

Perhaps you would answer "yes" if your previous encounters with Alan suggest that he is trustworthy.

Similarly, many mathematical ideas arise from observing a pattern in a series of examples. The "number magic" in LE (Lesson Exercise) 1 will lead your entire class to form a generalization based on a pattern in their individual results.

Discussion

LE 1   Opener

(a) Follow these instructions.
    Pick a number.
    Multiply it by 2.
    Add 10.
    Divide by 2.
    Subtract your original number.
(b) What number do you end up with?
(c) So why all the fanfare? Pick a different number and follow the instructions in part (a) again (or find out what everyone else in the class got). What number do you end up with this time? Are you mystified?
(d) Make a generalization based on your results in part (b) and part (c).
(e) Are you sure that your generalization will work for all starting numbers?

You may wonder how the instructions in LE 1 work. When we return to it later, in Section 1.3, you'll be able to prove that the pattern works for all starting numbers.

The four statements in the cartoon at the beginning of the lesson and the pattern in LE 1 illustrate inductive reasoning. Inductive reasoning works like this: You might be adding pairs of odd numbers and see a pattern in the results: $1 + 3 = 4$, $3 + 7 = 10$, and $35 + 31 = 66$. Then you make a generalization: The sum of any two odd numbers is an even number. This process of reasoning from the specific to the general is called inductive reasoning.

Definition: **Inductive Reasoning**

Inductive reasoning is the process of making a generalization based on a limited number of observations or examples.

Inductive reasoning is the basis of experimental science and a fundamental part of mathematics. A scientist uses inductive reasoning to formulate a hypothesis based on a pattern in experimental results. A mathematician uses inductive reasoning to form a reasonable generalization from a pattern in a series of examples.

In addition, many of our beliefs, prejudices, and theories are based on this process. This lesson will help you make intelligent use of inductive reasoning in mathematics and in everyday life. Use inductive reasoning with care, since your generalizations may turn out to be wrong. For example, your friend Alan may eventually turn out to be untrustworthy.

MAKING GENERALIZATIONS

The essence of inductive reasoning is making generalizations (also called "conjectures") that appear to be true. The following example and exercises will help you become familiar with the process of making generalizations.

• **EXAMPLE 1**

Does every number squared equal itself? Use inductive reasoning to make a generalization based on the examples given.

Examples: $0^2 = 0$ $1^2 = 1$
Generalization: _____

SOLUTION

In both examples, a number squared equals itself. Then, by inductive reasoning, I conclude that every number squared equals itself.

Generalization: Every number squared equals itself. (This will turn out to be a false generalization.) ●

In each of the following exercises, use inductive reasoning to make a generalization based on the examples given. Write each generalization as a complete sentence.

LE 2 Reasoning

The product of two odd numbers is what type of number?

Examples: $5 \cdot 9 = 45$ $3 \cdot 7 = 21$
Generalization: _____

LE 3 Reasoning

Examples: The two times I went on a picnic, it rained.
Generalization: _____
_____

(*Note:* The answers to many lesson exercises can be found at the end of the section, just before the homework exercises.)

Figure 1–1 shows how the first-grade *Harcourt Math* textbook asks the child to make a generalization after a series of subtraction problems.

Of course, we would prefer that our generalizations were always true. In everyday life, we settle for generalizations that are true *most* of the time. When we wake up each morning, we assume that the refrigerator will be working and that we will have hot water. Such generalizations are reasonable and usually true. They give our lives order and enable us to plan ahead. However, such generalizations would not meet mathematical standards. In mathematics, we seek generalizations that are *always* true.

Unfortunately, inductive reasoning does not always lead to true generalizations. Although the mathematical generalization you made in LE 2 is true, the mathematical generalization in Example 1 is false. Often a false generalization results when one selects a small or nonrepresentative set of examples. By observing a larger number of individual events, one can make more reasonable generalizations that are more likely to be true.

Classroom Connection

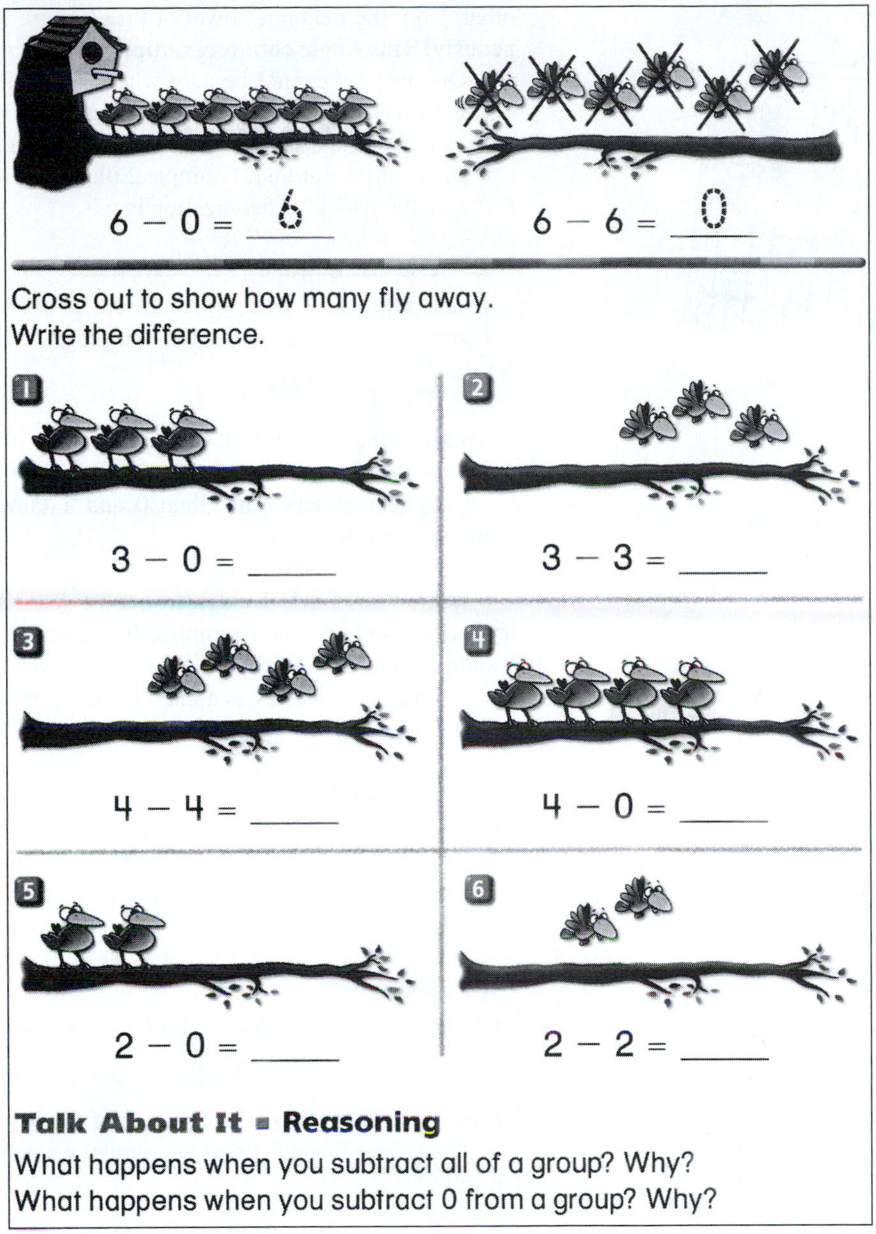

$6 - 0 = \underline{6}$ $6 - 6 = \underline{0}$

Cross out to show how many fly away.
Write the difference.

1 $3 - 0 = \underline{}$

2 $3 - 3 = \underline{}$

3 $4 - 4 = \underline{}$

4 $4 - 0 = \underline{}$

5 $2 - 0 = \underline{}$

6 $2 - 2 = \underline{}$

Talk About It ▪ Reasoning

What happens when you subtract all of a group? Why?
What happens when you subtract 0 from a group? Why?

From Harcourt Math, Grade 1 (Orlando: Harcourt School Publishers, 2002), p. 39.

Figure 1–1

USING COUNTEREXAMPLES TO DISPROVE FALSE GENERALIZATIONS

How can one recognize false generalizations? Often it is a matter of finding just one exception. For over 1900 years, people accepted Aristotle's (384–322 B.C.) generalization that heavier objects fall faster than lighter objects. To test this generalization, Galileo (1564–1643) is alleged to have dropped two metal objects, one much heavier than the

Figure 1–2

other, from the Leaning Tower of Pisa (Figure 1–2). The objects hit the ground simultaneously! This single **counterexample** disproved Aristotle's longstanding generalization.

One usually cannot be sure whether a generalization reached by induction is true, because one cannot examine every single possibility. Although one cannot *prove* a generalization is true by picking examples, one can show a generalization is false by finding just *one* counterexample! Example 2 illustrates how one counterexample is used to show that a mathematical generalization is false.

• EXAMPLE 2

In Example 1, it was conjectured that every number squared equals itself. Find a counterexample that disproves this generalization.

SOLUTION

This statement is false for other decimal numbers. For example, $2^2 = 4$, not 2. In making a generalization, it would have been wiser to use more than two examples, including some numbers other than 0 and 1 (since 0 and 1 have special properties for multiplication).   ●

In LE 4 and LE 5, decide whether the generalization is reasonable or false. If possible, try a few other specific examples. If the generalization is false, disprove it by giving a counterexample. (The in front of an exercise (such as LE 5) indicates that it is a good exercise to discuss as a class or in groups. Communicating your ideas orally or in writing helps to clarify them.)

LE 4   Reasoning

I take a number and add it to itself. Then I take the same number and multiply it by itself. I obtain the same answer either way in two examples: $0 + 0 = 0 \cdot 0$ and $2 + 2 = 2 \cdot 2$. So I generalize that this will always happen.

Discussion

LE 5   Reasoning

I add the first two, three, and four odd numbers.

$$1 + 3 = 2^2 \qquad 1 + 3 + 5 = 3^2 \qquad 1 + 3 + 5 + 7 = 4^2$$

From this, I generalize that the sum of the first N odd numbers is N^2, where $N = 2, 3, 4, \ldots$. (*Hint:* Write the next two equations that continue the pattern, and see if they are true.)

How can you make more *reasonable* generalizations and detect faulty ones? By checking more examples and a greater variety of them.

Classroom Connection

LE 6   Reasoning

A sixth grader says she has proved that a positive number plus a negative number is a positive number. She tried six examples and all the answers came out positive. Is her generalization correct? If not, what would you tell her?

APPLICATIONS OF INDUCTIVE REASONING

At its best, inductive reasoning leads to reasonable generalizations. In mathematics, these generalizations are then proved true (if possible) using deductive reasoning. (Deductive reasoning will be discussed in the next section.) In everyday life, however, a reasonable generalization based upon a large number of examples is often the best we can do. It is often impossible to prove that such a generalization is true.

Because nonmathematical generalizations are often impossible to prove, one should be careful about making them. Unfortunately, people are not always careful about making nonmathematical generalizations. Consider prejudice. One example of prejudice is a generalization about a whole group of people. Some people learn prejudices from their family or community, but prejudices can also be developed by using inductive reasoning.

After I meet a few people from a certain group and observe a particular common characteristic, I might assume that all people from that group share that characteristic. No matter what generalization I make about a group of people, there are bound to be some exceptions.

In LE 7, consider how some superstitions arise from inductive reasoning. (Note the use of the instruction "*Explain*" in the exercise. In this book, *Explain* means you should write an explanation that you might give to an elementary-school student. The explanation can be brief but should proceed step by step from start to finish.)

Discussion

LE 7   Connection
Explain how someone could use inductive reasoning to develop a superstition.

Discussion

LE 8   Connection
Give an example of how you have used inductive reasoning to form a general idea.

 ## AN INVESTIGATION: SUMS OF CONSECUTIVE WHOLE NUMBERS

An important part of mathematics is exploring new situations and making conjectures about them. The following investigation can be done in class (alone or in groups) or as homework.

LE 9   Reasoning
Use inductive reasoning to answer the following questions.

(a) Is the sum of any three consecutive whole numbers divisible by 3?
(b) Is the sum of any four consecutive whole numbers divisible by 4?
(c) Is the sum of any five consecutive whole numbers divisible by 5?
(d) Investigate further examples and complete the following generalization:
The sum of N consecutive whole numbers is divisible by N when
$N = $ _____

LE 10   Summary
Tell what you learned about inductive reasoning in this section.

ANSWERS TO SELECTED LESSON EXERCISES

2. The product of two odd numbers is an odd number.

3. Every time I go on a picnic it rains.

4. False; $3 + 3 \neq 3 \cdot 3$

5. Reasonable

6. No. Ask her if she has enough variety in her examples.

7. A person may associate events that have occurred in succession a few times, such as a black cat crossing the person's path and something bad happening soon after. After such experiences, the person generalizes that every time a black cat crosses one's path, something bad will happen soon after.

1.1   HOMEWORK EXERCISES

Basic Exercises

1. Consider the following number trick.

Pick any number at all.
Add 8.
Multiply by 2.
Subtract 10.
Subtract your original number.

(a) Try the trick with two or three different numbers.
(b) What is the general pattern in your results?

2. $3 \cdot 5 = 5 \cdot 3$ $6 \cdot 2 = 2 \cdot 6$ $1 \cdot 9 = 9 \cdot 1$
Write a generalization based upon these results.

3. A Martian met three people from Boston, and they all wore high heels. Now the Martian thinks everyone from Boston wears high heels. Clearly, this generalization is false. Did the Martian use inductive reasoning correctly?

4. Why is it usually impossible to show that a statement is true by checking examples?

5. How can a proposed generalization be disproved?

In Exercises 6–10(a), decide whether the generalization is reasonable or false. If it's false, give a counterexample.

6. *Examples:*

Generalization: All four-sided figures have four right angles.

7. *Examples:* José likes to drink beer and watch football on TV.
    Brad likes to drink beer and watch football on TV.
Generalization: All men like to drink beer and watch football on TV.

8. *Examples:* 5, 15, 25
Generalization: If a number is odd, then it ends in 5.

9. *Examples:* $1 + 2 = \dfrac{2(3)}{2}$

$1 + 2 + 3 = \dfrac{3(4)}{2}$

$1 + 2 + 3 + 4 = \dfrac{4(5)}{2}$

Generalization: The sum of the first N counting numbers is $N(N + 1)/2$ for $N = 2, 3, 4, \ldots$.

10. (a) A farmer uses fertilizer for 3 years in a row, and his corn crop yield improves each year. On the basis of these results, he generalizes that fertilizer improves corn crop yield.
(b) If the generalization in part (a) is true, what other factors should the farmer consider in deciding whether or not to use this fertilizer to grow corn?

11. (a) $12{,}345{,}679 \times 36 = $ _____
(b) $12{,}345{,}679 \times 45 = $ _____
(c) Write two more equations that extend this pattern.
(d) *Explain* how this pattern works. (*Hint:* What is $12{,}345{,}679 \times 9$?)

12. Ready for some math magic?

$$112 \times 124 = 13{,}888$$

No, of course that's not the whole trick! Read on.
(a) Reverse the digits in each factor and multiply.
$211 \times 421 = $ _____
(b) What is interesting about the result in part (a)?
(c) $312 \times 221 = $ _____
$213 \times 122 = $ _____
(d) Make a generalization based on the results to parts (a) and (c).
(e) Try some other examples, and decide whether your generalization is reasonable or false.

13. Galileo used inductive reasoning to develop hypotheses about the length of a pendulum and the period of a full swing.

Use the following data to develop your own hypothesis.

L (length of pendulum)	P (period of a full swing)
1 unit	1 second
2 units	1.41 seconds
3 units	1.73 seconds
4 units	2 seconds
9 units	3 seconds

(a) If the pattern continues, what is the *length* of a pendulum that has a *period* of 4 seconds?
(b) What is the general relationship between the length and the period?
(c) Write a formula relating L to P.
(d) Use your formula from part (c) to estimate the period of a pendulum 5 units long.

14. How would one use inductive reasoning to form a prejudice?

15. *Explain* (a) how each of the following generalizations can be formed by using inductive reasoning and (b) whether or not you agree with the generalization.
1. There will always be wars.
2. Boys are better than girls at mathematics.
3. A movie sequel is never as good as the original movie.

16. Give an example showing how each of the following groups might use inductive reasoning.
(a) Teachers
(b) College admissions officers

17. Give an example showing how each of the following groups might use inductive reasoning.
(a) Students   (b) Mathematicians

Extension Exercises

18.

E (weight on Earth)	100 lb	200 lb	300 lb
M (weight on Mars)	38 lb	76 lb	114 lb

(a) An astronaut in a space suit might weigh about 400 lb on Earth. If the pattern continued, what would be her weight on Mars?

(b) If someone's weight doubled on Earth, what do you think would happen to the person's weight on Mars?

(c) Make a conjecture about the general relationship between weight on Earth and weight on Mars, based upon these data. Tell how you developed your conjecture. (Compare each E value to the corresponding M value.)

(d) Write an equation relating E to M, based upon your conjecture.

19. Try some examples, and decide whether each of the following statements is probably true or definitely false.

(a) The product of any three consecutive whole numbers is divisible by 3.

(b) The product of any four consecutive whole numbers is divisible by 4.

(c) The product of any five consecutive whole numbers is divisible by 5.

(d) Investigate further examples, and state a generalization of your results.

20. (a) Is every number that is divisible by 4 also divisible by 3?

(b) Is every number that is divisible by 5 also divisible by 3?

(c) Is every number that is divisible by 6 also divisible by 3?

(d) Investigate further examples, and state a generalization of your results.

21. When people clasp their own hands, do they all prefer to put the right thumb on top?

22.

Mapmakers often shade adjacent states or countries in different colors. One of the most famous problems in mathematics is the map-coloring problem. It asks: How many different colors are needed to color any map drawn on a flat surface or a sphere so that adjacent regions are different colors?

In 1878, Arthur Cayley posed this problem to the London Mathematical Society. Ninety-eight years later, Wolfgang Haken and Kenneth Appel of the University of Illinois found and proved the answer using over 1000 hours of computer time.

(a) Find the minimum number of different colors needed to color the map of the United States so that no two adjacent regions are the same color. (You don't have to color in every state.)

(b) On the basis of your results, make a conjecture about the answer to the map-coloring question.

1.2 DEDUCTIVE REASONING

NCTM Standards
- develop and evaluate mathematical arguments and proofs (pre-K–12)
- recognize and apply mathematics in contexts outside of mathematics (pre-K–12)

It's the first day of class. You assume that the teacher will be older than most of the students and that the teacher will settle in at the front of the room. A middle-aged woman walks in the door and heads directly to the front of the room. Next, she lays her things on the teacher's desk. What would you conclude about this person?

Or how about this? I'm thinking of a one-digit odd number that is greater than 5 and divisible by 3. Can you deduce what number it is?

In drawing conclusions from information given in the preceding examples, one can use a second important reasoning process: deductive reasoning.

> **Definition: Deductive Reasoning**
>
> **Deductive reasoning** is the process of reaching a necessary conclusion from given facts or hypotheses.

One of the most famous deductive thinkers is Sherlock Holmes. Notice how he uses deductive reasoning in the following scene.

A FAMOUS APPLICATION OF DEDUCTIVE REASONING

LE 1   Opener

It's show time! Select two people from the class to play the parts of Holmes and Watson in the following scene, adapted from *The Adventures of the Dancing Men* by Sir Arthur Conan Doyle. Places! Action!

HOLMES: (*coolly*) So Watson, you were just discussing gold investments.

WATSON: (*astonished*) How on earth did you know that?!

HOLMES: In a few moments, you'll say it's so absurdly simple.

WATSON: (*in a huff*) I certainly will not.

HOLMES: After seeing chalk between your index finger and thumb, I feel sure that you were discussing gold investments.

WATSON: (*still confused*) What is the connection?

HOLMES: Let me explain and diagram my argument using step-by-step deductive reasoning.

1. You have chalk between your index finger and thumb.
2. Therefore, you must have played billiards.
3. You must have played billiards with Thurston, since you always play with him.
4. Thurston had an option to invest in gold no later than today and wanted you to invest with him, so you must have discussed it.

WATSON: How absurdly simple! You've done a proof like I used to do in high-school geometry class.

Curtain (and applause).

In reaching his conclusion, Sherlock Holmes used deductive reasoning. Each statement in his explanation necessarily leads to the next statement when combined with certain assumptions. For example, Holmes combined the fact that Watson had chalk between his index finger and thumb with the assumption that if Watson had chalk on his fingers, then he must have been playing billiards. The necessary conclusion is that Watson was playing billiards.

$$\text{Chalk on fingers} \rightarrow \text{Played billiards} \rightarrow \text{Was with Thurston} \rightarrow \text{Discussed gold investment}$$

Deductive reasoning is used by detectives solving crimes, mathematicians drawing conclusions, and philosophers thinking logically. When the assumptions are true, deductive reasoning produces certain results. Now it's your turn to solve a mystery using deductive reasoning.

LE 2   Reasoning

The painting "By Numbers" was stolen from Jane Dough's Illinois home last night. Today the painting was found in an alley. Can you find the thief?

The only suspects are Jane Dough (the owner of the house), Jeeves (the butler), Sharky (the pool man), and Fluffy (the pet Doberman pinscher). The police have gathered the following evidence:

1. The painting was stolen between 8:00 and 9:00 P.M. last night.
2. Fluffy can't carry a painting.
3. Jeeves has been visiting Mumsy in Liverpool all week.
4. Sharky's and Jane's fingerprints were on the canvas.
5. Jane was with three friends at the movie *Return of the Stepson of Darth Vader's Cousin* from 7:30 to 10:00 P.M. last night.

DRAWING CONCLUSIONS

The essence of deductive reasoning is drawing a necessary conclusion (from given hypotheses). You used this process repeatedly in the preceding exercise.

LE 3   Reasoning

Fill in the conclusion that follows from the given hypotheses.

Hypotheses: The painting was stolen from the house between 8:00 and 9:00 P.M.
Jane was at a movie from 7:30 to 10:00 P.M.
Conclusion: _____

Mathematicians often use this same process. The following lesson exercise illustrates how applying a geometry theorem to a particular figure requires deductive reasoning.

LE 4 Reasoning

Fill in a conclusion that follows from the given theorem and hypothesis.

> *Theorem:* The opposite sides of a parallelogram are equal in length.
> *Hypothesis:* ABCD is a parallelogram.

Conclusion: _____

In LE 4, you applied the theorem that the opposite sides of a parallelogram are equal in length in a particular parallelogram *ABCD*. Mathematicians also use deductive reasoning in applying a definition to a particular case, as is done in the following lesson exercise. The following exercise introduces a second format that is commonly used in deductive reasoning: the "if-then" format.

LE 5 Reasoning

Fill in a conclusion that follows from the given hypotheses.

If all numbers that are divisible by 2 are even, and 264 is divisible by 2,
then _____.

In LE 5, the definition that an even number is divisible by 2 was applied to the number 264. If we know that all numbers that are divisible by 2 are even and that 264 is divisible by 2, then we can conclude that 264 is even. In LE 5, the hypotheses are (1) "All numbers that are divisible by 2 are even numbers" and (2) "264 is divisible by 2." The conclusion is "264 is an even number." In the "if-then" statement, the hypotheses come after the word "if," and the conclusion is stated after the word "then." You can also think of hypotheses as assumptions.

LE 6 Concept

Consider the following statement: If $x = 4$ and $y = x + 3$, then $y = 7$.

(a) What are the hypotheses (assumptions)?
(b) What is the conclusion?

The mathematician, like the detective, uses deductive reasoning in a process of elimination. This is illustrated in the following exercise.

LE 7 Reasoning

(a) Fill in the blank and **(b)** identify the hypotheses and conclusion.

Two distinct lines in a plane are either parallel or intersecting. Suppose distinct lines *a* and *b* in a plane are not parallel. Therefore, _____.

Aristotle (384–322 B.C.), the father of deductive reasoning, was the first to study logic systematically. He stated the most famous example of deductive reasoning. It consists of two hypotheses and a necessary conclusion.

Example 1 models Aristotle's famous example of deductive reasoning with a Venn diagram (set picture). Aristotle didn't have the luxury of using a Venn diagram. John Venn (1834–1923) invented these diagrams, which are used for illustrating set relationships, more than 2000 years after Aristotle's death.

• EXAMPLE 1

Deduce the conclusion and represent it with a Venn diagram.

Hypothesis: All people are mortal.
            Socrates is a person.
*Conclusion:*_____

SOLUTION

Figure 1–3

Since Socrates is a person and all people are mortal, we can conclude that Socrates must be mortal. The same conclusion can be confirmed with Venn diagrams, as shown in Figure 1–3. The first hypothesis, "All people are mortal," is shown by drawing a circle or oval that represents the set of "all people" inside a circle or oval that represents the set of "all mortals." This Venn diagram shows that the set of "all mortals" contains the set of "all people." The second hypothesis, "Socrates is a person," is shown by drawing a point that represents Socrates inside the set of "all people," since the set of "all people" contains Socrates. A point is used to represent a single element such as "Socrates." A circle or oval is used to represent a group such as "all people."

The point representing Socrates is also contained in the circle representing mortals. So the set of "mortals" contains Socrates. We have our conclusion: Socrates is mortal. ●

LE 8   Reasoning

Deduce the conclusion to the following, and represent both hypotheses in a Venn diagram. (*Hint:* For the Venn diagram, identify a larger group [circle or oval] that contains another smaller group [circle or oval].)

Hypotheses: All doctors are college graduates.
            All college graduates finish high school.
Conclusion: _____

Figure 1–4, on the next page, shows how the first-grade *Harcourt Math* textbook asks children to make deductions based on two hypotheses about a set of coins.

DOES DEDUCTIVE REASONING ALWAYS WORK?

Are the conclusions that result from deductive reasoning always true? Find out what happens if we make a slight change in LE 8.

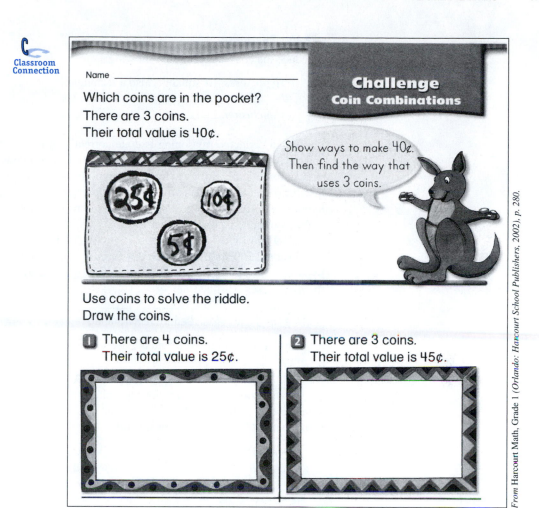

Figure 1–4

LE 9   Concept

(a) Fill in the blank and (b) identify the hypotheses and conclusion.

If all doctors are college graduates and all college graduates like roller skating, then _____.

LE 9 illustrates that valid (correct) deductive reasoning can sometimes lead to a false conclusion.

Valid

FALSE   reasoning?   FALSE

HYPOTHESIS   ——————→   CONCLUSION

LE 10 contains another example of this type.

LE 10   Reasoning

Deduce the conclusion to the following.

> *Hypotheses:* A square is also a rectangle.
>   All rectangles have six sides.
> *Conclusion:* _____

When deductive reasoning is done correctly, it is called valid deductive reasoning, and the conclusion is called a valid conclusion. In **valid** deductive reasoning, the conclusion automatically follows from the hypotheses. Deductive reasoning is said to be **invalid** (done incorrectly) when the conclusion does not automatically follow from the hypotheses. The use of the word "valid" or "invalid" does *not* indicate whether any hypothesis or conclusion is true or false.

LE 9 and LE 10 are examples of how valid deductive reasoning can lead to a false conclusion. LE 9 and LE 10 each have a false conclusion that results from having a false hypothesis. To guarantee that the conclusion reached by deductive reasoning is true, all the hypotheses must be true.

True Conclusions from Deductive Reasoning

If the hypotheses are true and the deductive reasoning is valid, then the conclusion must be true.

In analyzing a deductive sequence, check to see whether the hypotheses and conclusion are true or false, and check whether reasoning from the hypotheses to the conclusion is valid (done correctly) or invalid.

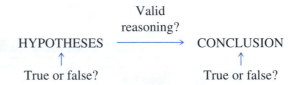

RULES OF LOGIC AND DEDUCTIVE REASONING

How can you tell if deductive reasoning is valid? First, check if the reasoning fits one of two basic rules of logic.

LE 11   Reasoning

Fill in the conclusion in parts (a) and (b).

(a) *Hypotheses:* If a person is a doctor, then the person is a college graduate.
          Maija is a doctor.
   Conclusion: _____

(b) *Hypotheses:* If a person is a doctor, then the person is a college graduate.
          Melissa is not a college graduate.
   Conclusion: _____

(c) Draw a Venn diagram that shows why your conclusion in part (a) is true. (*Hint:* The first statement is the same as "All doctors are college graduates.")

(d) Draw a Venn diagram that shows why your conclusion in part (b) is true. (*Hint:* Draw two circles or ovals for the first statement. Then say, a point representing Melissa is not in the . . . so it cannot be in the . . ..)

LE 11 illustrates two rules of logic that are often used in deductive reasoning. Part (a) is an example of **affirming the hypothesis**. We have a general if-then statement that is true and a specific example (for Maija) in which the hypothesis is true. We can deduce that the "then" part is true for our specific example. Maija is a college graduate.

Part (b) is an example of **denying the conclusion**. We have a general if-then statement that is true and a specific example (for Melissa) in which the "then" part is false. We can deduce that the hypothesis ("if" part) is false for our specific example. Melissa is not a doctor.

VENN DIAGRAMS AND DEDUCTIVE REASONING

How can you recognize invalid deductive reasoning? It occurs when someone attempts to use deductive reasoning even though it does not apply to a situation. Venn diagrams, which were used earlier to illustrate valid conclusions, are also helpful in detecting invalid deductive reasoning. If deductive reasoning is valid, it is impossible to draw a Venn diagram that satisfies the hypotheses but not the conclusion. If deductive reasoning is invalid, it *is* possible to draw a Venn diagram that satisfies all the hypotheses but not the conclusion.

• EXAMPLE 2

Is the following conclusion valid? If so, draw a Venn diagram that illustrates it. If not, draw a Venn diagram showing that the conclusion does not follow from the two hypotheses.

Hypotheses: All dogs are animals.
               All cats are animals.
Conclusion: All dogs are cats.

SOLUTION

The conclusion does not seem to follow from the two hypotheses. If the conclusion is invalid, there should be a Venn diagram that satisfies the hypotheses but not the conclusion (see Figure 1–5). Such a diagram would be a counterexample to the conclusion. The first hypothesis, "All dogs are animals," is shown by drawing an oval that represents the set of all dogs inside a circle that represents the set of all animals, since the group of animals contains all dogs (plus other animals).

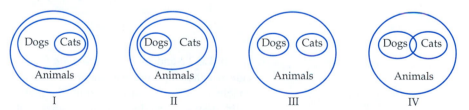

Figure 1–5

Similarly, the hypothesis "All cats are animals" is shown by drawing an oval that represents all cats inside the circle that represents all animals, since the group of all animals contains all cats.

It is possible to draw a diagram (I, III, or IV) that satisfies the hypotheses but contradicts the conclusion. The "Dogs" oval does not have to be inside the "Cats" oval as the conclusion suggests, so the conclusion does not necessarily follow from the hypotheses.   ●

The result of Example 2 can be illustrated as follows.

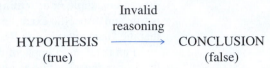

The error in Example 2 is obvious. However, advertisers sometimes use the same form of faulty reasoning, and many people do not notice it. Example 3 illustrates this technique.

• EXAMPLE 3

Does the conclusion follow from the two hypotheses? If so, draw a Venn diagram that illustrates it. If not, draw a Venn diagram showing that the conclusion does not follow from the two hypotheses.

Hypotheses: All beautiful women use Coverall lipstick.
              Sally uses Coverall lipstick.
Conclusion: Sally is a beautiful woman.

SOLUTION

According to the first hypothesis, Coverall users include all beautiful women. Draw an oval for beautiful women inside a circle for Coverall users. Sally is an individual represented by a point (not an oval) inside the circle for Coverall users (Figure 1–6). The point representing Sally does not have to be inside the "beautiful woman" circle. In fact, Sally may be a child! The Venn diagram in Figure 1–6 satisfies the hypotheses but not the conclusion. Therefore, the deductive reasoning must be invalid. ●

Figure 1–6

Note that Figure 1–6 and diagram III in Figure 1–5 have the same structure. This diagram results from one common form of invalid deductive reasoning. As shown in Figure 1–7, if *A* is in *C* and *B* is in *C,* then the deduction that *B* is in *A* is invalid.

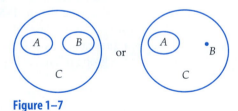

Figure 1–7

In LE 12 and LE 13, *assume the first two statements (hypotheses) are true,* and decide whether the third statement (conclusion) can be deduced from the first two. If the conclusion is valid, draw a Venn diagram that illustrates this. If the conclusion is invalid, draw a Venn diagram that disproves it.

Discussion

LE 12   Reasoning

A rectangle is a polygon. A triangle is a polygon. Therefore, a rectangle is a triangle.

Discussion

LE 13   Reasoning

Harold uses Some laundry detergent. People who use Some have clean clothes. Therefore, Harold has clean clothes.

A GAME: PICA-CENTRO

Pica-Centro is a game of deductive reasoning for two players that Douglas Aichele describes in his May 1972 article in *Arithmetic Teacher*. The object of the game is to determine the three-digit number your opponent has secretly selected by using deductive reasoning (and a little luck).

To begin the game, Player A secretly chooses a three-digit number with three different digits. Player B then keeps guessing three-digit numbers until Player B can determine Player A's number. For example, suppose Player A secretly picks the number 807. Player B records guesses as follows.

Guesses by Player B	Responses by Player A	
Digits	**Pica: Correct Digit, Wrong Position**	**Centro: Correct Digit, Correct Position**
7 4 2	1	0
4 2 5	0	0
3 8 7	1	1
8 0 7	0	3

For each of Player B's guesses, Player A tells how many digits are correct and whether they are in the correct positions, as shown in the preceding table. This process continues until Player B finds Player A's number. Then the players switch roles.

Discussion

LE 14   Reasoning
Find a partner and play a game of Pica-Centro.

LE 15   Summary
Tell what you learned about deductive reasoning in this section.

ANSWERS TO SELECTED LESSON EXERCISES

2. Sharky

3. Jane did not steal the painting.

4. $AB = CD$ and $AD = BC$.

5. 264 is an even number.

6. (a) $x = 4$ and $y = x + 3$ (b) $y = 7$

7. (a) a and b are intersecting.
(b) The first two sentences (about the lines) are the hypotheses, and the conclusion is "a and b are intersecting."

8. All doctors finish high school.

9. (a) All doctors like roller skating.
  (b) The phrase after "If" and before "then" contains the two hypotheses, and the conclusion is "all doctors like roller skating."

10. All squares have six sides.

11. (c)

College graduates

(d)

Melissa is not in the oval for the college graduates, so she cannot be in the oval for doctors.

12. Invalid

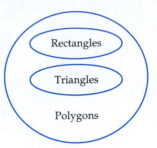

13. Valid

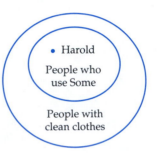

1.2   HOMEWORK EXERCISES

Basic Exercises

1. Use deductive reasoning to fill in a conclusion that follows from the given statements, and draw a Venn diagram (set picture).
   All rectangles are parallelograms. All parallelograms are quadrilaterals.
   Conclusion: _____

2. Use deductive reasoning to fill in the conclusion.
   Watson was playing billiards today. Watson plays billiards only with Thurston.
   Conclusion: _____

3. (a) Fill in the blank and (b) identify the hypotheses and conclusion.

If *ABCD* is a rectangle and the opposite sides of a rectangle are parallel, then

_____

4. (a) Fill in the blanks and (b) identify the hypotheses and conclusion.
   If $x = 4$ and $y = 2x - 3$, then _____

5. Write each of the following statements in if-then form.
   (a) All students want the option to earn extra credit.
   (b) All rectangles have four sides.

6. Write each of the following statements in if-then form.
   (a) All carrots contain vitamin A.
   (b) All two-digit whole numbers are greater than 9.

7. What will guarantee that the conclusion reached through valid deductive reasoning is true?

8. Make up an example of valid deductive reasoning that leads to a false conclusion.

9. Fill in the conclusion in each part.
   (a) *Hypotheses:* If a number is negative, then it is less than 10.
       -5 is a negative number.
   Conclusion: _____

(b) *Hypotheses:* If a number is negative, then it is less than 10.
12 is not less than 10.

Conclusion: _____

(c) Draw a Venn diagram that shows why your conclusion to part (a) is true. (*Hint:* The first statement is the same as "All negative numbers are also numbers less than 10.")

(d) Draw a Venn diagram that shows why your conclusion to part (b) is true.

10. Fill in the conclusion in each part.

(a) *Hypotheses:* If today is Tuesday, then tomorrow is Wednesday.
Tomorrow is not Wednesday.

Conclusion: _____

(b) *Hypotheses:* If today is Tuesday, then tomorrow is Wednesday.
Today is not Tuesday.

Conclusion: _____

11. Your teacher says, "Anyone who gets an A on the final will get at least a B in the course." What can you conclude about

(a) a student who gets an A on the final?
(b) a student who gets a B on the final?

12. If a figure is a square, then the figure is a rectangle. What can you can conclude about

(a) a figure that is not a rectangle?
(b) a figure that is not a square?

In Exercises 13–16, decide whether or not the conclusion can be deduced from the first two hypotheses. If so, draw a Venn diagram that supports your answer. If not, draw a Venn diagram that satisfies the hypotheses but does not support the conclusion.

13. *Hypotheses:* Two is a whole number.
All whole numbers can be written as fractions.

Conclusion: Two can be written as a fraction.

14. *Hypotheses:* All elephants have legs.
All chairs have legs.

Conclusion: All elephants are chairs.

15. *Hypotheses:* Models drink diet cola.
I drink diet cola.

Conclusion: I am a model.

16. *Hypotheses:* All cockroaches are young.
All young things are beautiful.

Conclusion: All cockroaches are beautiful.

In Exercises 17 and 18, decide whether or not the third statement can be deduced from the two hypotheses.

17. *Hypotheses:* Martina is taller than Gabriela.
Gabriela is taller than Steffi.

Conclusion: Martina is taller than Steffi.

18. *Hypotheses:* Martina beat Gabriela at tennis.
Gabriela beat Steffi at tennis.

Conclusion: Martina will beat Steffi at tennis.

19. Consider the following Pica-Centro Game.

Guesses	Responses	
Digits	Pica: Correct Digit, Wrong Position	Centro: Correct Digit, Correct Position
5 3 2	2	0
4 2 3	1	0

Which digit is definitely part of the secret number? Justify your answer.

20. Consider the following Pica-Centro Game.

Guesses	Responses	
Digits	Pica: Correct Digits, Wrong Position	Centro: Correct Digit, Correct Position
6 9 2	2	0
7 5 2	0	1
5 4 3	0	0

What is the secret number? Justify your answer.

21. Consider the following Pica-Centro game.

Guesses	Responses	
Digits	Pica: Correct Digit, Wrong Position	Centro: Correct Digit, Correct Position
5 2 3	2	0
0 1 2	0	0
6 7 8	0	1
9 7 6	0	0

(a) Which digits can be eliminated?
(b) Which digit is in the correct position in the third guess? Explain why.
(c) Which two digits are correct in the first guess?
(d) What are the correct positions of the two correct digits in the first guess? Explain why.
(e) What is the secret number?

22. Consider the following Pica-Centro game.

Guesses	Responses	
Digits	**Pica:** **Correct Digit,** **Wrong Position**	**Centro:** **Correct Digit,** **Correct Position**
3 0 7	0	0
2 6 4	1	1
6 8 4	0	1
1 5 3	1	0

Find the secret number.

In Exercises 23–26, use deductive reasoning to fill in a conclusion that follows from all the given statements.

23. If you have your teeth cleaned twice a year, you will have less tooth decay. If you have less tooth decay, you will lose fewer teeth.
Conclusion: _____

24. If my students don't like me, they will complain to the dean. If the dean hears complaints about me, then I won't get a raise in salary. If I give low grades, then my students won't like me.
Conclusion: _____

25. *Hypotheses:* Some adults watch television.
    People who watch television don't have time to read.
    Conclusion: _____

26. *Hypotheses:* Some people throw litter on the street.
    People who throw litter do not care about their surroundings.
    Conclusion: _____

27. Write a valid deduction from the following two assumptions.

    Some students like mathematics.
    All people who like mathematics are intelligent.

28. Write a valid deduction from the following two assumptions.

    No students like pop quizzes.
    Miguel likes pop quizzes.

29. Complete the following to create an example of valid deductive reasoning.
    Assumptions: All elephants are good dancers. _____

    Conclusion: _____

30. "Come to Bloated Burgers for a Glad Meal.™ You'll be done in a few moments and have plenty of money left over." Write three if-then statements that are suggested by this ad.

31. Add a hypothesis so that the given conclusion follows deductively from the two hypotheses.
    (a) *Hypotheses:* Ada is a mathematics teacher.
    _____
    Conclusion: Ada loves mathematics.
    (b) *Hypotheses:* You are having fun.
    _____
    Conclusion: Time flies.

32. Add a hypothesis so that the given conclusion follows deductively from the two hypotheses.
    (a) *Hypotheses:* $w = 2t - 8$
    _____
    Conclusion: $w = 12$
    (b) *Hypotheses:* Today is cold.
    _____
    Conclusion: You are wearing a wool sweater.

33. What conclusions can be drawn about Sandy on the basis of the following diagram?

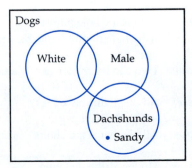

34. Joe and Sue are the same age. Joe is younger than Paul. Paul is older than Jane. Is Joe older than Jane, younger than Jane, or is it impossible to tell from the given information?

Extension Exercises

35. Everyone in the Smith family has dressed up for Halloween, but they didn't change their shoes or socks. Match each family member out of costume to the same family member in costume. Write a deductive explanation of your answer.

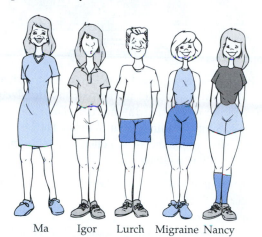

Ma   Igor   Lurch   Migraine   Nancy

36. (a) Draw a Venn diagram that represents the following statements.

All dolphins are swimmers.
All swimmers wear bathing suits.
No tigers wear bathing suits.

(b) Which of the following statements are confirmed by your diagram?
1. All dolphins wear bathing suits.
2. If you are not a tiger, then you wear a bathing suit.
3. All dolphins are tigers.
4. If you are not a swimmer, then you are not a dolphin.

37. Lewis Carroll (real name: Charles Dodgson), author of *Alice's Adventures in Wonderland,* was also a mathematics professor who devised logic puzzles. The following are from his book *Symbolic Logic.* In each part, draw a valid conclusion that follows from all the given statements.
(a) Everyone who is sane can do logic.
   No lunatics are fit to sit on a jury.
   None of your sons can do logic.
(b) No ducks waltz.
   No officers decline to waltz.
   All my poultry are ducks.
(c) No birds except ostriches are 9 feet high.
   There are no birds in this aviary that belong to anyone but me.
   No ostrich lives on mince pies.
   I have no birds less than 9 feet high.

38. The Three Stooges are a real pain. When you ask one of them a question, they all answer. Furthermore, one of them always lies, and the other two answer truthfully. I asked them who is the smartest. They replied:

CURLY: I am not the smartest.
MOE: Larry is the smartest.
LARRY: Curly is the smartest.

(a) Who is really the smartest, and who is the liar?
(b) Explain how you reached your conclusion.

39. Raymond Smullyan, a contemporary mathematician, devised the following puzzle:

The Politician Puzzle

One hundred politicians attended a convention. Each politician was either crooked or honest. Also:

1. At least one of the politicians was honest.
2. Given any two politicians, at least one of the two was crooked.

(a) How many politicians were honest, and how many were crooked?
(b) Explain how you reached your conclusion.

40. Mr. Barber is a barber who shaves every man in town who does not shave himself. Who shaves Mr. Barber?

41. Three boxes contain yellow and white tennis balls. One box has all yellow tennis balls; one has all white tennis balls; and one has some yellow and some white balls.

Unfortunately, all three boxes are labeled incorrectly! How can you determine the correct labeling after selecting just one ball?

| Y | W | W and Y |

42. You want to determine whether the following statement is true for the cards shown: "If a 1 is printed on one side, then a 2 is printed on the other side." Which of these cards *must* you turn over?

| 1 | 2 | 3 |

Puzzle Time

43. (a) Fill the numbers in the puzzle. Each number should be used once.

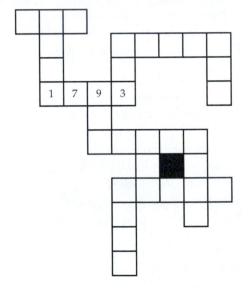

3 Digits	4 Digits	5 Digits
632	1562	12683
683	1793	14623
904	3081	98729
913	8421	
981		

(b) Give an example showing how you used deductive reasoning to fill in the numbers in this puzzle.

44. The following puzzle is adapted from the magazine *Murder Ink.*

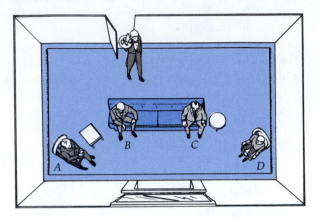

Boggs has been found dead. Four men, seated on the sofa and two chairs in front of the fireplace (as shown), are discussing the murder. Their names are Howell, Scott, Jennings, and Wilton. They are—not necessarily respectively—a general, a schoolmaster, an admiral, and a doctor.

1. The waiter brings a whiskey for Jennings and a beer for Scott.
2. In the mirror over the fireplace, the general sees the door close as the waiter departs. The general then speaks to Wilton, next to him.
3. Neither Howell nor Scott has any sisters.
4. The schoolmaster does not drink alcohol.
5. Howell is the admiral's brother-in-law. Howell is sitting in a chair, and the schoolmaster is next to him on his left.
6. The murderer suddenly moves to drop something in Jennings' whiskey. No one has moved from his seat.

(a) Tell the position of each man, give his profession, and name the murderer.
(b) Give an example showing how you used deductive reasoning to solve this puzzle.

1.3   INDUCTIVE AND DEDUCTIVE REASONING

NCTM Standards

- make and investigate mathematical conjectures (pre-K–12)

- develop and evaluate mathematical arguments and proofs (pre-K–12)

- use the language of mathematics to express mathematical ideas precisely (pre-K–12)

Mathematicians use inductive and deductive reasoning to create new mathematics. First they propose generalizations (using induction), and then they try to prove the logical validity of their generalizations (using deduction).

In this lesson, you'll first learn to tell the difference between induction and deduction. Then you'll see how mathematicians use these two processes to create new mathematics.

INDUCTIVE VERSUS DEDUCTIVE REASONING

LE 1   Opener

Are the men in the cartoon in Figure 1–8 discussing deductive or inductive reasoning?

Figure 1–8

How can one avoid confusing inductive reasoning and deductive reasoning? Remember, inductive reasoning leads to a generalization from specific examples; deductive reasoning draws a necessary conclusion from given assumptions. Induction involves an "inductive leap" from specific examples to a general idea. Deduction leads to a conclusion that *follows automatically* from the given assumptions.

Can you tell the difference? Find out in the following exercises.

In LE 2 through LE 4, identify whether induction or deduction is being used.

LE 2   Concept

The fingerprints on the wall match yours. I know that everyone has a unique set of fingerprints. I conclude that you touched the wall.

LE 3   Concept

I meet two people from France who wear berets. I conclude that all people from France wear berets.

| **LE 4   Concept**
Given: $m = n + 6$ and $n = -8$. I conclude that $m = -2$.

INDUCTIVE AND DEDUCTIVE REASONING

Many mathematical ideas are developed using a two-step process. First inductive reasoning is used to make a generalization about an observed pattern in a variety of examples. Then deductive reasoning demonstrates that the general statement *must* be true if certain assumptions are made.

Remember the mystifying number trick from LE 1 in Section 1.1? You observed that no matter what number you started with, you ended up with 5. Unfortunately, no matter how many examples you tried, you could not be sure you would *always* end up with 5.

Inductive reasoning carries the risk of an incorrect generalization, so the conclusion it produces lacks the certitude required in mathematics. A mathematician would use deductive reasoning to prove that the number trick always produces a 5.

• EXAMPLE 1

Prove that the following number trick always results in 5.

> Pick a number.
> Multiply by 2.
> Add 10.
> Divide by 2.
> Subtract your original number.

SOLUTION

Mathematicians use algebra and deductive reasoning to prove general results about numbers. In the case of the number puzzle, instead of starting with a number such as 3 or -2, start with a variable N that could represent any number. Then go through all the steps in the puzzle, and deduce what happens to the variable along the way.

1. Pick a number. N
2. Multiply by 2.   _____ (Deduction?)

What goes in the blank? If you take N and multiply by 2, you obtain $2N$.

Next, how do we deduce the result of step 3 from step 2?

2. Multiply by 2. $2N$
3. Add 10.   _____ (Deduction?)

Take $2N$ and add 10, and you obtain $2N + 10$. Continuing this series of steps and deductions yields the following proof.

1. Pick a number. N
2. Multiply by 2. $2N$
3. Add 10. $2N + 10$
4. Divide by 2. $N + 5$
5. Subtract your original number. 5

Deductive reasoning shows that no matter what N you start with, you will end up with 5.   ●

LE 5   Reasoning

Consider the following number trick.

> Pick any number at all.
> Subtract 3.
> Multiply by 2.
> Subtract your original number.
> Add 6.

(a) Try two or three different numbers, and use inductive reasoning to make a conjecture about what will happen with any number.

(b) Use deductive reasoning to prove your generalization from part (a).

Mathematics is our unique attempt to construct a system of ideas based on precise deductive reasoning. A mathematical system begins with a few simple assumptions and then, through step-by-step logical arguments, arrives at new conclusions. The conclusions of deductive reasoning are irrefutable if the assumptions are correct.

Ancient Greek philosophers required that all their mathematical theorems be based on deductive reasoning, and it has remained that way for over 2000 years! Mathematicians will not accept any statement as a theorem of mathematics unless it has been proved deductively.

LE 6   Summary

How can you tell the difference between inductive and deductive reasoning?

THE CONVERSE OF A STATEMENT

When you interchange the hypothesis and conclusion of an if-then statement, you obtain a new statement called the converse. The converse of "if A then B" is "if B then A."

Discussion

LE 7   Opener

Suppose a statement is true. Does its converse also have to be true?
Statement: "If a number is divisible by 10, then the number is even."

(a) is the statement true?
(b) Write the converse of the statement.
(c) Try some examples, and determine if the converse is probably true or definitely false.

LE 8   Concept

If $x + 3 = 5$, then $x = 2$.

(a) Is this statement true?
(b) Write its converse.
(c) Is the converse true or false?

LE 7 and LE 8 suggest the following.

Statements and Their Converses

The converse of a true "if-then" statement may be either true or false.

The extension (homework) exercises address two other statements that are related to an "if-then" statement: the contrapositive and the inverse.

"IF AND ONLY IF"

It would be simpler if all conditional statements were in an "If A, then B" format, but they are not. Two other common forms are "A if B," as in

$$X^2 = 4 \quad \text{if} \quad X = -2$$

and "A only if B," as in

You have mononucleosis *only if* you have a fever and a sore throat.

How are the "A if B" and "A only if B" formats related to the "If A, then B" format? Each of the two preceding statements can be rewritten using the "if-then" format. The first statement

$$X^2 = 4 \quad \text{if} \quad X = -2$$

is the same as

$$\text{If} \quad X = -2 \quad \text{then} \quad X^2 = 4$$

Both of these statements mean that whenever $X = -2$, it is also true that $X^2 = 4$. See if you can rewrite the "only-if" statement in the "If A, then B" format in the following lesson exercise.

LE 9   Reasoning

Which of the following are equivalent to the statement "You have mononucleosis only if you have a fever and a sore throat"?

(a) People who have mononucleosis must have a fever and a sore throat.
(b) If you have a fever and a sore throat, then you must have mononucleosis.
(c) If you have mononucleosis, then you have a fever and a sore throat.

The preceding exercise illustrates that "A only if B" means that whenever A is true, B is also true. The preceding example and exercise are special cases of the following.

Definition: **"If" Statements and "Only-If" Statements**

"A if B" means "If B, then A."
"A only if B" means "If A, then B."

Use these definitions to translate the following statements.

LE 10   Skill

Write each of the following statements in an "if-then" format.

(a) You may teach only if you have a college degree.
(b) N is an even number if N is divisible by 2.

Definitions and theorems sometimes contain the more powerful phrase "if and only if," which combines the "if" and "only-if" formats. Consider a compound statement that contains both "if" and "only if!"

$$x = 2 \quad \text{if and only if} \quad x + 3 = 5$$

An "if-and-only-if" statement is actually two statements in one. The statement "$x = 2$ if and only if $x + 3 = 5$" means

1. $x = 2$ if $x + 3 = 5$ and
2. $x = 2$ only if $x + 3 = 5$

These two statements can be rewritten in the "if-then" format as follows.

1. If $x + 3 = 5$, then $x = 2$ and
2. If $x = 2$, then $x + 3 = 5$

In symbols, $x = 2 \leftrightarrow x + 3 = 5$ means "If $x = 2$, then $x + 3 = 5$" *and* "If $x + 3 = 5$, then $x = 2$."

So an "if-and-only-if" statement describes a very strong connection between two conditions. It gives you two statements in one: a statement *and* its converse. It means that the two conditions (separated by "if and only if") are equivalent. For example, to say "$x + 3 = 5$" is equivalent to saying "$x = 2$." Whenever one is true, the other must be true; whenever one is false, the other must be false.

Definition: **"If-and-Only-If" Statements**

"A if and only if B" means

1. If A, then B, *and*
2. If B, then A.

In other words, "A if and only if B" means that A implies B and B implies A. This relationship is often written symbolically as $A \leftrightarrow B$. "If and only if" is a standard phrase used in many mathematical definitions and theorems. Look for it later in this book. Remember: "If and only if" indicates a theorem or definition that satisfies *two* "if-then" statements.

LE 11   Skill

Write the following statement as two "if-then" statements.

A whole number is divisible by 10 if and only if its ones digit is a 0.

LE 12   Summary

Tell what you have learned about "if-then" and "if-and-only-if" statements.

ANSWERS TO SELECTED LESSON EXERCISES

1. Inductive reasoning
(generalization based on over 800,000 examples)

2. Deduction

3. Induction

4. Deduction

5. (a) I always end up with my original number.
(b) Assume that N = my original number. Then the following deductions must be true.

1. Subtract 3, and I get $N - 3$.
2. Multiply by 2, and I get $2N - 6$.
3. Subtract my original number N, and I get $N - 6$.
4. Add 6, and I end up with N.

No matter what number N that I start out with, I end up with the same N. In short form, $N \rightarrow N - 3 \rightarrow 2N - 6 \rightarrow N - 6 \rightarrow N$.

7. (a) Yes
   (b) If a number is even, then it is divisible by 10.
   (c) No. The number 4 would be a counterexample.

8. (a) Yes
   (b) If $x = 2$, then $x + 3 = 5$.
   (c) True

9. (a) and (c)

10. (a) If you teach, then you have a college degree.
   (b) If N is divisible by 2, then N is even.

11. If a whole number is divisible by 10, then its ones digit is a 0. If its ones digit is a 0, then a whole number is divisible by 10.

1.3 HOMEWORK EXERCISES

Basic Exercises

1. What kind of reasoning involves a leap from given examples to a general conclusion?

2. Which type of reasoning leads to a true conclusion whenever the given statements are true?

In Exercises 3–7, identify whether induction or deduction is being used.

3. My teacher gave a quiz five Thursdays in a row. I conclude that she will give a quiz every Thursday.

4. Given that $AB = CD$ and $AB = 6$ cm, I conclude that $CD = 6$ cm.

5. On the last two Friday the 13ths, I had bad luck. I conclude that Friday the 13th is an unlucky day for me.

6. Since $4 \times 6 = 24$ and $2 \times 8 = 16$, I conclude that the product of two even numbers is an even number.

7. My mother and grandmother were homemakers. I conclude that all women are homemakers.

8. Our courts accept fingerprints as evidence because, after millions of comparisons, no two identical sets of fingerprints have been found. This is an example of _____ reasoning.

9. Consider the following number trick.

Pick a number.
Multiply by 3.
Subtract your original number.
Add 8.
Divide by 2.

(a) Try two or three different numbers, and use inductive reasoning to make a conjecture about what will happen with any number.
(b) Why don't your results to part (a) *prove* that the number trick adds 4 to any number?
(c) Use deductive reasoning to prove your generalization from part (a).
(d) *Explain* how you used deductive reasoning to derive the result of the fourth step using the result of the third step.

10. Consider the following number trick.

Pick a number.
Subtract 8.
Add your original number.
Divide by 2.
Add 5.

(a) Try two or three different numbers, and use inductive reasoning to make a conjecture about what will happen with any number.
(b) Use deductive reasoning to prove your generalization from part (a).
(c) *Explain* how you used deductive reasoning to derive the result of the third step using the result of the second step.

11. The number trick in Exercise 9 can also be verified with drawings of blocks (such as algebra tiles or algebra lab gear). Represent the number selected with an unknown block ▭, and represent ones with squares ▪. Begin with ▭, and show what blocks you would have after each step in the number puzzle.

12. Show how to verify Example 1 with drawings of blocks. (Refer to Exercise 11.)

13. How do mathematicians use inductive and deductive reasoning in sequence to develop a new idea?

14. How do you write the converse of an "if-then" statement?

15. Consider the following statement: If you have a fever, then you are sick.
(a) Is the statement true?
(b) Write the converse of the statement.
(c) Is the converse true?

16. Consider the following statement: If a figure is a square, then the figure has exactly four sides.
(a) Is the statement true?
(b) Write the converse of the statement.
(c) Is the converse true?

17. Write the following statements in an "if-then" format.
(a) You may be successful if you are smart.
(b) A triangle is equilateral only if it is isosceles.

18. Write the following statements in an "if-then" format.
(a) You can graduate only if you have taken a writing course.
(b) The number 5 is odd if 4 is even.

19. Write the following statement as two "if-then" statements: A triangle has two equal sides if and only if it has two equal angles.

20. Write the following statement as two "if-then" statements: A triangle is a right triangle if and only if it has a right angle.

21. Write the following two statements as one "if-and-only-if" statement.

If $2x = 4$, then $x = 2$.
If $x = 2$, then $2x = 4$.

22. Write the following two statements as one "if-and-only-if" statement.

If you are my parent, then I am your child.
If I am your child, then you are my parent.

Extension Exercises

23. Make up a number puzzle in which a person will always end up with the number 2. (Make it complicated enough so that someone else could not easily see how it works.)

24. Make up a number puzzle in which a person will always end up with a number that is 6 more than the person's original number.

25. Assume the following five statements are true.

Canaries like to sing. Blue jays like to sing. Tweety is a canary. Cardinals like to sing. Arnold is a pig.

(a) Use some of the preceding statements to create an example of valid inductive reasoning.
(b) Use some of the preceding statements to create an example of valid deductive reasoning.

26. (a) Try some examples, and answer the following question. What is the sum of two even numbers?
(b) Fill in the blanks in the following deductive proof.
  1. Assume your two even numbers are $2M$ and $2N$, where M and N are whole numbers.
  2. The sum of the two even numbers is

    _____.
  3. $2M + 2N = 2($_____$)$.
  4. $2(M + N)$ is an even number because

    _____.
  5. So the sum of two even numbers is an even number.

27. In this lesson, you studied the converse of a statement. Another related statement of this type is the inverse. To make the **inverse** of a statement, negate the hypothesis and the conclusion of the statement.

Statement: If it is raining, then I shall wear a raincoat.
Inverse: If it is not raining, then I shall not wear a raincoat.

(a) Write the inverse of the following statement:
   If $x + 3 = 5$, then $x = 2$.
(b) Write the inverse of the following statement:
   If I confess, then I am guilty.
(c) If possible, write a true "if-then" statement that has a false inverse.

28. Write the inverse of the following statements.
(a) If I drive a truck, then I need a driver's license.
(b) If a woman uses nose powder, then her nose looks beautiful.

29. The **contrapositive** of a statement combines the converse and the inverse. To make the contrapositive, interchange the hypothesis and conclusion *and* negate them both.

Statement: If a number is divisible by 100, then the number is even.
Contrapositive: If a number is not even, then the number is not divisible by 100.

(a) Write the contrapositive of the following statement: If $x + 3 = 5$, then $x = 2$.
(b) Write the contrapositive of the following statement: If I confess, then I am guilty.
(c) If possible, write a true "if-then" statement that has a false contrapositive.

30. Consider the following statement: If you are a U.S. Marine, then you are a real man.
(a) Write the converse.
(b) Write the contrapositive.
(c) Write the inverse.

31. Suppose the following statement is true: If it was raining, then I drove to work.
(a) Write the converse.
(b) Write the contrapositive.
(c) Write the inverse.
(d) Which of the three statements in parts (a), (b), and (c) is (are) true?

32. Suppose the following statement is true: If it rains today, then I will scream. Which of the following must also be true?
(a) If I do not scream today, then it is not raining.
(b) If I scream today, then it is raining.
(c) If it does not rain today, then I will not scream.

33. Suppose the following statement is true: If a four-sided figure is a square, then it is a rectangle. Which of the following must also be true?
(a) If a four-sided figure is a rectangle, then it is a square.
(b) If a four-sided figure is not a square, then it is not a rectangle.
(c) If a four-sided figure is not a rectangle, then it is not a square.

34. (a) On the basis of the preceding exercises, whenever an "if-then" statement is false, its *contrapositive*
(1) is true   (2) is false
(3) could be true or false
(b) On the basis of the preceding exercises, whenever an "if-then" statement is true, its contrapositive
(1) is true   (2) is false
(3) could be true or false

35. Suppose the following advertising claim is true: "If you rent a car from Mary, then you will be able to travel around with ease." Is it then true that "if you do not rent a car from Mary, you will not be able to travel around with ease"?

36. Your teacher tells you, "If you study hard, then you will do well on the test." Which of the following would contradict this statement?
(a) You do not study hard and do well on the test.
(b) You do not study hard, and you do not do well on the test.
(c) You study hard and do not do well on the test.

37. Which of the following are the same as "If A, then B," and which are the same as "If B, then A"?
(a) A implies B.   (b) A follows from B.
(c) A is a sufficient condition for B.
(d) A is a necessary condition for B.
(e) If A is false, then B is false.

Puzzle Time

38. Fill in the correct numbers in the square, using the following clues.

A	B	C	D
E	F	G	H
I	J	K	L
M	N	O	P **8**

1. Each number from 1 to 16 is used once.
2. Each row adds up to 34, and each column adds up to 34.
3. K is twice as large as G and three times E.
4. J and N add up to F.
5. H is 5 times C.
6. B and D are two-digit numbers.

1.4 PATTERNS

NCTM Standards

- describe, extend, and make generalizations about geometric and numeric patterns (3–5)

- express mathematical relationships using equations (3–5)

- recognize and use connections among mathematical ideas (pre-K–12)

Look out the window. At first glance, the world appears complex and confusing, but in reality it is full of patterns. Patterns give order to the world. In mathematics, we find patterns in shapes and quantities.

Human beings are better than other animals at finding mathematical patterns. The ability to find patterns is supposed to be a sign of intelligence.

Experimenters at Tulane University wanted to teach a rat the concept of "two." The rat had to choose among three doors. One door had one mark, one had two marks, and one had three marks. In each trial, the food was behind the door with two marks. How long would it take you to discover such a pattern? It took the rat 1,500 trials.

People are better at finding mathematical patterns than rats are. (Sometimes we find patterns even where they don't exist, as the cartoon in Figure 1–9 shows.)

Figure 1–9

 Discussion

LE 1 Opener

Describe some patterns you see in your classroom or in your everyday life.

The ability to see patterns is important in mathematical problem solving. Working with patterns also develops number sense. In this lesson, you will study two common types of mathematical pattern problems: sequences and patterns in sums and differences.

Discovering and extending patterns requires inductive and deductive reasoning. First, conjecturing a general pattern based on examples is inductive reasoning. Then, proving the generalization requires deductive reasoning. Writing new examples of a pattern based on a general rule is another use of deductive reasoning.

G. H. Hardy (1877–1947), a British mathematician, said, "A mathematician, like a painter or poet, is a maker of patterns. If his patterns are more permanent than theirs, it is because they are made with ideas. The mathematician's patterns, like the painter's or the poet's, must be beautiful . . . [and] fit together in a harmonious way."

ARITHMETIC SEQUENCES

The first set of numbers children study is the **counting** (or **natural**) **numbers** $\{1, 2, 3, \ldots\}$. A 3-year-old child who is 94 cm tall may grow about 6 cm per year until age 9, setting up the following pattern: 94 cm, 100 cm, 106 cm, 112 cm, 118 cm, 124 cm, 130 cm. These are two examples of number sequences. A **sequence** is an ordered arrangement of numbers, figures, or letters.

LE 2   Concept

What is similar about the pattern in the two sequences just discussed: 1, 2, 3, . . . and 94, 100, 106, . . . , 130?

The two sequences in LE 2 are arithmetic sequences. In an **arithmetic sequence,** the difference between each pair of successive terms is the same. In the counting numbers, the difference is 1, and in the growth sequence, the difference is 6 cm. The second-grade *Harcourt Math* textbook (Figure 1–10) shows three more examples of arithmetic sequences.

Classroom
Connection

LE 3   Concept

Tell whether the sequences in parts (a) and (b) are arithmetic. If a sequence is arithmetic, give the value of the difference.

(a) the even numbers: 2, 4, 6, 8, . . .   **(b)** cubes: 1, 8, 27, 64, . . .
(c) In part (a), what is a simple way to describe the pattern from one term to the next?
(d) If you have a graphing calculator, you can generate the sequence in part (a) as follows: Enter 2. Then enter Ans + 2. Continue to press enter. Write the sequence of numbers you obtain.

A child might describe the sequence in LE 3(a) by saying, "You add 2." This means you begin with 2, and then add 2 to each term to find the next one. Symbolically, you could write this sequence as INITIAL = 2 and CURRENT = PREVIOUS + 2. This is an example of a **recursive sequence,** in which (1) a starting value or values is given and (2) each successive term is found by applying a rule (such as "add 2") to the previous term or terms.

LE 4   Skill

A candle that is 12 cm tall decreases in height by 2 cm each hour.

(a) Write a sequence based on the height each hour.
(b) Is your sequence arithmetic?
(c) How many terms are in the sequence?
(d) Use the variables INITIAL, CURRENT, and PREVIOUS to write two equations that define the sequence.
(e) If you have a graphing calculator, use it to generate the sequence.

Classroom Connection

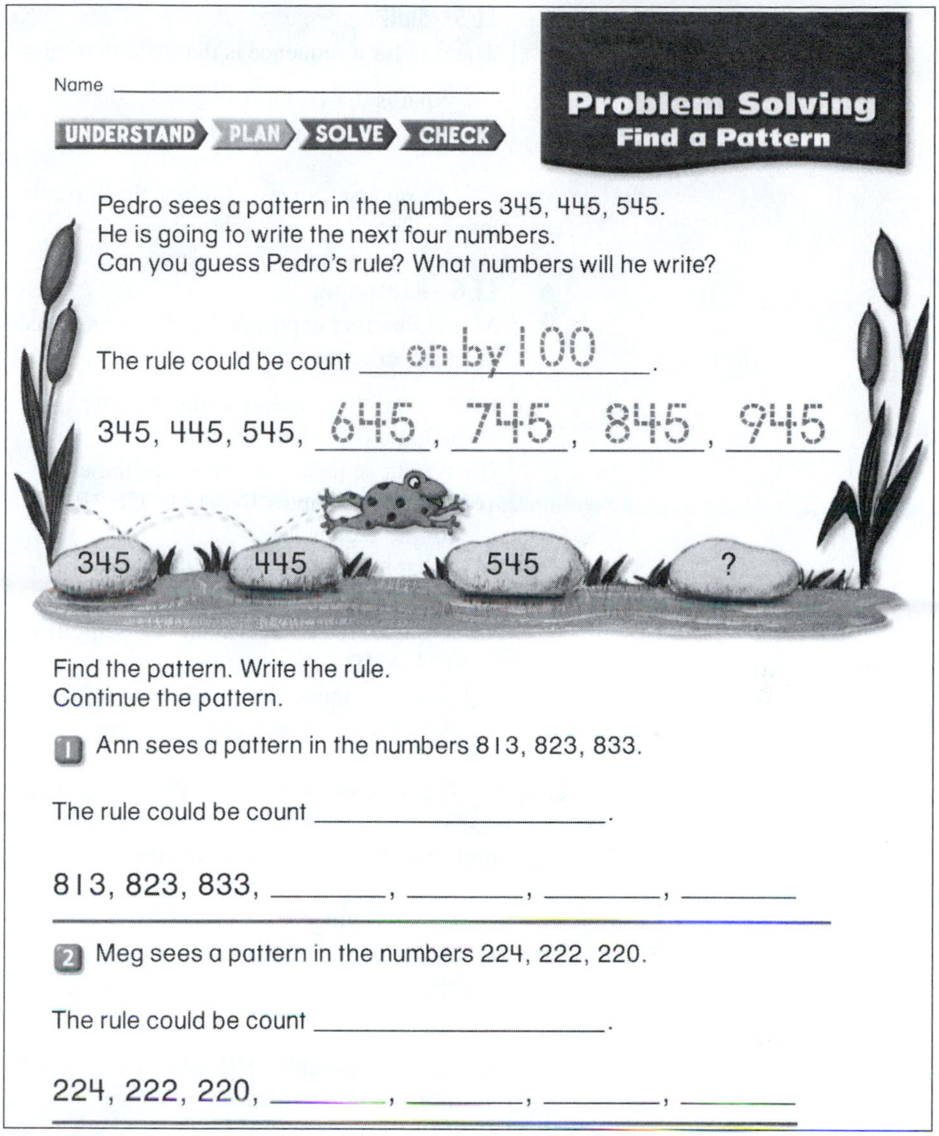

Name _____

UNDERSTAND 〉 PLAN 〉 SOLVE 〉 CHECK

Problem Solving
Find a Pattern

Pedro sees a pattern in the numbers 345, 445, 545.
He is going to write the next four numbers.
Can you guess Pedro's rule? What numbers will he write?

The rule could be count ____on by 100____.

345, 445, 545, __645__, __745__, __845__, __945__

345   445   545   ?

Find the pattern. Write the rule.
Continue the pattern.

1 Ann sees a pattern in the numbers 813, 823, 833.

The rule could be count _____.

813, 823, 833, _____, _____, _____, _____

2 Meg sees a pattern in the numbers 224, 222, 220.

The rule could be count _____.

224, 222, 220, _____, _____, _____, _____

From Harcourt Math, Grade 2 (Orlando: Harcourt School Publishers, 2002), p. 239.

Figure 1–10

In LE 4, the sequence 12, 10, 8, . . . , 0 has a difference of -2, since it decreases by 2 from one term to the next. Its general rule is nth term $= 12 - 2(n - 1)$ or $14 - 2n$. This rule can be used to find other terms in the sequence.

• **EXAMPLE 1**

The sequence 12, 10, 8, 6, . . . has the rule nth term $= 12 - 2(n - 1)$ or $14 - 2n$. Use the rule to deduce the 5th term.

SOLUTION

In the rule, $14 - 2n$, the n represents the position of the term. For example, the 1st term is $14 - 2 \cdot 1 = 12$. To find the 5th term, substitute 5 for n in the rule. The 5th term $= 14 - 2 \cdot 5 = 4$.

LE 5   Skill

The rule for a sequence is that the nth term $= 2n + 1$.

(a) What is the 1st term?
(b) What are the next three terms?

The reverse process, finding the formula for the nth term of a given sequence, is more challenging.

LE 6   Reasoning

Mario's mother deposited \$125 in a bank account to start a college fund for him. Mario has a paper route, and he plans to deposit \$10 into the bank account each week.

(a) Start with 125, and write the next three terms of a sequence that show the total amount deposited after each week.
(b) Is your sequence in part (a) arithmetic?
(c) Use the variables INITIAL, CURRENT, and PREVIOUS to write two equations for the sequence.
(d) What is the 10th term in the sequence?
(e) What is the 100th term in the sequence?
(f) What is the nth term in the sequence? (*Hint:* Think of the sequence as 125, $125 + 10$, $125 + 10 + 10$, ...)
(g) Compare the recursive formula in part (c) and the nth-term formula in part (f) and describe an advantage of each one.

The recursive formula is often easier to determine and easier to understand than the nth-term formula, but the nth-term formula enables you to find any term from its term number without having to know the terms that come just before it.

Discussion

LE 7   Reasoning

Consider a more general example. Suppose an arithmetic sequence starts with the number a and the difference between consecutive terms (called the **common differ- ence**) is d.

(a) Use the variables INITIAL, CURRENT, and PREVIOUS to write the two equa- tions for the sequence.
(b) What is the 10th term?
(c) What is the 100th term?
(d) What is the nth term?

The results of the preceding exercise suggest the general formula for the nth term of an arithmetic sequence. An arithmetic sequence with a first term a and common differ- ence d has the formula **nth term** $= a + (n - 1)d$.

LE 8   Skill

Consider the sequence 50, 47, 44, 41, ...

(a) What is the rule for the nth term?
(b) What is the 50th term?
(c) Does part (a) require inductive or deductive reasoning?
(d) Does part (b) require inductive or deductive reasoning?

Classroom Connection

LE 9 Concept

A seventh grader says that 5, 15, 35, 65, . . . is arithmetic because it increases by 10, then 20, then 30, and so on. Is the child right? If not, what would you tell her?

GEOMETRIC SEQUENCES

Another important type of sequence is a geometric sequence. Look at the sequences in the following exercise, and see if you can tell what makes them geometric.

LE 10 Concept

The powers of 3 are 3, 9, 27, 81, ... The mass of a 64-gram radioactive substance will decrease every 5 years (its half-life): 64 g, 32 g, 16 g, 8 g, ...
Both these sequences are geometric. What pattern do they have in common?

As you may have observed in the preceding exercise, you multiply each term by the same number to obtain the next term. This number is the ratio of the current term to the previous one. A **geometric sequence** has a **common ratio** (or multiplier) called r between each pair of consecutive terms. In the sequence 3, 9, 27, 81, . . . , the value of $r = 3$. In the sequence 64, 32, 16, 8, . . . , the value of $r = \frac{1}{2}$. The letter a is still used to refer to the first term.

LE 11 Skill

(a) Tell whether each sequence is arithmetic, geometric, or neither.
   (i) 8, 3, 6, 1, 4, −1 . . .   (ii) 12, 20, 28, . . .   (iii) 100, 20, 4, 0.8, . . .
(b) For each arithmetic sequence, give the values of a and d. For each geometric sequence, give the values of a and r.
(c) For each arithmetic or geometric sequence, use the variables INITIAL, CURRENT, and PREVIOUS to write two equations that define the sequence.

How do you find various terms in a geometric sequence? Because of the simple pattern, there is a shortcut.

LE 12 Concept

The ant population in a house is now 40, and you believe it will triple every 10 weeks.

(a) Start with 40, and write the next two terms of a sequence that shows the total ant population at 10-week intervals.
(b) Is your sequence in part (a) geometric?
(c) Use the variables INITIAL, CURRENT, and PREVIOUS to write the two equations for the sequence.
(d) If you have a graphing calculator, see if you can generate the sequence.
(e) What is the 5th term in the sequence? (Write the result using an exponent. You do not have to multiply it out.)
(f) What is the 50th term in the sequence? (Write the result using an exponent. You do not have to multiply it out.)
(g) What is the nth term in the sequence? (Write the result using an exponent.)

Discussion

LE 13   Concept

Consider a more general example. Suppose a geometric sequence starts with the number a and the common ratio is r.

(a) Write the first five terms using a and r.
(b) What is the 5th term?
(c) What is the 50th term?
(d) What is the nth term?

The results of the preceding exercise suggest the general formula for the nth term of a geometric sequence. A geometric sequence with a first term a and common ratio r has the formula **nth term** $= ar^{n-1}$. Children now study both arithmetic and geometric sequences in grades 7 and 8.

LE 14   Skill

(a) Is the sequence 2, 10, 50, 250, . . . arithmetic, geometric, or neither?
(b) What is a formula for the nth term?
(c) What is the 80th term? Do not simplify the answer.
(d) What kind of reasoning is used to guess the nth term formula for the sequence?
(e) What kind of reasoning is used to apply the general formula to find the 80th term?

The following chart summarizes the use of induction and deduction with sequences.

	Induction			*Deduction*	
Specific examples	$\longrightarrow$	General rule	General rule and given sequence	$\longrightarrow$	Apply rule to extended sequence

OTHER SEQUENCES

Not all sequences are arithmetic or geometric. One famous example is the Fibonacci sequence.

Daisy (13)

Buttercup (5)

Figure 1–11

LE 15   Reasoning

The number of petals in some varieties of flowers (see Figure 1–11), is found in the following sequence (called the **Fibonacci sequence** after a twelfth century Italian mathematician).

(a) Fill in the blanks: 1, 1, 2, 3, 5, ___, ___, ___, ___, ___, ___
(b) Tell how you obtained each new term.

Pythagoras and his followers were the first to study numbers and number sequences for their own sake. The Pythagoreans thought of quantities geometrically. They represented numbers with pebbles.

LE 16   Concept

(a) What is the common name for numbers such as the following?

1   4   9

(b) Is the sequence 1, 4, 9, 16, . . . arithmetic, geometric, or neither?

PATTERNS IN SUMS AND DIFFERENCES

Thinking of numbers geometrically reveals patterns in sums and establishes a connection between arithmetic and geometry.

LE 17   Reasoning

$$1 = (\ )^2$$
$$1 + 3 = (\ )^2$$
$$1 + 3 + 5 = (\ )^2$$

(a) Fill in the missing numbers. (Recall that $n^2 = n \cdot n$.)
(b) Draw geometric dot pictures of the three sums that show the pattern. (*Hint:* Use squares.)
(c) What would the next equation be if the pattern continued? Is this equation true?
(d) The sum of the first three odd numbers is _____ squared.
(e) The sum of the first four odd numbers is _____ squared.
(f) Write a generalization for any counting number N, based on parts (d) and (e).
(g) Part (f) involves _____ reasoning.
(h) Use your generalization to compute $1 + 3 + 5 + 7 + \cdots + 79$.

Square number patterns are part of a branch of mathematics called number theory, a subject you will study further in Chapter 4. Number theory also includes such familiar topics as factors, multiples, and prime numbers.

Other patterns in sums may be too complicated to show geometrically, but your experience with number sequences will help you find the patterns.

LE 18   Reasoning

$$2^2 - 1^2 = 3$$
$$3^2 - 2^2 = 5$$
$$4^2 - 3^2 = 7$$

(a) If the pattern continues, what is the next equation? Is the next equation true?
(b) Complete the following generalization for any counting number c.
   $c^2 -$ _____ $=$ _____.
(c) Show that your equation in part (b) is true.
(d) Part (b) involves _____ reasoning, and part (c) involves _____ reasoning.

AN INVESTIGATION: DIFFERENCES OF SQUARES

LE 19   Reasoning

Which counting numbers are the difference of two squared counting numbers? (For example, $5 = 3^2 - 2^2$.)

LE 20   Summary

Tell what you have learned about sequences in this section.

ANSWERS TO SELECTED LESSON EXERCISES

3. (a) Yes; $d = 2$ (b) No   (c) Add 2

4. (a) 12, 10, 8, 6, 4, 2, 0   (b) Yes   (c) 7
(d) INITIAL = 12, CURRENT = PREVIOUS $-$ 2

5. (a) 3   (b) 5, 7, 9

6. (a) 125, 135, 145, 155   (b) Yes
(c) INITIAL = 125, CURRENT = PREVIOUS + 10
(d) $125 + 9 \cdot 10 = 215$ (e) $125 + 99 \cdot 10 = 1115$
(f) $125 + (n - 1) \cdot 10$

7. (a) INITIAL = a, CURRENT = PREVIOUS + d
(b) $a + 9d$ (c) $a + 99d$

8. (a) $50 + (n - 1) \cdot (-3)$ or $53 - 3n$ (b) -97
(c) Inductive   (d) Deductive

9. It is not arithmetic. An arithmetic sequence changes by the same amount each time. If an arithmetic sequence started with 5, 15, what would the next number be?

11. (a)   (i) Neither   (ii) Arithmetic   (iii) Geometric
(b) (ii) $a = 12$, $d = 8$
(c) (ii) INITIAL = 12, CURRENT = PREVIOUS + 8
      (iii) INITIAL = 100, CURRENT = PREVIOUS/5

12. (a) 40, 120, 360   (b) Yes
(c) INITIAL = 40, CURRENT = PREVIOUS $\cdot$ 3
(e) $40 \cdot 3^4$ (f) $40 \cdot 3^{49}$ (g) $40 \cdot 3^{n-1}$

13. (a) a, ar, ar^2, ar^3, ar^4
(b) $a \cdot r^4$ (c) $a \cdot r^{49}$ (d) $a \cdot r^{n-1}$

14. (a) Geometric   (b) $2 \cdot 5^{n-1}$ (c) $2 \cdot 5^{79}$

15. (a) 8, 13, 21, 34, 55, 89
(b) By adding the previous two terms

16. (a) Square numbers   (b) Neither

17. (a) 1; 2; 3
(b) 1 $1 + 3$ $1 + 3 + 5$

1^2 2^2 3^2

(c) $1 + 3 + 5 + 7 = 4^2$; yes
(d) 3   (e) 4
(f) The sum of the first N odd numbers is N^2.
(g) Inductive   (h) $(40)^2 = 1600$

18. (a) $5^2 - 4^2 = 9$; yes
(b) $c^2 - (c - 1)^2 = (c + c - 1)$ (*Hint:* Look at the pattern as you read an equation from left to right.)
(c) $c^2 - (c - 1)^2 \overset{?}{=} (c + c - 1)$

$c^2 - (c^2 - 2c + 1) \overset{?}{=} 2c - 1$

$c^2 - c^2 + 2c - 1 \overset{?}{=} 2c - 1$
$2c - 1 = 2c - 1$

(d) Inductive; deductive

1.4   HOMEWORK EXERCISES

Basic Exercises

1. Tell whether each of the following sequences is arithmetic. If it is, give the value of the difference.
  (a) 5, 10, 20, 40, . . .   (b) 900, 870, 840, 810, . . .
  (c) In part (a), what is a simple way to describe the pattern from one term to the next?
  (d) If you have a graphing calculator, use it to generate the arithmetic sequence.

2. Tell whether each of the following sequences is arithmetic. If it is, give the value of the difference.
  (a) 8, 16, 24, 32, . . .   (b) 1, 8, 27, 64, . . .
  (c) In part (a), what is a simple way to describe the pattern from one term to the next?
  (d) If you have a graphing calculator, use it to generate the arithmetic sequence.

3. A plumber charges $20 for transportation and $40 for each hour of work.
  (a) Write a sequence based on the total charges for 1 hour, 2 hours, and so on.
  (b) Is the sequence arithmetic?
  (c) Use the variables INITIAL, CURRENT, and PREVIOUS to write two equations that define the sequence.
  (d) If you have a graphing calculator, use it to generate the sequence.

4. The 17-year cicada reappears every 17 years. Suppose the cicadas appeared in a town in 1963.
  (a) Write a sequence based on the years in which cicadas will appear starting with 1963.
  (b) Is your sequence arithmetic?
  (c) Use the variables INITIAL, CURRENT, and PREVIOUS to write two equations that define the sequence.
  (d) If you have a graphing calculator, use it to generate the sequence.

5. The rule for a sequence is that the nth term $= 10 - n$. Write the first five terms of the sequence.

6. The rule for a sequence is that the nth term $= n + 4$. Write the first four terms of the sequence.

7. Consider the following sequence.

$$2, 9, 16, 23, . . .$$

  (a) Is the sequence arithmetic?
  (b) Use the variables INITIAL, CURRENT, and PREVIOUS to write two equations that define the sequence.
  (c) What is the 10th term in the sequence?
  (d) What is the 100th term in the sequence?
  (e) What is the nth term in the sequence?

8. Consider the following sequence.

$$50, 49, 48, 47, . . .$$

  (a) Is the sequence arithmetic?
  (b) Use the variables INITIAL, CURRENT, and PREVIOUS to write two equations that define the sequence.
  (c) What is the 10th term in the sequence?
  (d) What is the 100th term in the sequence?
  (e) What is the nth term in the sequence?

9. Consider the sequence 580, 574, 568, 562, . . .
  (a) What is the rule for the nth term?
  (b) What is the 30th term?
  (c) Does part (a) require inductive or deductive reasoning?
  (d) Does part (b) require inductive or deductive reasoning?
  (e) Would 100 be a term in the sequence? Tell how you know.

10. Consider the sequence 11, 20, 29, 38, ...
  (a) What is the rule for the nth term?
  (b) What is the 40th term?
  (c) Does part (a) require inductive or deductive reasoning?
  (d) Does part (b) require inductive or deductive reasoning?
  (e) Would 100 be a term in the sequence? Tell how you know.

11. (a) Write the even numbers as a number sequence. Start with 2.
  (b) Write a rule for the nth term of the sequence.
  (c) Use the variables INITIAL, CURRENT, and PREVIOUS to write two equations that define the sequence.

12. (a) Write the odd numbers as a number sequence.
  (b) Write a rule for the *n*th term of the sequence.
  (c) Use the variables INITIAL, CURRENT, and PREVIOUS to write two equations that define the sequence.

13. Examine the following design.

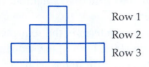

Row 1
Row 2
Row 3

  (a) If the design were to continue downward, how many squares would there be in the 50th row?
  (b) What would be the total number of small squares in the first 50 rows?

14. Examine the following designs.

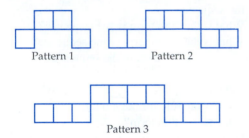

Pattern 1   Pattern 2

Pattern 3

  (a) How many squares would there be in Pattern 5?
  (b) How many squares are in Pattern *N,* where $N = 1, 2, 3, \ldots$?

15. Examine the following designs from the NCTM *Navigating Through Algebra* series.

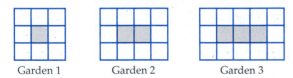

Garden 1   Garden 2   Garden 3

  (a) How many white squares are in garden 4?
  (b) How many white squares are in garden *N* where $N = 1, 2, 3, \ldots$?

16. Examine the following designs from the NCTM *Navigating Through Algebra* series.

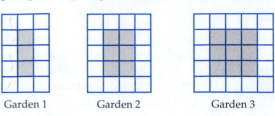

Garden 1   Garden 2   Garden 3

  (a) How many white squares are in garden 4?
  (b) How many white squares are in garden *N* where $N = 1, 2, 3, \ldots$?

17. (a) What is the formula for the *n*th term of an arithmetic sequence?
  (b) Explain why the formula makes sense.

18. (a) What is the formula for the *n*th term of a geometric sequence?
  (b) Explain why the formula makes sense.

19. (a) Tell whether each sequence is arithmetic, geometric or neither.
      (i) 50, 61, 72, 83, . . .
      (ii) 4, 40, 4,000, 4,000,000, . . .
      (iii) 600, 120, 24, 4.8, . . .
  (b) For each arithmetic sequence, give the values of *a* and *d.* For each geometric sequence, give the values of *a* and *r.*
  (c) For each arithmetic or geometric sequence, use the variables INITIAL, CURRENT, and PREVIOUS to write two equations that define the sequence.

20. (a) Tell whether each sequence is arithmetic, geometric or neither.
      (i) 7, 70, 700, 7,000, . . .   (ii) 14, 12, 10, 8, . . .
      (iii) 9, 16, 25, 36, . . .
  (b) For each arithmetic sequence, give the values of *a* and *d.* For each geometric sequence, give the values of *a* and *r.*
  (c) For each arithmetic or geometric sequence, use the variables INITIAL, CURRENT, and PREVIOUS to write two equations that define the sequence.

21. All living things contain carbon. Some scientists use radioactive carbon-14 to estimate the age of fossils. Carbon-14 has a half-life of 5,600 years, meaning that *half its mass decays every 5,600 years.*
(a) Complete the following table.

Time (years)	0	5,600	
Fraction of C-14 Left	1	$\frac{1}{2}$	$\frac{1}{4}$

(b) In 1947, prehistoric charcoal paintings in Lascaux, France, were found to have about $\frac{1}{5}$ of their original C-14. Estimate their age.

22. Strontium-90 is a common waste product of the production of nuclear weapons and nuclear energy. It has a half-life of 28 years. Some scientists estimate that it will be safe when about $\frac{1}{100}$ of it is left. About how long will this take?

23. A sequence begins 3, 15, ... What is the next term if the sequence is
(a) arithmetic?   (b) geometric?

24. A sequence begins 10, 40, ... What is the next term if the sequence is
(a) arithmetic?   (b) geometric?

25. (a) Is the sequence 5, 35, 245, . . . arithmetic, geometric, or neither?
(b) What is a formula for the nth term?
(c) What is the 40th term? Do not simplify the answer.

26. (a) Is the sequence 11, 25, 39, 53, . . . arithmetic, geometric, or neither?
(b) What is a formula for the nth term?
(c) What is the 20th term?

27. (a) Is the sequence 30, 20, 10, 0, . . . arithmetic, geometric, or neither?
(b) What is a formula for the nth term?
(c) What is the 60th term?

28. (a) Is the sequence 400, 800, 1600, 3200, . . . arithmetic, geometric, or neither?
(b) What is a formula for the nth term?
(c) What is the 34th term? Do not simplify the answer.

29. Examine these multiplication facts.

$$2 \times 9 = 18$$
$$3 \times 9 = 27$$
$$4 \times 9 = 36$$
$$5 \times 9 = 45$$
$$6 \times 9 = 54$$

(a) Name two patterns you see in these multiplication facts of 9.
(b) Do the patterns you mentioned in part (a) also work for 7×9, 8×9, and 9×9?

 30. Counting only blood relatives, state the total number of
(a) parents a person has.
(b) grandparents a person has.
(c) great-grandparents a person has.
(d) Great-grandparents are 3 generations back. How many direct ancestors would a person have in the 12th generation back?

31. (a) Determine how many numbers are in the following sequence.

$$3, 9, 15, 21, 27, \ldots, 159$$

(b) Tell how you worked out the answer.

 32. (a) Determine how many numbers in the following sequence are less than 10,000.

$$3, 9, 27, 81, \ldots$$

(b) Tell how you worked out the answer.

33. (a) Find two different reasonable answers for the next term in the sequence 1, 2, 4, _____.
(b) What rule did you use to obtain each of your answers?

34. (a) Find two different reasonable answers for the next term in the sequence 2, 6, 18, _____.
(b) What rule did you use to obtain each of your answers?

35. The Fibonacci sequence begins with two 1's. Each term after that is obtained by adding the preceding two terms.

$$1, 1, 2, 3, 5, 8, \ldots$$

(a) Write the first ten terms of the sequence.

(b) Compare the sum of the first 3 terms to the 5th term and the sum of the first 4 terms to the 6th term. What pattern do you see?

(c) Write a generalization of your results.

(d) Write another example that supports your generalization.

36. Examine the following pattern based on the Fibonacci sequence (see Exercise 35).

$$1^2 + 1^2 = 1 \times 2$$
$$1^2 + 1^2 + 2^2 = 2 \times 3$$
$$1^2 + 1^2 + 2^2 + 3^2 = 3 \times 5$$

What would the next equation be if the pattern continued? Is this equation true?

37. (a) Fill in the blank: 1, 16, 81, 256, _____

(b) Explain how you used both induction and deduction in solving part (a).

38. Fill in the blank: 4, 12, 14, 42, 44, _____

39. In 1772, an astronomer named Bode found a pattern in the distances of the six known planets from the sun (where the distance from the Earth to the sun is 10 units).

Planet	Actual Distance	Bode's Pattern		
Mercury	4	4		
Venus	7	$4 + 3$	=	7
Earth	10	$4 + (3 \times 2)$	=	10
Mars	15	$4 + (3 \times 2^2)$	=	16
??????		$4 + (3 \times 2^3)$	=	28
Jupiter	52	$4 + (3 \times 2^4)$	=	52
Saturn	96	$4 + (3 \times 2^5)$	=	100

(a) What might be an explanation for the extra equation between Mars and Jupiter?

(b) What would the equation after Saturn's be?

(c) In 1781, the next planet, Uranus, was discovered. It is 192 units from the sun. Is this close to Bode's prediction?

(d) In 1801, the asteroid Ceres was discovered 28 units from the sun. How does this relate to Bode's model?

(e) Neptune, the planet after Uranus, is 301 units from the sun. How close is this to Bode's prediction?

40. The sequence 6, 24, 54, 96, . . . comes from the surface area of cubes.

(a) What is the next number in the sequence?

(b) Tell how you figured out the next number.

41.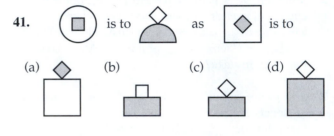

42.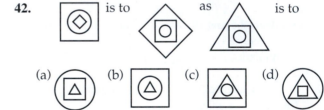

43.
$$2 = 2$$
$$2 + 4 = 6$$
$$2 + 4 + 6 = 12$$

(a) Draw geometric dot pictures of the three sums that show a pattern. (*Hint:* Use rectangles.)

(b) The sum of the first three even numbers is _____ times _____.

(c) The sum of the first four even numbers is _____ times _____.

(d) Write a generalization for any counting number N, based on parts (a) and (b).

(e) Part (d) involves _____ reasoning.

(f) Use your generalization to compute
$$2 + 4 + 6 + \cdots + 62.$$

44. The Pythagoreans called certain numbers such as 1, 3, and 6 "triangular."

1   3   6

(a) What is the next triangular number?
(b) Each triangular number can be written as the sum of consecutive numbers. Show how this works with a dot drawing.
(c) Make dot drawings that show how 4, 9, and 16 are each the sum of two triangular numbers.
(d) What generalization does part (c) suggest?

45. Examine the following pattern.
$$1 + 8 \cdot 1 = 3^2$$
$$1 + 8 \cdot 3 = 5^2$$
$$1 + 8 \cdot 6 = 7^2$$

(a) What would the next equation be if the pattern continued? Is this equation true?

(b) A general formula for these equations would be
$$1 + 8\left(\frac{n(n+1)}{2}\right) = \underline{\hspace{1cm}}.$$

(c) Show that the two sides of your equation in part (b) are equal.

46. Examine the following pattern.
$$3^2 + 4^2 \quad\quad = (\ \ )^2$$
$$3^3 + 4^3 + 5^3 = (\ \ )^3$$

(a) Fill in the missing numbers.
(b) What would the next equation be if the pattern continued? Is it a correct equation?

47. Examine the following pattern.
$$2^2 - 0^2 = 4$$
$$3^2 - 1^2 = 8$$
$$4^2 - 2^2 = 12$$

(a) What would the next equation be if the pattern continued? Is it a correct equation?
(b) Complete the following generalization. For any counting number c, $c^2 - \underline{\hspace{1cm}} = \underline{\hspace{1cm}}$.
(c) Prove that your equation in part (b) is correct by showing that both sides are equal.

48. Examine the following pattern.
$$1^2 + 2^2 + \ \ 2^2 = (\ \ )^2 \quad 9$$
$$2^2 + 3^2 + \ \ 6^2 = (\ \ )^2 \quad 49$$
$$3^2 + 4^2 + 12^2 = (\ \ )^2 \quad 185$$

(a) Fill in the missing numbers.
(b) What would the next equation be if the pattern continued? Is this equation true?
(c) Complete the following generalization of the pattern. For any counting number N,
$$N^2 + \underline{\hspace{1cm}} + \underline{\hspace{1cm}} = (\quad\quad)^2.$$
(d) Show that the two sides of your equation in part (c) are equal.

Extension Exercises

49. A finite sum has a definite answer, but infinite sums can be tricky.
(a) The sum of
$$(1 - 1) + (1 - 1) + (1 - 1) + \cdots$$
appears to be _____.
(b) The sum of
$$1 - (1 - 1) - (1 - 1) - (1 - 1)\cdots$$
appears to be _____.
(c) Write
$$(1 - 1) + (1 - 1) + (1 - 1) + \cdots$$
without parentheses.
(d) Write $1 - (1 - 1) - (1 - 1) - \cdots$ without parentheses.
(e) Is $(1 - 1) + (1 - 1) + (1 - 1) + \cdots$ the same as
$$1 - (1 - 1) - (1 - 1) - \cdots?$$

50. A set of ten cubical blocks has edges of 1 cm, 2 cm, 3 cm, . . . , 10 cm. Is it possible to build two 5-cube "towers" of the same height by stacking cubes one on top of the other?

51. (a) Two different-colored stripes are painted (equally spaced) on a stick, as shown below. How many different distinguishable sticks can be made by changing the positions of the colors?

(b) Three different-colored stripes are painted (equally spaced) on a stick. How many different distinguishable sticks can be made by changing the positions of the colors?

(c) Four different-colored stripes are painted (equally spaced) on a stick. How many different distinguishable sticks can be made by changing the positions of the colors?

52. (a) Three different-colored stripes are painted (equally spaced) on a ring, as shown. How many different distinguishable rings can be made by changing the positions of the colors?

(b) Four different-colored stripes are painted (equally spaced) on a ring. How many different distinguishable rings can be made by changing the positions of the colors?

53. If a fixed number is added to each term of a geometric sequence, is the resulting sequence geometric? Try some examples and decide.

54. If each term of a geometric sequence is multiplied by a fixed number, is the resulting sequence geometric? Try some examples and decide.

55. (a) Drop a ball from different heights and see how high it goes on the rebound. Describe any pattern in your results.

(b) Predict how high the ball will go on the *second* bounce in relation to its initial height. Check your prediction.

56. Select a current elementary-school mathematics textbook. What kind of pattern problems does it have?

Enrichment Topic

57. Carl Friedrich Gauss (1777–1855) was one of the greatest mathematicians who ever lived (Figure 1–12). According to one story, when Gauss was 10, his teacher, desiring to keep the children occupied, asked them to add a sum like $1 + 2 + 3 + \cdots + 98 + 99 + 100$. Expecting the students to be busy for a half hour or so, the teacher was astonished when Gauss came up with the answer in less than a minute. How did Gauss do it and what was the sum? (*Hint:* Gauss probably paired off the numbers.)

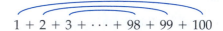

$$1 + 2 + 3 + \cdots + 98 + 99 + 100$$

Photo courtesy of Library of Congress.

Figure 1–12 Carl Friedrich Gauss

58. Use the method of the preceding exercise to find the sum of the following.

$$5 + 10 + 15 + \cdots + 290 + 295 + 300$$

59.

x	1	2	3	4	5	6	N
y	1	3	6	10	15		

(a) When $x = 6$, $y = $ _____.
(b) When $x = N$, $y = $ _____.

60. The following diagram shows that

$$1 + 2 = \frac{1}{2}(2 \times 3)$$

(a) Make a similar diagram showing that

$$1 + 2 + 3 = \frac{1}{2}(3 \times 4)$$

(b) Suggest a shortcut for computing

$$1 + 2 + 3 + \cdots + 98 + 99 + 100$$

using the same approach, and see if you obtain the correct sum.

  61. A 15-row auditorium seats 15 people in the first row, 16 in the second, 17 in the third, and so on. Use shortcuts to find the total number of seats in the auditorium.

62. (a) What is the 50th triangular number (see Exercise 44)?
(b) Explain how you figured out the answer.

1.5 PROBLEM SOLVING

NCTM Standards

- monitor and reflect on the process of mathematical problem solving (pre-K–12)

For too long, school mathematics has emphasized learning isolated facts and skills while devoting little time to mathematical reasoning. The most important part of school mathematics is not specific facts and skills; rather, it is learning how facts and skills are used by citizens analyzing data, consumers deciding what to buy, and workers dealing with technology or finance.

Admittedly, it is easier to teach children to perform routine computations and memorize facts and formulas, but as teachers, we must tackle a more significant task: developing children's problem-solving ability. The National Council of Teachers of Mathematics (NCTM) says, "Problem solving must be the focus of school mathematics." You have already studied two of the most important components of mathematical reasoning: induction and deduction. The next two sections will further develop your problem-solving ability.

What is problem solving? To understand what problem solving is, it is helpful to know what a problem is. A **problem** has two characteristics: (1) It requires a solution, and (2) the solution is not immediately obvious. The best classroom problems offer a situation of interest to the learner, in which the need for mathematics arises naturally.

TYPES OF PROBLEMS

What kinds of problems are studied in elementary school?

LE 1   Opener

What kinds of word problems do you remember solving in elementary school?

Read and solve the following three examples of elementary-school problems.

LE 2   Opener

Pierre has 21 oranges. He gives Jane 12. How many oranges does Pierre have left?

LE 3   Opener

The members of the Environment Club want to raise $50 by selling apples at $0.25 each. So far, they have sold 120 apples. How many more apples must they sell?

LE 4   Opener

Farmer Laura had 36 cows. She sold all but 10. How many cows does she have left?

Like the preceding three problems, many elementary-school problems fit into one of three categories: (1) one-step translation problems, (2) multistep translation problems, and (3) puzzle problems. These categories can be used to help organize the types of problems you present to children.

A **one-step translation** problem can be solved with a single arithmetic operation. LE 2 is an example of a one-step translation problem, the most common type in elementary school. One-step translation problems illustrate common applications of arithmetic and help reinforce arithmetic skills. Some mathematics educators would rather not call these "problems," since they become familiar and routine for most students who practice them.

A **multistep translation** problem can be solved with two or more arithmetic steps. LE 3 is an example of a multistep translation problem. Multistep translation problems also illustrate common applications of arithmetic and reinforce arithmetic skills, but they require higher-level thinking than one-step problems. Current elementary-school textbooks contain more multistep problems than less recent textbooks.

A **puzzle** problem is often solved with some unusual approach or insight. LE 4 is an example of a puzzle problem. Puzzle problems are nonroutine problems that develop flexible thinking.

These classifications are not precise; they depend on a student's background. LE 2 might be a routine one-step subtraction problem for a second-grader but nonroutine for a first-grader, who might solve it by counting back.

Read LE 5–LE 7, and tell whether each would usually be classified as a one-step translation problem, a multistep translation problem, or a puzzle problem.

LE 5   Concept

A classroom has 5 rows of desks. There are 6 desks in each row. If all but 2 desks are occupied, how many children are in the class?

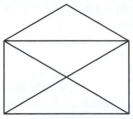

Figure 1–13

LE 6   Concept

Draw the figure shown in Figure 1–13 without lifting your pencil off the paper or retracing any line segments.

LE 7   Concept

Sidney has 8 incredibly good chocolate brownies. Four of us are waiting to eat them. If we all eat the same number of brownies, how many will we each eat?

PROBLEM SOLVING

Photo courtesy of G.L. Alexanderson, Santa Clara University.

Figure 1–14 George Polya

Solving problems is the specific achievement of intelligence and intelligence is the specific gift of mankind: solving problems can be regarded as the most characteristically human activity. . . . you can learn it only by imitation and practice.

(George Polya, *Mathematical Discovery,* J. Wiley, 1962, p. ix.)

George Polya (1887–1985; Figure 1–14), a brilliant teacher and mathematician, believed that people could learn to be better problem solvers. He explained how in his famous book *How To Solve It* (1945). More recently, organizations such as NCTM have promoted this idea, and it has become a standard topic in the elementary-school mathematics curriculum. Until recently, school mathematics focused on computational skills; now problem solving is considered to be of central importance in mathematics, and computational skills play a supporting role.

Good problem solvers take time to solve problems. After doing some thinking, they are not afraid to go ahead and try out some way of solving the problem. If one approach does not work, they are flexible and try another approach. They solve lots of problems and learn from their experiences.

Thanks to George Polya, we can learn specific methods for solving problems that increase our chances of being successful. First, Polya described four steps that can be used in analyzing many mathematical and nonmathematical problems:

Four Steps in Problem Solving

1. Understand the problem.
2. Devise a plan.
3. Carry out the plan.
4. Look back.

A mathematician could use these steps to solve a mathematics problem. A doctor could use them to treat an illness. A mechanic could use them to repair a car. Elementary-school mathematics textbooks utilize three-, four-, five-, or six-step variations of this approach.

People frequently solve problems without thinking about each step in Polya's plan. However, if one is having difficulty with a problem, the four-step plan is a useful guide. First, read the problem carefully and see if you understand it. What do you know, and what do you want to figure out? Second, in devising a plan, consider how the unknown is connected to the given in the problem. Also, how does the problem relate to concepts you know or other problems you have solved? Third, carry out the steps of your plan. Finally, look back, reviewing and checking your results. Have you answered the original question? Is there a way to check your answer to see if it is reasonable? Look back over the problem to improve your understanding of it. You can use this knowledge to solve related problems in the future.

Most elementary textbook series introduce the problem-solving steps early on. Figure 1–15 shows how the second-grade *Harcourt Math* textbook introduces Polya's steps.

Most elementary school textbooks present these four steps as the sequence:

$$\textbf{Understand} \rightarrow \textbf{Plan} \rightarrow \textbf{Solve} \rightarrow \textbf{Look Back}$$

Real problem solving is more dynamic; you might want to move backward and forward in this sequence. For example, the attempt to devise a plan might lead you to go back and try to understand the problem better.

Simply put, the four steps are understand, plan, solve, and check. How are these steps used in solving problems? Try them out as you work on LE 8.

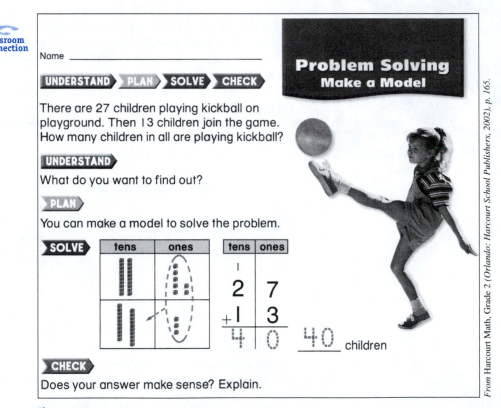

Classroom Connection

From Harcourt Math, Grade 2 (Orlando: Harcourt School Publishers, 2002), p. 165.

Figure 1–15

LE 8   Concept

Discussion

Twenty-two years ago, Jamaal's daughter was 1/4 his age, and his dog Yapper was 2. Today Jamaal's daughter is 1/2 his age. How old are Jamaal and his daughter now?

Understanding the Problem
(a) What are you supposed to figure out?
(b) Do you have enough information to do it?
(c) Is any information given that is not needed?

Devising a Plan
(d) You can solve this problem by using a guess-and-check (trial-and-error) approach or by using algebraic equations. Which will you try? (Guessing and checking is easier for most people.)
(e) Which of the following will you use in solving the problem: a calculator, paper and pencil, or mental computation?

Carrying Out the Plan
(f) Carry out your plan and obtain an answer.

Looking Back
(g) Check your answer.
(h) Make up a similar problem that can be solved in the same way.

AN INVESTIGATION: COUNTING SQUARES

LE 9   Reasoning

Discussion

(a) Solve the following problem using Polya's four steps. Make up one appropriate activity or question and answer it for each of the four steps. See LE 8 for ideas.

How many different squares (of all sizes) are in the following picture?

(b) How many different squares are in the following picture?

(c) How many different squares are in the following picture?

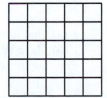

(Continued on next page)

(d) What pattern do you see in the numbers in parts (a), (b), and (c)?
(e) Solve a famous problem: How many different squares are there on an 8-by-8 checkerboard? (Use a shortcut.)

The four-step approach does not guarantee that you will be able to solve a problem, but it does provide some guidance. Previous experience in solving problems is also a great help in solving new problems.

OPEN-ENDED PROBLEMS

In Japan, mathematics teachers have used the "open approach" to problems for about 30 years. In a typical Japanese open-ended lesson, the teacher gives students a single problem to study. The teacher spends about 5 minutes asking students questions about the problem to make sure they understand it. Then the students work individually or with others on the problem for about 20 minutes. For the next 10 minutes, some students present their solutions and answer questions from others about their work. Finally, the teacher summarizes the lesson.

In **open-ended problems**, students are asked to find more than one answer or use more than one method. When they are finished, students might be asked to create or solve a problem like the one they just worked on.

The following problems come from the work of Jerry P. Becker of Southern Illinois University and his collaborators from Japanese schools. The first exercise has only one right answer, but can be solved in many ways.

LE 10   Reasoning

How many dots are in the picture at the right?
(a) Find as many different ways of grouping them to count them as you can.
(b) Make up a similar problem.

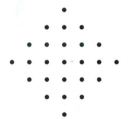

The next exercise starts with a question that can lead to different follow-up investigations.

LE 11   Reasoning

A row of connected squares can be made from toothpicks as follows.
How many toothpicks would be used to make a row of five connected squares?

(a) Find three different ways to count the toothpicks.
(b) Make up a series of related new problems and solve them. You could change the number of squares, use more than one row, or change the shape.

As a teacher, you may want to change a standard exercise into a more open-ended one.

LE 12   Concept

A standard exercise says, "The sum of four consecutive odd numbers is 176. Find the four numbers." Make a more open-ended problem. (*Hint:* Require more than one method or include a problem extension.)

LE 13   Summary

Tell what you learned about problem solving in this section.

ANSWERS TO SELECTED LESSON EXERCISES

2. 9

3. 80

4. 10

5. Multistep translation

6. Puzzle

7. One-step translation

8. (a) The current ages of Jamaal and his daughter
(b) Yes   (c) Yapper's age
(f) 66 years; 33 years

9. (a) *Understanding the Problem* What are you supposed to find? (The total number of squares of all sizes in the picture.)

Devising a Plan How will you do this? (Group the squares by size. Count all the 1-by-1 squares. Then count all the 2-by-2 squares. Finally, count all the 3-by-3 squares.)

Carrying Out the Plan Carry out your plan and obtain an answer.

Size	Number of Squares
1 by 1	9
2 by 2	4
3 by 3	1
Total	14

Looking Back Make up another problem like this one. How many different squares are in the following picture?

After solving this problem and the preceding one, develop a general solution for solving the same kind of problem with any size square.
(b) 30   (c) 55   (e) 204

1.5   HOMEWORK EXERCISES

Basic Exercises

Tell whether Exercises 1–3 would usually be classified as one-step translation, multistep translation, or puzzle problems.

1. There are 10 boys and 12 girls going on a field trip. Each car will hold 5 children and 1 parent. How many cars are needed?

2. A classroom has 4 rows with 5 desks in each row. How many desks are there in all?

3. About how many marbles would fit inside a basketball?

4. (a) Chick N. Little bought 3 dozen eggs, and 2 of the eggs were broken. How many good eggs were there?
(b) Write a problem about 12-year-old Bea Young and her father that involves computing $3 \times 12 - 2$ to find his age.
(c) Write a problem about the total number of people in rows of chairs that uses the same $3 \times 12 - 2$ computation.

5. Make up a word problem about the Country Kitchen's menu that involves multiplication and addition.

The Country Kitchen	
Roadside squirrel	$1.50
Twice-baked kelp	$1.05
Scalloped corn	$ .60
Pond water	$ .10

6. Describe with one word each of Polya's four steps for solving a problem.

7. Write a paragraph describing Polya's four steps for problem solving.

8. Explain how a doctor treating an illness could use Polya's four steps.

9. Describe a plan for solving the following problem. "The area of a square field is 49 m^2. What is the length of a fence that goes around the outside of the field?"

10. Seth Borgas buys two shirts for $12.98 each and a cap for $8.98. The sales clerk says the total cost is $48.94. Without computing the exact answer, tell whether the clerk's total seems reasonable. Why or why not?

11. Solve the following problem with Polya's four steps. Make up an appropriate question or activity and answer it for each of Polya's four steps. The problem is as follows: "You have 6 black socks, 8 white socks, 2 red socks, and 2 green socks in a drawer. On a dark morning, you pull out socks (without replacement) and stop when you obtain one matching pair. What is the greatest number of socks you would have to pull out?

12. Solve the following problem using Polya's four steps. Make up an appropriate question or activity and answer it for each of Polya's four steps. The problem is as follows. "You have 10 black socks, 10 white socks, and 4 red socks in a drawer. On a dark morning, you pull out socks (without replacement) and stop when you obtain one matching pair. What is the greatest number of socks you would have to pull out?"

13. Solve the following problem using Polya's four steps. Make up an appropriate question or activity and answer it for each of Polya's four steps. The problem is as follows. "How many cuts does it take to divide a log into five cross-sectional (cylindrical) pieces?"

14. How many cuts does it take to divide a log into
    (a) six equal cross-sectional pieces?
    (b) seven equal cross-sectional pieces?
    (c) N equal cross-sectional pieces?

15. How many squares are in the following picture?

16. (a) A 2-by-2-by-2 cube is built from eight 1-by-1-by-1 cubes. How many cubes of all sizes are there?
    (b) A 4-by-4-by-4 cube is built from sixty-four 1-by-1-by-1 cubes. How many cubes of all sizes are there?

17. (a) The Bathula family has 2 sons. Each son has 3 sisters. How many children are there?
    (b) The Dulfano family has T sons. Each son has N sisters. How many children are there?

18. A pizza restaurant has 10 different toppings for its cheese-and-tomato pizza: mushrooms, peppers, pepperoni, sausage, onion, anchovies, tuna, pickles, shredded wheat, and celery. How many different kinds of pizza can be made by varying the combination of toppings? (*Hint:* Try the same kind of problem with 1 topping, then 2 toppings, and so on, and look for a pattern.)

19. Move only three dots to make this design

look like this:

20. Remove two toothpicks so that you have two squares of different sizes.

21. An exercise says, "A square and a rectangle have equal areas. If the sides of the square are 4 ft and the rectangle has a width of 2 ft, what is the length of the rectangle?" Make a more open-ended problem. (*Hint:* Require more than one method or include a problem extension.)

22. An exercise says, "A school has a student-teacher ratio of 1 to 16. If there are 25 teachers, about how many students are there?" Make a more open-ended problem. (*Hint:* Require more than one method or include a problem extension.)

Extension Exercises

23. Consider the following dot patterns.

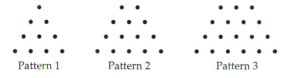

Pattern 1   Pattern 2   Pattern 3

(a) How many dots are in pattern 4? Find the answer as many different ways as you can.
(b) How many dots are in pattern N?

24. Consider the following problem: "Larry starts cycling at 20 mi/hr. One hour later, Rebecca starts cycling from the same place at 25 mi/hr. How long will it take Rebecca to catch up to Larry?"
(a) Find three different ways to solve the problem. (*Hint:* You might use arithmetic, a table, a number line, or algebra.)
(b) Make up a similar problem and describe a shortcut for solving problems of this type.

25. You have 16 coins and a balance. Fifteen of the coins are regular, and one is a lighter counterfeit coin. Show how you could identify the counterfeit coin after three weighings.

26. You have 24 coins and a balance. Twenty-three of the coins are regular, and one is a lighter counterfeit coin. Show how you could identify the counterfeit coin after three weighings.

27. Consider the entire set of problems that are like Exercises 25 and 26. What is the greatest number of coins you could start with and be sure to identify the counterfeit coin after
(a) three weighings?   (b) two weighings?

28. Connect the nine dots by drawing four lines, without lifting your pencil off the paper or retracing any lines.

• • •
• • •
• • •

Puzzle Time

29. A woman lines up five red checkers and five black checkers as follows.

R	R	R	R	R	B	B	B	B	B
1	2	3	4	5	6	7	8	9	10

She asks, "Can you switch the positions of only four checkers so that the black and red checkers alternate?"
(a) How would you do it?
(b) If you started with only 1 pair of checkers, how many checkers would you have to move to solve the same problem?
(c) If you started with 3 pairs of checkers, how many checkers would you have to move to solve the same problem?
(d) What is the general pattern in the numbers of checkers that must be moved in parts (a), (b), and (c)?
(e) Does this pattern work for 2 pairs of checkers?
(f) Does it work for 4 pairs of checkers?
(g) Can you think of a different rule that would work for parts (e) and (f) ?
(h) Make a conjecture about how many checkers must be moved in the same problem involving 25 pairs of checkers.
(i) Make a conjecture about how many checkers must be moved in the same problem involving 50 pairs of checkers.

1.6 PROBLEM-SOLVING STRATEGIES

NCTM Standards
• apply and adapt a variety of appropriate strategies to solve problems (pre-K–12)

Do you ever have difficulty solving mathematics problems? Devising a plan, the second step of Polya's procedure, is often the most difficult step. Elementary-school children now learn specific strategies that they can use to solve a variety of problems. Learning these strategies can help any student become a better problem solver.

Research shows that children who are good at solving problems are more likely to use certain problem-solving strategies. Children now do sets of problems that focus upon these strategies. The following problem-solving strategies are discussed in this book.

Some Problem-Solving Processes and Strategies

1. Using inductive reasoning
2. Using logical (deductive) reasoning
3. Guessing and checking
4. Making a table or list
5. Drawing a picture
6. Choosing an operation
7. Working backward
8. Using a graph
9. Using an equation
10. Taking a break and trying again

Problem-solving strategies are now included in most elementary-school textbooks. You have already studied induction and deduction. Choosing an operation and working backward will be introduced in Chapter 3. Chapter 11 discusses using graphs and equations to solve problems.

In this section, you will study three useful problem-solving strategies: guessing and checking, making a table, and drawing a picture.

LE 1 Opener

Tell what you know about the following problem-solving strategies:

(a) guessing and checking (trial and error)
(b) making a table or list
(c) drawing a picture

The best way for you or your students to learn a problem-solving strategy is to practice using it first by itself. After each of the three individual strategies becomes familiar, you can try to solve problems in which you must decide which of the three strategies to use.

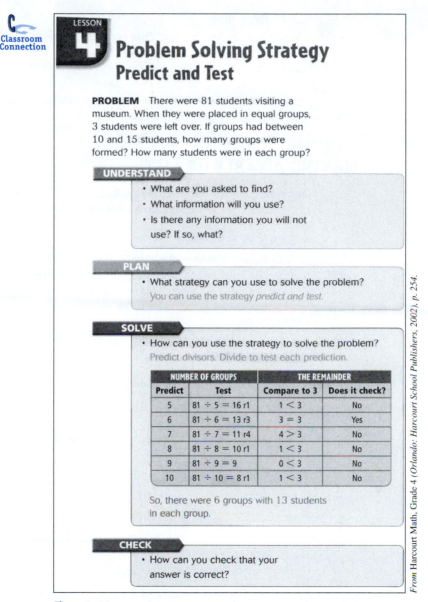

Figure 1–16

GUESSING AND CHECKING

Sometimes you have to be bold in mathematics! The guessing-and-checking strategy requires you to start by making a guess and then checking how close your answer is. Next, on the basis of this result, you revise your guess and try again. Figure 1–16 shows how the fourth-grade *Harcourt Math* textbook presents guessing and checking.

Try guessing and checking in the following exercise.

LE 2   Reasoning

Sandy bought 18 pieces of fruit (oranges and grapefruits), which cost $4.62. If an orange costs $0.19 and a grapefruit costs $0.29, how many of each did she buy? (*Guess and check.*)

Consider guessing and checking when you have a question with a limited number of possible answers and you can check how close a guess is to the correct answer.

MAKING A TABLE OR LIST

Organizing information often makes it easier to solve a problem. You may have used a table in the preceding lesson exercise. A table can also help you solve the problem in Example 1.

• EXAMPLE 1

A cashier wants to make change for $1 using nickels, dimes, and quarters. Furthermore, he will use at least 2 quarters and no more than 8 coins all together. How many ways are there to do this? (*Make a table.*)

SOLUTION

Understanding the Problem   The cashier must use 2, 3, or 4 quarters and no more than a total of 8 coins to make $1.

Devising a Plan   All the solutions can be organized into a table.

Carrying Out the Plan   Fill out the table in an organized way.

Quarters	Dimes	Nickels
4	0	0
3	2	1
3	1	3
3	0	5
2	5	0
2	4	2

There are 6 different ways.

Looking Back   Check the table to see that all combinations are included. They are.   ●

Tables enable one to organize information in a simple, clear way.

LE 3   Reasoning

It will take 11 quarts of paint to paint your new apartment. Quarts cost $3.50 and gallons cost $8.50. (*Note:* 1 gallon = 4 quarts.)

(a) What are the different ways you can purchase the paint for the job? (*Make a table or list.*)
(b) Which way is the least expensive?

Consider making a table or list when you have a limited number of options that can be listed in an organized way, and you want to examine all those options.

DRAWING A PICTURE

Sometimes a drawing or a diagram can help you analyze a problem. For example, you can use Venn diagrams to help check deductive reasoning. Solve LE 4 by *drawing a picture.*

Discussion

LE 4 Reasoning

A well is 30 ft deep. An athletic snail (Figure 1–17) at the bottom climbs up 3 ft each day and slips back 2 ft each night. On what day does the snail reach the top of the well? (*Hint:* Draw a picture. The answer is *not* the 30th day!)

Figure 1–17

Consider drawing a picture for any problem that involves measurements or geometric shapes. Besides using problem-solving strategies, successful problem solvers tend to (1) approach new problems with confidence, (2) analyze and explore the conditions of the problem, and (3) obtain a lot of experience in solving problems.

AN INVESTIGATION: PROBLEM ANALYSIS

The following lesson exercises are all related to the preceding puzzle problem. By doing this series of problems, you will deepen your understanding of this kind of puzzle problem.

LE 5 Reasoning

A well is 30 ft deep. A muscle-bound worm at the bottom climbs up 4 ft each day and slips back 1 ft each night. On what day does it reach the top of the well? (*Draw a picture.*)

LE 6 Reasoning

A well is 50 ft deep. A muscle-bound worm at the bottom climbs up 4 ft each day and slips back 2 ft each night. On what day does it reach the top of the well? (*Draw a picture.*)

Discussion

LE 7 Reasoning

A well is 80 ft deep. Complete a worm or snail problem so that the slimy animal reaches the top on the
(**a**) 20th day.   (**b**) 19th day.   (**c**) 18th day.

LE 8 Summary

Tell what you learned about problem-solving strategies in this section.

ANSWERS TO SELECTED LESSON EXERCISES

2. 6 oranges and 12 grapefruits

3. (a)

Number of Gallons	Number of Quarts	Total Cost
0	11	$38.50
1	7	$33.00
2	3	$27.50
3	0	$25.50

(b) Three gallons would be the least expensive.

4. 28th day (*Hint:* Where is the snail after 27 days and nights?)

5. 10th day

6. 24th day

7. *Hint:* Use an "up" number that is 4 more than the "down" number.

1.6   HOMEWORK EXERCISES

Basic Exercises

1. Describe the guess-and-check strategy in your own words.

2. Which step of Polya's four-step scheme involves the consideration of various problem-solving strategies?

3. Craig Chandler is 5 years older than Linda. The product of their ages is 1184. How old are they? (*Guess and check.*)

4. Find two consecutive even numbers whose product is 50,624. (*Guess and check.*)

5. Approximate a solution of $(x + 2)(x + 3) = 50$ to one decimal place. (*Guess and check.*)

6. Find $\sqrt{1444}$ without using a calculator $\boxed{\sqrt{}}$ key. (*Guess and check.*)

7. A baseball league has 6 teams: the Bats, the Diamonds, the Flies, the Goose Eggs, the Hot Dogs, and the Relish. If every team plays each of the other teams 4 times, how many games must be scheduled? (*Make a table.*)

8. A bank sells the following packets of traveler's checks: five $20 bills, three $50 bills, three $100 bills, and five $100 bills. How many different ways are there to buy $500 in traveler's checks? (*Make a table.*)

9. A well is 100 ft deep. A muscle-bound worm at the bottom climbs up 5 ft each day and slips back 2 ft each night. On what day does it reach the top of the well? (*Draw a picture.*)

10. A snail at the bottom of a well goes up 8 feet each day and slides back 4 feet at night. When will the snail reach the top of the well if it is
(a) 50 feet deep?
(b) 100 feet deep?
(c) Suppose the well is y feet deep. Describe a method for finding the answer.

11. A well is 60 ft deep. Complete a worm or snail problem so that the slimy animal reaches the top on the
(a) 20th day.   (b) 19th day.   (c) 18th day.

12. (a) A square has an area of 81 square units. What is its perimeter? Recall that the perimeter is the distance around the outside. (*Draw a picture and use the correct units.*)
(b) Computing the perimeter of a square when you know the length of each side involves what type of reasoning, inductive or deductive?

13. An office staff includes a manager (M), an assistant manager (AM), two secretaries (S), and two sales agents (SA). Construct a 7-week vacation schedule for them, taking into consideration the following constraints.

1. The manager is taking weeks 5, 6, and 7 off. Everyone else gets 2 consecutive weeks off.
2. No more than two people can be on vacation at one time.
3. At least one secretary and one sales agent must be present in the office each week.
4. When the manager is away, the assistant manager and both sales agents must be at work.

	M	AM	S1	S2	SA1	SA2
Week 1						
Week 2						
Week 3						
Week 4						
Week 5	X					
Week 6	X					
Week 7	X					

14. During 6 weeks in the summer, each of 4 workers will take 3 weeks off. No more than 2 workers may be off at one time, and everyone wants at least 2 consecutive weeks off. Complete the following schedule.

	Week 1	Week 2	Week 3	Week 4	Week 5	Week 6
Molly	X			X	X	
Paul	X		X	X		
Wong						
Clio						

In problems 15–17, tell which of the three strategies you would use to solve the problem. You do not have to solve the problem.

15. I took a certain two-digit number and reversed the digits. Then I added the numbers together. The sum was 131. What were the two numbers?

16. I have a piggy bank full of nickels, dimes, and quarters. What are all the possible ways to make $0.60?

17. How many 6-inch by 6-inch square tiles would it take to cover a floor that is 10 ft by 14 ft?

18. Solve the following problem and tell which of the three strategies you used. A 10-lb bag of mixed nuts contains 20% peanuts. How many pounds of peanuts should be added to change the mixture to 80% peanuts?

19. Solve the following problem and tell which of the three strategies you used. After throwing 25 darts at the following target, what is the *highest* score below 100 that it is *impossible* to score?

20. Solve the following problem and tell which of the three strategies you used. An apartment has two adjacent rectangular rooms, each 9 ft by 12 ft. They share a 9-ft-long wall. What is the perimeter of the apartment?

Extension Exercises

 21. A telephone company offers two plans for local calls. Plan A charges $0.09 per call. Plan B charges $4 for the first 50 calls and $0.06 for each call after that. Determine under what conditions each plan is cheaper. (*Make a table and guess and check.*) The table headings are as follows:

Number of Calls	Cost of Plan A	Cost of Plan B

 22. A travel agency is planning a one-week trip to Italy that has an initial cost of $2000 plus $1140 per person. At a price of $1600, 15 people will sign up. Past experience suggests that for each $20 decrease in the price, an additional person will sign up. Also, for each $20 increase in the price, one less person will sign up. Approximately what price will maximize the profit? (*Make a table and guess and check.*)

23. (a) A woman wants to enclose 100 yd^2 of field with a rectangular fence. What is the minimum perimeter of fencing she can use?

    (b) A woman wants to enclose N yd^2 of field with a rectangular fence. What is the minimum perimeter of fence she can use?

24. (a) A woman has 40 yards of fencing for her yard. What is the maximum rectangular area she can enclose? (*Draw a picture, make a table, and guess and check.*)

    (b) A woman has F yards of fencing for her yard. What is the maximum rectangular area she can enclose? (*Draw a picture, make a table, and guess and check.*)

25. A farmer wants to transport a fox, a goose, and a bag of corn across a river in a boat. He can take only one of the three across on each trip. He cannot leave the fox and the goose alone, since the fox will eat the goose. He cannot leave the goose and the corn alone, since the goose will eat the corn. How will he get the fox, the goose, and the corn across the river? (*Draw a picture and guess and check.*)

26. (a) A man and two small children want to cross the river in a small boat. The boat is big enough only to hold either the man or the two children, both of whom can row. How can all three get across the river in the boat?

    (b) Solve the same problem for four adults and two children, all of whom can row.

    (c) Solve the same problem for six adults and two children, all of whom can row.

    (d) Describe how to work out the problem for two children and any number of adults.

27. There are 20 children in a class, including exactly 10 girls and 12 eight-year-olds. How many eight-year-old girls could there be? Give *all* possible answers. (*Draw a picture and guess and check.*)

28. There are 26 children in a class, including exactly 12 girls and 20 eight-year-olds. How many eight-year-old girls could there be?

29. There are C children in a class, including exactly G girls and E eight-year-olds. Assume that $G < E < C$. How many eight-year-old girls could there be?

30. (a) I'm inviting 20 people to a party, and I want to seat them all at one long table. I'm going to put together a series of card tables that seat one person on each side and form one long table. If I arrange the card tables in a single row, how many card tables will I need?

    (b) Give a general solution for seating $2N$ people (N is a whole number greater than 3). How many tables are needed?

31. Another option for the party in the preceding exercise is to use rectangular tables, as shown.

What is a formula that relates the number of chairs (C) to the number of tables (T)?

32. In the preceding exercise, one student gives the answer $C = 2 + 4T$, and another gives the answer $C = 6 + 4(T - 1)$. Are both answers correct? Explain why or why not.

33. Another arrangement of rectangular tables (see the preceding exercise) gives more room on the table for serving platters.

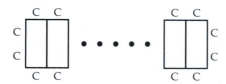

What is the formula that relates the number of chairs to the number of tables?

34. At a party where the guests have never met one another, everyone wants to shake hands with everyone else. How many handshakes will occur if there are
(a) 2 people at the party?
(b) 3 people at the party?
(c) 4 people at the party?
(d) 8 people at the party? Use the pattern from parts (a), (b), and (c).

35. Look at a current textbook series for Grades 1–6. What does the series teach about problem-solving strategies in each grade?

CHAPTER 1 SUMMARY

What methods of reasoning do mathematicians use that people also use to analyze problems in everyday life? Two methods commonly used to discover new ideas in mathematics and everyday life are inductive and deductive reasoning. Induction involves making a reasonable generalization from specific examples. Scientists and mathematicians use induction to develop general hypotheses or theorems. People may also develop prejudices and superstitions with induction.

To verify conjectures, one uses deductive reasoning, the process of drawing a necessary conclusion from given assumptions. Detectives use deduction in drawing conclusions, and mathematicians use deduction to prove theorems. Mathematicians often develop new ideas by using induction to make a generalization or state a theorem, followed by deduction to prove that the generalization or theorem must be true.

Both types of reasoning can lead to false conclusions. Induction is not totally reliable, because what happens a few times may not always happen. Superstitions and prejudices are examples of erroneous generalizations. Deductions are sometimes false because valid deductive reasoning based on false assumptions can lead to a false conclusion.

A superior ability to find patterns distinguishes human beings from other animals and machines. Mathematicians develop new ideas by recognizing and generalizing patterns. Finding a pattern makes use of induction, and extending the pattern based on a general rule requires deduction. Finding and extending patterns enables people to make predictions about the future and to develop classifications.

Problem solving is now a focus of school mathematics. Whenever possible, mathematical ideas should be developed from problem situations of interest to the learner. Three types of problems children study are one-step translation problems, multistep translation problems, and puzzle problems. In solving most of these problems, Polya's four steps can be followed: (1) Understand the problem, (2) devise a plan, (3) carry out the plan, and (4) check and assess your results.

Devising a plan or strategy is often the hardest step. Three strategies commonly used to solve mathematics problems are guessing and checking, making a table, and drawing a picture. Arithmetic, algebra, and geometry problems can sometimes be solved more easily with these strategies.

STUDY GUIDE

To review Chapter 1, see what you know about each of the following ideas and terms that you have studied. You can also use this list to generate your own questions about the chapter.

MATHEMATICAL REASONING IN GRADES 1–8

The following chart shows at what grade levels selected mathematical reasoning topics typically appear in elementary and middle-school mathematics textbooks. Underlined numbers indicate grades in which the most time is spent on the given topic.

Topic	Typical Grade Level in Current Textbooks
Using inductive reasoning	1, 2, 3, 4, 5, 6, 7, 8
Using deductive reasoning	1, 2, 3, <u>4</u>, <u>5</u>, <u>6</u>, <u>7</u>, <u>8</u>
Number sequences	1, 2, 3, 4, 5, 6, <u>7</u>, 8
Arithmetic and geometric sequences	7, 8
Finding patterns	1, 2, 3, 4, 5, 6, 7, 8
Problem-solving steps (Polya)	<u>1</u>, 2, <u>3</u>, <u>4</u>, <u>5</u>, <u>6</u>, <u>7</u>, <u>8</u>
Guess and check	2, 3, 4, 5, <u>6</u>, <u>7</u>, 8
Make a table or list	1, 2, <u>3</u>, 4, 5, 6, 7, 8
Draw a picture	1, 2, 3, 4, 5, 6, 7, 8

REVIEW EXERCISES

1. Tell what you learned about inductive reasoning in this chapter. Include what it is, how it works, and when it does not work. Give examples in mathematics and everyday life of how it is used.

2. *Explain* how a student might use inductive reasoning.

3. Is the sum of any three consecutive counting numbers divisible by 3?

4.
$$142857 \times 1 = 142857$$
$$142857 \times 2 = 285714$$
$$142857 \times 3 = 428571$$

(a) Write the next example, continuing the pattern.
(b) Make a generalization about the digits in the result.
(c) How long does the pattern continue? Try 5, 6, and 7 as factors.

5. What is deductive reasoning?

6. Make up an example of valid deductive reasoning that
(a) leads to a true conclusion.
(b) leads to a false conclusion.

7. Does the given conclusion follow from the two hypotheses? If so, draw a Venn diagram that illustrates it. If the conclusion is invalid, draw a Venn diagram that satisfies the hypothesis but does not support the conclusion.

Hypotheses: All squares have four sides. All rhombuses have four sides.
Conclusion: Therefore, all rhombuses are squares.

8. The rule is: "If you finish your fish, then I will give you some fruit."
(a) What can you conclude if you do not receive some fruit?
(b) What will happen if you do not finish your fish?

9. Consider the following Pica-Centro game.

Guesses	Responses	
Digits	Pica Correct Digits, Wrong Position	Centro Correct Digit, Correct Position
4 3 7	1	1
2 5 7	0	0
1 4 3	1	1
0 8 9	0	1

What is the secret number? Justify your answer.

In Exercises 10–12, identify whether induction or deduction is being used.

10. People have died in the past. I assume that everyone will die in the future.

11. I know Alicia is in the bedroom, the kitchen, or the bathroom. I do not find her in the bedroom or the kitchen. I conclude that she is in the bathroom.

12. I know that $x > y$ and $y > z$. I conclude that $x > z$.

13. Write a paragraph describing inductive reasoning and deductive reasoning and how to tell the difference between them.

14. Consider the following number trick.

Pick a number.
Add 5.
Multiply by 3.
Subtract 9.
Subtract your original number.
Divide by 2.

Prove that you will always end up with three more than you started.

15. (a) Write the converse of the following statement:
   If it is raining, then the ground gets wet.
   (b) Is the converse true?

16. Write the following statements in an "if-then" format.
   (a) A rectangle is a square if it has four congruent sides.
   (b) I will call only if there is a problem.

17. A sequence begins 6, 30, ... What is the next term if the sequence is
   (a) arithmetic?   (b) geometric?

18. Consider the following sequence.
$$20, 19, 18, 17, \ldots$$
   (a) Is the sequence arithmetic, geometric, or neither?
   (b) What is the rule for the nth term?
   (c) What is the 40th term?
   (d) Does part (b) require inductive or deductive reasoning?
   (e) Does part (c) require inductive or deductive reasoning?

19. Consider the following sequence.
$$27{,}000, 9{,}000, 3{,}000, 1{,}000, \ldots$$
   (a) Is the sequence arithmetic, geometric, or neither?
   (b) If it is arithmetic give the value of a and d. If it is geometric, give the value of a and r.
   (c) If it is arithmetic or geometric, use the variables INITIAL, CURRENT, and PREVIOUS to write the two equations for the sequence.
   (d) What is the nth term in the sequence?

20. Consider the following sequence.
$$3, 12, 27, 48, \ldots$$
   (a) Is the sequence arithmetic, geometric, or neither?
   (b) If it is arithmetic give the value of a and d. If it is geometric, give the value of a and r.
   (c) If it is arithmetic or geometric, use the variables INITIAL, CURRENT, and PREVIOUS to write the two equations for the sequence.
   (d) What is the 6th term in the sequence?

21. (a) Fill in the missing numbers.
$$1 = (\ )^3$$
$$3 + \ 5 = (\ )^3$$
$$7 + 9 + 11 = (\ )^3$$
   (b) What would the next equation be if the pattern continued? Is the equation true?

22. Examine the following pattern.
$$4^2 - 1^2 = 3 \cdot 5$$
$$5^2 - 2^2 = 3 \cdot 7$$
$$6^2 - 3^2 = 3 \cdot 9$$
   (a) Write the next example that follows the pattern.
   (b) Complete the following generalization. For any counting number N greater than two,
$$N^2 - \underline{\hspace{1.5cm}} = \underline{\hspace{1.5cm}}.$$
   (c) Prove that your equation in part (b) is true.

23. Make up an example of a multistep translation problem.

24. Explain how an auto mechanic repairing an engine could use Polya's four steps.

25. Solve the following problem using Polya's four steps. Make up an appropriate question and answer it for each of the four steps.

Examine the following figure.

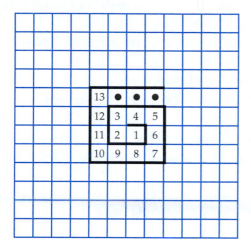

If the pattern in the square continues, what number will appear in the upper right-hand corner?

26. How many rectangles are there in the figure?

27. Consider the following problem. You took 11 orders for either a tuna salad sandwich ($1.90) or a buckwheat Newburg ($1.50) but forgot how many of each. If the bill comes to $17.70, how many of each were ordered?
    (a) What would be a good strategy for solving this problem?
    (b) Solve the problem using your strategy.

28. An employer wishes to select 2 people for a job from Alice, José, Mary, and Reggie. How many ways are there to do this?

29. A well is 40 ft deep. A snail climbs up 9 feet each day and slips back 5 feet at night. How long will the snail take to reach the top of the well?

30. A well is 40 ft deep. Make up a snail problem in which the snail reaches the top on the
    (a) 18th day.   (b) 19th day.

31. An exercise says, "You have $\frac{3}{4}$ lb of cottage cheese, and you buy $\frac{1}{2}$ lb more. How much do you have now?" Make a more open-ended problem. (*Hint:* Require more than one method or include a problem extension.)

32. You have 9 coins and a balance. Eight of the coins are regular, and one is a lighter, counterfeit coin. Show how you could identify the counterfeit coin after 2 weighings.

ALTERNATE ASSESSMENT—KEEPING A PORTFOLIO

Use a portfolio to display your best work. You can also show how your abilities developed during the course. Include topics that you found interesting or exercises that taught you something significant.

To start your portfolio, select one or more pieces of your work from Chapter 1. Write a description of each piece telling what it is, what mathematics is used, and why you think it is one of your most significant or best works. Add to your portfolio as you go through the course.

 ## SUGGESTED READINGS

Becker, J., and S. Shimda. *The Open-Ended Approach: A New Proposal for Teaching Mathematics.* Reston, VA: NCTM, 1997.

Burns, M. *About Teaching Mathematics.* 2nd ed. Parsippany, NJ: Pearson, 2000.

Brown, S., and M. Walter. *The Art of Problem Posing.* 2nd ed. Mahwah, NJ: Lawrence Erlbaum, 1990.

Charles, R., and F. Lester. *Teaching Problem Solving: What, Why, and How.* Palo Alto, CA: Dale Seymour, 1982.

Conference Board of Mathematical Sciences. *Issues in Mathematics Education.* Providence, RI: AMS, 2001.

Developing Mathematical Reasoning in Grades K–12. 1999 Yearbook. Reston, VA: NCTM, 1999

Erickson, T. *Get It Together.* Berkeley, Calif.: Equals, 1989.

Gardner, M. *The Colossal Book of Mathematics: Classic Puzzles, Paradoxes, and Problems.* New York: W. W. Norton, 2001.

Jacobs, Harold. *Mathematics: A Human Endeavor.* 3rd ed. New York: Freeman, 1994.

National Council of Teachers of Mathematics. *Principles and Standards for School Mathematics.* Reston, VA: NCTM, 2000.

National Council of Teachers of Mathematics. *Professional Development for Teachers of Mathematics.* 1994 Yearbook. Reston, Va.: NCTM, 1994.

Navigating Through Mathematical Reasoning and Proof (2 vols.). Reston, VA: NCTM, forthcoming.

Polya, G. *How To Solve It.* 2nd ed. Princeton, NJ: Princeton Univ. Press, 1973.

Stein, S. *How the Other Half Thinks: Adventures in Mathematical Reasoning.* Columbus, OH: McGraw-Hill, 2001.

Journals are another good source of information. Many of the most useful journals are published by NCTM. Two of the best are *Teaching Children Mathematics* (grades Pre-K–6) and *Mathematics Teaching in the Middle School* (grades 5–9). Thomson Learning provides Infotrac, a fully searchable on-line library of thousands of journals, including *Teaching Children Mathematics,* as well as many popular science, technology, social science, news, and business magazines.

Additional information and links to other useful Web sites can be found at the publisher's Web site: www.brookscole.com

2 | SETS AND FUNCTIONS

Every area of mathematics utilizes sets and functions in some way. For example, Chapters 3, 4, 5, and 7 each concern a particular set of numbers (e.g., whole numbers). You will also study different sets of shapes (e.g., rectangles and parallelograms) and the relationships among them. Graphing involves functions and relations (sets of ordered pairs). Functions are used to represent a variety of everyday situations that involve two variables in which each value of one variable determines a unique value for the second variable. An example would be the weight of some apples and their price. In statistics, one studies data sets; and in probability, one examines sets of possible outcomes. You have already used set pictures (Venn diagrams) to illustrate logical relationships.

This chapter introduces the basic ideas of sets, but this book as a whole does not place great emphasis on formal set theory. Neither does the current elementary-school mathematics curriculum. Functions are introduced in Section 2.3 and covered in more depth in Chapter 11.

2.1 SETS

NCTM Standards

- create and use representations to organize, record, and communicate mathematical ideas (pre-K–12)

George Boole (1815–1864; Figure 2–1) developed set theory. Born to an English working-class family, Boole taught himself six foreign languages and a good deal of mathematics. He believed that the essence of mathematics lies in its deductive organization.

Georg Cantor (1845–1918; Figure 2–2, on the next page), a German mathematician, extended set theory to include infinite sets, a topic so controversial that it caused one of his former teachers to turn against him. Partly as a result of being ridiculed by that teacher, Cantor suffered a nervous breakdown. He died in a mental hospital.

Today Boole's and Cantor's work with finite and infinite sets is an accepted part of mathematics. In this text, sets will be used to define whole-number addition and multiplication and will serve as a model for each of the four whole-number operations in Chapter 3.

Photo courtesy of Library of Congress

Figure 2–1 *George Boole*

Photo courtesy of Library of Congress

Figure 2–2 *Georg Cantor*

In the 1960s and early 1970s, some elementary-school teachers used sets to explain counting and arithmetic. Some of you might remember studying sets as children. Others might claim to be too young to remember 1970. (Others, like myself, can't believe it isn't still 1970.)

Teaching that was heavily based on set theory was not well received in elementary schools, and today sets are generally not studied as a separate topic at that level. However, we often talk about sets of numbers or shapes, about subsets, and about the elements two sets have in common. Venn diagrams are introduced in most middle-school programs as an aid to problem solving.

LE 1   Opener

What did you learn about sets in elementary or secondary school?

SETS

What education courses must you pass to receive state teacher certification? Who in your class likes to eat pizza? The answers to these questions would each comprise the members of a set.

A **set** is a collection of objects called **members** or **elements.** To define a set, one can list the members and enclose them in braces. For example, the set of primary colors is {red, yellow, blue}.

The expression "$x \in A$" means that x is a member of set A. The expression "$x \notin A$" means that x is *not* a member of set A.

LE 2   Concept

Fill in each blank with $\in$ or $\notin$.

(a) 4 _____ {2, 4, 6}
(b) Orange juice _____ the set of all junk foods.

A set with no members is called an **empty** or **null set.** For example, the set of all pink elephants in your math class is an empty set. The set of all whole numbers that are between 2 and 3 is another empty set. An empty set is denoted by the symbol { } or the Danish letter $\varnothing$.

LE 3   Concept

Make up another description that would represent an empty set.

People invented numbers so they could count objects. At first, people used only the first few counting numbers. The infinite set of counting or natural numbers is the result of thousands of years of work in expanding this initial set of numbers and developing more efficient notation.

We take the symbol 0 for granted, but at first people thought a symbol for "nothing" was unnecessary. Once 0 was included as a symbol for "nothing" and as a placeholder, numeration systems made significant advances (see Chapter 3).

By putting together 0 and the set of counting numbers, one obtains the set of whole numbers.

> **Definition:**   **Whole Numbers**
>
> The set of **whole numbers** $W = \{0, 1, 2, 3, \ldots\}$.

EQUAL, EQUIVALENT, FINITE, AND INFINITE SETS

One might compare two sets to see whether they are identical or whether they contain exactly the same number of elements. Two sets are **equal** if and only if they contain exactly the same elements. For example, if $A = \{2, 3\}$ and $B = \{3, 2\}$, then $A = B$.

To compare two sets, one can try to place their elements into a one-to-one correspondence. A **one-to-one correspondence** pairs the elements of two sets so that for each element of one set, there is exactly one element of the other. For example, suppose Bill and Sue are going to sit in seats 1 and 2 in a row. The sets $\{1, 2\}$ and $\{Bill, Sue\}$ can be placed into the two different one-to-one correspondences shown in Figure 2–3.

Figure 2–3

Two sets are **equivalent** if and only if there is a one-to-one correspondence between the sets. The following exercise examines the differences between equal and equivalent sets.

LE 4   Concept

$A = \{1, 2, 3\}$ $B = \{1, 3, 5\}$ $C = \{1, 2, 3, 4\}$

$D = \{Sara, John, Will\}$ $E = \{bat, ball, glove, cap\}$ $F = \{1, 3, 2\}$

(a) Which sets are equivalent?
(b) Which sets are equal?
(c) If two sets are equivalent, are they equal?
(d) If two sets are equal, are they equivalent?
(e) Show a one-to-one correspondence between A and D.
(f) How would you use equivalent sets from this exercise to show how many objects are in set E?

A set is a **finite set** if it is empty or if it can be placed into a one-to-one correspondence with a set of the form $\{1, 2, 3, \ldots, N\}$, where N is a whole number. The number of elements in a finite set must be a whole number. The set of days of the week is a finite set containing seven elements. The set of whole numbers is an **infinite set** because it does not contain a finite number of elements. A more formal definition of an infinite set appears in the extension exercises.

LE 5   Concept

Decide whether each of the following sets is a finite set or an infinite set.

(a) The set of whole numbers less than 6
(b) The set of all the pancakes in Arizona right now
(c) The set of counting numbers greater than 6

UNIVERSAL SETS AND SUBSETS

If you wanted to decide which days to exercise, you would choose from the 7 days of the week. Let

$$U = \{\text{Sunday, Monday, Tuesday, Wednesday, Thursday, Friday, Saturday}\}$$

U is the **universal set** that contains all elements being considered in a given situation.

Assume that you decide on a 3-day schedule, either A or B, or a very light exercise schedule, C.

$$A = \{\text{Monday, Wednesday, Friday}\}$$
$$B = \{\text{Tuesday, Thursday, Saturday}\}$$
$$C = \{\ \}$$

Under schedule A, the set of days on which you will not exercise is $\overline{A} = \{\text{Sunday, Tuesday, Thursday, Saturday}\}$, the complement of A.

Definition: Complement of a Set

The **complement** of a set D, written $\overline{D}$, is the set of elements in the universal set that are not in D.

Think of set $\overline{D}$ as "not D."

LE 6   Concept

(a) Write the set of elements in $\overline{B}$ in the exercise-scheduling example.
(b) What does $\overline{B}$ represent in relation to the days on which you will exercise?

Figure 2–4 shows that set A is an example of a subset of U, since each element of A is contained in U. For the same reason, B and C are also subsets of U.

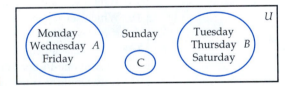

Figure 2–4

The definition of a subset is as follows.

Definition: Subset

Set A is a **subset** of B, written "$A \subseteq B$," if and only if every element of A is also an element of B.

When A is not a subset of B, written "$A \nsubseteq B$," it means that A contains an element that is not in B.

LE 7   Concept

Which of the following sets is a subset of the one-digit whole numbers that are even: $\{0, 2, 4, 6, 8\}$?

(a) $\{4, 8\}$ (b) $\{3\}$ (c) $\{\ \}$ (d) $\{4, 6, 8\}$ (e) $\{4, 6, 8, 10\}$

You may wonder why { } in the preceding exercise is a subset of {0, 2, 4, 6, 8}. For { } to be a subset, every element in { } must also be in {0, 2, 4, 6, 8}. Since there are no elements in { }, it is true that every element in { } is also an element of {0, 2, 4, 6, 8}!

A subset of {0, 2, 4, 6, 8} represents a possible choice. For example, you might choose {4, 6} or {0, 2, 4, 6, 8} from {0, 2, 4, 6, 8}. Another choice would be to select no elements, creating the subset { }.

You may be familiar with ⊂, the notation for a **proper subset.** The expression "$A \subset B$" means that every element of A is also an element of B *and* that B contains at least one element that is *not* in A. So {1, 2} is a proper subset of {1, 2, 3}, but {1, 2, 3} is not a proper subset of {1, 2, 3}. The symbols ⊆ and ⊂ are analogous to ≤ and < for relationships between numbers.

How does one distinguish between ⊆ and ∈? The symbol ⊆ shows a relationship between *sets*, as in "{8} ⊆ {8, 10}." It shows that a *set* containing 8 can be formed using elements of {8, 10}. The symbol ∈ shows that one object is a member of a set, as in "8 ∈ {8, 10}" or "{8} ∈ {{8}, {10}}."

Discussion

LE 8 Concept

Fill in each blank with ∈ or ⊆.

(a) 12 _____ {10, 11, . . . , 19}   **(b)** {2, 4} _____ {0, 2, 4, 6, 8}

AN INVESTIGATION: SUBSETS

LE 9 Reasoning

Is there a pattern in the number of subsets that different-sized sets have?

(a) *Understanding the Problem*   What are you supposed to do in this investigation?
(b) *Devising a Plan and Carrying Out the Plan*   Complete the following table. (*Hint:* The empty set is a subset of every set.)

Set	List of Subsets	Number of Subsets
{1}	{ } {1}	2
{1, 2}		
{1, 2, 3}		
{1, 2, 3, 4}		

(c) *Explain* why {1, 2, 3, 4} has twice as many subsets as {1, 2, 3}.
(d) On the basis of your response to part (c), a set with N elements appears to have _____ subsets.
(e) *Looking Back*   In part (d), you used _____ reasoning.

LE 10 Summary

Tell what new set symbols or terminology you learned in this lesson.

ANSWERS TO SELECTED LESSON EXERCISES

2. (a) $\in$ (b) $\notin$

3. The set of all positive numbers less than 0

4. (a) A, B, D, and F are equivalent, and C and E are equivalent.
(b) $A = F$ (c) No   (d) Yes
(e)

(f) Put set E into a one-to-one correspondence with set C.

5. (a) Finite   (b) Finite   (c) Infinite

6. (a) {Sunday, Monday, Wednesday, Friday}
(b) The days you do not exercise

7. (a), (c), (d)

8. (a) $\in$ (since 12 is not a set)   (b) $\subseteq$

2.1 HOMEWORK EXERCISES

Basic Exercises

1. Name five English words that refer to a set (for example, a coin *collection* or a *herd* of cattle).

2. List the members of the set of whole numbers less than 10 that are even.

3. Let A be the set of odd numbers. Are the following true or false?
(a) $11 \in A$ (b) $10 \in A$
(c) $6 \notin A$ (d) $\{3\} \in A$

4. Write a verbal description of each set.
(a) $\{4, 8, 12, 16, \ldots\}$
(b) $\{3, 13, 23, 33, \ldots\}$

5. Which of the following would be an empty set?
(a) The set of purple crows
(b) The set of odd numbers that are divisible by 2

6. What two symbols are used to represent an empty set?

7. Which of the following represent equal sets?

$D = \{\text{orange, apple}\}$ $F = \{\text{apple, orange}\}$
$A = \{1, 2\}$ $G = \{1, 2, 3\}$
$E = \{\ \}$ $N = \varnothing$ $H = \{a, b, c, d\}$

8. (a) Which of the following represent equivalent sets?

$D = \{\text{orange, apple}\}$ $F = \{\text{apple, orange}\}$
$A = \{1, 2\}$ $G = \{1, 2, 3\}$
$E = \{\ \}$ $N = \varnothing$ $H = \{a, b, c, d\}$

(b) How would you use sets and set concepts from this exercise to show a first grader how many objects are in set D?
(c) How would you use sets and set concepts from this exercise to show a first grader that set H has more than three members?

9. (a) True or false? If two sets are not equal, then they are not equivalent.
(b) If part (a) is true, give an example that supports it. If part (a) is false, give a counterexample.

10. Historical evidence reveals that some hunting tribes counted using equivalent sets. The tribal mathematician placed a rock in a pile for each hunter who left on an expedition. What do you think the mathematician did when the hunters returned (besides eating)?

11. Decide whether each set is a finite set or an infinite set.
(a) The set of whole numbers greater than 6
(b) The set of all the grains of sand on Earth

12. Decide whether each set is an infinite set or a finite set.
(a) the set of people named Lucky
(b) the set of all perfect square numbers

13. Suppose the universal set U = {math, science, English, Spanish, history, art}, and consider two possible subsets, A = {math, English, history, art} and B = {math, English, science, Spanish, art}. List the elements of $\overline{A}$ and $\overline{B}$.

14. Suppose the universal set U = {1, 2, 3, 4, 5, 6, 7, 8, 9, 10}, A = {1, 3, 5, 7, 9}, and B = {1, 2, 3, 4, 5}. List the elements of $\overline{A}$ and $\overline{B}$.

15. The following Venn diagram shows sets A, B, and U. Shade in the interior part of the rectangle that represents $\overline{A}$.

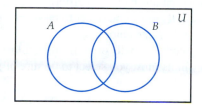

16. Suppose A contains all positive whole numbers. If the universal set is all whole numbers, describe $\overline{A}$.

17. Which of the following sets are subsets of {1, 3, 5, 7}?
    (a) {1, 3, 5, 7}   (b) {9, 11, 13, ...}
    (c) { }   (d) {5}

18. Rewrite the following expressions using symbols.
    (a) A is a subset of B.
    (b) The number 2 is not a member of set T.

19. Let A = {3, 4, 5, 6, 7, 8}. Are the following true or false?
    (a) $2 \notin A$ (b) {4} $\in A$
    (c) $4 \in A$ (d) {4} $\subseteq A$

20. Suppose S is the set of all squares, Q is the set of all quadrilaterals (four-sided figures), and T is the set of all triangles. Are the following true or false?
    (a) $T \subseteq S$ (b) $S \subseteq Q$ (c) $Q \subseteq S$

21. Fill in each blank with $\in$ or $\subseteq$.
    (a) { } _____ {1, 3}
    (b) Jean _____ {Tom, Jean}

22. Decide which symbol, $\in$, $\notin$, $\subseteq$, or $\not\subseteq$, is equivalent to the underlined word in each sentence.

(a) German shepherds <u>are</u> dogs.
(b) Juanita <u>is</u> Spanish.
(c) Jane <u>is not</u> a skydiver.

23. A committee C = {Bobbie, Rupert, Sly, Jenny, Melissa} requires a majority vote to pass any new rules. Each of the following two sets contains a winning coalition with enough voters to pass a new rule. Complete the list of winning coalitions, using just the first letter of each person's name. (Make an organized list.)

$$\{B, R, S\} \quad \{B, R, S, J\}$$

24. The United Nations Security Council has 15 members, of which 5 are permanent and 10 are elected for 2-year terms. No substantive measure can pass unless all 5 permanent members (the United States, the Russian Federation, the United Kingdom, France, and China) vote for it. Overall, nine votes are needed to pass a proposal.
    (a) How many votes are needed from temporary members?
    (b) You may refer to the 10 temporary members as $T_1, T_2, T_3, \ldots, T_{10}$. List five different winning coalitions that each contain exactly nine members.

Extension Exercises

25. How many one-to-one correspondences are possible between each of the following pairs of sets?
    (a) Two sets, each having 2 members
    (b) Two sets, each having 3 members
    (c) Two sets, each having 4 members
    (d) Two sets, each having N members

26. (a) Make up a nonempty set, and list all possible subsets.
    (b) How does the number of subsets with an even (whole) number of elements compare to the number of subsets with an odd number of elements? (*Note:* The even whole numbers start with 0.)
    (c) Repeat parts (a) and (b) for a different nonempty set.
    (d) Propose a generalization of your results.

27. Set A is **infinite** if and only if it can be put into a one-to-one correspondence with a proper subset of itself. For example, {0, 1, 2, 3, ...} is infinite because it can

be put into a one-to-one correspondence with its proper subset, {10, 11, 12, 13, . . .}.

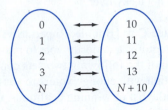

Show that the following sets are infinite.
(a) {0, 2, 4, 6, 8, . . .}
(b) {20, 21, 22, . . .}

28. Georg Cantor was the first to apply set theory to infinite sets. This led to some strange and surprising results.
(a) Make a conjecture about which set has more elements, $W = \{0, 1, 2, 3, . . .\}$ or $E = \{0, 2, 4, 6, . . .\}$.
(b) Cantor reasoned that two equivalent infinite sets (like finite sets) are those that can be put into a one-to-one correspondence. How did he match the members of W and E in a one-to-one correspondence to show that they are equivalent sets?

29. (a) A bag contains balls of 2 different colors. How many balls must you select to be sure of getting at least 2 of one color?

(b) A bag contains balls of 2 different colors. How many balls must you select to be sure of getting at least 3 of one color?
(c) A bag contains balls of 2 different colors. How many balls must you select to be sure of getting at least 4 of one color?
(d) A bag contains balls of 2 different colors. How many balls must you select to be sure of getting at least N of one color?

30. (a) A bag contains balls of 3 different colors. How many balls must you select to be sure of getting at least 2 of one color?
(b) A bag contains balls of 3 different colors. How many balls must you select to be sure of getting at least 3 of one color?
(c) A bag contains balls of 3 different colors. How many balls must you select to be sure of getting at least 4 of one color?
(d) A bag contains balls of 3 different colors. How many balls must you select to be sure of getting at least N of one color?

31. A bag contains balls of K different colors. How many balls must you select to be sure of getting at least N balls of one color?

2.2 OPERATIONS ON TWO SETS

NCTM Standards
- create and use representations to organize, record, and communicate mathematical ideas (pre-K–12)
- identify, compare, and analyze attributes of two- and three-dimensional shapes and develop vocabulary to describe the attributes (3–5)
- use geometric models to solve problems in other areas of mathematics, such as number and measurement (3–5)

Finding the complement of a set is an operation on one set that produces another set. Other set operations act on two sets to produce another set, just as the operation of addition on two numbers, such as 2 + 3, results in another number, 5. This lesson covers two set operations: intersection and union. A third operation, the Cartesian product, is presented in the homework exercises.

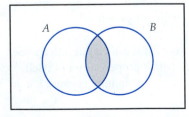

$A \cap B$

Figure 2–5

LE 1 Opener
In Figure 2–5, suppose A is the set of students in the math club and B is the set of students in the science club. What do you know about the students in the shaded region?

INTERSECTION AND UNION

Should Susan and Bert get married? Susan likes skydiving, bronco busting, and chess. Bert likes Sumo wrestling, mountain climbing, and chess. What interests do they have in common?

If A is the set of Susan's interests and B is the set of Bert's interests, the answer to the question is the intersection of sets A and B (members of A *and* B), written $A \cap B$. The only member of $A \cap B$ is "chess." "And" is a key word suggesting intersection. In everyday language and in mathematics, "and" indicates that *both* conditions must be true.

> **Definition: Intersection**
>
> The **intersection** of sets A and B, written $A \cap B$, is the set containing the elements that are in both A and B.

The shaded part of the set picture in Figure 2–5 is $A \cap B$. If A and B have no elements in common, then $A \cap B = \{\ \}$ or $\varnothing$.

LE 2   Concept

The math club membership $A = \{$Joe, Sam, Li, Clara$\}$, and the science club membership $B = \{$Juan, Li, Clara$\}$.

(a) What is $A \cap B$?

(b) Which region of Figure 2–5 would contain students who are in the math club but not in the science club?

(c) The region in part (b) is the intersection of which two sets? (*Hint:* One is the complement of a set.)

Discussion

LE 3   Reasoning

If $B \cap C = C$, what is the relationship between B and C?

When two or more sets are joined together, the new set they form is called the union of those sets. Getting back to Susan and Bert, Susan has a sports car and an airplane, and she has a bicycle that she and Bert received as an engagement gift. Set $C = \{$car, plane, bicycle$\}$. Bert has a motorcycle, a skateboard, and that same bicycle. Set $D = \{$motorcycle, skateboard, bicycle$\}$. If Susan and Bert marry, they will combine these vehicles to form the union of items belonging to either Susan *or* Bert *or both*, written $C \cup D$. The set $C \cup D = \{$car, plane, bicycle, motorcycle, skateboard$\}$.

> **Definition: Union**
>
> The **union** of sets A and B, written $A \cup B$, is the set containing all elements that are either in A or in B or in both A and B.

"Or" is a key word suggesting union. "Or" has a slightly different meaning in mathematics than it sometimes does in everyday speech. In everyday speech, "or" often means "one or the other (but not both)." In mathematics, "A or B" means "A or B or *both*." The shaded area in the set picture in Figure 2–6 is $A \cup B$.

Union and intersection are called **set operations** because they replace two sets with a third set, just as arithmetic operations replace two numbers with a third number. Two properties of set operations are discussed in the homework exercises.

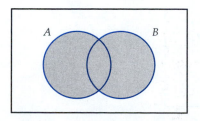
$A \cup B$

Figure 2–6

LE 4 refers back to LE 2.

LE 4   Concept

(a) In LE 2, what is $A \cup B$?

(b) Describe a situation in which the members of $A \cup B$ would be together after school.

Intersection and union are represented in words as "and" and "or," respectively.

LE 5   Reasoning

Decide whether each of the following is true or false. Use the logical meaning of "or" as *one or the other or both*. (This sort of question appears on some teachers' exams.)

(a) $5 + 7 = 8$ or $6 - 2 = 4$.

(b) $8 + 3 = 11$ and $9 - 5 = 7$.

(c) $7 + 2 = 9$ or $8 - 1 = 7$.

Now consider a problem that involves two or more set operations. How does grouping (parentheses) affect the results?

LE 6   Skill

$$A = \{1, 10, 100\}, \qquad B = \{10, 20, 30\}, \qquad \text{and} \qquad C = \{20, 25, 30\}$$

Which two of the following are equal? $(A \cup B) \cap C, A \cup (B \cap C), (A \cup B) \cap (A \cup C)$ Find out in parts (a)–(c).

(a) $(A \cup B) \cap C = $ _____   **(b)** $A \cup (B \cap C) = $ _____

(c) $(A \cup B) \cap (A \cup C) = $ _____

The results of LE 6 suggest that $A \cup (B \cap C) = (A \cup B) \cap (A \cup C)$. Is this equation true for all sets A, B, and C? Shade two Venn diagrams to find out.

LE 7   Reasoning

(a) Draw a three-set Venn diagram, and label the sets A, B, and C.

(b) Shade the region or regions representing $A \cup (B \cap C)$.

(c) Draw a second three-set Venn diagram, and shade the region or regions representing $(A \cup B) \cap (A \cup C)$. (*Hint:* Use the numbered regions shown below. First find $A \cup B$ and $A \cup C$. Then find their intersection. For example, $A \cup B$ would be regions 1, 2, 3, 4, 5, and 6.)

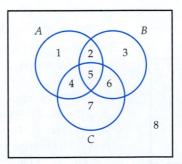

Now consider four different set symbols that you have studied in Sections 2.1 and 2.2. See if you can use them correctly.

Discussion

LE 8   Concept

Fill in each blank with ∈, ∪, ∩, or ⊆.

(a) {3, 5} _____ {6} = {3, 5, 6}
(b) 3 _____ {1, 2, 3}
(c) {3} _____ {1, 2, 3}

TWO-SET VENN DIAGRAMS

All teachers are college graduates. *No* dirty clothes smell nice. *Some* smart people are athletic. (In logic, "some" means "at least one.") Each of these statements can be illustrated with a two-set Venn diagram, as shown in Figure 2–7.

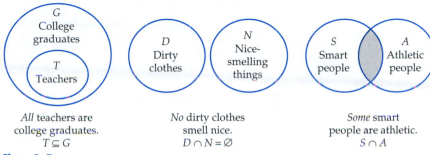

All teachers are college graduates.
$T \subseteq G$

No dirty clothes smell nice.
$D \cap N = \varnothing$

Some smart people are athletic.
$S \cap A$

Figure 2–7

Classroom Connection

LE 9   Concept

A child says the first diagram in Figure 2–7 shows that all college graduates are teachers. How would you convince the child that this is not correct?

LE 10   Concept

Draw a two-set Venn diagram illustrating each of the following statements, and write the relationship between the two sets, using symbols.

(a) All fish are good swimmers.
(b) No U.S. senators are teenagers.
(c) Some football fans are happy.

LE 11   Concept

(a) Write a sentence describing the set relationship shown in Figure 2–8.
(b) Use symbols to represent the relationship between sets A and P.

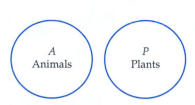

Figure 2–8

LE 12   Reasoning

Let U = {all students at Euclid College}, M = {students taking a math course}, and S = {students taking a science course.}.

The Venn diagram in Figure 2–9 shows student enrollment in mathematics and science courses. How many students are

(a) taking mathematics but not science?
(b) taking mathematics and science?
(c) taking neither mathematics nor science?

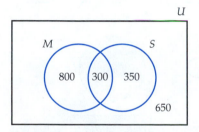

Figure 2–9

If you know some information about sets of people, you can sometimes deduce other information. The following exercise uses a grid model instead of a Venn diagram.

LE 13   Reasoning

A fourth-grade class has 25 students. Ten of them are ten-years-old, twelve of them are boys, and four of them are ten-year-old boys.

(a) Shade a square with a pencil for each student who is ten years old.
(b) Shade a square with a pen for each student who is a boy (and remember that four students are ten-year-old boys).
(c) On the basis of your diagram, how many students are girls who are not ten years old?

Next consider a situation in which you have only limited information about the sets.

Discussion

LE 14   Reasoning

A class has 28 students. There are 13 twelve-year-olds and 16 girls.

(a) What is the largest number of students that could be twelve-year-old boys?
(b) What is the smallest number of students that could be twelve-year-old boys?

ATTRIBUTE BLOCKS

As a teacher, you will sometimes use objects or models to introduce ideas to your students. Attribute blocks are manipulatives (objects) used in elementary school to study shapes, sets, and classification. Attribute blocks are so named because they have a variety of attributes: shape, color, size, and (sometimes) thickness. A typical set of attribute blocks is shown in Figure 2–10. You can buy a set or make one of your own.

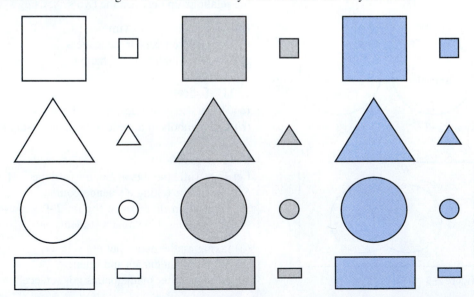

Figure 2–10

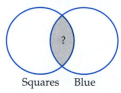

Squares   Blue

Figure 2–11

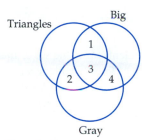

Triangles

Big

1

3

2   4

Gray

Figure 2–13

LE 15   Concept

Which attribute blocks in Figure 2–11 belong in the overlapping part? If you have a set of attribute blocks and some circular loops, use them to make a Venn diagram for Figure 2–11.

LE 16   Reasoning

In Figure 2–12, which attribute blocks belong to the left-hand part of the circle on the left? (Assume all the shapes shown are the larger size.)

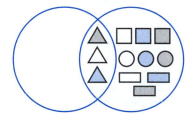

Figure 2–12

LE 17   Reasoning

Which attribute blocks in Figure 2–10 belong in Sections 1, 2, 3, and 4 of Figure 2–13?

LE 18   Summary

Tell what you learned about intersection and union in this section.

ANSWERS TO SELECTED LESSON EXERCISES

2. (a) {Li, Clara}
   (b) The unshaded region of the circle on the left.
   (c) $A \cap \overline{B}$

3. $C \subseteq B$

4. (a) {Joe, Sam, Li, Clara, Juan}
   (b) A joint meeting of the math and science clubs

5. (a) True   (b) False   (c) True

6. (a) {20, 30}   (b) {1, 10, 20, 30, 100}
   (c) {1, 10, 20, 30, 100}

7. Shade regions 1, 2, 4, 5, and 6 in parts (b) and (c).

8. (a) $\cup$ (b) $\in$ (c) $\subseteq$ ({3} is a set)

10. (a)

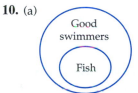

Good swimmers

Fish

(b)

U.S. senators   Teenagers

(c)

Football fans   Happy

11. (a) No animals are plants.
   (b) $A \cap P = \emptyset$

12. (a) 800   (b) 300   (c) 650

13. (b)

   (c) 7

14. (a) 12 (b) 0

15. The small and large blue squares

16. The large gray, blue, and white triangles

17. 1 = {big blue and white triangles},
   2 = {small gray triangles},
   3 = {big gray triangles},
   4 = {big gray squares, circles, and rectangles}

2.2   HOMEWORK EXERCISES

Basic Exercises

1. $A = \{1, 3, 5, 7, 9, 11\}$ and
   $B = \{3, 6, 9, 12, 15, 18\}$.
   (a) $A \cap B = $ _____
   (b) $A \cup B = $ _____
   (c) Is $A \subseteq B$?

2. Suppose U is the set of cards in a regular deck of 52
   cards, R is the set of red cards, and T is the set of
   twos.
   (a) What cards are in $R \cap T$?
   (b) What cards are in $R \cup T$?
   (c) What is $\bar{R}$?
   (d) Parts (a) and (b) involve sets R and T, set definitions, and _____ reasoning.

3. Explain how an intersection of two streets is like an
   intersection of two sets.

4. Tell the difference between the union of two sets and
   the intersection of two sets.

5. Decide whether each of the following is true or false.
   Use the logical meaning of "or" as *one or the other or
   both*.
   (a) $8 + 5 = 13$ or $6 - 2 = 4$
   (b) $7 - 4 = 1$ or $10 - 2 = 4$.
   (c) $5 + 7 = 8$ and $6 - 2 = 4$.

6. Decide whether each of the following is true or false.
   (a) $7 + 2 = 9$ and $8 - 1 = 7$
   (b) $4 - 3 = 1$ or $8 + 1 = 9$.
   (c) $4 + 4 = 8$ or $5 - 3 = 0$.

7. U is the set of all college students, M is the set of male
   college students, F is the set of female college students, E is the set of education majors, and Y is the
   set of college students under 22 years of age.
   (a) Describe $\bar{E} \cap \bar{F}$ in words.
   (b) Describe $\bar{E} \cap \bar{Y}$ in words.
   (c) $M \cup F = $ _____
   (d) $M \cap F = $ _____

8. W is the set of overweight people, S is the set of cigarette smokers, and E is the set of people who exercise
   regularly. Describe the following sets in words.
   (a) $W \cap S$ (b) $W \cup S$ (c) $W \cap S \cap E$
   (d) $\bar{W}$ (e) $\overline{W \cap S}$

9. A logician says, "I will eat fish or spinach for dinner."
   What will the logician possibly eat for dinner?

10. A logician says, "Tomorrow will be rainy or cold."
   What is the logician possibly predicting for
   tomorrow?

11. If $B \cup C = C$, then how are sets B and C related?

12. If $A \cap B \cap C = B$, then how are A and C related to B?

13. (a) The operation $\cap$ is **associative** because
   $(A \cap B) \cap C = A \cap (B \cap C)$ for all sets A and B.
   Write a comparable equation for $\cup$ and decide
   whether the operation $\cup$ is associative.
   (b) Support your conclusion in part (a) by shading two
   Venn diagrams and comparing them.

14. The operation ∩ is **commutative** because $A \cap B = B \cap A$ for all sets A and B. Is the operation ∪ commutative?

15. $A = \{1, 2, 3\}$, $B = \{3, 4, 5\}$, and $C = \{2, 4, 6, 8\}$.
(a) $A \cap (B \cup C) = $ _____
(b) $(A \cap B) \cup C = $ _____
(c) $(A \cap B) \cup (A \cap C) = $ _____
(d) Which two answers from parts (a)–(c) are the same?
(e) Shade two three-set Venn diagrams to show that the two equal expressions you selected in part (d) will be equal for all sets A, B, and C.

16. $A = \{1, 2, 3, 4, 5\}$, $B = \{3, 5, 7\}$, $C = \{2, 4, 8\}$, and $U = \{1, 2, 3, 4, 5, 6, 7, 8\}$.
(a) $\overline{A} \cap \overline{B} \cup C = $ _____   (b) $\overline{A \cup B} \cup C = $ _____
(c) $\overline{A} \cup \overline{B} \cup C = $ _____
(d) Which two answers from parts (a)–(c) are the same?
(e) Shade two three-set Venn diagrams to show that the two equal expressions you selected in part (d) will be equal for all sets A, B, and C.

17. A has 3 members, and B has 2 members.
(a) What is the largest number of members $A \cup B$ could have?
(b) What is the smallest number of members $A \cup B$ could have?

18. A has a members and B has b members, where $a \geq b$.
(a) What is the largest number of members $A \cup B$ could have?
(b) What is the smallest number of members $A \cup B$ could have?

19. A sixth-grade class has 24 students. Four of them are in the chess club, 6 are in the Spanish club, and 2 are in both clubs.
(a) Shade squares with a pencil to represent the students in the chess club.

(b) Shade squares with a pen to represent the students in the Spanish club (and remember that 2 are in both clubs).
(c) Based on your diagram, how many students are in neither club?

20. A class has 32 students. There are 17 boys and 22 eight-year-olds.
(a) What is the largest number of the girls that could not be eight years old?
(b) Tell how you solved the problem.

21. In an apartment complex of 40 apartments, 25 have porches and 22 have storage rooms.
(a) What is the largest number of apartments that could have a porch and a storage room?
(b) What is the smallest number of apartments that could have a porch and a storage room?

22. In an apartment complex of 30 apartments, 20 apartments receive the *New York Times* and 14 apartments receive the *Sporting News*.
(a) What is the largest number of apartments that could receive *both* publications?
(b) What is the smallest number of apartments that could receive *both* publications?

23. Assume $2 \in (A \cup B)$. Which of the following statements could be true?
(a) 2 is in A but not in B.
(b) 2 is in B but not in A.
(c) 2 is in both A and B.
(d) 2 is in neither A nor B.

24. True or false?
(a) If $A \subseteq B$, then $B \subseteq A$.
(b) $A \cup \varnothing = \varnothing$
(c) $A \cap \varnothing = \varnothing$
(d) $0 \in \varnothing$

25. Fill in each blank with $\in$, $\cup$, $\cap$, or $\subseteq$.
(a) 1 _____ $\{1, 2\}$
(b) $\{1, 3, 5\}$ _____ $\{5\} = \{5\}$
(c) $\{6, 7\}$ _____ $\{5, 6, 7\}$

26. Fill in each blank with $\in$, $\cup$, $\cap$, or $\subseteq$.
(a) $\{3\}$ _____ $\{2, 3, 6\}$
(b) 8 _____ $\{7, 8, 9\}$

27. If possible, make up two sets such that the number of elements in A plus the number of elements in B is
   (a) less than the number of elements in $A \cup B$.
   (b) equal to the number of elements in $A \cup B$.
   (c) greater than the number of elements in $A \cup B$.

28. Make up two sets A and B such that the number of elements in A equals the number of elements in $A \cap B$.

29. Draw a two-set Venn diagram to illustrate each of the following statements, and use symbols to represent the relationship between the two sets.
   (a) All parallelograms are quadrilaterals.
   (b) No man is an island.
   (b) Some college graduates are taxi drivers.

30. Draw a two-set Venn diagram to illustrate each of the following statements, and use symbols to represent the relationship between the two sets.
   (a) All musicians are creative people.
   (b) No square is a triangle.
   (c) Some vegetables are green.

31. (a) Write a sentence describing the set relationship shown in the figure.

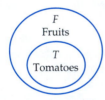

   (b) Use symbols to represent the relationship between sets F and T.

32. (a) Write a sentence describing the set relationship shown in the figure.

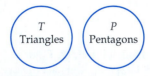

   (b) Use set symbols to represent the relationship between sets T and P.

33. The following type of question appears on a national teachers' exam. Which of the following is NOT consistent with the sentence:
   "Some values of N are greater than 10." (*Hint:* "NOT consistent" means it contradicts the given sentence.)

   (a) 5 is a value of N.
   (b) All values of N are greater than 10.
   (c) Some values of N are less than 10.
   (d) 40 is not a value of N.
   (e) No numbers greater than 10 are values of N.

34. Which of the following is NOT consistent with the sentence:
   "All values of x are greater than 4."
   (a) 5 is not a value of x.
   (b) 95 is a value of x.
   (c) Some values of x are less than 2.
   (d) All values of x are less than 100.
   (e) No numbers less than 10 are values of x.

35. A survey is taken of 500 adults. The results follow. Let $U = \{\text{all adults}\}$, $T = \{\text{adults who watch television}\}$, and $R = \{\text{adults who read books}\}$.

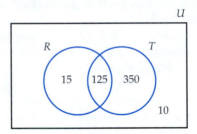

   How many adults
   (a) watch television and read?
   (b) watch television but do not read?
   (c) do not watch television and do not read?
   (d) Express the sets in parts (a), (b), and (c) with set notation.

36. A survey is taken of 300 college students about whether they eat breakfast and whether they have a job. Two hundred seventy eat breakfast, and 150 have a job. Twenty have a job and do not eat breakfast.
   (a) Draw a two-set Venn diagram that displays these results.
   (b) How many students eat breakfast but do not have a job?

37. A survey asked 100 students whether they play tennis or swim. Twenty-one play tennis, and 30 swim. Fourteen play tennis and swim.
   (a) Draw a two-set Venn diagram that displays these results.
   (b) How many students play neither sport?

38. A survey asked 50 students if they drive to school and if they work. Twenty-four drive to school, and 34 work. Eighteen drive to school and work.
(a) Draw a two-set Venn diagram that displays these results.
(b) How many students neither drive to school nor work?

39. Suppose T = {tennis players}, C = {chess players}, and A = {artists}. The regions of the Venn diagram of sets T, C, and A are labeled 1–8.

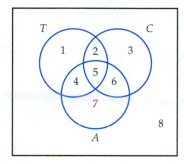

Which region or regions would contain
(a) your name?
(b) people who are artists and tennis players?
(c) people who are chess players but not artists?
(d) people who are tennis players?
(e) people who are artists and chess players but do not play tennis?

40. Suppose P = {people who like prunes}, G = {people who like grapes}, and R = {people who like raspberries}. The regions of a Venn diagram of sets P, G, and R are labeled 1–8.

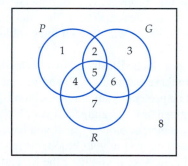

Which region or regions would contain
(a) your name?
(b) people who like raspberries and grapes?
(c) people who like raspberries and grapes but not prunes?
(d) people who like raspberries?
(e) people who like raspberries but not grapes?

41. Consider a set of attribute blocks; T = {triangles} and G = {gray}.
(a) What is $T \cup G$?
(b) What is $T \cap G$?

42. Using the set of attribute blocks described in the lesson, tell which ones would belong in the right-hand section of the circle on the right.

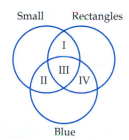

43. Which attribute blocks described in the lesson belong in Sections I, II, III, and IV?

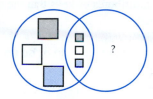

44.

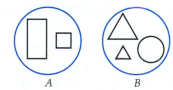

All the attribute pieces in set A have a certain attribute that the shapes in set B do not have. What attribute could it be?

45.

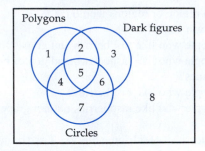

Which region would each figure go in?

(a)   (b)   (c)   (d)

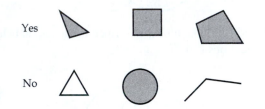

46. Teachers sometimes have children sort keys. Name two different attributes of keys that the children could use to sort them.

47. Following are figures that do or do not qualify for set *P*.

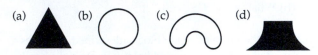

Which of the following is in set *P*?

(a)   (b)   (c)   (d)

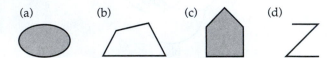

48. Following are words that do or do not qualify for set *P*.

Yes	LATE	PIE	ROAM
No	LOW	HOUSE	ENTREE

Which of the following is in set *P*?
(a) EITHER   (b) QUIT   (c) SHE   (d) I

Extension Exercises

49.

Reprinted with special permission of NAS, Inc. © Field Enterprises, Inc. 1976

(a) What do the percentages in the cartoon add up to?
(b) Why can the percentages add up to more than 100% and the data still be correct?

50. (a) At a college, 90% of all first-year students take English and 80% take mathematics. What is the minimum percentage of students who take both subjects?
(b) At the same college, 75% of the first-year students take social science. What is the minimum percentage of students who take all three subjects?

51. Blood types can be illustrated by a Venn diagram. There are three antigens, A, B, and Rh, that may or may not be present in any human's blood. If you have the A antigen or the B antigen in your blood, that letter appears in your blood type. If you have neither the A nor the B antigen in your blood, the letter O appears in your blood type. If you have the Rh antigen in your blood, a plus sign (+) appears in your blood type. If you don't have Rh, a minus sign (−) appears in your blood type. (You probably wonder how anyone ever made up this system to type blood. I wonder, too.)

(a) The blood types can be shown nicely in a Venn diagram. I started it for you. Fill in the types in the regions that have question marks.

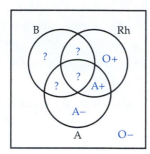

(b) The approximate percentages of each blood type in the world are as follows.

	A	O	B	AB
+	37%	32%	11%	5%
−	6%	6.5%	2%	0.5%

Fill in these percents in the appropriate regions of a Venn diagram.

(c) The following chart can be used for transfusions.

Blood Type	Can Receive
O	O
A	A, O
B	B, O
AB	A, B, AB, O

Which blood type is the "universal donor"?

52. Following are the results of a survey about preferred alcoholic beverages.

Age	Don't Drink (N)	Drink Only Beer or Wine (B)	Drink Hard Liquor (H)
18–30 (Y)	40	42	18
31–55 (M)	26	46	28
Over 55 (E)	21	43	36

Using the letters in parentheses from the table to represent each set, tell how many people are in each of the following sets.

(a) Y (b) $E \cap B$ (c) $M \cap N$
(d) $Y \cup M$ (e) N (f) $Y \cup M \cup E$

53. Another set operation, the **Cartesian product**, shows all the ways you can pair a member of one set with a member of a second set.
(a) Suppose you have 3 shirts, $S = \{$purple, yellow, green$\}$, and 2 pairs of pants, $P = \{$red, blue$\}$. How many possible shirt-pant outfits could you create?
(b) Complete the following. The Cartesian product $S \times P = \{$(purple, red), (purple, blue), (,   ), (,   ), (,   ), (,   )$\}$.

54. (a) If A has 3 elements and B has 1 element, then $A \times B$ has _____ elements.
(b) If A has 3 elements and B has 3 elements, then $A \times B$ has _____ elements.
(c) Using the results of the preceding exercise and parts (a) and (b), complete the following. If A has m elements and B has n elements, then $A \times B$ has _____ elements.

55. Apply the generalization from part (c) of the preceding exercise. Marissa has 10 blouses and 7 skirts. How many different blouse-skirt outfits can she create?

56. Lois has 2 different skirts and 6 different blouses. How many possible blouse-skirt outfits can she create?

57. $A = \{1, 2\}$ and $B = \{0, 2, 4, 6\}$
(a) $A \times B =$ _____
(b) How many ordered pairs are in $A \times B$?

58. Create sets A and B so that $A \times B$ has 6 members.

59. Which of the following is equal to $\overline{A \cup B}$? (*Hint:* Make up sets of elements for A, B, and U, or shade some Venn diagrams.)
(a) $\overline{A} \cup \overline{B}$ (b) $\overline{A} \cap \overline{B}$ (c) $A \cap B$

60. Which of the following is equal to $\overline{A \cap B}$?
(a) $\overline{A} \cup \overline{B}$ (b) $\overline{A} \cap \overline{B}$ (c) $A \cap B$

2.3 FUNCTIONS AND RELATIONS

NCTM Standards
- represent and analyze patterns and functions, using words, tables, and graphs (3–5)
- investigate how a change in one variable relates to a change in a second variable (3–5)

It is often useful to study two things that are changing at the same time.

LE 1 Opener

(a) List a quantity or measurement that changes.

(b) List something else that changes in relation to your answer to part (a).

(c) You have some hot soup sitting on the dining room table. Name two measurements that are changing, and tell how they are changing.

In the preceding exercise, you considered two quantities that are connected to one another. Functions can often be used to describe such relationships.

FUNCTIONS

The amount you learn in a mathematics class is a function of the amount of time you spend studying. The height of a candle is a function of the time it has been burning. A person's salary is a function of the number of hours worked.

A function can be represented with a table, an equation, a rule (in words), or a graph. Consider the following specific example of a function.

LE 2 Concept

A bank teller at Merger Savings Bank makes $14 an hour for working up to 8 hours each day.

(a) Complete the table.

number of hours (x)	0	2	4	6	8
pay in $ (y)	0	28			

(b) An equation that relates x and y is $y =$ _____.

(c) Give a rule in words that tells how to find y for a given value of x.

(d) Sketch a graph of the five points in part (a).

A function connects two sets of numbers such as the number of hours you work and your pay. A function matches each value of the first variable (called the **input**) to *exactly one* value of the second variable (called the **output**). Functions result from our observation and analysis of patterns.

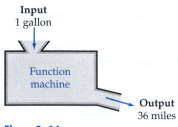

Input
1 gallon

Function
machine

Output
36 miles

Figure 2–14

Lejeune Dirichlet, a German mathematics professor, defined a function in 1837.

> **Definition: Function**
>
> A **function** from set *A* to set *B* assigns each member of set *A* to exactly one member of set *B*.

A function is often illustrated with a "function machine" (Figure 2–14). The set of input values is the **domain**, and the resulting set of output values is the **range**. Figure 2–15 shows how functions are introduced in a fifth-grade textbook.

**Classroom
Connection**

LESSON

3 **Functions**

▶ **Learn**

DOG WALKER   After school, Selena walks some of her neighbors' dogs. She earns $2 for each day she walks a dog. Last week, she walked Bandit on 4 days. How much money did Selena earn walking Bandit last week?

You can use a function table to show the number of dollars Selena earns for walking dogs for different numbers of days. The table matches each input value, *t* (days), with an output value, *d* (dollars). It shows a function.

days, *t*	1	2	3	4	5	8
dollars, *d*	2	4	6	8	10	16

So, Selena earned $8.

MATH IDEA   A **function** is a relationship between two variables in which one quantity depends on the other.

In this problem, the number of dollars earned, *d*, depends on the number of days walked, *t*. You can use an equation to represent this function.

$d = 2 \times t$, or $d = 2t$

From Harcourt Math Grade 5 (Orlando: Harcourt School Publishers, 2002) p. 155.

Figure 2–15

LE 3   Concept

What are the domain (input set) and range (output set) for the bank teller function in LE 2?

Does every table of values make a function? No. Consider a table that shows the square roots (y) of some perfect square numbers (x).

x	0	1	1	4	4
y	0	1	-1	2	-2

equation: $x = y^2$

This example is not a function because each x-value is not assigned to exactly one y-value. An x-value of 1, for example, is assigned to two different y-values.

When you match the members of one set to the members of another set, you define a **relation**. But not all relations are functions.

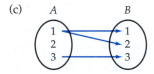 **LE 4   Concept**

Each arrow diagram matches the corresponding members of the input and output sets with arrows. Which of the diagrams illustrate a function from set A to set B?

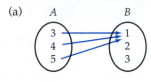

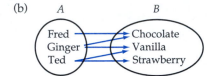

LE 5   Concept

Tell whether each of the following represents a function.

(a) input: person   output: zip code of residence
(b) input: country   output: capital
(c) input: month   output: number of days

GRAPH OF A FUNCTION

The tales in this lesson illustrate how the pay for the bank teller (a function) does not have two ordered pairs with the same x-value, but $x = y^2$ (not a function) does have two ordered pairs with the same x-value. How is this difference reflected in graphs?

LE 6   Concept

(a) Suppose a graph has two different points with the same x-value. How will the points appear in relation to one another?

(b) Part (a) describes a graph that is not a function. On the basis of this, how can you use a vertical line to tell when a graph is not a function?

Did you formulate the vertical line test in the preceding lesson exercise?

Vertical Line Test
If any vertical line intersects a graph at more than one point, the graph does not represent a function.

LE 7   Skill

State whether each graph represents a function.

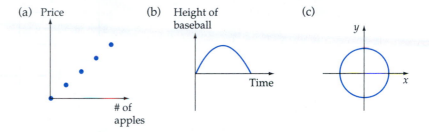

(a) Price

# of apples

(b) Height of baseball

Time

(c)

y

x

When a function involves a countable set such as apples, the graph (LE 7(a)) is a **discrete** set of points. When a function involves a measurement such as height (LE 7(b)), the graph is a **continuous** curve or set of curves.

FINDING A RULE

In Section 1.4, you found rules for the nth term of arithmetic and geometric sequences. This is related to finding an equation of a function.

LE 8   Concept

Consider the number sequence 30, 40, 50, . . .

(a) Is the sequence arithmetic, geometric, or neither?

(b) What is the rule for the nth term of the sequence?

(c) The relationship between n and each number in the sequence creates a function. The table shows three points.

n	1	2	3
y	30	40	50

Find the equation of the function.

(d) Suppose a repair person charges $\$y$ for n hours of labor using this function. Describe the repair rates in words.

In business and science, researchers often look for a pattern in a set of data points. One can sometimes find an equation that fits the data.

LE 9   Reasoning

Suppose you light a 10-inch candle when the lights go out, and you record the following data.

T (elapsed time in hours)	0	1	2	3
H (candle height in inches)	10	8	6	

(a) Fill in the last value for H.
(b) What is H when $T = 5$?
(c) Describe the relationship between H and T.
(d) Write a formula: $H = $ _____ .
(e) Graph each pair of values for T and H.

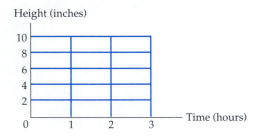

(f) What pattern do you see in the graph?

The game "guess my rule" is often used to introduce the concept of a function in elementary school. See if you can guess the rules in the following exercise.

LE 10   Reasoning

Suppose a teacher thinks of a rule. Then for each input given by a student, the teacher applies the rule and gives the result. In each part, state the teacher's rule in words and as an equation.

(a)

Student says (x)	0	1	10	20
Teacher says (y)	3	4	103	403

(b)

Student says (x)	0	1	5	10
Teacher says (y)	0	1	1	0

If you have the equation of a function, you can use a graphing calculator or spreadsheet to quickly generate a set of output values for a given set of input values. Spreadsheets are discussed in more detail in Section 7.2. Graphing calculators and spreadsheets are used with functions in Chapter 11.

LE 11   Summary

Tell all the different ways you can represent a function. Make up an example of an application of a function and show the different ways to represent it.

ANSWERS TO SELECTED LESSON EXERCISES

2. (a)

number of hours (x)	0	2	4	6	8
pay in $ (y)	0	28	56	84	112

(b) $y = 14x$
(c) y is 14 times x.
(d) The five points lie on a straight line that passes through the origin.

3. Domain: $\{0, 2, 4, 6, 8\}$ range: $\{0, 28, 56, 84, 112\}$

5. (a) No   (b) Yes   (c) No (February)

8. (a) Arithmetic
(b) nth term $= 20 + 10n$
(c) $y = 20 + 10n$
(d) The charge is $20 plus $10 for each hour, or the charge is $30 for the first hour and $10 for each additional hour.

10. (a) $y = x^2 + 3$
(b) If x is even, $y = 0$. If x is odd, $y = 1$.

2.3   HOMEWORK EXERCISES

1. A construction worker at Hard-Headed Builders makes $23 an hour for working up to 8 hours each day.
(a) Complete the table.

number of hours (x)	0	2	4	6	8
pay in $ (y)	0	46			

(b) An equation that relates x and y is $y =$ _____.
(c) Give a rule in words that tells how to find y for a given value of x.
(d) Sketch a graph of the five points in part (a).

2. A manufacturer of tennis balls packs 3 balls in each can.
(a) Complete the table.

number of cans (x)	1	2	3	4
number of balls (y)	3	6		

(b) An equation for y in terms of x is $y =$ _____.
(c) Give a rule in words that tells how to find y for a given value of x.
(d) Sketch a graph of the four points in part (a).

3. What are the domain (input set) and range (output set) for the function in Exercise 1?

4. What are the domain and range for the function in Exercise 2?

5. Measuring creates a relation between two sets. In a shoe store, a salesperson measures the length of each customer's left foot. This would create a measuring function.
(a) What could be the domain of the function?
(b) What numbers might be in the range?

6. A function has the rule $P = 8N - 50$. The range (output set) for P is $\{46, 62, 78\}$. What is the domain (input set)?

7. (a) Tell which of the following diagrams represents a function from set A to set B.
(b) Give a possible rule relating each set of ordered pairs.

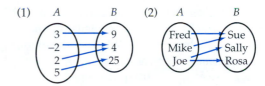

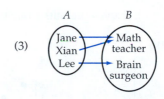

8. (a) Tell which of the following diagrams represents a function from set A to set B.

(b) Give a possible rule relating the ordered pairs.

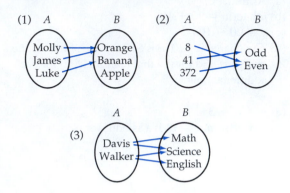

(1) A　B

Molly, James, Luke → Orange, Banana, Apple

(2) A　B

8, 41, 372 → Odd, Even

(3) A　B

Davis, Walker → Math, Science, English

9. Which of the following are functions from x to y? (Assume that the entire domain is given.)

(a) $\{(4, 2)\ (4, 3)\ (4, 4)\ (4, 5)\}$

(b) $\{(1, 7)\ (2, 8)\ (3, 9)\ (4, 10)\}$

(c) $\{(2, 4)\ (3, 4)\ (4, 4)\ (5, 4)\}$

10. Which of the following are functions from x to y? (Assume that the entire set of ordered pairs is given.)

(a) $\{(1, 2), (2, 2), (3, 4), (4, 5)\}$

(b) $\{(1, 3), (5, 1), (5, 2), (7, 9)\}$

(c) $\{(95, 15), (10, 30), (15, 45)\}$

11. Which of the following assignments creates a function?

(a) Each student in a school is assigned a teacher for each course.

(b) Each dinner in a restaurant is assigned a price.

(c) Each person is assigned a birth date.

12. A movie theater charges $2 for children ages 5–17, $6 for adults, and $4 for seniors (65 and up). There is no charge for children under 5. Which of the following sets of ordered pairs creates a function?

(a) (age, price)　　(b) (price, age)

13. Tell whether each graph represents a function.

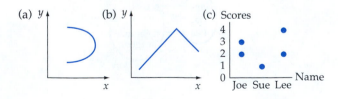

(a) y, x　(b) y, x　(c) Scores — 4, 3, 2, 1, 0 / Joe Sue Lee / Name

14. Tell whether each graph represents a function.

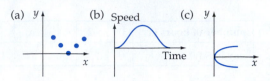

(a) y, x　(b) Speed, Time　(c) y, x

15. A construction company charges $60,000 plus $65 per square foot to build a house. The total cost $C = 60{,}000 + 65F$, in which F equals the floor space in square feet.

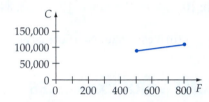

C: 150,000 / 100,000 / 50,000 / 0 — 0, 200, 400, 600, 800 F

(a) On the basis of the graph, what is the domain (input set)?

(b) What is the range (output set)?

(c) Explain why the graph exhibits a function from set F to set C.

16. Your car gets 36 miles per gallon. You fill up the tank with 15 gallons and drive. The number of gallons, g, left after you drive m miles is given by $g = 15 - \dfrac{m}{36}$.

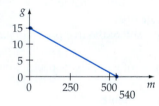

g: 15, 10, 5, 0 — 0, 250, 500, 540 m

(a) On the basis of the graph, what is the domain (input set)?

(b) What is the range (output set)?

(c) Explain why the graph exhibits a function from set m to set g.

17. Read the following table.

N (number sold)	1	2	3	4	5
P (profit)	1	3	5	7	

(a) Write a formula that relates each N value to its corresponding P value.

(b) Use your rule to fill in the last value of P.
(c) You found a rule in part (a) by using
_____ reasoning.
(d) Plot each pair of values for N and P on a graph.
(e) What pattern do you see in your graph?
(f) Write a rule for the nth term of
$$1, 3, 5, 7, \ldots$$

18.

x	0	1	2	3
y	7	9	11	

(a) Fill in the last value for y, continuing the pattern in the table.
(b) A formula relating y to x is $y = $ _____.
(c) Does part (b) involve induction or deduction?

19. Read the following table.

L (length)	1	2	3	4	5
V (volume)	1	8	27	64	

(a) A rule that relates L to V is $V = $ _____.
(b) Use your rule to fill in the last value of V.
(c) Write a rule for the nth term of
$$1, 8, 27, 64, \ldots$$

20. Read the following table.

X	1	2	3	4	5
Y	5	12	31	68	

(a) A rule that relates X to Y is $Y = $ _____.
(*Hint:* See the preceding exercise.)
(b) Use your rule to fill in the last value of Y.

 21. (a) Look at the following table and find an approximate formula for obtaining the flash number N from the ASA film speeds. (*Hint:* Find $\sqrt{S}$.)

ASA Film speed (S)	100	200	400	800
Flash Number (N)	120	170	240	340

(b) Use your formula to find the flash number for an ASA film speed of 64.

22. If you jump in the air, the length of time you remain in the air is related to how high you jump.

Height (ft)	1	4	9
Time in the Air (s)	0.5	1	1.5

(a) If the pattern continues, fill in the next height and time in the table.
(b) A formula relating time T to height H is $T = $ _____.

23. Examine the following designs.

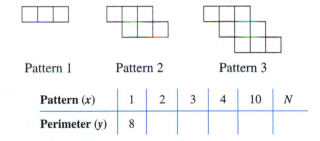

Pattern (x)	1	2	3	4	N
Perimeter (y)	4	6			

(a) Find the perimeter of Pattern 3 and write it in the table.
(b) Draw Pattern 4 and find its perimeter.
(c) Plot each pair of values for x and y.
(d) What pattern do you see in your graph?
(e) Figure out the formula for the perimeter of a Pattern N.

24. Examine the following designs.

Pattern (x)	1	2	3	4	10	N
Perimeter (y)	8					

(a) Find the perimeters of Patterns 2 and 3 and write them in the table.
(b) Draw Pattern 4 and find its perimeter.
(c) Graph each pair of values for x and y.
(d) What pattern do you see in your graph?
(e) Try to guess the perimeter for Pattern 10.
(f) Figure out a formula for the perimeter of Pattern N.

25. Find a rule that works for all of the following number pairs, and use that rule to fill in the blanks.

Expenses		Reserves
0	$\longrightarrow$	8
1	$\longrightarrow$	7
2	$\longrightarrow$	6
3	$\longrightarrow$	___
N	$\longrightarrow$	___

26. Fill in the blanks, finding a rule that works for each number pair.

Regular Price		Sale Price
$10	$\rightarrow$	$ 8
$20	$\rightarrow$	$16
$30	$\rightarrow$	$24
$40	$\rightarrow$	_____
$N	$\rightarrow$	_____

Extension Exercises

27. Some relations are **transitive.** The relations "is equal to" and "is less than" are transitive, as illustrated by the following. For real numbers x, y, and z, if $x = y$ and $y = z$, then $x = z$; and if $x < y$ and $y < z$, then $x < z$.

The relation "is the same age as" is also transitive. If person x is the same age as person y and person y is the same age as person z, then person x is the same age as person z. (x, y, and z are not necessarily different people.) Name two other relations that are transitive.

28. $y = x + 2$ and $z = 3y$.
(a) Describe (in words) the rule that relates x to y.
(b) Describe (in words) the rule that relates y to z.
(c) Find a function that relates z directly to x.
(d) Describe (in words) the rule that relates z to x.

29. $A = \{5, 10, 15, 20\}$, and $B = \{9, 18\}$. List the elements (x, y) of $A \times B$ such that $x < y$.

Labs and Projects

30. Drop a ball from different heights. Each time, measure the initial height and the height of the first bounce. Can you find an equation that relates the height of the first bounce to the initial height?

CHAPTER 2 SUMMARY

Sets are an organizing concept in mathematics. In elementary-school arithmetic, one examines the operations and properties of sets of numbers. Geometric shapes can also be grouped into sets according to various characteristics, such as the number of sides. In statistics, data are collected in sets. In probability, one studies sets of possible outcomes.

A set can be represented by a list of elements, a verbal description, or a symbol. In studying two sets, one may want to compare their elements. If one set is contained in the other, then it is a subset of the other set.

One can operate on two sets to create a third set that is related to the other two. The intersection of two sets contains all objects that are in both sets. The union of two sets contains all objects that are in one set or the other set or both. The Cartesian product of two sets creates ordered pairs from the elements of the two sets.

Attribute blocks help children develop their understanding of geometric concepts as they sort and classify shapes. Attribute blocks can also be used within Venn diagrams to give a visual representation of relationships among sets.

In many mathematics problems, one studies a function that assigns each member of a set to exactly one member of a second set. A function can usually be represented with a graph, a table, an equation, or words.

STUDY GUIDE

To review Chapter 2, see what you know about each of the following ideas or terms that you have studied. You can also use this list to generate your own questions about the chapter.

SETS AND FUNCTIONS IN GRADES 1–8

Although set language is used in all areas of mathematics, set theory is not studied in most elementary schools currently as it was in the 1960s and early 1970s. The following chart shows at what grade levels set topics typically appear in elementary- and middle-school mathematics textbooks.

Topic	Typical Grade Level in Current Textbooks
One-to-one correspondence	1
Venn diagrams	4, 5, 6, 7, 8
Functions	5, 6, 7, 8

REVIEW EXERCISES

1. (a) How many subsets does $\{3, 5, 7\}$ have?
   (b) If the universal set is $U = \{3, 4, 5, 6, 7\}$, create a set A that is equivalent to $\{3, 5, 7\}$.
   (c) Show a one-to-one correspondence between set A and $\{3, 5, 7\}$.
   (d) What is $\overline{A}$?

2. Give an example of two sets that are equivalent but not equal.

3. A partition of a set divides it into nonempty proper subsets. Two partitions of the set $\{1, 2, 3\}$ are shown.

   (a) How many different partitions are there of $\{1, 2, 3\}$?
   (b) How many different partitions are there of $\{1, 2, 3, 4\}$?

4. Fill in each blank with $\in$, $\cup$, $\cap$, or $\subseteq$.
   (a) $\{7, 9\}$ _____ $\{7, 9, 10\} = \{7, 9\}$
   (b) 8 _____ $\{3, 5, 8\}$

5. Let $R = \{31, 32, 33, 34, 35\}$, and let W be the set of whole numbers. Determine whether the following are true or false.
   (a) $R \subseteq W$ (b) $10 \notin R$ (c) $R \cup W = W$

6. Draw a Venn diagram illustrating the following statement, and use symbols to represent the relationship between the two sets: "All elementary-school teachers are college graduates."

7. This semester, 1000 students registered for classes at a college; 300 of them signed up for math.
   (a) What is the smallest number of students who could have registered for a course other than math?
   (b) What is the largest number of students who could have registered for an English class?
   (c) What is the largest number of students who could have registered for both math and English?

8. (a) Use the following information to determine how many households in a community have dogs.
   1. Seven households have only cats.
   2. Six households have no pet.
   3. Two households have only dogs.
   4. There are no pets in the community other than cats and dogs.
   5. There are 18 households.
   (b) Part (a) is an example of _____ reasoning.

9. In which region or regions does each of the following belong?

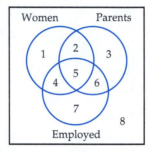

   (a) A seven-year-old boy who plays when he's not in school
   (b) A man with two young children who works
   (c) Women who are not parents
   (d) You

10. What attribute blocks belong in Section I?

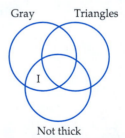

11. (a) Tell which of the following diagrams represents a function from set A to set B.
   (b) Give a possible rule relating each ordered pair.

(1) 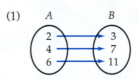   (2)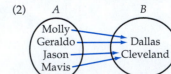

12. Does the following represent a function for the domain $\{0, 1, 2, \ldots, 90\}$?

x = Number of Students in a Section of Math 130	y = Number of Sections of Math 130
0–7	0
8–32	1
33–62	2
63–90	3

13. Suppose the daily profit P (in $) from selling x neckties is $P = 10x - 25$.

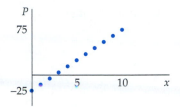

(a) According to the graph, what is the domain (input set)?

(b) What is the range (output set)?

(c) Explain why the graph exhibits a function from x to P.

14. Consider the following function for the speed and force of crash impact for a particular car.

Speed (mph)		Force
20	→	100
30	→	225
40	→	400
50	→	625
60	→	_____

(a) When the speed doubles, the force is multiplied by _____.

(b) Fill in the force for 60 mph that continues the pattern.

(c) Find a formula relating force to speed.

(d) What factors in addition to speed would affect the impact force of a car?

(e) What is a reasonable domain for this function?

15. Suppose you have to wait for a train to pass at a railroad crossing. The waiting time is related to the speed of the train.

Speed (mi/hr)	20	30	40	50	60
Waiting time (min)	6	4	3	2.4	2

A formula that relates waiting time T to speed S is $T = $ _____.

16. Which of the following is NOT consistent with the sentence: "All values of t are less than 25."

(a) 8 is a value of t.

(b) 20 is not a value of t.

(c) No numbers greater than 20 are values of t.

(d) Some values of t are greater than 50.

(c) All values of t are greater than 10.

ALTERNATE ASSESSMENT—KEEPING A JOURNAL

Use a journal to describe in your own words the major concepts and procedures that you learned in Chapter 2. You can also write your thoughts or feelings about what you have learned. What is something interesting that you learned? What topics are giving you difficulty? What questions do you have about the material?

Alternatively, you could add to your portfolio.

SUGGESTED READINGS

Eves, H. *Great Moments in Mathematics After 1650*. Washington, D.C.: Mathematical Association of America, 1983.

Goodnow, J. *Moving On with Attribute Blocks*. Oak Lawn, Ill.: Creative Publications, 1988.

3 | WHOLE NUMBERS

Many of us take whole numbers for granted, not fully appreciating that whole numbers have many significant applications and that they provide the basis for working with fractions, decimals, and integers. Whole numbers help us locate streets and houses. They help us keep track of how many bananas we have. And in a country with about 300 million people, they help us keep track of the federal budget and the amount of unemployment.

3.1 NUMERATION SYSTEMS

NCTM Standards

- Use multiple models to develop initial understandings of place value and the base-ten number system (K–2)
- Create and use representations to organize, record, and communicate mathematical ideas (pre-K–12)

How much is that dress?

Figure 3–1

Today most countries use the simple, efficient base-ten place-value system. Simple as it appears, it took thousands of years to develop. Using place value and a base, one can express amounts in the hundreds or thousands with only a few digits! Big deal, you say? Try writing a number such as 70 using tally marks (Figure 3–1).

LE 1 Opener

Tell how to represent 7 in another numeration system that you know.

THE TALLY SYSTEM

People first used numbers to count objects. Before people understood the abstract idea of a number such as "four," they associated the number 4 with sets of objects such as four cows or four stars. A major breakthrough occurred when people began to think of "four" as an abstract quantity that could measure the sizes of a variety of concrete sets.

No one knows for sure, but the first numeration system was probably a tally system. The first nine numbers would have been written as follows.

A system such as this, having only one symbol, is very simple, but large numbers are difficult to read and write. Computation with large numbers is also slow.

THE EGYPTIAN NUMERATION SYSTEMS

About five thousand four hundred years ago, even before your professors were born, the ancient Egyptians improved the tally system by inventing additional symbols for 10, 100, 1000, and so on. Symbols for numbers are called **numerals.** The Egyptians recorded numerals on papyrus with ink brushes. Most of our knowledge about Egyptian numerals comes from the Moscow Papyrus (1850? B.C.) and the Rhind Papyrus (1650 B.C.).

Egyptian numerals are rather attractive symbols called hieroglyphics (Figure 3–2). Using symbols for groups as well as for single objects was a major advance that made it possible to represent large quantities much more easily. The ancient Egyptian numeration system uses ten as a **base,** whereby a new symbol replaces each group of 10 symbols. The Egyptian system is **additive,** since the value of a number is the sum of the values of the numerals. For instance, 𝟡𝟡‖‖‖ represents $100 + 100 + 1 + 1 + 1$, or 203.

Egyptian Numerals	Value
/ (staff)	1
∩ (heel bone)	10
℗ (scroll or coil of rope)	100
🪷 (lotus flower)	1,000
⌐ (pointing finger)	10,000
⌒ (tadpole or fish)	100,000
👤 (astonished man)	1,000,000

Figure 3–2

LE 2 Skill

Discussion

(a) Speculate as to why the Egyptians chose some of the symbols shown in Figure 3–2.
(b) Write 672 using Egyptian numerals.

You can see from LE 2 that the early Egyptian system requires more symbols than our base-ten system. The Egyptians later shortened their notation by creating a different symbol for each number from 1 to 9 and for each multiple of 10 from 10 to 90, as shown in Figure 3–3 on the next page. For example, 7 changed from ‖‖‖‖ to **Z**. This change necessitated memorizing more symbols, but it significantly shortened the notation.

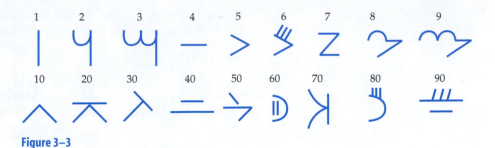

Figure 3–3

LE 3   Skill

Write 672 using the later Egyptian system.

THE BABYLONIAN NUMERATION SYSTEM

The Babylonian numeration system developed about the same time as the Egyptian system. However, more of the Babylonians' work has been preserved because they used clay tablets instead of papyrus.

The Babylonians made a significant improvement by developing a system based on **place value,** in which the value of a numeral changed according to its position. A place-value system reduces the number of different symbols needed.

The Babylonians used place value (for numbers greater than 59), a base of 60, and two symbols: **<** for 10 and **Y** for 1. The 3,800-year-old tablet in Figure 3–4 shows Babylonian numerals.

The University Museum, University of Pennsylvania (neg. #69434).

Figure 3–4

LE 4   Concept

To write 672 in Babylonian numerals, use eleven 60s and twelve 1s. (*Note:* 672 ÷ 60 = 11 R 12.) Since ⟨𝗬 represents 11 and ⟨𝗬𝗬 represents 12, the Babylonian numeral for 672 is ⟨𝗬⟨𝗬𝗬. How does the number of symbols needed to write 672 compare with the Egyptian system used in LE 2 and LE 3?

The Babylonians lacked a symbol for zero until 300 B.C. Before that time, it was impossible to distinguish between certain numerals. For example, 83 and 3623 are $1 \cdot 60 + 23$ and $1 \cdot 60^2 + 23$, both written as 𝗬⟨⟨𝗬𝗬𝗬. The Babylonians sometimes, but not always, put a space between groupings to indicate an empty place.

Zero serves two purposes: as a placeholder and as a symbol for nothing. Around 300 B.C., the Babylonians invented ▲ as a placeholder. The expression $1 \cdot 60^2 + 23$ was then written as 𝗬▲⟨⟨𝗬𝗬𝗬, where ▲ indicates that there are zero 60s. Even then, they did not use ▲ on the right, so 1 and 60 were both written as 𝗬.

LE 5   Skill

What base-ten numeral could represent the same number as ⟨▲⟨⟨⟨⟨𝗬?

Between them, the Egyptian and Babylonians developed most of the foundations of our numeration system: a base (ten), unique symbols for one through nine, place value, and a partial symbol for zero. The symbol ▲ worked like 0 within a number but was not used at the end of a number.

THE MAYAN NUMERATION SYSTEM

Between 300 and 900, the remarkable Mayans of Central America developed an accurate calendar, a system of writing, and a modified base-twenty numeration system with place value. While the Babylonians eventually developed a symbol for zero as a placeholder, the Mayan system developed a symbol for zero that also signified nothing. The system only required three symbols although the Mayans used different variations of the shell (zero) depending upon its position in a numeral.

Mayan numeral	🐚	•	___
Base-ten value	0	1	5

They used combinations of these three symbols to write numerals for 0 through 19. For example, 8 was written as •••, and 19 was written as ⬮⬮⬮⬮. Notation for whole numbers greater than 19 used vertical place value. The next vertical place value is 20. For example, ••• represents eight 1s and five 20s, or 108, in base ten.

After the 20s place comes the $18 \cdot 20$s place. Each subsequent place is 20 times the previous one: $18 \cdot 20^2$, $18 \cdot 20^3$, ...

LE 6   Skill

Write each Mayan numeral as a base-ten numeral.

(a) ••• ‾‾‾ (b) • •‾• (c) ‾‾ 🐚

THE ROMAN NUMERATION SYSTEM

Roman numerals can still be found today in books, outlines, and movies and on monuments.

LE 7 Concept

Where do you find Roman numerals

(a) in a book?
(b) in a movie?

The basic Roman numerals have the following base-ten values.

Roman numeral	I	V	X	L	C	D	M
Base-ten value	1	5	10	50	100	500	1000

To find the value of a Roman numeral, start at the left. Add when the symbols are alike or decrease in value from left to right. Subtract when the value of a symbol is less than the value of the symbol to its right. For example, XI is 11, and IX is 9. The Roman numeration system was developed between 500 B.C. and A.D. 100, but the subtractive principle was not introduced until the Middle Ages.

LE 8 Skill

(a) Write MCMLIX as a base-ten numeral.
(b) Write 672 (base ten) as a Roman numeral.

OUR BASE-TEN PLACE-VALUE SYSTEM

Between A.D. 200 and 1000, the Hindus and Arabs developed the base-ten numeration system we use today, which is appropriately known as the **Hindu-Arabic numeration system.** The Hindus (around A.D. 600) and the Mayans (date unknown) were the first to treat 0 not only as a placeholder but also as a separate numeral.

LE 9 Skill

Which numeration system generally uses the smallest total number of symbols to represent large numbers?

(a) The tally system
(b) The early Egyptian system
(c) The Babylonian system
(d) The Mayan system
(e) The Roman system
(f) Our Hindu-Arabic system

The Hindu-Arabic numerals 0, 1, 2, 3, 4, 5, 6, 7, 8, and 9 were developed from 300 B.C. to A.D. 1522. The numerals 6 and 7 were used in 300 B.C., and the numerals 2, 3, 4, and 5 were all invented by A.D. 1530.

We use these numerals in a base-ten place-value (Hindu-Arabic) system to represent any member of the set of whole numbers, $W = \{0, 1, 2, 3, \ldots\}$. The Hindu-Arabic system has three important features: (1) a symbol for zero, (2) a way to represent any whole number using some combination of ten basic symbols (called **digits**), and (3) **base-ten place value,** in which each digit in a numeral, according to its position, is multiplied by a specific power of ten. Each place has ten times the value of the place immediately to its right.

The following diagram shows the first four place values for whole numbers, starting with the "ones" place on the far right.

T h o u s a n d s (10^3)	H u n d r e d s (10^2)	T e n s (10^1)	O n e s (1)
3	4	2	6

3,426 can be decomposed in expanded notation as

3 thousands,	4 hundreds,	2 tens,	and 6 ones
or (3000)	+ (400)	+ (20)	+ 6
or (3×10^3)	+ (4×10^2)	+ (2×10^1)	+ 6

This **expanded notation** shows the value of each place.

Exponents offer a convenient shorthand. In base ten, $10^1 = 10$, $10^2 = 10 \cdot 10 = 100$, and $10^3 = 10 \cdot 10 \cdot 10 = 1000$. In the numeral 10^3, 10 is the base and 3 is the exponent. René Descartes invented the notation for an exponent in 1637. In an expression of numbers being multiplied such as $10 \cdot 10 \cdot 10$, each 10 is called a factor. (The term "factor" will be defined formally in Section 3.3.)

Definition: Base, Exponent, and a^n

If a is any number and n is a counting number, then

$$a^n = a \cdot a \cdot \ldots \cdot a$$

$$n \text{ factors}$$

where a is the **base** and n is the **exponent.**

Your calculator may have a special key for working with exponents. Scientific calculators usually have a powering key such as $\boxed{y^x}$. To compute 5^3, press $\boxed{5}\ \boxed{y^x}\ \boxed{3}\ \boxed{=}$. Graphing calculators also have a powering key. To compute 5^3, press $\boxed{5}\ \boxed{\wedge}\ \boxed{3}\ \boxed{\text{ENTER}}$.

LE 10   Skill

A base-ten numeral has digits A B C. Write this numeral in expanded notation.

LE 11   Communication

What is the difference between a place-value numeration system such as the Hindu-Arabic system and a system such as the Egyptian system that does not have place value?

You may be so accustomed to using place value that you don't give it a second thought. The Babylonians made some use of place value in their systems. Our own place-value system was first developed by the Hindus and was significantly improved by the Arabs.

Until the twelfth century, when paper became more readily available, the abacus was the most convenient tool for computations. Paper offered a new medium for doing arithmetic.

Another widely used place-value model is the abacus. The abacus was developed over 2,400 years ago as a calculating device. Today, many people still use it for computation. A simplified abacus like the one in Figure 3–5 may be used in elementary school after children have had experience with more concrete models, such as base-ten blocks. The abacus in Figure 3–5 shows 1233.

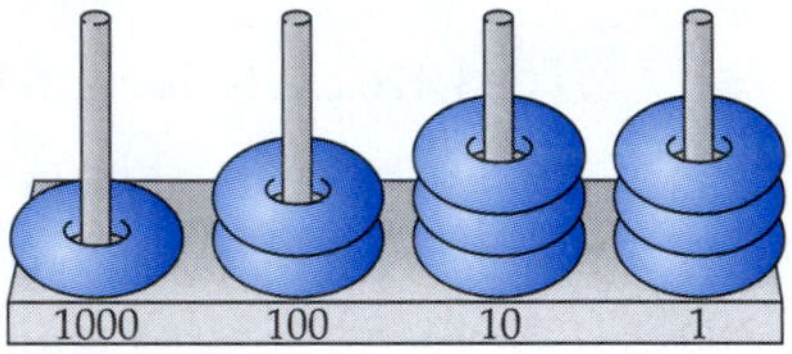

| 1000 | 100 | 10 | 1 |

Figure 3–5

A battle ensued from 1100 to 1500 between the *abacists,* who wanted to use Roman numerals and an abacus, and the *algorists,* who wanted to use Hindu-Arabic numerals and paper (Figure 3–6). By 1500, the algorists had prevailed.

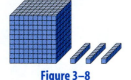

Abacists vs. Algorists

Figure 3–6

MODELS FOR PLACE VALUE

To understand and communicate mathematical concepts, one must be able to relate concrete and pictorial models to abstract ideas. For example, children learn about base ten using a variety of concrete materials such as Dienes blocks. Dienes blocks for base ten come in the shapes shown in Figure 3–7 on the next page.

Figure 3–8

LE 12   Skill

What number is represented in Figure 3–8?

LE 13   Communication

What makes the abacus more abstract than base-ten blocks?

Base-ten blocks are a **proportional model** because they show the proportional size differences in digits with different place values. For example, a ten is 10 times bigger than a one. The abacus is a **nonproportional model** because representations of digits with different place values look the same (except for the location). For example, the ring for a ten looks just like the ring for a one instead of being ten times bigger.

Block	Name	Value
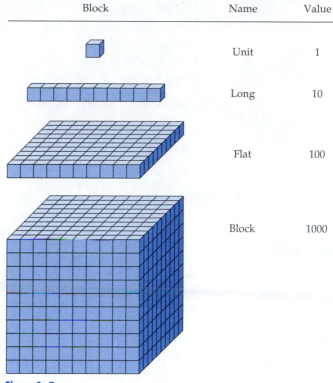	Unit	1
	Long	10
	Flat	100
	Block	1000

Figure 3–7

What sort of difficulties do children have with place value?

Classroom Connection

LE 14   Communication

A first grader says that the next number after 39 is "three and ten." Give an explanation that will help the child understand why the next number is 40.

COUNTS AND COUNTING UNITS

Suppose 386,212 people attended a concert. It would be possible to count the number of people at the concert one by one. The number 386,212 is a count, and "people" is the counting unit. The **count** (or **cardinal number**) of a set is the number of elements in the set. Every count has a counting unit.

LE 15   Skill

Name the count and counting unit for

(a) the number of letters in your first name.
(b) the number of seats in your classroom.

MEASURES AND MEASURING UNITS

Jackie weighs 108 pounds. The number 108 is a **measure,** and pounds are the unit of measure. Unlike counting units such as people, **units of measure** such as pounds can be split into smaller parts. Other examples of units of measure are hours, meters, and square feet.

LE 16   Skill

Name the measure and unit of measure for your height in inches.

LE 17   Concept

Tell whether the number in each part is a count or a measure.

(a) 26 cows   **(b)** 26 seconds   **(c)** 26 miles

Besides representing counts and measures, numbers are also used as identifiers (player No. 27) and to indicate relative position (4th).

ROUNDING

24,000 Attend U2 Concert

If you recently read about attendance at a concert or looked up the population of the world, you probably read an approximate, or rounded, figure. A number is always rounded to a specific place (for example, to the nearest thousand). The following exercise concerns the rounding of whole numbers between 24,000 and 25,000.

LE 18   Opener

(a) If a whole number between 24,000 and 25,000 is closer to 24,000, what digits could be in the hundreds place?
(b) If a whole number between 24,000 and 25,000 is closer to 25,000, what digits could be in the hundreds place?

As you saw in the preceding exercise, numbers between 24,000 and 24,500 are closer to 24,000 than to 25,000, and they have 0, 1, 2, 3, or 4 in the hundreds place. In rounding to the nearest thousand, these numbers would be rounded *down* to 24,000. However, numbers that are between 24,500 and 25,000, such as 24,589, would be rounded *up* to 25,000 because they are closer to 25,000 than to 24,000. All of these numbers have 5, 6, 7, 8, or 9 in the hundreds place.

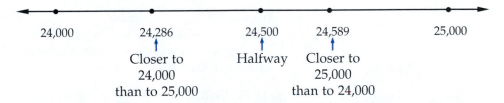

The number 24,500 is exactly halfway between 24,000 and 25,000, so it could be rounded either way. The rule used in our schools is to round up all such numbers that are exactly in the middle. All other whole numbers from 24,501 to 24,599 with 5 in the hundreds place should be rounded up to 25,000. Rounding 24,500 up to 25,000 enables children to learn one rule for all numbers with a 5 in the hundreds place.

> **Rounding Whole Numbers**
>
> The steps for rounding a whole number to a specific place are as follows.
>
> **1.** Locate the place and then check the digit one place to its right.
> **2.** Round *up* if the digit to the right is 5 or more, and round *down* if it is 4 or less.

LE 19 Skill

Exactly 386,212 people attended the Earaches' rock concert.

(a) Round this number to the nearest thousand.
(b) Use a number line to show that your answer is correct.

LE 20 Reasoning

Devise a plan and solve the following problem.

A newspaper headline says "83,000 SEE KITTENS DEFEAT DUST BUNNIES." If this is a rounded figure, the actual attendance could have been between what two numbers?

Whole-number rounding will be used in estimation later in this chapter.

Rounding is not always done to "the nearest." The other two types of rounding are rounding up and rounding down. In buying supplies, one often has to round up to determine how much to purchase.

LE 21 Concept

I need 627 address labels for a company mailing. They come in groups of 100.

(a) How many labels will I buy?
(b) How does this illustrate rounding up?

> ## LE 22 Summary
> Tell what you learned about place value in this section.

ANSWERS TO SELECTED LESSON EXERCISES

2. (a) The Egyptians would have used a staff in traveling. The scroll relates to their development of papyrus. The lotus flower, tadpoles, and fish would have been found in a culture that developed around the Nile River. The astonished man is amazed by the amount of 1,000,000.

(b) ꝯꝯꝯꝯꝯꝯ∩∩∩∩∩∩∩ΙΙ

3. ꝯꝯꝯꝯꝯꝯ𓏭

4. It takes fewer distinct symbols and a smaller total number of symbols.

5. $10 \cdot 60^2 + 31 = 36{,}031$

6. (a) 13   (b) 131   (c) 100

7. (a) Page numbers in the preface
    (b) Copyright date

8. (a) 1959   (b) DCLXXII

9. (d) or (f)

10. $100\,A + 10\,B + C$

11. In a place-value system, the value of a numeral changes depending upon its location. In a system that does not have place value, the value of a numeral is always the same.

12. 1030

13. The abacus distinguishes place value with position but not size. Base-ten blocks do use size.

17. (a) Count   (b) Measure   (c) Measure

18. (a) 0, 1, 2, 3, or 4   (b) 5, 6, 7, 8, or 9

19. (a) 386,000

(b)

20. 82,500 and 83,499 (inclusive) if attendance is rounded to the nearest thousand

21. (a) 700

(b) 627 is rounded up to the next 100.

3.1 HOMEWORK EXERCISES

Basic Exercises

1. Tell whether each word in quotation marks is used like a numeral (that is, as a symbol) or like a number (an idea).
(a) "Cat" is a three-letter word.
(b) A "cat" is a warm, furry animal.

2. What is the difference between a number and a numeral?

3. Write the following Egyptian numerals as Hindu-Arabic numerals.
(a) 𓏺𓏺𓍢𓃀𓃀𓃀𓃀   (b) 𓎆𓂀𓏭𓏭𓏭

4. Write the following Egyptian numerals as Hindu-Arabic numerals.

(a) ⎯Z   (b) ⌒𓏴𓏲⌐

5. (a) Do both of the following Egyptian numerals represent the same number?
⌒𓏭𓏭   𓏭⌒𓏭
(b) What difference between Egyptian and Hindu-Arabic numerals is suggested by part (a)?

6. Write the following Babylonian numerals as Hindu-Arabic numerals.
(a) ⟨⟨⟨𐏑𐏑   (b) ⟨𐏑𐏑⟨⟨𐏑𐏑𐏑

7. Write 324 as
(a) an early Egyptian numeral.
(b) a late Egyptian numeral.
(c) a Babylonian numeral.

8. Write 92 as
(a) an early Egyptian numeral.
(b) a late Egyptian numeral.
(c) a Babylonian numeral.

9. Write each Mayan numeral as a base-ten numeral.
(a) ● ● ● ●   (b) ●/⌒   (c) (two dots over bar over two dots)

10. Write each base-ten numeral as a Mayan numeral.
(a) 14   (b) 27   (c) 60

11. What are three important characteristics of our base-ten numeration system?

12. Name three places where you might find Roman numerals.

13. Give the Hindu-Arabic numeral for
(a) XIV   (b) XL
(c) MDCXIII   (d) MCMLXIV

14. Write a Roman numeral for
(a) 4   (b) 24   (c) 1998

15. How many symbols would be needed to write 642 using each of the following?
(a) Tallies
(b) Early Egyptian numerals
(c) Roman numerals

16. What was an advantage of the Babylonian system over the early Egyptian system?

17. Decompose the following base-ten numbers using expanded notation.
(a) 407   (b) 3125

18. A base-ten numeral has the digits A A A. The value of the A at the far left is _____ times the value of the A at the far right.

19. A base-ten numeral has the digits A B C D. Decompose this numeral using expanded notation.

20. Rewrite each of the following as a base-ten numeral.
(a) $(8 \times 10^3) + (6 \times 10) + 2$
(b) $(2 \times 10^4) + (3 \times 10^3) + (4 \times 10^2) + 5$

21. What number is represented in each part?
(a)

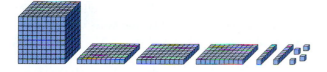

(b) 2 flats and 7 units

22. How would you represent 380 using base-ten blocks?

23. Chip trading is a common place-value model that uses colored chips and a mat. In base ten, the columns on the mat represent powers of ten. For example, 4523 is represented by 4 red, 5 green, 2 blue, and 3 yellow chips.

Red	Green	Blue	Yellow
4	5	2	3

(a) How would you represent 372?
(b) Ten blue chips can be traded in for one _____ chip.
(c) Which is more abstract, chip trading or base-ten blocks? Explain why.

24. What number is represented in the following picture?

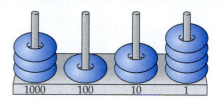

25. One way to represent 127 is with 12 tens and 7 ones. List three other ways.

26. I have 7 base-ten blocks. Some are longs and some are units. Their value is an even number between 40 and 60. What blocks do I have?

27. A second grader writes "two hundred and ten" as 20010.
(a) How might this child write "three hundred and seven"?
(b) What does the child understand about place value?
(c) Give an explanation that will help the child understand why the correct notation is 210.

28. A second grader knows that 34 is 3 tens and 4 ones. Then you ask, "How else could you represent 34? For example it would be 2 tens and _____ ones." The child cannot answer this question. What hint would you give the child?

29. Tell whether the number in each part is a count or a measure.
(a) 42 yards   (b) 42 pens   (c) 42 minutes

30. Tell whether the number in each part is a count or a measure.
(a) 7 cats   (b) 7 days   (c) 7 meters

31. Is money a counting unit or a unit of measure? Consider whether the definition of a unit of measure in the text would apply to 25 dollars and to 25 cents.

32. How can you distinguish between counting units and units of measure?

33. Round each of the following numbers to the nearest hundred.
    (a) $625, the price of lunch for two at La Vielle Chaussure
    (b) $36,412, the one-game salary of a superstar athlete

34. According to the 2000 U.S. Census, the population of Cleveland was 478,403. Round this number to the nearest hundred thousand.

35. Give the place to which each number has been rounded.

	Original Number	Rounded Number
(a)	418	420
(b)	418	400
(c)	368,272	370,000

36. Match the amount in each part with one of the following estimates.

    4   40   400   4000
    4,000,000   400,000,000

    (a) Number of students in a typical elementary school
    (b) Population of Chicago
    (c) Number of miles an adult can walk in an hour
    (d) Distance, in miles, from Boston to San Diego
    (e) Population of North America

37. You have $126 in a checking account. The automated cash machine will give you only $10 bills.
    (a) What is the most money you can withdraw?
    (b) Does this illustrate rounding up, rounding down, or rounding to the nearest?

38. Ballpoint pens come in packs of 10. You want 23 pens for your class.
    (a) How many pens will you order?
    (b) Does this illustrate rounding up, rounding down, or rounding to the nearest?

39. (a) About 8,000 people attended a baseball game. If this figure has been rounded to the nearest thousand, the actual attendance was between what two numbers?
    (b) Tell how you found the answer.

40. If the quarterback in the following cartoon is right, what is the greatest number of people that could be there watching?

© 1991, Carchria Biological Supply Company

Extension Exercises

41. Write a ten-digit numeral in which the first digit tells the number of zeroes in the numeral, the second digit tells the number of 1s, the third digit tells the number of 2s, and so on. (*Hint:* There are no 8s or 9s in the numeral.)

42. A number is called a palindrome if it reads the same forward and backward; 33 and 686 are examples. How many palindromes are there between 1 and 1000 (inclusive)?
    (a) Devise a plan and solve the problem.
    (b) Make up a similar problem and solve it.

Technology Exercise

43. If you have a calculator such as the TI-73 with a place-value feature, enter 1234 and find out how the calculator shows the place value of different digits.

Labs and Projects

44. (a) Find out what the five digits of the zip code tell the post office.
    (b) Find out what the first three digits of a Social Security number indicate.

3.2 ADDITION AND SUBTRACTION OF WHOLE NUMBERS

NCTM Standards

- understand various meanings of addition and subtraction of whole numbers and the relationship between the two operations (K–2)
- model situations that involve addition and subtraction of whole numbers, using objects, pictures, and symbols (K–2)
- create and use representations to organize, record, and communicate mathematical ideas (pre-K–12)

Addition and subtraction involve more than computing $3 + 6$ or $63 - 27$. Computational skills are not useful unless you know when to use them in solving problems. Children should learn how to recognize situations that call for addition or subtraction.

LE 1 Opener

How would you describe what addition is?

ADDITION DEFINITION AND CLOSURE

In elementary school, teachers explain the meaning of addition by giving examples. Later, one can formally define important ideas such as addition. The idea of combining sets (that is, union) is used to define addition. An example is combining 2 fresh bananas with 3 older bananas to obtain a total of 5 bananas.

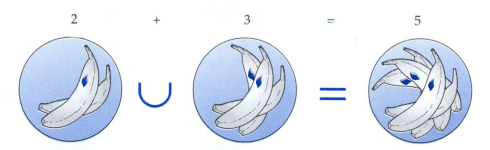

Definition: Whole-Number Addition

If set A contains a elements, set B contains b elements, and $A \cap B = \varnothing$, then $a + b$ is the number of elements in $A \cup B$.

In the addition equation $a + b = c$, a and b are called **addends,** and c is called the **sum.** The German mathematician Widmann used the $+$ sign for addition (and the $-$ sign for subtraction) in his book published in 1498.

LE 2 Concept

(a) If two whole numbers are added, is the result always a whole number?
(b) In part (a), did you use induction or deduction?
(c) Would a whole-number addition problem ever have more than one answer?

Your responses to the preceding exercise should confirm that the sum of two whole numbers is a unique whole number. For example, $3 + 5$ equals 8, a unique whole number. This is called the closure property of addition of whole numbers.

The Closure Property of Addition of Whole Numbers

If a and b are whole numbers, then $a + b$ is a unique whole number.

Closure requires that an operation on two members of a set results in a unique member of the same set, "unique" meaning that the result is the only possibility.

CLASSIFYING ADDITION APPLICATIONS

Research with children indicates that the most difficult aspect of word problems in arithmetic is choosing the correct operation. How do children learn which operation to use?

Consider addition. Many applications of addition fall into one of two categories. If children learn to recognize these two categories, they will recognize most real-world problems that call for addition. The following exercise will introduce you to the classification of addition applications.

Discussion

LE 3 Opener

What kind of situations call for addition? To illustrate this, make up two word problems that can be solved by adding 3 and 5. Compare your problems to others in your class.

Do your word problems fit into one of the following categories? Many addition applications do.

1. **Combine Sets** Combine two nonintersecting sets and find out how many objects are in the new set.
   Example: Tina has 3 cute beagle puppies, and Maddy has 5. How many puppies do they have altogether (Figure 3–9)?

Figure 3–9

2. **Combine Measures** Combine two measures of the same type and find the total measure.
   Example: Laura is going to Paris for 3 days and to London for 5 days. How long is the whole trip (Figure 3–10)?

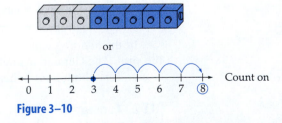

or

Figure 3–10

Numbers are used to count and to measure. The only difference between categories 1 and 2 is the difference between "sets" and "measures." Set problems involve objects that *can be counted* with whole numbers; objects such as apples, dogs, or people. Measure problems involve measures such as distance, weight, and time, which are not limited to having whole-number values. Most arithmetic applications can be classified as either set problems or measure problems.

Not all addition applications fit into the "combine sets (counts)" or "combine measures" categories, but these two categories identify specific characteristics of addition applications. Children learn to recognize these two addition categories in elementary school. Although your future students may not learn the terms "combine sets" and "combine measures," these categories will help you (the teacher) to identify more specifically what your students do or do not understand. For example, a child may know what 2 + 4 equals but may not be able to recognize combining sets as an addition situation.

LE 4   Concept

Classify the addition situation in the cartoon in Figure 3–11.

Figure 3–11

By permission of Johnny Hart and Creators Syndicate, Inc.

LE 5   Concept

Classify the following addition application. "Hy Climber scales some rocks to a height of 7 ft. If he climbs 3 more feet, he will reach the top. How high is the top?"

Teachers first model the concepts of combining sets and combining measures for children using manipulatives (objects) and pictures. Manipulatives and pictures establish connections between addition and everyday applications of it.

• EXAMPLE 1

Make a drawing that represents 2 + 3 = 5 (LE 4) with a set of counters.

SOLUTION

LE 6   Connection

Show how to solve LE 5.

(a) with squares (or snap cubes).
(b) by counting on with a number line.

Example 1 illustrates how a combine sets application can be represented with a set of counters.

Most elementary-school textbooks use pictures of snap cubes to illustrate combine measures applications. In whole-number addition and subtraction, textbooks use number lines only when a problem is solved by counting.

Classroom Connection

LE 7   Concept

A first grader works out $7 + 3$ by counting on. "I start at 7 and want 3 more. That's 7, 8, 9. The answer is 9."

(a) What is the child confused about?
(b) How would you help the child understand the correct procedure?

A major intellectual breakthrough occurred when people realized that addition applies to all kinds of objects. The equation $3 + 4 = 7$ describes a relationship that applies to sets of people, apples, and rocks. The power of mathematics lies in the way we can apply basic ideas such as addition to such a wide variety of situations.

Using addition with many different kinds of sets also has its drawbacks. Numbers emphasize similarities between situations. People must then make allowances for differences in the *qualities* of the living things or objects being analyzed. Although 3 peaches plus 4 peaches is the same mathematically as 3 bombs plus 4 bombs, peaches and bombs are rather different.

LE 8   Connection

Angela has 2 apples and buys 3 more. Now she has 5 apples. What information about the apples is ignored in this description?

SUBTRACTION DEFINITION

In school, children learn how subtraction is related to addition.

LE 9   Skill

Name an addition equation that is equivalent to $6 - 2 = 4$.

In answering LE 9, you used the following definition to convert a subtraction equation to an equivalent addition equation.

> **Definition:**   **Whole-Number Subtraction**
>
> For any whole numbers a and b, $a - b = c$ if and only if $a = b + c$ for a unique whole number c.

In the subtraction equation $a - b = c$, c is called the **difference.** You can illustrate the relationship between addition and subtraction with a diagram.

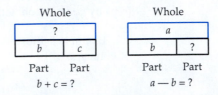

Because of this relationship, subtraction is called the inverse operation of addition. Although subtraction is defined in terms of addition, most children first understand subtraction as the process of removing objects from a set.

CLASSIFYING SUBTRACTION APPLICATIONS

Children can learn to recognize applications that call for subtraction by recognizing the common categories of these applications.

LE 10   Opener

What kind of situations call for subtraction? To illustrate this, make up two word problems that can be solved by subtracting 2 from 6. Try to create two different types of problems. Compare your problems to others in your class.

Do your problems fit into any of the following categories?

1. **Take Away Sets**   Take away some objects from a set of objects.
   Example: Dan had 6 pet fish. Two of them died. How many are left (Figure 3–12)?

Figure 3–12

2. **Take Away Measures**   Take away a certain measure from a given measure.
   Example: A lumberjack has 6 ft of rope. He uses 2 ft of rope to tie some logs. How much rope is left (Figure 3–13)?

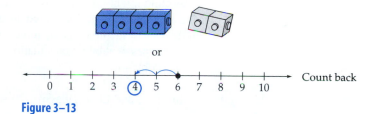

Figure 3–13

3. **Compare Sets**   Find how many more objects one set has than another.
   Example: Maria has 6 cuddly kittens and Pete has 2. How many more kittens does Maria have than Pete (Figure 3–14)?

Figure 3–14

4. Compare Measures   Find how much larger one measure is than another.
Example: Sue's worm is 6 inches long. Twelve months ago, it was 2 inches long. How much has the worm grown in the last year (Figure 3–15)?

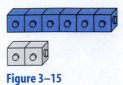

Figure 3–15

5. Missing Part (Sets)   Find how many objects are in one part of a set when you know how many are in the whole set and the other part.
Example: Reese needs 6 stamps for her letters. She finds 2 in a drawer. How many more does she need (Figure 3–16)?

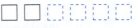

Figure 3–16

6. Missing Part (Measures)   Find one part of a measure when you know the whole measure and the other part.
Example: Luis wants to gain 6 pounds to qualify for the wrestling team. So far he has gained 2 pounds. How many more pounds does he need to gain (Figure 3–17)?

Figure 3–17

A child might initially solve a missing part (addend) problem by counting on or by using an addition fact. Solving equations such as $2 + \boxed{} = 6$ for the missing part (addend) is one of the first algebra topics in elementary school. Once they learn to recognize these six types of subtraction situations, children usually know when to use subtraction in solving problems.

In LE 11 and LE 12, classify the subtraction application.

LE 11   Concept
Jaclyn and Erica are working on a project. Jaclyn has worked 7 hours, and Erica has worked 4. How much longer has Jaclyn worked than Erica?

LE 12   Concept
The following is a true story. I used to have 3 umbrellas. I no longer have 2 of them. I lost 1 on the subway, and the other suffered an early internal breakdown. How many umbrellas do I have now?

The take away and compare models of subtraction are usually introduced with manipulatives and pictures in elementary school. Missing-part problems currently receive limited coverage in most elementary-school textbook series.

LE 13   Skill

Show how to solve LE 11 with squares (or snap cubes), and write the arithmetic equation it illustrates.

Classroom Connection

LE 14   Communication

A child in a first-grade class asks, "What does $2 - 3$ equal?" How would you respond?

AN INVESTIGATION: NUMBER CHAINS

LE 15   Reasoning

A **number chain** is created by adding and subtracting. The number in each square in Figure 3–18 is the sum of the numbers that are next to it on both sides.

Figure 3–18

What is the relationship between the far-left-hand and far-right-hand numbers on the number chain in Figure 3–19?

Figure 3–19

(a) *Understanding the Problem*   If you put 10 in the far-left-hand circle, what number results in the far-right-hand circle?
(b) *Devising and Carrying Out a Plan*   Try some other numbers in the far-left-hand circle, and fill in the remaining circles. Look for a pattern.
(c) Make a generalization of your results from part (b).
(d) *Looking Back*   Did you use inductive or deductive reasoning in part (c)?
(e) Start with X in the far-left-hand circle and fill in all the other circles to prove your conjecture in part (c).

LE 16   Reasoning

Consider the number chain in Figure 3–20.

(a) Fill in the circles with numbers that work.
(b) If possible, find a second solution.
(c) Start with X in the upper left-hand circle and use algebra to fill in all the circles.
(d) What does your answer in part (c) tell you about the relationship between the numbers in opposite corners?

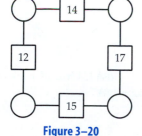

Figure 3–20

LE 17   Summary

Tell what you have learned about addition and subtraction in this section.

ANSWERS TO SELECTED LESSON EXERCISES

2. (a) Yes   (b) Induction   (c) No

4. Combine sets

5. Combine measures

6. (a) 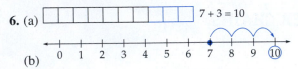 $7 + 3 = 10$

(b)

8. Are they fresh? What kind of apples are they? How much did she pay? What will she do with them? Were they sprayed with pesticides?

9. $2 + 4 = 6$

11. Compare measures

12. Take away sets

13.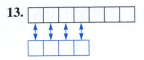

15. (a) 23   (d) Inductive

3.2   HOMEWORK EXERCISES

Basic Exercises

1. Which of the following sets are closed under addition?
   (a) $\{2, 4, 6, 8, \ldots\}$
   (b) $\{1, 3, 5, 7, \ldots\}$
   (c) $\{0, 1\}$

2. Make up a set containing one element that is closed under addition.

3. The Samuels built a house and lived in it for 8 years. They sold it to the Wangs, who have now lived there for 11 years. How old is the house?
   (a) Which category of addition, combine sets or combine measures, is illustrated?
   (b) Name some other significant information that is not included in the problem.

4. Which category of addition, combine sets or combine measures, is illustrated in the following problem? "Mary sold 61 apples yesterday and 48 apples today. How many apples did she sell in the last two days?"

5. (a) What addition fact is illustrated in the diagram?

   (b) How would you use a number line to show a child that $4 + 2 = 6$?

6. (a) What addition fact is illustrated by placing the rods together as shown?

   (b) Which category of addition is illustrated?

7. Make up a word problem that illustrates the combine sets category.

8. Make up a word problem that illustrates the combine measures category.

9. Use the definition of subtraction to rewrite each of the following subtraction equations as an addition equation.
   (a) $7 - 2 = ?$
   (b) $N - 72 = 37$
   (c) $x - 4 = N$

10. Is whole-number subtraction closed? (*Hint:* Check the two parts of the closure property.)

In Exercises 11–14, classify each application as one of the following: take away sets, take away measures, compare sets, compare measures, missing part (sets), or missing part (measures).

11. Mary is 4 feet, 8 inches tall and Sidney is 4 feet, 3 inches tall. How much taller is Mary?

12. Hy Climber hikes to an altitude of 3,600 ft. The summit has an altitude of N feet. How much higher does he have to climb?

13. I need to make 7 sandwiches for a party. I have made 4 so far. How many more do I have to make?

14. Kong had 8 bananas. Kong ate 3. How many are left?

15. Show how to solve Exercise 13 with blocks, and tell what arithmetic equation it illustrates.

16. Show how to solve Exercise 14 with blocks, and tell what arithmetic equation it illustrates.

17. (a) What subtraction fact is shown?

(b) How would you use a number line to show a child that $5 - 2 = 3$?

18. (a) What subtraction fact is illustrated?

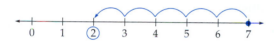

(b) Which category of subtraction is illustrated?

19. (a) What subtraction fact is illustrated?

(b) Which category of subtraction is illustrated?

20. How would you act out each of the following with counters?
(a) Cedric has 8 apples. He gives three to a friend. How many are left?
(b) Cedric has 8 apples. Dominique has 3. How many more does Cedric have than Dominique?
(c) Cedric wants 8 apples for a party. He has 3. How many more does he need?

21. Make a drawing that shows $8 - 2 = 6$, using
(a) take away sets.   (b) compare sets.
(c) compare measures.

22. The following two third-grade problems require subtracting 12 from 30. How do they differ in tone?
(a) Jody Winer had 30 records. She gave 12 to her brother. How many records does she have left?
(b) Jody Winer has 30 records. Her brother has 12. How many more records does she have than her brother?

23. Make up a word problem that illustrates the missing-part (measures) category.

24. Make up a realistic problem that illustrates the compare-sets category.

25. Make up a word problem that illustrates the take-away measures category.

26. Write a word problem that requires computing $10 - (3 + 5)$.

27. A first grader works out $8 - 3$ by counting back on a number line. "I start at 8 and go back 3. That's 8, 7, 6. The answer is 6."
(a) What is the child confused about?
(b) How would you help the child understand the correct procedure?

28. A first grader works out $6 + \underline{\hspace{1cm}} = 10$ by counting forward. "I start at 6 and count up to 10. That's 6, 7, 8, 9, 10. The answer is 5."
(a) What is the child confused about?
(b) How would you help the child understand the correct procedure?

29. Represent the following with algebraic expressions, using the variable x.
    (a) The difference between 10 and a number
    (b) A number decreased by 2
    (c) The sum of a number and 6

30. Represent the following with algebraic expressions, using the variable n.
    (a) The sum of 4 and a number
    (b) 3 less than a number
    (c) A number increased by 6

31. Solve the following using addition and subtraction: "Margarita starts off with A. She buys food for F and clothes for C, and then she receives a paycheck for P. Write an expression representing the total amount of money she has now."

32. Consider the following problem. "Last year, your salary was A and your expenses were B. This year, you received a salary increase of X, but your expenses increased by Y. Write an expression showing how much of your salary will be left this year after you have paid all your expenses."
    (a) Devise a plan and solve the problem.
    (b) Make up a similar problem.

33. Consider the following problem. "Under what conditions for whole numbers a, b, and c would $(a - b) - c$ be a whole number?"
    (a) Devise a plan and solve the problem.
    (b) Make up a similar problem.

34. If $a - b = c$, then $a - (b + 1) =$ _____

35. Have you ever seen a combining problem that requires subtraction? These problems are usually taught in problem-solving units.
    (a) Consider the following problem. "Dad just hung up 3 shirts from the laundry in the closet next to his other shirts. Now there are 10 shirts in the closet. How many were there before?" What category (from the section) and operation are illustrated?
    (b) Write another word problem that involves combining and requires subtraction.

36. Have you ever seen a take-away or comparison problem that requires addition? Consider the following

problem. "A clerk takes 24 cartons of orange juice out of the storeroom. Now 10 cartons are left. How many were there to start with?" Write a word problem that involves a comparison and requires addition.

37. You are familiar with the "greater than" ($>$) symbol. "Greater than" can be defined using addition. For whole numbers a and b, $a > b$ when $a = b + k$ for some counting number k.
    (a) Use the definition to show why $8 > 6$.
    (b) Write a definition for "less than" ($<$) using subtraction.

38. "Less than" and "greater than" can be defined in terms of a number line.

    (a) For whole numbers a and b, $a > b$ means that a is located to the _____ of b on the
        (left, right)
        number line.
    (b) Write a similar definition of $a < b$.

Extension Exercises

39. (a) If possible, write each of the following numbers as the sum of two or more consecutive counting numbers. For example, $13 = 6 + 7$ and $14 = 2 + 3 + 4 + 5$.

1 =	6 =
2 =	7 =
3 =	8 =
4 =	9 =
5 =	10 =

    (b) Propose a hypothesis regarding what kinds of counting numbers can be written as the sum of two or more consecutive counting numbers.
    (c) The hypothesis in part (b) is based on _____ reasoning.

40. (a) For each number from 1 to 20, find *all* the ways to write it as the sum of two or more consecutive counting numbers.
    (b) Find a pattern in your results to part (a), and see if it applies to other counting numbers.
    (c) On the basis of your results, tell whether 27 and 38 can be written as the sum of three consecutive counting numbers. How about four consecutive counting numbers?

41. Last night, three women checked in at the Quantity Inn and were charged $30 for their room. Later, the desk clerk realized he had charged them a total of $5 too much. He gave the bellperson five $1 bills to take to the women. The bellperson wanted to save them the trouble of splitting $5 three ways, so she kept $2 and gave them $3.

Each woman had originally paid $10 and received $1 back. So the room cost them $9 apiece. This means they spent $27 plus the $2 "tip." What happened to the other dollar?

42. Consider the following problem. "How could you measure 1 oz of syrup using only a 4-oz container and a 7-oz container?"
(a) Devise a plan and solve the problem.
(b) Make up a similar problem.

43. All the pairs of whole numbers that add up to 6 comprise a **fact family.**

$$0 + 6, \quad 1 + 5, \quad 2 + 4, \quad 3 + 3,$$
$$4 + 2, \quad 5 + 1, \quad 6 + 0$$

(a) Plot the points $(0, 6)$, $(1, 5)$, $(2, 4)$, $(3, 3)$, $(4, 2)$, $(5, 1)$, and $(6, 0)$ on an xy graph.
(b) What geometric pattern do you see in the points?
(c) On a separate graph, plot the points for the fact family of 7.
(d) What geometric pattern do you see in the points in part (c)?
(e) Make a generalization based on your results to parts (b) and (d).
(f) Is part (e) an example of induction or deduction?

44. (a) Find six pairs of solutions to $x - y = 3$ for whole numbers x and y.
(b) Graph $x - y = 3$ for whole numbers x and y.
(c) Repeat parts (a) and (b) with $x - y = 4$.
(d) Propose a generalization about $x - y = k$ in which x, y, and k are whole numbers.
(e) Is part (d) an example of induction or deduction?

45. A function has exactly one output for each input or set of inputs. Explain why the operation of addition is a function.

46. Each letter represents a digit. Find a possible solution.

$$
\begin{array}{r}
A \\
A \\
A \\
+\ B \\
\hline
BA
\end{array}
$$

47. Consider the following problem. "Place the digits 1 to 9 in the squares so that all horizontal, vertical, and diagonal sums of three numbers equal 15."

(a) Devise a plan and solve the problem.
(b) Make up a similar problem.

48. Fill in the digits from 1 to 9 in the circles so that the sum on each side of the triangle is the same *and* as large as possible.

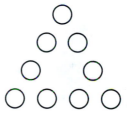

Questionnaire

49. By quantifying characteristics and adding them together, you can estimate your risk of cardiac arrest (not a cheery topic, but useful).
(a) Fill in the following questionnaire, which is adapted from one by the Executive Health Group.

Risk of Heart Disease
1. **Cigarette smoking**
 (0) Never
 (1) None in past year
 (2) Smoked in past year or under 10 per day
 (3) 20 per day
 (5) 40 or more per day

2. **Family history (parents, grandparents)**
   (0) No heart disease before age 75
   (1) One relative with heart disease between age 60 and 75
   (2) Two relatives with heart disease between age 60 and 75
   (3) One relative with heart disease under age 60
   (4) Two relatives with heart disease under age 60
3. **Diabetes (parents, grandparents, siblings)**
   (0) None
   (1) One relative
   (2) Two relatives
   (4) You have diabetes between ages 20 and 60.
4. **Exercise**
   (Subtract 1) Regular aerobic exercise
   (1) Moderate exercise during and after work
   (3) Sedentary work and moderate recreation
   (5) No exercise
5. **Disposition**
   (0) Always easy-going
   (2) Frequently impatient
   (4) Constantly hard-driving
6. **Age/gender**
   (0) Under 45, F
   (1) Under 40, M

   (2) 45–55, F, or 40–50, M
   (3) Over 55, F
   (4) 50–60, M
   (5) Over 60, M
7. **Blood pressure**
   (0) Low   (3) Medium   (6) High
8. **HDL cholesterol (if known)**
   (0) Under 3   (3) 6 to 7.9
   (1) 3 to 4.5   (4) 8 to 9.9
   (2) 4.6 to 5.9   (5) 10 or over
9. **Weight**
   (0) Not much overweight
   (1) About 10 lb overweight
   (2) About 23 lb overweight
   (3) About 38 lb overweight
   (4) About 53 lb overweight

To obtain your score, add up the numbers of all your choices. The results are as follows: 0–17, low risk to acceptable; 18–22, average; 23–32, above average—try to lower some risk factors; 33 and up, high—consult with a physician.

(b) Which factors in the questionnaire are most under your personal control?

3.3 MULTIPLICATION AND DIVISION OF WHOLE NUMBERS

NCTM Standards

- understand various meanings of multiplication and division (3–5)
- create and use representations to organize, record, and communicate mathematical ideas (pre-K–12)

To understand multiplication and division better, you will classify the kinds of situations that call for multiplication or division. Sound familiar?

 LE 1   Opener

How would you describe what multiplication is?

MULTIPLICATION DEFINITION

Historically, people developed multiplication as a shortcut for certain addition problems, such as 4 + 4 + 4. Research suggests that children first understand multiplication as repeated addition, but this has not been clearly established.

LE 2   Skill

Write $4 + 4 + 4$ as a multiplication problem.

This **repeated-addition** model of multiplication can be used to define multiplication of whole numbers.

Definition: Whole-Number Multiplication

For any whole numbers a and b where $a \neq 0$,

$$a \times b = \underbrace{b + b + \cdots + b}.$$
$$a \text{ terms}$$

If $a = 0$, then $0 \times b = 0$.

According to the definition, $3 \times 4 = 4 + 4 + 4$. However, because multiplication is commutative (see Section 3.4), $4 \times 3 = 4 + 4 + 4$ is also an acceptable equation in elementary school. In $a \times b = c$, a and b are called **factors** and c is called the **product.**

An English minister, William Oughtred (1574–1660), created over 150 mathematical symbols, only three of which are still in use, the most important being the symbol $\times$ for multiplication. Some seventeenth-century mathematicians objected, saying that the symbol $\times$ for multiplication would be confused with the letter x. One of them, Gottfried Wilhelm von Leibnitz (1646–1716), preferred the dot symbol $\cdot$ for multiplication. Multiplication of two variables such as m and n can be written as mn without the use of either multiplication symbol.

LE 3   Concept

(a) If you multiply any two whole numbers, is the result always a whole number?
(b) In part (a), did you use induction or deduction?
(c) Would a whole-number multiplication problem ever have more than one answer?

Your responses to LE 3 should suggest that whole-number multiplication is closed.

The Closure Property of Multiplication of Whole Numbers

If a and b are whole numbers, then $a \cdot b$ is a unique whole number.

CLASSIFYING MULTIPLICATION APPLICATIONS

Multiplication is used in a variety of applications, which usually fall into one of five categories.

Discussion

LE 4   Opener

Make up two word problems that can be solved by computing 3×4. Try to create two different types of problems. Compare your problems to others in the class.

Do your problems fit into any of the following categories?

1. **Repeated Sets**   Find the total number of objects, given a certain number of equivalent sets, each of which has the same number of objects.
   Example: I bought 3 packages of tomatoes. Each package contained 4 tomatoes. How many tomatoes did I buy (Figure 3–21)?

Figure 3–21

2. **Repeated Measures**   Find the total measure that results from repeating a given measure a certain number of times.
   Example: On a long car trip, I average 50 miles per hour. How far will I travel in 6 hours (Figure 3–22)?

Figure 3–22

0   50   100   150   200   250   300 miles

Figure 3–23

3 ft

4 ft

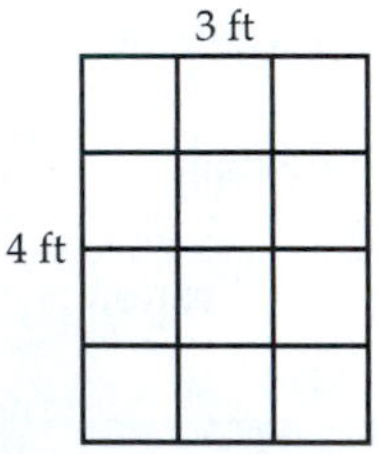

Figure 3–24

3. **Array (Sets)**   Find the total number of objects needed to occupy a given number of rows and columns.
   Example: A class has 4 rows of desks with 3 desks in each row. How many desks are there (Figure 3–23)?

4. **Area (Measures)**   Find the total measure in square units, given the width (number of rows of squares) and length (number of columns of squares). This category establishes an important connection between multiplication and area.
   Example: A rug is 4 feet by 3 feet. What is its area in square feet (Figure 3–24)?

5. **Counting Principle (Pairings)**   Find the total number of different pairs formed by pairing any object from one set with any object from a second set.
   Example: Ella has 3 blouses and 4 skirts. How many different blouse-skirt outfits can she make (Figure 3–25)? In each case, the first item is chosen from $B = \{B_1, B_2, B_3\}$ and the second item from $S = \{S_1, S_2, S_3, S_4\}$. The counting principle states that we can count the outfits (ordered pairs) by multiplying 3 by 4.

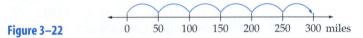

$$
\begin{array}{cccc}
B_1S_1 & B_1S_2 & B_1S_3 & B_1S_4 \\
B_2S_1 & B_2S_2 & B_2S_3 & B_2S_4 \\
B_3S_1 & B_3S_2 & B_3S_3 & B_3S_4
\end{array}
$$

Organized List of Ordered Pairs

	S_1	S_2	S_3	S_4
B_1	X	X	X	X
B_2	X	X	X	X
B_3	X	X	X	X

Figure 3–25   Array

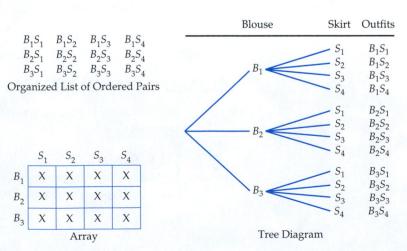

Tree Diagram

Anyone who can recognize these five situations will usually know when to multiply to solve a problem. The units in multiplication problems present a new kind of difficulty for children. In many cases, the two factors have *different kinds* of units (such as "tomatoes" and "packages of tomatoes"), unlike addition and subtraction which use the same units throughout. LE 5 and LE 6 will give you practice in recognizing different types of problems that require multiplication.

LE 5   Connection

Classify the following multiplication application. "A juice pack contains 3 cartons of juice. How many cartons are in 4 juice packs?"

LE 6   Connection

Classify the following multiplication application. "How many square feet is the surface of a rectangular tabletop that is 5 feet long and 4 feet wide?"

In elementary school, repeated sets, repeated measures, arrays, area, and the counting principle are usually introduced with manipulatives and pictures.

LE 7   Skill

Show how to solve LE 5 using a set of counters, and write the arithmetic equation it illustrates.

LE 8   Skill

Show how to solve LE 6 using paper squares, and write the arithmetic equation it illustrates.

DIVISION DEFINITION

Just as subtraction is defined in terms of addition, division is defined in terms of multiplication.

LE 9   Skill

What multiplication equation corresponds to $12 \div 3 = 4$?

In answering LE 9, you used the following relationship between multiplication and division.

Definition: Whole-Number Division

If x, y, and q are whole numbers and $y \neq 0$, then $x \div y = q$ if and only if $x = y \cdot q$.

The definition tells us that a problem such as $18 \div 6 = q$ can be viewed as $6 \cdot q = 18$. This means we can solve a division problem by looking for a **missing factor.** Children can learn basic division facts by thinking about related multiplication facts. This relationship is also useful for solving equations.

In $x \div y = q$, x is called the **dividend,** y is called the **divisor,** and q is called the **quotient.** The Swiss mathematician Johann Rahn introduced the division ($\div$) symbol in 1659. Some countries preferred Leibniz's division symbol, the colon, which is still used today. The definition of division establishes that division is the inverse operation of multiplication. The connections among whole-number operations are summarized in the following diagram.

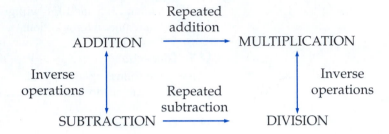

All of these connections have already been discussed except for division as **repeated subtraction.** Imagine that a child (who does not know what "division" is) has 8 cookies to distribute. She serves 2 cookies per person, and she wants to figure out how many people she can serve. She could use repeated subtraction to determine the answer.

$$8 - 2 = 6$$
$$6 - 2 = 4$$
$$4 - 2 = 2$$
$$2 - 2 = 0$$

Whole-number division problems such as $8 \div 2$ can be thought of as asking how many times 2 can be subtracted from 8 until nothing remains. The answer is 4. Some children use addition and go up instead of down. A child might add 2s until the total equals 8 and conclude that it takes four 2s to make 8. So it is possible to relate division to addition, subtraction, or multiplication.

The preceding definition of division applies to examples with whole-number quotients and no remainder. For other whole-number division problems with nonzero divisors, one can always find a whole-number quotient and a whole-number remainder. For example, $13 \div 4 = 3\,R1$.

LE 10   Skill

Using multiplication and addition, write an equation that is equivalent to $13 \div 4 = 3\,R1$.

The preceding exercise illustrates the division algorithm.

The Division Algorithm (for $a \div b$)

If a and b are whole numbers with $b \neq 0$, then there exist unique whole numbers q and r such that $a = bq + r$, in which $0 \leq r < b$.

The division algorithm equation $a = bq + r$ means

$$\text{Dividend} = \text{divisor} \cdot \text{quotient} + \text{remainder}$$

The quotient q is the greatest whole number of b's that are in a. The remainder r is how much more than bq is needed to make a.

CLASSIFYING DIVISION APPLICATIONS

Have you ever needed to divide 12 by 3? What types of situations require computing 12 ÷ 3?

Discussion

LE 11   Opener

Make up two word problems that can be solved by dividing 12 by 3. Try to create two different types of problems. Compare your results to others in your class.

Do your problems fit into any of the following categories?

1. **Repeated Sets**   Find how many sets of a certain size can be made from a group of objects.
   Example: I have a dozen eggs. How many 3-egg omelets can I make (Figure 3–26)?

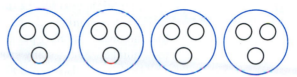

Figure 3–26

2. **Repeated Measures**   Find how many measurements of a certain size equal a given measurement.
   Example: A walk is 12 miles long. How long will it take if a walker averages 3 miles per hour (Figure 3–27)?

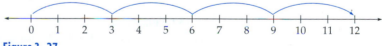

Figure 3–27

3. **Partition (Share) a Set**   Find how many are in each group when you divide a set of objects equally into a given number of groups.
   Example: We have 12 yummy strawberries for the 3 of us. How many will each person get if we are fair about it (Figure 3–28)?

Figure 3–28

4. **Partition (Share) a Measure**   Find the measure of each part when you divide a given measurement into a given number of equal parts.
   Example: Let's divide this delicious 12-inch submarine sandwich equally among the three of us. How long a piece will each of us receive (Figure 3–29)?

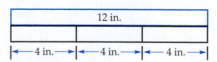

Figure 3–29

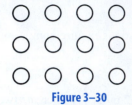

Figure 3–30

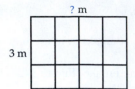

Figure 3–31

5. Array   Find the number of rows (or columns), given an array of objects and the number of columns (or rows).
Example: Twelve people are seated in 3 rows. How many people are there in each row (Figure 3–30)?

6. Area   Find the length (or width) of a rectangle, given its area and its width (or length).
Example: A rug has an area of 12 m² and a width of 3 m. What is its length (Figure 3–31)?

Students who recognize these types of situations will usually know when to divide to solve a problem. The hardest distinction for people to make is between the "repeated" categories (1 and 2) and the "partition" categories (3 and 4). For 12 ÷ 3, a "partitioning" problem would say, "Divide 12 into 3 equal parts," and a "repeating" problem would say, "How many 3's make 12?"

You might also see the difference between repeated and partitioning problems by acting out a similar example of each type. To put it another way, in repeated sets, total number ÷ number per group = how many groups. In partitioning (sharing) a set, total number ÷ how many groups = number per group.

Division is also related to fractions and ratios. These topics are discussed in Chapters 6 and 7, respectively.

In LE 12, LE 13, and LE 14, classify the division applications.

LE 12   Connection
Judy, Joyce, and Zdanna split a box of 6 pears equally. How many pears does each woman receive?

LE 13   Connection
The Arcurris drove 1,400 km last week. What was the average distance they drove each day?

LE 14   Connection
My National Motors Gerbil gets 80 miles per gallon. How many gallons would I use on a 400-mile trip?

Repeated sets and measures, partitioning a set or measure, arrays, and area are usually introduced with manipulatives and pictures in elementary school.

LE 15   Skill
Show how to solve LE 12 using a set of counters, and write the arithmetic equation it illustrates.

LE 16   Skill
Show how to solve LE 14 using a number line, and write the arithmetic equation it illustrates.

The following chart summarizes the most common classifications of arithmetic applications.

Classifying Arithmetic Applications	
Addition	**Subtraction**
1. Combine sets	**1.** Take-away sets
2. Combine measures	**2.** Take-away measures
	3. Compare sets
	4. Compare measures
	5. Missing part (sets)
	6. Missing part (measures)
Multiplication	**Division**
1. Repeated sets	**1.** Repeated sets
2. Repeated measures	**2.** Repeated measures
3. Array	**3.** Partition (share) a set
4. Area	**4.** Partition (share) a measure
5. Counting principle	**5.** Array
	6. Area

Classroom Connection

LE 17   Concept

A third grader solves a word problem. While most of the class divides (15 ÷ 3), she uses subtraction to find the correct answer. How do you think she does it?

DIVISION BY ZERO

Division problems involving zero are a source of confusion. What is 4 ÷ 0? 0 ÷ 0? 0 ÷ 4? According to the definition of division, 4 ÷ 0 and 0 ÷ 0 are undefined because you cannot have 0 as a divisor. However, 0 ÷ 4 is defined. Children wonder why one problem is defined, but the other two are not.

Discussion

LE 18   Opener

Why are problems such as 4 ÷ 0 undefined?

Expressions involving division by zero are most easily examined by converting the expression to a related multiplication equation. They can also be analyzed using any of the six division categories. The repeated-sets model is usually the easiest to use.

• EXAMPLE 1

(a) What is 4 ÷ 0?

(b) *Explain* why, using multiplication or the division category of your choice.

SOLUTION

Method 1 uses the definition of division to rewrite the division problem as a multiplication equation with a missing factor.

Method 1 $4 \div 0 = ?$ would mean $0 \times ? = 4$. There is no solution to $0 \times ? = 4$, so we make $4 \div 0$ undefined.

Method 2 Using repeated sets, $4 \div 0$ would mean "How many sets of 0 will make 4?" There is no solution. No matter how many sets of 0 you have, you can never make 4. So we make $4 \div 0$ undefined. ●

Discussion

LE 19 Communication

(a) What is $0 \div 4$?

(b) *Explain* why, using multiplication or the division category of your choice.

Discussion

LE 20 Communication

(a) What is $0 \div 0$?

(b) *Explain* why, using multiplication or the division category of your choice.

In elementary school, we simply teach children that they cannot divide a number by zero, after doing some examples such as $4 \div 0$ and $2 \div 0$. The difference between $0 \div 0$ and these other examples is not discussed. You are unlikely to find $0 \div 0$ in an elementary-school textbook, but a child may ask you about it.

LE 21 Summary

Tell what you have learned about multiplication and division in this section.

WORKING BACKWARD

In Chapter 1, you studied problem-solving strategies such as induction, guess and check, and drawing a picture. In the previous two sections, you have seen how children can learn to choose the correct operation to solve a problem. Now you can add another strategy, working backward, to your repertoire. Sometimes you can solve a problem by **working backward** from the final result to the starting point. Some working-backward problems involve the four operations of arithmetic you just studied. The following example illustrates this strategy.

● **EXAMPLE 2**

Hy Roller got his weekly paycheck. He spent half of it on a gift for his mother. Then he spent $8 on a pizza. Now he has $19. How much was his paycheck?

SOLUTION

Understanding the Problem You know that he ended up with $19, and you know how he spent his money. How much did he start with?

Devising a Plan Draw a diagram and then work backward.

Carrying Out the Plan The diagram shows the sequence from the beginning to the end of the problem.

Now work backward and fill in the question marks. He ended up with $19. What did he have before he spent $8 on the pizza? $27.

So Hy had $27 just after he spent $\frac{1}{2}$ on a gift. How much did he start with? He started with 2 × $27, or $54.

Looking Back After you obtain an answer, start with $54 and work forward to check the answer. Start with $54. Spend $\frac{1}{2}$ on a gift. That leaves $27. Spend $8 on a pizza. That leaves $19. It checks! ●

One could solve the preceding example using inverse operations.

Original Problem	Same Problem Working Backward
↓ ÷ 2	↑ × 2
↓ − 8	↑ + 8
$19	$19

Now you try one.

LE 22   Reasoning

(a) Hy Roller received a raise in his weekly pay. After he received his first new paycheck, he spent $6 on tacos. He spent $\frac{1}{2}$ of what was left on a gift for his father. Now he has $25. How much is his weekly pay? (Draw a diagram and work backward.)

(b) Does working backward involve inductive reasoning or deductive reasoning?

Consider working backward when a problem involves a series of reversible steps, the final result is known, and the initial situation is unknown.

ORDER OF OPERATIONS

What is $7 + 3 \times 6$? When you do a computation, the order in which you add and multiply may affect the answer.

LE 23   Reasoning

(a) Everyone in the class should do 7 $\boxed{+}$ 3 $\boxed{\times}$ 6 $\boxed{=}$ on a calculator. Collect the results.
(b) *Explain* how you can obtain 60 as an answer.
(c) *Explain* how you can obtain 25 as an answer.
(d) Which answer is right?

Because problems such as $7 + 3 \times 6$ are potentially ambiguous, mathematicians have made rules for the order of operations. Normally, we use parentheses to indicate the order, by writing either $(7 + 3) \times 6$ or $7 + (3 \times 6)$. However, when using a calculator, parentheses are often omitted. All computers and most calculators automatically use the following rules for order of operations with and without parentheses. These rules tell how to handle grouping symbols (usually parentheses or brackets), exponents, and the four arithmetic operations.

Order of Operations

Work within the innermost parentheses or brackets,

1. Evaluate all exponents.
2. Multiply and divide, working from left to right.
3. Add and subtract, working from left to right.

Repeat this process until all calculations in parentheses and brackets are done.

4. Follow rules 1–3 for the remaining computations.

The rules for the order of operations are also useful in working out a complicated set of computations with a calculator. You might find the answer to the first few computations, and then use that answer to continue your calculations. These same rules will also apply to negative numbers, fractions, and decimals later in the course. Apply the order of operations in the following exercise.

LE 24   Skill

Compute the following using the correct order of operations.

(a) $6 - 18 \div 3 + 7 \times (2 + 1)$
(b) $12 - 3^2 + 2 \times 5$

🔍 AN INVESTIGATION: OPERATING ON 3S

LE 25   Reasoning

Using four 3s and any of the four operations, try to obtain all the numbers from 1 to 10. Here, 0 is done as an example.

$$0 = 3 - 3 + 3 - 3$$

ANSWERS TO SELECTED LESSON EXERCISES

2. 3×4

3. (a) Yes   (b) Induction   (c) No

5. Repeated sets (4 is repeated 3 times)

6. Area

7. ◯◯◯ ◯◯◯ ◯◯◯ ◯◯◯

$4 \times 3 = 12$

8.

$5 \times 4 = 20$

9. $12 = 3 \times 4$

10. $4 \times 3 + 1 = 13$

12. Partition (share) a set

13. Partition (share) a measure (the divisor 7 partitions the distance)

14. Repeated measures (how many 80s make 400)

15.

$6 \div 3 = 2$

16.

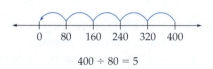

$400 \div 80 = 5$

17. $15 - 3 = 12, 12 - 3 = 9, 9 - 3 = 6, 6 - 3 = 3,$
$3 - 3 = 0.$ You can take away three 5 times.
So $15 \div 3 = 5.$

19. (a) $0 \div 4 = 0$
(b) $0 \div 4 = ?$ is the same as $4 \times ? = 0.$ So $? = 0.$

20. (a) $0 \div 0$ is undefined.
(b) $0 \div 0 = ?$ is the same as $0 \times ? = 0.$ So ? could stand for any number! Since each division problem must have one definite answer, we say $0 \div 0$ is undefined.

22. (a) $56 (double $25 and add $6)
(b) Deductive

24. (a) 21   (b) 13

3.3   HOMEWORK EXERCISES

Basic Exercises

1. (a) Rewrite $7 + 7 + 7$ as a multiplication problem.
   (b) Children may first compute 3×7 by **skip counting** (counting by some whole number greater than 1). How might a child skip count to find the answer to 3×7?
   (c) Skip counting is sometimes shown by a series of jumps along a number line. Show how to do this for 3×7.

2. (a) How might a child skip count to find $7 + (4 \times 5)$?
   (b) Illustrate part (a) with jumps along a number line.

3. Tell which of the following sets are closed under multiplication. If a set is not closed, show why not.
   (a) $\{1, 2\}$ (b) $\{0, 1\}$ (c) $\{2, 4, 6, 8, \ldots\}$

4. Explain what it means to say that whole numbers are closed under multiplication.

In Exercises 5–8, classify each application as one of the following: repeated sets, repeated measures, array, area, or counting principle.

5. Each person in the United States creates about 6 pounds of solid garbage per day. How much solid garbage does a person dispose of in a year?

6. In an election, 5 people are running for president and 3 people are running for vice president. How many different pairs of candidates can be elected to the two offices?

7. Tacos come 10 to a box. How many tacos are in 3 boxes?

8. An ice cream store sells 3 types of cones (cake, sugar, waffle) and 12 different flavors of ice cream. How many different one-scoop cones can they make?

9. How would you explain to a child why the area of the rectangle shown is 6 ft²?

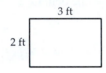

10. The following two word problems require multiplying 8 by 3. Which problem illustrates a more realistic use of mathematics?
    (a) Bill has 8 pencils. Courtney has 3 times as many. How many pencils does Courtney have?
    (b) Concert tickets cost $8 each. How much would 3 tickets cost?

11. (a) What multiplication problem is shown in the following diagram?

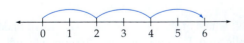

    (b) How would you use a number line to show a child that $3 \times 4 = 12$?

12. (a) What multiplication fact is illustrated:

    (b) What category of multiplication is illustrated? (area, array, counting principle, repeated)

13. Make a drawing that shows $2 \times 5 = 10$, with
    (a) repeated measures.
    (b) an array.

14. Make a drawing that shows $5 \times 3 = 15$ with
    (a) repeated sets.   (b) the counting principle.

15. Make up a word problem that illustrates the array category for multiplication.

16. Make up a word problem that illustrates the repeated measures category for multiplication.

17. Write a realistic word problem that requires computing $(3 + 6) \times 7$.

18. Write a realistic word problem that requires computing $(4 \times 8) - 5$.

19. A second grader wants to work out 5×2 with counters. She gets a set of 5 and a set of 2, and then she is stuck.
    (a) Where would she get the idea for this approach?
    (b) What would you tell the child to help her?

20. A sixth grader doesn't understand why you can make 20 outfits from 5 blouses and 4 skirts. What would you tell the child?

21. Fill in the blanks in each part by following the same rule used in the completed examples.
    (a) $4 \to 26$ (b) $3 \to 10$
        $5 \to 35$ $4 \to 14$
        $6 \to 46$ $5 \to 18$
        $8 \to$ _____ $10 \to$ _____
        $X \to$ _____ $X \to$ _____
    (c) The process of making a conjecture about the general formula in the last blank in parts (a) and (b) is an example of _____ reasoning.

22. What is the next number in each sentence?
(a) 7, 15, 31, _____
(b) 20, 37, 71, 139, _____

23. (a) Write the number of small squares using an exponent other than 1.

(b) Write the number of small cubes using an exponent other than 1.

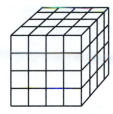

 24. (a) $11^2 =$ _____ (b) $111^2 =$ _____
(c) Without a calculator, guess the values of 1111^2 and $11,111^2$.
(d) Use a calculator to check your answers to part (c).

25. Use the definition of division to rewrite each division equation as a multiplication equation.
(a) $63 \div 7 = ?$ (b) $0 \div N = C$

26. Use the definition of division to rewrite each division equation as a multiplication equation.
(a) $40 \div 8 = ?$ (b) $32 \div N = 6$

In Exercises 27–30, classify each application as one of the following: repeated sets, repeated measures, partition a set, partition a measure, array, or area.

27. A car travels 180 miles in 4 hours. What is its average speed?

28. If I deal a deck of 52 cards to 4 people, how many cards does each person receive?

29. A car travels 180 miles, averaging 45 miles per hour. How long does this trip take?

30. A teacher wants to seat 30 students evenly in 5 rows. How many children sit in each row?

31. (a) What division fact is illustrated in the diagram?

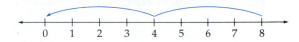

(b) How would you use a number line to show a child that $10 \div 5 = 2$?

32. (a) What division fact is illustrated in the diagram?

(b) Use a set-partition picture to show that $6 \div 3 = 2$.

33. Make a drawing that illustrates $10 \div 5 = 2$, with
(a) an array.
(b) partition a measure.
(c) repeated sets.

34. Make a drawing that illustrates $12 \div 4 = 3$ with
(a) partition a set. (b) repeated measures. (c) area.

35. Make up a word problem that illustrates the partition (share) a set category.

36. Make up a word problem that illustrates the repeated measures category for division.

37. Make up a word problem that illustrates the repeated sets category for division.

38. Make up a word problem that illustrates the area category for division.

39. (a) How might a child skip count (forward or back) to find $12 \div 3$?
(b) Illustrate part (a) with jumps along a number line.

40. (a) How might a child skip count (forward or back) to find $18 \div 6$?
(b) Illustrate part (a) with jumps along a number line.

41. Write a realistic word problem for $19 \div 4$ for which the answer is
(a) 4. (b) 5. (c) 3.

42. Write a realistic word problem for $82 \div 10$ for which the answer is
(a) 9. (b) 8. (c) 2.

43. Helmer Junghans has 37 photographs to put in a photo album. Write a question about Helmer's task that has an answer of
    (a) 10.   (b) 1.   (c) 9.

44. (a) When whole numbers are divided by 5, how many different possible remainders are there?
    (b) Repeat part (a) for division by 6.
    (c) Repeat part (a) for division by a counting number N.

45. The Hersheys buy a 6-slice pizza for the 4 members of their family. They each eat an equal number of slices and give any leftover slices to their pet goat Dainty.
    (a) How many slices does each eat?
    (b) How many slices does Dainty get?
    (c) Is the answer to part (a) a quotient, a remainder, or neither?
    (d) Is the answer to part (b) a quotient, a remainder, or neither?

46. (a) Ninety-seven children want to go on a field trip in school buses. Each bus holds 40 children. How many buses are needed?
    (b) Is the answer a quotient, a remainder, or neither?

47. A class has 26 students. The teacher forms teams with 6 students on them. The students who are left over will be substitutes.
    (a) How many substitutes are there?
    (b) Is the answer to part (a) a quotient, a remainder, or neither?

48. The number $N = 15 \times 12 + 3$.
    (a) What is the remainder when N is divided by 15?
    (b) What is the quotient when N is divided by 15?

49. (a) What is $7 \div 0$?
    (b) Explain why, using multiplication or the division category of your choice.

50. (a) What is $0 \div 7$?
    (b) Explain why, using multiplication or the division category of your choice.

 51. Try computing $4 \div 0$ on a calculator. What is the result? What does the result mean?

52. (a) What is $0 \div 0$?
    (b) Explain why with multiplication or the division category or your choice.

53. A fourth grader says that "3 divided into 6 is the same as 3 divided by 6." Is that right? If not, what would you tell the child?

54. You are teaching third grade. One of your colleagues comes over and tells you that one of his students was solving a word problem in which she was supposed to divide $(20 \div 5)$ but she found the correct answer with addition. What did the child do?

55. Tell how to solve $40 \div 8$ with
    (a) addition.   (b) subtraction.   (c) multiplication.

56. Tell how to solve $12 \div 4$ with
    (a) addition.   (b) subtraction.   (c) multiplication.

57. (a) Fill in the missing numbers.

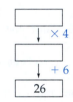

    (b) What kind of reasoning, inductive or deductive, do you use to fill in the blanks?

58. Fill in the missing numbers.

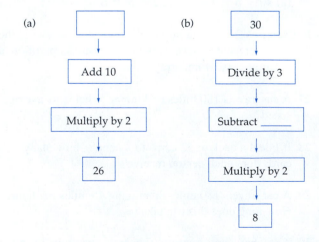

59. Frank picked some daisies. He gave half of them to his wife. Then he divided what was left evenly among his brother, his daughter, and his horse. They each got 6 daisies. How many daisies did he pick?

60. Carroll Matthews went to a gambling casino. He did not look in his wallet beforehand, but he remembers how he spent his money. He paid $5 for parking. Then he spent $3 on an imitation orange juice drink. Then he lost half of his remaining money gambling. Now he has $14. How much did Carroll start with?

61. In a two-person game of Last Out, the players take turns removing either one or two chips. The person who removes the last chip loses the game. If you go first and the game starts with 11 chips, describe a winning strategy. (Work backward.)

62. Sharon takes a number, adds 2 to it, and then multiplies the result by 3. She ends up with N. What was the original number in terms of N?

63. (a) Complete each table.

x		
$x + 3$	5	7

x		
$4x$	8	12

x		
$4x + 3$	27	51

(b) Tell how you completed each table.

64. (a) Complete each table.

t		
$t - 5$	3	7

t		
$t/3$	4	6

t		
$t/3 - 5$	2	5

(b) Tell how you completed each table.

65. Compute the following.
 (a) $(3 + 5) - 4 + 2 \times 3$
 (b) $7^2 + 2 \times 3^2$
 (c) $8 - (7 - 2) + 3 \times 2$

66. Compute the following.
 (a) $10 \times 2 - 6 \div 2 + 1$
 (b) $18 - 2 \times (4 + 1)$
 (c) $8^2 - (2 + 4 \times 3)$

 67. Without using a graphing calculator, tell the result of each of the following. Then check your answer.
 (a) To compute $\sqrt{9 + 16}$, a child enters: $9 + 16$
 (b) To compute $2(3^4)$, a child enters: $2 \times 3 \wedge 4$

68. Without using a graphing calculator, tell the result of each of the following. Then, check your answer.
 (a) To compute $\sqrt{8 \cdot 4 + 3 \cdot 2}$, a child enters:
   $(8 \times 4) + (3 \times 2)$
 (b) To compute $(3 + 6) \div (1 + 2)$, a child enters:
  $3 + 6 \div (1 + 2)$

69. Represent the following expressions with the variable t.
 (a) Twice a number
 (b) 5 less than a number
 (c) The product of a number and 5

70. Represent the following expressions using the variable y.
 (a) The product of 8 and a number
 (b) A number divided by 3
 (c) A number decreased by 7

71. Solve the following problems mentally.
 (a) $\dfrac{r}{5} = 20$ (b) $\dfrac{35}{t + 2} = 7$
 (c) $8(y - 2) = 40$

72. Solve the following problems mentally.
 (a) $4 \times \square = 28$ (b) $\dfrac{8}{n} = 2$
 (c) $7(y + 2) = 35$

73. Is whole-number division closed? (*Hint:* Check the two parts of the closure property.)

74. Because whole number division is not closed, what kind of numbers had to be invented?

75. For each of the following, tell if it is a sets or measures problem and tell what two categories are illustrated in the problem. (area, array, combine, compare, counting principle, missing part, partition, repeated, take away)
 (a) At a carnival, Puneet tried to knock down 4 rows of 3 pins. On his first throw, he hit 7 pins. How many were left?
 (b) Wally ate 2 pears. Juanita ate twice as many. How many pears did the two of them eat all together?

76. For each of the following, tell if it is a sets or measures problem and tell what two categories are illustrated in the problem. (area, array, combine, compare, counting principle, missing part, partition, repeated, take away)
   (a) How many miles must you travel each day to complete a 3260-mile car trip in 2 weeks?
   (b) A National Motors Capon travels 276 miles on 12 gallons of gas. How far does it travel on a full 20-gallon tank of gas?

77. For each of the following, tell if it is a sets or measures problem and tell what two categories are illustrated in the problem. (area, array, combine, compare, counting principle, missing part, partition, repeated, take away)
   (a) You have 10 library books. Then you return 2 of them and lend 3 to a friend. How many books do you have left?
   (b) An auditorium has 20 rows with 28 seats in each row. If 16 seats are empty at a concert, how many people are seated there?

78. (from NAEP, 1996, grade 4) Kevin has 5 rows of flowerpots, and there are 4 pots in each row. He needs to plant 3 seeds in each pot. How many seeds does he need?
   (a) Is this a sets or measures problem?
   (b) What two operations and categories are illustrated in the problem? (area, array, combine, compare, counting principle, missing part, partition, repeated, take away)

79. Gwen Cyrkiel buys A shirts for $\$B$ each and C skirts for $\$D$ each. What is the total cost of all her purchases?

80. For whole numbers a, b, and c, with a an even number and $b \neq 0$, if $a \div b = c$, then $a \div 2b =$ _____.

81. A function has exactly one output for each input or set of inputs. Explain why the operation of multiplication is a function.

82. A function has exactly one output for each input or set of inputs. Explain why the operation of division (with a nonzero divisor) is a function.

Extension Exercises

83.

5	10	20	5
4	8	16	4
6	12	24	6
9	18	36	9

   (a) Copy the table and circle four numbers, making sure that only one number is in each row and only one number is in each column.
   (b) Multiply your four numbers together.
   (c) Repeat parts (a) and (b) after picking a different group of four numbers.
   (d) Repeat parts (a) and (b) after picking a different group of four numbers.
   (e) Explain the pattern in the table.
   (f) Make your own three-row-by-three-column table that does the same thing.

84. $25 \times 25 = 625$
   $26 \times 24 = 624$
   $27 \times 23 = 621$
   (a) Predict the answer to the next multiplication problem that extends the pattern. (*Hint:* Look at the last digit of each factor.)
   (b) Check your guess in part (a).
   (c) Repeat parts (a) and (b).
   (d) Start with $40 \times 40 = 1600$ and repeat parts (a) and (b) three times.
   (e) Make a generalization about your results.
   (f) Use algebra to explain why your generalization is true. (*Hint:* Use $(x + a)(x - a)$.)

85. A 5×5 square grid has 16 squares on the border.

How many squares would an $n \times n$ grid have on the border? (Simplify your answer.)

86. At a restaurant, you can order a chicken, turkey, cheese, or roast beef sandwich on rye, wheat, or white bread. How many different kinds of sandwiches are possible?

87. You need to measure 7 oz of medicine, and all you have are two transparent 2-oz measuring cups and a glass. How can you do it?

88. The square of a number is 9 less than 10 times the number. What is the number? (Guess and check.)

 89. Every new book now has an ISBN (International Standard Book Number) such as 0-86576-009-8 (*Precalculus Mathematics in a Nutshell*). The first digit indicates the language of the country in which the book is published (for example, 0 for English). The digits 86576 represent the publisher (William Kaufmann, Inc.). The digits 009 identify the book for the publisher. Finally, the last digit is the **check digit,** which is used to check that the rest of the number is recorded correctly. To obtain an ISBN check digit:

Step 1 Multiply the first nine digits by 10, 9, 8, 7, 6, 5, 4, 3, and 2, respectively. For 0-86576-009-8, $(0 \times 10) + (8 \times 9) + (6 \times 8) + (5 \times 7) + (7 \times 6) + (6 \times 5) + (0 \times 4) + (0 \times 3) + (9 \times 2) = 245$.

Step 2 Divide the sum by 11 and find the remainder.

$$245 \div 11 = 22 \text{ R } 3$$

Step 3 Subtract the remainder from 11 to match the check digit.

$$11 - 3 = 8 \qquad \text{Yes, it checks!}$$

The check digit is used to check ISBN numbers that are copied onto order forms. Check the following ISBN numbers to see if they seem to be correct.
(a) ISBN   0-03-008367-2
(b) ISBN   0-7617-1326-8

90. The Universal Product Code (UPC) appears on many grocery items. The first digit identifies the type of product. The next five digits identify the manufacturer. The next five digits identify the specific product. The 12th digit (not always printed) is the check digit, which will reveal some errors in the first 11 digits.

For the product 0-16300-15114 (a brand of orange juice), the check digit is obtained as follows.

1. Add the digits in positions 1, 3, 5, 7, 9, and 11 and triple the sum.

2. Add the sum of the remaining digits to the result from step 1.
3. Create a check digit that makes the sum resulting from step 2 end in 0.

In this example, $0 + 6 + 0 + 1 + 1 + 4 = 12$ and $12 \times 3 = 36$. Then $1 + 3 + 0 + 5 + 1 + 36 = 46$. Make the check digit 4 so that $46 + 4$ ends in 0.

Find the check digit for
(a) 0-14019-02747 (*Consumer Reports* magazine)
(b) 0-74333-47052 (a brand of peanut butter)
(c) Find a product at home with a UPC and verify that the check digit is correct.

91. (a) Suppose you know how many groups of equal size there are and the number of items in each group. You are asked to find how many items there are in all. What operation and category does this illustrate?
(b) Suppose you know how many items there are in all and how many equal-sized groups there are. You are asked to find the number in each group. What operation and category does this illustrate?

92. Whole-number multiplication can also be defined using the Cartesian product. For whole numbers a and b, if set A contains a elements and set B contains b elements, then $a \cdot b$ is the number of elements in $A \times B$. Use this definition with $A = \{1, 2\}$ and $B = \{1, 2, 3\}$ to explain why $2 \cdot 3 = 6$.

Puzzle Time

93. In the game Krypto, you are dealt five numbers on cards and must combine them using any of the four arithmetic operations to obtain the value on a sixth card. For example, if you are dealt 3, 4, 6, 8, and 10, and the sixth card is a 2, you could write: $(10 - 3 - 6) \times 8 \div 4 = 2$. Parentheses are needed so that the subtraction is done before the multiplication and division. Try the following Krypto exercises.
(a) Use 5, 7, 9, 10, and 12 to obtain 3.
(b) Use 4, 8, 12, 15, and 20 to obtain 15.

Magic Time

94. Here is some number "magic."
(a) Pick a number between 50 and 100. Now add 58 to the number. Next, cross out the hundreds digit of your number and add that to the remaining two-digit number. Subtract this number from your original number. I bet you ended up with 41.
(b) Show why you will always end up with 41.

3.4 PROPERTIES OF WHOLE-NUMBER OPERATIONS

NCTM Standards

- illustrate general principles and properties of operations, such as commutativity, using specific numbers (K–2)

- understand and use properties of operations, such as the distributivity of multiplication over addition (3–5)

- identify such properties as commutativity, associativity, and distributivity and use them to compute with whole numbers (3–5)

Properties of whole-number operations make it easier to memorize basic facts and do certain computations. If you learn $7 \times 9 = 63$, then you know what 9×7 equals. The sum $(24 + 2) + 8$ is more easily computed as $24 + (2 + 8)$. These properties are also the basis for the efficient procedures we use to add and multiply larger numbers. The properties have names such as "commutative," "associative," "identity," and "distributive." Do you feel your memory being jarred?

THE COMMUTATIVE AND ASSOCIATIVE PROPERTIES

Is a "night light" the same as a "light night"? Is 7×9 the same as 9×7? These questions involve the commutative property. In mathematics, the commutative property says that you can change the *order* of two numbers in certain arithmetic operations and still obtain the same answer. Which of the four whole-number operations are commutative? Investigate this question in LE 1 and LE 2.

Classroom Connection

LE 1 Opener

Answer the following questions to help determine whether whole-number addition is commutative.

(a) How would you convince a first-grader that $2 + 6$ is the same as $6 + 2$?
(b) Does $6 + 8 = 8 + 6$?
(c) For any two whole numbers x and y, do you think $x + y = y + x$?
(d) Your answer to part (c) is based on _____ reasoning.

LE 2 Reasoning

Try some examples and see whether you think the following whole-number operations are commutative. Give a counterexample for any operation that is not commutative.

(a) Subtraction   **(b)** Multiplication   **(c)** Division

By now, you should be convinced that whole-number addition and multiplication are commutative.

The Commutative Property of Addition for Whole Numbers

For any whole numbers x and y, $x + y = y + x$.

> ### The Commutative Property of Multiplication for Whole Numbers
>
> For any whole numbers x and y, $xy = yx$.

The term "commutative" was first used by Francois Servois in 1814. The commutative properties can be stated in words as follows: When you add two whole numbers, you may add them in either order, and when you multiply two whole numbers, you may multiply them in either order. Figure 3–32 illustrates these properties with sets and measures.

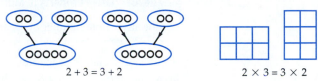

$$2 + 3 = 3 + 2 \qquad\qquad 2 \times 3 = 3 \times 2$$

Figure 3–32

Note that the commutative property applies to expressions involving *one* operation, either addition or multiplication. But how do human beings use this property? It makes learning the addition and multiplication tables a lot easier.

LE 3 Concept

How does the commutative property reduce the number of one-digit addition facts (for example, $3 + 5 = 8$) that must be memorized?

What about the associative (grouping) property? This property says that the grouping of numbers for an arithmetic operation will not change the answer. So which whole-number operations are associative? The following exercises will help you decide.

LE 4 Reasoning

Answer the following questions, and see whether you think whole-number addition is associative.
(a) Does $(6 + 3) + 5 = 6 + (3 + 5)$?
(b) Does $(24 + 2) + 18 = 24 + (2 + 18)$?
(c) Do you think $(x + y) + z = x + (y + z)$ for all whole numbers?

LE 5 Reasoning

Try some examples, and see whether you think the following whole-number operations are associative. Give a counterexample for any operation that is not associative.

(a) Subtraction   **(b)** Multiplication   **(c)** Division

LE 6 Reasoning

If A, B, and C are whole numbers, under what conditions is

(a) $(A - B) - C > A - (B - C)$? Tell why.
(b) $(A - B) - C = A - (B - C)$? Tell why.
(c) $(A - B) - C < A - (B - C)$? Tell why.

Are you now entirely convinced that whole-number addition and multiplication are commutative and associative? They are.

The Associative Property of Addition for Whole Numbers

For any whole numbers x, y, and z, $(x + y) + z = x + (y + z)$.

The Associative Property of Multiplication for Whole Numbers

For any whole numbers x, y, and z, $(xy)z = x(yz)$.

The associative properties can be stated in words as follows: When you add a series of whole numbers, parentheses have no effect on the results, and when you multiply a series of whole numbers, parentheses also have no effect on the results. In these kinds of problems, you can move parentheses around or remove them altogether.

Figure 3–33 illustrates these properties with sets.

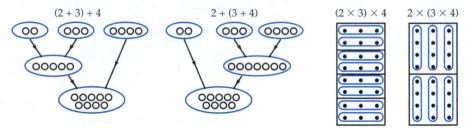

Figure 3–33

Like the commutative property, the associative property applies to expressions involving *one* operation, either addition or multiplication. The commutative property allows you to change the order of the numbers being added (or multiplied). The associative property allows you to change the order in which the operations are performed. We use the associative property in algebra to simplify an expression such as $8 \cdot (3x)$.

LE 7 Concept

(a) Does $8 \cdot (3 \cdot x) = 8 \cdot 3 \cdot 8 \cdot x$? (If you're not sure, try substituting some numbers for x.)
(b) What property says that for a whole number x, $8 \cdot (3 \cdot x) = (8 \cdot 3) \cdot x$?
(c) Part (b) confirms that $8 \cdot (3 \cdot x)$ simplifies to _____.

Together, the commutative and associative properties allow us to reorder and regroup numbers in an addition problem in any way we want! We can put the addends in any order, and we can remove or put them in parentheses. For example, we can change $46 + (28 + 4) + 2$ to $(46 + 4) + (28 + 2)$ using the commutative and associative properties to create easier computations. The same reordering and regrouping may be done to any multiplication problem.

LE 8 Concept

(a) Rewrite $8 \cdot (4 \cdot 7) \cdot 25$ to make it easier to compute mentally.
(b) What properties justify your answer to part (a)?

THE IDENTITY PROPERTY

Some whole-number operations have a unique number called an *identity element.* The following exercises will guide you through an investigation of identity elements.

LE 9   Concept

What whole number, if any, can go in both blanks in the following statement?

For every whole number w, $w +$ _____ = _____ $+ w = w$.

LE 10   Concept

What whole number, if any, can go in both blanks in the following statement?

For every whole number w, $w \cdot$ _____ = _____ $\cdot w = w$.

LE 11   Concept

What whole number, if any, can go in both blanks in the following statement?

For every whole number w, $w -$ _____ = _____ $- w = w$.

LE 12   Concept

What whole number, if any, can go in both blanks in the following statement?

For every whole number w, $w \div$ _____ = _____ $\div w = w$.

In LE 9 through LE 12, did you find that only addition and multiplication have a unique whole number that goes in both blanks and works for all whole numbers? These special numbers are called **identity elements.**

Identity Elements for Whole-Number Addition and Multiplication

Zero is the unique additive identity such that, for every whole number w,
$w + 0 = 0 + w = w$.
One is the unique multiplicative identity such that, for every whole number w,
$w \cdot 1 = 1 \cdot w = w$.

There is no identity element for whole-number subtraction or division. Don't let this upset you too much. Let's return to addition and multiplication and enjoy the fact that these operations do have identities.

LE 13   Connection

(a) The equation $3 + 5 = 8$ is an example of a one-digit addition fact. What one-digit addition facts use the identity property?
(b) What one-digit multiplication facts use the identity property?

Together, the commutative and identity properties of addition significantly reduce the number of separate addition facts that need to be memorized. In multiplication, the commutative and identity properties, along with the fact that $0 \times a = 0$ for every whole number a, greatly reduce the number of separate multiplication facts that children must memorize. You will learn additional strategies children can use to memorize addition and multiplication facts if you take a course in methods of teaching mathematics.

THE DISTRIBUTIVE PROPERTY

The distributive property plays an important part in the procedure for multiplying numbers such as 34×2. The distributive property is the only property in this lesson that involves two different operations at the same time. The distributive property of multiplication over addition is

$$F \cdot (G + H) = (F \cdot G) + (F \cdot H)$$

Discussion

LE 14   Reasoning

(a) Try some examples and see whether you think the distributive property of multiplication over addition holds for all whole numbers.

(b) The conclusion is based on _____ reasoning.

LE 14 suggests that the whole numbers possess the distributive property of multiplication over addition.

The Distributive Property of Whole-Number Multiplication Over Addition

For any whole numbers x, y, and z, $x(y + z) = (xy) + (xz)$.

Figure 3–34 illustrates the distributive property using arrays.

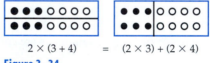

$2 \times (3 + 4)$ $=$ $(2 \times 3) + (2 \times 4)$

Figure 3–34

LE 15   Skill

Draw a set diagram showing that $3 \times (2 + 5) = (3 \times 2) + (3 \times 5)$.

People use the distributive property in algebra to combine like terms. Consider the following exercise.

LE 16   Skill

If x is a whole number, the distributive property of multiplication over addition says that $7 \cdot x + 3 \cdot x = $ _____.

Do any other distributive properties work for whole-number operations?

LE 17   Reasoning

(a) $F + (G \times H) = (F + G) \times (F + H)$ would be called the distributive property of _____ over _____.

(b) Try some whole-number examples, and see whether you think the property in part (a) is true.

LE 18   Reasoning

(a) $F \times (G - H) = (F \times G) - (F \times H)$ would be called the distributive property of
_____.

(b) Try some whole-number examples, and see whether you think the property in part
(a) is true.

(c) The conclusion in part (b) is based on _____ reasoning.

As LE 18 suggests, another distributive property holds for whole numbers.

The Distributive Property of Whole-Number Multiplication Over Subtraction

For any whole numbers x, y, and z, $x(y - z) = (xy - xz)$.

LE 19   Connection

A storekeeper buys 24 televisions for $99 each. A method for mentally multiplying
24×99 is to compute $(24 \times 100) - (24 \times 1)$. Use a distributive property to show
why these two expressions are equal.

The following chart summarizes all the properties of whole-number operations presented in this section.

Properties of Whole-Number Operations

Whole-number addition is commutative and associative.

Whole-number multiplication is commutative and associative.

The additive identity for whole numbers is 0.

The multiplicative identity for whole numbers is 1.

Multiplication is distributive over addition and subtraction for whole numbers.

LE 20   Summary

Tell what you learned about properties of whole-number operations in this section.

ANSWERS TO SELECTED LESSON EXERCISES

1. (a) *Hint:* Use counters. (b) Yes (d) Inductive

2. (a) No, $2 - 1 \neq 1 - 2$
(c) No, $2 \div 1 \neq 1 \div 2$

3. After memorizing any fact with two different addends
(e.g., $3 + 5 = 8$), you will also know a companion
fact with the addends reversed (e.g., $5 + 3 = 8$). This
reduces the number of basic facts that need to be
learned from 100 to 55.

+	0	1	2	3	4	5	6	7	8	9
0	0	1	2	3	4	5	6	7	8	9
1		2	3	4	5	6	7	8	9	10
2			4	5	6	7	8	9	10	11
3				6	7	8	9	10	11	12
4					8	9	10	11	12	13
5						10	11	12	13	14
6							12	13	14	15
7								14	15	16
8									16	17
9										18

4. (a) Yes   (b) Yes

5. (a) No, $(3 - 2) - 1 \neq 3 - (2 - 1)$
(c) No, $(8 \div 4) \div 2 \neq 8 \div (4 \div 2)$

6. (a) This is the same as $A - B - C > A - B + C$. Since $-C$ is never greater than C, there is no solution.
(b) $-C = C$ when $C = 0$.
(c) The remaining possibility is $C \neq 0$.

7. (a) No
(b) Associative property of $\times$

8. (a) $(4 \cdot 25) \cdot 8 \cdot 7$
(b) Commutative and associative properties

9. 0

10. 1

11. None

12. None

13. (a) $0 +$ any number or any number $+ 0$
(b) $1 \times$ any number or any number $\times 1$

14. (b) inductive

15.

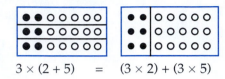

$$3 \times (2 + 5) \quad = \quad (3 \times 2) + (3 \times 5)$$

16. $(7 + 3) \cdot x$

17. (a) addition over multiplication
(b) No

18. (a) multiplication over subtraction
(b) Yes
(c) inductive

19. $24 \times 99 = 24 \times (100 - 1) = (24 \times 100) - (24 \times 1)$

3.4   HOMEWORK EXERCISES

Basic Exercises

1. Write a sentence telling the difference between the commutative property of addition and the associative property of addition for whole numbers.

2. (a) Four sets of 3 is the same as 3 sets of 4 because of the _____ property of _____.
(b) Draw a picture that shows that the two amounts in part (a) are the same.

3. A second grader figures out $(19 + 2) + 8$ as follows. "First, I add $2 + 8$ to get 10. Then $19 + 10$ is 29." What property is the child using?

4. A first grader figures out $2 + 5$ by starting at 5 and counting on 2 more to get 7. What property is the child using?

5. What whole-number operations are commutative?

6. What whole-number operations are associative?

7. (a) Give a counterexample showing that whole-number subtraction is not associative.
(b) Explain why one side of your equation results in a smaller number than the other side.

8. Some people confuse $8 \div 2$ and $2 \div 8$.
(a) $8 \div 2 \neq 2 \div 8$ is a counterexample disproving what property?
(b) Explain the difference between $8 \div 2$ and $2 \div 8$.

9. How does the commutative property reduce the number of one-digit multiplication facts that must be memorized?

10. Tell whether each sequence of activities is commutative.
(a) putting on your socks, putting on your shoes
(b) brushing your teeth, combing your hair.

11. In English, the meaning of a phrase may change depending upon which words are associated. "Slow-motion picture" and "slow motion picture" have different meanings. Therefore, the phrase "slow-motion picture" is not associative. Which of the following phrases is associative?
    (a) Man eating shark
    (b) Smart handsome stranger
    (c) Hot dog salesperson
    (d) High-school student

12. In English, the meaning of a phrase may change depending upon the order of the words. Does changing the order change the meanings of the following phrases?
    (a) Hall light   (b) Killer shark

13. A carpet that is 6 ft by 9 ft costs $5 per square foot. Explain an easy way to compute mentally the total cost of the carpet.

14. Three items cost $246, $38, and $4. Explain an easy way to compute mentally the total cost.

15. You buy a simulated Astroturf carpet for $38, a talking wastebasket for $57, and a half-pound chocolate moose for $2. What is the easiest way to compute your total bill?

16. Tell all the different ways to compute $3 \cdot 2 \cdot 4$ by multiplying.

17. $6 + 0 = 0 + 6 = 6$
    $8 + 0 = 0 + 8 = 8$
    These examples illustrate that _____ is the
    _____ for _____.

18. What one-digit multiplication facts use the identity property?

19. Name the property illustrated.
    (a) $3 \cdot (2 \cdot r) = (3 \cdot 2) \cdot r$
    (b) $4 \cdot x + 3 \cdot x = (4 + 3)x$
    (c) $n \cdot 2 = 2 \cdot n$
    (d) $(8 + 2w) + 5w = 8 + (2w + 5w)$

20. Match each property name with an equation.
    Associative property of addition
    Commutative property of addition
    Identity property of addition
    Identity property of multiplication
    Distributive property of multiplication over addition
    (a) $n \cdot 1 = 1 \cdot n = n$
    (b) $a(b + c) = ab + ac$
    (c) $a + 0 = a$
    (d) $a + (b + c) = (a + b) + c$

21. Until fourth or fifth grade, many mathematics textbooks refer to some of the properties by alternative names. Match each name in column 1 with its corresponding name in column 2.

Lower Elementary Grades	Upper Elementary Grades
Order property	Associative
Zero property	Commutative
Grouping property	Identity for addition
Property of one	Identity for multiplication

22. If x is a whole number, justify each of the following steps.
$$\begin{aligned} (x + 3)(x + 6) &= x(x + 6) + 3(x + 6) \ \underline{\hspace{1cm}} \\ &= x^2 + 6x + 3x + 18 \ \underline{\hspace{1cm}} \\ &= x^2 + (6 + 3)x + 18 \ \underline{\hspace{1cm}} \\ &= x^2 + 9x + 18 \quad \text{Basic addition fact} \end{aligned}$$

23. A fourth grader says that
    $3 \times (4 \times 5) = 3 \times 4 \times 3 \times 5$. Is this correct? If not, why would the child think it is?

24. A fourth grader says that 0 is the identity number for subtraction. Is this correct? If not, what would you tell the child?

25. You ask two children if $2 + 5 = 5 + 2$. Margaret says, "I counted some blocks for each problem and got the same answer." Nino says, "They are equal because one combines 2 blocks and 5 blocks, and the other combines 5 blocks and 2 blocks. It doesn't matter which set comes first. You will get the same answer." What is different about each child's understanding of the problem?

26. A child says that $8 - 2$ equals $2 - 8$. Is this right? If not, what would you tell the child?

27. (a) Find the total area of the yard shown in the following picture.
(b) Find the total area a different way.
(c) If possible, show that your two methods are related by the distributive property.

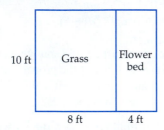

28. You drove your car on two different days for business purposes, going 120 miles and 62 miles. The company will reimburse you at the rate of 20 cents per mile. Show two ways of computing the total cost that illustrate the distributive property.

29. A, B, and C are whole numbers, and $A \neq 0$ and $B \neq 0$. For each equation, find one set of numbers that makes the equation true and one set of numbers that makes the equation false.
(a) $A \div B = B \div A$
(b) $C \div (A \div B) = (C \div A) \div B$
(c) $A - (B - C) = (A - B) - C$
(d) $A - B$ is a whole number.
(e) $A \div B$ is a whole number.
(f) $A \times (C \div B) = (A \times C) \div (A \times B)$
(g) $A - (C \div B) = (A - C) \div (A - B)$

30. A, B, and C are whole numbers with $C \neq 0$. The following represent three fairly common algebra errors. For each equation, find one set of numbers that makes the equation true and one set of numbers that makes the equation false.
(a) $(A - B)^2 = A^2 - B^2$
(b) $(A + B)^2 = A^2 + B^2$
(c) $A(BC) = (AB)(AC)$

Extension Exercises

31. (a) When you double both addends, what happens to the sum?
(b) Use the distributive property to prove your answer to part (a).
(c) What kind of reasoning did you use in part(a)?
(d) What kind of reasoning did you use in part(b)?

32. (a) When you double both numbers in a subtraction problem, what happens to the difference?
(b) Use the distributive property to prove your answer to part (a).

Enrichment Topic

33. Consider the set of whole numbers and the operation ∇, which takes the larger of any two numbers as the result. For example, $3 \nabla 6 = 6$ and $8 \nabla 2 = 8$.
(a) Is ∇ commutative for whole numbers?
(b) Is there an identity number for ∇ in W?
(c) Is ∇ associative for whole numbers?

34. Consider the set of numbers $A = \{5, 6, 7, 8\}$ with a made-up operation called $*$. The results for $*$ are shown in the table.

$*$	5	6	7	8
5	8	7	6	5
6	7	6	5	8
7	6	5	8	7
8	5	8	7	6

(a) Is $*$ commutative for set A?
(b) Is there a number I (an identity element) in A such that $I * a = a * I = a$ for all numbers a in A?
(c) Does $(5 * 6) * 8 = 5 * (6 * 8)$?
(d) Is $*$ associative for set A?

3.5 ALGORITHMS FOR WHOLE-NUMBER ADDITION AND SUBTRACTION

NCTM Standards

- develop and use strategies for whole-number computations, with a focus on addition and subtraction (pre-K–2)
- develop fluency in adding, subtracting, multiplying, and dividing whole numbers (3–5)
- analyze and evaluate the mathematical thinking and strategies of others (pre-K–12)

People have developed procedures to perform paper-and-pencil computations involving large numbers more easily. Despite having such procedures, people in the Middle Ages considered whole-number arithmetic a college-level subject! Today children in elementary school study some of the best of these computational procedures, called **algorithms.** We typically use algorithms to perform computations with two- and three-digit numbers.

"Algorithm" is a term you can use to impress friends at parties when they ask, "What *are* you studying in that math class, anyway?" Tell them, "We're analyzing a few algorithms." Actually, the word "algorithm" comes from al-Khowarizmi, an eighth-century Persian mathematician, whose books described procedures for arithmetic.

ADDITION ALGORITHMS

The *Principles and Standards for School Mathematics* recommend that rather than simply teaching students the most efficient algorithm, students first have the opportunity to develop their own algorithms. Research suggests that children develop a better understanding of base-ten arithmetic after devising their own procedures. After a discussion of various methods, students should learn a standard algorithm.

But what algorithms do children invent? After exploring problems such as 34 + 2, 30 + 40 and 34 + 40, children would be ready for a problem such as 24 + 32. How would they work it out?

Classroom Connection

LE 1 Opener

Suppose you ask a group of second graders to compute

$$\begin{array}{r} 24 \\ + 32 \\ \hline \end{array}$$

before they have studied the standard algorithm. Write down all the ways you think a child might find the correct answer. Assume they have base-ten blocks (Activity Card 1) to help them but no calculators.

When a teacher asks children to devise their own addition algorithms, many children will use compensation or breaking apart to compute the answer. In **compensation,** the numbers in a problem are adjusted in a way that makes computation easier but leaves the final result the same. For example, a child might change 24 + 32 to 26 + 30.

In **breaking apart,** you break apart one *or* both addends often by using the place value of the number. For example, a child might compute 20 + 30 = 50 and 4 + 2 = 6 and then combine 50 + 6 = 56. This method is called the **partial sums algorithm.** Whichever method children use, they are more likely to add from left to right instead of right to left.

The partial sums algorithm is an **expanded algorithm,** which is longer and easier to understand. In the standard addition algorithm, children line up digits according to place value and proceed from right to left, adding the digits in each column. Figure 3–35 shows how to compute 42 + 26 with the partial sums and standard algorithms.

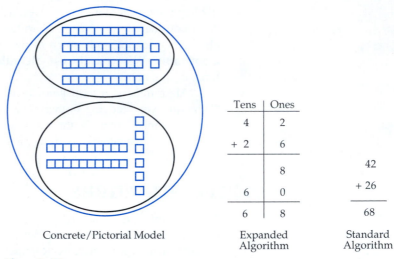

Concrete/Pictorial Model   Expanded Algorithm   Standard Algorithm

Figure 3–35

While the partial sums algorithm does not really change for a problem such as 38 + 54, the standard algorithm becomes more complicated because it requires regrouping. Children usually learn about regrouping with the standard algorithm by using base-ten blocks to work out the steps.

• EXAMPLE 1

Suppose you want to introduce a second grader to the standard algorithm for computing

$$\begin{array}{r} 38 \\ + 54 \\ \hline \end{array}$$

Explain how to work out the 38 + 54 with base-ten blocks.
Follow the same sequence of steps as the standard algorithm.

SOLUTION

Step 1   Show 38 and 54 with base-ten blocks.

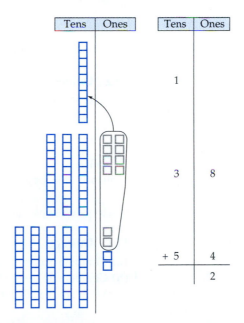

Step 2   Add the ones.
8 ones + 4 ones = 12 ones.
Regroup 12 ones as 1 ten 2 ones.

Step 3   Add the tens.
1 ten + 3 tens + 5 tens = 9 tens.
The sum is 9 tens 2 ones = 92.

Tens	Ones
	1
3	8
+ 5	4
9	2

Now it's your turn

LE 2   Connection

Suppose you want to introduce a second grader to the standard algorithm for computing

$$\begin{array}{r} 26 \\ + 35 \\ \hline \end{array}$$

Explain how to find the sum with base-ten blocks (Activity Card 1) following the same sequence of steps as the standard algorithm. (Drawings are optional.)

How do the partial sums and standard algorithms compare?

LE 3   Concept

(a) Compute 562 + 275 with the partial sums algorithm.
(b) Compute 562 + 275 with the standard algorithm.
(c) What are the advantages and disadvantages of each algorithm?

Did you notice that the process of regrouping is easier to understand in the partial sums algorithm? It is also easier to understand place value in the partial sums algorithm. The standard algorithm is shorter to write and faster once it has been mastered.

The partial sums algorithm can serve two different purposes. Some teachers use it as a transitional algorithm to help students progress from base-ten blocks to the standard algorithm. Other instructors teach it to students who have been unable to learn the standard algorithm. In that case, the procedure is usually done from left to right rather that right to left. Since calculators are available for more complicated computations, it is reasonable to consider teaching children a somewhat less efficient procedure. Another alternative algorithm some children prefer is scratch addition (see Exercise 29).

One can use the properties of whole-number operations to show how the standard addition algorithm works. For example, in computing

$$\begin{array}{r} 56 \\ + 32 \\ \hline \end{array}$$

why are we allowed to add 2 + 6 and then add 50 + 30 to that result?

LE 4   Reasoning

Discussion

Fill in the properties that justify the last three steps.

$$
\begin{aligned}
56 + 32 &= (50 + 6) + (30 + 2) & \text{Expanded notation} \\
&= 50 + (6 + 30) + 2 & \text{+ is associative} \\
&= 50 + (30 + 6) + 2 & \underline{\hspace{3cm}} \\
&= (50 + 30) + (6 + 2) & \underline{\hspace{3cm}} \\
&= (6 + 2) + (50 + 30) & \underline{\hspace{3cm}}
\end{aligned}
$$

As LE 4 demonstrates, the commutative and associative properties are the basis for the standard addition algorithm.

SUBTRACTION ALGORITHMS

In studying subtraction algorithms, children first solve problems that do not require regrouping. Children line up digits according to place value and proceed from right to left, subtracting each digit from the one above it. This process is sufficient to compute the answer to a problem such as 35 − 23.

LE 5   Opener

Classroom Connection

Suppose you ask a group of second graders to compute 35 − 23 before they have studied the standard algorithm. Write down all the ways you think children might find the correct answer. Assume they have base-ten blocks (Activity Card 1) to help them but no calculators.

A useful expanded subtraction algorithm is the **partial differences algorithm.** You may have thought of something like it in the preceding lesson exercise. Figure 3–36 shows how to compute 35 − 23 with the partial differences (expanded) algorithm and the standard algorithm.

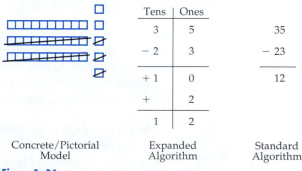

Figure 3–36

Like the partial sums algorithm, the partial differences algorithm can be used as a transition to the standard algorithm or as an alternative to the standard algorithm. Subtraction is more difficult when the top digit in any column is less than the bottom digit. While the partial differences algorithm is somewhat different for a problem such as 54 − 38, the standard algorithm is more complicated because it requires regrouping. Children usually learn about regrouping with the standard algorithm by using base-ten blocks to work out the steps.

Discussion

LE 6   Connection

Suppose you want to introduce a second grader to the standard algorithm for computing

$$54$$
$$\underline{-\ 38}$$

Explain how to find the difference with base-ten blocks (Activity Card 1) following the same sequence of steps as the standard algorithm.
(*Hint:* Step 1 shows 54 and asks if you can take away 8 ones. In step 2, regroup and then subtract the ones. Subtract the tens and state the difference in step 3.)

How does the partial differences algorithm change in a problem such as 54 − 38?

Tens	Ones	
5	4	
− 3	8	
+ 2	0	(a plus sign is used when the larger digit is in the minuend (top number))
−	4	(a minus sign is used when the larger digit is in the subtrahend (the 38))
1	6	(then compute following the signs: 20 − 4)

Now see for yourself how the two algorithms compare.

LE 7   Concept

(a) Compute 562 − 275 with the partial differences algorithm.
(b) Compute 562 − 275 with the standard algorithm.
(c) What are the advantages and disadvantages of each algorithm?

Which algorithm do you think would be easier for most children to learn? It is easier to understand place value in the partial differences algorithm. The standard algorithm is shorter to write and faster once it has been mastered.

COMMON ERROR PATTERNS IN ALGORITHMS

The bad news is that not all students learn the algorithms just the way you teach them. Confused students may initially develop their own erroneous procedures. The good news is that you get to play detective in trying to uncover students' error patterns.

Children's written work provides the first evidence of many learning difficulties. Try your luck at finding the error patterns in the following exercises.

Classroom Connection

In LE 8 and LE 9, (a) complete the last example, repeating the error pattern in the completed examples, (b) write a description of the error pattern, and (c) write what you would tell the child about his or her error. (You may use base-ten blocks.)

LE 8   Reasoning

$$
\begin{array}{r} 49 \\ + \ 37 \\ \hline 76 \end{array}
\qquad
\begin{array}{r} 67 \\ + \ 43 \\ \hline 100 \end{array}
\qquad
\begin{array}{r} 92 \\ + \ 39 \\ \hline \end{array}
$$

LE 9   Reasoning

$$
\begin{array}{r} 92 \\ - \ 39 \\ \hline 67 \end{array}
\qquad
\begin{array}{r} 408 \\ - \ 322 \\ \hline 126 \end{array}
\qquad
\begin{array}{r} 921 \\ - \ 376 \\ \hline \end{array}
$$

LE 10   Summary

Tell what you learned about the addition and subtraction algorithms in this section.

ANSWERS TO SELECTED LESSON EXERCISES

2. Step 1   Show 26 as 2 tens and 6 ones and 35 as 3 tens and 5 ones.

   Step 2   Add the ones.
   6 ones + 5 ones = 11 ones.
   Regroup 11 ones as 1 ten 1 one.

   Step 3   Add the tens.
   1 ten + 2 tens + 3 tens = 6 tens
   The sum is 6 tens 1 one = 61.
   26 + 35 = 61

3. (a)
$$
\begin{array}{r} 562 \\ + \ 275 \\ \hline 7 \\ 130 \\ 700 \\ \hline 837 \end{array}
$$
(b)
$$
\begin{array}{r} 1 \\ 562 \\ + \ 275 \\ \hline 837 \end{array}
$$

4. + is commutative; + is associative; + is commutative

6. Step 1   Show 54 as 5 tens and 4 ones. Can you take away 8 ones?

Step 2   Regroup 1 ten as 10 ones.
   Subtract the ones.
   14 ones − 8 ones = 6 ones.

Step 3   Subtract the tens.
   4 tens − 3 tens = 1 ten
   The difference is 1 ten 6 ones = 16.

8. (a) 121
   (b) The child does not regroup the 1 to the tens column.

9. (a) 655
   (b) The child subtracts the smaller digit from the larger in each place-value column.

3.5   HOMEWORK EXERCISES

Basic Exercises

1. Four children work out $76 + 38$. Tell whether each solution is correct. If so, what does the child understand about addition? If the answer is wrong, what would you tell the child about how to solve the problem?

(a) $\begin{array}{r} 76 \\ +\ 38 \\ \hline 100 \\ 14 \\ \hline 114 \end{array}$ (b) $\begin{array}{r} 76 \\ +\ 38 \\ \hline 104 \end{array}$

(c) 76 plus 30 is 106. Then add 8 to get 114.
(d) $76 + 38 = 74 + 40 = 114$

2. Three second graders are asked to solve a problem in which they compute $64 + 76$. For each method, tell what the child understands about addition.
(a) $64 + 70$ is 134. Then add 6 to get 140.
(b) $64 + 76$ is the same as $60 + 80$ which is 140.
(c) $60 + 70$ is 130 and $4 + 6$ is 10. Then $130 + 10 = 140$.

3. Suppose you want to introduce a third grader to the standard algorithm for computing $182 + 336$. Explain how to find the sum with base-ten blocks following the same sequence of steps as the standard algorithm. (Drawings are optional.)

4. Suppose you want to introduce a second grader to the standard algorithm for computing $57 + 36$. Explain how to find the sum with base-ten blocks following the same sequence of steps as the standard algorithm. (Drawings are optional.)

5. (a) Compute $357 + 529$ with the partial sums algorithm.
(b) Compute $357 + 529$ with the standard algorithm.
(c) What are the advantages and disadvantages of each algorithm?

6. (a) Compute $89 + 34$ with the partial sums algorithm.
(b) Compute $89 + 34$ with the standard algorithm.

7. When you add $39 + 48$ using the standard algorithm, you add $9 + 8 + 30 + 40$ and regroup. The following shows why this method is correct. Fill in the blanks with properties you studied in this lesson.

$$
\begin{aligned}
39 + 48 &= (30 + 9) + (40 + 8) && \text{Expanded} \\
&&& \text{notation} \\
&= 30 + (9 + 40) + 8 && \underline{\hspace{2cm}} \\
&= 30 + (40 + 9) + 8 && \underline{\hspace{2cm}} \\
&= (30 + 40) + (9 + 8) && \underline{\hspace{2cm}} \\
&= (9 + 8) + (30 + 40) && \underline{\hspace{2cm}} \\
&= 17 + (30 + 40) && \text{Addition fact} \\
&= 7 + 10 + (30 + 40) && \text{Expanded} \\
&&& \text{notation} \\
&= 7 + (10 + 30 + 40) && \underline{\hspace{2cm}}
\end{aligned}
$$

8. Fill in the blank with the property that justifies the step.

$$
\begin{aligned}
36 + 8 &= (30 + 6) + 8 && \text{Expanded notation} \\
&= 30 + (6 + 8) && \underline{\hspace{2cm}}
\end{aligned}
$$

9. The sum of 4 three-digit whole numbers is greater than 1200. Tell whether each of the following is correct and explain why.
(a) Each of the numbers is greater than 300.
(b) Two of the numbers have a sum greater than 600.

10. The sum of 4 two-digit whole numbers is less than 200. Tell whether each of the following is correct and explain why.
(a) At least one of the numbers is less than 50.
(b) Two of the numbers have a sum less than 100.

11. A group of second-grade children are asked to compute $62 - 37$ (CBMS, pp. 59–60). For each method, tell what the student understands about subtraction.
(a) Take 37. Then 3 more makes 40. Then add 20 more to get to 60. Then 2 more to get to 62. I added $3 + 20 + 2 = 25$. That's the answer.
(b) I took 3 tens from 6 tens which leaves 3 tens. Then I took 7 from 1 of those tens, which leaves 3. So I am left with $3 + 2$ tens $+ 5 = 25$.
(c) I counted from 37 up to 57 which is 20 and then 5 more to 62. So the answer is $20 + 5 = 25$.

12. Three third graders are working out $135 - 87$. Tell whether each solution is correct. If so, what does the child understand about subtraction? If the answer is wrong, what would you tell the child about how to solve the problem?

(a) $\begin{array}{r} 1^13^15 \\ -\ \ 8\ 7 \\ \hline 1\ 5\ 8 \end{array}$

(b) 135 minus 80 is 55. Now 55 minus 7 is 48.

(c) 135 minus 87 is the same as 138 minus 90 which is 48.

13. In subtracting 462 from 827, the 827 must be regrouped as _____ hundreds, _____ tens, and _____ ones.

14. In subtracting 87 from 325, 325 must be regrouped as _____ hundreds, _____ tens, and _____ ones.

15. Suppose you want to introduce a third grader to the standard algorithm for computing $336 - 182$. Explain how to find the difference with base-ten blocks following the same sequence of steps as the standard algorithm. (Drawings are optional.)

16. Suppose you want to introduce a second grader to the standard algorithm for computing $82 - 49$. Explain how to find the difference with base-ten blocks following the same sequence of steps as the standard algorithm. (Drawings are optional.)

17. (a) Compute $814 - 391$ with the partial differences algorithm.
(b) Compute $814 - 391$ with the standard algorithm.
(c) What are the advantages and disadvantages of each algorithm?

18. (a) Compute $765 - 329$ with the partial differences algorithm.
(b) Compute $765 - 329$ with the standard algorithm.

19. (a) What is an addition problem a child could solve with counting by tens?
(b) What is a subtraction problem a child could solve with counting by tens?

20. (a) How could a child compute $35 + 21$ by counting on from 35?
(b) How could a child compute $35 - 21$ by counting back from 35?

In Exercises 21–24, (a) complete the last example, repeating the error pattern in the completed examples, (b) write a description of the error pattern, and (c) write what you would tell the child about his or her error.

21.
$\begin{array}{r} 1 \\ 76 \\ +\ 6 \\ \hline 142 \end{array}$
$\begin{array}{r} 1 \\ 98 \\ +\ 7 \\ \hline 175 \end{array}$
$\begin{array}{r} 87 \\ +\ 8 \\ \hline \end{array}$

22.
$\begin{array}{r} 1 \\ 62 \\ +\ 57 \\ \hline 110 \end{array}$
$\begin{array}{r} 1 \\ 52 \\ +\ 84 \\ \hline 37 \end{array}$
$\begin{array}{r} 57 \\ +\ 76 \\ \hline \end{array}$

23.
$\begin{array}{r} 1 \\ 86 \\ -\ 48 \\ \hline 48 \end{array}$
$\begin{array}{r} 1 \\ 72 \\ -\ 37 \\ \hline 45 \end{array}$
$\begin{array}{r} 93 \\ -\ 28 \\ \hline \end{array}$

24.
$\begin{array}{r} 40 \\ -\ 27 \\ \hline 20 \end{array}$
$\begin{array}{r} 306 \\ -\ 215 \\ \hline 101 \end{array}$
$\begin{array}{r} 809 \\ -\ 763 \\ \hline \end{array}$

25. The **adding up algorithm** for subtraction is invented by some children in first or second grade. It is based upon the comparison model. For example, think of $87 - 39$ as "How much more is 87 than 39?" Find out what you add to 39 to obtain 87.

$\begin{array}{ll} 39 & +\ 1 \\ 40 & +\ 40 \\ 80 & +\ 7 \\ 87 & +\ 48 \qquad \text{So } 87 - 39 = 48 \end{array}$

Show how you would compute the following with the adding up algorithm.
(a) $62 - 29$ (b) $212 - 137$

26. Show how you would compute the following with the adding up algorithm.
(a) $83 - 37$ (b) $306 - 189$

Extension Exercises

27. Show two other ways besides the standard algorithm to compute each of the following.
(a) $35 + 29$ (b) $41 - 26$

28. (a) Select a three-digit number whose first and third digits are different.
(b) Reverse the digits of your number and subtract the smaller of the two numbers from the larger.
(c) Select another three-digit number and do the same thing.
(d) Select another three-digit number and do the same thing.
(e) What pattern do you see in your answer?
(f) Finding the general pattern from examples involves _____ reasoning.

29. Scratch addition requires only addition of single digits. To compute $57 + 86 + 39$, write the example vertically.

```
  57     Add the numbers in the units place, starting
  86₃    at the top. If the sum is 10 or more, "scratch"
+ 39     a line through the last digit added and write
         the number of units just below it.

  57
  86₃    Continue adding units.
+ 39
─────
   2

 1  2
  5 7    Count the number of scratches in the
 8₅6₃    column, and write it at the top of the
+ 3 9    next column. Repeat the procedure
─────
 1 8 2   for each successive column.
```

(a) Compute $38 + 97 + 246$ with scratch addition.
(b) What is easier about scratch addition?

30. Compute the following with scratch addition.
(a) 386
      97
      58
    + 87
(b) 4679
      345
    + 276

31. Suppose you add the same amount to both numbers in a subtraction problem. What happens to the answer? Try the following.
(a) What is $86 - 29$?
(b) Add 1 to both numbers in part (a) and subtract. Do you obtain the same answer?
(c) Add 11 to both numbers in part (a) and subtract. Do you obtain the same answer?

32. The **equal-additions algorithm** has been used in some U.S. schools in the past 60 years. The property developed in the preceding exercise is the basis for the equal-additions algorithm. For example, in computing $563 - 249$, one needs to add 10 to the 3. To compensate, one adds 10 to 249. Then the subtraction can be done without regrouping.

```
   563        5 6¹3       5 6¹3
 - 249      - 2⁵4 9     - 2⁵4 9
                        ───────
                          3 1 4
```

(a) Compute $86 - 29$ using the equal-additions algorithm.
(b) How do you think this algorithm compares to the standard algorithm?

33. Compute the following using the equal-additions algorithm.
(a)   72
    −47
(b)   821
    − 376

Magic Time

34. (a) Try the following trick on a friend. Before you start the trick, write 44 on a sheet of paper and put it face down in front of you. Now tell your friend to do each of the following steps. Write down any number between 50 and 100. Then add 55 to the number. Next, cross out the hundreds digit of the result, and add that digit to the remaining two-digit number. Finally, subtract the answer from the number you started with. Now turn over the paper so your friend can see it.
(b) Let x represent the number your friend selected. See if you can show why your friend will end up with 44.

3.6 ALGORITHMS FOR WHOLE-NUMBER MULTIPLICATION AND DIVISION

NCTM Standards

- develop fluency in adding, subtracting, multiplying, and dividing whole numbers (3–5)

- analyze and evaluate the mathematical thinking and strategies of others (pre-K–12)

MULTIPLICATION ALGORITHMS

After exploring problems such as 30×2 and 300×2, a child is ready for a problem such as 34×2. How would the child work it out?

Classroom Connection

LE 1   Opener

Suppose you asked a group of fourth graders to compute

$$\begin{array}{r} 34 \\ \times\, 2 \\ \hline \end{array}$$

before they had studied the standard algorithm. Write down all the ways you think children might find the correct answer. Assume they have blocks to help them but no calculators.

A useful expanded multiplication algorithm is the **partial products algorithm.** You may have thought of something like it in LE 1. Figure 3–37 shows how to compute 34×2 with the partial products (expanded) algorithm and the standard algorithm.

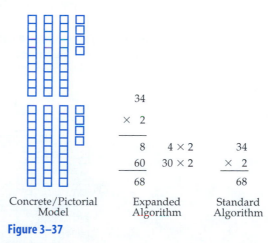

| Concrete/Pictorial Model | Expanded Algorithm | Standard Algorithm |

Figure 3–37

The partial products algorithm can be used as a transition to the standard algorithm or as an alternative to the standard algorithm. To introduce regrouping with the standard algorithm, it helps to use base-ten blocks.

Discussion

LE 2 Connection

Suppose you want to introduce a fourth grader to the standard algorithm for computing

$$26$$
$$\times 3$$

Explain how to find the product with base-ten blocks (Activity Card 1) following the same sequence of steps as the standard algorithm. (*Hint:* Step 1 shows 3 groups of 26. In step 2, multiply the ones and regroup them. In step 3, multiply the tens, total them up, and state the product.)

Whole-number properties help justify the standard procedure. For example, why are we allowed to multiply 2×4 and 2×30 and add the results together (Figure 3–37)? Try the following exercise.

Discussion

LE 3 Reasoning

Fill in the properties that justify the last two steps.

$$34 \times 2 = (30 + 4) \times 2 \qquad \text{Expanded notation}$$
$$= (30 \times 2) + (4 \times 2) \qquad \underline{\hspace{3cm}}$$
$$= (4 \times 2) + (30 \times 2) \qquad \underline{\hspace{3cm}}$$

The preceding exercise illustrates how the distributive property is used in the standard multiplication algorithm. Multiplying larger numbers is merely an extension of this process.

Classroom Connection

LE 4 Opener

Suppose you asked a group fourth graders to compute

$$35$$
$$\times 26$$

before they had studied the standard algorithm. Write down all the ways you think children might find the correct answer.

Figure 3–38 shows how a child might work out 35×26 with a square grid diagram, the partial products algorithm, and the standard algorithm.

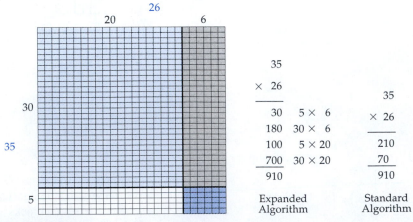

Figure 3–38

LE 5   Connection

Make a square grid diagram for 23×14 on a piece of graph paper as follows.

(a) Outline a rectangle that is 23 units long and 14 units wide.
(b) Break apart each factor by place value (20 + 3) and (10 + 4). Outline the four rectangles that correspond to the partial products.
(c) Multiply each partial product and add to find out how many squares are inside the original rectangle.

Now see for yourself how the partial products algorithm and the standard algorithm compare.

LE 6   Concept

(a) Compute 83×47 with the partial products algorithm.
(b) Compute 83×47 with the standard algorithm.
(c) What are the advantages and disadvantages of each algorithm?

It is easier to understand place value in the partial products algorithm. However, the partial products algorithm requires extended skill with the basic facts, such as $30 \times 6 = 180$ and $30 \times 20 = 600$. The standard algorithm is shorter to write and faster once it has been mastered. Another alternative algorithm some children prefer is lattice multiplications (see Exercise 25).

DIVISION ALGORITHMS

Next, consider the most difficult of the four algorithms, the division algorithm. Children first study examples such as $38 \div 3$ that do not require regrouping. After that, they would learn about problems with a one-digit divisor that requires regrouping the tens.

LE 7   Opener

Suppose you asked a group of fourth graders to compute $53 \div 4$ before they had studied the standard algorithm. Write down all the ways you think children might find the correct answer. Assume they have blocks to help them but no calculators.

You may have devised something that resembles the repeated subtraction algorithm. The **repeated subtraction (or scaffold) algorithm** is another useful expanded algorithm. Figure 3–39 shows how to compute $53 \div 4$ with the expanded and standard algorithms.

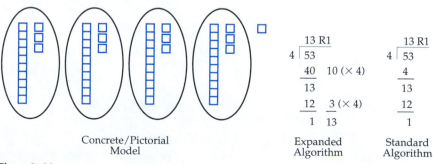

Concrete/Pictorial Model   Expanded Algorithm   Standard Algorithm

Figure 3–39

So $53 \div 4 = 13$ R1.

• **EXAMPLE 1**

Suppose you want to introduce a child to the procedure for computing 53 ÷ 4. Explain how to find the quotient with base-ten blocks following the same sequence of steps as the standard algorithm.

SOLUTION

Step 1   You want to divide 53 into 4 equal groups. Show 53 and try to divide the tens into 4 equal groups. Put 1 ten in each group. There is 1 ten left over.

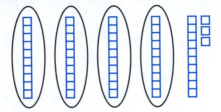

Step 2   Regroup the leftover 10 as 10 ones. This makes 10 ones + 3 ones = 13 ones.

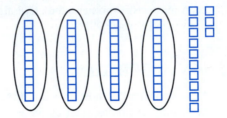

Step 3   Try to divide the 13 ones into 4 equal groups. Put 3 ones in each group. There is 1 one left over.

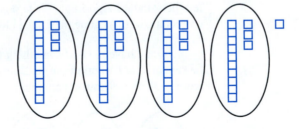

Step 4   The quotient is 13 remainder 1.

$$53 ÷ 4 = 13 \text{ R1}.$$

Now it's your turn.

LE 8   Connection

Suppose you want to introduce a fourth grader to the procedure for computing 71 ÷ 3. Explain how to find the quotient with base-ten blocks (Activity Card 1) following the same sequence of steps as the standard algorithm. (Drawings are optional.)

The standard long-division algorithm for whole numbers is more complicated than, and somewhat unlike, the algorithms for the other three operations. Instead of lining up the digits of the numbers as in the other operations, one writes the divisor to the left of the dividend. Unlike the other three operations, one finds the quotient from *left to right*. And while one simply computes one-digit basic facts to perform the other algorithms, the division algorithm may require rounding the divisor (if it is more than one digit) and the dividend and then using estimation (Section 3.7) to obtain digits in the quotient.

How do the repeated subtraction and standard algorithms compare?

LE 9   Concept

(a) Compute 394 ÷ 6 with the repeated subtraction algorithm.
(b) Compute 394 ÷ 6 with the standard algorithm.
(c) What are the advantages and disadvantages of each algorithm?

It is easier to understand place value in the repeated subtraction algorithm. Most people find that the repeated subtraction algorithm is longer and easier to understand than the standard long-division algorithm. The standard algorithm is simply a more efficient and compact version of the repeated subtraction algorithm. Both algorithms require some facility with estimation especially for larger divisors (see Section 3.7).

The repeated subtraction algorithm relies on successive approximations rather than having to guess the exact digits of the quotient. It also allows students to work at different levels according to how skilled they are at approximating. The repeated subtraction algorithm is useful as either a transitional algorithm to the standard algorithm or an alternative algorithm for students who have been unable to learn the standard algorithm.

COMMON ERROR PATTERNS IN ALGORITHMS

What kind of errors might children make with the multiplication and division algorithms?

Classroom Connection

In LE 10 and LE 11, (a) complete the last example repeating the error pattern in the completed examples, (b) write a description of the error pattern, and (c) write what you would tell the child about his or her error.

LE 10   Reasoning

$$
\begin{array}{r} 36 \\ \times\ 8 \\ \hline 568 \end{array}
\qquad
\begin{array}{r} 42 \\ \times\ 6 \\ \hline 302 \end{array}
\qquad
\begin{array}{r} 72 \\ \times\ 9 \\ \hline \end{array}
$$

LE 11   Reasoning

$$
\begin{array}{r} 121 \\ 4\overline{)623} \end{array}
\qquad
\begin{array}{r} 184 \\ 8\overline{)912} \end{array}
\qquad
\begin{array}{r} \\ 3\overline{)782} \end{array}
$$

LE 12   Summary
Tell what you learned about the multiplication and division algorithms in this lesson.

ANSWERS TO SELECTED LESSON EXERCISES

2. *Step 1*   Shows 3 groups of 26.

Step 2   Combine the ones. You have 18 ones. Trade 10 ones for 1 ten leaving 8 ones.

Step 3   Combine the tens. You have 6 tens plus 1 ten for a total of 7 tens. The product is 7 tens 8 ones or 78. So $26 \times 3 = 78$.

3. Distributive $\times$ over $+$; $+$ is commtative.

5. (b)

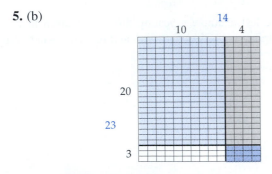

(c) $12 + 80 + 30 + 200 = 322$

6. (a) $\begin{array}{r} 83 \\ \times\ 47 \\ \hline 21 \\ 560 \\ 120 \\ 3200 \\ \hline 3901 \end{array}$ (b) $\begin{array}{r} 83 \\ \times\ 47 \\ \hline 581 \\ 332 \\ \hline 3901 \end{array}$

8. *Step 1*   You want to divide 71 into 3 equal groups. Show 71 and try to divide the tens into 3 equal groups. Put 2 tens in each group. There is 1 ten left over.

Step 2   Regroup the leftover ten as 10 ones. This makes 10 ones + 1 one = 11 ones.

Step 3   Try to divide the 11 ones into 3 equal groups. Put 3 ones in each group. There are 2 ones left over.

Step 4   The quotient is 23 remainder 2. So $71 \div 3 = 23$ R2.

9. (a) $\begin{array}{r} 65\ \text{R4} \\ 6\overline{)394} \\ 300 \quad 50 \\ \hline 94 \\ 60 \quad 10 \\ \hline 34 \\ 30 \quad 5 \\ \hline 4 \quad 65 \end{array}$ (b) $\begin{array}{r} 65\ \text{R4} \\ 6\overline{)394} \\ 36 \\ \hline 34 \\ 30 \\ \hline 4 \end{array}$

10. (a) 728.
(b) In regrouping the ones to the tens column, the child adds the number to the tens digit *before* multiplying.

9. (a) 221.
(b) The child divides the smaller digit into the larger and discards all remainders.

3.6   HOMEWORK EXERCISES

Basic Exercises

1. Three fourth graders are asked to compute 38×7. In each case, tell what the child understands about multiplication.

(a) $30 \times 7 = 210$ and $8 \times 7 = 56$. Then $210 + 56 = 266$.
(b) $20 \times 7 = 140$ and $10 \times 7 = 70$ and $8 \times 7 = 56$. Then $140 + 70 + 56 = 266$.
(c) $40 \times 7 = 280$. Then take away $2 \times 7 = 14$. The answer is $280 - 14 = 266$.

2. Four fourth graders work out 32×15. Tell whether each solution is correct. If so, what does the child understand about multiplication? If the answer is wrong, what would you tell the child about how to solve the problem?

   (a) 32×10 is 320. Add half of 320, which is 160. You get 480.

   (b)
   $$\begin{array}{r} 32 \\ \times\ 15 \\ \hline 160 \\ 32 \\ \hline 480 \end{array}$$

   (c)
   $$\begin{array}{r} 32 \\ \times\ 15 \\ \hline 160 \\ 32 \\ \hline 192 \end{array}$$

   (d) 32×15 is the same as 16×30, which is 480.

3. Suppose you want to introduce a fourth grader to the standard algorithm for computing 24×4. Explain how to find the product with base-ten blocks following the same sequence of steps as the standard algorithm. Draw at least one picture.

4. Suppose you want to introduce a fourth grader to the standard algorithm for computing 47×2. Explain how to find the product with base-ten blocks following the same sequence of steps as the standard algorithm. Draw at least one picture.

5. When you multiply 39×48 using the standard algorithm, you multiply and add partial products $(9 \times 8) + (30 \times 8) + (9 \times 40) + (30 \times 40)$. The following proof shows why this is correct. Fill in the blanks with properties you studied in this lesson.

   $$39 \times 48$$
   $$= (30 + 9) \times (40 + 8) \qquad \text{Expanded notation}$$
   $$= [(30 + 9) \times 40] \qquad \underline{\hspace{3cm}}$$
   $$\quad + [(30 + 9) \times 8]$$
   $$= [(30 \times 40) + (9 \times 40)] \qquad \underline{\hspace{3cm}}$$
   $$\quad + [(30 \times 8) + (9 \times 8)]$$
   $$= [(30 \times 8) + (9 \times 8)] \qquad \underline{\hspace{3cm}}$$
   $$\quad + [(30 \times 40) + (9 \times 40)]$$
   $$= [(9 \times 8) + (30 \times 8)] \qquad \underline{\hspace{3cm}}$$
   $$\quad + [(9 \times 40) + (30 \times 40)]$$

6. In multiplying 62×3, we use the fact that $(60 + 2) \times 3 = (60 \times 3) + (2 \times 3)$. What property does this equation illustrate?

7. Make a square grid model for 13×26 on a piece of graph paper as follows.

   (a) Outline a rectangle for the product.

   (b) Outline the four rectangles that correspond to the partial products.

   (c) Multiply each partial product and add to find out how many squares are in the original rectangle.

8. Make a square grid model for 34×23 on a piece of graph paper.

   (a) Outline a rectangle for the product.

   (b) Outline the four rectangles that correspond to the partial products.

   (c) Multiply each partial product and add to find out how many squares are in the original rectangle.

9. (a) Compute 49×62 with the partial products algorithm.

   (b) Compute 49×62 with the standard algorithm.

   (c) What are the advantages and disadvantages of each algorithm?

10. (a) Compute 123×56 with the partial products algorithm.

    (b) Compute 123×56 with the standard algorithm.

11. Children can more easily see connections among different multiplication problems by studying a series of related computations such as the following: 6×3, 20×3, 20×30, 26×30, 26×300, 26×34, 26×340. What are two other multiplication problems that would be related to 52×40?

12. What are two other multiplication problems that would be related to 83×7?

13. Two fourth graders work out $96 \div 8$. Tell whether each solution is correct. If so, what does the child understand about division? If the answer is wrong, what would you tell the child about how to solve the problem?

    (a) Ten 8s make 80. That leaves 16 which is 2 more 8s. It takes 12 of those 8s to make 96.

    (b) The divisor 8 goes into 9 one time. Then 8 into 6 goes 0 times with remainder 6. The answer is 10 R6.

14. Two fourth graders work out $56 \div 3$. Tell whether each solution is correct. If so, what does the child understand about division? In each case, tell what the child understands about division.

(Continued on the next page)

(a) How many 3s make 56? Ten 3s make 30. That leaves 26. That will take 8 more 3s, and 2 are left over. The answer is 18 R2.

(b) Twenty times 3 is 60. That is 4 too much. Take off two 3s. That makes eighteen 3s and 2 extra. The answer is 18 R2.

15. Suppose you want to introduce a fourth grader to the standard algorithm for computing 246 ÷ 2. Explain how to find the quotient with base-ten blocks following the same sequence of steps as the standard algorithm. (Drawings are optional.)

16. Suppose you want to introduce a fourth grader to the standard algorithm for computing 43 ÷ 3. Explain how to find the quotient with base-ten blocks following the same sequence of steps as the standard algorithm. (Drawings are optional.)

17. (a) Compute 217 ÷ 4 with the repeated subtraction algorithm.
    (b) Compute 217 ÷ 4 with the standard algorithm.
    (c) What are the advantages and disadvantages of each algorithm?

18. (a) Compute 870 ÷ 21 with the repeated subtraction algorithm.
    (b) Compute 870 ÷ 21 with the standard algorithm.

In Exercises 19–22, (a) complete the last example, repeating the error pattern in the completed examples, (b) write a description of the error pattern, and (c) write what you would tell a child about his or her error.

19. 36   42   72
    × 8   × 6   × 9
    ―――   ―――   ―――
    248   242

20. 82   41   27
    × 37   × 79   × 32
    ―――   ―――   ―――
    574   369
    246   287
    ―――   ―――
    820   656

21.   36   57
   6)378   5)375   8)512
     36   35
     ――   ――
     18   25
     18   25

22.   26   15
    4)824   5)525   6)1254

23. A fourth grader works out 26 × 15 as follows. "Since 20 × 10 = 200 and 6 × 5 = 30. The answer will be 230."
    (a) What is wrong with this procedure?
    (b) Make a square grid picture that shows what the child is leaving out.

24. A fourth grader works out 117 ÷ 6 as follows. She finds 100 ÷ 6 and 17 ÷ 6. She gets 16 + 2 = 18 with 9 left over. Then 9 ÷ 6 = 1 with 3 left over. That makes a total of 19 sixes with remainder 3. So 117 ÷ 6 = 19 R3.
    (a) Tell how to find 159 ÷ 7 with the same method.
    (b) How do you think this method compares to the standard algorithm?

25. **Lattice multiplication** was passed along from the early Hindus and Chinese to the Arabs to medieval Europe. It appeared in the earliest known arithmetic book, which was published in Italy in 1478. Today it is sometimes taught as an enrichment topic in the upper elementary grades. The algorithm for 27 × 34 is shown here.

  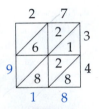

Step 1: Write the numbers.

Step 2: Multiply.

Step 3: Sum the numbers on each diagonal, beginning at the right. On the second diagonal, 1 + 2 + 8 = 11. Put down the 1 and carry the other 1 to the third diagonal. Then 1 + 2 + 6 = 9.

So 27 × 34 = 918.
    (a) Compute 38 × 74 with lattice multiplication.
    (b) Compute 123 × 35 with lattice multiplication.
    (c) How do you think this algorithm compares to the standard one?

26. (a) Compute 46 × 29 with lattice multiplication.
    (b) Compute 234 × 76 with lattice multiplication.

Extension Exercises

27. Show two other ways besides the standard algorithm to compute each of the following.

(a) 56×7 (b) 41×26 (c) $86 \div 20$

28. The **Russian peasant multiplication algorithm** was used thousands of years ago by the Egyptians and until recently by Russian peasants. The Russian peasant algorithm for multiplying employs halving and doubling. Remainders are ignored when halving. The algorithm for 33×47 is shown here.

	Halving	Doubling	
Smaller →	33	⟨47⟩	← Larger
factor	16	94	factor
	8	188	
	4	376	
	2	752	
	1	⟨1504⟩	

Circle and add all the numbers in the doubling column that are paired with *odd* numbers in the halving column. $47 + 1504 = 1551$, so $33 \times 47 = 1551$.

(a) Compute 28×52 with this algorithm.

(b) Compute 18×127 with this algorithm.

(c) How does it compare to the standard algorithm?

29. Find the quotient and *remainder* of $8569 \div 23$, using a calculator.

Puzzle Time

30. Fill in the missing digits.

$$
\begin{array}{r}
2?? \\
6\overline{)?2??} \\
?? \\
\hline
3 \\
? \\
\hline
?? \\
?? \\
\hline
0
\end{array}
$$

3.7 WHOLE NUMBERS: MENTAL COMPUTATION AND ESTIMATION

NCTM Standards

- use a variety of methods and tools to compute, including objects, mental computation, estimation, paper and pencil, and calculators (pre-K–2)

- develop and use strategies to estimate the results of whole-number computations and judge the reasonableness of such results (3–5)

Believe it or not, your ancestors used to do all their arithmetic without electronic calculators! To cope with this hardship, they developed a repertoire of computational shortcuts. Today, most adults solve complicated arithmetic problems with a calculator. But what happens if you don't have a calculator with you? Would you be limited to doing one-digit arithmetic?

Mental computation is used to obtain an *exact* answer to a computation that can be done easily without paper and pencil or a calculator. **Estimation** (usually done mentally) is used to obtain an *approximate* answer to a more difficult computation.

MENTAL ADDITION

Do you enjoy doing something differently from the way it's "normally" done? Sometimes that's a good idea!

LE 1   Opener
You buy a sweater for $46 and a coat for $79. How would you mentally compute the (exact) total cost? Tell what steps you use.

Classroom
Connection

LE 2   Reasoning
Four second graders were asked to compute 46 + 79. Here's what each child did.

Abdul: 40 + 70 = 110 and 6 + 9 = 15. Then 110 + 15 = 125.
Rachel: 46 + 79 = 45 + 1 + 79 = 45 + 80 = 125.
Callista: 46 + 70 = 116 and 116 + 9 = 125.
Ramon: 46 + 79 = 45 + 80 = 125.

(a) Tell what each child knows about addition.
(b) A man buys a jacket for $58 and a CD player for $36. Show how to compute the total cost with each child's method of mental addition.
(c) What property makes 50 + 8 + 30 + 6 = 50 + 30 + 8 + 6?
(d) Describe a way to compute 58 + 36 with the following chart.

50	51	52	53	54	55	56	57	58	59
60	61	62	63	64	65	66	67	68	69
70	71	72	73	74	75	76	77	78	79
80	81	82	83	84	85	86	87	88	89
90	91	92	93	94	95	96	97	98	99

In LE 2, the children used breaking apart or compensation to compute the answer. Abdul and Callista use breaking apart based on place value. Rachel had a different way of breaking apart. Ramon used compensation to change the numbers in a way that makes the computation easier but leaves the final result the same.

MENTAL SUBTRACTION

LE 3 Opener

You have $93 in your checking account, and you write a check for $38. How would you mentally compute the balance?

Classroom Connection

LE 4 Reasoning

Four third graders were asked to compute $93 - 38$. Here's what each child did.

> Abdul: $93 - 30 = 63$ and $63 - 8 = 55$.
> Rachel: $90 - 30 = 60$ and $8 - 3 = 5$. Then $60 - 5 = 55$.
> Callista: $38 + 2$ makes 40. Then $40 + 53 = 93$. The answer is $2 + 53 = 55$.
> Ramon: $93 - 38 = 95 - 40 = 55$.

(a) Tell what each child knows about subtraction.
(b) Which children used breaking apart?
(c) Which child used compensation?
(d) A woman starts out with $62 and spends $29. Show how to compute how much money she has left with each of the four methods of mental subtraction.

Callista did not use breaking apart or compensation. Her method is called **adding on.** Add on to $38 until you reach $93. The answer is the total amount you added on. Another way to add on would be as follows. First, how many tens can you add to $38 without going over 93? Compute $38 + 50 = 88$. Then add ones to $88 to reach $93. So $88 + 5 = 93$. How much was added? $50 + 5 = 55$.

MENTAL MULTIPLICATION

LE 5 Opener

Do each of the following computations in your head. Write down the steps that you used.

(a) 68×7 **(b)** 42×13 **(c)** 800×30 **(d)** 28×25

Did you do parts (a) and (b) of LE 5 by breaking apart? In part (a), $68 \times 7 = 60 \times 7 + 8 \times 7 = 420 + 56 = 476$. Similarly, in part (b), $42 \times 13 = 42 \times 10 + 42 \times 3 = 420 + 126 = 546$. The familiar shortcut for part (c) is based on breaking apart, too! It comes from $800 \times 30 = 8 \times 3 \times 100 \times 10 = 24 \times 1,000 = 24,000$.

LE 6 Concept

Show the steps to compute 28×25 mentally by breaking apart.

How does compensation work with multiplication? Consider the following.

Classroom Connection

LE 7 Reasoning

A child computes 13×28 by using compensation as follows. 13×28 is the same as 11×30 which equals 330.

(a) Is this correct?
(b) How can you find 13×28 by using multiplication involving 30?

ESTIMATION

You were probably expecting "mental division" next. However, it is harder to obtain an exact answer with mental division than with addition, subtraction, or multiplication. For example, although one might do $410 + 7$, $410 - 7$, and 410×7 mentally, one would be more inclined to estimate $410 \div 7$. This leads us to the next topic: estimation. Division will be addressed as part of this topic.

LE 8   Opener

Name a situation in which you have used estimation.

LE 9   Opener

In each situation, tell why an estimate might be preferred over an exact answer.

(**a**) You want to tell how many people attended a political rally.
(**b**) You want to decide how much money to take on a summer trip.
(**c**) You want to figure out how long it will take to drive from Chicago to Cleveland.

An estimate may be preferred over an exact answer when (1) an exact answer is difficult or impossible to obtain, (2) an estimate is easier to use in computations, or (3) an overestimate or underestimate includes a safety factor. An estimate may also be used to check if an answer is reasonable.

Estimation shows another side of mathematics, in which one seeks a reasonable result rather than an exact one. People use estimation either to obtain an approximate answer without having to use a calculator, or to check work done with a calculator or by another person.

To estimate, devise an easier problem that can be computed mentally but still has approximately the same answer as your original problem. The three most useful strategies for whole-number estimation are rounding, the compatible-numbers strategy, and the front-end strategy.

ROUNDING STRATEGY

How many calories should you eat each day? If you are a young adult, as I used to be, you could multiply your weight in pounds by 19 to get a rough calorie estimate.

LE 10   Skill

May Wong is a 14-year-old who weighs 93 pounds. How would you estimate how many calories she should consume to maintain her weight if the recommended number of calories is 18 times the number of pounds she weighs?

Did you use rounding to estimate the answer to LE 10? You could have estimated 93×18 by rounding the factors to 90 and 20, multiplying 9 by 2 mentally, and adding two zeroes. So the answer is about 1,800 calories. In symbols, this is written $93 \times 18 \approx 1,800$, in which the symbol $\approx$ means "is approximately equal to."

The rounding strategy involves two steps.

Rounding Strategy

1. Round the numbers to obtain a problem you can compute mentally.
2. Add, subtract, or multiply the rounded numbers to obtain an estimate.

Use the rounding strategy to estimate the answers to LE 11 and LE 12.

LE 11   Skill

A stadium seats 58,921 people. If only 3,426 tickets are left for next Sunday's match between the Poodles and the Goulash, show the steps you would use to estimate the number of tickets that have been sold.

LE 12   Reasoning

In estimating $A - C$, in which $A > C$, you round A up and C down. Your estimate

(a) is too high
(b) is too low
(c) could be too high or too low

THE COMPATIBLE-NUMBERS STRATEGY

In division, rounding to the "nearest" does not always yield a simpler problem.

LE 13   Opener

A group of milking cows produces 712 oz of milk one fine day. How would you estimate the number of quarts produced? (One quart contains 32 oz.)

In LE 13, you needed to estimate $712 \div 32$. You could have rounded it to $700 \div 30$ or $710 \div 30$, but a better choice would be $600 \div 30$, $750 \div 30$, or $700 \div 35$. For example, $712 \div 32 \approx 750 \div 30 = 25$. The average per cow is a little less than 25 quarts.

The computation $750 \div 30$ is an example of **compatible numbers,** a set of numbers whose sum, difference, product, or quotient is easy to compute mentally. Numbers that do not divide evenly, such as $700 \div 30$, are not compatible.

The compatible-numbers strategy involves two steps.

The Compatible-Numbers Strategy

1. Round the numbers to nearby compatible numbers.
2. Perform the computation with the compatible numbers, and use the answer as an estimate.

Children also use compatible numbers to perform the standard division algorithm. For example, in doing $32\overline{)712}$ with paper and pencil, one first estimates $71 \div 32$. Use the compatible-numbers strategy to estimate in the following exercises.

LE 14   Skill

You want to pay for an $19,847 National Motors Ulcer in 24 easy, interest-free monthly payments.

(a) How could you estimate the cost of a monthly payment?
(b) If you were going to work out $19{,}847 \div 24$ with the standard long division algorithm, how would you estimate to find the first digit in the quotient?

Discussion

LE 15   Reasoning

A and C are whole numbers. In estimating $A \div C$, you round A up and C down. If you assume that $C \neq 0$, your estimate

(a) is too high
(b) is too low
(c) could be too high or too low

You can also use compatible numbers to estimate when adding three or more numbers, by adding groups of numbers that total approximately 100 or 1000 (or some other round number).

• EXAMPLE 1

A movie theater had six shows of "Return of the Dodecahedron" on a Saturday. The numbers of tickets sold at the six shows were 64, 59, 32, 43, 27, and 77. How can you estimate the total number of tickets sold, using the compatible-numbers strategy?

SOLUTION

Add compatible numbers.

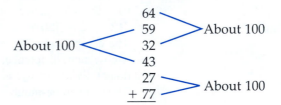

The sum is about 300 tickets.   ●

LE 16   Skill

(a) The next day, the movie theater mentioned in the preceding example had six more shows. The total numbers of tickets sold at the six shows were 36, 52, 51, 98, 71, and 96. Explain how to estimate the total number of tickets sold, using the compatible-numbers strategy.
(b) What properties allow you to reorder and regroup the numbers?

THE FRONT-END STRATEGY

Front-end (**leading digit**) **estimation** is especially useful in addition. It is based on adding a column of numbers from left to right. The estimate is made by adding the digits in the left-hand column and is then adjusted by considering the digits in the next column to the right.

• EXAMPLE 2

The Desir family went on a three-day trip. They drove 462 miles the first day, 385 miles the second day, and 447 miles the third day. Explain how you would use front-end estimation with adjustment to estimate the total distance they drove in three days.

SOLUTION

First, add the front-end or leading digits.

$$
\begin{array}{rcr}
462 & \rightarrow & 400 \\
385 & \rightarrow & 300 \\
+\,447 & \rightarrow & +\,400 \\
\hline
& & 1100
\end{array}
$$

Second, use the next digit in each column (tens place) to make an adjustment in the hundreds place. In the tens column, $6 + 8 + 4$ will make about 200 more.

$$
\begin{array}{rcl}
1100 & \leftarrow & \text{front end sum} \\
+\,200 & \leftarrow & \text{adjustment} \\
\hline
1300 & &
\end{array}
$$

So the Desirs traveled about 1,300 miles   ●

Apply this method in the following exercise.

LE 17   Skill

A salesperson sells three used cars for $3793, $4391, and $2807. *Explain* how you would use front-end estimation with adjustment to estimate the salesperson's total sales.

A GAME: MAXIMIZE IT

LE 18   Reasoning

This game is for two to four players. You will need a regular die and a calculator.

(a) Each player tries to complete the following arithmetic problem and obtain the highest answer.

_____ _____ ÷ _____ _____ + _____ _____

One player rolls the die six times. After each roll, each player must put the number that is rolled in one of the blanks. At the end of the six rolls, everyone works out his or her computation. The winner(s) will have the highest result.

(b) Discuss strategies that are helpful in this game.

LE 19   Summary

(a) Describe a method of mental computation that you learned about in this section, preferably one you did not know about before.

(b) Give an example of how you would use this method.

ANSWERS TO SELECTED LESSON EXERCISES

2. (b) Abdul: $50 + 30 = 80$ and $8 + 6 = 14$. Then
$80 + 14 = 94$.
Rachel: $58 + 36 = 54 + 4 + 36 = 54 + 40 = 94$.
Callista: $58 + 30 = 88$ and $88 + 6 = 94$.
Ramon: $58 + 36 = 54 + 40 = 94$.
(c) Commutative property of addition.

4. (d) Abdul: $62 - 20 = 42$ and $42 - 9 = 33$.
Rachel: $60 - 20 = 40$ and $9 - 2 = 7$. Then
$40 - 7 = 33$.
Callista: $29 + 1$ makes 30. Then $30 + 32 = 62$.
The answer is $1 + 32 = 33$.
Ramon: $62 - 29 = 63 - 30 = 33$.

6. $28 \times 25 = 28 \times 20 + 28 \times 5 = 560 + 140 = 700$

7. (a) no (b) $13 \times 28 = 13 \times 30 - 13 \times 2 =$
$390 - 26 = 364$

9. (a) It is not possible to find the exact answer.
(b) You use an overestimate to be safe.
(c) You cannot know the exact time in advance.

11. Round the problem to $59,000 - 3,000 = 56,000$
tickets.

12. (a) *Hint:* Try $29 - 11$.

14. (a) $19,847 \div 24 \approx 20,000 \div 20 = 1,000$
(b) $198 \div 24 \approx 200 \div 25 = 8$

15. (a) *Hint:* Try $29 \div 11$.

16. (a) $36 + 71, 52 + 51, 98$ and 96 are each about 100.
The sum is about 400.
(b) Commutative and associative properties of
addition

17. Add $3 + 4 + 2$ thousand $= \$9000$. Then
$\$793 + \$391 + \$807$ is about another $\$2000$.
Compute $\$9000 + \$2000 = \$11,000$.

3.7 HOMEWORK EXERCISES

Basic Exercises

1. The distance from Washington, D.C., north to Baltimore is 39 miles; the distance from Baltimore north to Philadelphia is 97 miles.
(a) Show the steps for three different ways to compute mentally the total distance from Washington, D.C., to Philadelphia.
(b) What property justifies the equation
$30 + 9 + 90 + 7 = 30 + 90 + 9 + 7$?

2. Show the steps for three different ways to compute mentally $93 + 59$.

3. Show the steps for three different ways to compute mentally $134 - 58$.

4. A book is 86 pages long. Mariel has read 47 pages. How many pages does she have left? Show the steps for three different ways to compute mentally

5. Show the steps to use compensation to compute mentally
(a) $236 + 89$. (b) $82 - 58$.

6. Show the steps to use adding on to compute mentally
(a) $74 - 57$. (b) $822 - 298$.

7. A watermelon costs 39¢ per pound. Show the steps to compute mentally the cost of a 6-pound watermelon by
(a) breaking apart. (b) compensation.
(c) What property justifies the equation
$39 \times 6 = (30 \times 6) + (9 \times 6)$?

8. A restaurant serves lunch to 90 people per day. Show the steps to mentally compute the number of people served lunch in 31 days.

9. Show the steps to compute mentally $(500)^3$.

10. You plan to drive 1,000 miles from Memphis to Albuquerque at an average speed of 50 mph. Show the steps to compute mentally how long the drive will take.

11. In each situation, tell why an estimate might be used instead of an exact answer.
   (a) You want to predict the U.S. population in 2010.
   (b) You want to decide how many people should ride on an elevator that carries up to 2,000 pounds.
   (c) You want to tell how long it took you to finish a mathematics test.

12. What is an everyday situation in which you recently used mental computation or estimation?

13. Use estimation to tell whether the following calculator answers are reasonable. Explain why or why not.
   (a) $657 + 542 + 707 = \boxed{543364}$
   (b) $26 \times 47 = \boxed{1222}$

14. A child computes 4364×38 with a calculator and gets 17,632. Tell the child how to check this answer (which is wrong!) by estimating.

15. *Explain* how you would use rounding to estimate mentally $5692 + 8091 + 3721$.

16. An auditorium has 56 rows, each seating 23 people.
   (a) Explain how you would use rounding to estimate mentally the total number of seats by rounding.
   (b) Which category of multiplication is illustrated? (area, array, counting principle, repeated)

17. The distance from Boston to Buffalo is about 437 miles.
   (a) Explain how you would mentally estimate the time the drive would take if you averaged 55 mph.
   (b) Which category of division is illustrated? (area, array, partition, repeated)

18. A college has 4832 students and 324 teachers.
   (a) Explain how you would mentally estimate the number of students per teacher (called the student-teacher ratio).
   (b) Which category of division is illustrated? (area, array, partition, repeated)

19. A and C are two-digit whole numbers. In estimating $A \div C$, you round A up and C down. Is your estimate too high or too low, or is it impossible to tell?

20. Estimate each quotient with compatible numbers.
   (a) $8\overline{)549}$ (b) $21\overline{)152}$ (c) $74\overline{)4692}$

21. If you were going to work out $4,692 \div 74$ with the standard long division algorithm, how would you estimate to find the first digit in the quotient?

22. If you were going to work out $2,158 \div 47$ with the standard long division algorithm, how would you estimate to find the first digit in the quotient?

23. *Explain* how you would estimate the following with compatible numbers.
$$\begin{array}{r} 59 \\ 32 \\ 42 \\ + \ 97 \\ \hline \end{array}$$

24. Show the steps to estimate the following using compatible numbers.
$$87 + 45 + 37 + 22 + 98 + 51$$

25. Compatible numbers can be used to compute mentally exact answers to some addition and multiplication problems. Show the steps to compute mentally the following and name the property of whole numbers that justifies your shortcut.
   (a) $8 \times 22 \times 5$ (b) $2 + (78 + 43)$

26. Tell how you would mentally compute the exact answers to the following with compatible numbers.
   (a) $2 \times 9 \times 6 \times 5$ (b) $46 + 28 + 32 + 4$

27. The Department of Education spent the following amounts on three projects: \$3,462,871, \$830,212, and \$21,172,806. Explain how to estimate the total expense to the nearest *million* dollars, using the front-end or rounding strategy.

28. Three towns with populations of 3692, 1527, and 4278 make up a voting district. *Explain* how to estimate the total population of the voting district using front-end estimation.

29. Consider the following problem. "A company orders 32 boxes of lightbulbs. Each box contains 48 lightbulbs. What is the total number of lightbulbs?"
    (a) Show how to use one step of front-end estimation to obtain an answer.
    (b) Show how to use rounding to obtain an estimate.
    (c) Which estimate is closer to the exact answer?

30. Crusty's Pizza sold 2,621 pizzas in March, 1,522 pizzas in April, and 2,218 pizzas in May. Explain how to estimate the total number of pizzas sold for those 3 months, using front-end estimation.

31. Show the steps to estimate
    (a) $4872 - 3194$.   (b) $3279 \div 65$.

32. Show the steps to estimate
    (a) $8327 \div 36$.   (b) 427×62.

33. Tell whether an estimate or an exact answer is more appropriate in each problem.
    (a) You want to buy two books that cost $7.95 and $8.95. You have $20. Is that enough money?
    (b) You want to buy two books that cost $7.95 and $8.95. How much change will you receive back from a $20 bill?

34. In each situation, tell whether it makes more sense to compute mentally or to use a calculator. Assume the calculator is stored somewhere nearby.
    (a) You want to estimate the total revenue at a baseball game with 38,542 people who paid an average of about $5 per ticket.
    (b) You just purchased 264 scientific calculators for your school at $18 each and you want to check the total bill.

35. For each computation, tell which computation method you would use (mental computation, paper and pencil, or calculator) and why.
    (a) $87,347 \times 144$ (b) $750 + 422 + 250$
    (c) $782 - 246$

36. For each computation, tell which method you would use (mental computation, paper and pencil, or calculator) and give the result.
    (a) $146 + 225 + 307$
    (b) $85,432 - 67,679$
    (c) 5000×20

37. Sometimes, you need to know only a range (or interval) of possible answers. For example, 23×46 is between $20 \times 40 = 800$ and $30 \times 50 = 1500$. "Between 800 and 1,500" is a **range estimate** for 23×46. Give a range estimate for
    (a) 59×32.   (b) $8627 + 2432$.   (c) $385 \div 12$.

38. Give a range estimate for
    (a) 86×64.   (b) $627 - 243$.   (c) $864 \div 37$.

39. You ask your fourth grade class to compute 28×5. A child says that multiplication by 5 is the same as dividing by 2 and adding a zero at the end. "I find $28 \div 2 = 14$ and put a 0 at the end. The answer is 140." Is this correct? If so, why does it work? If not, what would you tell the child?

40. A fifth grader computes 29×12 as follows: $30 \times 12 = 360$ and $360 - 12 = 348$. On what property is the child's method based?

41. Some children do mental multiplication by **building up.** For example, to find 74×22, a child computed $74 \times 10 = 740$. Then $740 + 740 = 1480$. Then $1480 + 148 = 1628$.
    (a) Explain what the child is doing.
    (b) Show possible steps for computing 31×33 with building up.

42. Show possible steps for computing 56×21 with building up.

43. **Clustering** is a method of estimating a sum when the numbers are all close to one value. For example, $3648 + 4281 + 3791 \approx 3 \cdot (4000) = 12,000$. Show how to estimate the following using the clustering strategy.
    (a) $897 + 706 + 823 + 902 + 851 \approx$ _____
    (b) $36,421 + 41,362 + 40,987 + 42,621 \approx$ _____

44. How could you use a calculator *and* mental computation to figure the cost of buying 20 carpets that are 18 ft by 24 ft, if each square foot costs $5?

 45. In each exercise, estimate the second factor so that the product falls in the range given. Check your guess on a calculator and revise it as needed.
(a) 300 × _____ = (between 6000 and 6500)
(b) 46 × _____ = (between 700 and 750)
(c) 67 × _____ = (between 2500 and 2600)

 46. In each exercise, estimate the divisor so that the quotient falls in the range given. Check your guess on a calculator and revise it as needed.
(a) 463 ÷ _____ = (between 80 and 90)
(b) 3246 ÷ _____ = (between 200 and 300)
(c) 4684 ÷ _____ = (between 65 and 70)

 47. Consider the following problem. "Use each of the digits from 1 to 6 once to obtain the largest possible product."

$$\begin{array}{r} ?\,?\,? \\ \times\,?\,?\,? \\ \hline \end{array}$$

(a) Devise a plan and solve the problem.
(b) Make up a similar problem.

 48. Select 6 different digits. Use each of the digits once and obtain the largest possible quotient.

49. Solve mentally.
(a) $34 - n = 20$ (b) $\dfrac{x + 3}{9} = 20$
(c) $30(y - 2) = 150$

50. Solve mentally.
(a) $\square - 10 = 26$ (b) $\dfrac{100}{r + 2} = 5$
(c) $4(x + 8) = 200$

Extension Exercises

51. There is a shortcut for multiplying a whole number by 99. For example, consider 15 × 99.
(a) Why does 15 × 99 = (15 × 100) − (15 × 1)?

(b) Compute 15 × 99 mentally, using the formula in part (a).
(c) Compute 95 × 99 mentally, using the same method.

52. (a) Develop a shortcut for multiplying 8 × 999 mentally. (*Hint:* See the preceding exercise.)
(b) Compute 6 × 999 mentally, using the same shortcut.
(c) Show how this shortcut is based on the distributive property.

53. You can mentally divide by 25 if you think of four 25s making 100 (or four quarters making a dollar, if you prefer). In other words, each 100 has four 25s. Try the following.
(a) 200 ÷ 25 (Think of four 25s per 100.)
(b) 700 ÷ 25
(c) How many quarters in $9?
(d) 650 ÷ 25
(e) How many quarters in $4.25?

54. (a) Develop a shortcut for multiplying by 25 mentally in a computation such as 24 × 25.
(b) Compute 44 × 25 using the same shortcut.

55. Estimate the total number of restaurants in the United States.

 56. 33 × 37 = 1221
46 × 44 = 2024
58 × 52 = 3016
(a) Make up another example of this type.
(b) How can these examples be computed mentally?
(c) Prove that the shortcut always works. (*Hint:* Represent the two digits of the two numbers by *a* *b* and *a* 10 − *b*, and write them in expanded notation.)

Labs and Projects

57. Look at a current series of mathematics textbooks for Grades 1–6. What do they teach about estimation and mental computation with whole numbers in each grade?

3.8 PLACE VALUE AND ALGORITHMS IN OTHER BASES

NCTM Standards

- create and use representations to organize, record, and communicate mathematical ideas (pre-K–12)
- recognize and use connections among mathematical ideas (pre-K–12)

You are well acquainted with base-ten place value and arithmetic. This familiarity will make it harder for you to understand the difficulties your students will encounter in learning about place value and arithmetic for the first time.

Studying place value and the algorithms in less familiar number systems will enable you to reexperience learning these concepts. Working in other bases will also deepen your understanding of place value and the algorithms.

OTHER NUMERATION SYSTEMS

LE 1 Opener

Devise your own numeration system using only the symbols A, B, C, D, and E. Show how you would represent some different base-ten numbers, including 0, 1, 2, 3, 4, 5, 10, 30, 50, and 100.

Today, most countries use base-ten numeration systems. The basis for this is the tendency to count on our fingers.

In fact, any counting number greater than 1 could be used as a base. The Babylonians used base sixty, the Mayans used base twenty, and some ancient tribes used base two. Today computers work with bases two, eight, and sixteen, and some goods, such as eggs and pencils, are grouped in dozens and grosses (groups of 12 and 12^2).

In representing 19 items, one could group them in different ways, as shown in Figure 3–40.

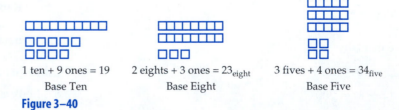

1 ten + 9 ones = 19   2 eights + 3 ones = 23_{eight} 3 fives + 4 ones = 34_{five}

Base Ten   Base Eight   Base Five

Figure 3–40

Consider base eight, the **octal** system, one base that is used in computers. Just as base ten has ten digits (0, 1, 2, 3, 4, 5, 6, 7, 8, 9), base eight has eight digits (0, 1, 2, 3, 4, 5, 6, 7).

LE 2 Concept

What digits would base five have?

Counting with the eight digits of base eight goes like this: 0, 1, 2, 3, 4, 5, 6, 7, 10, 11, 12, ... To indicate a numeral in a base other than ten, one writes the base as a subscript, as in 12_{eight} (read "one-two, base eight").

Figure 3–41 shows how to represent some base-eight numerals, with base-eight blocks.

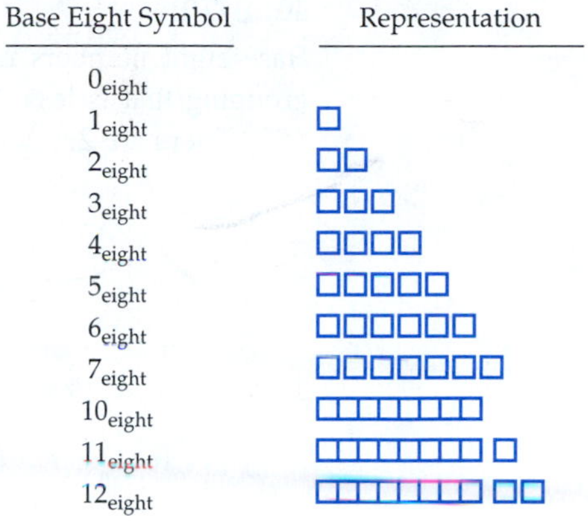

Base Eight Symbol	Representation
0_{eight}	
1_{eight}	□
2_{eight}	□□
3_{eight}	□□□
4_{eight}	□□□□
5_{eight}	□□□□□
6_{eight}	□□□□□□
7_{eight}	□□□□□□□
10_{eight}	▭▭▭▭▭▭▭▭
11_{eight}	▭▭▭▭▭▭▭▭ □
12_{eight}	▭▭▭▭▭▭▭▭ □□

Figure 3–41

LE 3   Skill

What base-ten numeral is the same as 32_{eight}?

Base-eight place value works like base-ten place value. In base ten, each place value is ten times greater than the place value to its right. In base eight, each place value is eight times greater than the place value to its right, with the right-hand place for whole numbers being the ones place.

100	10	1	← **Place values expressed** →	64	8	1
3	1	4	**using base-ten numerals**	3	1	4

$$314 = 3(100) + 1(10) + 4(1)$$

$$314_{eight} = 3(64) + 1(8) + 4(1)$$
$$= 204$$

Thus, 314_{eight} is the same as 204.

LE 4   Skill

Convert 426_{eight} to base ten.

LE 5   Skill

Convert 2134_{five} to base ten. (*Hint:* First, write place-value columns for base five.)

One can also convert base-ten numerals to any other base.

LE 6   Opener

You visit another country where they only have quarters, nickels, and pennies. How would you make each of the following amounts with the smallest number of coins?

(a) 12 cents   **(b)** 83 cents

The following example shows a method for converting a base-ten numeral to a different base.

• EXAMPLE 1

Convert 150 to base eight.

SOLUTION

Base-eight numbers use groups of 1, 8, 64, 512, and so on. Start with the largest grouping that is less than or equal to 150—namely, 64. How many 64s are there in 150? There are 2.

$$
\begin{array}{r}
2 \\
64\overline{)150} \\
128 \\
\hline
22
\end{array}
$$

This leaves 22. Proceed to the next lower grouping, 8s. How many 8s are in 22? There are 2.

$$
\begin{array}{r}
2\,\text{R}6 \\
8\overline{)22}
\end{array}
$$

This leaves 6 ones. So $150 = (2 \times 64) + (2 \times 8) + 6$, or 226_{eight}. •

> **LE 7   Skill**
> **(a)** Convert 302 to base eight.
> **(b)** Convert 302 to base five.

ALGORITHMS IN BASE FIVE

The base-ten arithmetic algorithms also work in other bases. Studying algorithms in other bases will increase your understanding of them. First, consider addition in base five. How would you construct an addition table?

> **LE 8   Skill**
> **(a)** Start with 0, and count off the first eleven base-five numerals.
> **(b)** Complete the following base-five addition table.

+	0	1	2	3	4
0	0	1	2		
1	1	2	3		
2	2	3	4		
3	3	4	10		
4	4	10	11		

You can also work out basic addition facts with a number line. Figure 3–42 shows $4_{\text{five}} + 2_{\text{five}}$ on a number line.

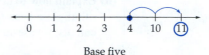

Base five

Figure 3–42

How does one compute $34_{\text{five}} + 22_{\text{five}}$? How is it similar to base-ten addition?

• EXAMPLE 2

(a) Compute $34_{\text{five}} + 22_{\text{five}}$ with the partial sums algorithm.

(b) Explain how to compute $34_{\text{five}} + 22_{\text{five}}$ with base-five blocks and the standard algorithm.

SOLUTION

(a) Use the addition table as needed for computations.

$$
\begin{array}{r}
34_{\text{five}} \\
+\ 22_{\text{five}} \\
\hline
11 \\
100 \\
\hline
111_{\text{five}}
\end{array}
$$

(b) Follow the same steps as in the base-ten algorithm, but remember that the base-five digits are 0, 1, 2, 3, and 4 and that regrouping (carrying) involves groups of 5 rather than groups of 10.

Step 1

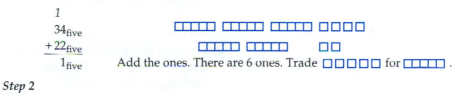

$$
\begin{array}{r}
1 \\
34_{\text{five}} \\
+\ 22_{\text{five}} \\
\hline
1_{\text{five}}
\end{array}
$$
Add the ones. There are 6 ones. Trade ▢▢▢▢▢ for ▭▭▭▭▭ .

Step 2

$$
\begin{array}{r}
1 \\
34_{\text{five}} \\
+\ \ 22_{\text{five}} \\
\hline
111_{\text{five}}
\end{array}
$$

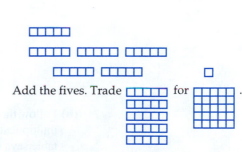

Add the fives. Trade ▭▭▭▭▭ for ▦ .

So $34_{\text{five}} + 22_{\text{five}} = 111_{\text{five}}$, or ▦ ▭▭▭▭▭ ▢

LE 9 Connection

(a) Compute $24_{five} + 14_{five}$ with the partial sums algorithm.

(b) Explain how to compute $24_{five} + 14_{five}$ with base-five blocks and the standard algorithm. (Use base-five blocks if you have them. It may also be possible to subdivide cutouts of base-ten blocks.)

LE 10 Connection

(a) Compute $32_{five} - 14_{five}$ with the partial differences algorithm.

(b) Explain how to compute $32_{five} - 14_{five}$ with base-five blocks and the standard algorithm. Use the addition table or a number line as needed. Remember that regrouping involves 5s and not 10s.

Next, consider the multiplication algorithm. A base-five multiplication table will be helpful.

LE 11 Skill

Complete the base-five multiplication table.

×	0	1	2	3	4
0	0	0	0	0	0
1	0	1	2	3	4
2	0	2	4	11	13
3					
4					

Example 3 utilizes this table to compute $23_{five} \times 14_{five}$.

• EXAMPLE 3

(a) Compute $23_{five} \times 14_{five}$ with the partial products algorithm.

(b) Compute $23_{five} \times 14_{five}$ with the standard algorithm.

SOLUTION

(a) There are four partial products. Use the base-five multiplication table to compute them. Compute $3_{five} \times 4_{five}$, $20_{five} \times 4_{five}$, $3_{five} \times 10_{five}$, and $20_{five} \times 10_{five}$. Then add the four partial products.

$$
\begin{array}{r}
23_{five} \\
\times\ 14_{five} \\
\hline
22_{five} \\
130_{five} \\
30_{five} \\
200_{five} \\
\hline
432_{five}
\end{array}
$$

(b) Follow the same steps as in the standard base-ten algorithm. Refer to the base-five multiplication table to obtain the basic facts. First multiply $4_{five} \times 3_{five}$, which the table says is 22_{five}. Put down the 2 and carry the 2.

$$
\begin{array}{r}
\overset{2}{2}3_{five} \\
\times\ 14_{five} \\
\hline
2_{five}
\end{array}
$$

Then $4_{\text{five}} \times 2_{\text{five}} = 13_{\text{five}}$, and $13_{\text{five}} + 2_{\text{five}} = 20_{\text{five}}$

$$
\begin{array}{r}
\overset{2}{2}3_{\text{five}} \\
\times\ 14_{\text{five}} \\
\hline
202_{\text{five}}
\end{array}
$$

Next, compute $10_{\text{five}} \times 23_{\text{five}}$ and add the partial products.

$$
\begin{array}{r}
23_{\text{five}} \\
\times\ 14_{\text{five}} \\
\hline
202 \\
230 \\
\hline
432_{\text{five}}
\end{array}
$$

(c)
$$
\begin{array}{r}
23_{\text{five}} \\
\times\ 14_{\text{five}} \\
\hline
202 \\
23 \\
\hline
432_{\text{five}}
\end{array}
$$

LE 12   Connection

(a) Compute $44_{\text{five}} \times 22_{\text{five}}$ with the partial products algorithm.

(b) Compute $44_{\text{five}} \times 22_{\text{five}}$ with the standard algorithm.

Long division in base five can be done with a long-division algorithm analogous to the base-ten algorithm.

• EXAMPLE 4

(a) Compute $1442_{\text{five}} \div 4_{\text{five}}$ with the repeated subtraction algorithm.

(b) Compute $1442_{\text{five}} \div 4_{\text{five}}$ with the standard products algorithm.

SOLUTION

(a) Use the multiplication table as needed. Repeatedly subtract off the largest group of 4 that you can find. For example, you might subtract off 200 sets of 4. The rest of a possible solution is shown below.

$$
\begin{array}{r}
4_{\text{five}} \overline{)1442_{\text{five}}} \\
1300 \qquad 200_{\text{five}} (\times 4) \\
\hline
142 \\
130 \qquad 20_{\text{five}} (\times 4) \\
\hline
12 \\
4 \qquad 1_{\text{five}} (\times 4) \\
\hline
3 \qquad 221_{\text{five}}
\end{array}
$$

$$1442_{\text{five}} \div 4_{\text{five}} = 221_{\text{five}} R3_{\text{five}}$$

(b) Using long division, follow the same steps as in base ten. First, how many times can 4_{five} go into 14_{five}? Two times. (You can refer to the base-five multiplication table.) $2_{\text{five}} \times 4_{\text{five}} = 13_{\text{five}}$. Subtract this from 14_{five}, and bring down the next digit in the next dividend.

$$
\begin{array}{r}
2 \\
4_{\text{five}} \overline{)1442_{\text{five}}} \\
13 \\
\hline
14
\end{array}
$$

Next, divide 4_{five} into 14_{five}. It goes in 2 times, as it did before. Continuing the algorithm in this manner yields the following results.

$$
\begin{array}{r}
221_{\text{five}}\ \text{R}3_{\text{five}} \\
4_{\text{five}})\overline{1442_{\text{five}}} \\
\underline{13} \\
14 \\
\underline{13} \\
12 \\
\underline{4} \\
3
\end{array}
$$

LE 13 Connection

(a) Compute $1343_{\text{five}} \div 3_{\text{five}}$ with the repeated subtraction algorithm.
(b) Compute $1343_{\text{five}} \div 3_{\text{five}}$ with the standard algorithm.

The homework exercises include problems involving alternative algorithms in other bases.

LE 14 Summary

How did studying other bases help you understand base ten better?

ANSWERS TO SELECTED LESSON EXERCISES

2. 0, 1, 2, 3, 4

3. $3(8) + 2 = 26$

4. $4(64) + 2(8) + 6 = 278$

5. $2(125) + 1(25) + 3(5) + 4 = 294$

6. (a) Two nickels and two pennies
(b) Three quarters, one nickel, and three pennies

7. (a) 456_{eight}
(b) 2202_{five}

8. (a) 0, 1, 2, 3, 4, 10, 11, 12, 13, 14, 20

(b)

+	0	1	2	3	4
0	0	1	2	3	4
1	1	2	3	4	10
2	2	3	4	10	11
3	3	4	10	11	12
4	4	10	11	12	13

9. (a)
$$
\begin{array}{r}
24_{\text{five}} \\
+\ 14_{\text{five}} \\
\hline
13 \\
30 \\
\hline
43_{\text{five}}
\end{array}
$$

(b) *Step 1* Show 24_{five} as 2 fives and 4 ones and 14_{five} as one five and four ones.

Step 2 Add the ones.
4 ones + 4 ones = 8 ones.
Regroup 8 ones as 1 five 3 ones.

Step 3 Add the fives.
2 fives + 1 five + 1 five = 4 fives
The sum is 4 fives 3 ones = 43_{five}
$24_{\text{five}} + 14_{\text{five}} = 43_{\text{five}}$

10. (a)
$$
\begin{array}{r}
32_{\text{five}} \\
-\ 14_{\text{five}} \\
\hline
20 \\
-\ 2 \\
\hline
13_{\text{five}}
\end{array}
$$

(b) *Step 1* Show 32_{five} as 3 fives and 2 ones. Can you take away 4 ones?

Step 2 Regroup 1 five as 5 ones.
Subtract the ones.
12_{five} ones − 4 ones = 3 ones.

Step 3   Subtract the fives.

2 fives − 1 five = 1 five

The difference is 1 five 3 ones = 13_{five}.

11.

×	0	1	2	3	4
0	0	0	0	0	0
1	0	1	2	3	4
2	0	2	4	11	13
3	0	3	11	14	22
4	0	4	13	22	31

12. (a)

$$\begin{array}{r} 44_{\text{five}} \\ \times\ 22_{\text{five}} \\ \hline 13 \\ 130 \\ 130 \\ 1300 \\ \hline 2123_{\text{five}} \end{array}$$

(b)

$$\begin{array}{r} 44_{\text{five}} \\ \times\ 22_{\text{five}} \\ \hline 143 \\ 143 \\ \hline 2123_{\text{five}} \end{array}$$

13. The quotient is $244_{\text{five}}\ \text{R}1_{\text{five}}$

3.8   HOMEWORK EXERCISES

Basic Exercises

1. How would you group

X X X X X X X X X X X X

in each of the following bases?
(a) Base eight   (b) Base five

2. How many different digits are needed for base twelve?

3. Write the first 12 counting numbers in base three.

4. What base-eight numeral follows 377_{eight}?

5. Convert each of the following to base ten.
(a) 75_{eight} (b) 423_{five} (c) 213_{eight}

6. Which is larger, 1011_{five} or 72_{eight}?

7. Convert each of the following base-ten numerals to numerals in the indicated base.
(a) 46 to base eight
(b) 26 to base five
(c) 324 to base five

8. Change 36_{eight} to base five.

9. Consider the following problem. "How can one recognize a base-five numeral that is divisible by 5?" Devise a plan and solve the problem.

10. How can one distinguish even whole numbers from odd ones in the following systems?
(a) Base eight   (b) Base five

11. Write a base-five numeral represented by the base-five blocks shown. (Make all possible trades first.)

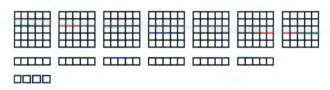

12. Write a numeral for the following set, using the base indicated by the groupings.

X X X X X
X X X X X
X X X X X
X X X X X
X X X X X

| X X X X X | | X X X X X | | X |

13. (a) Compute $13_{\text{five}} + 22_{\text{five}}$ with the partial sums algorithm.
(b) Explain how to compute $13_{\text{five}} + 22_{\text{five}}$ with base-five blocks and the standard algorithm.

14. (a) Compute $31_{\text{five}} + 22_{\text{five}}$ with the partial sums algorithm.
(b) Explain how to compute $31_{\text{five}} + 22_{\text{five}}$ with base-five blocks and the standard algorithm.

15. (a) Compute $43_{\text{five}} - 24_{\text{five}}$ with the partial differences algorithm.
(b) Explain how to compute $43_{\text{five}} - 24_{\text{five}}$ with base-five blocks and the standard algorithm.

16. (a) Compute $32_{\text{five}} - 14_{\text{five}}$ with the partial differences algorithm.
   (b) Explain how to compute $32_{\text{five}} - 14_{\text{five}}$ with base five-blocks and the standard algorithm.

17. Perform the following computations.
   (a) $\begin{array}{r} 23_{\text{five}} \\ + \ 34_{\text{five}} \\ \hline \end{array}$
   (b) $\begin{array}{r} 324_{\text{five}} \\ + \ 132_{\text{five}} \\ \hline \end{array}$
   (c) $\begin{array}{r} 432_{\text{five}} \\ + \ 233_{\text{five}} \\ \hline \end{array}$

18. Perform the following computations.
   (a) $\begin{array}{r} 41_{\text{five}} \\ - \ 23_{\text{five}} \\ \hline \end{array}$
   (b) $\begin{array}{r} 312_{\text{five}} \\ - \ 133_{\text{five}} \\ \hline \end{array}$
   (c) $\begin{array}{r} 432_{\text{five}} \\ - \ 143_{\text{five}} \\ \hline \end{array}$

19. (a) Compute $23_{\text{five}} \times 14_{\text{five}}$ with the partial products algorithm.
   (b) Compute $23_{\text{five}} \times 14_{\text{five}}$ with the standard algorithm.

20. (a) Compute $33_{\text{five}} \times 42_{\text{five}}$ with the partial products algorithm.
   (b) Compute $33_{\text{five}} \times 42_{\text{five}}$ with the standard algorithm.

21. (a) Compute $121_{\text{five}} \div 3_{\text{five}}$ with the repeated subtraction algorithm.
   (b) Compute $121_{\text{five}} \div 3_{\text{five}}$ with the standard algorithm.

22. (a) Compute $324_{\text{five}} \div 4_{\text{five}}$ with the repeated subtraction algorithm.
   (b) Compute $324_{\text{five}} \div 4_{\text{five}}$ with the standard algorithm.

23. Perform the following computations.
   (a) $\begin{array}{r} 34_{\text{five}} \\ \times \ 23_{\text{five}} \\ \hline \end{array}$
   (b) $\begin{array}{r} 412_{\text{five}} \\ \times \ 321_{\text{five}} \\ \hline \end{array}$

24. Perform the following computations.
   (a) $213_{\text{five}} \div 4_{\text{five}}$
   (b) $4123_{\text{five}} \div 3_{\text{five}}$

25. Complete the following base-eight addition and multiplication tables.

26. Perform the following computations.
   (a) $\begin{array}{r} 32_{\text{eight}} \\ + \ 66_{\text{eight}} \\ \hline \end{array}$
   (b) $\begin{array}{r} 132_{\text{eight}} \\ - \ 66_{\text{eight}} \\ \hline \end{array}$
   (c) $\begin{array}{r} 24_{\text{eight}} \\ \times \ 35_{\text{eight}} \\ \hline \end{array}$

27. Perform the following computations.
   (a) $124_{\text{eight}} \div 4_{\text{eight}}$
   (b) $756_{\text{eight}} \div 3_{\text{eight}}$

28. Fill in the missing digits.
   (a) $\begin{array}{r} 1 \ 3 \ \_\_\_{\text{eight}} \\ + \ 2 \ \_\_ \ 4_{\text{eight}} \\ \hline \_\_ \ 2 \ 1_{\text{eight}} \end{array}$
   (b) $\begin{array}{r} 2 \ 0 \ \_\_\_{\text{five}} \\ - \ 1 \ 0 \ 2 \ 2_{\text{five}} \\ \hline \_\_ \ \_\_ \ 1 \ 3_{\text{five}} \end{array}$

Extension Exercises

29. Computers use base two since it contains two digits, 0 and 1, that correspond to electronic switches in the computer being off or on. For example,

Base-Two Place Value			
Eights 2^3	**Fours** 2^2	**Twos** 2^1	**Ones** 1
	1	0	1

So $101_{\text{two}} = 5$.
   (a) Write 1101_{two} as a base-ten numeral.
   (b) Write 17 as a base-two numeral.
   (c) $1011_{\text{two}} + 111_{\text{two}} = $ _____ (base two).

30. (a) Construct addition and multiplication tables for base two.
   Use your tables to compute the following.
   (b) $1101_{\text{two}} + 101_{\text{two}}$
   (c) $1101_{\text{two}} \times 101_{\text{two}}$

31. Computers sometimes take numbers in base eight or sixteen rather than base ten, because they are easier to convert to base two. Can you figure out why?
   (a) Convert 555_{eight} to base two.
   (b) Describe an easier way for you or a computer to do part (a). (*Hint:* Work with each digit separately.)
   (c) Use the shortcut from part (b) to convert 642_{eight} to base two.

32. In base sixteen (**hexadecimal**), the digits are 0 through 9 and *A, B, C, D, E*, and *F* for 10 through 15.
   (a) Convert $B6_{\text{sixteen}}$ to base ten.
   (b) Convert $B6_{\text{sixteen}}$ to base two.
   (c) How can a computer convert $B6_{\text{sixteen}}$ to base two without first converting it to base ten?

33. In base twelve, the **duodecimal** system, one uses 12 digits: 0, 1, 2, 3, 4, 5, 6, 7, 8, 9, *T*, and *E* (*T* and *E* represent ten and eleven).
    (a) Convert $E6_{\text{twelve}}$ to base ten.
    (b) Convert 80 to base twelve.

34. Pencils are often packed by the dozen or by the gross (144).
    (a) Inventory shows that a college bookstore has 3 gross, 2 dozen, and 8 pencils. How many pencils does the store have?
    (b) Write a mathematics problem involving bases that is equivalent to the problem in part (a).

35. Perform the following computations.
    (a) $\begin{array}{r} 321_{\text{four}} \\ + 233_{\text{four}} \\ \hline \end{array}$ (b) $\begin{array}{r} 1001_{\text{two}} \\ - 110_{\text{two}} \\ \hline \end{array}$

36. Perform the following computations.
    (a) $\begin{array}{r} 35_{\text{six}} \\ \times 42_{\text{six}} \\ \hline \end{array}$ (b) $123_{\text{six}} \div 3_{\text{six}}$

37. Consider the following problem. "An inspector wants to check the accuracy of a scale in weighing any whole-number amount from 1 to 15 pounds. Find the smallest number of weights the inspector needs, and tell how heavy each of these weights is."
    (a) What amounts must the inspector be able to weigh?
    (b) Devise a plan and solve the problem.
    (c) Make up a similar problem.

38. A magician asks a volunteer to think of a number from 1 to 15. The magician then shows the following four cards and asks the volunteer which cards contain the volunteer's number.

1	3
5	7
9	11
13	15

2	3
6	7
10	11
14	15

4	5
6	7
12	13
14	15

8	9
10	11
12	13
14	15

By adding the numbers in the upper left corner of the selected cards, the magician finds the secret number.
(a) Pick a number from 1 to 15 and show that the trick works.
(b) Explain how the trick works, using base-two place value.

39. Devise a number system using the symbols *A*, *B*, *C*, and 0.
    (a) What do *A*, *B*, and *C* represent?
    (b) How would you count to 10 in your system?
    (b) How would you represent 50 in your system?

40. Consider the following problem. "If $37_{\text{eight}} = 133_b$, what base is *b*?"
    (a) Devise a plan and solve the problem.
    (b) Make up a similar problem.

41. For what bases *b* would $32_b + 25_b = 57_b$?

42. For what bases *a* and *b* would $12_a = 22_b$?

Labs and Projects

43. Businesses that use the ZIP + 4 (nine-digit) bar code on postage-paid reply envelopes use the binary system. The tall bars represent 1s and the short bars represent 0s. The first and last bars are used to mark the beginning and end. Each group of five bars besides the first and last represents a digit. The last group before the end bar is a check digit that is not part of the zip code. The five-bar codes are *not* the binary equivalents of the digits. Find some postage-paid reply envelopes and try to figure out what five-bar code is used for each digit in a zip code. For example, ⁙ represents 1.

CHAPTER 3 SUMMARY

Our Hindu-Arabic numeration system took a long time to develop. By contrasting the Hindu-Arabic system with older systems such as ancient Egyptian and Babylonian numeration, one gains a greater appreciation of the time it took to develop a base-ten place-value system that has a symbol for zero. Children today study place value with concrete materials such as the base-ten blocks and the abacus.

To know when to add, subtract, multiply, or divide in everyday life, one must recognize common application categories for each operation. Although educators do not agree on the exact number or names of the categories, it is clear that each operation has anywhere from two to seven different classifications. These categories can be illustrated with pictures or objects.

Whole-number operations possess properties that simplify certain computations as well as memorization of the addition and multiplication tables. Addition and multiplication are commutative and associative, and the distributive property holds for multiplication over addition and multiplication over subtraction. The whole numbers also possess identity elements: 0 for addition and 1 for multiplication.

Algorithms are faster and more efficient methods for paper-and-pencil computation involving larger numbers. Elementary-school children learn an efficient algorithm for each operation. There are alternative algorithms for children who have trouble learning the standard ones. Manipulatives such as base-ten blocks clarify the steps in the algorithms. The addition and multiplication algorithms are based on the commutative, associative, and distributive properties.

Adults frequently use mental computation and estimation. The elementary-school curriculum now devotes more time to teaching children about whole-number computations that are especially easy to compute mentally. Children also learn how to estimate with rounding, compatible numbers, and front-end strategies.

One can develop a deeper understanding of place value and algorithms for the four operations by studying other bases. Computers are well suited to base two, with its two digits that can correspond to "on" and "off."

STUDY GUIDE

To review Chapter 3, see what you know about each of the following ideas or terms that you have studied. You can also use this list to generate your own questions about the chapter.

WHOLE NUMBERS IN GRADES 1–8

The following chart shows at what grade levels selected whole-number topics typically appear in elementary- and middle-school mathematics textbooks. Underlined numbers indicate grades in which the most time is spent on the given topic.

Topic	Typical Grade Level in Current Textbooks
Place value	1, 2, 3, 4, 5
Addition concepts	1, 2
Addition algorithm	1, 2, 3, 4
Subtraction concepts	1, 2
Subtraction algorithm	1, 2, 3, 4
Multiplication concepts	2, 3, 4
Multiplication algorithm	3, 4, 5
Division concepts	3, 4
Division algorithm	3, 4, 5, 6
Working backward	3, 4, 5, 6, 7, 8
Order of operations	5, 6, 7, 8
Estimation	2, 3, 4, 5, 6

REVIEW EXERCISES

1. What is the total number of symbols needed to write 37 with each of the following?
(a) Tallies
(b) Late Egyptian numerals
(c) Roman numerals

2. Describe the difference between a numeration system that has place value and one that does not.

3. About 130,000 people attended an outdoor concert. If this figure has been correctly rounded to the nearest ten thousand, the actual attendance was between what two numbers?

4. Name a set of numbers that is closed under addition.

5. Make a drawing that shows $7 - 3 = 4$ using compare measures.

6. "My math class has 36 students, and your class has 27 students. How many more students are there in my class?"
(a) Is this a sets or measures problem?
(b) What category is illustrated? (compare, missing part, take away)

7. "A box holds 10 diskettes. How many boxes does Diane Lorenz need to hold 40 diskettes?"
(a) Is this a sets or measures problem?
(b) What category is illustrated? (area, array, partition, repeated)

8. Write a paragraph describing the difference between the repeated-sets and partition-sets classifications for division.

9. "Ira Reed bought 6 cans of tennis balls. Each can contained 3 tennis balls. Since then, he has lost 2 balls. How many tennis balls does he have left?"
(a) Is this a sets or measures problem?
(b) What two operations and categories are illustrated in the problem? (area, array, combine, compare, counting principle, missing part, partition, repeated, take away)

10. "A man has agreed to transport 1600 lb of boxes. He has already transported 600 lb of them. If each box weighs 40 lb, how many boxes does he have left to transport?"
(a) Is this a sets or measures problem?
(b) What two operations and categories are illustrated in the problem? (area, array, combine, compare, counting principle, missing part, partition, repeated, take away)

11. (a) What is $3 \div 0$?
(b) *Explain* why, using multiplication or the division category of your choice.

12. Make up a word problem that illustrates the combine measures category.

13. Make up a word problem that illustrates the array category for multiplication.

14. Make up a word problem that illustrates the missing part (sets) category.

15. A man earns D dollars per hour. If he works H hours and then spends S dollars for taxes and expenses, how much money does he have left?

16. Joy spent $30 on groceries. Then she spent half of her remaining money on a book. She now has $8. How much money did she start out with? Explain how you found the answer.

17. Name the property illustrated.
(a) $3 + 5 = 5 + 3$ (b) $8 \times 1 = 1 \times 8 = 8$
(c) $8 \times (6 \times 3) = (8 \times 6) \times 3$

18. (a) Is whole-number division associative?
(b) If so, give an example. If not, give a counterexample.

19. Find one set of whole numbers A, B, and C, in which $B \neq 0$ and $C \neq 0$, that makes the following equation true; then find one set that makes the equation false.
$$(A \div B) + (A \div C) = A \div (B + C)$$

20. (a) What is an easy way to compute
$(8 \times 24) + (8 \times 16)$?
(b) What property justifies your answer to part (a)?

21. A child says that $(4 + 5)^2 = 4^2 + 5^2$. Is this correct? If not, what would you tell the child?

22. Suppose you want to introduce a third grader to the standard algorithm for computing $326 + 293$. Explain how to find the sum with base-ten blocks following the same sequence of steps as the standard algorithm.

23. Compute $127 + 246$ with the partial sums algorithm.

24. Two second-grade children are asked to compute $74 - 26$. For each method, tell what the student understands about subtraction.
(a) I counted from 26 up to 30 which is 4 and then 44 more to 74. So the answer is $4 + 44 = 48$.
(b) I changed $74 - 26$ to $78 - 30 = 48$.

25. A fourth grader computes 46×7 as follows. "First $40 \times 7 = 280$ and $6 \times 7 = 42$. The answer is $280 + 42 = 322$." How would this child work out 324×5?

26. Suppose you want to introduce a child to the standard algorithm for computing 26×3. Explain how to find the product with base-ten blocks following the same sequence of steps as the standard algorithm.

27. (a) Compute $312 \div 14$ with the repeated subtraction algorithm.
(b) Compute $312 \div 14$ with the standard algorithm.
(c) What are the advantages and disadvantages of each algorithm?

In Exercises 28–30, (a) complete the last example, repeating the error pattern in the completed examples, (b) write a description of the error pattern, and (c) write what you would tell the child about his or her error.

28.

82	46	89
+ 79	+ 37	+ 74
1511	*713*	

29.

37	426	371
− 10	− 302	− 130
20	*104*	

30.

$\overset{5}{3}7$	$\overset{2}{2}7$	48
$\times\ 48$	$\times\ 93$	$\times\ 57$
296	*81*	
178	*203*	
2076	*2111*	

31. Compute $239 - 167$ with the adding up algorithm.

32. Compute 457×28 with lattice multiplication.

33. Show the steps for three different ways to compute mentally $81 - 43$.

34. You are planning a 2270-mile trip. Your car gets 37 miles per gallon.

    (a) Explain how you could mentally estimate how much gas you will need. (area, array, partition, repeated)

    (b) Is this a sets or a measures problem?

    (c) Which category is illustrated?

35. Write a sentence or two describing the difference between the rounding strategy for multiplication and the compatible-numbers strategy for division.

36. Show the steps to compute $458 + 327 + 118$ by

    (a) rounding.

    (b) a front-end estimate with adjustment.

37. Convert 100 to base six.

38. (a) Compute $641_{\text{eight}} - 235_{\text{eight}}$ with the partial differences algorithm.

    (b) Explain how to compute $641_{\text{eight}} - 235_{\text{eight}}$ with base eight blocks and the standard algorithm.

39. (a) Compute $1324_{\text{seven}} \div 6_{\text{seven}}$ with the repeated-subtraction algorithm.

    (b) Compute $1324_{\text{seven}} \div 6_{\text{seven}}$ with the standard algorithm.

40. Find bases a and b such that $11_a = 22_b$.

ALTERNATE ASSESSMENT—WRITING A UNIT TEST

Create your own test for the material your class covered in Chapter 3. Try out the items on a classmate to see if they are clearly worded and the difficulty level is appropriate. Or you could add to your portfolio or your journal.

SUGGESTED READINGS

Ashlock, R. *Error Patterns in Computation.* 8th ed. Englewood Cliffs, N.J.: Prentice-Hall, 2001.

Bresser, R., and Holtzman, C. *Developing Number Sense.* Sausalito, CA: Math Solutions, 1999.

Chapin, S., and Johnson, A. *Math Matters: Grades K–6.* Sausalito, CA: Math Solutions, 2000.

Copley, J., ed. *Mathematics in the Early Years.* Reston, VA: NCTM, 1999.

Cuoco, A., ed. *The Role of Representations in School Mathematics.* 2001 Yearbook. Reston, VA: NCTM, 2001.

Edwards, R. *Operation Magic Tricks.* Pacific Grove, CA: Critical Thinking Press, 1995.

Eves, H. *Introduction to the History of Mathematics.* 6th ed. Stamford, CT: Thomson, 1990.

Morrow, L. J., ed. *The Teaching and Learning of Algorithms in School Mathematics.* 1998 Yearbook. Reston, VA: NCTM, 1998.

Navigating Through Number and Operations (4 vols). Reston, VA: NCTM, 2003.

National Council of Teachers of Mathematics. *Historical Topics for the Mathematics Classroom.* Reston, VA: NCTM, 1989.

Ostrow, J. *Making Problems, Creating Solutions: Challenging Young Mathematicians.* Portland, ME: Stenhouse, 1999.

Schifter, D., et al. *Building a System of Tens.* Parsippany, NJ: Dale Seymour, 1999.

Use Infotrac to find articles in *Teaching Children Mathematics* on teaching children about whole numbers.

4 | NUMBER THEORY

Written records indicate that until the Pythagoreans came along around 550 B.C., people used numbers primarily in practical applications. The Pythagoreans believed that numbers revealed the underlying structure of the universe, so they studied patterns of counting numbers. This was probably the first significant work in the field of mathematics that we now call number theory.

Number theory is useful in certain computations with fractions and in algebra. Divisibility checks are used to verify codes on merchandise and food, identification numbers on tickets and books, signals from compact discs and TV transmitters, and data from space probes. Coded messages sometimes utilize factors of very large numbers. People also study number theory because number patterns fascinate them, and they enjoy the challenge of determining whether or not the patterns apply to all whole or counting numbers.

4.1 FACTORS

NCTM Standards

- use factors, multiples, prime factorization, and relatively prime numbers to solve problems (6–8)
- develop and evaluate mathematical arguments and proofs (pre-K–12)
- communicate their mathematical thinking coherently and clearly to peers, teachers, and others (pre-K–12)

Pythagoras (580?–500? B.C., Figure 4–1) was the charismatic leader of a group of 300 called the Pythagoreans. One of the female Pythagoreans was Theano, a talented student of Pythagoras, who later married him. After Pythagoras died, Theano continued his work. The Pythagoreans observed some rather strange rules. They (1) never ate meat or beans except after a religious sacrifice, (2) never walked on a highway, and (3) never let swallows sit on their roofs.

They were also mathematicians. Believing that numbers were the basis of all things, they associated numbers with ideas, just as some people today believe that 7 is lucky or 13 is unlucky. The Pythagoreans associated the number 1 with reason (a consistent whole), 2 with opinion (two sides), 4 with justice (balanced square and product of equals), and 5 with marriage (unity of their first odd or masculine number, 3, and the first even or feminine number).

Courtesy of Brown University.

Figure 4–1

The Pythagoreans' interest in numbers also led them to investigate factors. They labeled numbers as "perfect," "abundant," or "deficient," depending upon what factors they possessed (see Exercise 33).

 LE 1   Opener

Suppose 4 is a factor of x and 4 is a factor of y.

(a) What other counting numbers are factors of x?
(b) 4 is also a factor of what other numbers related to x or y?

DEFINITION OF A FACTOR

Before finding out which numbers are perfect or amicable, you'll need to review factors. As you learned in Chapter 3, the term "factor" is used in describing the parts of any multiplication problem such as $2 \times 5 = 10$.

$$2 \times 5 \quad = \quad 10$$
$$\uparrow \; \uparrow \qquad\qquad \uparrow$$
$$\text{Factors} \qquad \text{Product}$$

LE 2   Concept

Write the two division equations that are equivalent to $2 \times 5 = 10$.

In the equation $10 \div 5 = 2$, 5 is the divisor, and in the equation $10 \div 2 = 5$, 2 is the divisor. So 2 and 5 are called either divisors or factors of 10. Although the word "factor" generally refers to multiplication and "divisor" to division, factors and divisors are the same in number theory.

LE 3   Concept

If someone asked you why 3 is a factor of 21, how would you respond

(a) using division?
(b) using multiplication?

The formal definition of "factor" or "divisor" uses multiplication rather than division.

Definition: Factor or Divisor

If A and B are whole numbers, with $A \neq 0$, then A is a **factor** or **divisor** of B, written $A \mid B$, if and only if there is a whole number C such that $A \cdot C = B$.

The statement "A is a factor of B" means that A divides B evenly. If C is the quotient for $B \div A$, the relationship between A and B can be shown in the following three ways.

$$A\overline{\smash{)}B}^{\,C} \qquad \text{or} \qquad A \cdot C = B \qquad \text{or } A \mid B$$

If asked to explain why 6 is a factor of 30, you might say that it is because 6 divides 30 evenly ($30 \div 6 = 5$). Using the *definition* of a factor to explain why 6 is a factor of 30 requires using a *multiplication* equation. By the definition, 6 is a factor of 30 because there is a whole number 5 such that $6 \cdot 5 = 30$.

One advantage of defining "factor" using multiplication instead of division is that a multiplication equation is easier to work with in proofs involving factors.

LE 4   Concept

(a) Use the definition of a factor to explain why 2 is a factor of 40.
(b) If 2 is a factor of R, then there must be a whole number N such that _____. (Write a multiplication equation.)
(c) Applying the definition to 2 and R in part (b) to fill in the blank involves _____ reasoning.

The notation $2 \mid 12$ means "2 is a factor of 12" or "2 divides 12." In $a \mid b$, the first number, a, represents the divisor, or factor. Note that $a \mid b$ is not a number. It is a relation between two numbers. To symbolize that 5 is not a factor of 12, one writes $5 \nmid 12$.

LE 5   Concept

(a) Name all the factors of 12.
(b) True or false? $12 \mid 3$.
(c) Show the factor pairs of 12 using rectangles. The rectangles for 2 and 6 are shown below. You can outline these rectangles on graph paper if you have it.

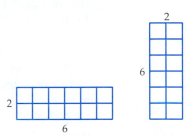

(d) The number 12 is divisible by _____. (List all possibilities.)
(e) Use the definition of a factor to explain why $8 \nmid 12$.

AN INVESTIGATION: FACTORS

LE 6   Reasoning

(a) Find three examples of counting numbers that have exactly three different factors.
(b) Propose a generalization about counting numbers that have exactly three different factors.
(c) Does part (b) involve induction or deduction?

THEOREMS ABOUT FACTORS

Patterns they noted in the factors of particular numbers led the Pythagoreans and other Greek mathematicians to investigate more general questions about factors. For example, if A is a divisor of B, then A must also be a divisor of what other numbers?

In LE 7–LE 9, try numerical examples and use inductive reasoning to make an educated guess regarding whether each statement is true or false. Remember that in mathematics and logic, a "true" statement must *always* be true. If a mathematical statement is false in even one instance, it is "false." Give an example if you think a statement is true, and give a counterexample if you discover a statement that is false.

LE 7 Reasoning

True or false? If A is a factor of an even number, then A is even.

LE 8 Reasoning

True or false? If $A \mid 1200$, then $A \mid 2400$.

LE 9 Concept

In LE 8, why isn't $A = 7$ a counterexample?

When you see the phrase "there exist(s)," you need to find only one instance of it, as in the following exercise.

LE 10 Concept

True or false? There exist counting numbers A, B, and C such that $A \mid (B+C)$ and $A \mid BC$. (*Hint:* "There exist" is the same as "It is possible to find.")

For a "there exists" statement, one true example makes it true. For a generalization, one false example makes it false.

You may have guessed that the general statement in LE 8 is true. But how can you be sure? In this same situation, the classical Greeks, whenever possible, proved their conjectures using deductive reasoning or found a counterexample to disprove their conjectures. Euclid organized and published many of the Greeks' proofs in his work *Elements*.

Deductive proofs about factors usually employ the definition of a factor. The following exercises will help prepare you for these proofs.

LE 11 Reasoning

Suppose A, B, and C are counting numbers and $A \cdot B = C$. What can you deduce about factor relationships among A, B, and C?

LE 12 Concept

Suppose $A \mid B$. Using the definition of a factor, one can deduce that _____
_____.

• EXAMPLE 1

Prove that if $A \mid 1200$, then $A \mid 2400$.

SOLUTION

In LE 8, you probably found a few different examples that worked. For example, when $A = 2$, $2 \mid 1200$ and $2 \mid 2400$.

To show that the statement is true for any counting number A when $A \mid 1200$, use the definition of a factor on the hypothesis $A \mid 1200$.

1. $A \mid 1200$
2. $A \mid 1200$ means that there is a whole number W such that $A \cdot W = 1200$.
3. ?
4. ?

The last (fifth?) step of the proof is

5. $A \mid 2400$

Now try to work backward from the last step. If step 5 says $A \mid 2400$, how would we show that in step 4? By finding a whole number ? such that $A \cdot ? = 2400$.

1. $A \mid 1200$
2. $A \cdot W = 1200$, in which W is a whole number
3. ?
4. $A \cdot \underline{\quad} = 2400$, in which $\underline{\quad}$ is a whole number
5. $A \mid 2400$

Now, how do you get from step 2 to step 4? To make $A \cdot W = 1200$ look like $A \cdot ? = 2400$, multiply both sides by 2. This is step 3. Since we know that $2W$ is a whole number, this shows that $A \mid 2400$, because $A \cdot 2W = 2400$.

1. $A \mid 1200$
2. $A \cdot W = 1200$, in which W is a whole number
3. $A \cdot 2W = 2400$
4. $A(2W) = 2400$, in which $2W$ is a whole number
5. $A \mid 2400$ ●

The preceding proof and the one in the following exercise can be carried out as follows.

1. Write the first (hypothesis) and last (conclusion) steps.
2. Write a multiplication equation for each factor statement to obtain step 2 and the next-to-last step.
3. Use algebra to go from the equation(s) in step 2 to the equation in the next-to-last step.

Try to prove the statement in the following exercise.

LE 13   Reasoning

Prove that if $3 \mid A$, then $3 \mid 5A$, in which A is a whole number. (*Hint:* Start by saying $3 \mid A$, and then write an equation that follows from this fact. At the end of the proof, consider what equation will show that $3 \mid 5A$.)

The statements in LE 14–LE 17 are possible theorems about divisors. Try numerical examples to determine which of these statements might be true. Give a counterexample for any statement that is false. If you believe that a statement is true, prove that it is true with deductive reasoning. In all statements, A, B, and C are whole numbers, with $A \neq 0$.

LE 14   Reasoning

Discussion

True or false? If $A \mid B$, then $B \mid A$.

LE 15   Reasoning

Discussion

True or false? If $A \mid B$ and $A \mid C$, then $A \mid (B + C)$.

LE 16   Reasoning

Discussion

True or false? If $A \mid B$, then $A \mid BC$.

LE 17   Reasoning

Discussion

True or false? If $A \mid BC$, then $A \mid B$ and $A \mid C$.

Were you able to prove the statements in LE 15 and LE 16? LE 15 is a new theorem called the Divisibility-of-a-Sum Theorem. LE 16 is called the Divisibility-of-a-Product Theorem. Both these theorems have the word "divisibility" in them. For counting numbers A and B, A is **divisible** by B if and only if B is a factor of A. The Divisibility-of-a-Sum-Theorem is as follows.

The Divisibility-of-a-Sum Theorem

For any whole numbers A, B, and C, with $A \neq 0$, if $A \mid B$ and $A \mid C$, then $A \mid (B + C)$.

Figure 4–2 shows an example of the Divisibility-of-a-Sum Theorem. If 9 and 15 can be arranged in arrays with 3 rows, then so can $9 + 15$.

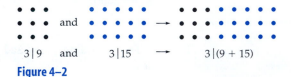

$3 \mid 9$ and $3 \mid 15$ → $3 \mid (9 + 15)$

Figure 4–2

This theorem is useful in explaining why the divisibility tests in the next lesson work. See if you can apply the theorem in the following exercise.

LE 18   Reasoning

According to the Divisibility-of-a-Sum Theorem, if $5 \mid 4000$ and $5 \mid 200$, then _____.

A similar theorem works for subtraction. The proof is left for the homework exercises.

The Divisibility-of-a-Difference Theorem

For any whole numbers A, B, and C, with $A \neq 0$ and $C > B$, if $A \mid B$ and $A \mid C$, then $A \mid (C - B)$

Figure 4–3 shows an example of the Divisibility-of-a-Difference Theorem. If 9 and 15 can be arranged in arrays with 3 rows, then so can 15 − 9.

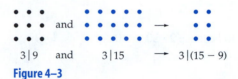

$$3 \mid 9 \quad \text{and} \quad 3 \mid 15 \quad \longrightarrow \quad 3 \mid (15 - 9)$$

Figure 4–3

The Divisibility-of-a-Product Theorem is as follows.

The Divisibility-of-a-Product Theorem

A, B, and C are whole numbers with $A \neq 0$. If $A \mid B$, then $A \mid BC$.

See whether you can apply the Divisibility-of-a-Product Theorem in the following exercises.

LE 19 Reasoning

According to the Divisibility-of-a-Product Theorem, if A and C are whole numbers, with $A \neq 0$ and $A \mid 3$, then _____.

LE 20 Communication

Explain how you could use the Divisibility-of-a-Product Theorem to recognize that $6 \mid (12 \cdot 37)$.

A GAME: FACTOR OUT

LE 21 Skill

Factor Out is a game for two players.

1. Use the chart of numbers from 1 to 50, or copy it on a piece of paper.
2. Flip a coin to decide who goes first and marks the numbers with circles. The second player will mark numbers with squares.
3. Players take turns picking a number and marking the number and all its factors that have not already been marked. The game is over when all the numbers are marked with a square or circle.
4. The winner is the player whose marked numbers have the greater sum.

1	2	3	4	5	6	7	8	9	10
11	12	13	14	15	16	17	18	19	20
21									30
31									40
41									50

LE 22 Summary

Tell what you learned about factors in this section.

ANSWERS TO SELECTED LESSON EXERCISES

2. $10 \div 5 = 2$; $10 \div 2 = 5$

4. (a) 2 is a factor of 40 because $2 \cdot 20 = 40$.
 (b) $2N = R$ (c) deductive

5. (a) 1, 2, 3, 4, 6, 12   (b) False
 (c)

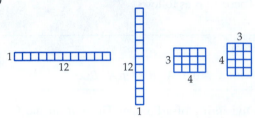

 (d) 1, 2, 3, 4, 6, 12
 (e) 8 is not a factor of 12 because there is no whole number C such that $8C = 12$.

6. (a) 4, 9, 25
 (b) They are perfect squares.
 (c) Induction

7. False for $A = 3$

8. True; $A = 2$ is an example.

9. Because $7 \nmid 1200$.

10. True for $A = 1$ and any B and C

11. A and B are factors of C.

12. A times some whole number equals B.

13. 1. $3 \mid A$
 2. $3C = A$, in which C is a whole number
 3. $3(5C) = 5A$, in which $5C$ is a whole number
 4. $3 \mid 5A$

14. False; $A = 2$, $B = 4$

15. True
 1. $A \mid B$ and $A \mid C$
 2. $A \cdot D = B$ and $A \cdot E = C$, in which D and E are whole numbers
 3. $A \cdot D + A \cdot E = B + C$
 4. $A \cdot (D + E) = B + C$, in which $D + E$ is a whole number
 5. $A \mid (B + C)$

16. True
 1. $A \mid B$
 2. $AD = B$, in which D is a whole number
 3. $A(DC) = BC$, in which DC is a whole number
 4. $A \mid BC$

17. False; $A = 2$, $B = 3$, $C = 4$

18. $5 \mid 4200$

19. $A \mid 3C$

20. $6 \mid 12$, so 6 is a factor of 12 times any whole number such as $12 \cdot 37$.

4.1   HOMEWORK EXERCISES

Basic Exercises

1. (a) Name all the factors of 28.
 (b) Use the definition of a factor to explain why 6 is not a factor of 28.

2. Use the definition of a factor to explain why $11 \mid 0$.

3. True or false? $8 \nmid 2$.

4. True or false? $5 \mid 21$.

5. (a) You want to arrange 20 blocks into a rectangular shape. What are all the ways this can be done? (Outline the rectangles on graph paper if you have it.)
 (b) What mathematical concept does part (a) illustrate?

6. Show all the factor pairs for 14 by drawing rectangles or outlining rectangles on graph paper.

Decide whether the statements in Exercises 7–12 are true or false. Assume that A, B, C, and K are whole numbers, with $A \neq 0$. If a statement is true, explain why or give an example. If it is false, explain why or give a counterexample.

7. True or false? If $A \mid B$, then $A \mid KB$.

8. True or false? If $2 \nmid B$, then $4 \nmid B$.

9. True or false? If $A \mid B$ and $A \mid C$, then $A \mid (B + C)$.

10. True or false? If $A \mid B$, then $B \mid A$.

11. True or false? There exist A, B, and C such that if $A \mid BC$, then $A \mid B$ and $A \mid C$.

12. True or false? There exist A, B, and C such that if $A \mid B$, then $AC \mid B$.

13. If $3 \cdot A = B$, then _____ is a factor of _____, and _____ is a factor of _____.

14. If $2 \mid B$, then there exists a whole number C such that _____.

15. The statement "If $5 \mid 20$, then $5 \mid 200$" is an instance of what theorem?

16. The statement "If $5 \mid 20$ and $5 \mid 40$, then $5 \mid 60$" is an instance of what theorem?

17. According to the Divisibility-of-a-Sum Theorem, if C is a whole number such that $8 \mid C$ and $8 \mid 16$, then _____.

18. According to the Divisibility-of-a-Sum Theorem, if $10 \mid 1000$ and $10 \mid 20$, then _____.

19. The number $C = 12 \times 71 + 48$. Without computing the value of C, explain whether or not C is divisible by 12.

20. According to the Divisibility-of-a-Difference Theorem, if $7 \mid 49$ and $7 \mid 91$, then _____.

21. (a) According to the Divisibility-of-a-Product Theorem, if A and C are whole numbers, with $A \neq 0$, and $A \mid 8$, then _____.
(b) Does part (a) involve induction or deduction?

22. If $3 \mid A$, then $3 \mid 2A$. This is an instance of what theorem?

Extension Exercises

In Exercises 23–30, decide whether each statement is true or false. If it is true, prove it. If it is false, give a counterexample. Assume that A, B, C, and D are whole numbers, with $A \neq 0$.

23. True or false? If $5 \mid B$, then $10 \mid B$. (Read the instructions that precede Exercise 23.)

24. True or false? $1 \mid B$ for all B. (Read the instructions that precede Exercise 23.)

25. True or false? If $A \mid B$ and $C \mid D$, then $AC \mid BD$.

26. True or false? If $A \mid B$, then $A \mid (B + 1)$.

27. True or false? If $A \neq 0$, $B \neq 0$, and $A \mid B$ and $B \mid C$, then $A \mid C$. (*Hint:* You will have to write equations based on $A \mid B$ and $B \mid C$.)

28. True or false? If B and C are divisible by 2, then BC is divisible by 2.

29. True or false? The product of two even whole numbers is even. (*Hint:* C is **even** if $C = 2W$ for some whole number W.)

30. True or false? The product of two odd whole numbers is odd. (*Hint:* C is **odd** if $C = 2W + 1$ for some whole number W.)

31. Use the Divisibility-of-a-Product Theorem to show why, for whole numbers x and y with $x \neq 0$, $x \mid x^2 y^2$.

32. Prove the Divisibility-of-a-Difference Theorem. If $B > C$ and B and C are divisible by A, then $B - C$ is divisible by A.

33. What makes a counting number perfect? According to the Pythagoreans, a **perfect number** equals the sum of all its factors that are less than itself. For example, the factors of 6 are 1, 2, 3, and 6. All of the factors that are less than 6—namely 1, 2, and 3—add up to 6. How perfect! For most numbers, the sum of the corresponding set of factors is either less than the number, as for 9, where $1 + 3 < 9$ (called a **deficient number**), or greater than the number, as for 12, where $1 + 2 + 3 + 4 + 6 > 12$ (called an **abundant number**). Test each of the following numbers to see if it is perfect, deficient, or abundant.
(a) 20   (b) 28   (c) 38

34. Test each of the following numbers to see if it is perfect, deficient, or abundant.
(a) 120   (b) 315   (c) 496

35. Tell why for any whole number $N \geq 2$, $3 \mid (N^3 - N)$. (*Hint:* Factor $N^3 - N$.)

36. Consider the following problem. "Investigate the result of adding 1 to the product of four consecutive whole numbers."

Understanding the Problem
(a) Give an example of adding 1 to the product of four consecutive whole numbers.

Devising a Plan and Carrying Out the Plan
(b) Repeat part (a) for four different consecutive whole numbers.
(c) Propose a generalization.

Looking Back
(d) Prove your generalization using N as the first whole number.

4.2 DIVISIBILITY

NCTM Standards

- describe classes of numbers according to their characteristics such as the nature of their factors (3–5)
- use factors, multiples, prime factorization, and relatively prime numbers to solve problems (6–8)
- develop and evaluate mathematical arguments and proofs (pre-K–12)
- recognize reasoning and proof as fundamental aspects of mathematics (pre-K–12)

LE 1   Opener

List all the divisibility tests you remember from the following: divisibility by 2, 3, 4, 5, 6, 9, and 10.

Three drugstore owners pool their resources to buy 3,456 bottles of aspirin. Can they divide the bottles evenly among themselves?

What is a shortcut for determining if a number such as 3,456 is divisible by 3? The shortcut can be discovered by looking at the multiples of 3.

MULTIPLES

Do different numbers that are divisible by 3 have anything in common? Consider 0, 3, 6, 9, 12, 15, 18, 21, 24, 27, ...

To study divisibility by 3, you would look at the multiples of 3, {0, 3, 6, 9, 12, 15, 18, 21, 24, 27, . . . }. Multiples are generated when you count by a number starting at 0. The multiples of 3 would be written as $3 \cdot 0, 3 \cdot 1, 3 \cdot 2, 3 \cdot 3, 3 \cdot 4$, or 3 times any whole number.

The formal definition of a multiple is based on the definition of a factor. A number *B* is a **multiple** of *A* if and only if *A* is a factor of *B*. For example, 20 is a multiple of 5 because 5 is a factor of 20.

LE 2   Concept

List the multiples of 8.

LE 3   Concept

According to the definition, 50 is a multiple of 10 because there is a whole number _____ such that _____ · _____ = _____.

Students sometimes confuse factors and multiples because they are closely related ideas. Try the following.

LE 4   Concept

Fill in each blank with "factor(s)" or "multiple(s)."

(a) 7 is a _____ of 14.
(b) 30 is a _____ of 10.
(c) 8*x* is a _____ of *x* when *x* is a whole number.
(d) Every counting number has an infinite set of _____.

DIVISIBILITY TESTS FOR 2, 5, AND 10

Is 77,846,379,820 divisible by 2, 5, or 10? The longer way to find out is to divide this 11-digit number by 2, 5, and then 10. There is a shorter way.

The divisibility tests for 2, 5, and 10 are all based on checking the last digit of the number. Do you see the patterns in the last digits of each of the following sets of numbers?

Numbers divisible by 2: {0, 2, 4, 6, 8, 10, 12, 14, 16, 18, . . . }
Numbers divisible by 5: {0, 5, 10, 15, 20, 25, . . . }
Numbers divisible by 10: {0, 10, 20, 30, 40, 50, . . . }

See whether you can describe the divisibility tests in the following exercise.

LE 5   Reasoning

Fill in the blanks with the divisibility test for each number.

(a) A whole number is divisible by 2 if and only if _____.
(b) A whole number is divisible by 5 if and only if _____.
(c) A whole number is divisible by 10 if and only if _____.

The divisibility tests for 2, 5, and 10 are as follows.

Divisibility Tests for 2, 5, and 10

• A whole number is divisible by 2 if and only if its ones digit is divisible by 2.
• A whole number is divisible by 5 if and only if its ones digit is 0 or 5.
• A whole number is divisible by 10 if and only if its ones digit is 0.

These divisibility tests are grouped together because they all require checking the last digit of the whole number.

LE 6 Skill

Without dividing, determine whether each number in parts (a)–(c) is divisible by 2, 5, and/or 10.

(a) 8,479,238   (b) 1,046,890   (c) 317,425

(d) Applying the divisibility rules to answer parts (a), (b), and (c) is an example of _____ reasoning.

Discussion

LE 7 Reasoning

Complete the number so that it is divisible by 2 but not by 5 or 10. Place one digit in each blank.

$$8\ 6\ 3,1\ \_\!\_\ \_\!\_$$

Divisibility tests are useful in studying number theory and fractions. Later in this chapter, divisibility tests will be used to factor a number and to test whether a number is prime. In Chapter 6, divisibility tests will be used to simplify fractions and to find common denominators.

DIVISIBILITY TESTS FOR 3 AND 9

Do you know how to tell if a number is divisible by 3 or 9? Both divisibility tests require adding up the digits of the number.

Numbers divisible by 3: {0, 3, 6, 9, 12, 15, 18, . . . }
Numbers divisible by 9: {0, 9, 18, 27, 36, 45, . . . }

LE 8 Opener

(a) State the divisibility tests for 3 and 9, if you recall them. Use the sets just listed to check your guesses.

(b) What is the relationship between divisibility by 3 and divisibility by 9?

The rules are as follows.

Divisibility Tests for 3 and 9

• A whole number is divisible by 3 if and only if the sum of all its digits is divisible by 3.

• A whole number is divisible by 9 if and only if the sum of all its digits is divisible by 9.

These divisibility tests are grouped together because they both require computing the sum of the digits.

LE 9 Skill

Use the divisibility tests to determine whether each number is divisible by 3 or 9.

(a) 468,172   (b) 32,094

LE 10   Reasoning

Fill in the two missing digits so that the number is divisible by 3 and 9. Place one digit in each blank.

$$1\,0,8\,2\,1,7\,\underline{\ \ }\,\underline{\ \ }$$

(a) How will you know if your answers are correct?
(b) Devise a plan for solving the problem.
(c) Solve the problem.
(d) Do you have all possible answers?

DIVISIBILITY TEST FOR 4

The divisibility test for 4 is similar to the divisibility tests for 2, 5, and 10.

LE 11   Reasoning

(a) Write down two three-digit numbers not ending with two zeroes that are divisible by 4.
(b) Write down two four-digit numbers not ending with two zeroes that are divisible by 4.
(c) What pattern do you see in the last two digits of all the numbers you found in parts (a) and (b)?
(d) State a divisibility test for 4 based on looking at the last two digits of a number.

The rule is as follows.

Divisibility Test for 4

A whole number is divisible by 4 if and only if the last two digits represent a number divisible by 4.

DIVISIBILITY TEST FOR 6

The divisibility test for 6 makes use of some of the divisibility tests in this section.

LE 12   Reasoning

(a) What patterns do you see in numbers divisible by 6, such as 6, 12, 18, 24, 30, and 36?
(b) Devise a divisibility test for 6.

The rule is as follows.

Divisibility Test for 6

A whole number is divisible by 6 if and only if the number is divisible by both 2 and 3.

LE 13   Skill

Without dividing, complete the following chart.

| Number | Divisible by | | | | | | |
	2	3	4	5	6	9	10
8,172	X	X				X	
403,155							
800,002							
68,710							

Figure 4–4 shows the divisibility tests in the fifth-grade textbook for *Harcourt Math*.

Classroom Connection

LESSON

Divisibility

▶ **Learn**

BRANCH OUT   Duncan and Leila are planning this year's Arbor Day celebration. They have 1,764 seedlings that they want to divide evenly among the 9 schools in the county. Will they have any seedlings left over?

If 1,764 is divisible by 9, there will not be any seedlings left over. A number is **divisible** by another number when the quotient is a whole number and there is a remainder of zero.

Some numbers have a divisibility rule. A number is divisible by 9 if the sum of the digits is divisible by 9.

1,764 → 1 + 7 + 6 + 4 = 18   Think: 18 is divisible by 9, so 1,764 is divisible by 9.

So, there will not be any seedlings left over.

Look at the divisibility rules in the table.

	A number is divisible by	Divisible	Not Divisible
2	if the last digit is an even number.	324	753
3	if the sum of the digits is divisible by 3.	612	7,540
4	if the last two digits form a number divisible by 4.	6,028	861
5	if the last digit is 0 or 5.	5,875	654
6	if the number is divisible by 2 and 3.	3,132	982
9	if the sum of the digits is divisible by 9.	9,423	6,032
10	if the last digit is 0.	450	7,365

 MATH IDEA   Knowing the rules of divisibility will help you determine if one number is divisible by another number.

From Harcourt Math Grade 5 (Orlando: Harcourt School Publishers, 2002), p.258.

Figure 4–4

In LE 14–LE 16, do the following. (a) Try some examples, and decide whether the statements are true or false. (b) If a statement is true, give an example that supports it. If a statement is false, give a counterexample. (c) Draw a Venn diagram showing the relationship between the sets of numbers in each statement.

LE 14   Reasoning
Discussion

True or false? If a number is divisible by 20, then it is divisible by 10.

LE 15   Reasoning
Discussion

True or false? If a number is divisible by 2 and 4, then it is divisible by 8.

LE 16   Reasoning
Discussion

True or false? If a number is not divisible by 2, then it is not divisible by 4.

WHAT MAKES THE DIVISIBILITY TESTS WORK?

Why can we tell that 3,640 is divisible by 10 just by looking at the last digit? The numbers of thousands, hundreds, and tens do not matter, because any number of thousands, hundreds, or tens is divisible by 10. Only the ones digit matters!

LE 17   Opener

(a) In 3,640, the 3 represents _____, the 6 represents _____, and the 4 represents _____.

(b) The number 3,640 has place values of 3000, 600, 40, and 0. Is each of these place values divisible by 10?

(c) Then 3000 + 600 + 40 + 0 (or 3,640) is divisible by 10 because of the _____ Theorem.

The following example gives a more general proof of the divisibility test for 10.

• EXAMPLE 1

Prove that if A B C 0 is a four-digit number, then A B C 0 is divisible by 10. (A, B, and C represent the first three digits of the number.) In other words, a four-digit number that ends in 0 is divisible by 10.

SOLUTION

Understanding the Problem   Assume that A B C 0 is a four-digit number. Show that it must be divisible by 10.

Devising a Plan   The first step is the hypothesis, and the last step is the conclusion. Let's estimate that it will take five steps.

1. A B C 0 is a four-digit number.
2.
3.
4.
5. A B C 0 is divisible by 10.

Use expanded notation and the Divisibility-of-a-Sum Theorem to work from step 1 to step 5.

Carrying Out the Plan

1. $\underline{A}\ \underline{B}\ \underline{C}\ \underline{0}$ is a four-digit number.
2. $\underline{A}\ \underline{B}\ \underline{C}\ \underline{0} = 1000A + 100B + 10C$
3. 1000A, 100B, and 10C are each divisible by 10.
4. By the Divisibility-of-a-Sum theorem, $1000A + 100B + 10C$ is divisible by 10.
5. $\underline{A}\ \underline{B}\ \underline{C}\ \underline{0}$ is divisible by 10.

Looking Back   In a whole number ending in 0, there are no ones. All other place values (tens, hundreds, and so on) are divisible by 10 no matter what digit is in that place. Therefore, all whole numbers ending in 0 are divisible by 10.   ●

Now it's your turn.

LE 18   Reasoning
Prove that if a three-digit number ends in 5 or 0 (denoted $\underline{A}\ \underline{B}\ \underline{0}$ or $\underline{A}\ \underline{B}\ \underline{5}$), then the number is divisible by 5.

LE 19   Reasoning
Fill in the steps in the proof of the divisibility test for 3: If $\underline{A}\ \underline{B}\ \underline{C}\ \underline{D}$ is a four-digit number with $A + B + C + D$ divisible by 3, then $\underline{A}\ \underline{B}\ \underline{C}\ \underline{D}$ is divisible by 3.

(a) Write the hypothesis as step 1.
(b) Write $\underline{A}\ \underline{B}\ \underline{C}\ \underline{D}$ in expanded notation as step 2. $\underline{A}\ \underline{B}\ \underline{C}\ \underline{D} = $ _____.
(c) In step 3, regroup the terms in expanded notation so that each term is divisible by 3. For example, 1000A may not be divisible by 3, but 999A is, so pull out a 999A. Fill in the blanks to complete step 3.

$$1000A + 100B + 10C + D = 999A + 99B + 9C + _____$$

(d) Step 4 says that 999A, 99B, 9C, and $A + B + C + D$ are each _____.
(e) Step 5 says that $999A + 99B + 9C + A + B + C + D$ is _____.
(f) Step 6 says, therefore, $\underline{A}\ \underline{B}\ \underline{C}\ \underline{D}$ is divisible by 3.

Classroom Connection

LE 20   Communication
After studying the divisibility rules for 2, 3, 5, 9, and 10, a child asks why you cannot check divisibility for 3 and 9 by checking the last digit. How would you explain this?

LE 21   Summary
Discuss similarities and differences among the divisibility rules that you studied in this section.

ANSWERS TO SELECTED LESSON EXERCISES

2. 0, 8, 16, 24, 32, ...

3. 5; $10 \cdot 5 = 50$

4. (a) factor
(b) multiple
(c) multiple
(d) multiples

5. See the box that follows the exercise.

6. (a) 2
   (b) 2, 5, 10
   (c) 5
   (d) deductive

7. The last digit is 2, 4, 6, or 8. The other digit can be any number.

8. (b) Any whole number that is divisible by 9 is also divisible by 3.

9. (a) Neither
   (b) 3, 9

10. (c) Any two digits with a sum of 8

13.

Number	Divisible by						
	2	3	4	5	6	9	10
8,172	X	X	X		X	X	
403,155		X		X		X	
800,002	X						
68,710	X			X			X

14. True; 20 is an example

15. False for 4

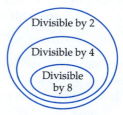

16. True; 7 is an example

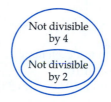

17. (a) 3000; 600; 40
    (b) Yes   (c) Divisibility-of-a-Sum

18. 1. A B 0 and A B 5 are three-digit numbers.
    2. A B 5 = (A · 100) + (B · 10) + 5 and
       A B 0 = (A · 100) + (B · 10)
    3. (A · 100), (B · 10), and 5 are all divisible by 5.
    4. (A · 100) + (B · 10) + 5 and (A · 100) + (B · 10)
       are divisible by 5.
    5. A B 0 and A B 5 are divisible by 5.

19. (a) A B C D is a four-digit number with
        A + B + C + D divisible by 3.
    (b) (A · 1000) + (B · 100) + (C · 10) + D
    (c) A + B + C + D
    (d) Divisible by 3
    (e) Divisible by 3

4.2   HOMEWORK EXERCISES

Basic Exercises

1. What are the multiples of 7?

2. X is a whole number. What are the multiples of X?

3. Explain why 3 is not a multiple of 6.

4. Name a number that is a factor and a multiple of 15.

5. Fill in each blank with "multiple" or "factor."
   (a) 1 is a _____ of every counting number.
   (b) 3 is a _____ of 12.
   (c) 25 is a _____ of 5.
   (d) $2X$ is a _____ of $8X^3$ when X is a whole number not equal to 0.

6. Write a sentence telling the difference between a factor of x and a multiple of x.

7. A sixth grader writes the divisibility test for 4 as "A number is divisible by 4 if the last two digits are divisible by 4." Is this correct? If not, how could it be improved?

8. A sixth grader writes the divisibility test for 6 as "A number is divisible by 6 if the ones digit is divisible by 2 and 3." Is this correct? If not, give a counterexample.

9. Use shortcuts to test each number for divisibility by 2, 3, 4, 5, 6, 9, and 10.
(a) 7,533   (b) 1,344   (c) 410,330

10. Use shortcuts to test each number for divisibility by 2, 3, 4, 5, 6, 9, and 10.
(a) 5,260   (b) 8,197   (c) 345,678

11. There will be 219 children in next year's third grade. If the school has 9 teachers, can the school assign each teacher the same number of students?

12. Three sisters earn a reward of $37,500 for solving a mathematics problem. Can they divide the money equally?

13. What three-digit numbers are less than 130 and divisible by 6?

14. What two-digit numbers greater than 80 are divisible by 2 and 3?

15. Fill in the missing digit(s) so that each number is divisible by the indicated number or numbers. Determine all possible answers.
(a) 4 1, ___ 7 2 (by 3 but not by 9)
(b) 8 2 6, 3 ___ ___ (by 4 and 5)
(c) 4 1 7, 2 ___ ___ (by 6 and 10)

16. Fill in the missing digit(s) so that each number is divisible by the indicated number or numbers. Determine all possible answers.
(a) 2 3, 5 ___ 4 (by 4 and 6)
(b) 5 4 6, 7 ___ ___ (by 2 and 5 but not by 3)
(c) 1 2 4, 5 ___ ___ (by 4 and 9)

17. (a) Suppose you want to mail a package that requires $1.96 in postage, using $0.25 stamps and $0.15 stamps. Is it possible to find stamps totaling the exact amount needed? Tell why.
(b) Explain the result of part (a) using divisibility.

18. A grocery store sells eggs in cartons of 6 or 12. Last week's inventory report says the store sold 5,002 eggs. Your job is to check the inventory figures. What would you tell your boss?

19. Consider the numbers 0, 1, 4, 8, and 64.
(a) Tell which number does not belong with the others, and why.
(b) Give a reason why each of the other numbers could be given as the answer to part (a).

20. Tell whether it would be quicker to use a calculator or a divisibility test to answer each of the following questions.
(a) is 34,865,345 divisible by 2?
(b) Is 34,865,345 divisible by 9?

In Exercises 21–24, do the following. (a) Tell whether the statement is true or false. (b) If the statement is true, tell why or give an example that supports it; if the statement is false, give a counterexample. (c) Draw a Venn diagram showing the correct relationship among the sets of numbers in each statement.

21. True or false? If a number is divisible by 5, then it is divisible by 10.

22. True or false? If a number is not divisible by 5, then it is not divisible by 10.

23. True or false? If a number is divisible by 6 and 8, then it is also divisible by 48.

24. True or false? If a number is divisible by 8 and 10, then it is also divisible by 40. (Use inductive reasoning.)

In Exercises 25 and 26, do the following. (a) Tell whether the statement is true or false. (b) If the statement is true, tell why or give an example that supports it; if the statement is false, give a counterexample.

25. True or false? If A is divisible by 2, then $4 + A$ is divisible by 2. (Instructions follow Exercise 24.)

26. True or false? The sum of any five consecutive whole numbers is divisible by 5. (Instructions follow Exercise 24.)

27. If $15 \mid N$, then what other counting numbers must be factors of N?

28. If M is a multiple of 20, then M must also be a multiple of what other whole numbers?

29. Fill in the blanks to complete the following proof. A three-digit even number has the form $\underline{A}\,\underline{B}\,\underline{C}$ where $C = 0, 2, 4, 6,$ or 8. In expanded notation, the number is _____.

100A, _____, and _____ are each divisible by 2. Therefore, _____ is divisible by 2.

30. Prove that if $\underline{A}\,\underline{B}\,\underline{0}$ is a three-digit number, then it is divisible by 10.

31. The number 729 can be written all of the following ways:

$$9^3, \quad 27^2, \quad 2 \times 364 + 1, \quad 3 \cdot 243$$

What does each of the four expressions tell you about 729?

32. The number 1,296 can be written all of the following ways:

$$6^4, \quad 2 \times 648, \quad 3 \times 432, \quad 5 \times 259 + 1$$

What does each of the four expressions tell you about 1296?

Extension Exercises

33. Prove that if the sum of the digits of a four-digit number $\underline{A}\,\underline{B}\,\underline{C}\,\underline{D}$ is divisible by 9, then the number itself is divisible by 9.

34. Prove that if $\underline{A}\,\underline{B}\,\underline{C}$ is a three-digit number and $A + B + C$ is divisible by 3, then $\underline{A}\,\underline{B}\,\underline{C}$ is divisible by 3.

35. Prove that if $\underline{A}\,\underline{B}\,\underline{C}\,\underline{D}$ is a four-digit number and $\underline{C}\,\underline{D}$ is divisible by 4, then $\underline{A}\,\underline{B}\,\underline{C}\,\underline{D}$ is divisible by 4.

36. Devise a divisibility test for 8. (*Hint:* It is an extension of the divisibility test for 4.) Tell how you found the answer.

37. (a) What is the divisibility test for 100?
(b) A leap year must be divisible by 4. Furthermore, if a leap year is divisible by 100, then it must also be divisible by 400. Which of the following are leap years?
(1) 1776　　(2) 1994　　(3) 1996
(4) 2000　　(5) 2010

38. Devise a divisibility test for 25.

39. Devise a divisibility test for 15. (*Hint:* Look at the divisibility test for 6.)

40. (a) Use the divisibility tests for 6 and 15 to complete the following generalization. If x is the product of two primes a and b, then a whole number y is divisible by x if and only if _____.
(b) Try some examples, and conjecture whether the following more general hypothesis is true. If x is the product of two relatively prime numbers a and b, then a whole number y is divisible by x if and only if y is divisible by a and b. (*Hint:* Start with $a = 4$ and $b = 5$.)
(c) Does part (b) involve induction or deduction?

41. When the entire fifth grade is divided into groups of 4, there is 1 child left over. When the fifth grade is divided into groups of 3, there are 2 children left over. How many children could be in the fifth grade? (Give all possible answers.)

42. Consider the following problem. "The ages of a woman and her granddaughter have a surprising property. First of all, they were born on the same day of the year. And for the last 6 years in a row, the grandmother's age has been divisible by her granddaughter's age! How old are they today?"
(a) Devise a plan and solve the problem.
(b) Check your answer.

43. (a) Choose a two-digit number and reverse the digits. Subtract the smaller number from the larger. Is the difference divisible by 9?
(b) Repeat part (a) with a different number.
(c) Prove that the difference will always be divisible by 9.

44. $10^2 - 8^2$ is divisible by 18.
$20^2 - 3^2$ is divisible by 23.
  (a) Write another statement that fits this pattern, and see if it is true.
  (b) Use algebra to explain why this pattern works for $a^2 - b^2$ in which a and b are counting numbers and $a > b$.

45. (a) Devise a divisibility test for 4 in base eight.
  (b) Devise divisibility tests for 2 and 7 in base eight.

46. Find the remainder when $3^{888,888}$ is divided by
  (a) 4.   (b) 5.

Puzzle Time

 47. (a) Think of any three-digit number.
  (b) Write it down twice to form a six-digit number (e.g., 382,382).
  (c) Is your number divisible by 7?
  (d) Is your number divisible by 11?
  (e) Is your number divisible by 13?
  (f) Repeat parts (a)–(e) with a new number.
  (g) What is $7 \times 11 \times 13$?
  (h) What is 382×1001?
  (i) Can you explain why your six-digit numbers were divisible by 7, 11, and 13?

4.3 PRIME AND COMPOSITE NUMBERS

NCTM Standards

- describe classes of numbers according to their characteristics such as the nature of their factors (3–5)
- use factors, multiples, prime factorization, and relatively prime numbers to solve problems (6–8)

The Greeks were the first to study numbers systematically. They discovered a subset of the whole numbers that are the building blocks of all whole numbers greater than 1.

PRIME NUMBERS AND COMPOSITE NUMBERS

The prime numbers are the essential components of counting numbers greater than 1.

LE 1 Opener

Try to write a definition that tells what property of a counting number makes it a prime number. (*Hint:* Say something about factors.)

Compare your definition to the first part of the following.

Definitions: Prime and Composite Numbers

A counting number greater than 1 is **prime** if it has exactly two distinct factors: 1 and itself. A counting number is **composite** if it has more than two distinct factors.

The number 11 is prime because it has exactly two factors: 1 and 11. The number 10 is composite (and not prime) because it has four factors: 1, 2, 5, and 10. A composite number has a factor other than 1 and itself.

LE 2   Concept
(a) How many different factors does 1 have?
(b) Is 1 prime, composite, or neither?

Since 1 has only one factor, it is neither prime nor composite. The Pythagoreans, who studied prime numbers 2500 years ago, called 1 the unity that generates all prime and composite numbers. The Pythagoreans studied prime and composite numbers geometrically.

 LE 3   Reasoning
In this exercise, you will draw or outline all possible rectangles using the given number of squares. For example, with 4 squares, you could draw rectangles that are 2 by 2, 1 by 4, and 4 by 1.

In parts (a)–(d), draw all possible rectangles using

(a) 2 squares.
(b) 6 squares.
(c) 7 squares.
(d) 1 square.
(e) On the basis of your results, how is the number of different rectangles related to whether or not the number is prime?
(f) Does part (e) involve induction or deduction?

In geometric terms, you can represent a prime number with exactly two rectangles with counting-number dimensions. You can represent a composite number with more than two such rectangles. (Note that 1, which is neither prime nor composite, has exactly one rectangle with counting-number dimensions.)

 LE 4   Concept
Name all prime numbers that are less than 20.

Children use prime numbers to find common denominators and to simplify fractions. This will be discussed in Chapter 6.

THE PRIME NUMBER TEST

It's fairly easy to see that 2, 3, 5, 7, and 11 are primes. But how can you tell if a larger number is prime?

Discussion   **LE 5   Opener**
How would you check to see if 367 is prime?

In the preceding exercise, one would know that 367 is prime if it were not divisible by 2, 3, 4, 5, . . . , 365, 366. However, it is not necessary to try all these numbers as divisors. This procedure can be shortened. Consider the following questions.

LE 6   Reasoning

(a) If 2 is not a divisor of a number (such as 367), then what other numbers could not possibly be divisors of the number?

(b) If 3 is not a divisor of a number (such as 367), then what other numbers could not possibly be divisors of the number?

The approach of LE 6 would go as follows for 367.

$2 \nmid 367$. Therefore, 4, 6, 8, 10, . . . , 366 are not divisors of 367. (The Divisibility-of-a-Product Theorem says that 4, 6, 8, . . . cannot be divisors unless 2 is a divisor.)

$3 \nmid 367$. Therefore, 6, 9, 12, . . . , 366 are not divisors of 367. The divisor 4 is already eliminated.

$5 \nmid 367$. Therefore, 10, 15, 20, 25, . . . , 365 are not divisors of 367. The divisor 6 is already eliminated.

$7 \nmid 367$. Therefore, 14, 21, 28, 35, . . . , 364 are not divisors of 367. The divisors 8, 9, and 10 are already eliminated.

What numbers do we have to check as divisors? Only primes. If *they* are not divisors, other whole numbers greater than 1 (which are all multiples of primes) could not be divisors either.

Do we need to check all primes less than 367? No. There's an additional shortcut. It results from the fact that every composite number x has a prime factor less than or equal to $\sqrt{x}$. Recall that $\sqrt{x}$ means the nonnegative or **principal square root** of x, which is a number $y \geq 0$ such that $y^2 = x$.

We need to check only primes less than or equal to $\sqrt{x}$. Consider 360. Any larger factor of 360, such as 180, has a smaller companion factor (in this case, 2).

$$360 = 180 \cdot 2$$
$$= 120 \cdot 3$$
$$= 90 \cdot 4$$
$$= 72 \cdot 5$$
$$= 60 \cdot 6$$
$$= 40 \cdot 9$$
$$= 36 \cdot 10$$
$$= 30 \cdot 12$$
$$= 24 \cdot 15$$
$$= 20 \cdot 18$$

In fact, any factor larger than $\sqrt{360} \approx 19$ has a companion factor smaller than $\sqrt{360} \approx 19$. So if a number does not have a factor less than or equal to its square root, then it is prime. Putting the two ideas together, one needs to check only prime numbers less than or equal to the square root of the number.

The Prime Number Test

To determine if a number N is prime, find out if any prime number less than or equal to $\sqrt{N}$ is a divisor of N.

• EXAMPLE 1

You want to determine if 367 is prime.

(a) What is the minimum set of numbers you must try as divisors?
(b) Use the Prime Number Test to determine if 367 is prime.

SOLUTION

(a) A calculator shows that $\sqrt{367} \approx 19.2$. So we need to test all prime numbers less than 19.2 as divisors. This would include 2, 3, 5, 7, 11, 13, 17, and 19.
(b) By divisibility tests, $2 \nmid 367$, $3 \nmid 367$, and $5 \nmid 367$.

$$367 \div 7 = 52 \text{ R3, so } 7 \times 367$$
$$367 \div 11 = 33 \text{ R4, so } 11 \times 367$$
$$367 \div 13 = 28 \text{ R3, so } 13 \times 367$$
$$367 \div 15 = 21 \text{ R10, so } 17 \times 367$$
$$367 \div 17 = 19 \text{ R6, so } 19 \times 367$$

(You could also check the divisors 11, 13, 17, and 19 with a calculator.)
367 is not divisible by any prime number $\leq \sqrt{367}$. This means that 367 cannot have any prime factors less than 367. Therefore, 367 is a prime number.   ●

LE 7   Skill

Discussion

Use the Prime Number Test to determine if the following numbers are prime.

(a) 347   (b) 253

PRIME FACTORIZATION OF COMPOSITE NUMBERS

Is it possible to factor any composite number using only prime numbers {2, 3, 5, 7, 11, 13, 17, 19, 23, . . . }? For example, $8 = 2 \cdot 2 \cdot 2$ and $264 = 2 \cdot 2 \cdot 2 \cdot 3 \cdot 11$. In other words, is it possible to write any composite number as a product of prime numbers?

LE 8   Opener

Discussion

(a) Select a composite number. Can you write it as a product of primes?
(b) Can you find a different set of prime factors for this same number?
(c) Repeat parts (a) and (b) for two other composite numbers.
(d) What generalization can you make from parts (a)–(c)?
(e) Part (d) is an example of _____ reasoning.

After studying many examples, Greek mathematicians conjectured that every composite number could be written as a product of a unique set of prime numbers. Euclid, a mathematics professor at the University of Alexandria, proved this in Book 9 of his *Elements*.

The prime numbers are the building blocks for all counting numbers greater than 1. Every whole number greater than 1 is prime or can be expressed using prime factors. The prime factorization is like a signature for each composite number. For example, $12 = 2 \cdot 2 \cdot 3 = 2^2 \cdot 3$, and $26 = 2 \cdot 13$.

It is customary to write the prime factors in increasing order and to use exponents as a shorthand for repeated factors. The possibility of writing any composite number as a unique product of numbers is so important that it is called the Fundamental Theorem of Arithmetic.

The Fundamental Theorem of Arithmetic

Every composite number has exactly one prime factorization.

Two methods are commonly used to find prime factorizations. Most middle-school textbooks teach the **factor-tree method.** The other method I call the "prime-divisor method."

• EXAMPLE 2

Find the prime factorization of 84 using both the factor-tree and prime-divisor methods.

SOLUTION

The Factor-Tree Method Start with 84 and find any two factors (for example, 2 and 42).

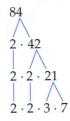

Then continue finding prime factors of these factors until all the factors are prime.

The result, $2 \cdot 2 \cdot 3 \cdot 7$, is the prime factorization of 84. Repeated factors are usually written with exponents.

$$84 = 2^2 \cdot 3 \cdot 7$$

The Prime-Divisor Method In this method, try all prime numbers in increasing order as divisors, beginning with 2. Use each prime number as a divisor as many times as possible.

Try 2.	$\longrightarrow$	$2\lfloor\overline{84}$
It works twice.	$\longrightarrow$	$2\lfloor\overline{42}$
Next, try 3.	$\longrightarrow$	$3\lfloor\overline{21}$
7 is prime.		7

The divisors and final quotient give the prime factorization of 84.

$$84 = 2 \cdot 2 \cdot 3 \cdot 7$$

Try the following exercise using both methods from Example 2.

LE 9 Skill

Find the prime factorization of 120 using both methods.

Classroom Connection

LE 10 Communication

A fifth grader finds the prime factorization of 12 as follows.

$$12 = 2 \cdot 3$$

What would you tell the child to help him?

FAMOUS UNSOLVED PROBLEMS

Some mathematics problems have remained unsolved for centuries. Since number theory deals only with numbers, its unsolved problems are the simplest to describe and the most well known. Do you want to be famous? Just solve one of these problems!

One famous unsolved problem concerns twin primes. **Twin primes** are any two consecutive odd numbers, such as 3 and 5, that are prime.

LE 11 Concept

(a) Find all twin primes less than 100.
(b) Look at the number between each pair of twin primes greater than 3. What property do all these numbers have?
(c) Do you think there is an infinite number of twin primes?

Euclid proved that there is an infinite number of primes. Number theorists believe that there is also an infinite number of twin primes, but no one has been able to prove it!

Another famous unsolved problem is Goldbach's conjecture. Christian Goldbach (1690–1764) said that every even number greater than 2 is the sum of two primes. No one knows for sure whether this is true or false.

LE 12 Concept

Show that Goldbach's conjecture is true for the following numbers.

(a) 8 (b) 22 (c) 120

LE 13 Reasoning

Goldbach's conjecture has been checked for all even numbers up to 100 million (using a computer), and it works! Why isn't this a sufficient proof?

LE 14 Summary

Tell what you learned about prime and composite numbers in this section.

ANSWERS TO SELECTED LESSON EXERCISES

2. (a) One

3. (e) A prime number can be represented by exactly two different rectangles with lengths that are counting numbers.
(f) Induction

4. 2, 3, 5, 7, 11, 13, 17, 19

6. (a) Multiples of 2   (b) Multiples of 3

7. (a) Yes   (b) No, divisible by 11

8. (a) Yes   (b) No   (e) inductive

9. $120 = 2^3 \cdot 3 \cdot 5$

10. Does 12 equal $2 \cdot 3$? What number did the $2 \cdot 3$ come from in your factor tree? See if you can correct your work.

11. (a) 3 and 5, 5 and 7, 11 and 13, 17 and 19, 29 and 31, 41 and 43, 59 and 61, 71 and 73
(b) They are all divisible by 6.

12. (a) $8 = 3 + 5$ (b) $22 = 11 + 11$
(c) $120 = 47 + 73$ or $7 + 113$ or $11 + 109$ or $13 + 107$ or $19 + 101$, and so on

13. This is inductive reasoning. Some even number greater than 100 million may turn out to be a counterexample.

4.3 HOMEWORK EXERCISES

Basic Exercises

1. Classify the following numbers as prime, composite, or neither, and draw all possible rectangles with counting-number dimensions for each number.
(a) 11   (b) 10   (c) 1

2. Classify the following numbers as prime, composite, or neither and draw all possible rectangles with counting-number dimensions for each number.
(a) 14   (b) 13

3. One can often obtain a prime number by performing the following steps. Begin with each number from 1 through 10, and see how many prime numbers you end up with.
Step 1. Choose a counting number.
Step 2. Multiply it by the next highest counting number.
Step 3. Add 17 to your result.

4. The formula $2^n - 1$ often produces prime numbers when n is a counting number. In 2001, Cameron, Woltman, and Kurowski used a computer to show that $2^{13,466,917} - 1$ is prime.
(a) Compute $2^n - 1$ for $n = 2, 3, 4, 5,$ and 6.
(b) Which of your resulting numbers are prime?

5. Many mathematicians have tried to find formulas that produce only prime numbers. Fermat conjectured that $2^{2^n} + 1$ would be prime for $n = 0, 1, 2, 3, \ldots$. The resulting numbers, called **Fermat numbers,** are $2^{2^0} + 1 = 3$, $2^{2^1} + 1 = 5$, $2^{2^2} + 1 = 17$, and so on.
(a) The first five Fermat numbers are prime. Find the fourth Fermat number.
(b) The sixth Fermat number is a composite number divisible by 641. Find this number, and show that it is divisible by 641.

6.
$$2 \qquad\quad + 1 = 3$$
$$2 \times 3 \quad\ \, + 1 = 7$$
$$2 \times 3 \times 5 + 1 = 31$$

(a) Write the next two equations that continue the pattern.
(b) The numbers 3, 7, and 31 are prime. Are the numbers on the right sides of your equations in part (a) prime?
(c) Make a generalization based on the results of parts (a) and (b).
(d) Does part (c) involve induction or deduction?
(e) Show that the next equation that continues the pattern results in a composite number divisible by a prime number between 50 and 60.

7. (a) Choose any prime number greater than 3. Square it. Add 17. Find the remainder that results when you divide by 12.
  (b) Repeat part (a) for two other prime numbers greater than 3.
  (c) Make a table of numbers from 1 to 102 with six numbers in each row. Circle all prime numbers. Where are all prime numbers other than 3 located?
  (d) All prime numbers greater than 3 have the form $6n + 1$ or $6n - 1$, in which n is a counting number. Show why the procedure in part (a) always results in a remainder of 6.

8. The number 5 can be written as $1 + 2^2$. Find three other prime numbers that have the form $1 + W^2$ for some whole number W.

9. Without computing the results, explain why each of the following problems will result in a composite number.
  (a) $3 \times 5 \times 7 \times 11 \times 13$
  (b) $(3 \times 4 \times 5 \times 6 \times 7 \times 8) + 2$
  (c) $(3 \times 4 \times 5 \times 6 \times 7 \times 8) + 5$

10. In 1978, Hugh Williams discovered the following prime number with 317 ones. $11,111,111, \ldots ,111$. Explain why this number is not divisible by 2, 3, or 5.

11. To determine if 431 is prime, what is the minimum set of numbers you must *try* as divisors?

12. To determine if 817 is prime, what is the minimum set of numbers you must *try* as divisors?

13. Use the Prime Number Test to classify the following numbers as prime or composite.
  (a) 71   **(b)** 697

14. Use the Prime Number Test to classify the following numbers as prime or composite.
  (a) 91   **(b)** 577

 15. The following numbers are prime.

31   331   3331   33,331   333,331
3,333,331   33,333,331

  (a) Write a generalization of this pattern.
  (b) Does part (a) involve inductive or deductive reasoning?
  (c) Use the Prime Number Test to classify 333,333,331 as prime or composite.

 16. Use the Prime Number Test to classify 1,523 as prime or composite.

17. According to the Fundamental Theorem of Arithmetic, every composite number has

_____.

18. Consider a property similar to the Fundamental Theorem of Arithmetic. Show that every composite number greater than 1 cannot be *uniquely* expressed as a sum of prime numbers.

19. Find the prime factorization of each number using both methods.
  (a) 495   **(b)** 320

20. Write the prime factorizations of the following numbers.
  (a) 90   **(b)** 3155

21. (a) How many different divisors does $2^5 \cdot 3^2 \cdot 7$ have?
  (b) Show how to use the prime factorization to determine how many different factors 148 has.

22. (a) How many different factors does 120 have?
  (b) List all the divisors of 84. Use the prime factorization of 84 to confirm that your list is complete.

23. Find all twin primes between 101 and 140.

24. Triplet primes are any three consecutive odd numbers that are prime.
  (a) Find a set of triplet primes.
  (b) Why can't there be any other sets of triplet primes?

25. Show that Goldbach's conjecture is true for the following numbers.
  (a) 12   **(b)** 30   **(c)** 108

26. In how many ways can 28 be expressed as the sum of two prime numbers?

Extension Exercises

27. Complete the following table for counting numbers from 1 to 25. The numbers 1 through 6 have already been placed in the appropriate columns for their numbers of divisors.

Total Number of Divisors for Counting Numbers									
1	2	3	4	5	6	7	8	9	10
1	2	4	6						
	3								
	5								

Describe any pattern you see in the numbers in any particular column.

28. (a) The number $162 = 2 \cdot 3^4$. First, find all the divisors of 2.
    (b) Next, find all the divisors of 3^4.
    (c) How many different divisors does 162, or $2 \cdot 3^4$, have?
    (d) Try the same process for $225 = 3^2 \cdot 5^2$. First, find all divisors of 3^2.
    (e) Next, find all divisors of 5^2.
    (f) How many different divisors does 225, or $3^2 \cdot 5^2$, have?
    (g) Based on your results in parts (a)–(f), if p and q are prime, how many different divisors does $p^m \cdot q^n$ have?

29. Here is a way to find five consecutive composite numbers. First, compute $2 \times 3 \times 4 \times 5 \times 6 = 720$. Then, using the Divisibility-of-a-Sum Theorem, fill in the blanks.
    (a) $2 \mid 720$ and $2 \mid 2$, so _____.
    (b) $3 \mid 720$ and $3 \mid 3$, so _____.
    (c) $4 \mid 720$ and _____.
    (d) Complete the pattern of parts (a), (b), and(c).

30. Use the method of the preceding exercise to find 10 consecutive composite numbers.

31. The mathematician Sophie Germain (1776–1831) lived in a time when women were not permitted to attend French universities. She learned advanced mathematics by collecting lecture notes from the professors at the École Polytechnique in Paris. Germain became fascinated with divisibility problems such as the following.

    Suppose we want $x^3 \div p$ to have a remainder of 2 for a counting number x and an odd prime p. For example, $2^3 \div 3$ has a remainder of 2. If possible, in parts (a)–(c), find an x such that:
    (a) $x^3 \div 5$ has a remainder of 2.
    (b) $x^3 \div 7$ has a remainder of 2.
    (c) $x^3 \div 11$ has a remainder of 2.
    (d) What pattern do you see in the values of x?

Technology Exercise

32. Some calculators such as the TI-73 find prime factorizations of numbers. To find the prime factorization of 18, enter the fraction $\frac{18}{18}$. Now press [Simp]. The first prime factor 2 and a simplified fraction will appear: fac = 2 $\frac{9}{9}$. Continue pressing [Simp] until the fraction is simplified to $\frac{1}{1}$. Use the same method to find the prime factorization of 2,233.

Enrichment Topic

33. Eratosthenes, a Greek mathematician, developed the Sieve of Eratosthenes about 2200 years ago as a method for finding all prime numbers less than a given number. Follow the directions to find all prime numbers less than or equal to 50.

1	2	3	4	5	6
7	8	9	10	11	12
13	14	15	16	17	18
19	20	21	22	23	24
25	26	27	28	29	30
31	32	33	34	35	36
37	38	39	40	41	42
43	44	45	46	47	48
49	50				

(a) Copy the list of numbers.
(b) Cross out 1 because 1 is not prime.
(c) Circle 2. Count by 2s from there, and cross out 4, 6, 8, . . . , 50 because all these numbers are divisible by 2 and therefore are not prime.
(d) Circle 3. Count by 3s from there, and cross out all numbers not already crossed out because these numbers are divisible by 3.
(e) Circle the smallest number not yet crossed out. Count by that number, and cross out all numbers that are not already crossed out.
(f) Repeat part (e) until there are no more numbers to circle. The circled numbers are all the prime numbers.
(g) List all the primes between 1 and 50.

34. At Hexagon High, the students play mathematical pranks. There are exactly 400 lockers, numbered 1 to 400. One day, the 400 students filed into school one by one. The first student opened every locker. The second student closed every even-numbered locker. The third student changed (opened or closed) every third locker (3, 6, 9, . . .). The fourth student changed every fourth locker, and so on. After all the students were finished, which lockers were open?

4.4   COMMON FACTORS AND MULTIPLES

NCTM Standards

- use factors, multiples, prime factorization, and relatively prime numbers to solve problems (6–8)
- recognize and use connections among mathematical ideas (pre-K–12)

After studying factors and multiples of individual whole numbers, it is natural to examine the factors and multiples that two whole numbers have in common. With the exception of 0, any two whole numbers have a largest common factor and a smallest nonzero common multiple.

THE GREATEST COMMON FACTOR (GCF)

You can use greatest common factors to simplify both numerical and algebraic fractions.

LE 1   Opener

(a) What methods do you know for finding the greatest common factor of 20 and 30? If you don't know any, continue on to part (b).

(b) What are the common factors of 20 and 30?

(c) What is the greatest common factor of 20 and 30?

(d) Fill in the factors of 20 and 30 in the appropriate regions in the Venn diagram and confirm your answer for the greatest common factor of 20 and 30.

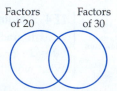

Factors of 20   Factors of 30

The **greatest common factor (GCF)** of counting numbers is the largest number that is a factor of both numbers. For example, 20 and 30 have common factors of 1, 2, 5, and 10. So the greatest common factor of 20 and 30 is 10, or GCF(20, 30) = 10. Note that the definition of the greatest common factor excludes 0. The following exercise concerns what would happen if 0 were included in the definition.

LE 2   Concept

The definition of GCF is limited to counting numbers. Suppose 0 were included in the definition. What would the greatest common factor of 0 and any counting number be?

What follows are two methods for finding the greatest common factor: the factor-list method and the prime-factorization method. (A third method, the Euclidean algorithm, is presented later in the section.)

• EXAMPLE 1

Find the greatest common factor of 60 and 140.

SOLUTION

Two methods are as follows.

The Factor-List Method	*The Prime-Factorization Method*
1. List all factors.	**1.** Write the prime factorizations.
60: 1, 2, 3, 4, 5, 6, 10, 12, 15, 20, 30, 60	$60 = 2 \cdot 2 \cdot 3 \cdot 5$ $140 = 2 \cdot 2 \cdot 5 \cdot 7$
140: 1, 2, 4, 5, 7, 10, 14, 20, 28, 35, 70, 140	(*Note:* Since 2, 2, and 5 are on both lists, 2 is a common factor, $2 \cdot 2$ is a common factor, and $2 \cdot 5$ is a common factor, but we want the greatest common factor.)
2. List all common factors.	**2.** Multiply all common prime factors.
1, 2, 4, 5, 10, 20	$2 \cdot 2 \cdot 5$
3. So GCF(60, 140) = 20.	**3.** So GCF(60, 140) = 20.

The factor-list method is less abstract, so children can more easily understand why it works. For this reason, it is the first method children study in school. Although both methods work well for small numbers, the prime-factorization method is usually superior for larger numbers that have many factors.

LE 3   Skill

Compute the GCF of 546 and 650 using whichever method you think is easier.

Classroom Connection

LE 4   Communication

A sixth grader works out GCF(60, 140) as follows.
$60 = 2 \cdot 2 \cdot 3 \cdot 5$ and $140 = 2 \cdot 2 \cdot 5 \cdot 7$. The GCF is 5.
What would you tell the child to help her?

When two counting numbers have no common factors other than 1, they are said to be relatively prime. If the numerator and denominator of a fraction are relatively prime, the fraction is in simplest form. For example, 7/10 is in simplest form because 7 and 10 are relatively prime. In general, A and B are **relatively prime** if and only if GCF(A, B) = 1. For example, 7 and 10 are relatively prime because GCF(10, 7) = 1. Note that A and B need not be prime numbers to be relatively prime.

LE 5 Skill

Which of the following pairs of numbers are relatively prime?

(a) 8 and 6   (b) 8 and 25   (c) 13 and 21

LE 6 Reasoning

(a) If A and B are two different prime numbers, are A and B relatively prime? Try some examples and make an educated guess.
(b) Your guess in part (a) is based on _____ reasoning.

EUCLIDEAN ALGORITHM FOR FINDING THE GCF

Suppose you want to find the GCF of two numbers that are difficult to factor, such as 374 and 1173. There is a third method for finding the GCF that is more efficient.

LE 7 Opener

(a) According to the Divisibility-of-a-Difference Theorem from Section 4.1, if $x \mid 374$ and $x \mid 1173$, then _____.
(b) If $y \mid 374$ and $y \mid 1173$, then _____.

The preceding exercise suggests a method of simplifying a GCF problem. Any number (such as x or y in the preceding exercise) that is a common factor of 374 and 1173 is also a common factor of 374 and 799. If x is GCF(374, 1173), then x is also GCF(374, 799).

Continuing this way, x is GCF(374, 799 − 374), or GCF(374, 425). Then x is also GCF(374, 425 − 374), or GCF(374, 51). This simplifies the original problem from GCF(374, 1173) to GCF(374, 51).

LE 8 Concept

51 is simply the remainder after you select as many 374s as possible from 1173. How can you find 51 from 374 and 1173 more quickly?

The method of simplifying GCF(374, 1173) with division is called the Euclidean algorithm. Once we reach GCF(374, 51), we repeat the process on those two numbers. Here is the method from start to finish.

• EXAMPLE 2

Find GCF(374, 1173) using the Euclidean algorithm.

SOLUTION

$$\text{GCF}(374, 1173) \rightarrow 374 \overline{)1173}^{3} \\ \underline{1122} \\ 51$$

$$\text{GCF}(374, 51) \rightarrow 51 \overline{)374}^{7} \\ \underline{357} \\ 17$$

$$\text{GCF}(17, 51) \rightarrow 17 \overline{)51}^{3} \\ \underline{51} \\ 0$$

This shows that GCF(17, 51) = 17. Therefore, GCF(374, 1173) = 17.

Try one yourself.

LE 9   Skill
Use the Euclidean algorithm to find GCF(253, 322).

THE LEAST COMMON MULTIPLE (LCM)

Least common multiples of denominators are the least common denominators used to add and subtract fractions. Note that 40, 80, 120, 160, . . . are all common nonzero multiples of 10 and 8. Any two counting numbers have an infinite set of common multiples.

LE 10   Opener
(a) What methods do you know for finding the least common multiple of 8 and 10? If you don't know any, continue on to part (b).
(b) Name three common multiples of 8 and 10.
(c) What is the least common multiple of 8 and 10?

The **least common multiple (LCM)** of two counting numbers is the smallest counting number that is a multiple of both numbers. For example, the nonzero multiples of 10 are 10, 20, 30, 40, 50, . . . , and the nonzero multiples of 8 are 8, 16, 24, 32, 40, 48, ... So the least common multiple (LCM) of 10 and 8 is 40, or LCM(10, 8) = 40.

LE 11   Reasoning
(a) Is 18 a common multiple of 3 and 6?
(b) Is 40 a common multiple of 5 and 8?
(c) On the basis of parts (a) and (b), what number would you expect to be a common multiple of counting numbers M and N?
(d) What kind of reasoning did you use in part (c)?
(e) Use the definition of a multiple to prove that your answer to part (c) is true.

LE 11 verifies the following theorem.

> **Products as Common Multiples**
>
> MN is a common multiple of M and N for all counting numbers M and N.

Computing the product of two counting numbers is an easy way to find a common multiple. However, the result may not be the *least* common multiple. Two methods for finding the LCM are the multiple-list method and the prime-factorization method. Both methods are used in the following example to find the LCM of 10 and 8.

• EXAMPLE 3
Find the LCM of 10 and 8.

SOLUTION
Two methods are as follows.

The Multiple-List Method

1. List the multiples of each number.

    10:   10, 20, 30, 40, 50, . . .

    8:    8, 16, 24, 32, 40, 48, . . .

2. Find the first common multiple.

    10, 20, 30, 40, 50, . . .

    8, 16, 32, 40, 48, . . .

3. So the LCM = 40.

The Prime-Factorization Method

1. Write the prime factorizations.

$$10 = 2 \cdot 5$$
$$8 = 2 \cdot 2 \cdot 2$$

2. Multiply all prime numbers that appear in either factorization. Use each prime number the greatest number of times it appears in either factorization. The LCM of 10 and 8 must contain $2 \cdot 5$ (which is 10) and $2 \cdot 2 \cdot 2$ (which is 8). So it is $2 \cdot 5 \cdot$ _____. For the LCM to contain 8, we need two more 2s as factors. $2 \cdot 5 \cdot \underline{2 \cdot 2} = 40$.

3. So the LCM = 40. (*Note:* $2 \cdot 2 \cdot 2 \cdot 5$ contains $2 \cdot 2 \cdot 2$, or 8, and $2 \cdot 5$, or 10, so it is a multiple of both 8 and 10. ●

The multiple-list method is less abstract, so children can more easily understand why it works. For this reason, it is the first method taught in school. Although both methods work well for smaller numbers, the prime-factorization method is usually superior for larger LCMs.

Discussion

LE 12   Skill
Compute the LCM of 120 and 72 using both methods.

LE 13   Skill
Compute the LCM of 62 and 80 using either method.

AN INVESTIGATION: LEAST COMMON MULTIPLES

LE 14   Reasoning

Consider the following problem. "When is *MN* the least common multiple of counting numbers *M* and *N*?"

(a) Restate the question in your own words.
(b) Devise a plan.
(c) Solve the problem.

LE 15   Summary
(a) What are the different methods for finding the GCF?
(b) What are the different methods for finding the LCM?

ANSWERS TO SELECTED LESSON EXERCISES

1. (d)

Factors of 20   Factors of 30

4, 20 | 1, 2, 5, 10 | 3, 6, 15, 30

2. The counting number

3. 26

4. The GCF is the largest number that is a factor of both 60 and 140. What numbers besides 5 are common factors of 60 and 140? Is 2 · 5 a common factor? What is the greatest common factor?

5. (b), (c)

6. (a) Yes   (b) inductive

7. (a) $x \mid 799$ (b) $y \mid 799$

8. Find the remainder for 1173 ÷ 374.

9. 23

11. (a) Yes   (b) Yes   (c) *MN*
    (d) Inductive
    (e) *MN* is a whole number times *M*, so it is a multiple of *M*.
       MN is a whole number times *N*, so it is a multiple of *N*.

12. 360

13. 2,480

4.4   HOMEWORK EXERCISES

Basic Exercises

1. (a) Draw a Venn diagram showing the factors and common factors of 10 and 24.
    (b) What is the greatest common factor of 10 and 24?

2. (a) Draw a Venn diagram showing the factors and common factors of 15 and 21.
    (b) What is the greatest common factor of 15 and 21?

3. (a) Compute the GCF of 42 and 120 using factor lists.
    (b) Draw a Venn diagram showing the factors and common factors of 42 and 120.
    (c) Compute the GCF of 42 and 120 using prime factorizations.

4. (a) Compute the GCF of 45 and 130 using factor lists.
    (b) Draw a Venn diagram showing the factors and common factors of 45 and 130.
    (c) Compute the GCF of 45 and 130 using prime factorizations.

5. (a) Find the GCF of 4 with each of the following: 5, 6, 7, 8, 9, 10, 11, and 12.
(b) Write a generalization of your results in part (a).
(c) Does part (b) involve induction or deduction?

6. If A is a counting number, then GCF(A, A^2) = _____.

7. $a = 3^2 \cdot 5 \cdot 11^4$, and $b = 3 \cdot 5^4 \cdot 7 \cdot 11$. Why isn't 11 the GCF?

8. $a = 5 \cdot 7 \cdot 11^3$, and $b = 2^3 \cdot 5^2 \cdot 7 \cdot 11$. What is GCF($a$, b)?

9. $a = 2^2 \cdot 3^4 \cdot 5^7$, and $b = 2^3 \cdot 3 \cdot 5^6$. What is GCF($a$, b)? (You may write your answer as a prime factorization.)

10. $a = 2 \cdot 3^2 \cdot 7^3$, and GCF($a$, b) = $2 \cdot 3^2 \cdot 7$. Give two possible values for b.

11. Which of the following pairs of numbers are relatively prime?
(a) 11 and 12   (b) 34 and 51
(c) 157 and 46

12. Which of the following pairs of numbers are relatively prime?
(a) 28 and 33   (b) 14 and 15
(c) 108 and 333

13. Find the GCF for each of the following, using the Euclidean algorithm.
(a) 627 and 665   (b) 851 and 2035
(c) 551 and 609

14. Find the GCF for each of the following, using the Euclidean algorithm.
(a) 583 and 795   (b) 602 and 1330
(c) 1705 and 527

15. (a) Name three common multiples of 10 and 12.
(b) How many common multiples do 10 and 12 have?
(c) What is the LCM of 10 and 12?

16. Suppose 0 were included as a possible multiple in the definition of the least common multiple. What would be the LCM of any two whole numbers?

17. (a) Find the LCM of 20 and 32 using multiple lists.
(b) Find the LCM of 20 and 32 using prime factorizations.

18. (a) Find the LCM of 14 and 20 using multiple lists.
(b) Find the LCM of 14 and 20 using prime factorizations.

19. Loanne wants to lay three rows of floor tiles of different sizes and colors. One row will contain white tiles, each 8 cm long. One row will contain blue tiles, each 10 cm long. One row will contain gray tiles, each 24 cm long.
(a) At what lengths will the tiles first line up across all three rows?
(b) At what other lengths will the tiles line up?

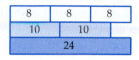

20. Two buses leave the terminal at 8 A.M. Bus 36 takes 60 minutes to complete its route; bus 87 takes 75 minutes. When is the next time the two buses will arrive together at the terminal (if they are on time)?

21. Consider the following problem. "John has a cloth that is 30 in. by 48 in. He wants to cut out the largest possible squares of the same size and use all the material. How big can the squares be?"
(a) Select a problem-solving strategy.
(b) Solve the problem.

22. You have a balance and an unlimited supply of 3-oz and 5-oz weights. It is possible to weigh any whole-number amount of ounces using these weights on one or both sides of the balance!
(a) How would you weigh the following amounts of a substance with the 3- and 5-oz weights?
(1) 13 oz   (2) 2 oz   (3) 1 oz
(b) Find another set of weights (measured in ounces) of two different denominations other than 1 oz that you could use to weigh any whole-number amount of ounces.

23. (a) Is $3^2 \cdot 2^4$ a factor of $3^4 \cdot 2^7$?
(b) Is $3^2 \cdot 2^5$ a factor of $3 \cdot 2^7$?

24. Suppose that x and y are whole numbers.
   (a) Is x^3y^5 a factor of x^2y^8?
   (b) Is x^2y^2 a factor of x^3y^5?

25. $a = 2 \cdot 3^2 \cdot 5^7 \cdot 7 \cdot 11$, and $b = 2^3 \cdot 3^4 \cdot 5 \cdot 11$.
   (a) What is LCM(a, b)?
   (b) What is GCF(a, b)?

26. $a = 2^5 \cdot 7 \cdot 11^3$, and $b = 2^3 \cdot 3 \cdot 7^2 \cdot 11$.
   (a) What is LCM(a, b)?
   (b) What is GCF(a, b)?

27. $a = 2^3 \cdot 5^2 \cdot 7^3$, GCF($a$, b) $= 2 \cdot 5^2 \cdot 7$, and LCM(a, b) $= 2^3 \cdot 3^2 \cdot 5^4 \cdot 7^3$. Find b.

28. Let x, y, and z be counting numbers. Tell whether each of the following is always true, sometimes true, or never true. Give a numerical example or examples to support your answer.
   (a) GCF(x, y) $= x$
   (b) LCM(x, y) $< x$
   (c) If LCM(x, y) $= y$, then GCF(x, y) $=$ ___.

29. In simplifying $\dfrac{30}{45}$, divide the numerator and denominator by the _____ of 30 and 45.
   (a) LCM   (b) GCF   (c) Neither

30. The lowest common denominator for two fractions is the same as the _____ of the denominators.
   (a) LCM   (b) GCF   (c) Neither

31. Find the greatest common factor of $3x^2y$ and $6y^2$, in which x and y are counting numbers.

32. Find three common multiples of x^2 and xy, in which x and y are counting numbers.

In Exercises 33 and 34, A and B are counting numbers. If you think a statement is true, give an example. If a statement is false, give a counterexample.

33. True or false? If A and B are relatively prime, then A is prime and B is prime.

34. True or false? The LCM of two prime numbers is their product.

Extension Exercises

35. You want to pack new "Fiber 'n Wood Chips" cereal boxes standing up in a carton. The cereal boxes and all possible cartons are 10 in. high. The cereal boxes are 8 in. long and 3 in. wide. Which of the following cartons could be used to pack them without any wasted space between boxes?

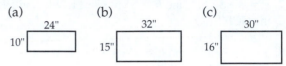

(Top view: All boxes are 10" high.)

36. Suppose that in the preceding exercise, you want to use a box with a square base (since it would use less material). What are the smallest possible dimensions of the base?

37. How are GCF(a, b) and LCM(a, b) related?
   (a) Select three different pairs of counting numbers greater than 2, and fill in the table.

a	b	GCF(a, b)	LCM(a, b)	GCF(a, b) $\cdot$ LCM(a, b)

   (b) Make a generalization based on your results in part (a).

38. (a) Suppose LCM(10, x) $= 120$ and GCF(10, x) $= 2$. Find x.
   (b) LCM(15, x) $= 45$ and GCF(15, x) $= 3$. Find x.

Technology Exercise

39. Some calculators such as the TI-73 compute the LCM or GCF of two numbers. On the TI-73, find LCM(on the MATH menu. To compute LCM (22, 40), complete the expression on the regular screen and press enter. Find GCF(242, 792).

CHAPTER 4 SUMMARY

The classical Greeks were the first to study patterns in the set of counting numbers in detail, breaking down the counting numbers greater than 1 into their basic components. This analysis led to the study of factors (divisors) and multiples (by reversing factor-number relationships). The Divisibility-of-a-Sum Theorem and the Divisibility-of-a-Product Theorem describe the relationships between factors of different numbers. These theorems can be used with expanded notation to show how divisibility tests work.

The crowning achievement of ancient number theory was the discovery of the basic building blocks of counting numbers greater than 1: prime numbers. All composite numbers can be uniquely factored using prime numbers. This result is called the Fundamental Theorem of Arithmetic.

The Greeks also studied the relationship between the factors and multiples of counting numbers. Any two counting numbers have a greatest common factor and a least common (nonzero) multiple. In simplifying fractions, one divides the numerator and denominator by their greatest common factor. To find the lowest common denominator for adding or subtracting two fractions, one can compute the least common multiple of the denominators.

Number theory continues to fascinate mathematicians because it contains seemingly simple conjectures that have remained unproved for hundreds of years. People have tested these conjectures for many examples, but no one has been able to prove these statements.

STUDY GUIDE

To review Chapter 4, see what you know about each of the following ideas or terms that you have studied. You can also use this list to generate your own questions about the chapter.

NUMBER THEORY IN GRADES 1–8

The following chart shows at what grade levels selected topics in number theory typically appear in elementary- and middle-school mathematics textbooks. Underlined numbers indicate grades in which the most time is spent on the given topic.

Topic	Typical Grade Level in Current Textbooks
Divisibility tests	5, <u>6</u>, <u>7</u>
Factors	<u>4</u>, <u>5</u>, 6
Prime and composite numbers	4, <u>5</u>, 6
Multiples	<u>4</u>, <u>5</u>, 6
GCF and LCM	<u>5</u>, <u>6</u>, 7

REVIEW EXERCISES

1. A, B, and C are whole numbers. The number $A \cdot (B + C)$ is even whenever _____ is even. Select all correct choices and tell why they are correct.
  (a) A (b) B (c) C

2. Write a paragraph describing how to construct a proof of a theorem about factors.

3. By the definition of a multiple, N is a multiple of 8 if and only if _____.

4. Fill in the two missing digits so that the number is divisible by 3, 5, and 10.

$$3\,2,0\,0\,5,\underline{}\,2\,\underline{}$$

Determine all possible answers.

5. (a) Use expanded notation and the Divisibility-of-a-Sum Theorem to prove that if a three-digit numeral ends in 0 or 5 ($\underline{A}\ \underline{B}\ \underline{0}$ or $\underline{A}\ \underline{B}\ \underline{5}$), then the number is divisible by 5.
  (b) Each step of the proof in part (a) involves _____ reasoning.

6. Suppose that you want to determine if 577 is prime. Using the Prime Number Test, what is the minimum set of numbers you must try as divisors?

7. Which of the following could describe set A, set B, set C, and set D?

Multiples of 3   Multiples of 10
Divisors of 20   Prime numbers
Odd numbers

(a)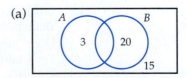

(b)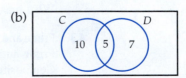

8. Write the prime factorization of 312.

In Exercises 9–11, assume that A, B, and C are counting numbers. If a statement is true, give an example that illustrates it. If a statement is false, give a counterexample.

9. True or false? If $A \mid B$ and $A \nmid C$, then $A \nmid (B + C)$.

10. True or false? If $A \mid B$, then $A \mid B^2$.

11. True or false? The GCF of any two prime numbers is 1.

In Exercises 12 and 13, if the statement is true, prove it. If the statement is false, give a counterexample.

12. X, Y, and Z are whole numbers, with $X \neq 0$ and $Y \neq 0$. True or false? If $XY \mid Z$, then $XY \mid (XY + Z)$.

13. True or false? The LCM of any prime number and any composite number is their product.

14. (a) Compute the GCF of 35 and 56 using factor lists.

(b) Draw a Venn diagram showing the factors and common factors of 35 and 56.
(c) Compute the GCF of 35 and 56 using prime factorizations.

15. The numbers A and 10 are relatively prime. Give two possible values of A.

16. $a = 2 \cdot 5^2 \cdot 11^3$, and $b = 2^3 \cdot 3 \cdot 5^4 \cdot 11^2$.
(a) What is GCF(a, b)?
(b) What is LCM(a, b)?

17. LCM(24, x) = 168 and GCF(24, x) = 2. Find x.

18. To simplify $\frac{28}{40}$, you divide the numerator and denominator by _____, which is the _____ of 28 and 40.

19. Find the smallest number that is divisible by 5, 6, 7, and 8.

ALTERNATE ASSESSMENT: SELF-ASSESSMENT

Evaluate your work in class so far. How well do you understand the main ideas of the course? How well have you done on homework and tests? Also rate your effort in class, your effort outside of class, and how well you work with others in the class. Or you could add to your portfolio, add to your journal, or write a unit test.

SUGGESTED READINGS

Brown, S. *Some Prime Comparisons.* Reston, VA: NCTM, 1979.

Burton, D. *The History of Mathematics.* 5th ed. Columbus, OH: McGraw-Hill, 2003.

Mathematics Teaching in the Middle School. April, 2002 Focus Issue on the Magic of Numbers. Reston, VA: NCTM, 2002.

National Council of Teachers of Mathematics. *Historical Topics for the Mathematics Classroom.* Reston, VA: NCTM, 1989.

Use Infotrac to find articles in *Teaching Children Mathematics* on teaching children about factors, multiples, primes, and divisibility.

5 | INTEGERS

As far back as 200 B.C., Chinese accountants used black (negative) rods for debts and red (positive) rods for credits. Around A.D. 900, the Hindus invented negative numbers to solve equations such as $x + 5 = 2$. But it wasn't until the 1500s that Giralumo Cardano (1501–1576) gave the first detailed description of negative numbers and their properties.

Cardano was a physician who did mathematics and astrology in his spare time. After publishing a horoscope of the life of Jesus Christ, Cardano was imprisoned for heresy. In spite of this scandal, he was later appointed as an astrologer to the papal court! Some sources claim that Cardano killed himself in 1576 so that his prediction of the date of his death would be correct.

Not everyone accepted Cardano's negative numbers. René Descartes called them "false" numbers. However, by the eighteenth century, negative numbers were widely used.

Today, negative numbers retain their importance in accounting. If you have $80 and spend $100, you have a net worth of $-\$20$. People also use negative numbers to measure temperatures, golf scores, and changes in stock prices.

In mathematics, the set of integers results from enlarging the set of whole numbers to create solutions for every whole-number subtraction problem (for example, $6 - 10$). The set of integers retains many properties and patterns of whole-number operations.

In this course, negative whole numbers (integers) are introduced before fractions and decimals to illustrate more clearly the connections among different sets of numbers in elementary school. However, in elementary-school mathematics, most fraction and decimal topics precede most topics involving negative numbers.

5.1 ADDITION AND SUBTRACTION OF INTEGERS

NCTM Standards

- explore numbers less than 0 by extending the number line and through familiar applications (3–5)

- develop meaning for integers and represent and compare quantities with them (6–8)

- develop and analyze algorithms for computing with fractions, decimals, and integers, and develop fluency in their use (6–8)

 LE 1 Opener

(a) Give some examples of everyday uses of negative numbers.
(b) What mathematical problems necessitated the development of negative numbers?

First, consider the mathematical problems. When you add or multiply any two whole numbers, the result is a whole number. However, some whole-number subtraction problems, such as 2 − 5, do not have whole-number answers. These subtraction problems provided the mathematical motivation for the creation of negative integers.

To compute 2 − 5 on a number line, we would start at 2 and move 5 to the left. The whole-number line goes only 2 units to the left of 2, to reach 0. We mark units to the left of 0 and create negative integers.

$$-3 \quad -2 \quad -1 \quad 0 \quad 1 \quad 2 \quad 3$$

For example, 3 is 3 units to the right of 0, and "negative 3," written −3, is 3 units to the left of 0. For each positive number to the right of 0 on a number line, there is a corresponding negative number the same distance to the left of 0. The set $\{-1, -2, -3, \ldots\}$ is called the set of **negative integers.**

By combining the set of whole numbers with the set of negative integers, one obtains the set of integers.

Definition: Integers

The union of the set of whole numbers and the set of negative integers is called the set of **integers.** The set of integers is denoted by $I = \{ \ldots -3, -2, -1, 0, 1, 2, 3, \ldots \}$.

The integers are an extension of the whole numbers. With whole numbers, we can only count forward from 0, or move to the right from 0 on a number line. With integers, we can count forward or backward from 0 and move to the right or left from 0 on a number line. A greater number is always to the right of a lesser number on the standard integer number line. For integers m and n, $m > n$ if and only if m is located to the right of n on the number line.

 LE 2 Reasoning

What number is 2 units to the left of −130 on the standard number line?

Using numbers greater than or equal to 0 is sufficient for measurements such as weight, area, and the number of raisins in a bowl of cereal. Such measurements have a lower limit of 0. In some measurement scales, such as temperature, the scale extends above *and* below 0.

Figure 5–1 illustrates common uses of negative numbers. Investors do not like negative numbers, but golfers love them!

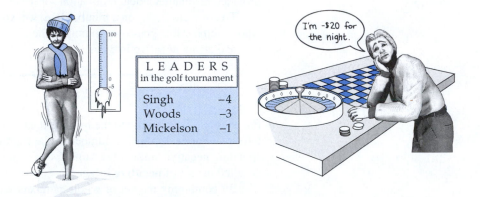

Figure 5–1

 LE 3 Opener

Tell in words what the − sign means in

(a) a temperature of −5°C.
(b) a golf score of −4.
(c) an account balance of −$400.
(d) a change in a stock price of −3 points.

As a measure, −5 (called "negative 5") may represent an amount shown by a point on a number line. It may also represent a change (decrease of 5) shown by a move of 5 to the left.

Using a set model, -5 can be represented by 5 signed (white) counters.

The number -5 could represent a debt of $5. Positive numbers such as 2 would be represented by differently colored positive counters.

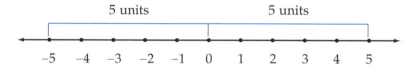

Many middle-school textbooks use yellow signed counters for positives and red signed counters for negatives.

LE 4 Skill

Represent -2 in three ways: as a point on a number line, as a move on a number line, and as a set of signed counters.

The number -5 is also "the opposite of 5." A number and its **opposite** (for example, 5 and -5) are the same distance from 0, but on opposite sides.

The opposite of -5, written $-(-5)$, is 5. If x is an integer, x can be positive, negative, or zero; consequently, $-x$ can be negative, positive, or zero. Thus, calling $-x$ "the opposite of x" instead of "negative x" may be less confusing. The use of the $-$ sign to indicate "the opposite of" is one of the three uses of the minus sign: as a subtraction symbol, as a negative sign, and as an opposite sign.

The distance between 0 and an integer on the integer number line is called the **absolute value** of x and written $|x|$. The absolute values of both 5 and -5 are 5. In symbols,

$$|5| = 5 \qquad \text{and} \qquad |-5| = 5$$

LE 5 Communication

Using the word "distance," explain why $|-13| = |13|$.

Integer arithmetic is more abstract than whole-number arithmetic. People are sometimes surprised by the answers to integer arithmetic problems. The remainder of this lesson shows how to explain the results of integer addition and subtraction.

INTEGER ADDITION

If you are $3 in debt and you receive a bill for $2, what is your new net worth? If a football player loses 3 yards on one play and 2 yards on the next, what is the player's overall yardage on the two plays? Both these questions are applications of $-3 + (-2)$. (*Note:* Any time a signed number is written after an operation symbol, parentheses are placed around that signed number.)

The first question is best modeled by a set of signed counters. The second question is best modeled by a number line. Example 1 shows how to compute $-3 + (-2)$ with signed counters or a number line.

• EXAMPLE 1

Explain how to compute $-3 + (-2)$ with

(a) signed counters.
(b) a number line.

SOLUTION

(a) Show -3 as 3 negative counters and -2 as 2 negative counters.

Combine these counters to obtain 5 negative counters, which would represent -5. So $-3 + (-2) = -5$.
(b) Go to -3. Add the second number, -2, by moving 2 units to the left from -3.

You end up at -5. So $-3 + (-2) = -5$.   ●

Next, consider some examples involving a positive number and a negative number. What will happen when you add a positive counter and a negative counter? The sum is 0. In other words, 1 positive counter cancels out 1 negative counter just as $1 cancels out a debt of $1. These two counters form a **zero pair.**

The simplest addition of a positive integer and a negative integer has the form $a + (-a)$.

LE 6   Communication

(a) Explain how to compute $3 + (-3)$ with a number line.
(b) Explain how to compute $3 + (-3)$ with signed counters.
(c) The general result for a whole number a is $a + (-a) =$ _____.

Next, consider addition of the form $a + (-b)$ for whole numbers a and b, in which $a > b$ or $a < b$. Suppose you reach an intersection and are unsure about whether to turn right or left. So you turn right and drive for 2 miles. Uh oh! This can't be the right way, so you turn around and head 3 miles in the opposite direction. Where are you now in relation to the intersection? This trip is an application of $2 + (-3)$.

• EXAMPLE 2

(a) Explain how to compute $2 + (-3)$ with a number line.
(b) Explain how to compute $2 + (-3)$ with signed counters.
(c) In adding 2 and -3, why does the answer come out negative?

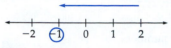

Figure 5–2

SOLUTION

(a) Go to 2. Add the second number, -3, by moving 3 to the left from 2 (Figure 5–2). You end up at -1. So $2 + (-3) = -1$. This process corresponds to going 2 miles to the right and 3 miles to the left. You end up 1 mile to the left of where you started.

(b) Represent 2 with 2 positive counters and -3 with 3 negative counters.

Combine the counters. Form two zero pairs.

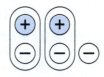

This leaves one negative counter. So $2 + (-3) = -1$.

(c) The answer comes out negative because -3 is larger in absolute value than 2.   ●

LE 7   Communication

(a) Explain how to compute $4 + (-1)$ with a number line.
(b) Explain how to compute $4 + (-1)$ with signed counters.
(c) In adding 4 and -1, why does the answer come out positive?

What are the general rules for addition involving negative integers? First consider the sum of two negative integers.

LE 8   Communication

(a) The general rule for whole numbers a and b is $-a + (-b) =$ _____.

        (1) $-(a + b)$ (2) $b - a$ (3) $a + b$ (4) $a - b$

(b) To add two negative integers, add their _____ and make the result _____.

It is more difficult to express the rule for adding a positive integer and a negative integer in words.

LE 9   Communication

(a) In adding a positive integer and a negative integer, the sign of the sum is determined by _____.
(b) To add a positive integer and a negative integer, find the _____ of their absolute values. The sum has the sign of the integer with the _____.

LE 8 and LE 9 suggest the rules for addition involving negative integers. Combined with whole-number addition rules, these rules comprise the *definition* of integer addition.

> **Definition: Addition of Integers**
>
> **1.** For any integer a, $a + 0 = 0 + a = a$.
> **2.** To add two positive integers, add them as whole numbers.
> **3.** To add two negative integers, add their absolute values and make the result negative.
> **4.** To add a positive integer and a negative integer, compute the larger absolute value minus the smaller absolute value. The sum has the sign of the integer with the larger absolute value. If both integers have the same absolute value, the sum is 0.

Consider the following applications of integer addition.

LE 10   Connection

(a) In today's mail you will receive a check for $282 and a bill for $405. Write an integer addition equation that gives your overall gain or loss.
(b) What type of addition application is this?

Although most applications fall into the set and measure categories, money does not fit clearly into either category. Money is taught as a measurement in school, since it measures the value of goods and services. Yet money is more easily modeled by counters (sets) in units of dollars or cents. In this text, applications involving money will be classified as "measures/sets," indicating that money has characteristics of both measures and sets.

Can you write a realistic application of a given integer addition problem? You might use temperature, money, or a journey. Try the following.

LE 11   Connection

Write an application problem that represents $-10 + (-14)$.

INTEGER SUBTRACTION

In elementary school, a child who is learning whole-number subtraction may ask, "What is $3 - 5$?" The child is asking a question that motivates the creation of negative numbers. The four approaches in LE 12 can be used to develop subtraction. See which of them you already know about. Then continue reading to learn about the others.

LE 12   Opener

Show how to determine the result of $3 - 5$

(a) with a number line.
(b) with a temperature or money application.
(c) by adding the opposite.
(d) with signed counters.

Like integer addition, integer subtraction can be illustrated with a number line or with signed counters. First consider a number line. To show $3 - 5$, first go to 3. Then move 5 to the left. You end up at -2. So $3 - 5 = -2$ (Figure 5–3).

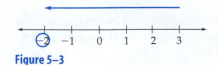

Figure 5–3

The following exercise concerns addition and subtraction on an integer number line.

LE 13   Reasoning

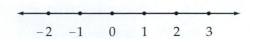

-2   -1   0   1   2   3

(a) On the standard number line, adding a positive integer is shown as a move to the _____.

(b) On the standard number line, adding a negative integer is shown as a move to the _____.

(c) On the standard number line, subtracting a positive integer is shown as a move to the _____.

(d) On the basis of parts (a)–(c), it would make sense for subtraction of a negative integer to be shown as a move to the _____.

Apply the ideas of LE 13 to three numerical examples.

LE 14   Communication

Explain how to compute each of the following with a number line.

(a) $-2 - 1$ **(b)** $2 - (-3)$ **(c)** $-3 - (-2)$

Many elementary-school textbooks now use signed counters as a model for integer subtraction. The easiest type of computation to work out with signed counters is one like $-3 - (-2)$.

LE 15   Communication

To compute $-3 - (-2)$ with signed counters, first show -3. Then take away -2 and see what is left. Write a more complete explanation, and include a drawing of the counters.

Many integer subtraction computations are more difficult to work out with counters. They require changing the way you represent the first number in the subtraction problem.

LE 16   Concept

(a) Which of the following shows -4?

(1) ⊖ ⊖ ⊖ ⊖   (2) ⊖ ⊖ ⊖ ⊖ ⊖

(3) ⊖ ⊖ ⊖ ⊖ ⊖   (4) ⊖ ⊖ ⊖ ⊖ ⊖ ⊖

(b) Show -4 with signed counters in another way.

The sixth-grade textbook for *Harcourt Math* (Figure 5–4) shows how to model $^-7 - 5$ with colored counters. Consider the following similar example.

Classroom Connection

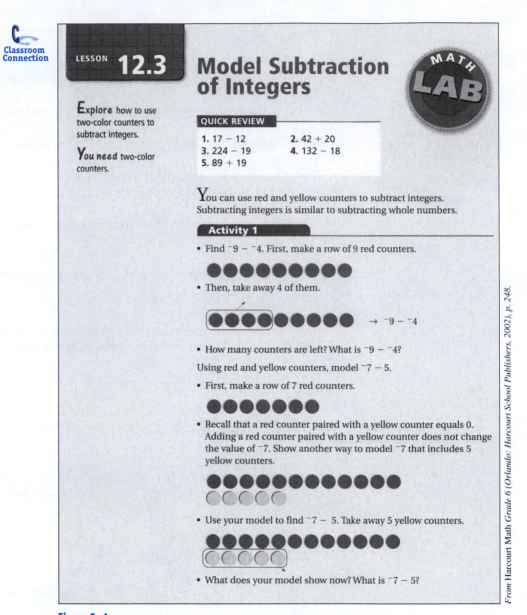

LESSON 12.3

Model Subtraction of Integers

MATH LAB

Explore how to use two-color counters to subtract integers.

You need two-color counters.

QUICK REVIEW

1. $17 - 12$ 2. $42 + 20$
3. $224 - 19$ 4. $132 - 18$
5. $89 + 19$

You can use red and yellow counters to subtract integers. Subtracting integers is similar to subtracting whole numbers.

Activity 1

• Find $^-9 - ^-4$. First, make a row of 9 red counters.

• Then, take away 4 of them.

→ $^-9 - ^-4$

• How many counters are left? What is $^-9 - ^-4$?

Using red and yellow counters, model $^-7 - 5$.

• First, make a row of 7 red counters.

• Recall that a red counter paired with a yellow counter equals 0. Adding a red counter paired with a yellow counter does not change the value of $^-7$. Show another way to model $^-7$ that includes 5 yellow counters.

• Use your model to find $^-7 - 5$. Take away 5 yellow counters.

• What does your model show now? What is $^-7 - 5$?

From Harcourt Math Grade 6 (Orlando: Harcourt School Publishers, 2002), p. 248.

Figure 5–4

• EXAMPLE 3

Explain how to find $3 - 5$ with signed counters.

SOLUTION

First show 3.

To subtract 5, you must take away 5 positive counters. Since there are only 3 positive counters, add 2 zero pairs.

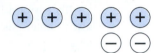

Now take away 5.

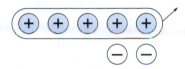

Two negative chips are left. So $3 - 5 = -2$.   •

LE 17   Communication

Explain how to find the following using signed counters.

(a) $-2 - 1$ **(b)** $-2 - (-3)$

Using a number line or signed counters, one finds that $3 - 5 = -2$, $-2 - 1 = -3$, $2 - (-3) = 5$, and $-3 - (-2) = -1$. But there is a shorter way: the add-the-opposite rule. The rule tells how to rewrite an integer subtraction problem as an equivalent integer addition problem. You probably remember this rule, but do you know why it works?

The rule can be explained with money or a number line.

LE 18   Concept

Suppose you are $10 in debt.

(a) Would you rather lose $5 or gain another $5 debt?
(b) Part (a) shows that subtracting _____ is the same as adding _____.
(c) $-10 - 5 =$ _____ and $-10 + (-5) =$ _____.
(d) Would you rather have someone remove a $5 debt, or would you rather receive $5?
(e) Part (d) shows that subtracting _____ is the same as adding _____.
(f) $-10 - (-5) =$ _____ and $-10 + 5 =$ _____.

LE 19   Reasoning

(a) On a number line, why is subtracting a positive number (such as 3) the same as adding its opposite (-3)?
(b) On a number line, why is subtracting a negative number (such as -5) the same as adding its opposite (5)?
(c) Make a generalization based on parts (a) and (b).

The preceding two exercises suggest the add-the-opposite rule for subtracting integers, which will be our definition of integer subtraction.

Definition: **Integer Subtraction**

Subtracting an integer is the same as adding its opposite. If x and y are integers, $x - y = x + (-y)$.

For example, $-2 - (+1) = -2 + (-1) = -3$. So any subtraction problem can be rewritten as an addition problem. To compensate for changing the subtraction sign to an addition sign, the sign of the number being subtracted must also be changed.

To perform integer arithmetic such as $-2 - 1$ on a calculator, you need to know how to enter a negative number. To enter -2 into most graphing calculators, push the negative key $\boxed{(-)}$ followed by 2. For a scientific calculator, try pushing 2 followed by the $\boxed{+/-}$ change-of-sign key.

 LE 20   Skill

Use the integer subtraction rule to compute the following. Then check your results with a calculator.

(a) $420 - (-506)$ **(b)** $-208 - 80$

The familiar categories fit applications of integer arithmetic.

LE 21   Connection

At 2 P.M., the temperature was 14°C. Since then, it has dropped 38°.

(a) Write an integer subtraction equation for this situation.
(b) What is the temperature now?
(c) Is this a sets or measures problem?
(d) What subtraction category does this illustrate? (compare, missing part, take away)

LE 22   Communication

A sixth grader says, "How is $-2 - 6$ used in everyday life? Also, I am not sure if my answer is right." Help the child by making up a temperature or money problem for $-2 - 6$ and giving the result.

LE 23   Summary

Tell what you learned about modeling integer addition and subtraction with number lines and signed counters.

ANSWERS TO SELECTED LESSON EXERCISES

2. -132

3. (a) 5 degrees below 0   (b) 4 under par
(c) A debt of $400
(d) A decrease of $3/share

4.

5. -13 and 13 are both equal distances (13) from 0.

6. (a) Go to 3. To add -3, move 3 to the left.

You end up at 0. So $3 + (-3) = 0$.

(b) Show 3 as 3 positive counters and -3 as 3 negative counters. Form three zero pairs There are no other counters left. So $3 + (-3) = 0$.

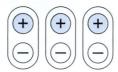

(c) 0

7. (a) Go to 4. Then move 1 to the left.

You end up at 3. So $4 + (-1) = 3$.

(b) Show 4 as 4 positive counters and -1 as 1 negative counter.

Form one zero pair. This leaves 3 positive counters. So $4 + (-1) = 3$.

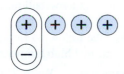

(c) The answer comes out positive because 4 has a larger absolute value than -1.

8. (a) (1)   (b) absolute values; negative

9. (a) The number with the larger absolute value
(b) difference; larger absolute value

10. (a) $\$282 + (-\$405) = -\$123$
(b) See the discussion that follows the exercise.

13. (a) right   (b) left
(c) left   (d) right

14. (a) Go to -2. Move 1 to the left. You end up at -3. So $-2 - 1 = -3$.

(b) Go to 2. Adding -3 would move to the left so subtracting -3 moves 3 to the right. You end up at 5. So $2 - (-3) = 5$.

(c) Go to -3. Adding -2 is a move to the left, so subtracting -2 is a move of 2 to the right. You end up at -1. So $-3 - (-2) = -1$.

15. Show -3 as 3 negative counters.

Now take away -2.

One negative counter is left. So $-3 - (-2) = -1$.

16. (a) (1), (2), (4)

17. (a) Show -2 as 2 negative counters. To subtract 1, you must take away 1 positive counter. Since there are no positive counters, add 1 zero pair.

Now take away 1 positive counter.

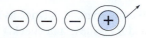

Three negative counters are left. So $-2 - 1 = -3$.

(b) Show -2 as 2 negative counters. To subtract -3, you must take away 3 negative counters. Since there are only 2 negative counters, add 1 zero pair.

Now take away 3 negative counters.

One positive counter is left. So $-2 - (-3) = 1$.

18. (a) No difference
(b) 5; -5
(c) -15; -15
(d) No difference
(e) -5; 5
(f) -5; -5

19. (a) In subtracting a positive integer a, you move a units to the left, and in adding $-a$, you move a units to the left.
(b) In subtracting a negative integer $-a$, you move a units to the right, and in adding a, you move a units to the right.

20. (a) 926   (b) -288

21. (a) $14 - 38 = -24$
(b) $-24°C$
(c) Measures
(d) Take away

5.1 HOMEWORK EXERCISES

Basic Exercises

−1. Which of the following are integers?

(a) -4 (b) 0   (c) $\frac{2}{3}$ (d) $\frac{-8}{4}$

0. What is the largest negative integer?

1. $I = \{ \ldots, -2, -1, 0, 1, 2, \ldots\}$
$P = \{1, 2, 3, \ldots\}$
$N = \{-1, -2, -3, \ldots\}$
$W = \{0, 1, 2, 3, \ldots\}$
(a) $N \cup W = $ _____
(b) $N \cap P = $ _____

2. True of false? All whole numbers are integers.

3. What number is 3 to the right of -100 on the standard number line?

4.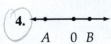

A 0 B

A and B are 9 units apart. A is twice as far from 0 as B. What are A and B? (Guess and check.)

5. Explain the meaning of the $-$ sign in
(a) making -2 yards on a football play.
(b) a federal budget balance of $-\$140$ billion (called a deficit).
(c) an altitude of -50 ft.

6. Write an integer to represent
(a) a debt of \$40.   (b) a fever of 102°.
(c) a loss of 4 yards on a football play.

7. A newspaper lists the following information on "departure from normal high temperature."

Boston -6 Wash. D.C. $+7$

Explain what these numbers mean.

8. Find an example of a place where negative numbers appear in the newspaper.

9. Represent -4 in three ways: as a point on a number line, as a move on a number line, and as a set of signed counters.

10. (a) Use signed counters to represent a debt of \$3.
(b) Use a vertical number line to represent a temperature of 4° below 0°C.

11. Give the opposite idea and the integer that represents it.
(a) 20 seconds after liftoff   (b) A loss of $3
(c) 3 floors higher

12. $a < 0$. Then $-a$ is
(a) positive   (b) zero   (c) negative

13. Using the word "distance," explain why $|-5| = 5$.

14. (a) For what integers x is $|x| < x$?
(b) For what integers x is $|x| = x$?
(c) For what integers x is $|x| > x$?

15. What integer *addition* problem is shown on the number line?

16. What integer addition problem is shown with the signed counters?

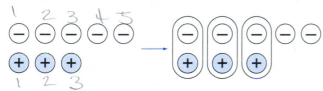

17. (a) Explain how to compute $-5 + 3$ with a number line.
(b) What addition category is illustrated by part (a)?
(c) Explain how to compute $-5 + 3$ with signed counters.

18. Explain how to compute $6 + (-2)$ with
(a) signed counters.   (b) a number line.

19. If a and b are whole numbers, and $a > b$, then a formula relating $-a + b$ to whole-number subtraction is $-a + b =$
(a) $a - b$ (b) $a + b$
(c) $-(a - b)$ (d) $-(a + b)$

20. State the rule for adding two negative integers.

21. In today's mail, you will receive a check for $86, a bill for $30, and a bill for $20.
(a) Write an integer addition equation that gives the overall gain or loss.
(b) What application category is this?

22. An ion contains 42 protons, each with a single positive charge, and 44 electrons, each with a single negative charge.
(a) What is the overall positive or negative charge of the ion?
(b) Write an addition equation for this situation.
(c) What addition category is illustrated?

23. A sixth grader says that -10 is greater than -5 because 10 is greater than 5. How would you explain the correct solution to the child?

24. A seventh grader says that $-2 + (-3) = 5$ because a negative plus a negative makes a positive. How would you explain the correct solution to the child?

25. Compute the following without a calculator.
(a) $-54 + 25$ (b) $400 + (-35)$
(c) $-98 + (-10)$

26. Compute the following without a calculator.
(a) $-23 + 30$ (b) $-8 + (-17)$
(c) $10 + (-25)$

27. Which of the following is a correct way to say $-(-6)$?
(a) "Minus minus 6"
(b) "The opposite of negative 6"
(c) "Minus negative 6"

28. Which of the following are correct ways to say $-3 - (-2)$?
(a) "Minus 3 minus negative 2"
(b) "Negative 3 minus negative 2"
(c) "Minus 3 minus minus 2"
(d) "The difference between minus 3 and minus 2"
(e) "The difference between negative 3 and negative 2"

29. *Explain* how to compute each of the following with a number line.
(a) $4 + (-6)$ (b) $5 - (-2)$

30. *Explain* how to compute each of the following with a number line.
(a) $-2 + 5$ (b) $-6 - (-3)$

31. Show two ways to represent each integer with signed counters.
(a) 3   (b) -2

32. Show two ways to represent each integer with signed counters.
(a) -3 (b) 4

33. Tell what subtraction problem each picture illustrates.

(a)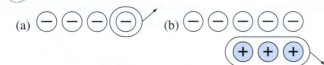

(b)

(c)

34. Tell what subtraction problem each picture illustrates.

(a) (b)

(c)

35. *Explain* how to compute the following with signed counters.
(a) $-6 - (-2)$ (b) $2 - 6$

36. *Explain* how to compute the following with signed counters.
(a) $-2 - 3$ (b) $4 - (-2)$

37. Tell whether each of the following is easier to compute with a number line or with signed counters.
(a) $2 - 5$ (b) $-3 - (-2)$
(c) $-1 - 3$ (d) $2 - (-4)$

38. A football team runs four plays with results of $+6$ yd, $+10$ yd, $+3$ yd, and -5 yd.
(a) What is the net yardage?
(b) How could you use this example to explain why $14 - (-5) = 19$?

39. (a) On a standard number line, subtracting 3 is the same as moving _____ units to the _____.
(b) On a standard number line, adding -3 is the same as moving _____ units to the _____.
(c) What conclusion is suggested by parts (a) and (b)?
(d) Make a broader generalization based upon part (c).

40. (a) Taking away a debt of $10 is the same as receiving _____.
(b) Part (a) illustrates that subtracting _____ is the same as adding _____.

41. Compute the following without a calculator. Then check your result with a calculator.
(a) $-51 - 22$ (b) $-32 - (-70)$

42. Compute the following without a calculator. Then check your result with a calculator.
(a) $21 - 48$ (b) $34 - (-10)$

43. Evaluate each expression if $x = -2$, $y = 3$, and $z = -5$. Do not use a calculator.
(a) $x + y - z$ (b) $-x - y + z$
(c) $x - (-y) + z$

44. Compute the following without a calculator.
(a) $-3 - 4 + 2$ (b) $-8 + (-2) - (-5)$
(c) $4 - (-10) + (-2)$ (d) $-5 + 7 - (-3)$

45. Solve mentally. (Guess and check.)
(a) $\Box + 8 = -6$ (b) $r - 10 = -2$
(c) $4 - x = -6$

46. Solve mentally. (Guess and check.)
(a) $9 - \Box = -3$ (b) $r + (-2) = -7$
(c) $10 - y = 14$

47. An elevator is at an altitude of -10 ft. The elevator goes down 30 ft.
(a) Write an integer equation for this situation.
(b) What is the new altitude?
(c) What subtraction category does this illustrate? (compare, missing part, take away)

48. Euclid was born around 360 B.C.
(a) If he lived for 50 years, when did he die?
(b) Write an integer equation for this situation.
(c) What operation and category does this illustrate?

49. Consider the following problem. "A football team gains 5 yards on their first play and loses 12 yards on the second play."
   (a) What is their net gain or loss?
   (b) What operation and category does this illustrate? (Give two possible answers.)

50.

	Continent	
	North America	**Europe**
Highest point	Mt. McKinley	Mt. Elbrus
Altitude	6194 m	5642 m
Lowest point	Death Valley	Caspian Sea
Altitude	-86 m	-28 m

   (a) What is the difference between the highest and lowest elevations in North America?
   (b) What is the difference between the highest and lowest elevations in Europe?
   (c) What operation and category are illustrated in parts (a) and (b)? (combine, compare, missing part, take away)

51. Make up a temperature or money problem for $-6 - 4$, and give the result.

52. Make up a temperature or money problem for $4 - 8$, and give the result.

53. Fill in the chart.

Temperature at 8 A.M.	Temperature at 6 P.M.	Change
7		-10
-6	-2	
5	-3	
	-4	-2
	3	11

54.

Wind-Chill Temperature (°F)						
Wind Speed	**Actual Temperature (°F)**					
	50°	**40°**	**30°**	**20°**	**10°**	**0°**
10 mph	40	28	16	4	-8	
20 mph	32	18	4	-10	-24	
30 mph	28	13	-2	-17	-32	
40 mph	26	10	-6	-22	-38	

   (a) The temperature is 40°F, and there is a 30-mph wind. What is the wind-chill temperature?
   (b) The weather report said that the temperature is 10°F, and it feels like -25°F. What is the wind speed?
   (c) Use the pattern in each row to fill in the last column of the chart.
   (d) The temperature is 50°F, and the wind speed is 15 mph. Estimate the wind-chill temperature.

Extension Exercises

55. One can find the results to integer subtraction by extending patterns in whole-number subtraction. For example, to find $3 - 5$, we work our way down from more familiar problems in which a smaller positive number is subtracted from 3.

$$3 - 1 = 2$$
$$3 - 2 = 1$$
$$3 - 3 = 0$$
$$3 - 4 = \underline{}$$
$$3 - 5 = \underline{}$$

   (a) Fill in the blanks, continuing the pattern.
   (b) In part (a), assuming the answer decreases by 1 and using that to fill in the next blank involves _____ reasoning.
   (c) Find the answer to $3 - (-2)$ by creating a similar pattern, beginning with three whole-number subtraction examples. (*Hint:* Try 3 minus some small positive numbers.)
   (d) Find the result of $-2 - 1$ by extending a pattern in three whole-number subtraction examples.

56. (a) Find the result of $4 - 7$ by extending a pattern in three whole-number subtraction examples. (*Hint:* Start with $9 - 7$ or $4 - 2$.)
   (b) Find the result of $5 - (-2)$ by extending a pattern in three whole-number subtraction examples.

57. Whole-number subtraction such as $6 - 2 = n$ can be rewritten in the form $n + 2 = 6$ using the definition of subtraction. The same definition can be used with integers.
  (a) Using this approach, $6 - (-2) = n$ is the same as _____ and $n =$ _____.
  (b) Using this approach, $-3 - 2 = n$ is the same as _____ and $n =$ _____.

58. Use the method of the preceding exercise to complete the following.
  (a) Using this approach, $7 - 10 = n$ is the same as _____ and $n =$ _____.
  (b) Using this approach, $-4 - (-5) = n$ is the same as _____ and $n =$ _____.

59. Which integers can be written as a sum of each of the following?
  (a) Two consecutive integers
  (b) Three consecutive integers

60. The sum of two integers is 4. The difference of the two integers is 10. What are they? (Guess and check.)

61. Decide whether each statement is true or false for all integers x and y. If a statement is true, give an example that supports it. If it is false, give a counterexample.
  (a) True or false? $|x - y| = |y - x|$.
  (b) True or false? $|x - y| = |x| - |y|$.

62. If possible, find integers m and n such that
  (a) $|m + n| < |m| + |n|$.
  (b) $|m + n| = |m| + |n|$.
  (c) $|m + n| > |m| + |n|$.

63. Fill in each of the numbers $-8, -6, -4, -2, 0, 2, 4, 6, 8$ in one square so that every row, column, and diagonal has the same sum.

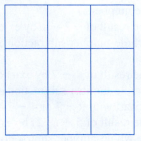

64. Fill in each of the numbers $-15, -12, -9, -6, -3, 0, 3, 6, 9$ in one square so that every row, column, and diagonal has the same sum.

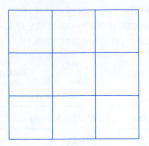

65. Are there more integers than whole numbers? Show that the sets are equivalent by establishing a one-to-one correspondence between the whole numbers and the integers.

Technology Exercise

66. Some calculators have the capability of downloading a program that shows addition and subtraction of integers on a number line. Set the window to MIN $= -10$, MAX $= 10$, START $= 0$ and compute the following:
  (a) $-7 + 2$ (b) $-3 - 5$.

Questionnaire

67. How can you increase your chance of living a longer life? The following predictor of life expectancy is adapted from Wallechinsky and Wallace's *People's Almanac #2*.
  (a) Start with 72, and add the integers for all applicable descriptors.

• Male	-3
Female	$+4$
• Urban residence over 2,000,000	-2
Rural residence under 10,000	$+2$
• Job with regular heavy labor	$+3$
Exercise five times per week	$+2$
• Alone for each 10 years since 25	-1
Live with spouse/friend	$+5$
• Easily angered, very aggressive	-3
Easygoing, a follower	$+3$
• Happy	$+1$
Unhappy	-2
• College graduate	$+1$
Graduate degree	$+2$
• One grandparent lived to 85	$+2$
All grandparents lived to 80	$+6$
• Parent died of stroke, heart attack before 50	-4
• Immediate family under 50 has cancer, heart disease, diabetes	-4

• Smoke > 2 packs/day	−8	• 30 to 39 years old	+2
Smoke 1 to 2 packs	−6	40 to 49	+3
Smoke 1/2 to 1 pack	−3	50 to 69	+4
• Drink at least 1/4 bottle liquor/day	−1		
• 50 or more pounds overweight	−8		
30 to 49 pounds overweight	−4		
10 to 29 pounds overweight	−2		

The number you obtain is your life expectancy.

(b) According to this scoring system, what are the three most important factors affecting your life expectancy, besides your current age?

5.2 MULTIPLICATION AND DIVISION OF INTEGERS

NCTM Standards

- understand the meanings and effects of operations with fractions, decimals, and integers (6–8)

- develop and analyze algorithms for computing with fractions, decimals, and integers, and develop fluency in their use (6–8)

- create and use representations to organize, record, and communicate mathematical ideas (pre-K–12)

Integer multiplication and division are similar to whole-number multiplication and division. In multiplying and dividing integers, the one new issue is whether the result is positive or negative. This lesson shows how to explain the sign of an integer product or quotient using patterns, applications, and definitions.

INTEGER MULTIPLICATION

Classroom Connection

LE 1   Opener

How would you convince a child that $3 \times (-2) = -6$?

The topic of integer multiplication begins with a positive times a negative such as $3 \times (-2)$. Why would someone want to multiply $3 \times (-2)$ in everyday life? Suppose the temperature drops 2° per hour for 3 hours. How much does it change altogether? −6°. This example suggests that $3 \times (-2) = -6$.

You can find the product of a positive and a negative with repeated addition, repeated sets, or repeated measures.

LE 2   Concept

(a) A sixth grader knows how to change 3×2 into a repeated addition problem. Change $3 \times (-2)$ into repeated addition, and compute the answer.

(b) *Explain* how to compute $3 \times (-2)$ with repeated sets of signed counters.

(c) Show how to compute $3 \times (-2)$ with a number line.

(d) On the basis of these results, a positive integer times a negative integer results in a _____.

After establishing that a positive times a negative is a negative, children next consider a negative times a positive, such as $(-2) \times 3$. They are told that integer multiplication, like whole number multiplication, is commutative. (Other basic properties of the integers will be covered in the next section.)

LE 3   Reasoning

(a) How could you use the result of the previous exercise and the commutative property to compute $(-2) \times 3$?

(b) This suggests more generally that a negative integer times a positive integer results in a _____.

One can use the knowledge that a negative integer times a positive equals a negative to establish the result of a negative integer times a negative. The next exercise shows how to do this by extending a pattern.

LE 4   Reasoning

Suppose a child knows that a negative times a positive is a negative and wants to use a pattern to determine what $-2 \cdot (-3)$ equals. Fill in the blanks, continuing the same pattern.

(a) $-2 \cdot 3 \quad = -6$
$\quad -2 \cdot 2 \quad = -4 \qquad$ Negative $\cdot$ positive
$\quad -2 \cdot 1 \quad = -2$
$\quad -2 \cdot 0 \quad = 0$
$\quad -2 \cdot (-1) = $ _____
$\quad -2 \cdot (-2) = $ _____
$\quad -2 \cdot (-3) = $ _____

(b) On the basis of part (a), a negative integer times a negative integer results in a _____.

It is also possible to use whole-number multiplication extending a pattern to show that a negative times a positive is a negative (see Exercise 5). An effective way to find the results of all three new types of multiplication with the same method is to use an application such as temperature, money, or a walker on a number line.

• EXAMPLE 1

Describe a temperature application that suggests why

(a) $3 \times (-2) = -6$ (b) $-3 \times 2 = -6$ (c) $-3 \times (-2) = 6$

SOLUTION

Assume the first factor is time and the second factor represents a change in temperature. To model the second and third equations, it is much easier to state that the temperature is now 0° and move forward or back from here.

(a) A temperature application for $3 \times (-2)$ was described earlier in the section. It is now 0°F. The temperature drops 2° per hour. What will the temperature be in 3 hours? -6°F. Therefore, $3 \times (-2) = -6$.

(b) It is now 0°F. The temperature has been increasing 2° per hour. What was the temperature 3 hours ago? -6°F. Therefore, $-3 \times 2 = -6$.

(c) It is now 0°F. The temperature drops 2° per hour. What was the temperature 3 hours ago? 6°F. Therefore, $-3 \times (-2) = 6$.   ●

Another good model for all three types of equations is a walker on a standard number line. If you want to use that, start your walker at 0, make the first factor the time, and the second factor the speed. A positive speed moves to the right (east) on the number line, and a negative speed moves to the left (west). Time works the same way as in the temperature model.

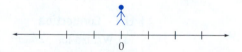

An example for $3 \times (-2)$ would be: A walker is now at 0 on the number line. If she walks 2 miles per hour to the left (west), where will she be in 3 hours? At -6 miles. Therefore, $3 \times (-2) = -6$.

The following chart compares three different applications of $-3 \times (-2)$. Consider which one you prefer, and apply it in the following exercise.

	Temperature	**Walker**	**Money**
Now 0	It is now 0°C	She is now at 0 on a number line.	I have $0.
-2	The temperature has been dropping 2° per hour.	She has been walking 2 miles per hour to the left.	I have been spending/ losing $2 per hour.
-3 (back 3 in time)	What was the temperature 3 hours ago?	Where was she 3 hours ago?	How much did I have 3 hours ago?
Answer	6°C	At $+6$ miles	$6
	So $-3 \times (-2) = 6$.	So $-3 \times (-2) = 6$.	So $-3 \times (-2) = 6$.

LE 5   Connection

Describe an application using temperature, a walker, money, or your own model that suggests why

(a) $5 \times (-6) = -30$ **(b)** $-5 \times 6 = -30$ **(c)** $-5 \times (-6) = 30$

The results of LE 2–LE 5 support the rules for multiplication involving negative integers. Each rule relates integer multiplication to whole-number multiplication. These rules comprise the definition of integer multiplication (when combined with the rules for whole-number multiplication).

Definition: Multiplication of Integers

If a and b are whole numbers, then	For *nonzero a* and *b,* this means that
$-a \cdot (-b) = +(ab)$	The product of two integers with the same sign is positive.
$-a \cdot b = a \cdot (-b) = -(ab)$	The product of two integers with different signs is negative.

INTEGER DIVISION

When, in everyday life, would someone do division involving negative numbers? Suppose that a town's population drops by 400 in 5 years. What is the average population change per year? -80 people. This problem suggests that $-400 \div 5 = -80$. Population, money, and temperature problems offer useful models of integer division.

LE 6   Connection

(a) Write a temperature or money problem for $-8 \div 4$, and give the result.
(b) Explain how to compute $-8 \div 4$ by partitioning a set of signed counters.
(c) The results of parts (a) and (b) suggest that a negative integer divided by a positive integer results in a _____.

Just as whole-number division is defined as the inverse of whole-number multiplication, integer division is defined as the inverse of integer multiplication.

Definition: **Integer Division**

If x, y, and q are integers and $y \neq 0$, then $x \div y = q$ if and only if $x = y \cdot q$.

This definition enables one to relate integer multiplication and integer division problems.

LE 7   Concept

(a) Write two division equations that are equivalent to $3 \times (-2) = -6$. (*Hint:* Use the definition of integer division and the commutative property of multiplication.)
(b) Write two division equations that are equivalent to $-3 \times (-2) = 6$.
(c) Would the quotient of two integers with the same sign be positive or negative?
(d) Would the quotient of two integers with different signs be positive or negative?

The following chart summarizes the rules for division of integers.

Division of Integers

The quotient of two integers with the same sign is positive.
The quotient of two integers with different signs is negative.

LE 8   Skill

Use one of the division rules to compute $-24 \div (-8)$.

The introductory applications of division modeled a negative integer divided by a positive. It is more difficult to describe applications of a positive integer divided by a negative and a negative integer divided by a negative.

LE 9   Connection

The temperature dropped 8°F in the last 2 hours. What was the average change in temperature each hour?

(a) Write a division equation for this application.
(b) What division category does this illustrate? (area, array, partition, repeated)

LE 10   Summary

Tell what you learned about integer multiplication and division in this section.

ANSWERS TO SELECTED LESSON EXERCISES

2. (a) $3 \times (-2) = -2 + (-2) + (-2) = -6$
(b) $3 \times (-2)$ means 3 sets of -2.

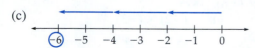

This makes -6. So $3 \times (-2) = -6$.

(c)

(d) Negative integer

3. (a) $(-2) \times 3 = 3 \times (-2) = -6$
(b) Negative integer

4. (a) 2; 4; 6
(b) positive

6. (a) The temperature fell 8°C during a 4-hour period. What was the average change in temperature each hour? -2°C.

(b) Take a set of 8 negative counters and divide them into four equal groups.

How many in each group? -2. So $-8 \div 4 = -2$.
(c) Negative integer

7. (a) $-6 \div 3 = -2$ and $-6 \div (-2) = 3$.
(b) $6 \div (-3) = -2$ and $6 \div (-2) = -3$.
(c) Positive
(d) Negative

8. 3

9. (a) $-8 \div 2 = -4$
(b) Partition measures

5.2   HOMEWORK EXERCISES

Basic Exercises

1. (a) Change $4 \times (-3)$ into repeated addition, and compute the answer.
(b) Explain how to compute $4 \times (-3)$ with repeated sets of signed counters.
(c) Show how to compute $4 \times (-3)$ with a number line.

2. (a) Change $2 \times (-4)$ into repeated addition, and compute the answer.
(b) Explain how to compute $2 \times (-4)$ with repeated sets of signed counters.
(c) Show how to compute $2 \times (-4)$ with a number line.

3. (a) If integer multiplication is commutative and we know that a positive times a negative is a negative, what can we conclude?
(b) Part (a) illustrates _____ reasoning.

4. If $5 \times (-7) = -35$ and multiplication is commutative, then we also know that _____.

5. Suppose a child knows how to multiply a negative by a positive. Fill in the blanks, continuing the same pattern, to show the result of a negative times a negative.

$$-5 \cdot 2 = -10$$
$$-5 \cdot 1 = -5$$
$$-5 \cdot 0 = 0$$
_____
_____

6. Show how -2×4 can be solved by extending a pattern in whole-number multiplication. (*Hint*: Start with 2×4.)

7. Describe an application using temperature, a walker, or your own model that suggests why
   (a) $4 \times (-3) = -12$
   (b) $-4 \times 3 = -12$
   (c) $-4 \times (-3) = 12$

8. Describe an application using temperature, a walker, or your own model that suggests why
   (a) $3 \times (-5) = -15$
   (b) $-3 \times 5 = -15$
   (c) $-3 \times (-5) = 15$

9. Mike lost 3 pounds each week for 4 weeks.
   (a) What was the total change in his weight?
   (b) Write an integer equation for this situation.
   (c) What application category does this illustrate?

10. Write a money or temperature problem for $4 \times (-5)$.

11. (a) Write a temperature or money problem for $-10 \div 2$, and give the result.
   (b) Explain how to compute $-10 \div 2$ by partitioning a set of signed counters.

12. (a) Write a temperature or money problem for $-6 \div 3$, and give the result.
   (b) Explain how to compute $-6 \div 3$ by partitioning a set of signed counters.

13. Write two division equations that are equivalent to $-2 \times 5 = -10$.

14. Write two division equations that are equivalent to $-3 \times (-6) = 18$.

15. Rewrite each problem as an equivalent multiplication problem, and give the solution.
   (a) $-54 \div (-6) = $ _____
   (b) $32 \div (-4) = $ _____

16. Rewrite each problem as an equivalent multiplication problem, and give the solution.
   (a) $0 \div (-3) = $ _____
   (b) $-3 \div 0 = $ _____

17. A store lost \$480,000 last year.
   (a) What was the average net change per month?
   (b) Write an integer equation for this situation.
   (c) What application category does this illustrate? (area, array, partition, repeated)

18. A school population has been dropping 15 students per year.
   (a) How many more students were at the school 2 years ago?
   (b) Write an integer equation for this situation.
   (c) What application category does this illustrate? (area, array, counting principle, repeated)

19. Compute the following without a calculator.
   (a) $3 \times (-8)$
   (b) $-5 \times (-8) \times (-2) \times (-3)$
   (c) $-24 \div 8$
   (d) $(-1)^{20}$

20. Compute the following without a calculator.
   (a) -5×3 (b) $-42 \div (-7)$
   (c) $-3 \times 4 \times (-2) \times (-6)$ (d) $(-1)^{45}$

21. Compute the following, using the correct rules for order of operations.
   (a) $-2^2 - 3$ (b) $-5 + (-4)^2 \times (-2)$

22. Compute the following, using the correct rules for order of operations.
   (a) $-2 - (-3)^3$ (b) $-12 + 3 \times (-2)$

23. Solve mentally.
   (a) $-5t = 0$ (b) $y^2 = 4$ (c) $\dfrac{30}{n} = -6$

24. Solve mentally.
   (a) $-10 \div \square = 2$ (b) $-4 \times \square = -32$
   (c) $9(r + 3) = -45$

25. Suppose a and b are positive and c and d are negative. Determine the sign of each expression.
   (a) $\dfrac{a - c}{b - d}$ (b) $\dfrac{-a - b}{c}$ (c) $ac - b$

26. Evaluate each expression if $x = 4$, $y = -8$, and $z = -2$. Do not use a calculator.
   (a) $\dfrac{x - y}{z}$ (b) $x + yz$ (c) $\dfrac{y^2}{z}$

27. A stock changes as follows for 5 days: $-2, 4, 6, 3, -1$. What is the average daily change in price?

28. One week, the daily temperatures (in °C) in a town were $-18, -13, 5, 2, -10, -8$, and -7. What was the average temperature for the week?

Extension Exercises

29. Use the following five integers to fill in the blanks in the box. You may use the same number more than once.

$$-4 \quad 2 \quad 1 \quad -8 \quad -2$$

1	×	___	×	___	=	-8
+		+		×		
___	−	___	−	___	=	4
+		+		×		
___	÷	___	÷	___	=	1
=		=		=		
1		-10		4		

30. If x is a member of $\{-3, -2, -1, 0, 1, 2\}$ and y is a member of $\{-6, -4, -2, 0, 2, 4\}$, find the largest and smallest possible values of each of the following.

(a) $|x + y|$ (b) $x - y$ (c) xy (d) $\frac{x}{y}$

31. If x and y are integers and $x < y$, what values of x and y would make $x^3 < y^3$?

32. If x and y are integers, and $x < y$, what values of x and y would make $x^2 < y^2$? Devise a plan and solve the problem.

Puzzle Time

33. *Four negative twos.* Using only four -2s and any combination of arithmetic symbols, write expressions equal to 1, 2, 3, . . . , 9. The first one has been done for you. Some of them are impossible.

$$-2 \div (-2) + (-2) - (-2) = 1$$

5.3 PROPERTIES OF INTEGER OPERATIONS

NCTM Standards

- use the associative and commutative properties of addition and multiplication and the distributive property of multiplication over addition to simplify computations with integers, fractions, and decimals (6–8)
- develop and analyze algorithms for computing with fractions, decimals, and integers, and develop fluency in their use (6–8)

Does $-3 \cdot (4 \cdot 5) = -3 \cdot 4 \cdot 5$, or does $-3 \cdot (4 \cdot 5) = -3 \cdot 4 \cdot (-3) \cdot 5$? Why does $-5x + 2x = -3x$? You can answer these questions if you understand integer properties.

First, let your mind relax. Now, let those whole-number properties you studied in Chapter 3 re-enter your consciousness.

LE 1 Opener

What properties does the set of whole numbers have?

Why recall these properties at this particular time? Well, wouldn't it be wonderful if integer arithmetic had these properties, too? It does!

WHOLE-NUMBER PROPERTIES RETAINED!

Integer operations retain the same commutative, associative, identity, and distributive properties as whole-number operations. The following list summarizes these properties.

> **Properties of Integer Operations**
>
> 1. Integer addition and multiplication are closed. For any integers x and y, $x + y$ is a unique integer and xy is a unique integer.
>
> 2. Integer addition and multiplication are commutative. For any integers x and y, $x + y = y + x$ and $xy = yx$.
>
> 3. Integer addition and multiplication are associative. For any integers x, y, and z, $(x + y) + z = x + (y + z)$ and $(xy)z = x(yz)$.
>
> 4. The unique additive identity for integers is 0, and the unique multiplicative identity for integers is 1. For any integer x, $x + 0 = 0 + x = x$ and $x \cdot 1 = 1 \cdot x = x$.
>
> 5. Integer multiplication is distributive over addition, and integer multiplication is distributive over subtraction. For any integers x, y, and z, $x(y + z) = xy + xz$ and $x(y - z) = xy - xz$.

The following exercises utilize these properties.

LE 2   Connection
(a) What is the easiest way to add $(3 + (-5)) + 5$?
(b) What property does part (a) illustrate?

LE 3   Skill
I is an integer. According to the distributive property of multiplication over addition, $-5I + 2I = $ _____.

What about integer subtraction and division? Are they commutative or associative?

LE 4   Skill
(a) Since 2 and 3 are integers, $2 - 3 \neq 3 - 2$ shows that integer subtraction is not _____.

(b) Explain why any commutative or associative property that does not hold for all whole numbers could not possibly hold for the set of integers.

INVERSES

So far, the integers have had all the same properties as the whole numbers. Big deal, you say? Well, it is nice to have some consistency. How would you like it if gravity stopped working?

Something different is also nice once in a while. The integers do have one additional property for addition that the whole numbers do not have. This additional property concerns the fact that people use integers to solve problems such as $3 + $ ____ $= 0$ and $4 + $ ____ $= 0$. The numbers that go in the blanks are the additive inverses of 3 and 4.

> **Definition: Additive Inverse**
> The integer y is an **additive inverse** of the integer x if and only if $x + y = y + x = 0$ (the additive identity).

Any integer added to its additive inverse should result in the additive identity, 0. Do all integers have unique additive inverses that are also integers? The results of LE 5 and LE 6 will help you decide.

LE 5   Skill

Fill in each blank with all possible answers.

(a) $3 + \underline{\hspace{1cm}} = 0$, and $\underline{\hspace{1cm}} + 3 = 0$. Therefore, $\underline{\hspace{1cm}}$ is an additive inverse of 3.
(b) $-8 + \underline{\hspace{1cm}} = 0$, and $\underline{\hspace{1cm}} + -8 = 0$. Therefore, $\underline{\hspace{1cm}}$ is an additive inverse of -8.
(c) Did each part have a unique (exactly one) answer?

LE 6   Skill

What is the additive inverse of each of the following?

(a) 7   **(b)** -2 **(c)** 0

LE 5 and LE 6 should convince you that every integer has a unique additive inverse that is an integer. This is an additional property for integers that does not work for the set of whole numbers.

Additive Inverses for Integers

For each integer x, there is a unique integer $-x$ such that $x + (-x) = -x + x = 0$.

CLOSURE

One reason for creating integers is to provide answers to problems such as $2 - 3$. Do all integer subtraction problems have integer answers?

LE 7   Opener

If x and y are integers, is $x - y$ always a unique integer?

LE 7 may have led you to the following conclusion.

The Closure Property for Integer Subtraction

Integer subtraction is closed. For any two integers x and y, $x - y$ is a unique integer.

COMMON ERROR PATTERNS

As a teacher, you will encounter certain common errors in your students' integer arithmetic. The following problems will give you practice in detecting them.

Classroom Connection

In LE 8–LE 10, (a) complete the last two examples, repeating the error pattern in the completed examples, and (b) describe the error pattern.

LE 8   Reasoning

$4 - 6 =$ __2__ $8 - 11 =$ __3__ $2 - 7 =$ __5__

$6 - 10 =$ _____ $10 - 12 =$ _____

LE 9   Reasoning

$-6 \times (-5) =$ __-30__ $-4 \times (-2) =$ __-8__ $-3 \times (-2) =$ __-6__

$-5 \times (-4) =$ _____ $-3 \times (-6) =$ _____

LE 10   Reasoning

$-8 + 5 \;\; =$ __-13__ $-4 + 4 =$ __-8__ $-2 + 6 =$ __-8__

$3 + (-6) =$ _____ $-2 + 8 =$ _____

AN INVESTIGATION: ORDER IN SUBTRACTION

Integer subtraction is not commutative, but $x - y$ is related to $y - x$ for all integers x and y.

LE 11   Skill

If $x - y = 2$, then $y - x =$ _____. (*Hint:* Try some numbers for x and y.)

LE 12   Reasoning

For integers a and b, how does $a - b$ compare to $b - a$?

(a) What is meant by "compare" in the question?
(b) Devise a plan.
(c) Carry out the plan.
(d) Make a generalization based on your results.
(e) What kind of reasoning is used to make a generalization from examples in part (d)?

LE 13   Reasoning

For what integer values of m and n does $m - n = n - m$?

LE 14   Summary

Tell what you have learned about the properties of integer operations in this section.

ANSWERS TO SELECTED LESSON EXERCISES

2. (a) Add $-5 + 5$, and then add 3.
   (b) Associative property of addition

3. $(-5 + 2)I$

4. (a) commutative
   (b) If a rule does not apply to *all* whole numbers, then it cannot apply to *all* integers, since every whole number is also an integer. Whatever was a counterexample for whole numbers is also a counterexample for the set of integers.

5. (a) $-3; -3; -3$ (b) 8; 8; 8   (c) Yes

6. (a) -7 (b) 2   (c) 0

7. Yes

8. (a) 4; 2
(b) The child computes the larger number minus the smaller.

9. (a) $-20; -18$
(b) The child thinks a negative number times a negative number is a negative number.

10. (a) $-9; -10$
(b) The child adds a negative number and a positive number by adding their absolute values and placing a negative sign in front of the result.

11. -2

13. When $m = n$

5.3 HOMEWORK EXERCISES

Basic Exercises

1. (a) What integer operations are commutative?
(b) What integer operations are associative?

2. Name all the properties of whole number operations that you studied that also apply to the set of integers.

3. What is an easy way to multiply $-5 \times (7 \times -8)$?

4. During 3 consecutive years, a man gains 14 pounds, loses 37 pounds, and then loses 14 pounds.
(a) Write an integer *addition* expression that represents his overall change.
(b) What is an easy way to add the numbers?

5. In adding a series of integers, it is often easier to add all the negative numbers and positive numbers separately and then add the results together.
(a) Compute $(-6 + 4) + (4 + (-3)) + (-7 + 5)$ mentally, using this method.
(b) Compute $-5 + (-2) + 6 + (-3) + 8 + 7$ mentally, using this method.
(c) What two properties enable you to add integers in a different order and still obtain the same answer?

6. What property guarantees that for an integer m, $4 \cdot (-3m) = (4 \cdot -3)m$?

7. The equation $-5 \times (-3) = -3 \times (-5)$ illustrates the _____ property of _____.

8. For an integer n, the distributive property of multiplication over addition states that $-6n + (-3n) =$ _____.

9. A seventh grader thinks that $-2 \cdot (3 \cdot 4) = -2 \cdot 3 \cdot (-2) \cdot 4.$
(a) What properties is she confusing?
(b) What would you tell the child to help her?

10. How can you compute -27×6 using a distributive property?

11. Give a counterexample that shows that integer division is not associative.

12. Give a counterexample that shows that integer subtraction is not commutative.

13. (a) Explain why a person who has studied whole-number operations would know that integer division is not associative.
(b) The conclusion in part (a) is an example of _____ reasoning.

14. Explain why a person who has studied whole-number operations would know that integer subtraction is not commutative.

15. The examples $-3 \times 1 = -3$ and $-5 \times 1 = -5$ illustrate that _____ is the _____ for _____.

16. What is the integer identity element for addition?

17. Any number added to its additive inverse equals _____.

18. What is the value of -0? Explain why.

19. What property guarantees that $6 - 12$ has a unique integer answer?

20. What does the closure property for subtraction of integers say?

In Exercises 21 and 22, (a) complete the last two examples, repeating the error pattern in the completed examples, and (b) write a description of the error pattern.

21. $-3 + (-4) = \underline{\quad 7 \quad}$ $-8 + (-2) = \underline{\quad 10 \quad}$

$-2 + (-3) = \underline{\qquad}$ $-6 + (-1) = \underline{\qquad}$

22. $3 - (-6) = \underline{\quad -3 \quad}$ $4 - (-5) = \underline{\quad -1 \quad}$

$6 - (-2) = \underline{\qquad}$ $-3 - (-2) = \underline{\qquad}$

23. Why might the error pattern in Exercise 21 occur?

24. Describe two errors a child might make in computing $-28 - 65$.

Extension Exercises

25. Consider the following problem. "For nonzero integers a and b, how does $a \div b$ compare to $b \div a$?"
(a) Devise a plan and solve the problem.
(b) For what integer values of a and b does $a \div b = b \div a$?

26. Fill in the blanks, following the rule from the completed examples.

$$-2 \rightarrow \quad 3$$
$$-4 \rightarrow \quad 5$$
$$6 \rightarrow \quad -5$$
$$-5 \rightarrow \underline{\quad}$$
$$8 \rightarrow \underline{\quad}$$
$$N \rightarrow \underline{\quad}$$
$$\underline{\quad} \rightarrow -17$$

27. Suppose you did not know that $-2 \cdot 4 = -8$. You could prove that $-2 \cdot 4 = -8$ using the additive inverse property and whole-number multiplication.
(a) Since $2 \cdot 4$ is the additive inverse of $-2 \cdot 4$, write an addition equation that shows their relationship.
(b) How does this show that $-2 \cdot 4 = -8$?

28. Suppose you did not know that $-2 \cdot (-4) = 8$. How could you use the result of the preceding exercise to prove it?

29. An **even** integer is any integer that can be written in the form $2m$, in which m is an integer. Show that the product of two even integers is even.

30. Show that the sum of two even integers is even.

CHAPTER 5 SUMMARY

People use negative integers to measure stock prices, golf scores, and altitudes. Integers were also developed as solutions to whole-number subtraction problems, such as $2 - 3$, that have no whole-number solution.

With integers, as with whole numbers, addition and multiplication are defined, and then subtraction is defined as the inverse of addition, and division is defined as the inverse of multiplication. The results of integer arithmetic can be illustrated with definitions, with applications such as temperature and money, or with models such as a number line, or signed counters.

The set of integers retains the commutative, associative, and distributive properties of whole-number operations. Both integers and whole numbers have the same identity elements for addition (0) and multiplication (1). The set of integers has two important additional properties that whole numbers do not have: additive inverses and closure for subtraction.

STUDY GUIDE

To review Chapter 5, see what you know about each of the following ideas or terms that you have studied. You can also use this list to generate your own questions about the chapter.

INTEGERS IN GRADES 1–8

The following chart shows at what grade levels selected integer topics typically appear in elementary- and middle-school mathematics textbooks.

Topic	Typical Grade Level in Current Textbooks
Integer concepts	4, 5, 6
Adding and subtracting integers	5, 6, 7, 8
Multiplying and dividing integers	6, 7, 8

REVIEW EXERCISES

1. Write a paragraph defining whole numbers and integers and telling how these two sets of numbers are related.

2. Explain how to compute $2 - 4$ with
(a) a number line.
(b) signed counters.

3. Explain how to compute $-4 - (-2)$ with
(a) signed counters.
(b) a number line.

4. An elevator is at an altitude of -30 ft. If it goes up 20 ft, what is its new altitude?
(a) Write an integer equation and a solution for this situation.
(b) What operation and category does this illustrate?

5. (a) An army that loses 300 soldiers is how much better off than an army that loses 800 soldiers?
(b) Write an integer equation for this situation.
(c) What subtraction category does this illustrate? (compare, missing part, take away)

6. Make up a temperature or money problem for $-3 - 6$ and give the result.

7. (a) Explain how to compute $3 \times (-2)$ with repeated sets of signed counters.
   (b) Show how to compute $3 \times (-2)$ with a number line.

8. A sixth grader knows how to multiply a negative by a positive. Start with $-4 \times 2 = -8$ and show how to find a negative times a negative by extending a pattern.

9. Describe an application using temperature, a walker, or your own model that suggests why
   (a) $3 \times (-4) = -12$ (b) $-3 \times 4 = -12$
   (c) $-3 \times (-4) = 12$

10. A sixth grader knows integer multiplication. Explain how to find the answer to $-18 \div (-3)$ by writing an equivalent multiplication question.

11. Sandy lost 18 pounds in 9 months. What was the average monthly change in her weight?
    (a) Write an integer equation and a solution for this situation.
    (b) What operation and category does this illustrate? (area, array, partition, repeated)

12. Compute the following without a calculator.
    (a) $-54 + 17$ (b) $-30 - (-28)$
    (c) $-8 \times 4 \times (-5) - 3$ (d) $72 \div (-8)$

13. What integer operations are commutative?

14. Give an example showing that integer subtraction is not associative.

15. For an integer n, the distributive property of multiplication over addition states that $-5n + 8n = $ _____.

16. Name a *property* that the integers have that whole numbers do not.

ALTERNATE ASSESSMENT—MAKE A PRESENTATION

Demonstrate one procedure and explain one major concept of the chapter to a classmate, your teacher, a friend, or family member. Or you could add to your portfolio, add to your journal, or write a unit test.

 ## SUGGESTED READINGS

Ashlock, R. *Error Patterns in Computation.* 8th ed. Englewood Cliffs, N.J.: Prentice-Hall, 2001.

Leutzinger, Larry, ed. *Mathematics in the Middle.* Reston, VA: NCTM, 1998.

Navigating Through Number and Operations, (4 vols.). Reston, VA: NCTM, 2003.

Use Infotrac to find articles in *Mathematics Teaching in the Middle School* on teaching children about integers.

6 | RATIONAL NUMBERS AS FRACTIONS

The word "fraction" comes from the Latin verb for "to break." The ancient Egyptians were using fractions in the marketplace to determine fair exchanges of goods long before the invention of negative integers. The Rhind Papyrus (1600 B.C.) gives a systematic treatment of unit fractions $\left(\dfrac{1}{\text{counting number}}\right)$. Teachers follow this historical order when they teach fractions before integers in elementary school.

Practical applications that led to the development of fractions include sharing problems (for example, 4 loaves of bread for 10 people) and measuring problems (for example, $2\frac{1}{2}$ ft). Today, we also talk about $\frac{3}{4}$ of a tank of gas or a $\frac{2}{3}$ majority to override a veto in Congress.

This chapter examines a subset of the set of fractions called the rational numbers: fractions that have an integer numerator and a nonzero integer denominator. In a mathematical development of number systems, one expands the whole numbers to the integers to provide answers to all subtraction problems. Still lacking answers to many whole-number and integer division problems (such as $2 \div 3$), one can expand the integers to form the rational number system, which provides answers to all integer division problems with nonzero divisors.

6.1 RATIONAL NUMBERS

NCTM Standards

- develop understanding of fractions as part of unit wholes, as parts of a collection, as locations on number lines, and as divisions of whole numbers (3–5)
- use models, benchmarks, and equivalent forms to judge the size of fractions (3–5)
- create and use representations to organize, record, and communicate mathematical ideas (pre-K–12)

Figure 6–1

The Babylonians and Egyptians used fractions in agriculture and business over 3500 years ago. Today, the increased use of computers has reduced the importance of fractions relative to decimals. However, fractional notation is still used in algebra, geometry, and calculus as well as for certain sharing and measuring applications. In this section, you'll examine the uses of fractions and methods for comparing the sizes of two fractions.

LE 1   Reasoning

Why do you think a fraction was used in the speed limit sign in Figure 6–1?

ELEMENTARY FRACTIONS

When you hear the word "fractions," how do you react?

Fractions are challenging, since they are the first numbers children study that represent a relationship between two quantities rather than a single quantity. A **fraction** is a symbol $\frac{a}{b}$, in which a and b are numbers and $b \neq 0$. In elementary school, children study fractions in which a and b are whole numbers, which will be referred to as **elementary fractions** in this text. Elementary fractions include $\frac{2}{3}$, $\frac{1}{2}$, and $\frac{11}{4}$ $\left(\text{or } 2\frac{3}{4}\right)$, whereas $\frac{-2}{3}$ and $\frac{\sqrt{2}}{3}$ are not elementary fractions. In a fraction $\frac{a}{b}$, a is called the **numerator**, and b is called the **denominator**.

Why do we use the terms "numerator" and "denominator"? Well, suppose I cut an apple pie into 8 equal pieces. The size of any serving is expressed in eighths (the denomination). The number of pieces that someone eats determines the numerator. If you eat 3 (number) pieces that are eighths (denomination), then you have eaten $\frac{3}{8}$ of the pie. (You also must have been rather hungry.)

FOUR MEANINGS OF AN ELEMENTARY FRACTION

What are the different meanings of elementary fractions?

Discussion

LE 2   Opener
What are some different things the fraction $\frac{2}{3}$ can represent in mathematics or in every-day life?

Did you come up with the following uses of $\frac{2}{3}$?

Four Meanings of Elementary Fractions

1. Part of a whole (or region): $\frac{2}{3}$ means to count 2 parts out of 3 equal parts.

2. Part of a set (or group): $\frac{2}{3}$ means to divide a set into 3 equal groups and find the number in 2 groups.

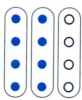

3. Point on a number line: $\frac{2}{3}$ is a number between 0 and 1. Divide the interval from 0 to 1 into 3 equal parts, and count 2 parts over from 0 to 1. In other words, go $\frac{2}{3}$ of the way from 0 to 1.

4. Division: $\frac{2}{3}$ means $2 \div 3$. Divide 2 into 3 equal parts.

Note that $\frac{2}{3}$ can also represent a ratio. This meaning of $\frac{2}{3}$ will be addressed in Chapter 7.

LE 3   Concept

Describe four common meanings of $\frac{3}{4}$. For each meaning, give its name, make a drawing, and describe how it works for $\frac{3}{4}$.

LE 4   Connection

Match each application with a fraction meaning.

(**a**) 2 desserts split equally among 4 people   (1) Part of a whole

(**b**) A ribbon $\frac{3}{4}$ ft long   (2) Point on a number line

(**c**) 2 slices of an 8-slice pizza   (3) Division

(**d**) $\frac{3}{5}$ of a group of 20 prefer juice over soda.   (4) Part of a set

LE 5   Connection

What are two different ways to divide 5 brownies equally between 2 people?

The part-of-a-whole meaning is not as simple as it seems. Many children have difficulty understanding how the numbers in the numerator and denominator have different meanings. The following exercise focuses on this.

LE 6   Reasoning

(**a**) If [ ] represents $\frac{2}{3}$ of a unit, what does the unit look like? (*Hint:* First represent $\frac{1}{3}$.)

(**b**) If [ ] represents $\frac{3}{4}$ of a unit, what does the unit look like?

Another difficulty in the part-of-a-whole meaning is the relationship between parts of two different wholes.

Classroom Connection

LE 7   Communication

A third grader says $\frac{1}{4}$ is more than $\frac{1}{2}$ in the following drawings. How would you explain why $\frac{1}{2}$ is greater than $\frac{1}{4}$?

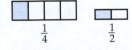

$\frac{1}{4}$ $\frac{1}{2}$

Numbers greater than 1 can also involve fractions. Suppose there are $2\frac{1}{4}$ waffles left to eat (Figure 6–2). A **mixed number** such as $2\frac{1}{4}$ is made up of an integer and a fraction.

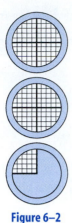

Figure 6–2

LE 8   Concept

(a) In Figure 6–2, how many fourths are in $2\frac{1}{4}$?

(b) Do you remember the fast way to convert $2\frac{1}{4}$ to $\frac{9}{4}$? Why does the method of computing $4 \times 2 + 1 = 9$ and putting it over 4 work?

(c) Start a number line at 0, and show where $2\frac{1}{4}$ is on the number line.

(d) Use division to show that $\frac{9}{4} = 2\frac{1}{4}$.

The mixed number $2\frac{1}{4}$ can be written as the improper fraction $\frac{9}{4}$. An **improper fraction** is an elementary fraction $\frac{a}{b}$, in which $a \geq b$.

Classroom Connection

LE 9   Communication

A fourth grader says she can show $\frac{3}{4}$ of a circle, but does not know how to show $\frac{4}{3}$ with a circle. What would you tell her?

RATIONAL NUMBERS

Children in elementary school study elementary fractions. In secondary-school mathematics, students also study the negatives of elementary fractions. The union of the set of elementary fractions and their negatives is the rational numbers. Secondary-school mathematics also includes other types of fractions, such as fractions that contain square roots.

This chapter focuses on rational numbers.

Definition: Rational Numbers

Rational numbers are all numbers that can be written as a quotient (ratio) of two integers $\frac{p}{q}$, in which $q \neq 0$.

The set of rational numbers includes all integers, since any integer can be written in rational form. For example, $-4 = \frac{-4}{1}$. Many decimals such as 0.3 are also rational numbers, since $0.3 = \frac{3}{10}$. Percents such as 42% are rational, since $42\% = \frac{42}{100}$. Decimals and percents will be discussed further in Chapter 7.

What numbers are *not* rational? It is impossible to write $\sqrt{2}$ and $\sqrt{3}$ as an integer over a nonzero integer! This will be proved for $\sqrt{2}$ in Chapter 7.

LE 10   Concept

Which of the following are rational numbers?

(a) $\frac{-3}{4}$ (b) 5   (c) $\sqrt{2}$ (d) 0   (e) 0.37   (f) $\sqrt{25}$

Both elementary fractions and the set of rational numbers written as fractions are subsets of the set of fractions. Examples of fractions include $\frac{1}{3}, \frac{-1}{3}, \frac{11}{4}, \frac{-3}{8}, \frac{\sqrt{5}}{7}, \frac{\sqrt{-3}}{2}$, and $-\frac{\pi}{2}$.

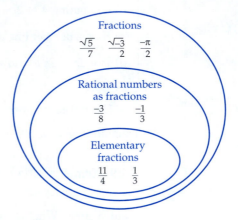

LE 11   Concept

Where in the preceding Venn diagram would each of the following go?

(a) $\frac{3}{1}$ (b) $\frac{0}{1}$ (c) $\frac{-2}{1}$ (d) $\frac{\sqrt{3}}{2}$

All properties and rules in this chapter will be stated for elementary fractions or rational numbers, although they usually apply to all fractions.

EQUIVALENT FRACTIONS FOR RATIONAL NUMBERS

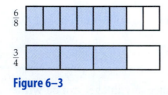

Figure 6–3

A friend of mine went on a diet. Instead of cutting a cake into 8 equal pieces and eating 6, he cut it into 4 equal pieces and ate only 3. Fractions such as $\frac{3}{4}$ and $\frac{6}{8}$ that look different may represent the same rational number (Figure 6–3).

Equivalent fractions are two fractions that represent the same *rational* number. Equivalent elementary fractions name the same amount. Many children first study equivalent fractions such as $\frac{1}{2} = \frac{2}{4} = \frac{3}{6} = \frac{4}{8}$ with fraction bars (strips).

You can also show equivalent fractions with number lines.

LE 12   Concept

(a) Draw a number line from 0 to $\frac{2}{2}$ and show where $\frac{1}{2}$ is.

(b) Underneath your number line, draw a second number line of the same length from 0 to $\frac{4}{4}$ and show where $\frac{2}{4}$ is.

(c) What patterns do you see in fractions that are equivalent, such as $\frac{1}{2}, \frac{2}{4}, \frac{3}{6}$, and $\frac{4}{8}$?

(d) If you multiply the numerator and denominator of $\frac{1}{2}$ by the same nonzero number, what happens? (Be sure to try both positive and negative numbers.)

The Fundamental Law of Fractions describes the general relationship between equivalent fractions.

The Fundamental Law of Fractions

For any rational number $\frac{a}{b}$ and any integer $c \neq 0$, $\frac{ac}{bc} = \frac{a}{b}$.

The Fundamental Law works in two ways. We can divide the numerator and denominator by the same nonzero integer c to simplify $\frac{ac}{bc}$ to $\frac{a}{b}$. We can also multiply both numerator and denominator by the same nonzero integer c to change $\frac{a}{b}$ to $\frac{ac}{bc}$. In most elementary-school texts, fractions are simplified with division symbols in accordance with the Fundamental Law.

$$\frac{12}{15} = \frac{12 \div 3}{15 \div 3} = \frac{4}{5}$$

LE 13   Concept

Use the Fundamental Law of Fractions to show why $\frac{6}{8} = \frac{3}{4}$.

LE 14   Reasoning

Can you *add* the same counting number to the numerator and denominator of an elementary fraction $\frac{a}{b}$ without changing its value? That is, does $\frac{a}{b} = \frac{a + c}{b + c}$ for counting numbers a, b, and c?

SIMPLIFYING ELEMENTARY FRACTIONS

In simplifying an elementary fraction such as $\frac{12}{20}$, one divides the numerator and denominator by the same counting number. The following exercise illustrates this process.

LE 15   Reasoning

(a) To write $\frac{12}{20}$ in simplest form in one step, divide the numerator and denominator

by _____.

(b) To write $\frac{28}{42}$ in simplest form in one step, divide the numerator and denominator

by _____.

(c) In parts (a) and (b), we divide the numerator and the denominator by the number known as the _____ of the numerator and denominator.

(d) Give a general procedure for simplifying an elementary fraction, based on your response to part (c).

A fraction $\frac{a}{b}$ is in **simplest form** if GCF(a, b) = 1. When the numerator and denominator of an elementary fraction have a greatest common factor (GCF) greater than 1, you can divide the numerator and denominator by the GCF, and the resulting fraction will be in simplest form! Try this method in the following exercise.

Discussion

LE 16   Skill

Find the GCF of 148 and 260, and use it to simplify $\frac{148}{260}$.

In simplifying fractions that have relatively large numerators and denominators, such as $\frac{148}{260}$, the GCF may make the work easier. Most fractions children encounter in elementary school $\left(\text{for example, } \frac{8}{16}\right)$ can be simplified just as easily without finding the GCF. But when children end up computing $\frac{8}{16} = \frac{4}{8} = \frac{2}{4} = \frac{1}{2}$ and ask why it took so many steps, you can tell them about the GCF. It always gets the simplification done in *one* step!

COMPARE AND ORDER FRACTIONS

The easiest fractions to compare are those with the same denominator.

LE 17   Skill

Show why $\frac{3}{4} > \frac{1}{4}$ with a fraction bar drawing.

Someone with a basic understanding of elementary fractions will have little difficulty determining which of two elementary fractions with the same denominator is larger $\left(\text{for example, } \frac{5}{6} > \frac{2}{6}\right)$. In general, for whole numbers a, b, and c with $c \neq 0$, if $a > b$, then $\frac{a}{c} > \frac{b}{c}$. On a standard number line, $\frac{a}{c}$ will be to the right of $\frac{b}{c}$.

When denominators are unequal, it is sometimes possible to compare the fractions with fraction bars.

LE 18   Skill

Compare $\frac{2}{3}$ and $\frac{5}{6}$ with a fraction bar drawing and state the result.

One can always compare two fractions by renaming them both with a common denominator.

LE 19 Skill

(a) In one group, 18 out of 30 people $\left(\dfrac{18}{30}\right)$ prefer butter to guns, and in a second group, 24 out of 40 people $\left(\dfrac{24}{40}\right)$ prefer butter to guns. How do the group preferences compare?

(b) Compare $\dfrac{2}{3}$ and $\dfrac{3}{5}$ by writing both fractions in terms of the least common denominator (which is the LCM of the two denominators.)

DENSENESS

LE 19(b) shows that $\dfrac{2}{3} > \dfrac{3}{5}$. Are there any fractions between $\dfrac{2}{3}$ and $\dfrac{3}{5}$? If so, how many are there?

LE 20 Reasoning

(a) Write $\dfrac{2}{3}$ and $\dfrac{3}{5}$ with a larger common denominator than 15. Then find a rational fraction that is between them.

(b) Write $\dfrac{2}{3}$ and $\dfrac{3}{5}$ with an even larger common denominator, and find more rational fractions between them.

(c) How many rational numbers are between $\dfrac{2}{3}$ and $\dfrac{3}{5}$?

This "betweenness" property that works for the set of rational numbers is called the denseness property.

The Denseness Property of Rational Numbers
Between any two rational numbers, there is another rational number.

There are infinitely many rational numbers *between* any two rational numbers. It's amazing, isn't it? If you pick any two rational numbers, you can keep finding numbers between them—forever.

LE 21   Summary

Tell what you learned about the different meanings of fractions in this section.

ANSWERS TO SELECTED LESSON EXERCISES

1. To catch people's attention

3. $\frac{3}{4}$ represents:

  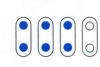

Part of a whole   Part of a set
(shade 3 of 4 equal   (shade 3 of 4 equal
parts)   groups)

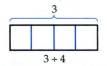   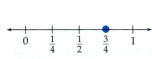

Division   Number line location
(divide 3   (go $\frac{3}{4}$ of the way
into 4   from 0 to 1)
equal parts)

4. (a) (3)   (b) (2)   (c) (1)   (d) (4)

5. Divide each brownie into 2 equal parts, or give each person 2 brownies and divide the remaining 1 into 2 equal parts.

6. (a)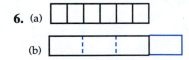

(b)

7. A fraction represents a relationship between equal parts and a whole. It does not compare the size of parts from two different-sized wholes.

8. (a) 9

(b) 2 wholes are 2 × 4 quarters plus 1 more quarter makes 9 quarters or $\frac{9}{4}$.

(c)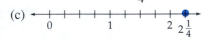

(d) $\frac{9}{4} = 9 \div 4 = 2\frac{1}{4}$

9. You will need to draw thirds of a circle. How many thirds in one whole circle? 3. Draw that. But that's not enough. Draw a second circle of the same size, and mark the 4th third in that circle.

10. (a), (b), (d), (e), (f)

11. (a), (b) In the elementary-fractions oval
(c) In the rational-numbers oval and outside the elementary-fractions oval
(d) In the fractions oval and outside the other two ovals

12. (d) You obtain an equivalent fraction.

13. $\frac{6}{8} = \frac{6 \div 2}{8 \div 2} = \frac{3}{4}$

14. No

15. (a) 4   (b) 14
(c) greatest common factor
(d) The answer follows the exercise.

16. GCF = 4; $\frac{148}{260} = \frac{148 \div 4}{260 \div 4} = \frac{37}{65}$

17.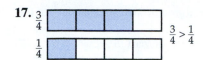
$\frac{3}{4}$
$\frac{1}{4}$
$\frac{3}{4} > \frac{1}{4}$

18. $\frac{2}{3}$
$\frac{5}{6}$
$\frac{2}{3} < \frac{5}{6}$

19. (a) $\frac{18}{30} = \frac{3}{5}$ and $\frac{24}{40} = \frac{3}{5}$ so $\frac{18}{30} = \frac{24}{40}$.

(b) $\frac{2}{3} = \frac{10}{15}$ and $\frac{3}{5} = \frac{9}{15}$ so $\frac{2}{3} > \frac{3}{5}$.

20. (a) $\frac{19}{30}$ (b) $\frac{37}{60}$ and $\frac{39}{60}$ $\left(\text{or } \frac{13}{20} \right)$

(c) An infinite number

6.1 HOMEWORK EXERCISES

Basic Exercises

1. Explain how to complete each diagram so that it shows $\frac{3}{5}$.

(a)

(b)

0   1

2. Explain how to complete each diagram so that it shows $\frac{2}{3}$.

(a)

(b)

0   1

3. Suppose you have two brownies to share equally.

(a) Two people would each get _____.

This shows 2 ÷ _____ = _____.

(b) Three people would each get _____.

This shows _____ ÷ _____ = _____.

(c) Four people would each get _____.

This shows _____ ÷ _____ = _____.

4. Suppose you have three grapefruits to share equally.

(a) Two people would each get _____.

This shows 3 ÷ _____ = _____.

(b) Three people would each get _____.

This shows _____ ÷ _____ = _____.

(c) Four people would each get _____.

This shows _____ ÷ _____ = _____.

5. A child shows $\frac{4}{5}$ as .

What is wrong with the diagram?

6. Divide the triangle into 3 equal regions.

7. Describe four common meanings of $\frac{1}{3}$. For each meaning, give its name, make a drawing, and describe how it works for $\frac{1}{3}$.

8. Describe four common meanings of $\frac{5}{8}$. For each meaning, give its name, make a drawing, and describe how it works for $\frac{5}{8}$.

9. Represent each situation with a diagram.

(a) A screw is $\frac{3}{8}$ in. long.

(b) You ate $\frac{3}{8}$ of a pizza.

(c) A group of 16 people is $\frac{3}{8}$ men.

(d) Eight people want to share a 3-ft sub sandwich equally.

10. Match each application with a fraction model.

Application	*Fraction Model*
(a) Your height in inches	(1) Part of a whole
(b) Fraction of those surveyed who jog	(2) Point on a number line
(c) The fraction of the day for which you are asleep	(3) Division
(d) Evenly sharing 3 pizzas among 5 people	(4) Part of a set

11. Describe two different ways to divide 3 brownies equally between 2 people.

12. Describe two different ways to divide 4 sandwiches equally among 3 people.

13. Explain how you would do each of the following. You may use a drawing.
(a) Divide 3 pizzas equally among 6 people.
(b) Divide 3 pizzas equally among 8 people.

14. Suppose you asked some fourth graders how they would divide 3 jumbo cookies equally among 4 people.
(a) How do you think they might work out the answer?
(b) What part of a cookie would each person receive?

15. Two children are figuring out the speed for someone who drives 425 miles at 50 miles an hour. One child gives the answer as 8 R25 The other says $8\frac{1}{2}$. They don't understand why their answers are the same. Explain why the 25 in 8 R25 and the $\frac{1}{2}$ in $8\frac{1}{2}$ mean the same thing.

16. Two third graders are figuring out how to divide 7 cookies equally among 3 people. One child gives the answer as 2 R1. The other says $2\frac{1}{3}$. They don't understand why their answers are the same. Explain why the 1 in 2 R1 and the $\frac{1}{3}$ in $2\frac{1}{3}$ mean the same thing.

17. *Explain* why $\frac{0}{5} = 0$.

18. *Explain* why $\frac{0}{0}$ is undefined.

19. (a) Draw 3 circles of the same size, and shade $\frac{1}{6}$ of the 3 circles.
(b) Show a second way to do part (a).
(c) What is $\frac{1}{6}$ of 3?

20. (a) Draw 2 circles of the same size, and shade $\frac{3}{4}$ of the 2 circles.
(b) Show a second way to do part (a).
(c) What is $\frac{3}{4}$ of 2?

21. Three seventh graders solved a problem. One got an answer of $\frac{-5}{12}$, one got $\frac{5}{-12}$, and one got $-\frac{5}{12}$.
(a) Which answers are the same?
(b) How would you use division rules for integers to convince the children which answers are the same?

22. Three seventh graders solved a problem. One got an answer of $\frac{-2}{-9}$, one got $\frac{2}{9}$, and one got $-\frac{2}{9}$.
(a) Which answers are the same?
(b) How would you use division rules for integers to convince the children which answers are the same?

23. If ▢▢▢▢▢ represents $\frac{3}{4}$ of a unit, what does a unit look like?

24. If ▢▢▢▢▢ represents $2\frac{1}{2}$ of a unit, what does a unit look like?

25. Six cats represent $\frac{3}{5}$ of all the cats in a house. Tell how to use a drawing to find how many cats are in the house.

26. If ▢▢▢ represents $\frac{1}{3}$, what does $1\frac{1}{6}$ look like?

27. If ▢ represents $\frac{3}{8}$ of a unit, what does $\frac{1}{2}$ look like?

28. If ▢ represents $2\frac{1}{3}$ of a unit, what does a unit look like?

29. Locate 1 on the number line below.

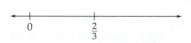

30. Locate 1 on the number line below.

31. Use fractions to explain the error made by the man in this cartoon.

"CUT MY PIZZA INTO FOUR PIECES...
NO WAY I COULD EAT EIGHT."

BLAIR

32. Give an example in which $\frac{1}{2}$ of something could be more than $\frac{3}{4}$ of something else.

33. (a) Use the part-of-a-whole meaning to show why
$3\frac{1}{2} = \frac{7}{2}$.

(b) Do you remember the fast way to convert $3\frac{1}{2}$ to $\frac{7}{2}$? Why does the method of computing $3 \times 2 + 1 = 7$ and putting it over 2 work?

(c) Start a number line at 0, and show where $3\frac{1}{2}$ is on the number line.

(d) Use division to show that $\frac{7}{2} = 3\frac{1}{2}$.

34. (a) Use the part-of-a-whole meaning to show why
$3\frac{2}{3} = \frac{11}{3}$.

(b) Do you remember the fast way to convert $3\frac{2}{3}$ to $\frac{11}{3}$? Why does the method of computing $3 \times 3 + 2 = 11$ and putting it over 3 work?

(c) Start a number line at 0, and show where $3\frac{2}{3}$ is on the number line.

(d) Use division to show that $\frac{11}{3} = 3\frac{2}{3}$.

35. Show that each number is rational by writing it as a quotient of two integers.

(a) 3   (b) -3 (c) $4\frac{1}{2}$ (d) -5.6

(e) 25%

36. Which of the following are rational numbers?

(a) -7 (b) $\frac{2}{3}$ (c) $\sqrt{3}$ (d) 0.2

37. W is the set of whole numbers, I is the set of integers, and Q is the set of rational numbers.
(a) Is $I \subseteq Q$?
(b) $W \cap Q = $ _____

38. Give an example of a number that is
(a) rational but not an integer.
(b) a fraction but not a rational number.

39. Show why $\frac{3}{5} = \frac{6}{10}$ with
(a) a part-of-the-whole diagram.
(b) two number lines.
(c) the Fundamental Law of Fractions.

40. Show why $\frac{2}{8} = \frac{1}{4}$ with
(a) a part-of-the-whole diagram.
(b) two number lines.
(c) the Fundamental Law of Fractions.

41. Two companies conduct surveys asking people if they favor stronger controls on air pollution. The first company asks 1,500 people, and the second asks 2,000 people. In the first group, 1,200 say yes. Make up results for the second group that would be considered equivalent.

42. The third grade is voting on whether to go to a movie or a play. In Ms. Chan's class, 12 out of 20 students prefer going to the movie. In Mr. Brussat's class, 15 out of 25 students prefer going to the movie. Explain in what sense both classes equally prefer the movie over the play.

43. Write each fraction in simplest form. (Assume that x, y, and z are counting numbers.)
(a) $\dfrac{168}{464}$ (b) $\dfrac{xy^2}{xy^3z}$

44. Write each fraction in simplest form.
(a) $\dfrac{48}{264}$ (b) $\dfrac{2^8 - 2^7}{2^7 - 2^6}$

45. Show why $\dfrac{1}{3} < \dfrac{2}{3}$ with a fraction-bar drawing.

46. Show why $\dfrac{5}{8} > \dfrac{3}{8}$ with a fraction-bar drawing.

47. Compare $\dfrac{7}{8}$ and $\dfrac{3}{4}$ with a fraction-bar drawing, and state the result.

48. Compare $\dfrac{1}{3}$ and $\dfrac{2}{9}$ with a fraction-bar drawing, and state the result.

49. Compare $\dfrac{4}{9}$ and $\dfrac{7}{15}$ by writing both fractions in terms of the least common denominator.

50. Compare $\dfrac{7}{10}$ and $\dfrac{3}{4}$ by writing both fractions in terms of the least common denominator.

 51. In one class, 14 out of 23 students would rather go to an aquarium than a circus. In another class, 17 out of 30 students would rather go to an aquarium than a circus. Compare their preferences by writing two fractions with a common denominator.

52. You have two different recipes for making orange juice from concentrate. The first says to mix 2 cups of concentrate with 6 cups of water. The second says to mix 3 cups of concentrate with 8 cups of water. Which recipe will have a stronger orange flavor?

53. Use a number line to show which is greater, $-2\dfrac{1}{4}$ or $-\dfrac{1}{2}$.

54. A third grader says $\dfrac{1}{4}$ is less than $\dfrac{1}{5}$ because 4 is less than 5. What would you tell the child?

55. You can sometimes compare fractions by relating them to $\dfrac{1}{2}$.
(a) Tell whether each of the following fractions is less than $\dfrac{1}{2}$ or greater than $\dfrac{1}{2}$.
(1) $\dfrac{4}{9}$ (2) $\dfrac{7}{12}$
(b) How did you know how each fraction compared to $\dfrac{1}{2}$?

56. (a) Tell whether each of the following fractions is less than 1 or greater than 1.
(1) $\dfrac{4}{5}$ (2) $\dfrac{5}{4}$
(b) How did you know how each fraction compared to 1?

57. Show how to use common denominators to find three fractions between $\dfrac{3}{4}$ and $\dfrac{7}{8}$.

58. Show how to use common denominators to find three fractions between $\dfrac{1}{2}$ and $\dfrac{2}{3}$.

59. Is the set of whole numbers dense? (If you pick any two whole numbers, is there always a whole number between them?)

60. Is the set of integers dense?

Extension Exercises

61. Consider the following regions, which represent manipulatives called pattern blocks (Activity Card 6). (There is also a parallelogram block.)

Hexagon   Trapezoid   Triangle

(a) How many triangular regions would it take to cover the hexagonal region?

(b) What fractional part of the hexagonal region is the triangular region?

(c) What fractional part of the hexagonal region is the trapezoidal region?

(d) Two triangular regions would be what fractional part of the trapezoidal region?

62. If a trapezoid represents $\frac{1}{3}$, what is the value of

(a) a triangle?   (b) two hexagons?

63. A sixth grader wants to find a fraction between $\frac{5}{8}$ and $\frac{5}{9}$. They say $\dfrac{5}{8\frac{1}{2}}$, which is $\dfrac{5 \times 2}{8\frac{1}{2} \times 2} = \dfrac{10}{17}$.

Is this correct? If not, what would you tell the child?

64. Prove: If a and b are counting numbers and $a > b$, then $\frac{1}{b} > \frac{1}{a}$.

Puzzle Time

65. Copy the picture, and cut out all nine squares. Fit the nine pieces together into one large square so that all the edges that touch have equivalent fractions.

1 0	$1\frac{1}{3}$ $\frac{1}{4}$	$\frac{1}{2}$ $\frac{7}{7}$ $\frac{2}{5}$
$\frac{2}{16}$ $\frac{4}{3}$ $\frac{3}{4}$	$\frac{1}{3}$ $\frac{2}{6}$ $\frac{2}{8}$	$\frac{1}{3}$ $\frac{6}{8}$ $\frac{3}{6}$ $\frac{10}{12}$
$\frac{1}{3}$ $\frac{4}{10}$	$\frac{5}{6}$ $\frac{2}{4}$	$\frac{1}{2}$ $\frac{0}{3}$ $\frac{1}{8}$

6.2   ADDITION AND SUBTRACTION OF RATIONAL NUMBERS

NCTM Standards

- use visual models, benchmarks, and equivalent forms to add and subtract commonly used fractions and decimals (3–5)

- understand the meanings and effects of operations with fractions, decimals, and integers (6–8)

- develop and analyze algorithms for computing with fractions, decimals, and integers, and develop fluency in their use (6–8)

You walked $\frac{1}{5}$ of a mile to school and then $\frac{3}{5}$ of a mile from school to your friend's house. How far did you walk altogether? How much farther was the second walk than the first?

To solve the first problem, you would add $\frac{1}{5}$ and $\frac{3}{5}$. To solve the second, you would subtract $\frac{1}{5}$ from $\frac{3}{5}$. Adding and subtracting rational numbers are the subjects of this lesson.

ADDING AND SUBTRACTING FRACTIONS THAT HAVE LIKE DENOMINATORS

Children first add and subtract fractions that have the same denominators. They would learn to solve the two introductory problems with manipulatives, pictures, and symbols (Figure 6–4).

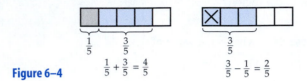

Figure 6–4

$$\frac{1}{5} + \frac{3}{5} = \frac{4}{5} \qquad \frac{3}{5} - \frac{1}{5} = \frac{2}{5}$$

Pictures and manipulatives help clarify why one should not add or subtract the denominators. The general addition and subtraction rules for like denominators can be given algebraically—by you! Complete the following.

LE 1   Skill

(a) For rational numbers, $\frac{a}{c} + \frac{b}{c} =$ _____.

(b) For rational numbers, $\frac{d}{f} - \frac{e}{f} =$ _____.

The rules for adding and subtracting rational numbers that have the same denominator are as follows.

Addition of Rational Numbers That Have Like Denominators
$$\frac{a}{c} + \frac{b}{c} = \frac{a + b}{c}$$

Subtraction of Rational Numbers That Have Like Denominators
$$\frac{d}{f} - \frac{e}{f} = \frac{d - e}{f}$$

Next, consider adding fractions with unlike denominators.

ADDING ELEMENTARY FRACTIONS THAT HAVE UNLIKE DENOMINATORS

Does $\frac{1}{2} + \frac{1}{3} = \frac{2}{5}$? According to the National Assessment of Educational Progress, about 30% of the seventh graders in the United States think so.

Classroom Connection

LE 2   Opener

A fifth grader thinks that $\frac{1}{2} + \frac{1}{3} = \frac{2}{5}$. How would you convince the child that $\frac{2}{5}$ is not the right answer?

Did you use pictures of fraction bars or circles? Pictures of fraction bars make it clear why a common denominator is needed. Then, by trading for fraction bars with the common denominator, one can find the answer.

• EXAMPLE 1

(a) Draw a fraction bar for each fraction in $\frac{1}{2} + \frac{1}{3}$.

(b) Draw a fraction bar for the sum, and explain why a common denominator is needed.
(c) Draw a fraction bar for each fraction with the least common denominator.

(d) Show how to compute $\frac{1}{2} + \frac{1}{3}$ using a picture.

SOLUTION

(a)

(b) Try to determine the sum of $\frac{1}{2}$ and $\frac{1}{3}$ without a common denominator.

What part of the whole is shaded to show the sum in Figure 6–5? There is no name for the sum, since it is made out of two different units (denominations): halves and thirds. The answer cannot be determined without a *common denominator*.

$\frac{1}{2} + \frac{1}{3} =$

Sum

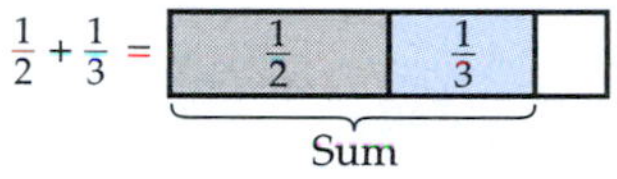

Figure 6–5

(c)

(d) After rewriting both fractions with a common denominator (sixths), one can use a combine-measures pictures to show the sum.

$$\frac{1}{2} + \frac{1}{3} = \frac{3}{6} + \frac{2}{6} = \qquad = \frac{5}{6}$$

Aren't common denominators wonderful? Without them, we could not name sums or differences of fractions that have unlike denominators.

LE 3   Connection

(a) Draw a fraction bar for each fraction in $\frac{1}{2} + \frac{1}{4}$.

(b) Draw a fraction bar for the sum, and explain why a common denominator is needed.
(c) Draw a fraction bar for each fraction with the least common denominator.

(d) Show how to compute $\frac{1}{2} + \frac{1}{4}$ using a picture.

Before you can add rational numbers that have unlike denominators, you must rewrite them with a common denominator. As you saw in Section 6.1, you can find the least common denominator (LCD) using the least common multiple (LCM) of all the denominators. For example, the LCD for $\frac{1}{6}$ and $\frac{7}{8}$ is the LCM of 6 and 8, which is 24.

$$\frac{1}{6} + \frac{7}{8}$$

$$\text{LCM} = 24$$

$$\frac{1 \cdot 4}{6 \cdot 4} + \frac{7 \cdot 3}{8 \cdot 3} = \frac{4}{24} + \frac{21}{24} = \frac{25}{24} = 1\frac{1}{24}$$

Negative rational numbers can be added in the same way. To use only positive denominators, a negative rational can be written as $\frac{-3}{5}$ rather than $\frac{3}{-5}$.

The general rule for adding rational numbers that have unlike denominators is as follows.

Addition of Rational Numbers That Have Unlike Denominators

To add rational numbers $\frac{a}{b} + \frac{c}{d}$ in which $b > 0$, $d > 0$, and $b \neq d$:

1. Rename each fraction with the least common denominator, that is, LCM(b, d).
2. Add the fractions using the addition rule for like denominators.

Use this rule in LE 4.

LE 4   Skill

(a) What is the least common denominator of $\frac{3}{4x}$ and $\frac{5}{6}$?

(b) Compute $\frac{3}{4x} + \frac{5}{6}$.

Are you aware that fractions can be added with any common denominator? However, if you do not use the *least* common denominator, additional simplification will be required at the end.

$$\frac{5}{12} + \frac{1}{6} = \frac{5}{12} + \frac{2}{12} = \frac{7}{12} \qquad \text{or} \qquad \frac{5}{12} + \frac{1}{6} = \frac{30}{72} + \frac{12}{72} = \frac{42}{72} = \frac{7}{12}$$

$$\text{(LCD)} \qquad\qquad\qquad \left(\begin{array}{c}\text{Other common}\\\text{denominator}\end{array}\right) \quad \text{(Extra step)}$$

Until the seventeenth century, most people used the *product* of all the denominators as the common denominator. For the last 300 years, most people have used the least common denominator in textbook examples, and they rarely need prime factorizations to do it.

SUBTRACTING ELEMENTARY FRACTIONS THAT HAVE UNLIKE DENOMINATORS

Suppose a family spends $\frac{1}{3}$ of their income on taxes and $\frac{1}{4}$ of their income on rent. How much more of their income do they spend on taxes than on rent? This example requires subtracting fractions that have unlike denominators.

Fraction pictures clarify subtraction rules for unlike denominators. The following example explains why one needs a common denominator to compute $\frac{1}{3} - \frac{1}{4}$.

• EXAMPLE 2

(a) Draw a fraction bar for each fraction in $\frac{1}{3} - \frac{1}{4}$.

(b) Explain why a common denominator is needed.

(c) Draw a fraction bar for each fraction with the least common denominator.

(d) Show how to compute $\frac{1}{3} - \frac{1}{4}$ using a picture.

SOLUTION

(a)

$\frac{1}{3}$ $\qquad\qquad$ $\frac{1}{4}$

(b) How much is left when we take $\frac{1}{4}$ away from $\frac{1}{3}$? We can't tell. A common denominator is needed.

(c)

$\frac{4}{12}$ $\qquad\qquad$ $\frac{3}{12}$

(d) $\frac{1}{3} - \frac{1}{4} = \frac{4}{12} - \frac{3}{12} =$  $= \frac{1}{12}$ ●

One can also use compare-measures subtraction with two separate fraction bars and see how much longer $\frac{4}{12}$ is than $\frac{3}{12}$.

LE 5   Connection

(a) Draw a fraction bar for each fraction in $\frac{1}{2} - \frac{1}{3}$.

(b) Explain why a common denominator is needed.

(c) Draw a fraction bar for each fraction with the least common denominator.

(d) Show how to compute $\frac{1}{2} - \frac{1}{3}$ using a picture.

The general rule for subtracting rational numbers that have unlike denominators is as follows.

> **Subtraction of Rational Numbers That Have Unlike Denominators**
>
> To subtract rational numbers $\dfrac{a}{b} - \dfrac{c}{d}$ in which $b > 0$, $d > 0$, and $b \neq d$:
>
> **1.** Rename each fraction with the least common denominator, that is, LCM(b, d).
>
> **2.** Subtract the fractions using the subtraction rule for like denominators.

Use this rule in the next exercise.

LE 6   Skill

Compute $\dfrac{3}{20} - \dfrac{1}{12}$.

The classifications for whole-number operations apply to some addition and subtraction applications involving rational numbers.

LE 7   Connection

What operation and category are illustrated in the following problem? "Last week you worked $37\frac{1}{2}$ hours, and this week you worked 45 hours. How many more hours did you work this week?" (compare, missing part, take away)

ADDING AND SUBTRACTING MIXED NUMBERS

After working with proper fractions, children learn to add and subtract mixed numbers in problems that do not involve regrouping. For example, you buy $1\frac{1}{4}$ lb of low-fat Swiss cheese and $2\frac{3}{8}$ lb of provolone. How much cheese do you have altogether? How much more provolone than Swiss do you have?

$$1\frac{1}{4}\text{ lb} + 2\frac{3}{8}\text{ lb} = 1\frac{2}{8} + 2\frac{3}{8} = 3\frac{5}{8}\text{ lb} \qquad 2\frac{3}{8}\text{ lb} - 1\frac{1}{4}\text{ lb} = 2\frac{3}{8} - 1\frac{2}{8} = 1\frac{1}{8}\text{ lb}$$

Then children study problems that involve regrouping.

• EXAMPLE 3

(a) Compute $1\frac{3}{4} + 4\frac{3}{8}$.   **(b)** Compute $4\frac{3}{8} - 1\frac{3}{4}$.

SOLUTION

(a) *Method 1 (Regrouping)*
Elementary school textbooks use method 1.

$$1\frac{3}{4} + 4\frac{3}{8} = 1\frac{6}{8} + 4\frac{3}{8} = 5\frac{9}{8} = 6\frac{1}{8}$$

Method 2 (Improper Fractions)
Some adults prefer this method.

$$1\frac{3}{4} + 4\frac{3}{8} = \frac{7}{4} + \frac{35}{8} = \frac{14}{8} + \frac{35}{8} = \frac{49}{8} = 6\frac{1}{8}$$

(b) *Method 1 (Regrouping)*
Elementary school textbooks use method 1.

$$4\frac{3}{8} - 1\frac{3}{4} = 4\frac{3}{8} - 1\frac{6}{8} = 3\frac{11}{8} - 1\frac{6}{8} = 2\frac{5}{8}$$

Method 2 (Improper Fractions)

$$4\frac{3}{8} - 1\frac{3}{4} = \frac{35}{8} - \frac{7}{4} = \frac{35}{8} - \frac{14}{8} = \frac{21}{8} = 2\frac{5}{8}$$ ●

LE 8   Skill

Compute $8\frac{2}{3} - 3\frac{5}{6}$

(a) with regrouping.
(b) with improper fractions.
(c) What are the advantages of each method?

FRACTION CALCULATORS

Fraction calculators have special keys for entering and simplifying fractions. The $\boxed{\text{Simp}}$ key can be used to simplify fractions. Suppose you display 12/16. To specify what number to divide into the numerator and denominator, type that number between $\boxed{\text{Simp}}$ and $\boxed{=}$. For example, 12 $\boxed{/}$ 16 $\boxed{\text{Simp}}$ 2 $\boxed{=}$ will display 6/8.

LE 9   Skill

If you have a fraction calculator, enter $\frac{50}{60}$ and see if you can simplify it in

(a) one step.   (b) two steps.

Fraction calculators also give results of fraction arithmetic in fraction form. To compute $\frac{1}{2} + \frac{3}{8}$, press 1 $\boxed{/}$ 2 $\boxed{+}$ 3 $\boxed{/}$ 8 $\boxed{=}$. The answer will be displayed as 7/8 or $\frac{7}{8}$. If the keys on your calculator are different, consult with your instructor or a manual.

LE 10   Skill

If you have a fraction calculator, compute $\frac{5}{8} - \frac{1}{2}$.

Fraction calculators also convert fractions to decimals, and they convert improper fractions to mixed numerals. If you have a fraction calculator, you can experiment with these features.

LE 11   Summary
Tell what you learned about adding and subtracting fractions with unlike denominators.

ANSWERS TO SELECTED LESSON EXERCISES

1. (a) $\dfrac{a+b}{c}$ (b) $\dfrac{d-e}{f}$

3. (a)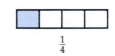

$\frac{1}{2}$ $\frac{1}{4}$

(b) $\dfrac{1}{2} + \dfrac{1}{4}$ would equal the shaded region shown,

 but we cannot name the sum

unless we use a common denominator (fourths).

(c) [diagram] $\frac{2}{4}$ [diagram] $\frac{1}{4}$

(d) $\dfrac{1}{2} + \dfrac{1}{4} = \dfrac{2}{4} + \dfrac{1}{4} =$ [diagram] $= \dfrac{3}{4}$

4. (a) $12x$ (b) $\dfrac{9+10x}{12x}$

5. (a) [diagram] $\frac{1}{2}$ [diagram] $\frac{1}{3}$

(b) How much is left when we take $\dfrac{1}{3}$ away from $\dfrac{1}{2}$?

We can't tell. A common denominator is needed.

(c)

$\frac{3}{6}$ $\frac{2}{6}$

(d) $\dfrac{1}{2} - \dfrac{1}{3} = \dfrac{3}{6} - \dfrac{2}{6} =$ [diagram] $= \dfrac{1}{6}$

6. $\dfrac{1}{15}$

7. Subtraction; compare measures

8. (a) $8\dfrac{2}{3} - 3\dfrac{5}{6} = 8\dfrac{4}{6} - 3\dfrac{5}{6} = 7\dfrac{10}{6} - 3\dfrac{5}{6} = 4\dfrac{5}{6}$

(b) $8\dfrac{2}{3} - 3\dfrac{5}{6} = \dfrac{26}{3} - \dfrac{23}{6} = \dfrac{52}{6} - \dfrac{23}{6} = \dfrac{29}{6} = 4\dfrac{5}{6}$

(c) The regrouping method takes fewer steps and

avoids very large numerators $\left(\text{for example, the}\right.$

improper fraction for $\left. 20\dfrac{3}{8}\right)$. The improper fractions

method is easier to understand, since it does not
require regrouping and is more like the methods
for multiplying and dividing mixed numbers.

6.2   HOMEWORK EXERCISES

Basic Exercises

1. Assume that the following fractions are rational.

(a) $\dfrac{3}{y} + \dfrac{1}{y} =$ _____

(b) $\dfrac{x}{n} - \dfrac{2x}{n} =$ _____

2. Draw a diagram that shows why $\dfrac{1}{8} + \dfrac{3}{8} = \dfrac{1}{2}$.

3. A child thinks $\dfrac{1}{2} + \dfrac{1}{8} = \dfrac{2}{10}$. Draw a fraction bar for

each fraction, and explain why $\dfrac{2}{10}$ cannot be the answer.

4. A child thinks $\dfrac{1}{2} + \dfrac{1}{4} = \dfrac{2}{6}$. Draw a fraction bar for

each fraction, and explain why $\dfrac{2}{6}$ cannot be the answer.

5. (a) Draw a fraction bar for each fraction in $\dfrac{1}{3} + \dfrac{1}{4}$.

(b) Draw a fraction bar for the sum, and explain why a
common denominator is needed.

(c) Draw a fraction bar for each fraction with the least
common denominator.

(d) Show how to compute $\dfrac{1}{3} + \dfrac{1}{4}$ using a picture.

6. (a) Draw a fraction bar for each fraction in $\frac{1}{2} + \frac{2}{5}$.

(b) Draw a fraction bar for the sum, and explain why a common denominator is needed.

(c) Draw a fraction bar for each fraction with the least common denominator.

(d) Show how to compute $\frac{1}{2} + \frac{2}{5}$ using a picture.

7. A fifth grader says $\frac{1}{3} + \frac{1}{3}$ is $\frac{2}{6}$. She draws two sets of three circles and shades one out of the three circles in each set. What would you tell the child?

8. What is the difference between what the 3 and the 4 represent in $\frac{3}{4}$?

9. Compute the following without a calculator.

(a) $\frac{5}{12} + \frac{3}{8}$ (b) $\frac{1}{a} + \frac{2}{b}$

10. Compute the following without a calculator.

(a) $\frac{2}{15} + \frac{1}{21}$ (b) $\frac{3}{2n} + \frac{4}{5n}$

11. (a) Draw a fraction bar for each fraction in $\frac{1}{4} - \frac{1}{6}$.

(b) Explain why a common denominator is needed.

(c) Draw a fraction bar for each fraction with the least common denominator.

(d) Show how to compute $\frac{1}{4} - \frac{1}{6}$ using a picture.

12. (a) Draw a fraction bar for each fraction in $\frac{1}{3} - \frac{1}{5}$.

(b) Explain why a common denominator is needed.

(c) Draw a fraction bar for each fraction with the least common denominator.

(d) Show how to compute $\frac{1}{3} - \frac{1}{5}$ using a picture.

13. (a) Find the least common denominator of $\frac{7}{20}$ and $\frac{3}{28}$.

(b) Give two other common denominators.

(c) Compute $\frac{7}{20} - \frac{3}{28}$ without a calculator.

14. Compute the following without a calculator.

(a) $\frac{5}{12} - \frac{1}{20}$ (b) $\frac{3}{4c} - \frac{5}{6c}$

15. You planned to work on a project for about $4\frac{1}{2}$ hours today.

(a) If you have been working on it for $1\frac{3}{4}$ hours, how much more time will it take?

(b) What operation and classification are illustrated here? (compare, missing part, take away)

16. In an apartment complex, $\frac{7}{8}$ of the people speak English as their native language, and $\frac{1}{16}$ speak Spanish as their native language.

(a) What fraction speak a language other than English or Spanish as their native language?

(b) What operations and classifications are illustrated here? (combine, compare, missing part, take away)

17. Compute $5\frac{3}{4} + 2\frac{5}{8}$

(a) with regrouping.   (b) with improper fractions.

18. Compute $3\frac{5}{12} + 7\frac{2}{3}$

(a) with regrouping.   (b) with improper fractions.

19. Compute $10\frac{1}{6} - 5\frac{2}{3}$

(a) with regrouping.   (b) with improper fractions.

20. Compute $90\frac{1}{3} - 32\frac{7}{9}$

(a) with regrouping.   (b) with improper fractions.

21. Make up a compare measures problem for $5\frac{1}{4} - 2\frac{1}{2}$, and give the answer as a fraction.

22. Make up a take away measures problem for $6\frac{1}{2} - 3\frac{3}{4}$ and give the answer as a fraction.

23. Solve mentally.

(a) $\frac{1}{8} + \square = \frac{5}{8}$ (b) $3\frac{9}{10} - r = 1\frac{3}{10}$

24. Solve mentally.

(a) $4\frac{1}{8} + x = 10\frac{3}{8}$ (b) $t - \frac{1}{12} = \frac{7}{12}$

25. Do the following computations in your head. Tell how you did each one.

(a) $4\frac{1}{4} + 3\frac{3}{4}$ (b) $8 - 3\frac{1}{3}$

26. Fill in each square with either a + sign or a − sign to complete each equation correctly.

(a) $1\frac{1}{4} \,\square\, \frac{1}{4} \,\square\, \frac{3}{4} = \frac{3}{4}$

(b) $1\frac{7}{8} \,\square\, \frac{1}{4} \,\square\, \frac{3}{8} = 1\frac{1}{4}$

27. One can avoid regrouping in some fraction subtraction problems by using equal addition.

For example, $3\frac{1}{5} - 1\frac{2}{5} = 3\frac{4}{5} - 2 = 1\frac{4}{5}$.

Show how to use this method to compute

(a) $8\frac{1}{3} - 3\frac{2}{3}$.   (b) $5\frac{1}{4} - 2\frac{3}{4}$.

28. One can solve subtraction problems that would involve regrouping by changing them to addition. For example, $3\frac{1}{5} - 1\frac{2}{5} = x$ is equivalent to $1\frac{2}{5} + x = 3\frac{1}{5}$.

What would you add to $1\frac{2}{5}$ to obtain $3\frac{1}{5}$?

$1\frac{2}{5} + \frac{3}{5} + 1 + \frac{1}{5} = 3\frac{1}{5}$. So $x = \frac{3}{5} + 1 + \frac{1}{5} = 1\frac{4}{5}$.

Show how to use this method to compute the following.

(a) $8\frac{1}{3} - 3\frac{2}{3}$ (b) $5\frac{1}{4} - 2\frac{3}{4}$

29. Use a fraction calculator to

(a) simplify $\dfrac{125}{300}$.   (b) compute $\dfrac{9}{10} - \dfrac{3}{4}$.

30. Use a fraction calculator to

(a) simplify $\dfrac{20}{48}$.   (b) compute $\dfrac{3}{8} + \dfrac{3}{4}$.

Extension Exercises

31. (a) What will the next equation be if the pattern continues? Is the equation true?

$$\frac{1}{2} = \frac{1}{3} + \frac{1}{6}$$
$$\frac{1}{3} = \frac{1}{4} + \frac{1}{12}$$

(b) Complete the general equation showing the pattern in part (a).

$$\frac{1}{N} = \underline{\hspace{1.5cm}}$$

(c) Is part (b) an example of induction or deduction?

(d) Show that the equation in part (b) is true.

(e) Does part (d) involve induction or deduction?

32. (a) What will the next equation be if the pattern continues?

$$\frac{1}{2} + \frac{1}{6} = \frac{8}{12}$$
$$\frac{1}{6} + \frac{1}{12} = \frac{18}{72}$$

(*Hint:* Find two factors of each denominator.)

(b) Complete the general equation showing the pattern in part (a).

$$\frac{1}{N(N-1)} + \underline{\hspace{1.5cm}} = \underline{\hspace{1.5cm}}$$

33. The ancient Egyptians represented every elementary fraction other than 0 and $\frac{2}{3}$ as the sum of unequal **unit fractions,** fractions that have a numerator of 1. For example, $\frac{2}{7} = \frac{1}{4} + \frac{1}{28}$. They did this to avoid certain computational difficulties. Write each of the following as the sum of unequal unit fractions.

(a) $\dfrac{3}{4}$ (b) $\dfrac{3}{26}$ (c) $\dfrac{5}{8}$ (d) $\dfrac{7}{9}$

34. In the December 1991 *Mathematics Teacher*, Arthur Howard describes situations in which the standard algorithm for adding fractions does not yield the correct answer. For example, suppose two third-grade classes go on a field trip. One class has 10 girls out of a class of 25, and the other has 12 girls out of a class of 24. What fraction of the whole group is girls?

$$\frac{10}{25} \oplus \frac{12}{24} = \frac{10 + 12}{25 + 24} = \frac{22}{49}$$

This example illustrates nonstandard fraction addition, denoted by ⊕, in which the "reference unit" (denominator) of the sum is the *sum* of the reference units of the addends. In other words, one adds the denominators of the addends!

(a) Make up a problem about fruit that is solved with ⊕.

(b) Make up a problem that is solved with an analogous ⊖ operation.

6.3 MULTIPLICATION AND DIVISION OF RATIONAL NUMBERS

NCTM Standards

- understand the meanings and effects of operations with fractions, decimals, and integers (6–8)
- develop and analyze algorithms for computing with fractions, decimals, and integers, and develop fluency in their use (6–8)
- create and use representations to organize, record, and communicate mathematical ideas (preK–12)

You are planning a chicken dinner. Each person will eat about $\frac{1}{4}$ of a chicken. How many chickens will you need for 3 people? To solve this problem, you would multiply $\frac{1}{4}$ by 3. Multiplying rational numbers is the first topic of this lesson.

MULTIPLYING RATIONAL NUMBERS

Now about that chicken dinner. You need 3 servings, each consisting of $\frac{1}{4}$ lb chicken. Using repeated addition, $3 \times \frac{1}{4} = \frac{1}{4} + \frac{1}{4} + \frac{1}{4} = \frac{3}{4}$.

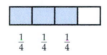

 LE 1 Concept

Show how to compute $5 \times \frac{1}{8}$ using repeated addition.

After studying how to multiply a whole number and a fraction, children study the product of two fractions. Consider the following application. Suppose $\frac{3}{4}$ of a field is plowed. Then $\frac{1}{2}$ of the plowed field is planted with tomatoes. Just as 2 sets of $\frac{3}{4}$ is the same as $2 \times \frac{3}{4}$, a $\frac{1}{2}$ of a set of $\frac{3}{4}$ is the same as $\frac{1}{2} \times \frac{3}{4}$. When you find $\frac{1}{2}$ of $\frac{3}{4}$, the fraction $\frac{1}{2}$ is known as an operator. Example 1 shows how to find the result of $\frac{1}{2}$ of $\frac{3}{4}$ or $\frac{1}{2} \times \frac{3}{4}$ using two diagrams.

• EXAMPLE 1

Suppose a child does not know the multiplication rule for fractions. *Explain* how to compute $\frac{1}{2} \times \frac{3}{4}$ using two diagrams.

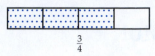

$\frac{3}{4}$

Figure 6–6

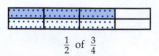

$\frac{1}{2}$ of $\frac{3}{4}$

Figure 6–7

SOLUTION

$\frac{1}{2} \times \frac{3}{4}$ is the same as $\frac{1}{2}$ of $\frac{3}{4}$. Show $\frac{3}{4}$ and then take $\frac{1}{2}$ of it. Place dots in $\frac{3}{4}$ of a rectangular diagram to show $\frac{3}{4}$, the part of the field that is plowed (Figure 6–6).

To show the part with tomatoes, $\frac{1}{2}$ of $\frac{3}{4}$, darken $\frac{1}{2}$ of the *dotted* part, as in Figure 6–7. What part of the whole figure is darkened? $\frac{3}{8}$. So $\frac{1}{2}$ of $\frac{3}{4} = \frac{1}{2} \times \frac{3}{4} = \frac{3}{8}$. You can also work out $\frac{1}{2} \times \frac{3}{4}$ with an area model.   ●

Discussion

LE 2   Communication

Explain how to compute $\frac{1}{2} \times \frac{3}{5}$ using two diagrams.

The results of Example 1 and LE 2 suggest the multiplication rule for fractions.

LE 3   Reasoning

Find the pattern in the preceding results.

(a) Complete the chart, showing results from Example 1 and LE 2.

Factors	Product
$\frac{1}{2} \times \frac{3}{4}$	
$\frac{1}{2} \times \frac{3}{5}$	

(b) The results from part (a) suggest that the general rule for multiplying rational numbers is: $\frac{a}{b} \times \frac{c}{d} = $ _____

(c) Is part (b) an example of induction or deduction?

LE 3 suggests the rule for multiplying rational numbers. This rule is the definition of multiplication of rational numbers.

Definition: Multiplication of Rational Numbers

$$\frac{a}{b} \times \frac{c}{d} = \frac{ac}{bd}$$

Because the definition is written for fractions, we must *rewrite mixed numbers as improper fractions* before applying this rule.

Another interesting process in multiplying fractions is simplifying the resulting fraction before completing the multiplication. This is no longer referred to as "canceling" in textbooks; it is now called a "shortcut" or "simplifying by dividing by a common factor."

LE 4 Opener

In multiplying $\frac{3}{7} \times \frac{10}{21}$, why is one allowed to simplify as follows?

$$\frac{\overset{1}{\cancel{3}}}{7} \times \frac{10}{\underset{7}{\cancel{21}}}$$

Why does the method of LE 4 work?

● **EXAMPLE 2**

Explain why, in multiplying $\frac{3}{7} \times \frac{10}{21}$, one can simplify the computation by changing the 3 and 21 to 1 and 7, respectively. Then state the result.

SOLUTION

$$\frac{3}{7} \times \frac{10}{21} = \frac{3 \times 10}{7 \times 21}$$

Now that it's all one fraction, you can divide the numerator and denominator by a common factor, 3. $\left(\text{Then } \frac{3}{21} \text{ becomes } \frac{1}{7}. \right)$

$$= \frac{\overset{1}{\cancel{3}}}{7} \times \frac{10}{\underset{7}{\cancel{21}}}$$

Then multiply to obtain the answer, $\frac{10}{49}$.   ●

When children use the shortcut in school, they may simplify before combining the two fractions into one, which makes the process faster but more mysterious. It may not be clear to them why the top of one fraction and the bottom of a *different* fraction can be divided by the same nonzero number.

LE 5 Concept

Explain why, in multiplying $\frac{4}{9} \times \frac{5}{8}$, one can simplify the computation by changing the 4 and 8 to 1 and 2, respectively. Then state the result.

To multiply mixed numbers, change them to improper fractions and follow the same rules. Some children will try to multiply mixed numbers without changing them to improper fractions.

LE 6 Communication

A sixth grader thinks that $4\frac{1}{3} \times 5\frac{1}{2} = 20\frac{1}{6}$. Use an area diagram to show why this is in-correct, and then obtain the correct answer.

DIVIDING RATIONAL NUMBERS

"Joe buys 4 lb of low-fat Swiss cheese. He and his wife eat a total of $\frac{1}{2}$ lb of low-fat Swiss cheese each day. For how many days will the cheese last?" Before learning the division rule, children solve problems such as this that model a whole number divided by an elementary fraction. This kind of problem can be solved using pictures and the repeated-measures model of division.

• EXAMPLE 3

A child has not yet learned the rule for dividing fractions.

(a) *Explain* how to compute $4 \div \frac{1}{2}$ using a measurement picture.

(b) *Explain* how to compute $4 \div \frac{1}{2}$ with a common denominator.

SOLUTION

(a) The expression $4 \div \frac{1}{2}$ means "How many $\frac{1}{2}$s does it take to make 4?" It takes two $\frac{1}{2}$s to make each whole.

Therefore, it takes eight $\frac{1}{2}$s to make 4. So $4 \div \frac{1}{2} = 8$.

(b) $4 \div \frac{1}{2} = \frac{8}{2} \div \frac{1}{2}$. Now $\frac{1}{2}$ goes into $\frac{8}{2}$ eight times. So $4 \div \frac{1}{2} = 8$.   ●

LE 7   Connection

A child has not yet learned the rule for dividing fractions.

(a) *Explain* how to compute $3 \div \frac{1}{4}$ using a measurement picture.

(b) *Explain* how to compute $3 \div \frac{1}{4}$ with a common denominator.

Division involving fractions with whole-number quotients can be done in a similar way.

LE 8   Connection

Explain how to compute $\frac{3}{4} \div \frac{1}{8}$ using a part-of-a-whole diagram.

Another way to confirm that $\frac{3}{4} \div \frac{1}{8}$ is 6 is to try multiplying $\frac{1}{8} \times 6$. In other words, $\frac{3}{4} \div \frac{1}{8} = 6$ means $\frac{1}{8} \times 6 = \frac{3}{4}$. This method of checking uses the definition of division. Like division of whole numbers and integers, division of rational numbers is defined in terms of multiplication.

> **Definition: Division of Rational Numbers**
>
> If $\frac{a}{b}$ and $\frac{c}{d}$ are rational numbers and $\frac{c}{d} \neq 0$, then $\frac{a}{b} \div \frac{c}{d} = \frac{e}{f}$ if and only if $\frac{a}{b} = \frac{c}{d} \times \frac{e}{f}$.

LE 9   Concept

How would you check the result of $3 \div \frac{1}{4}$ by writing it as multiplication?

The results from Example 3, LE 7, and LE 8 suggest the shortcut procedure for dividing fractions.

LE 10   Reasoning

Can you find the pattern in the preceding results? Fill in the blanks in parts (a)–(c) with a number written as a fraction.

(a) In Example 3, where $4 \div \frac{1}{2} = 8$, dividing by $\frac{1}{2}$ is the same as multiplying by _____.

(b) In LE 7, dividing by $\frac{1}{4}$ is the same as multiplying by _____.

(c) In LE 8, dividing by $\frac{1}{8}$ is the same as multiplying by _____.

(d) Parts (a)–(c) suggest that for rational numbers $\frac{a}{b}$ and $\frac{c}{d}$ with $\frac{c}{d} \neq 0$,

$$\frac{a}{b} \div \frac{c}{d} = \underline{\hspace{2cm}}.$$

LE 10 suggests the general invert-and-multiply rule.

Division of Rational Numbers

$$\frac{a}{b} \div \frac{c}{d} = \frac{a}{b} \times \frac{d}{c} = \frac{ad}{bc} \qquad (c \neq 0)$$

To use this rule with mixed numbers, first rewrite them as improper fractions.

LE 11   Skill

Compute $1\frac{2}{3} \div 1\frac{1}{2}$ using the invert-and-multiply rule. Put the answer in simplest form.

The invert-and-multiply rule is a shortcut for division that uses the reciprocal of a rational number. The **reciprocal** of a rational number $\frac{a}{b}$ in which $a \neq 0$ is $\frac{b}{a}$.

LE 12   Concept

Any nonzero rational number multiplied by its reciprocal equals _____.

You are familiar with the invert-and-multiply rule for dividing rational numbers and have seen how it can be developed from examples. One can show how the invert-and-multiply rule works in any rational-number division problem. For example, consider $\frac{2}{5} \div \frac{7}{9}$.

• EXAMPLE 4

A child wants to see why the invert-and-multiply rule works. Use the division model of fractions and the Fundamental Law to show that $\frac{2}{5} \div \frac{7}{9}$ is the same as $\frac{2}{5} \times \frac{9}{7}$.

SOLUTION

$$\frac{2}{5} \div \frac{7}{9} = \frac{\dfrac{2}{5}}{\dfrac{7}{9}} = \frac{\dfrac{2}{5} \times \dfrac{9}{7}}{\boxed{\dfrac{7}{9} \times \dfrac{9}{7}}} = \frac{\dfrac{2}{5} \times \dfrac{9}{7}}{1} = \frac{2}{5} \times \frac{9}{7}$$

Division rewritten as a fraction

Fundamental Law of Fractions

Omitted when shortcut is used

So $\frac{2}{5} \div \frac{7}{9} = \frac{2}{5} \times \frac{9}{7}$. It's the invert-and-multiply rule! This rule is simply a shortcut for the longer procedure just shown. ●

Discussion

LE 13 Skill

A child wants to see why the invert-and-multiply rule works. Use the division model of fractions and the Fundamental Law to show that $\frac{3}{5} \div \frac{2}{7}$ is the same as $\frac{3}{5} \times \frac{7}{2}$.

The categories of whole-number operations also apply to some situations involving multiplication and division of rational numbers. In LE 14 and LE 15, tell what operation and classification are illustrated.

LE 14 Connection

A computer printer produces a page in $\frac{1}{2}$ minute. How many pages would it print in 30 minutes? (area, array, partition, repeated)

LE 15 Connection

Last year, 20 people signed up for a course called "Mathematics Is Awesome." The word got around, and this year $3\frac{1}{2}$ times as many people signed up. How many people signed up this year? (area, array, counting principle, repeated)

🔍 AN INVESTIGATION: MULTIPLYING ELEMENTARY FRACTIONS

Discussion

LE 16 Reasoning

Consider the following problem. "In multiplying an elementary fraction $\frac{a}{b}$ by an elementary fraction $\frac{c}{d}$, tell what conditions would make the product (1) greater than $\frac{a}{b}$, (2) equal to $\frac{a}{b}$, and (3) less than $\frac{a}{b}$."

(a) Devise a plan and solve the problem.

(b) Use the results of this problem to analyze $\frac{a}{b} \div \frac{c}{d}$ (with $c \neq 0$) in a similar way.

Instead of computing the answers, apply your generalizations from the preceding exercise to answer the following.

LE 17 Concept

$\frac{3}{16} \times \frac{3}{5}$ is   **(a)** greater than $\frac{3}{16}$ **(b)** equal to $\frac{3}{16}$ **(c)** less than $\frac{3}{16}$

LE 18   Concept

$9\frac{1}{2} \div \frac{3}{4}$ is   **(a)** greater than $9\frac{1}{2}$ **(b)** equal to $9\frac{1}{2}$ **(c)** less than $9\frac{1}{2}$

LE 19   Summary

Tell what you learned about multiplying and dividing fractions in this section.

ANSWERS TO SELECTED LESSON EXERCISES

1. $5 \times \frac{1}{8} = \frac{1}{8} + \frac{1}{8} + \frac{1}{8} + \frac{1}{8} + \frac{1}{8} = \frac{5}{8}$

2. $\frac{1}{2} \times \frac{3}{5}$ is $\frac{1}{2}$ of $\frac{3}{5}$. Show $\frac{3}{5}$.

Darken $\frac{1}{2}$ of $\frac{3}{5}$.

$\frac{3}{10}$ of the figure is darkened. So $\frac{1}{2} \times \frac{3}{5} = \frac{3}{10}$.

3. (a) $\frac{3}{8}; \frac{3}{10}$ (b) $\frac{ac}{bd}$ (c) Induction

5. $\frac{4}{9} \times \frac{5}{8} = \frac{4 \times 5}{9 \times 8}$. Now that it's all one fraction, you can

divide the numerator and denominator by a common

factor, 4. $\left(\text{Then } \frac{4}{8} \text{ becomes } \frac{1}{2}.\right)$

$= \frac{\overset{1}{4} \times 5}{9 \times \underset{2}{8}} = \frac{5}{18}$

6.

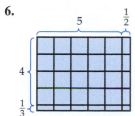

Area $= (4 \times 5) + \left(4 \times \frac{1}{2}\right) + \left(\frac{1}{3} \times 5\right) + \left(\frac{1}{3} \times \frac{1}{2}\right)$

$= 20 + 2 + \frac{5}{3} + \frac{1}{6} = 23\frac{5}{6}$ sq. units

7. (a) $3 \div \frac{1}{4}$ means how many $\frac{1}{4}$s does it take to make 3?

It takes 4 quarters to make 1.

So it takes 12 quarters to make 3. Therefore,

$3 \div \frac{1}{4} = 12$.

(b) $3 \div \frac{1}{4} = \frac{12}{4} \div \frac{1}{4}$. Now $\frac{1}{4}$ goes into $\frac{12}{4}$ twelve

times. So $3 \div \frac{1}{4} = 12$.

8. $\frac{3}{4} \div \frac{1}{8}$ means "How many $\frac{1}{8}$s make $\frac{3}{4}$?" The answer is 6.

So $\frac{3}{4} \div \frac{1}{8} = 6$.

9. $3 = \frac{1}{4} \times 12$

10. (a) 2   (b) 4   (c) 8   (d) $\frac{ad}{bc}$

11. $1\frac{1}{9}$

12. 1

13. $\dfrac{3}{5} \div \dfrac{2}{7} = \dfrac{\frac{3}{5}}{\frac{2}{7}} = \dfrac{\frac{3}{5} \times \frac{7}{2}}{\frac{2}{7} \times \frac{7}{2}} = \dfrac{\frac{3}{5} \times \frac{7}{2}}{1} = \dfrac{3}{5} \times \dfrac{7}{2}$

Note: The second step in LE 13 can also be justified as multiplication by 1 in the following form:

$$\dfrac{\frac{7}{2}}{\frac{7}{2}}$$

14. Division; repeated measures

15. Multiplication; repeated sets

17. (c)

18. (a)

6.3   HOMEWORK EXERCISES

Basic Exercises

1. Show how to compute $4 \times \dfrac{1}{5}$ using repeated addition.

2. Show how to compute $3 \times \dfrac{2}{7}$ using repeated addition.

3. Celine ate $\dfrac{3}{8}$ of a pizza. Roberto ate $\dfrac{2}{5}$ of the pizza that was left.

   (a) The $\dfrac{3}{8}$ is $\dfrac{3}{8}$ of what unit?

   (b) The $\dfrac{2}{5}$ is $\dfrac{2}{5}$ of what unit?

4. In $\dfrac{1}{2}$ of 8, $\dfrac{1}{2}$ operates on 8. Explain how the fraction as an operator is a function. (*Hint:* A function has exactly one output for each input.)

5. A child who does not know the multiplication rule for fractions wants to compute $\dfrac{1}{5} \times \dfrac{1}{3}$. *Explain* how to compute $\dfrac{1}{5} \times \dfrac{1}{3}$ using two diagrams.

6. *Explain* how to compute $\dfrac{1}{4} \times \dfrac{2}{3}$ using two diagrams.

7. *Explain* how to compute $\dfrac{3}{4} \times \dfrac{4}{5}$ using two diagrams.

8. *Explain* how to compute $\dfrac{2}{3} \times \dfrac{5}{6}$ using two diagrams.

9. What fraction product is illustrated by the following grid?

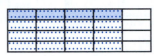

10. What fraction product is illustrated by the following grid?

11. How would you diagram and explain

   (a) $6 \times \dfrac{1}{3}$?   (b) $\dfrac{1}{3} \times 6$?

12. How would you diagram and explain

   (a) $4 \times \dfrac{1}{2}$?   (b) $\dfrac{1}{2} \times 4$?

13. *Explain* why, in multiplying $5\dfrac{1}{2} \times 2\dfrac{2}{3}$, one can simplify the computation by changing a 2 and 8 to 1 and 4, respectively.

14. *Explain* why, in multiplying $\dfrac{7}{10} \times \dfrac{5}{9}$, one can simplify the computation by changing the 10 and 5 to 2 and 1, respectively.

15. Show how to work out $3\dfrac{1}{2} \times 4\dfrac{1}{8}$ with an area diagram.

16. Show how to work out $2\dfrac{1}{6} \times 3\dfrac{1}{3}$ with an area diagram.

17. Consider the following diagram.

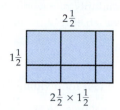

$2\frac{1}{2} \times 1\frac{1}{2}$

(a) Find $2\frac{1}{2} \times 1\frac{1}{2}$ from the diagram.

(b) Make a similar diagram for $3\frac{1}{2} \times 2\frac{1}{2}$ and give the answer.

(c) Repeat part (b) for $4\frac{1}{2} \times 3\frac{1}{2}$.

(d) Find a shortcut for computing $a\frac{1}{2} \times b\frac{1}{2}$, in which $a = b + 1$ for counting numbers a and b.

(e) Use your shortcut to compute $18\frac{1}{2} \times 19\frac{1}{2}$.

18. Consider the following diagram.

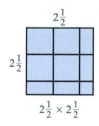

$2\frac{1}{2} \times 2\frac{1}{2}$

(a) Find $2\frac{1}{2} \times 2\frac{1}{2}$ from the diagram.

(b) Make a similar diagram for $3\frac{1}{2} \times 3\frac{1}{2}$ and give the result.

(c) Repeat part (b) for $4\frac{1}{2} \times 4\frac{1}{2}$.

(d) Find a shortcut for computing $a\frac{1}{2} \times a\frac{1}{2}$ for a counting number a.

(e) Use your shortcut to compute $29\frac{1}{2} \times 29\frac{1}{2}$.

(f) Show algebraically why the shortcut works.

19. (a) What multiplication is the same as $\frac{1}{2}$ of 8?

(b) What division is the same as $\frac{1}{2}$ of 8?

20. (a) What multiplication is the same as $\frac{1}{4}$ of 20?

(b) What division is the same as $\frac{1}{4}$ of 20?

21. Compute the following without a calculator and simplify.

(a) $\frac{2}{3} \times \frac{3}{4}$ (b) $2\frac{1}{3} \times \frac{2}{3}$

22. Compute the following without a calculator and simplify.

(a) $\frac{7}{8} \times \frac{4}{5}$ (b) $-\frac{1}{4} \times 4\frac{1}{6}$

23. A child has not yet learned the rule for dividing fractions.

(a) *Explain* how to compute $2 \div \frac{1}{4}$ using a measurement picture.

(b) *Explain* how to compute $2 \div \frac{1}{4}$ with a common denominator.

24. (a) *Explain* how to compute $3 \div \frac{1}{2}$ using a measurement picture.

(b) *Explain* how to compute $3 \div \frac{1}{2}$ with a common denominator.

25. *Explain* how to compute $\frac{2}{3} \div \frac{1}{6}$ using a part-of-a-whole diagram.

26. *Explain* how to compute $\frac{1}{2} \div \frac{1}{8}$ using a part-of-a-whole diagram.

27. *Explain* how to compute $\frac{3}{4} \div 2$ using a part-of-a-whole diagram.

28. *Explain* how to compute $\frac{1}{2} \div 3$ using a part-of-a-whole diagram.

29. *Explain* how to compute $1\frac{1}{4} \div \frac{1}{2}$ using a part-of-a-whole diagram.

30. *Explain* how to compute $2\frac{1}{3} \div \frac{1}{3}$ using a part-of-a-whole diagram.

31.

(a) Where do you see $\frac{3}{4}$ of something in the diagram?

(b) Where do you see $\frac{4}{3}$ of something in the diagram?

(c) Where can you see $\frac{1}{3}$ of $\frac{3}{4}$ in the diagram?

(d) Where can you see $1 \div \frac{3}{4}$ in the diagram?

32.

(a) Where do you see $\frac{2}{5}$ of something in the diagram?

(b) Where do you see $\frac{5}{2}$ of something in the diagram?

(c) Where can you see $\frac{1}{2}$ of $\frac{2}{5}$ in the diagram?

(d) Where can you see $1 \div \frac{2}{5}$ in the diagram?

33. Use the definition of division to rewrite $\frac{10}{21} \div \frac{2}{3} = \frac{n}{m}$ as a multiplication equation, and find the answer by inspection.

34. Use the definition of division to solve $\frac{1}{2} \div ? = 4$.

35. Compute the following without a calculator and simplify.

(a) $\frac{5}{8} \div \frac{1}{2}$ (b) $-8 \div 3\frac{1}{9}$ (c) $\frac{x}{5} \div \frac{x}{7}$

36. Compute the following without a calculator and simplify.

(a) $\frac{3}{5} \div \frac{7}{10}$ (b) $10 \div 4\frac{2}{3}$ (c) $\frac{x^2}{y} \div \frac{x}{3}$

37. Solve mentally.

(a) $\frac{1}{2}x = 10$ (b) $4 \div \frac{1}{2} = y$

38. Solve mentally.

(a) $\frac{1}{4}t = 8$ (b) $9 \div \frac{1}{3} = r$

 39. If you have a fraction calculator, compute $\frac{3}{8} \times \frac{1}{6}$ and simplify it.

 40. If you have a fraction calculator, compute $\frac{3}{4} \div \frac{1}{2}$, simplify it, and convert it to a mixed numeral.

41. A child wants to see why the invert-and-multiply rule works. Use the division model of fractions and the Fundamental Law to show that $\frac{1}{2} \div \frac{3}{4}$ is the same as $\frac{1}{2} \times \frac{4}{3}$.

42. Use the division model of fractions and the Fundamental Law to show that $\frac{9}{10} \div \frac{2}{3}$ is the same as $\frac{9}{10} \times \frac{3}{2}$.

43. It is preferable to use numbers to show children how the invert-and-multiply rule works. However, a proof requires using variables. Follow the steps of Example 4 and show that $\frac{a}{b} \div \frac{c}{d} = \frac{ad}{bc}$.

44. (a) What is a shortcut for dividing fractions with equal denominators such as $\frac{3}{8} \div \frac{1}{8}$ or $\frac{10}{3} \div \frac{2}{3}$?

(b) A general rule would be $\frac{a}{c} \div \frac{b}{c} =$ _____.

(c) This rule can be used to derive the invert-and-multiply rule. First, rewrite each fraction in $\frac{a}{b} \div \frac{c}{d}$ with a common denominator. Then apply the rule from part (b). Finally, write the result as a fraction.

45. A recipe calls for $\frac{1}{4}$ cup of flour per person. If you are cooking for 6 people, how much flour should you use?
(a) Solve the problem with a drawing.
(b) Solve the problem with an arithmetic computation.
(c) What operation and category does part (a) illustrate?

46. Last year a farm produced 1360 oranges. This year they produced $2\frac{1}{2}$ times as many oranges.

(a) How many oranges did they produce?
(b) What operation and category does part (a) illustrate?

47. A wall is $82\frac{1}{2}$ inches high. It is covered with $5\frac{1}{2}$-inch square tiles.
(a) How many tiles are in a vertical row from the floor to the ceiling? (Give an exact answer.)
(b) What operation and category does part (a) illustrate?

48. Norma and Irving want to split a delicious $10\frac{1}{2}$-inch submarine sandwich equally. What division category does this illustrate?

49. Four identical cartons are stacked one on top of the other. The stack is 10 feet high.
(a) If the room is 18 feet high, how many more identical cartons can be stacked on top of the pile?
(b) What operations and categories are illustrated here? (area, array, compare, missing part, partition, repeated, take away)

50. A baker takes $\frac{1}{2}$ hour to decorate a cake.

    (a) How many cakes can she decorate in H hours?

    (b) What operation and category does part (a) illustrate? (area, array, partition, repeated)

51. Four-fifths of a class bring in food for a charity drive. Of those who brought food, one-fourth brought canned soup. Make a drawing, and tell what fraction of the whole class brought canned soup.

52. Rainey puts 19 gallons of gas into the empty tank of her Marauder SUV. Now the tank is $\frac{5}{8}$ full. How much gas does the tank hold?

53. A recipe for 4 people calls for $\frac{1}{3}$ cup of olive oil. If you make the same dish for 2 people, how much olive oil should you use?

54. Pam has 8 yards of fabric to make shirts. Each shirt requires $1\frac{3}{4}$ yards of fabric. How many shirts can she make?

55. An adult pays $231 for plane tickets for himself and a child. If the child goes for half-price, how much is the adult fare?

56. A recipe calls for $\frac{3}{4}$ cup of flour. You have only $\frac{1}{2}$ cup of flour. What fraction of the recipe can you make?

57. Carol Burbage spent $\frac{1}{2}$ of her money at the movies. Then she spent $\frac{1}{3}$ of what was left at the store. Now she has $4 left. How much did she start with? (*Hint:* Draw a part-of-a-whole picture.)

58. Bill ate $\frac{1}{3}$ of the cookies in the cookie jar. Maria ate $\frac{1}{2}$ of what was left. Then Sidney ate 2 cookies. Now 4 are left. How many were in the jar to start with? (*Hint:* Draw a part-of-a-whole picture.)

59. Write a word problem that can be solved by computing

    (a) $\frac{2}{3} \times \frac{4}{5}$ (b) $\frac{5}{8} \div \frac{1}{4}$

60. Write a word problem that can be solved by computing

    (a) $1\frac{1}{2} \times \frac{3}{4}$ (b) $\frac{3}{4} \div \frac{1}{2}$

61. Some children think $A \div B$ will always result in an answer that is smaller than A. Give an example with positive fractions in which the answer is greater than A.

62. A fifth grader says you find $\frac{1}{2}$ of a number by dividing the number by $\frac{1}{2}$. Is this correct? If not, how would you explain what is wrong with this approach?

63. (a) Tell how to find what fraction of the figure is shaded.

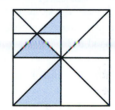

    (b) Describe a second way of finding the answer.

64. (a) $1 + \frac{1}{5} =$ _____

    (b) $1 + \dfrac{1}{1 + \frac{1}{5}} =$ _____

    (c) $1 + \dfrac{1}{1 + \dfrac{1}{1 + \frac{1}{5}}} =$ _____

    (d) What pattern do you see in the results to parts (a)–(c)?

    (e) Without computing, guess the answer to the following.

$$1 + \dfrac{1}{1 + \dfrac{1}{1 + \dfrac{1}{1 + \frac{1}{5}}}}$$

Extension Exercises

65. Crazy King Loopy just died. Loopy's will instructs his attorney, Ward E. Claus, to divide up his prize collection of 17 hogs as follows: $\frac{4}{9}$ of the hogs go to his eldest daughter Wacky, $\frac{1}{3}$ go to his son Harpo, and $\frac{1}{6}$ go to young Loopy II. Ward has no idea how he is going to carry out the will.

Fortunately, the court sage, Wiggy, tells Ward to borrow another hog, and then he will be able to carry out the will. Ward tries it, and it works! Ward returns the extra hog when he is done.
(a) How many hogs does each child receive?
(b) Why does Wiggy's approach seem to work?
(c) In the end, were the conditions of the will fulfilled, or did everyone receive more hog than he or she was supposed to?

66. Suppose a pollution control bill is supported by $\frac{3}{5}$ of all voters who are Democrats, $\frac{2}{5}$ of all Republicans, and $\frac{1}{2}$ of all Independents. If $\frac{2}{5}$ of the population is Democratic, $\frac{3}{10}$ is Republican, and $\frac{3}{10}$ is Independent, what fraction of all voters support the bill?

67. For $46 \div 6$, make up a word problem for which the answer is
(a) 7   (b) 8   (c) 4   (d) $7\frac{2}{3}$

68. Units of measurement can be treated like fractions. For example, if a man travels 10 miles per hour for 6 hours, how far does he travel?

$$10\,\frac{\text{miles}}{\text{hour}} \times 6 \text{ hours} = 10 \times 6 \,\frac{\text{miles}}{\text{hour}} \times \text{hours}$$
$$= 60 \text{ miles}$$

Carry the units along throughout each of the following exercises.
(a) Swiss cheese costs 4 dollars per pound. How much would $3\frac{1}{2}$ pounds cost?
(b) An ad claims that a car travels 495 miles on a full tank of gas. If the car gets 30 miles to the gallon, how many gallons does the gas tank hold?

69. Sometimes the difference of two fractions equals their product. For example,

$$\frac{3}{7} - \frac{3}{10} = \frac{3}{7} \times \frac{3}{10} \quad \text{and} \quad \frac{2}{3} - \frac{2}{5} = \frac{2}{3} \times \frac{2}{5}$$

(a) What is the relationship between the two fractions in each example?
(b) Make up two more examples that work.
(c) Show algebraically that the product and difference of two fractions of this type will always be equal. (*Hint:* It takes only two variables to write all the numerators and denominators.)

70.
$$\frac{1}{3} + \frac{1}{2 \cdot 3} = \underline{\hspace{2cm}}$$
$$\frac{1}{4} + \frac{1}{3 \cdot 4} = \underline{\hspace{2cm}}$$
$$\frac{1}{5} + \frac{1}{4 \cdot 5} = \underline{\hspace{2cm}}$$

(a) Fill in the blanks.
(b) If the pattern continues, what will the next equation be? Is the equation true?
(c) Write the general formula suggested by parts (a) and (b):
$$\underline{\hspace{1.2cm}} + \underline{\hspace{1.2cm}} = \underline{\hspace{1.2cm}}$$
(d) Writing a general formula based upon the examples in parts (a) and (b) requires $\underline{\hspace{2cm}}$ reasoning.
(e) Prove that your formula in part (c) is correct. (*Hint:* Find the common denominator on the left side of the equation, and add the two fractions.)

71. Fill in the blanks, following the rule in the completed examples.

(a)	(b)
$9 \to 6$	$4 \to 14$
$12 \to 8$	$8 \to 28$
$15 \to \underline{\ \ }$	$10 \to \underline{\ \ }$
$18 \to \underline{\ \ }$	$5 \to \underline{\ \ }$
$\underline{\ \ } \to 20$	$\underline{\ \ } \to 70$
$N \to \underline{\ \ }$	$N \to \underline{\ \ }$

72. Assume that b, c, and d are not 0. The statement
$$\frac{a}{b} \div \frac{c}{d} = \frac{a \div c}{b \div d} \text{ is}$$
(a) always true.
(b) sometimes true.
(c) never true.

6.4 RATIONAL NUMBERS: PROPERTIES, ESTIMATION, AND ERROR PATTERNS

NCTM Standards

- develop and use strategies to estimate computations involving fractions and decimals in situations relevant to students' experience (3–5)

- use the associative and commutative properties of addition and multiplication and the distributive property of multiplication over addition to simplify computations with integers, fractions, and decimals (6–8)

- develop and use strategies to estimate the results of rational-number computations and judge the reasonableness of the results (6–8)

The properties of whole-number and integer operations discussed in Chapters 3 and 5 hold for rational-number operations. However, the rational-number system possesses some additional properties.

INTEGER PROPERTIES RETAINED?

Do rational-number operations retain the same commutative, associative, identity, inverse, closure, and distributive properties as integer operations?

LE 1 Opener

Using examples, decide if you think rational-number

(a) addition is commutative.
(b) multiplication is associative.
(c) multiplication is distributive over addition.
(d) Are your conclusions in parts (a)–(c) based on induction or deduction?

The preceding exercise concerns some possible properties of rational numbers. The rational-number operations do retain all the properties of integer operations!

Some Properties of Rational-Number Operations

- Addition, subtraction, and multiplication of rational numbers are closed.

- Addition and multiplication of rational numbers are commutative.

- Addition and multiplication of rational numbers are associative.

- The unique additive identity for rationals is 0, and the unique multiplicative identity for rationals is 1.

- All rational numbers have a unique additive inverse that is rational. For any rational number $\frac{a}{b}$, there is a unique rational number $-\frac{a}{b}$ such that $\frac{a}{b} + -\frac{a}{b} = -\frac{a}{b} + \frac{a}{b} = 0$.

- Multiplication is distributive over addition, and multiplication is distributive over subtraction in the rational number system.

The following exercises make use of these properties.

LE 2   Concept

What is the additive inverse of $\frac{3}{4}$?

The properties also justify some procedures used in algebra.

LE 3   Concept

If x is rational, what property guarantees that $\left(x + \frac{1}{2}\right) + 2 = x + \left(\frac{1}{2} + 2\right)$?

LE 4   Skill

You can use the distributive property of multiplication over addition to compute $8 \times 5\frac{1}{2}$. Show how this is done. $\left(Hint: 5\frac{1}{2} = 5 + \frac{1}{2}.\right)$

What about rational-number subtraction and division? Are they commutative or associative?

LE 5   Reasoning

Explain how you know, after studying whole numbers and integers, that rational-number subtraction and division are neither commutative nor associative.

MULTIPLICATIVE INVERSES

Add any rational number to its additive inverse, and the result is the additive identity (0). Try to apply the same idea to a multiplicative inverse in the next lesson exercise.

LE 6   Concept

Any rational number multiplied by its multiplicative inverse should result in what number?

Do all rational numbers have multiplicative inverses that are rational numbers?

LE 7   Skill

What is the multiplicative inverse of each of the following?

(a) -4 **(b)** $\frac{2}{3}$

LE 8   Reasoning

(a) Do you think all rational numbers have multiplicative inverses that are rational numbers? Try some examples.
(b) Only one rational number does not have a multiplicative inverse. What is it?

LE 7 and LE 8 should convince you that every nonzero rational number has a unique rational multiplicative inverse.

> **Multiplicative Inverses for Nonzero Rational Numbers**
>
> Every rational number except 0 has a unique multiplicative inverse that is rational. That is, for each nonzero rational number $\frac{a}{b}$ there is a unique rational number $\frac{b}{a}$ such that $\frac{a}{b} \times \frac{b}{a} = \frac{b}{a} \times \frac{a}{b} = 1$.

THE ROUNDING STRATEGY

In the rare instance that you should make an error in arithmetic with elementary fractions, estimation might help you detect your error. For example, if you computed that $3\frac{1}{4} + 2\frac{2}{3} = \frac{11}{12}$ (the error pattern in LE 15), estimating would tell you that the answer could not be right.

The rounding strategy is often the best for estimating in addition, subtraction, and multiplication problems involving elementary fractions. Round the numbers to create a problem that you can compute mentally. In most cases, fractions are rounded to the nearest integer. In addition or subtraction, round fractions between 0 and 1 to 0, $\frac{1}{2}$, or 1 which are called **benchmarks**.

Use rounding to estimate in the following exercises.

LE 9 Skill

Celia plans to bake $4\frac{1}{2}$ pans of lasagna. Each pan requires $2\frac{3}{4}$ cups of tomato sauce.

(a) How can you use rounding to estimate the amount of tomato sauce needed?
(b) Is your estimate too high or too low?
(c) Round both numbers down, and give a range estimate.

LE 10 Skill

Estimate $\frac{5}{12} + \frac{7}{8}$ by rounding each fraction to 0, $\frac{1}{2}$, or 1.

THE COMPATIBLE-NUMBERS STRATEGY

Are you a flexible person? If so, you'll like using the compatible-numbers strategy with elementary fractions. Rounding to the *nearest* whole number is not always the best approach in fraction multiplication or division.

Rounding may not work in multiplication when at least one factor is close to or less than $\frac{1}{2}$. This factor can be rounded to the closest fraction that has 1 in the numerator and a counting number in the denominator, such as $\frac{1}{2}, \frac{1}{3}$, or $\frac{1}{4}$. Then change the other factor to a compatible number.

• EXAMPLE 1

A new electric eyeglass defogger regularly sells for $820, but Nutty Mike's is offering it for $\frac{2}{5}$ off the regular price. How can you estimate the amount of the discount?

SOLUTION

We need to estimate $\frac{2}{5}$ of $820. We cannot round $\frac{2}{5}$ to the nearest whole number, 0. Doing so would make the estimate 0 for $\frac{2}{5}$ of any dollar amount! Instead, round $\frac{2}{5}$ to $\frac{1}{2}$ or $\frac{1}{3}$, which is easier to use than $\frac{2}{5}$, and round $820 to a compatible number.

$$\frac{2}{5} \times \$820 \approx \frac{1}{2} \times \$800 = \$400$$

So $\frac{2}{5}$ of $820 is about $400. Or

$$\frac{2}{5} \times \$820 \approx \frac{1}{3} \times \$900 = \$300$$

So $\frac{2}{5}$ of $820 is about $300.   ●

LE 11   Skill

Nutty Mike's is offering $\frac{3}{5}$ off the price of a $604 electric bookmark. How can you estimate the discount?

COMMON ERROR PATTERNS

Your future students will amaze you with their own ingenious procedures for fraction arithmetic (which often do not work). Repetition of incorrect procedures results in error patterns in children's work.

Classroom Connection

In LE 12–LE 15, (a) complete the last two examples, repeating the error pattern in the completed examples; (b) write a description of the error pattern; and (c) state how one of the errors might be detected using estimation.

LE 12   Reasoning

$$\frac{1}{3} + \frac{2}{5} = \underline{\frac{3}{8}} \qquad \frac{2}{5} + \frac{1}{5} = \underline{\frac{3}{10}} \qquad \frac{2}{3} + \frac{4}{7} = \underline{\frac{6}{10}}$$

$$\frac{1}{5} + \frac{3}{4} = \underline{\qquad} \qquad \frac{1}{2} + \frac{2}{3} = \underline{\qquad}$$

LE 13   Reasoning

$$\frac{2}{3} \div \frac{1}{4} = \underline{\frac{3}{8}} \qquad 8 \div \frac{1}{2} = \underline{\frac{1}{16}} \qquad 8\frac{1}{2} \div 2\frac{2}{3} = \underline{\frac{16}{51}}$$

$$\frac{4}{7} \div \frac{2}{3} = \underline{\qquad} \qquad \frac{7}{8} \div 2 = \underline{\qquad}$$

LE 14   Reasoning

$$\frac{8}{4} = \underline{\quad 2 \quad} \qquad \frac{3}{6} = \underline{\quad 2 \quad} \qquad \frac{3}{9} = \underline{\quad 3 \quad} \qquad \frac{12}{3} = \underline{\qquad}$$

$$\frac{1}{4} = \underline{\qquad}$$

LE 15   Reasoning

$$3\frac{1}{4} = \frac{3}{12} \qquad\qquad 5\frac{1}{4} \qquad\qquad 6\frac{1}{6}$$
$$\underline{+ \; 2\frac{2}{3}} = \frac{8}{12} \qquad\qquad \underline{+ \; 2\frac{1}{2}} \qquad\qquad \underline{+ \; 4\frac{1}{4}}$$
$$\qquad\qquad \frac{11}{12}$$

LE 16   Summary
Tell what you learned about properties of rational-number operations in this section.

ANSWERS TO SELECTED LESSON EXERCISES

2. $\dfrac{-3}{4}$

3. Associative property of addition

4. $8 \times 5\frac{1}{2} = 8 \times \left(5 + \frac{1}{2}\right) = \left(8 \times 5\right) + \left(8 \times \frac{1}{2}\right) =$
$40 + 4 = 44$

5. The counterexamples for whole-number and integer subtraction and division would also be counterexamples for rational numbers. If a rule does not apply to *all* whole numbers, then it cannot apply to *all* rational numbers, since every whole number is also a rational number.

6. 1

7. (a) $-\dfrac{1}{4}$ (b) $1\dfrac{1}{2}$

8. (b) 0

9. (a) $4\frac{1}{2} \times 2\frac{3}{4} \approx 5 \times 3 = 15$ cups
   (b) Too high
   (c) $4\frac{1}{2} \times 2\frac{3}{4} \approx 4 \times 2 = 8$; between 8 and 15 cups

10. $\dfrac{5}{12} + \dfrac{7}{8} \approx \dfrac{1}{2} + 1 = 1\dfrac{1}{2}$

11. $\dfrac{3}{5}$ of $604 \approx \dfrac{1}{2} \times \$600 = \$300$

12. (a) $\dfrac{4}{9}; \dfrac{3}{5}$
   (b) The child is adding the denominators.
   (c) $\dfrac{4}{9}$ is less than $\dfrac{3}{4}$. How can $\dfrac{1}{5} + \dfrac{3}{4} = \dfrac{4}{9}$?

13. (a) $\dfrac{14}{12}$ or $1\dfrac{1}{6}$; $\dfrac{16}{7}$ or $2\dfrac{2}{7}$
   (b) The child is inverting the first fraction and multiplying it by the second.
   (c) $8\dfrac{1}{2} \div 2\dfrac{2}{3} \approx 9 \div 3 = 3$. How can the answer be $\dfrac{16}{51}$?

14. (a) 4; 4
   (b) The child always divides the smaller number into the larger.
   (c) Fractions with denominators that are larger than their numerators cannot be greater than 1.

15. (a) $\dfrac{3}{4}; \dfrac{5}{12}$
   (b) The child is adding only the fractional parts of the mixed numbers.
   (c) $3\dfrac{1}{4} + 2\dfrac{2}{3} \approx 3 + 3 = 6$. How can the answer be $\dfrac{11}{12}$?

6.4 HOMEWORK EXERCISES

Basic Exercises

1. (a) What rational-number operations are commutative?
   (b) What rational-number operations are associative?

2. $\frac{2}{3} + 0 = 0 + \frac{2}{3} = \frac{2}{3}$

   $-3\frac{5}{8} + 0 = 0 + -3\frac{5}{8} = -3\frac{5}{8}$

   These examples illustrate that _____ is the
   _____ for _____.

3. According to the associative property of multiplication

   for a rational number x, $\frac{1}{2} \cdot (4 \cdot x) =$ _____.

4. For any rational number x, what property guarantees that

   $$\frac{5}{2}x + (2x + 7) = \left(\frac{5}{2}x + 2x\right) + 7?$$

5. What is the additive inverse of $-2\frac{1}{2}$?

6. What property guarantees that

   $$8 \times \left(3 + \frac{1}{4}\right) = (8 \times 3) + \left(8 \times \frac{1}{4}\right)?$$

7. Use the distributive property of multiplication over
   addition for a rational number n to simplify

   $$-\frac{4}{3}n + \frac{2}{3}n = \text{_____} = \text{\_\_\_\_}.$$

8. Show the steps in mentally computing each of the
   following with the distributive property.
   (a) $40 \times 6\frac{1}{8}$ (b) $\frac{1}{4} \times 8\frac{1}{2}$

9. On a car trip, the Durst family averages 50 miles per
   hour for $6\frac{1}{2}$ hours. How can you mentally compute the
   total distance traveled?

10. A job pays $40 per hour for $3\frac{3}{4}$ hours. Mentally com-
    pute the exact total pay. Show how you did it.

11. Draw a Venn diagram that includes the following sets:
    rational numbers, whole numbers, and integers.

12. True or false? Every rational number is an integer.

13. *Explain* how you would know, after studying whole
    numbers and integers, that rational-number subtrac-
    tion is not commutative.

14. Give a counterexample showing that rational-number
    division is not associative.

15. What is the multiplicative inverse of $-2\frac{1}{2}$?

16. Any nonzero number multiplied by its multiplicative
    inverse equals _____.

17. (a) What property do rational numbers have that
        whole numbers and integers do not have?
    (b) What property do all nonzero rational numbers
        have that whole numbers and integers do not have?

18. One reason that we need rational numbers is to provide
    answers to whole-number and integer division problems.
    (a) Are the rational numbers closed under division?
        (That is, if x and y are rational numbers, is $x \div y$ a
        rational number?)
    (b) Would the set of rational numbers, excluding 0, be
        closed under division?

19. A recipe calls for $1\frac{3}{4}$ cups of sugar to make 1 lb of
    belly busters. How can you estimate how much sugar
    is needed to make $8\frac{1}{4}$ lb of belly busters?

20. The Westwood Dribblers have a trip of $836\frac{1}{10}$ miles
    to their away basketball game. So far, they have trav-
    eled $381\frac{7}{10}$ miles. How can you estimate the distance
    they have left to travel?

21. $26\frac{7}{3600} \times 32\frac{9}{60}$ is about
    (a) 7000   (b) 100   (c) 60   (d) 900

22. A, B, C, D, E, and F are counting numbers. You estimate
    $A\frac{B}{C} - D\frac{E}{F}$ by rounding $A\frac{B}{C}$ up and $D\frac{E}{F}$ down. Is your
    estimate too high or too low, or is it impossible to tell?

23. Suppose you nail together two boards that are $\frac{7}{8}$ in. thick and $\frac{1}{16}$ in. thick.

(a) Estimate the total thickness by rounding each measurement to either 0, $\frac{1}{2}$, or 1.

(b) Some children estimate this answer as 8 in. or 24 in. Why do they make these errors and what don't they understand about fractions?

24. Estimate $\frac{11}{12} + \frac{1}{6} + \frac{7}{12}$ by rounding each number to either 0, $\frac{1}{2}$, or 1.

25. Tell whether each fraction is close to 0, close to $\frac{1}{4}$, close to but less than $\frac{1}{2}$, close to but greater than $\frac{1}{2}$, or close to $\frac{3}{4}$.

(a) $\frac{2}{9}$ (b) $\frac{18}{23}$ (c) $\frac{5}{11}$ (d) $\frac{2}{35}$ (e) $\frac{9}{17}$

26. Tell whether each fraction is close to 0, close to $\frac{1}{4}$, close to but less than $\frac{1}{2}$, close to but greater than $\frac{1}{2}$, or close to $\frac{3}{4}$.

(a) $\frac{8}{15}$ (b) $\frac{4}{75}$ (c) $\frac{75}{99}$ (d) $\frac{47}{100}$ (e) $\frac{19}{41}$

27. How can you tell whether a fraction is

(a) close to $\frac{1}{4}$?   (b) close to but less than $\frac{1}{2}$?

28. How can you tell whether a fraction is

(a) close to 0?   (b) close to but greater than $\frac{1}{2}$?

29. Suppose the fifth grade has 126 children, and 41 of them bring lunch to school. Write a simple fraction that approximates the fraction of children who bring lunch.

30. In a school survey, 34 out of 164 students say they would like to wear school uniforms everyday. Write a fraction in simplest form that represents the fraction of children who want to wear school uniforms.

31. In a national survey, $\frac{1}{4}$ of all fifth graders say vanilla is their favorite flavor of ice cream. In a class of 23 fifth graders, show how to estimate how many would pick vanilla as their favorite.

32. A \$327 stereo is selling for $\frac{2}{5}$ off during a sale. How can you estimate the sale price?

33. You want to lay $17\frac{1}{4}$ in. wide panels across a wall that is $384\frac{1}{2}$ in. wide. How can you estimate how many panels you will need?

34. $286\frac{31}{900} \div 6\frac{1}{4}$ is about

(a) 280   (b) 1800   (c) 30   (d) 50

35. $872\frac{7}{16} \div \frac{3}{8}$ is approximately

(a) $\frac{1}{1,000,000}$ (b) 40   (c) 320   (d) 2200

36. The product of A, B, and C is approximately

(a) $\frac{1}{10}$ (b) $1\frac{1}{4}$ (c) 3   (d) 4   (e) 5

37. Using mental computation, one can see that $20\frac{1}{2} \times 2\frac{1}{2}$ is

(a) less than 44
(b) between 44 and 50
(c) more than 50

38. $\dfrac{\square}{\square} \div \dfrac{\square}{\square}$

Place the numbers 2, 3, 7, and 9 in the four boxes to make
(a) the largest possible quotient.
(b) the smallest possible quotient.

In Exercises 39-44, (a) complete the last two examples, repeating the error pattern in the completed examples, and (b) write a description of the error pattern.

39. $\dfrac{1}{2} + \dfrac{3}{4} = \dfrac{2+3}{4+4} = \dfrac{5}{8}$ $\dfrac{2}{3} + \dfrac{1}{6} = $ _____

$\dfrac{1}{5} + \dfrac{9}{10} = $ _____

40. $\dfrac{13}{36} = \underline{\dfrac{1}{6}}$ $\dfrac{16}{64} = \underline{\dfrac{1}{4}}$ $\dfrac{15}{25} = \underline{\dfrac{1}{2}}$

$\dfrac{21}{42} = \underline{\hphantom{xxx}}$ $\dfrac{17}{27} = \underline{\hphantom{xxx}}$

41. $\begin{array}{r} 4\frac{1}{5} = 3\frac{11}{5} \\ -2\frac{2}{5} = 2\frac{2}{5} \\ \hline \end{array}$ $\begin{array}{r} 3\frac{1}{3} \\ -1\frac{2}{3} \\ \hline \end{array}$ $\begin{array}{r} 8\frac{2}{5} \\ -3\frac{3}{5} \\ \hline \end{array}$

$1\frac{9}{5} = 2\frac{4}{5}$

42. $\begin{array}{r} 14\frac{3}{5} \\ -8\frac{1}{2} \\ \hline 6\frac{2}{3} \end{array}$ $\begin{array}{r} 12\frac{4}{5} \\ -6\frac{1}{3} \\ \hline \end{array}$ $\begin{array}{r} 8\frac{1}{12} \\ -5\frac{1}{2} \\ \hline \end{array}$

43. $\dfrac{4 + 3}{2 + 3} = \dfrac{4}{2} = 2$ $\dfrac{5 + 2}{5 + 3} = \underline{\hphantom{xxxxx}}$

$\dfrac{y + 3}{y + 1} = \underline{\hphantom{xxxxx}}.$

44. $\dfrac{3 + \overset{2}{\cancel{4}}}{2} = \dfrac{5}{1} = 5$ $\dfrac{6 + 5}{3} = \underline{\hphantom{xxxxx}}$

$\dfrac{\overset{1}{\cancel{7}} + 8}{4} = \underline{\hphantom{xxxxx}}.$

45. Why might the error pattern in Exercise 43 occur?

46. Why might the error pattern in Exercise 44 occur?

Extension Exercises

47. Does 1 = 4? Does 10 = 15? Find the incorrect step in each of the following.

(a) $1 = \dfrac{2}{1 + 1} = \dfrac{2}{1} + \dfrac{2}{1} = 4.$ So 1 = 4.

(b) $10 = \dfrac{20}{2} = \dfrac{10 + 10}{2} = \dfrac{10}{2} + 10 = 15.$

So 10 = 15.

48. People mentally multiply whole numbers by 25 by changing a problem such as 25×44 into $100 \times \dfrac{44}{4}$ or 100×11.

(a) Show why $25 \times 44 = 100 \times \dfrac{44}{4}$.

(b) How can you use this same approach to compute 25×84?

49. Show that multiplication of rational numbers $\dfrac{u}{v}, \dfrac{w}{x},$ and $\dfrac{y}{z}$ is associative.

50. Show that multiplication of rational numbers $\dfrac{x}{w}$ and $\dfrac{y}{z}$ is commutative.

51. Tell whether each of the following is true or false. If the equation is true, prove it. If it is false, give a counter-example. Assume x and y are nonzero rational numbers.

(a) $\dfrac{2}{x} + \dfrac{2}{y} = \dfrac{2}{x + y}$ (b) $\dfrac{x + y}{y} = x$

52. Tell whether each of the following is true or false. If the equation is true, prove it. If it is false, give a counter example. Assume that $\dfrac{w}{z}, \dfrac{x}{z},$ and $\dfrac{y}{z}$ are rational numbers.

(a) $\left(\dfrac{w}{z} - \dfrac{x}{z} \right) + \dfrac{y}{z} = \dfrac{w}{z} - \left(\dfrac{x}{z} + \dfrac{y}{z} \right)$

(b) $\left(\dfrac{w}{z} + \dfrac{x}{z} \right) \div \dfrac{y}{z} = \left(\dfrac{w}{z} \div \dfrac{y}{z} \right) + \left(\dfrac{x}{z} \div \dfrac{y}{z} \right)$ if $y \neq 0$

53. A common error pattern in addition of fractions is to add the numerators and the denominators, as in $\dfrac{1}{2} + \dfrac{2}{3} = \dfrac{3}{5}$. In fact, if A, B, C, and D are counting numbers, then $\dfrac{A}{B} + \dfrac{C}{D} \underset{(< \text{ or } >)}{\underline{\hphantom{xxxx}}} \dfrac{(A + C)}{(B + D)}$. Tell why.

(*Hint:* Rewrite the right side as a sum of two fractions.)

CHAPTER 6 SUMMARY

The set of integers is expanded to form the set of rational numbers so that nonzero integer division problems with nonzero divisors will have answers. In elementary school, children study the nonnegative subset of rational numbers that I have called elementary fractions.

Elementary fractions are quite versatile. An elementary fraction can represent a part of a whole, a division problem, a point on a number line, or a part of a set.

Arithmetic rules for fractions use these models. Fraction pictures help children understand the results of fraction arithmetic. The common categories of the four whole-number operations fit many word problems involving rational numbers. Estimation enables students to use their facility with whole numbers to develop their intuition about elementary fraction arithmetic.

Rational numbers include all the numbers most children study in elementary school. Whole numbers, elementary fractions, and integers are all rational numbers. Whole numbers, integers, and rational numbers all possess the commutative and associative properties for addition and multiplication and the distributive properties for multiplication over addition and for multiplication over subtraction. The integers and rational numbers have unique additive inverses as well, and nonzero rational numbers have unique multiplicative inverses. The rational numbers are also dense.

Children now study estimation and mental computation with fractions in school. They learn how to recognize problems that are easy to compute and how to use the rounding and compatible-numbers strategies to estimate.

STUDY GUIDE

To review Chapter 6, see what you know about each of the following ideas or terms that you have studied. You can also use this list to generate your own questions about the chapter.

ELEMENTARY FRACTIONS IN GRADES 1–8

The following chart shows at what grade levels selected elementary fraction topics typically appear in elementary- and middle-school mathematics textbooks.

Topic	Typical Grade Level in Current Textbooks
Fraction concepts	1, 2, 3, 4, 5, 6
Fraction addition and subtraction	3, 4, 5, 6, 7
Fraction multiplication and division	5, 6, 7
Fraction estimation	4, 5, 6, 7

REVIEW EXERCISES

1. Describe four common meanings of $\frac{5}{6}$. For each meaning, give its name, make a drawing, and describe how it works for $\frac{5}{6}$.

2. Show why $\frac{7}{4} = 1\frac{3}{4}$ using
   (a) the part-of-a-whole model.
   (b) a number line.
   (c) division.

3. If [grid figure] represents $\frac{1}{2}$, make a drawing that represents
   (a) $\frac{3}{4}$ (b) $\frac{1}{6}$

4. *Explain* why $\frac{5}{0}$ is undefined.

5. Give an example of a number that is rational but not an integer.

6. Write a paragraph that defines whole numbers, integers, and rational numbers and tells how the three sets of numbers are related.

7. Find the GCF of 135 and 162, and use it to simplify $\frac{135}{162}$.

8. (a) Compare $\frac{3}{4}$ and $\frac{2}{3}$ with a fraction-bar drawing, and state the result.
   (b) Compare $\frac{3}{4}$ and $\frac{2}{3}$ by writing both fractions in terms of the least common denominator, and state the result.

9. Eight math professors share five medium pizzas equally. At the next table, six science professors share four medium pizzas equally. Which department has slightly larger shares of pizza?

10. Suppose $\frac{a}{b} > \frac{c}{d}$ and $a = c$ for counting numbers a, b, and c.
    (a) What can you conclude about b and d?
    (b) Does part (a) involve induction or deduction?

11. (a) Name two fractions between $\frac{5}{7}$ and $\frac{6}{7}$.
    (b) What property of rational numbers does part (a) illustrate?

12. A child thinks $\frac{1}{4} + \frac{1}{3} = \frac{2}{7}$. Draw a fraction bar for each fraction, and explain why $\frac{2}{7}$ cannot be the sum.

13. (a) Draw a fraction bar for each fraction in $\frac{2}{3} + \frac{1}{6}$.
    (b) Draw a fraction bar for the sum, and explain why a common denominator is needed.
    (c) Draw a fraction bar for each fraction with the least common denominator.
    (d) Show how to compute $\frac{2}{3} + \frac{1}{6}$ using a picture.

14. (a) Draw a fraction bar for each fraction in $\frac{1}{2} - \frac{1}{5}$.
    (b) Explain why a common denominator is needed.
    (c) Draw a fraction bar for each fraction with the least common denominator.
    (d) Show how to compute $\frac{1}{2} - \frac{1}{5}$ using a picture.

15. Compute and simplify the following without a calculator.

(a) $\dfrac{3}{8} + \dfrac{2}{9}$ (b) $42\dfrac{5}{12} - 24\dfrac{2}{3}$ (c) $-3\dfrac{1}{2} - 2\dfrac{1}{4}$

16. (a) *Explain* how to compute $4 \div \dfrac{1}{3}$ with a measurement picture.

(b) *Explain* how to compute $4 \div \dfrac{1}{3}$ with a common denominator.

17. A child who does not know the multiplication rule for fractions wants to compute $\dfrac{2}{3} \times \dfrac{3}{5}$. *Explain* how to compute $\dfrac{2}{3} \times \dfrac{3}{5}$ using two diagrams.

18. Explain why, in multiplying $\dfrac{1}{4} \times \dfrac{8}{9}$, one can simplify the computation by changing the 8 and 4 to 2 and 1, respectively.

19. A sixth grader thinks $2\dfrac{1}{2} \times 3\dfrac{1}{4} = 6\dfrac{1}{8}$. Use an area diagram to show why this is incorrect, and then obtain the correct answer.

20. Explain how to compute $\dfrac{2}{3} \div 4$ with a part-of-the-whole diagram.

21. A child wants to see why the invert-and-multiply rule works. Use the division model of fractions and the Fundamental Law to show that $\dfrac{2}{5} \div \dfrac{7}{9} = \dfrac{2}{5} \times \dfrac{9}{7}$.

22. (a) You have 5 ounces of peanuts. Exactly how many $\dfrac{3}{4}$-ounce servings can you make?

(b) What operation and category does this problem illustrate? (area, array, partition, repeated)

23. Suppose you give $\dfrac{1}{10}$ of your earnings to charity and pay $\dfrac{1}{3}$ of your earnings in taxes.

(a) What fraction of your earnings is left for other expenses?

(b) What operations and classifications does part (a) illustrate? (combine, compare, missing part, take away)

24. Patty bought $1\dfrac{3}{4}$ pounds of low-fat Swiss cheese for 5 lunches during the work week.

(a) How much cheese will she have for each day? (Give an exact answer.)

(b) What operation and category does this problem illustrate? (area, array, partition, repeated)

25. Write a realistic word problem for $4 \div \dfrac{1}{2}$.

26. Compute and simplify the following without a calculator.

(a) $\dfrac{3}{5} \times 3\dfrac{5}{12}$ (b) $8 \div \dfrac{2}{3}$ (c) $-\dfrac{2}{3} \div \left(-\dfrac{4}{5}\right)$

27. Give an example illustrating the associative property of multiplication for rational numbers.

28. *Explain* how you would know, after studying whole numbers and integers, that rational number division for nonzero numbers is not associative.

29. Show the steps you would use to compute the exact answer to $24 \times 2\dfrac{1}{2}$ mentally.

30. Name a property that the nonzero rational numbers have that the integers do not have.

31. $472\dfrac{1}{4} \div \dfrac{3}{16}$ is approximately

(a) 2500   (b) 100   (c) 9   (d) $\dfrac{1}{10,000}$

32. (a) Complete the last problem, repeating the error pattern from the completed examples.

$$\dfrac{1}{4} + \dfrac{2}{3} = \dfrac{4}{7} + \dfrac{6}{7} = \dfrac{10}{7} = 1\dfrac{3}{7}$$

$$\dfrac{3}{5} + \dfrac{1}{9} = \dfrac{12}{14} + \dfrac{6}{14} = \dfrac{18}{14} = 1\dfrac{4}{14} = 1\dfrac{2}{7}$$

$$\dfrac{2}{5} + \dfrac{1}{4} = \underline{\hspace{3cm}}$$

(b) Write a description of the error pattern.

(c) Explain how the error pattern in the first example could be detected with estimation. (*Hint:* Why should the answer be less than 1?)

ALTERNATE ASSESSMENT—RUBRICS

Questions that require students to do more thinking and writing are an essential part of assessment, but they are much harder to score than multiple-choice or short-answer items. Rubrics give structured guidelines for uniform scoring of performance tasks. Showing the standards for the rubrics to the children helps them understand what distinguishes a good response from a poor one. While rubrics must be adapted to the type of question and the expectations of the teacher, the following example will give you some idea how rubrics work.

Suppose your class had to answer the following question. Three friends order 4 small pizzas. Describe how you would divide up the pizzas equally among the 3 people. What part of a pizza does each person receive? Score each of the answers below using the following rubrics (1–4 points).

4 - Excellent The solution is complete and correct or has one very trivial error.

3 - Good The solution is mostly correct. It shows most of the steps and has few errors.

2 - Partially Proficient The solution is partially correct, but it has a number of errors or incomplete work, and the final answer may be incorrect.

1 - Not Proficient The solution has errors throughout, or a significant part of the solution is missing.

> *Student # 1:* I would take 4 pizzas. Cut them into 3 slices each. That makes 12 slices. Each person gets 4 slices. They get 1 pizza plus an extra slice.
>
> *Student # 2:* I would give each person one pizza. Then cut the remaining pizza into 3 equal parts and give one to each person. So each person gets $1 + \frac{1}{3} = 1\frac{1}{3}$ pizzas.
>
> *Student # 3:* Each person gets $\frac{3}{4}$ of a pizza since $3 \div 4 = \frac{3}{4}$.
>
> *Student # 4:* I would cut each pizza into 6 pieces and then pass out slices to everyone.

Or you can do one of the following assessment activities: add to your portfolio, add to your journal, write another unit test, do another self-assessment, or give a presentation.

SUGGESTED READINGS

Ashlock, R. *Error Patterns in Computation.* 8th ed. Englewood Cliffs, N.J.: Prentice Hall, 2001.

Barnett, C., et al. *Fractions, Decimals, Ratios, and Percents.* Portsmouth, NH: Heinemann, 1994.

National Council of Teachers of Mathematics. *2002 Yearbook: Making Sense of Fractions, Ratios, and Proportions.* Reston, VA: NCTM, 2002.

Navigating Through Number and Operations (4 vols). Reston, VA: NCTM, 2003.

Use Infotrac to find articles in *Teaching Children Mathematics* on teaching children about fractions.

7 | DECIMALS, PERCENTS, AND REAL NUMBERS

All rational numbers can be written as fractions or as decimals. Each notation has its advantages.

The ancient Egyptians developed fraction notation over 3,500 years ago for measuring and accounting, but fraction notation is often awkward for comparing the sizes of two numbers or for doing computations. Decimal notation is one of the great labor-saving inventions of mathematics. Simon Stevin (1548–1620), a Flemish engineer, was the first to discuss decimal notation and decimal arithmetic in some detail.

Decimal notation uses an extension of our whole-number place-value system to represent numbers. As a result, decimal arithmetic algorithms use whole-number/integer algorithms, with additional rules for shifting and placing decimal points.

Decimals are also significant in mathematics because, as the Pythagoreans discovered over 2,000 years ago, rational numbers alone are insufficient for measuring all lengths. The first irrational (that is, "not rational") decimal numbers that the Pythagoreans found were certain square roots.

Although the set of rational numbers is dense, it does not represent every point on the standard number line. The fact that decimal numbers include both rational and irrational numbers makes it possible to label every point on a number line and to measure any length. The union of the set of rational and irrational numbers is called the set of real numbers.

7.1 DECIMALS: PLACE VALUE, ESTIMATION, AND MENTAL COMPUTATION

NCTM Standards

- understand the place-value structure of the base-ten number system and be able to represent and compare whole numbers and decimals (3–5)
- develop and use strategies to estimate computations involving fractions and decimals in situations relevant to students' experience (3–5)
- develop an understanding of large numbers and recognize and appropriately use exponential, scientific, and calculator notation (6–8)

315

> **Sally's Car Clinic**
> **Parts and Labor $362.25**   **GNP Is $3.4 Billion**

Whether it is a Gross National Product (GNP) of $3.4 billion, an atom of diameter 0.00000004 cm, or car repairs costing $362.25, people usually report statistics in decimal notation. In most everyday applications, decimal notation is easier to use than fractions, and decimal computations are easier than fractional computations. Furthermore, calculators and computers generally give output in decimal form, and the metric system employs decimal notation.

PLACE VALUE

Classroom Connection

LE 1 Opener

A fourth grader says that 0.1 is the same as 0.01 because adding the zero doesn't change the value. Is this right? If not, what would you tell the child?

Children in fourth grade learn about decimal place value for tenths and hundredths by connecting it to decimal-square pictures and fractions (Figure 7–1). Decimal-square pictures make it clear why $0.1 > 0.01$.

Decimal place value is an extension of whole-number place value, and it has a symmetry as shown in the following diagram. Each place value for whole numbers (extending to the left) corresponds to a place value for decimal places (extending to the right). For example, tens correspond to tenths, and hundreds correspond to hundredths.

3	4	2	6	·	5	1	7
thousands	hundreds	tens	ones	·	tenths	hundredths	thousandths

The decimal point is read "and." The number 3426.517 is read "three thousand, four hundred, twenty-six and five hundred seventeen thousandths."

LE 2 Concept

(a) In our numeration system, each place you move to the left multiplies place value by _____.

(b) Each place you move to the right divides place value by _____, which is the same as multiplying by _____.

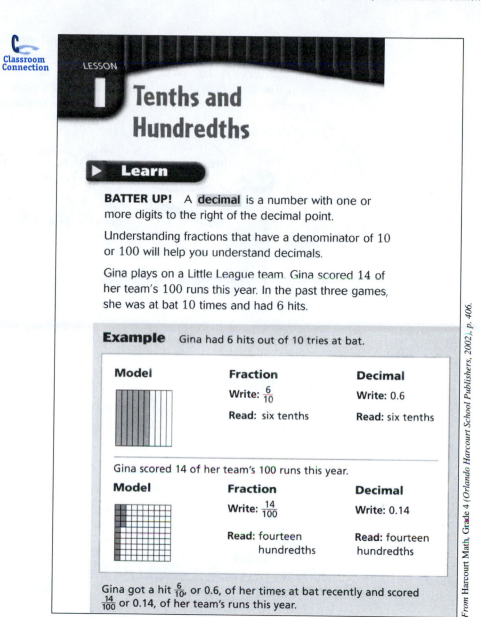

Classroom Connection

LESSON

Tenths and Hundredths

▶ **Learn**

BATTER UP! A **decimal** is a number with one or more digits to the right of the decimal point.

Understanding fractions that have a denominator of 10 or 100 will help you understand decimals.

Gina plays on a Little League team. Gina scored 14 of her team's 100 runs this year. In the past three games, she was at bat 10 times and had 6 hits.

Example   Gina had 6 hits out of 10 tries at bat.

Model	Fraction	Decimal
	Write: $\frac{6}{10}$	Write: 0.6
	Read: six tenths	Read: six tenths

Gina scored 14 of her team's 100 runs this year.

Model	Fraction	Decimal
	Write: $\frac{14}{100}$	Write: 0.14
	Read: fourteen hundredths	Read: fourteen hundredths

Gina got a hit $\frac{6}{10}$, or 0.6, of her times at bat recently and scored $\frac{14}{100}$ or 0.14, of her team's runs this year.

From Harcourt Math, Grade 4 (Orlando Harcourt School Publishers, 2002), p. 406.

Figure 7–1

COMPARE AND ORDER DECIMALS

To compare decimal numbers, you can use decimal-square pictures, a number line, or place-value columns. As with whole numbers and fractions, if a decimal number u is located to the right of a decimal number v on this number line, $u > v$.

LE 3   Connection

(a) Why do some children think that $0.32 > 0.4$?
Explain why $0.4 > 0.32$
(b) with a decimal-square picture (Activity Card 3).
(c) with a number line from 0 to 1.
(d) by putting the numbers in place-value columns of ones, tenths, and hundredths.

The preceding exercise contains examples of terminating decimals. **Terminating decimals** can be written with a finite number of digits to the right of the decimal point. The numbers 8.0, 0.6, and 0.317 are terminating decimals, but 0.3444 . . . is not (Figure 7–2). The ". . ." in 0.3444 . . . indicates that the 4 repeats infinitely. The number can also be written 0.34̄, in which a line is placed over the repeating block of digits.

Figure 7–2

Nonterminating decimals are discussed later in the chapter.

TYPES OF ROUNDING

In preceding chapters, almost all rounding was done to the nearest appropriate number. Actually, there are three kinds of rounding: rounding up, rounding down, and rounding to "the nearest."

• EXAMPLE 1

An office-supply store sells pens for $0.1225 each. Round this price

(a) down to the preceding hundredth.   **(b)** up to the next hundredth.
(c) to the nearest hundredth.

SOLUTION

The hundredths place contains a 2. This means that $0.1225 is between $0.12 and $0.13 and closer to $0.12.

(a) $0.12   **(b)** $0.13   **(c)** $0.12

$0.1200 $0.1225 $0.1300   ●

Calculators and computers may round down, or truncate, all long decimal numbers. To **truncate** means to discard digits to the right of a particular decimal place. Note that discarding the 25 in $0.1225 (truncating) is the same as rounding down to the preceding hundredth.

LE 4   Skill

A kilometer is 0.621371 mile. Round this distance

(a) down to the preceding thousandth.   **(b)** up to the next thousandth.
(c) to the nearest thousandth.

THE ROUNDING AND FRONT-END STRATEGIES

Decimal estimation, not surprisingly, is very similar to fraction estimation. The rounding strategy is usually best for estimating in decimal addition, subtraction, and most multiplication problems. Decimals are usually rounded to the nearest whole number to create a problem that can be computed mentally. The front-end strategy can also be used for addition, subtraction, and some multiplication problems.

Figure 7-3 shows how the fifth-grade *Harcourt Math* textbook compares the rounding and front-end strategies.

Classroom
Connection

LESSON

2 Estimate Decimal Sums and Differences

▶ **Learn**

TO THE POINT   Janet bought a ticket to a gymnastics meet. The ticket cost $8.19 and Janet paid with a $10 bill. About how much change did she receive?

Estimate by using front-end estimation. Then subtract.

$$\begin{array}{r} \$10.00 \rightarrow \quad \$10 \\ -8.19 \rightarrow \quad -8 \\ \hline \$2 \end{array}$$

So Janet received about $2.00 in change.

To estimate Ilene's all-around score, Janet rounds her individual scores to the nearest tenth. What is the estimate of Ilene's all-around score for the meet, based on rounding?

$$\begin{array}{rcl} 9.20 & \rightarrow & 9.2 \\ 9.575 & \rightarrow & 9.6 \\ 9.40 & \rightarrow & 9.4 \\ +9.575 & \rightarrow & +9.6 \\ \hline & & 37.8 \end{array}$$

Event	Ilene's Score
Vault	9.20
Uneven Bars	9.575
Balance Beam	9.40
Floor Exercise	9.575

So the estimate of Ilene's all-around score for the meet is 37.8.

 **MATH IDEA**   You can round decimals or use front-end estimation to estimate sums and differences. The method that you use depends on the situation.

 From Harcourt Math, Grade 5 (Orlando: Harcourt School Publishers, 2002), p. 48.

Figure 7–3

LE 5   Skill

You ask four fifth graders to solve the following problem: Estimate the price of 8.6 m of material that costs $7.79 per meter. Here are the children's answers. What does each child understand about decimal estimation?

(a) Change it to $9 \times 8 = \$72$.
(b) On my calculator, $8.6 \times 7.79 = 66.994$. My estimate is $67.
(c) Change it to $8 \times 7 = \$56$.
(d) Round 8.6 down to 8 and 7.79 up to 8. Then $8 \times 8 = \$64$.

THE COMPATIBLE-NUMBERS STRATEGY

The compatible-numbers strategy works for estimating in decimal division just as it does for whole-number and fraction division. Rounding to the *nearest* whole number is not always the best approach.

LE 6   Skill

Suppose that 6.8 lb of crabmeat costs $58.20. How could you estimate the cost per pound?

EXPONENTS

Exponents are often used in expressing large and small positive numbers in a briefer format. Decimal place values can be expressed using powers of 10.

   Counting-number exponents were used for place value in Chapter 3. Zero and negative-integer exponents are useful in decimal place value. Zero and negative-integer exponents are defined so that certain properties of positive exponents extend to all integer exponents.

LE 7   Reasoning

Suppose that a student knows how to compute 10^c, when c is a counting number.

(a) How could you suggest what 10^0, 10^{-1}, and 10^{-2} should equal by extending a pattern? (*Hint*: Start with $10^3 = 1000$.)
(b) Describe the pattern.

   LE 7 suggests an appropriate definition of zero and negative-integer exponents.

Zero and Negative-Integer Exponents

For all $a > 0$ and integers n:

$$a^n = 1$$

$$a^{-n} = \frac{1}{a^n}$$

The powers of 10 have a pattern in decimal place value, as follows.

$$\ldots \ 10^3 \qquad 10^2 \qquad 10^1 \qquad 10^0 \qquad 10^{-1} \qquad 10^{-2} \ldots$$

These definitions are consistent with exponent properties with which you are familiar.

Discussion

LE 8 Concept

Complete the following exercise, to show how the value of 10^0 is obtained, by using the rule for adding exponents. Parts (a) and (b) review the rule.

(a) $(10^3) \cdot (10^4) = (10 \cdot 10 \cdot 10) \cdot (\underline{\hspace{1.5cm}}) = 10^?$.
(b) What is the shortcut for multiplying $10^3 \cdot 10^4$?
(c) Suppose a child asks what 10^0 is. Using the same rule (shortcut),
$10^3 \cdot 10^0 = \underline{\hspace{1.5cm}}$.
(d) If $10^3 \cdot 10^0 = 10^3$, then 10^0 must equal $\underline{\hspace{1.5cm}}$.

How nice that letting $10^0 = 1$ and $10^{-1} = 0.1$ makes the pattern in LE 7 and the rule in LE 8 both work! The general addition rule for exponents states that $a^m a^n = a^{m+n}$, in which a is a nonzero rational number and m and n are integers. The subtraction rule for exponents is reviewed in the homework exercises.

MENTALLY MULTIPLYING OR DIVIDING BY POWERS OF 10

The shortcut for multiplying by a counting-number power of 10 is suggested by the pattern in the following lesson exercise.

Discussion

LE 9 Reasoning

Fill in the blanks.

(a) 9.52×10 or 9.52×10^1 $= \underline{\hspace{1.5cm}}$
(b) 9.52×100 or 9.52×10^2 $= \underline{\hspace{1.5cm}}$
(c) 9.52×1000 or $\underline{\hspace{2.5cm}}$ $= \underline{\hspace{1.5cm}}$
(d) Propose a general rule for multiplying by 10^n based on your answers to parts (a)–(c).
(e) Explain how you used inductive reasoning to answer part (d).

LE 9 suggests the following shortcut.

Multiplying a Decimal Number by a Power of 10

Multiplying a decimal number by 10^n ($n = 1, 2, 3, \ldots$) is the same as moving the decimal point n places to the right.

LE 10   Skill

Explain how to compute 4.6×10^3 mentally.

In 1987, Van Gogh's painting *Irises* sold for $53.9 million. The amount 53.9 million means "53.9 times 1 million," or 53.9×10^6. So $53.9 million = $53,900,000, which would buy a lot of real flowers. Figure 7–4 shows how 53.9 million is related to 53 million and 54 million.

Figure 7–4

LE 11   Skill

The cost of motor vehicle accidents in 2002 was $150.5 billion.

(a) Write this number in decimal notation (without the word "billion").
(b) If there were about 185 million drivers, what was the average cost per driver?

Now consider dividing a number by a counting-number power of 10. What is a shortcut for dividing by a power of 10 such as 100 or 1000?

LE 12   Reasoning

Complete the following.

(a) $9.52 \div 10$ or $9.52 \div 10^1 =$ _____
(b) $9.52 \div 100$ or $9.52 \div 10^2 =$ _____
(c) $9.52 \div 1000$ or $9.52 \div 10^3 =$ _____
(d) On the basis of parts (a)–(c), dividing a decimal number by 10^n, in which $n = 1, 2, 3, \ldots$, appears to be the same as moving the decimal point _____ places to the _____.
(e) The generalization in part (d) is formed with _____ reasoning.

LE 12 suggests the following shortcut.

Dividing a Decimal Number by a Power of 10

Dividing a decimal number by 10^n ($n = 1, 2, 3, \ldots$) is the same as moving the decimal point n places to the left.

LE 13   Skill

A store buys 1000 candy bars for $75. How much did it pay for each candy bar?

(a) *Explain* how to compute the exact answer mentally.
(b) What operation and category does part (a) illustrate? (area, array, counting principle, repeated)

The shortcut for multiplying or dividing by a power of 10 can be extended to exponents that are negative integers. Consider the following examples.

$$3.47 \times 10^2 \ = 347$$
$$3.47 \times 10^1 \ = 34.7$$
$$3.47 \times 10^0 \ = 3.47$$
$$3.47 \times 10^{-1} = 0.347$$
$$3.47 \times 10^{-2} = 0.0347$$

LE 14   Reasoning

Describe a shortcut you can use to multiply by negative-integer powers of 10.

The shortcut for multiplying a decimal number by an integer power of 10 is as follows.

Multiplying a Decimal Number by an Integer Power of 10

To multiply $a \times 10^c$, in which c is an integer, the decimal point in a is moved c places to the right if $c \geq 0$ or $|c|$ places to the left if $c < 0$.

LE 15   Skill

Write 5.67×10^{-6} using decimal notation.

SCIENTIFIC NOTATION

Archimedes (287–212 B.C.) was one of the first to use very large numbers. He supposedly computed the number of grains of sand needed to fill the entire universe (10^{63}). But why did he do this?

According to my wife, he was at the beach with some of his friends, who taunted him. "If you're so smart Archi, how many grains of sand would fill the universe?" Submitting to peer pressure, Archimedes proceeded to find out.

It is easier to write large numbers such as 10^{63} in shorthand notation. How do calculators deal with such large numbers? Try the following and find out.

LE 16   Opener

(a) Compute $400,000 \times 360,000$ by hand.
(b) Compute $400,000 \times 360,000$ on your calculator. Did you obtain something like
$\boxed{1.44 \quad 11}$? If you got an error message or "E," your calculator cannot deal with very small or very large numbers. Try this computation on a classmate's calculator that does show the correct answer.

The answer to LE 16(b) is in scientific notation. It means 1.44×10^{11}. Some calculators and all computers use scientific notation to abbreviate very large positive numbers and positive numbers that are close to 0. A computer might write 1.44×10^{11} as $\boxed{1.44 \quad E\,11}$. Scientific notation shows a number as an integer power of 10 multiplied by a number between 1 and 10 (including 1 but not 10).

> **Definition:** **Scientific Notation**
>
> Any positive decimal number can be written in **scientific notation**, $n \times 10^p$, in which $1 \leq n < 10$ and p is an integer.

Scientific notation is a useful shorthand for numbers that have many digits. Scientists, calculators, and computers each employ slightly different forms of scientific notation.

People using scientific notation also need to know how to convert numbers in standard form to scientific notation.

• EXAMPLE 2

Write the 2003 world population, 6,300,000,000, in

(a) scientific notation.   **(b)** billions.

SOLUTION

(a) First, move the decimal point to obtain a number between 1 and 10.

$$6.300000000$$

How many places was the decimal point shifted? Nine. The 9 gives the *magnitude* of the exponent of 10. Should the exponent be 9 or -9 (6.3×10^9 or 6.3×10^{-9})? Use estimation to decide. Which would equal 6,300,000,000?

$$6{,}300{,}000{,}000 = 6.3 \times 10^9$$

(b) Since 1 billion $= 10^9$, 6.3×10^9 is 6.3 billion.   ●

LE 17   Skill

A snail moves at a rate of about 0.00036 mile per hour. Write this rate in scientific notation.

> ### LE 18   Summary
> Tell what you learned about decimal place value in this section.

ANSWERS TO SELECTED LESSON EXERCISES

2. (a) 10   (b) 10; $\frac{1}{10}$

3. (a) With whole numbers such as 32 and 4, the number with more digits is larger.

(b) To show 0.4, shade 4 of the 10 columns. To show 0.32, shade 32 of the 100 squares. The area shaded for 0.4 is larger, so 0.4 > 0.32.

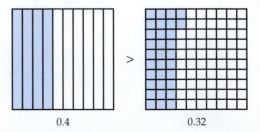

0.4   0.32

(c)

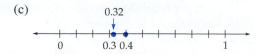

Locate 0.4 and 0.32 on a number line. Since 0.4 is to the right of 0.32, 0.4 > 0.32.

(d)

ones		tenths	hundredths
0	·	4	
0	·	3	2

Start at the place on the left. Both numbers have the same number of ones, but 0.4 has more tenths than 0.32, so 0.4 > 0.32.

4. (a) 0.621   (b) 0.622   (c) 0.621

5. (a) The child knows how to estimate by rounding.
   (b) The child does not understand that estimation is done separately from an exact computation.
   (c) The child knows front-end estimation.
   (d) Rounding one factor up and the other down may yield a more accurate estimate.

6. $58.20 ÷ 6.8 ≈ $56 ÷ 7 = $8/lb

7. (a) $10^3 = 1000, 10^2 = 100, 10^1 = 10, 10^0 = 1,$
   $10^{-1} = 0.1, 10^{-2} = 0.01$
   (b) When the exponent decreases by 1, the result is divided by 10.

8. (a) The final answer is 10^7.
   (b) Add the exponents.
   (c) 10^3
   (d) 1

9. (a) 95.2   (b) 952   (c) 9520
   (e) A generalization based on a pattern in some examples

10. Move the decimal point in 4.6 three places to the right. The answer is 4,600.

11. (a) $150,500,000,000   (b) $814

12. (a) 0.952   (b) 0.0952
    (c) 0.00952   (d) n; left
    (e) Inductive

13. (a) Dividing by 1000 requires moving the decimal point in $75 three places to the left to obtain $0.075.
    (b) Partition sets/measures

15. 0.00000567

16. (a) 144,000,000,000

17. $3.6 × 10^{-4}$ mph

7.1   HOMEWORK EXERCISES

Basic Exercises

1. If you move the decimal point in a number two places to the left, the value of the number is divided by _____ or multiplied by _____.

2. What is the smallest positive number you can display on your calculator?

3. Write each of the following as a decimal number.
   (a) Forty-one and sixteen hundredths
   (b) Seven and five thousandths

4. Write each of the following terminating decimals as a fraction in simplest form.
   (a) 0.27   (b) 3.036   (c) 8.0004

5. You ask a fourth grader to add 4.21 + 18. The child asks, "Where is the decimal point in the 18?" How would you respond?

6. A fourth grader asks whether you can write 0.2 as .2. How would you respond?

7. A sign in a store mistakenly says that apples are selling for .89¢/lb.
   (a) What should the sign say?
   (b) If the store manager insists that the sign is correct, would you want to buy the apples at that price?

8. How many cents is $0.043?

9. (a) Why do some children think that $0.11 > 0.2$?
   Explain why $0.11 < 0.2$
   (b) with a decimal-square picture (Activity Card 3).
   (c) with a number line from 0 to 1.
   (d) by putting the numbers in place-value columns of ones, tenths, and hundredths.

10. (a) Why do some children think that $-2.3 > -2.1$?
    Explain why $-2.3 < -2.1$
    (b) with a number line.
    (c) by putting the numbers in place-value columns of ones and tenths.

11. *Explain* how to use two decimal-square pictures to show that $0.40 = 0.4$.

12. (a) Why might a student think that $-0.4 > -0.17$?
    (b) Use a number line to *explain* why $-0.4 < -0.17$.

13. Round $0.86
    (a) up to the next tenth.
    (b) down to the preceding tenth.
    (c) to the nearest tenth.

14. Round 0.3678
    (a) up to the next hundredth.
    (b) down to the preceding hundredth.
    (c) to the nearest hundredth.

15. Suppose labels are sold in packs of 100.
    (a) If you need 640 labels, how many labels would you have to buy?
    (b) Does this application require rounding up, down, or to "the nearest"?

16. In everyday life, not all rounding of decimals is done to the nearest whole number. For example, if you mail a 1.1-oz letter, the post office charges you the 2-oz rate. Name another situation in which all fractional amounts are rounded up.

17. Mount Everest has an altitude of 8847.6 m, and Mount Api has an altitude of 7132.1 m. How much higher is Mount Everest than Mount Api?
    (a) Estimate the answer using rounding.
    (b) Estimate the answer using the front-end strategy.

18. A job pays $4.35 per hour. How can you estimate how much the job pays for a 32-hour work week?
    (a) Estimate the answer using rounding.
    (b) Estimate the answer using the front-end strategy.

19. A 46-oz can of apple juice costs $1.29. How can you estimate the cost per ounce?

20. You can plan to divide a wall that is 22.7 ft long into 6 equal sections. Estimate the length of each section.

21. I did the following computations on my calculator. Determine by estimating which answers could not be correct.
    (a) $2.4 \times 8.6 = 14.4$
    (b) $2.13 - 0.625 = 1.505$

22. Determine by estimating which of the following answers could not be correct.
    (a) $374 \times 1.1 = 41.14$
    (b) $43.74 \div 2.2 = 19.88181818$

23. A 3-line ad in a newspaper costs $1.79 per line per day. The cost to run such an ad for 4 days is about
    (a) $5   (b) $8   (c) $12
    (d) $22   (e) $54

24. The answer to 3.74×42.8125 has the digits 16011875. How can you use estimation to determine where the decimal point goes?

25. The length of a table to the nearest tenth of a centimeter is 153.7 cm. The exact length of the table is between _____ cm and _____ cm.

26. To estimate the positive decimal problem $a \div b$, a is rounded up and b is rounded down. Will the estimate be too high or too low? Or is it impossible to tell?

27. Show how to find the value of 10^0 by extending a pattern in positive exponents.

28. Show how to find the value of 10^{-3} by extending a pattern in positive exponents.

29. Consider the following pattern.
$$5^3 = 125$$
$$5^2 = 25$$
$$5^1 = 5$$

(a) Each result equals the previous result divided by
_____.

(b) Continuing this pattern,
$5^0 = $ _____ $5^{-1} = $ _____ $5^{-2} = $ _____

30. (a) How can you write 10^{-n} (n is a positive integer) with a positive exponent?

(b) How can you write X^{-n} (X and n are positive integers) with a positive exponent? (See the preceding exercise.)

31. Do you recall the shortcut for dividing numbers with the same base, such as $\dfrac{2^6}{2^4}$?

(a) $\dfrac{2^6}{2^4} = \dfrac{2 \cdot 2 \cdot 2 \cdot 2 \cdot 2 \cdot 2}{2 \cdot 2 \cdot 2 \cdot 2} = 2^?$

(b) What is the shortcut for dividing $\dfrac{2^6}{2^4}$?

Use the shortcut from part (b) in parts (c), (d), and (e). It works with all integer exponents. Assume that X is a positive number.

(c) $\dfrac{10^7}{10^4} = $ _____

(d) $\dfrac{5^7}{5^{-3}} = $ _____

(e) $\dfrac{X^6}{X^2} = $ _____

32. (a) Simplify $4^5 \cdot 4^{-3}$.

(b) $4^5 \cdot 4^{-3} = \dfrac{4^5}{?}$

(c) Simplify the answer to part (b).

33. Write the following in standard form.
(a) 3.62×10^7 (b) 0.056×10^5

34. Multiplying a decimal number by 10^n, in which $n = 1, 2, 3, \ldots$, is the same as moving the decimal _____ places to the _____.

35. The 2003 U.S. federal budget included a debt of about \$6.5 trillion. Write this number without the word "trillion."

36. Write each of the following population figures in millions (using the word "million").
(a) United States: 290,720,000
(b) Paris: 2,772,000
(c) World: 6,400,000,000

37. Write the following in standard form.
(a) $4{,}268 \div 10^6$ (b) $3.62 \div 10^3$

38. Write the following in standard from.
(a) $537 \div 10^4$ (b) $0.0167 \div 10^2$

39. Rosa bought 100 board feet of walnut board at \$3.29 per board foot. What was the total cost? *Explain* how to compute the exact answer mentally.

40. A store buys 1000 "Honk if you love quiet" bumper stickers for \$43. How much did they pay for each bumper sticker? *Explain* how to compute the exact answer mentally.

41. A television signal travels 1 mile in 5.4×10^{-6} second. Write this time interval in standard decimal form.

42. Which is greater, 3.2×10^{-6} or 3.2×10^{-5}?

43. A computer display shows $\boxed{3.4 \quad E\ 12}$. Write this number in scientific notation.

 44. Compute 5^{10} on your calculator and write the answer in scientific notation.

45. Each day, the earth picks up approximately 1.2×10^7 kg of dust from outer space. Does it seem like most of it falls in your house or apartment?
(a) Write this number in standard form.
(b) Name this number.

46. Many scientists believe that the earth is about 5×10^9 years old.
(a) Write this number in standard form.
(b) Name this number.

47. The mass of an oxygen atom is 0.000 000 000 000 000 000 000 013 g.
(a) Write this number in scientific notation.
(b) What advantage does scientific notation have over standard decimal form in this case?
(c) Write this number as it would appear on a calculator display.

48. It would take 3,000,000,000,000,000,000,000,000,000 candles to give off as much light as the sun. Write this number in scientific notation.

49. The average person in the United States discards 6 pounds of trash each day. The population of the United States is about 291 million people. How much trash does the U.S. population discard in a year? Write your answer in scientific notation.

50. The U.S. government has a large debt. The *interest* on the debt in 2004 was about $189 billion.
(a) Write this number in standard form.
(b) Write this number in scientific notation.

51. Why is 48×10^3 not correct scientific notation?

52. Change 386×10^{-5} to correct scientific notation.

53. The following chart gives the distance in meters of each planet from the sun.

Planet	Distance from Sun (meters)
Mercury	5.8×10^{10}
Venus	1.1×10^{11}
Earth	1.5×10^{11}
Mars	2.3×10^{11}
Jupiter	7.8×10^{11}
Saturn	1.4×10^{12}
Neptune	2.9×10^{12}
Uranus	4.5×10^{12}
Pluto	5.9×10^{12}

(a) Which planet is about 10 times as far from the sun as the earth?
(b) Pluto is about _____ times as far from the sun as the earth is.
(c) Mercury is about _____ times as far from the sun as the earth is.

54. The distance from Earth to Saturn is about 800 million miles. About how long would it take to reach Saturn traveling 35,000 mph?

55. Large numbers are difficult to grasp. How long is a million seconds? Is it a week? A month? A year? 5 years?
(a) Guess.
(b) Figure out the answer.

56. How long is a billion seconds? Is it a week? A month? A year? 5 years?
(a) Guess.
(b) Figure out the answer.

57. You can use the clustering strategy to estimate total costs. For example, in estimating the total cost of grocery items at $1.29, $0.45, $2.45, and $1.09, you can group items that add up to about $1.00 or $2.00. The first two items cost about $2; if you add $2 for the third plus $1 for the fourth, the total is about $5. Mentally estimate the total of each of the following groups of prices.
(a) $1.59, $0.30, $3.10, $1.15, $0.72, $2.00, $1.59, $0.89, $2.29
(b) $3.71, $2.62, $0.51, $0.30, $26.95, $9.98, $4.25

58. (a) Without using a calculator, tell which of the following are less than 42 and which are greater.
   (1) $42 \div 2.7$ (2) $42 \div 0.1$
   (3) $42 \div 1.01$ (4) $42 \div 0.999$
(b) Check your answers with a calculator.
(c) If $42 \div a$ is greater than 42, then a is _____.

59. A job pays $9.25 an hour. *Explain* how to compute the exact pay for 20 hours mentally.

$$\left(\text{Hint: Think of 5.25 as } 5\frac{1}{4}. \right)$$

60. A job pays $9.75 an hour. *Explain* how to compute the exact pay for 40 hours mentally.

61. You order 12 shirts for $10.98 each. Show how to compute the exact total cost mentally.

62. You order 8 books costing $7.99 each. Show how to compute the exact total cost mentally.

Extension Exercises

63. In some computer languages, you can truncate a decimal numeral. The TRUNC command takes the integer part of a real number and discards the decimal part. TRUNC(6.8) = 6, and TRUNC(−236.715) = −236. Truncate the following.
(a) 46.81792   (b) −278.987   (c) 4.325
(d) When does truncating produce a different result from rounding?

64. (a) In base five, what would be the place value of the first two digits to the right of the decimal point?
(b) Write 24.21_{five} in expanded notation.

Technology Exercise

 65. If you have a calculator such as the TI-73 with a place-value feature, enter 123.45 and find out how the calculator shows the place value of different digits.

Game Time

 66. The following game requires two people and a calculator. Play a game of "100 POINT" with a partner. The rules of 100 POINT are as follows.
1. Player 1 keys any number into the calculator.
2. Then each player in turn multiplies the number on the calculator by another number, trying to obtain 100 or "100 point something" (e.g., 100.54). The first player to succeed wins the round.

7.2 DECIMAL ARITHMETIC AND ERROR PATTERNS

NCTM Standards

- use visual models, benchmarks, and equivalent forms to add and subtract commonly used fractions and decimals (3–5)

- understand the meanings and effects of operations with fractions, decimals, and integers (6–8)

- develop and analyze algorithms for computing with fractions, decimals, and integers, and develop fluency in their use (6–8)

"Where do you want the decimal point?"

Where does the decimal point go in the answer? For someone who understands whole-number arithmetic, decimal points are the major issue in decimal arithmetic. Decimal arithmetic is closely related to both whole-number and fraction arithmetic. Rewriting decimal numbers as fractions helps clarify how decimal points are placed.

ADDING AND SUBTRACTING DECIMALS

You buy two packages of cheese weighing 0.36 lb and 0.41 lb, and you want to compute the total weight. Or you find a CD for $16.42 at one store and for $16.98 at another store, and you want to know the difference in price. These situations call for addition and subtraction of decimal numbers.

Classroom Connection

LE 1   Opener

Suppose, a fourth grader has not yet learned the rule for adding decimals, and you ask her to compute 0.3 + 0.4. Tell all the ways she might figure it out. (*Hint*: She has used decimal squares and already studied addition of fractions.)

After studying decimal addition in money problems, children learn to model simple decimal addition with decimal squares or base-ten blocks.

LE 2   Connection

(a) Show how to compute 0.3 + 0.4 in a single decimal square (Activity Card 3) by shading in two different colors.
(b) How is this problem similar to 3 + 4?
(c) How is this problem different than 3 + 4?

The decimal point in the sum is lined up with the decimal points in the addends. Note that lining up the decimal points corresponds to getting a common denominator in addition of fractions. For example, 3 tenths + 4 tenths = 7 tenths.

Adding Terminating Decimal Numbers Vertically

Line up the decimal points, add the numbers (ignoring the decimal points), and insert the decimal point in the sum directly below those in the addends.

Like addition, simple decimal subtraction can be modeled with decimal squares or base-ten blocks.

LE 3   Connection

This decimal-square picture (Activity Card 3) shows 0.3.

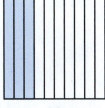

0.3

Use this decimal square to show the result of 0.3 − 0.1, using a take-away approach. Write the result as an equation.

The preceding exercise shows the following connection between decimals and whole-number subtraction.

$$
\begin{array}{ll}
3 \text{ tenths} & 0.3 \\
\underline{-1 \text{ tenth}} & \underline{-0.1} \\
2 \text{ tenths} & 0.2
\end{array}
\qquad \rightarrow
$$

The procedure for decimal subtraction is just like the procedure for decimal addition. Lining up the decimal corresponds to finding a common denominator in subtraction of fractions.

> **Subtracting Terminating Decimal Numbers Vertically**
>
> Line up the decimal points, subtract the numbers (ignoring the decimal points), and insert the decimal point in the difference directly below those in the other two numbers.

The classifications of whole-number operations also apply to many decimal word problems. Consider one of the problems from the beginning of this section.

LE 4   Connection

What subtraction category is illustrated in the following problem? "A CD costs $16.42 at one store and $16.98 at another. What is the difference in price?" (compare, missing part, take away)

MULTIPLYING DECIMALS

What is the cost of 0.6 lb of cole slaw that sells for $0.89 per pound? About how much do 5 bananas weigh if 1 banana weighs 0.3 lb? These applications call for multiplication of decimals.

Classroom
Connection

LE 5   Opener

Suppose, a fourth grader has not yet learned the rule for multiplying decimals and you ask him to compute 3×0.12. Tell all the ways he might figure it out. (*Hint*: He has used decimal squares and has already studied multiplication of fractions.)

Simple decimal multiplication can be modeled with decimal squares or converted into fraction multiplication. Try shading a decimal square to compute 0.6×0.4 in the next exercise.

LE 6   Connection

A child knows fraction multiplication but does not know the rule for decimal multiplication.

(a) Explain how to compute 0.6×0.4 using decimal-square pictures. First, dot 0.4 of a decimal square. Then, in a new picture, darken 0.6 (6 tenths) of each column of the 0.4.

(b) Work out the same problem using fractions.

(c) Write out the same problem using words for place value.

LE 5 and LE 6 indicate how decimal multiplication is just like whole-number multiplication except for having to place the decimal point in the answer. Some elementary-school textbooks now use estimation to place the decimal point in a product. Try it.

LE 7   Skill

The total cost of 12 shirts that are $7.95 each is

(a) $95.40   (b) $954   (c) $9.54   (d) $0.954

How could you use estimation to select the correct choice?

The results of LE 5–LE 7 suggest the rule for placing the decimal point in the product.

LE 8   Reasoning

(a) Complete the following chart.

Numbers			No. of Decimal Places		
Factor 1	Factor 2	Product	Factor 1	Factor 2	Product
3	0.12	0.36	0	2	2
0.6	0.4	0.24			
12	7.95	95.40			

(b) Use the results to write a rule for determining the number of decimal places in the product.

LE 8 suggests the familiar rule for multiplying terminating decimal numbers.

Multiplying Terminating Decimal Numbers

Multiply the two numbers, ignoring the decimal points. The number of decimal places in the product is the sum of the numbers of decimal places in the two factors.

We do not line up the decimal points in multiplication as we do in addition and subtraction. This corresponds to the fact that multiplication of fractions does not require a common denominator.

The categories for whole-number multiplication extend to some decimal applications.

LE 9   Connection

What operation and category are illustrated in the following problem? "About how much do 5 bananas weigh if 1 banana weighs 0.3 lb?" (area, array, counting principle, repeated)

DIVIDING DECIMALS

Suppose 4 laps around a track is a distance of 0.8 mile, and you want to know how long each lap is. Children begin decimal division by studying examples like this one that have whole-number divisors and terminating decimals as dividends and quotients.

Decimal squares can be used to compute $0.8 \div 4$.

LE 10   Connection

In computing $0.8 \div 4$ with a decimal square (Activity Card 3), first show 8 tenths.

Show how to use this picture to compute $0.8 \div 4$ by partitioning, and write the result as an equation.

In LE 10, each of the 8 columns represents 1 tenth, so the 2-column result represents 2 tenths. This exercise illustrates how $0.8 \div 4$ is computed by dividing 8 by 4 and then placing the decimal point in the quotient above the decimal point in the dividend.

$$\begin{array}{r} 2 \text{ tenths} \\ 4\overline{)8 \text{ tenths}} \end{array} \quad \rightarrow \quad \begin{array}{r} 0.2 \\ 4\overline{)0.8} \end{array}$$

This suggests the following procedure.

> **Dividing a Decimal Number by a Whole Number**
>
> Divide, ignoring the decimal point, and place the decimal point in the quotient directly over the decimal point in the dividend.

Now that you have studied problems involving whole-number divisors, what about problems with decimal divisors? These problems are converted to problems with *whole-number* divisors by moving the decimal points in the divisor and the dividend the same number of places to the right. But why does this method work?

$$0.4\overline{)0.08} \qquad \text{becomes} \qquad 4\overline{)0.8}$$

(Unfamiliar, new)   (Hey, we just
                                      studied these!)

What allows us to move the decimal points in the dividend and the divisor the same number of places? It had better not affect the answer to the division problem. If it does, we can't go around teaching children to do it!

LE 11   Reasoning

Consider $8 \div 2$.

(a) Multiply both original numbers by 10 and divide the first number by the second. Is the quotient still the same?
(b) Multiply both original numbers by 100 and divide. Is the quotient still the same?
(c) Multiply both original numbers by 1000 and divide. Is the quotient still the same?
(d) Part (a) works because $8 \div 2 = \dfrac{8}{2} = \dfrac{8}{2} \times \dfrac{10}{10} = \dfrac{80}{20} = 80 \div 20$. Show why part (b) works.
(e) For any two decimal numbers a and b where $b \neq 0$, why does $a \div b = (ac) \div (bc)$, in which $c \neq 0$? (*Hint:* Use the same approach as in part (d).)

LE 11 verifies the following conclusion.

> **Equivalent Division**
>
> If a, b, and c are decimal numbers, with $b \neq 0$ and $c \neq 0$, then
> $a \div b = (a \cdot c) \div (b \cdot c)$.

This means that we can multiply both numbers in a division problem by the same nonzero number without affecting the quotient. The equivalent-division property is merely a different form of the Fundamental Law of Fractions! The expression $a \div b = (a \cdot c) \div (b \cdot c)$ is the same as $\frac{a}{b} = \frac{a \cdot c}{b \cdot c}$. This conclusion can also be justified using multiplication by 1 in the form $\frac{c}{c}$. For nonzero b and c,

$$a \div b = \frac{a}{b} = \frac{a}{b} \cdot \boxed{\frac{c}{c}} = \frac{ac}{bc} = ac \div bc$$

LE 12   Concept

A child wants to know why we can move the decimal point to change $0.4\overline{)0.08}$ to $4\overline{)0.8}$.

(a) By what do you multiply both the divisor and dividend when you change $0.4\overline{)0.08}$ to $4\overline{)0.8}$?

(b) Write $0.08 \div 0.4$ as a fraction, and show how it is converted to $0.8 \div 4$.

(c) Compute $0.08 \div 0.4$.

As LE 12 indicates, the equivalent-division property (i.e., the Fundamental Law of Fractions) can be used to convert any decimal divisor problem to a whole-number divisor problem by moving the decimal points in the divisor and the dividend the same number of places to the right.

Decimal division is also used to change fractions to decimals.

LE 13   Connection

(a) $\frac{3}{11}$ means _____ $\div$ _____.

(b) Use paper and pencil to complete the division in part (a), and tell why the decimal representation of $\frac{3}{11}$ is called a repeating decimal.

Decimal division is also useful in certain applications. Can you correctly classify the following division application?

LE 14   Connection

"A National Motors Finite SST travels 460.8 miles on 16.2 gallons of gas. How many miles does the Finite SST get per gallon?" What division category does this illustrate? (area, array, partition, repeated)

AN INVESTIGATION: MULTIPLYING AND DIVIDING DECIMALS

Discussion

LE 15   Reasoning

Consider the following problem. "In multiplying positive decimals x and y, determine what would make the product greater than x, equal to x, or less than x.

(a) Devise a plan.
(b) Solve the problem

Discussion

LE 16   Reasoning

In dividing positive decimal x by positive decimal y, determine what would make the quotient $x \div y$ greater than x, equal to x, or less than x.

Instead of computing the answers, apply your generalizations from LE 15 and LE 16 to do LE 17 and LE 18.

LE 17   Connection

Peaches cost $0.69/lb. You buy 0.8 lb. The price you pay is

(a) greater than $0.69   (b) $0.69   (c) less than $0.69

LE 18   Connection

You need 12 kg of soil, and each bag holds 0.8 kg. How many bags do you need?

(a) More than 12   (b) 12   (c) Fewer than 12

COMMON ERROR PATTERNS

Students regularly make certain errors in decimal arithmetic. The following exercises will help you recognize some common error patterns.

Classroom
Connection

In LE 19–LE 21 (a) complete the last two examples, repeating the error pattern of the completed examples; (b) write a description of the error pattern; and (c) if possible, explain how estimation could be used to detect errors.

LE 19   Reasoning

0.6	0.8	0.7	0.9	0.5
+ 0.3	+ 0.9	+ 0.6	+ 0.2	+ 0.8
0.9	*0.17*	*0.13*		

LE 20   Reasoning

0.6	0.3	0.7	0.8	0.6
× 0.9	× 0.2	× 0.3	× 0.7	× 0.4
5.4	*0.6*	*2.1*		

LE 21   Reasoning

$$\overset{17}{0.3\overline{)3.21}} \qquad \overset{3.5}{5\overline{)15.25}} \qquad 8\overline{)5.672} \qquad 4\overline{)36.16}$$

SPREADSHEETS

Many elementary-school classrooms now make use of spreadsheets. A **spreadsheet** arranges data, labels, and formulas in a table with rows and columns. Spreadsheets were originally designed for accountants. Today, all sorts of people use spreadsheets at home or at work for recordkeeping. When you use a spreadsheet for computations, you have to figure out the algebraic formulas. Then the computer does all the arithmetic. Figure 7–5 shows part of a discussion about spreadsheets in the fifth-grade *Harcourt Math* textbook.

Each location in the spreadsheet is called a **cell**. In the lower spreadsheet in Figure 7–5, "July" is in cell A1. The cell that is highlighted by the computer is called the **active cell.** You can move from one cell to another using the four arrow keys.

Classroom Connection

Data ToolKit • Order and Add Numbers on a Spreadsheet

The table shows the money Susan earned babysitting last year. Susan uses a spreadsheet to order the amounts from greatest to least and to find the total amount she earned.

Month	Amount	Month	Amount	Month	Amount
January	$75.25	May	$61.75	September	$61.00
February	$87.75	June	$71.00	October	$56.00
March	$65.50	July	$89.75	November	$66.25
April	$54.75	August	$79.25	December	$71.50

• Enter the months into Column A and the amounts into Column B.

• Highlight both columns. Click ⓐ↓.

• Choose *Descending* to sort Column B from greatest to least.

• Click in cell B13. Click *fx* .

• Highlight *SUM* and click *OK*.

• Type *B1:B12*. This tells the spreadsheet to add the cells from B1 to B12.

• Press *return*.

So, Susan earned a total of $839.75 last year.

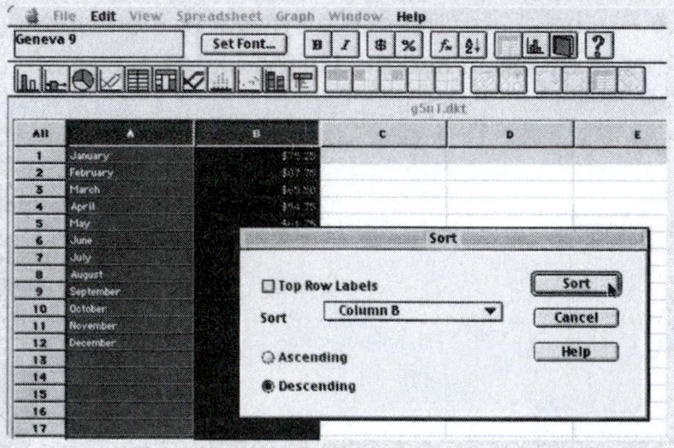

Practice and Problem Solving

Use Data ToolKit to order from greatest to least and find the total amount.

1. Chuck earned $55.25 last winter, $72.50 last spring, $95.25 last summer, and $62.50 last fall.

2. Stephanie earned $54.25 in May, $67.75 in June, $51.75 in July, $65.00 in August, and $71.50 in September.

All	A	B
1	July	$89.75
2	February	$87.75
3	August	$79.25
4	January	$75.25
5	December	$71.50
6	June	$71.00
7	November	$66.25
8	March	$65.50
9	May	$61.75
10	September	$61.00
11	October	$56.00
12	April	$54.75
13		$839.75

From Harcourt Math, Grade 5 (Orlando: Harcourt School Publishers, 2002), p. 65.

Figure 7–5

LE 22   Skill

Refer to Figure 7–5 on page 336.

(a) What does the number in cell B2 represent?
(b) How was the amount in cell B13 calculated?
(c) To compute the average monthly earnings in cell C13, you would enter "=" followed by a formula for C13. What is the formula?
(d) If you have a spreadsheet program, enter the data from the first table and arrange it from least to greatest.

LE 23   Summary

Tell what you learned about decimal arithmetic in this section.

ANSWERS TO SELECTED LESSON EXERCISES

2. (a)

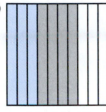

0.3 + 0.4 = 0.7

(b) You add 3 units and 4 units to obtain 7 units.
(c) The units are different (tenths rather than ones).

3.

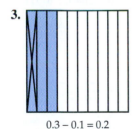

0.3 − 0.1 = 0.2

4. Compare sets/measures

5. Here are three ways.
To show 3 × 0.12, shade 0.12 of a decimal square 3 times. A total of 36 squares are shaded. So 3 × 0.12 = 0.36.

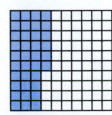

$$\frac{3}{1} \times \frac{12}{100} = \frac{36}{100}$$

$$\begin{array}{r} 12 \text{ hundredths} \\ \times\ \ 3 \text{ ones} \\ \hline 36 \text{ hundredths} \end{array}$$

6. (a) 0.6 × 0.4 means 0.6 (6 tenths) of 0.4. Dot 0.4 of a decimal square.

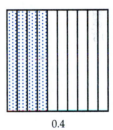

0.4

Now darken 6 tenths of the 0.4.

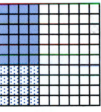

0.6 of 0.4

0.24 of the decimal square is darkened.
So 0.6 × 0.4 = 0.24.

(b) $0.6 \times 0.4 = \frac{6}{10} \times \frac{4}{10} = \frac{24}{100} = 0.24$

(c) 6 tenths × 4 tenths = 24 hundredths

7. $12 \cdot 7.95 \approx 12 \cdot 8 = 96$. The answer is (a).

8. (a)

No. of Decimal Places		
Factor 1	**Factor 2**	**Product**
0	2	2
1	1	2
0	2	2

9. Multiplication, repeated measures

10.

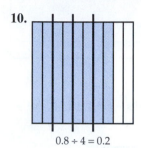

$$0.8 \div 4 = 0.2$$

11. (a) Yes   (b) Yes   (c) Yes

(d) $8 \div 2 = \dfrac{8}{2} = \dfrac{8}{2} \times \dfrac{100}{100} = \dfrac{800}{200} = 800 \div 200$

(e) $(a \cdot c) \div (b \cdot c) = \dfrac{ac}{bc} = \dfrac{a}{b} = a \div b$

12. (a) 10

(b) $0.08 \div 0.4 = \dfrac{0.08}{0.4} = \dfrac{0.08 \times 10}{0.4 \times 10} = \dfrac{0.8}{4} = 0.8 \div 4$

(c) 0.2

13. (a) 3; 11

(b) $0.272727\ldots$; the digits 2 and 7 repeat over and over.

14. Partition, measures

17. (c)

18. (a)

19. (a) 0.11; 0.13

(b) The sum from the tenths column is all placed to the right of the decimal point in the answer.

(c) $0.8 > 0.17$. How can $0.8 + 0.9 = 0.17$?

20. (a) 5.6; 2.4

(b) The product is given the same number of decimal places as each of the factors.

(c) $0.6 \times 0.9 \approx 0.6 \times 1 = 0.6$.
How can $0.6 \times 0.9 = 5.4$?

21. (a) 0.79; 9.4

(b) Zeroes are omitted from the quotient.

(c) $0.3 \times 17 \approx \dfrac{1}{3}$ of $17 \neq 3.21$

22. (a) Susan's monthly earnings for February

(b) It's the sum of the numbers from cells B1 to B12 (inclusive).

(c) (B13)/12

7.2   HOMEWORK EXERCISES

Basic Exercises

1. A child does not know the rule for adding decimals. Show how to compute $0.2 + 0.3$ in a single decimal square by shading in two different colors and state the result in an equation.

2. Show how to compute $0.18 + 0.24$ in a single decimal square by shading in two different colors. State the result in an equation.

3. Consider the following addition problem.

$$\begin{array}{r} \overset{1}{0.36} \\ +\ 0.27 \\ \hline 0.63 \end{array}$$

(a) When you add $6 + 7$ and separate the 13 into 3 and 10, this 10 represents 10 _____.

(b) The 1 that is regrouped represents 1 _____.

(c) Are the amounts in parts (a) and (b) equal?

4. (a) How is decimal addition like whole-number addition?

(b) How is decimal addition different from whole-number addition?

5. A fourth grader computes $0.4 + 0.8 = 0.12$. What would you tell the child?

6. A fourth grader computes 0.25 + 0.06 and obtains a sum of 0.211, which he reads as "2 tenths and 11 hundredths." Explain the logic in what the child is doing. Explain what the child does not understand.

7. Draw a decimal-square picture (Activity Card 3) that shows the result of 0.23 − 0.08. State the result in an equation.

8. Draw a decimal-square picture that shows the result of 0.4 − 0.3, using a take-away approach. State the result in an equation.

9. What operation and category are illustrated in the following problem? "Joe bought a bag of gourmet plant food for $9.89. How much change did he receive back from a $100 bill?"

10. Make up a realistic decimal problem illustrating the missing measure category of subtraction.

11. (a) Show how to compute 2 × 0.18 with a decimal-square picture (Activity Card 3).
   (b) Work out the same problem using fractions.
   (c) Work out the same problem using words for place value.

12. (a) Show how to compute 3 × 0.07 with a decimal-square picture.
   (b) Work out the same problem using fractions.
   (c) Work out the same problem using words for place value.

13. A child knows fraction multiplication but does not know the rule for decimal multiplication.
   (a) *Explain* how to compute 0.5 × 0.6 using two decimal-square pictures (Activity Card 3).
   (b) Work out the same problem using fractions.
   (c) Work out the same problem using words for place value.

14. (a) *Explain* how to compute 0.3 × 0.2 using two decimal-square pictures.
   (b) Work out the same problem using fractions.
   (c) Work out the same problem using words for place value.

15. As a result of studying whole numbers, many children expect that multiplying two numbers usually results in an answer that is larger than either factor.
   (a) Is 1.2 × 0.4 less than 1.2?
   (b) Use two decimal squares to show 1.2 × 0.4. State the result in an equation.

16. What decimal multiplication problem does the following decimal-square picture illustrate?

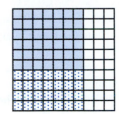

17. The product of 34.56 × 6.2 has the digits 214272.
   (a) Explain how to place the decimal point with estimation.
   (b) Explain how to place the decimal point by counting decimal places.

18. The product of 512.3 × 0.821 has the digits 4205983.
   (a) Explain how to place the decimal point with estimation.
   (b) Explain how to place the decimal point by counting decimal places.

19. A runner burns about 0.12 calorie per minute per kilogram of body mass.
   (a) How many calories does a 60-kg runner burn in a 10-minute run?
   (b) What category does this illustrate? (area, array, counting principle, repeated)

20. Make up a realistic decimal problem illustrating the area category of multiplication.

21. A fifth grader says 50 × 4.44 is the same as 0.50 × 444, which is 222. Is this right? If not, what would you tell the child?

22. A fifth grader says 0.2 × 0.3 = 0.6.
   (a) Why do you think the child did the problem this way?
   (b) What would you tell the child?

23. Show how to work out $0.6 \div 3$ with a decimal square picture (Activity Card 3). State the result in an equation.

24. Show how to work out $1.2 \div 0.4$ with decimal square pictures. State the result in an equation.

25. Show how to work out $0.2 \div 0.04$ with a decimal square picture. (*Hint*: Use repeated measures.) State the result in an equation.

26. Show how to work out $0.8 \div 0.2$ with a decimal square picture. State the result in an equation.

27. A child wants to know why we can move the decimal point to change $6.4 \div 0.32$ to $640 \div 32$. Write $0.32 \overline{|6.4}$ as a fraction, and show how the process works.

28. What do you multiply both numbers by when you change $72 \div 3.6$ to $720 \div 36$?

29. Which of the following are equal?
    (a) $8 \div 0.23$ (b) $800 \div 0.0023$
    (c) $80 \div 2.3$ (d) $0.8 \div 0.023$
    (e) $80 \div 0.023$

30. If you were asked to compute $5000 \div 12$, you could solve a simpler equivalent problem such as $2500 \div 6$ using the equivalent-division property. Change each problem to a simpler one.

    (a) $6000 \div 24$ (b) $8 \div \frac{2}{3}$

31. A sixth grader divides 16 by 3 and gets 5.1
    (a) How did the child obtain this answer?
    (b) What concept doesn't the child understand?

32. (a) $\frac{2}{9}$ means _____ $\div$ _____.

    (b) Complete the division, and explain why the decimal representation of $\frac{2}{9}$ is called a repeating decimal.

33. The quotient of $44.237 \div 3.1$ has the digits 1427. Explain how to place the decimal point with estimation.

34. The quotient of $8.2 \div 0.4$ has the digits 205. Explain how to place the decimal point with estimation.

35. What operation and category are illustrated in the following problem? "A baker needs 2 kg of flour to make a loaf of bread. How many loaves can he bake from 5 kg of flour?" (area, array, partition, repeated)

36. What operations and categories are illustrated in the following problem? "At the beginning of a 4-day car trip, an odometer read 58,427.7. At the end of the trip, it read 59,271.5. What was the average number of miles driven per day?" (area, array, compare, missing part, partition, repeated, take away)

37. What operations and categories are illustrated in the following problem? "Molly bought 4 tires for $37.28 each, an air pump for $11.43, and 2 windshield wipers for $3.10 each. How much did these items cost altogether if there was no tax?" (area, array, combine, counting principle, repeated)

38. At a Chinese restaurant, Szechuan chicken costs $7.50, Hunan shrimp costs $9.25, and moo shi pork costs $6.95. Write two multistep mathematics problems that could be answered using this information.

39. Compute the following without a calculator.
    (a) $4 - 0.03$ (b) $0.2 \times 0.3 \times 0.5$ (c) $0.5 \div 0.02$

40. Compute the following without a calculator
    (a) $0.07 - 0.2$ (b) $0.4 \times 0.3 \times 0.6$ (c) $0.084 \div 0.3$

 41. A store sells salmon for $4.99 a pound.
    (a) How much would 2.5 pounds cost?
    (b) How much would 0.82 pound cost?
    (c) While most adults recognize that part (a) requires multiplication, many adults use division or subtraction to incorrectly solve part (b). What makes part (b) more difficult to see as multiplication?

42. Children sometimes think $A \div B$ will result in an answer that is smaller than A. Give an example with positive decimals where the answer is greater than A.

43. Suppose that $0 < x < 1 < y$.
    (a) How does xy compare to y?
    (b) How does $\frac{y}{x}$ compare to y?
    (c) How does x^2 compare to x?

44. Suppose $-1 < x < 0$ and $y > 1$.

(a) How does xy compare to x?

(b) How does $\frac{x}{y}$ compare to x?

(c) How does x^3 compare to x?

45. Use two of the following numbers in a realistic word problem for which the third number is the answer: 0.3, 1.5, 5.

46. Use two of the following numbers in a realistic word problem for which the third number is the answer: 0.08, 0.2, 0.4.

In Exercises 47–49, (a) complete the last two examples, repeating the error pattern in the completed examples, and (b) write a description of the error pattern.

47.

```
   42        8.1        63        4.2
 − 3.71    −3.71     − 5.29    −3.17
 ──────    ──────    ──────    ──────
  39.71     4.41
```

48.

```
   6.2       3.5
 4⟌26      8⟌29     7⟌36     5⟌93
```

49.

```
  16.2       14.1      12.3       8.2
− 3.7      − 2.5     − 6.7     −4.8
─────      ─────     ─────     ─────
 13.5       12.4
```

50. Describe two errors children might make in computing 0.7×0.8.

51. The following table shows the 2002 annual energy consumption and population of three industrialized countries in tons of oil equivalent.

	Energy Consumption	**Population**
United Kingdom	2.5×10^8 tons	60 million
United States	2.3×10^9 tons	287 million
France	2.9×10^8 tons	59 million

Compute the annual energy consumption per person in each country.

52. In 2003, a new submarine cost about $2.6 billion, and the Head Start program cost about $5,000 per child for a year. How many children could be served by Head Start for the cost of the submarine?

53. (a) Fill in the parentheses.

$$1 \times 2 + 0.25 = (\quad)^2$$
$$2 \times 3 + 0.25 = (\quad)^2$$

(b) What is the next equation if the pattern continues?

(c) Is your equation in part (b) true?

54. (a) Fill in the blanks.

$$0.1089 \quad \times 9 = \underline{\qquad}$$
$$0.10989 \quad \times 9 = \underline{\qquad}$$
$$0.109989 \times 9 = \underline{\qquad}$$

(b) What is the next equation if the pattern continues?

(c) Is your equation in part (b) true?

55. An SUV gets about 6 mpg less than a family car. Suppose you drive 10,000 miles per year and pay $1.60 per gallon of gas.

(a) How many more gallons of gas will the SUV use each year than the family car?

(b) How much higher is the annual gas bill for the SUV than for the family car?

56. A 22-year-old woman owns a 2-year-old car. Consider the following facts. She drives it about 8,000 miles per year. Maintenance on the car costs $0.08 per mile. Gas costs an average of $1.30 per gallon. The car averages 26 miles per gallon. Insurance costs $480 per year.

(a) About how much does maintenance cost for a year?

(b) About how much does gas cost for a year?

(c) Excluding car loan payments, about how much does it cost to operate the car for a year?

Extension Exercises

57. Consider the following problem. "In charging a customer, a cashier interchanges dollars and cents in the price of an item and gives the customer an extra $11.88 in change for a $20 bill. What is the correct price of the item?"

(a) Devise a plan and solve the problem.

(b) Find all possible solutions.

58. Nate bought some oranges for $0.30 each. He ate two of them and sold the rest for $0.50 each. He made a profit of $1.40. How many oranges did he buy?

59. Make up two different decimal multiplication problems that have answers of 0.2. Do not use 1 as a factor.

60. Consider the following problem. "The sum of 2 two-digit decimal numerals is 0.42. Their product is 0.0297. What are the numbers?" Devise a plan and solve the problem.

Technology Exercise

61. A checkbook can be recorded in a spreadsheet. Enter the following on a computer spreadsheet if you have one. Otherwise, you can do the exercise without a computer.

	A	B	C	D	E	F
1	Checkbook					
2	Check #	Date	For	Withd.	Deposit	Balance
3	Starting					$3426.10
4	123	3-12	Rent	$950		
5		3-14			$875.40	
6	124	3-16	Electric	$85.11		
7	125	3-17	Shoes	$62.25		

(a) What does the number in cell D4 represent?
(b) What is a formula to determine the amount in F4?
(c) Complete column F. Use a spreadsheet program if you have it.

62.

	A	B	C	D	E
1	Item	Retail Price	Discount Rate	Discount	Sale Price
2	DVD player	$79	0.15		
3	Television	$325	0.1		
4	Microwave oven	$67	0.05		

(a) What does the number in cell C2 represent?
(b) What are the formulas to determine the numbers in cells D2 and E2?
(c) Compute the values of cells D2 and E2 with a spreadsheet or a calculator.
(d) If you have a spreadsheet program, enter the data from the table and find all the missing values.

63. To compute $\dfrac{8}{2+2}$ on a calculator, a child enters $8 \div 2 + 2$. What answer will the child obtain? Is it correct?

64. To compute $\dfrac{1}{2} + \dfrac{1}{4}$ on a calculator, a child enters $1 \div 2 + 1 \div 4$. What answer will the child obtain? Is it correct?

65. (a) Write $\dfrac{1}{3}$ as a decimal.

(b) Write $\dfrac{1}{3}$ in base three.

(c) Write $\dfrac{1}{3}$ in base six.

(d) Name another base in which $\dfrac{1}{3}$ is a terminating decimal, and write it as a decimal in that base.

66. Compute $3.12_{\text{five}} + 2.33_{\text{five}}$.

Special Exercises

67. The following chart gives the energy consumption of some common electric appliances.

Appliance	Usage	Kilowatt-Hours (kwh) Used
Oven	1 hour	1.1
TV (color)	1 hour	0.3
Refrigerator	1 month	120
Frost-free refrigerator	1 month	200
Dishwasher	1 cycle	0.6
Fan	1 hour	0.2
Room air conditioner	1 hour	1.5
Vacuum cleaner	15 minutes	0.2
Clock	1 month	1.8
Iron	1 hour	0.8
Light bulb (100 W)	1 hour	0.1
Baseboard heater	1 hour	2
Clothes washer	1 cycle	0.2
Clothes dryer	1 cycle	3
Water heater	1 month	350

Using a rate of $0.08 per kilowatt-hour, find the cost of the following.
(a) Watching color TV for 2 hours
(b) Doing one load of laundry in the washer and dryer
(c) Running a refrigerator for 1 month
(d) Running a water heater for 1 month
(e) Running a room air conditioner for 8 hours

68. Refer to the chart in Exercise 67, and consider the following. A couple owns an oven, a refrigerator, two clocks, two room air conditioners, a vacuum cleaner, a color TV, a water heater, an iron, and eight 100-watt bulbs. Assume that they run the air conditioner for 100 hours one month.

Estimate how much they will use each other appliance in a month, and then estimate their total electric bill for the month at a rate of $0.09 per kilowatt-hour.

7.3 RATIO AND PROPORTION

NCTM Standards

- represent and use ratios and proportions to represent quantitative relationships (6–8)
- develop, analyze, and explain methods for solving problems involving proportions, such as scaling and finding equivalent ratios (6–8)

Ratios and proportions are useful for representing relationships between two measures or counts. Common applications include ratios of men to women, students to teachers, miles to hours, and pounds to cubic feet. Ratios and proportions offer a new kind of challenge for children who have previously studied individual counts or measures. Isn't it harder to understand 40 miles per hour than 40 miles?

RATIOS

LE 1   Opener

(a) How many male and female students are in your class today?
(b) What is the ratio of male students to female students?
(c) What is the ratio of female students to all students in your class?

A ratio compares two numbers using division. In some ratios, you can compare a part to a part or a part to a whole. How did you write the ratios in LE 1? There are three common ways of writing a ratio: with the word "to," with a colon, or as a fraction.

Definition: **Ratio**

A **ratio** is a comparison of two numbers a and b, with $b \neq 0$. It can be written a to b, $a : b$, or $\frac{a}{b}$.

See if you can connect ratios and fractions in the following exercise.

LE 2   Connection

Use a rectangle to represent a submarine sandwich.

(a) Divide the sandwich into two sections A and B so that the ratio of A to B is 1:3.
(b) What fraction of the whole sandwich is section A?
(c) Section B is _____ times as long as section A.
(d) Section A is _____ times as long as section B.

When might someone want to compare two ratios? In selecting a college, you may have been interested in comparing the number of students and the number of professors or the number of men and the number of women.

LE 3   Concept

Tom Cruise College has 775 students and 25 professors. Compare the number of students to the number of professors in two ways.

(a) There are _____ more students than professors.
(b) There are _____ times as many students as professors.
(c) Suppose Brain Strain College has 800 students and 50 professors. Use part (a) or part (b) and compare the two colleges.

Both colleges have 750 more students than professors, but Brain Strain has significantly fewer students per professor. The number of students per professor (called the "student-professor ratio") gives the more useful figure.

At Brain Strain College, the student-professor ratio is 16 to 1. At Tom Cruise College, it is 31 to 1. These ratios could also be written using colons (16:1 and 31:1) or as fractions $\left(\frac{16}{1} \text{ and } \frac{31}{1}\right)$. The ratios tell us that Tom Cruise College has about twice as many students per teacher as Brain Strain College. A ratio such as $\frac{16 \text{ students}}{1 \text{ professor}}$ is more difficult to understand than a count such as 16 students or a measure such as 4 feet because a ratio is a relationship between two counts or measures.

LE 3(a) and (b) ask for additive and multiplicative comparisons of students and professors. The difference in two quantities is an **additive comparison.** A ratio is a **multiplicative comparison** because it tells how many times one quantity the other quantity is. In grades 6–8, many children begin to grasp the difference between additive and multiplicative comparisons. They need to see a variety of situations and decide whether or not it is appropriate to use ratios.

In many cases, the most useful comparison between two sets or two measures is how many times larger one set or measure is than the other. That is what a ratio of two sets shows.

To have a correspondence between ratios and fractions, the second number in a ratio is not allowed to be 0, as in the ratio of men to women on the New York Yankees, 25 to 0, since $\frac{25}{0}$ is undefined. One can compare the same two groups in accordance with the definition by looking at the ratio of women to men on the New York Yankees $\left(0 \text{ to } 25, \text{ or } \frac{0}{25}\right)$.

Since ratios can be represented as fractions, you can find equivalent ratios, write a ratio in simplest form, or write a ratio as a decimal.

LE 4   Skill

Kristin bicycles 33 miles in 6 hours.

(a) What is the ratio of miles to hours, and what does this measure?
(b) Write two ratios that are equivalent to the ratio in part (a).
(c) Write the ratio from part (a) in simplest form.
(d) Write the ratio as a decimal.
(e) Bill bicycles 42 miles in 8 hours. How does his average speed compare to Kristin's?

The ratios in LE 4 are also rates. A **rate** is a ratio that compares two quantities with different units, such as 33 miles/6 hours. A ratio with no units (that is, in which units cancel out), such as 8 inches/1 inch = 8 : 1, is not a rate.

In part (d), the ratio is simplified to 5.5 miles/hour, a unit rate. A rate in which the denominator is 1 is a **unit rate.** In part (e), unit rates could be used to make a comparison.

LE 5   Skill

Use the unit rate for Kristin's bicycle from LE 4 to complete the following rate table.

Rate Table

Time (hours)	1	2	3	4
Distance Traveled (miles)				

PROPORTIONS

Suppose you're driving one of those cars with automatic cruise control. You set it at 50 mph and sit back and relax. ZZZZ. Wake up! You still have to steer the car.

The car travels 50 miles/hour. You will travel 100 miles in 2 hours, 150 miles in 3 hours, and so on. When you double the driving time, the miles traveled also double. When you triple the driving time, the miles traveled also triple.

The ratios of miles to hours—for example, 100 to 2 and 150 to 3—are equal because $\frac{100}{2} = \frac{150}{3}$. The equation $\frac{100}{2} = \frac{150}{3}$ is a proportion. A **proportion** states that two ratios are equal. In this application, one would say that the distance traveled "is proportional to" the driving time.

LE 6   Concept

Two different classes are asked whether they would prefer a field trip to an amusement park or a sewage treatment plant. In class A, 10 out of 15 students prefer the amusement park. In class B, 20 out of 30 students prefer the amusement park.

(a) Write a ratio for the results for each class.

Consider three ways to show that the pair of ratios in part (a) can form a proportion.
(b) Write each ratio in simplest form.
(c) Use the Fundamental Law of Fractions to show why the two ratios in part (a) are equivalent.
(d) Write the ratios in part (a) as fractions. Multiply the numerator of one fraction times the denominator of the other and compare the two products.

Two ratios can form a proportion if they are equivalent. You can determine this by writing both ratios in simplest form, by finding a common multiplier (or scale factor) between the ratios (Fundamental Law of Fractions), or by comparing cross products. Part (d) is an example of how comparing cross products works. By using cross products, $\frac{10}{15} = \frac{20}{30}$ since $10 \times 30 = 15 \times 20$. The following exercise shows why this works.

LE 7   Reasoning

Suppose $\dfrac{a}{b} = \dfrac{c}{d}$.

(a) Use the Fundamental Law of Fractions to write $\dfrac{a}{b}$ and $\dfrac{c}{d}$ in terms of the common denominator bd. Fill in the resulting numerators in the following equation.

$$\frac{\overline{}}{bd} = \frac{\overline{}}{bd}$$

(b) Now if $\dfrac{ad}{bd} = \dfrac{bc}{bd}$, then $\underline{} = \underline{}$.

LE 7 shows that $\dfrac{a}{b} = \dfrac{c}{d}$ is equivalent to $ad = bc$. Therefore, by comparing cross products ad and bc, one can tell if $\dfrac{a}{b} = \dfrac{c}{d}$.

Cross Product Property

If a, b, c, and d are integers and $b \neq 0$ and $d \neq 0$, then $\dfrac{a}{b} = \dfrac{c}{d}$ if and only if $ad = bc$.

Cross products are both efficient and widely applicable. Children are taught to use cross products to solve proportions. However, some children have difficulty using cross products because the method does not make sense to them. Some schools also teach children to use mental math (finding a multiplier) or unit rates to solve proportions. Try these different methods in the following exercise.

Classroom Connection

LE 8   Connection

The weight of a pile of bricks is proportional to its volume. Suppose that 10 ft³ of bricks weigh 30 lb. How much would 40 ft³ of bricks weigh?

(a) What are three ways a child might solve this problem?
See which of the following you used in part (a).
(b) How could you find the answer by finding a multiplier (scale factor)?
(c) Write a proportion. Tell how you set up the proportion. Then solve it by using cross products.
(d) Find a unit rate for the weight of the bricks per ft³, and use that to find the answer. (Make a unit rate table if you wish.)

The method of part (b) is referred to as "mental math" in some middle-school texts. The process involves a multiplier (or scale factor) with the two given numbers that have the same units (10 ft ³ and 40 ft³). You find what you multiply by to go from one volume to the other (4 in Figure 7–6). Then the weight is also multiplied by 4. Compute $4 \cdot 30$ lb $= 120$ lb.

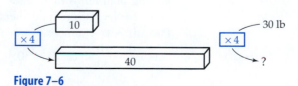

Figure 7–6

In part (c), note that there are four correct ways to write the proportion! They all have the same cross products. They are $\frac{30}{10} = \frac{x}{40}$, $\frac{10}{30} = \frac{40}{x}$, $\frac{40}{10} = \frac{x}{30}$, and $\frac{10}{40} = \frac{30}{x}$. The equal cross products of all four proportions are $1200 = 10x$. So $x = 120$ lb. If you write the x in the numerator, the proportion can also be solved by multiplying both sides by the same number (40 for the first proportion and 30 for the third proportion) instead of using cross products.

In part (d), the unit-rate method uses the two given numbers that form a complete rate measurement (30 lb/10 ft^3). Rewrite that ratio with a denominator of 1, and you have a unit rate (3 lb/ft^3). Then 40 ft^3 weighs $40 \cdot 3 = 120$ lb.

LE 9   Skill

You want to cook a new dish called "seaweed surprise" for 7 people. The recipe for 4 people calls for 18 ounces of seaweed.

(a) Estimate how much seaweed you will need.
(b) Find the exact answer by using a proportion. Tell how you set up the proportion. (Make sure that your units match up.)
(c) Find the unit rate, and solve the problem with it.
(d) Use the multiplier (scaling) method. Find a multiplier (or scale factor) to go from one portion of seaweed to the other and solve the problem.
(e) Estimate how many friends would accept a subsequent dinner invitation.

In some cases, there isn't a simple proportional relationship among the quantities you are given and the quantity you are looking for.

LE 10   Skill

A college has a male-female ratio of 3 to 2. If there are 1100 students, how many men and women are there?

 ## AN INVESTIGATION: CAPTURE-RECAPTURE

LE 11   Reasoning

People in your town are worried that the fish in the town's lake are dying out. You are hired to estimate how many fish are in the lake. Simulate how this is done using beans (or whatever else your instructor chooses) to represent the fish.

(a) You are given a bag of white beans. Each white bean represents a fish. It is not practical to capture every fish in order to count them. Instead, take a small handful of fish (white beans) out of the bag and count them.
(b) Suppose you tagged the fish (white beans) that you removed. Represent each tagged fish with a black bean. Put the tagged fish (black beans) into the bag instead of the white beans you removed.
(c) Mix up the beans in the bag. Now remove a handful of beans, and count how many white and black beans you obtained. Figure out a way to estimate the total number of beans in the bag.

LE 12   Summary

Tell what you learned about solving proportions in this section.

ANSWERS TO SELECTED LESSON EXERCISES

2. (a) Show a rectangle in four equal parts. A is one part, and B is three parts.

(b) $\dfrac{1}{4}$ (c) 3   (d) $\dfrac{1}{3}$

3. (a) 750   (b) 31   (c) Use part (b).

4. (a) 33 to 6; speed
(b) 11 to 2 and 66 to 12
(c) 11 to 2
(d) 5.5
(e) 42 to 8 = 5.25; less than Kristin's

5.

Time (hours)	1	2	3	4
Distance Traveled (miles)	5.5	11	16.5	22

6. (a) 10 : 15; 20 : 30   (b) 2 : 3; 2 : 3   (c) $\dfrac{10}{15} = \dfrac{10 \cdot 2}{15 \cdot 2} = \dfrac{20}{30}$

(d) $\dfrac{10}{15}$ and $\dfrac{20}{30}$; 10 · 30 and 20 · 15 both equal 300

7. (a) $\dfrac{ad}{bd} = \dfrac{bc}{bd}$ (b) $ad = bc$

9. (b) $4x = 7 \cdot 18$, so $x = 31.5$ ounces

(c) $\dfrac{18}{4} = 4.5$ oz/person. Then $4.5 \cdot 7 = 31.5$ oz.

(d) The scale factor is $\dfrac{7}{4} = 1.75$.

Then $18 \cdot 1.75 = 31.5$ oz.

10. *Hint*: First, find the ratio of women to all students.

7.3   HOMEWORK EXERCISES

Basic Exercises

1. A class has 12 boys and 17 girls.
(a) What is the ratio of girls to boys?
(b) What is the ratio of boys to all children in the class?

2. A fruit basket has 5 apples and 3 oranges.
(a) What is the ratio of oranges to fruit in the basket?
(b) What is the ratio of oranges to apples?

3. An all-female college has 600 students.
(a) What is the ratio of men to women?
(b) Why is the ratio of women to men undefined?

4. A mathematician tells you that the ratio of golf courses to baseball fields in a town is 0. What would you conclude?

5. Use a rectangle to represent a submarine sandwich.
(a) Divide the sandwich into two sections A and B so that the ratio of A to B is 2 : 3.
(b) What fraction of the whole sandwich is section A?
(c) Section B is _____ times as long as section A.
(d) Section A is _____ times as long as section B.

6. Use a rectangle to represent a submarine sandwich.
(a) Divide the sandwich into two sections A and B so that the ratio of A to B is 2 : 1.
(b) What fraction of the whole sandwich is section A?
(c) Section B is _____ times as long as section A.
(d) Section A is _____ times as long as section B.

7. One-fifth of a college class is men. What is the ratio of men to women in the class?

8. One-third of the vehicles in a business are cars, and the rest are trucks. What is the ratio of cars to trucks?

9. A mother and daughter are ages 40 and 20, respectively.
(a) Make an additive comparison of their ages.
(b) Make a multiplicative comparison of their ages.

10. In a survey, 24 out of 30 fifth graders said they liked music class. The rest said that they did not.
(a) Make an additive comparison of the number of students who liked music class and the number that did not.
(b) Make a multiplicative comparison of the number of students who liked music class and the number that did not.

11. The following table lists the numbers of students and teachers in five major public school systems, in the 2000–01 school year (National Center for Education Statistics).

School District	Number of Students	Number of Teachers	Student-Teacher Ratio
San Francisco	59,979	3,261	
Atlanta	58,230	3,950	
San Antonio	57,273	3,560	
Minneapolis	48,834	3,314	
Buffalo	45,721	3,471	

(a) Complete the last column of the chart, showing the ratio of students to teachers.
(b) Which city has the lowest student-teacher ratio?
(c) Which city has the highest student-teacher ratio?
(d) Would you rather go to a school with a higher or lower student-teacher ratio?

12. The *pitch* of a roof is the ratio of its rise (height) to its span (width).

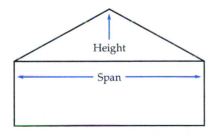

A roof has a rise of 8 ft and a span of 30 ft.
(a) What is its pitch?
(b) Write the pitch as a fraction in simplest form.
(c) What does the pitch tell you about the roof?

13. Is 7:4 the same as 6:3?

14. Is 4:3 the same as 8:6?

15. Two sets have a ratio of 10 to 3. Determine whether the ratio of the two sets will change if the number of members in each set is
(a) doubled.   (b) increased by 10.

16. Molly is 50, and her grandson Robert is 5. As they get older, the *ratio* of Molly's age to Robert's age will
(a) increase   (b) decrease
(c) stay the same   (d) do none of these

17. At a picnic, one table seating 10 people has 3 platters of food. A second table seats 14 people and has 4 platters of food. Which table is allotting more food per person?

18. The math department uses 3 tablespoons of coffee with 5 cups of water to make coffee. The science department uses 5 tablespoons of coffee with 9 cups of water. Which coffee is slightly stronger? Explain why.

19. In 2003, New York had a population of 19.2 million and an area of 47,214 square miles. California had a population of 35.7 million and a land area of 155,959 square miles.
(a) Find the population density for each state.
(b) Which state has a higher population density?
(c) Find information on the Web about how your own state compares to New York and California in population density. If you live in New York or California, find out how your state's population density ranks nationally.

20.

	Town A	Town B
Population	346,812	182,312
Number of restaurants	641	311

(a) If you had only this information, which town would appear to be a better place to open a restaurant?
(b) What other information should you obtain before deciding?

21. Perry is offered two jobs. One pays $500 for a 40-hour week. The second job pays $450 for a 36-hour week.
(a) Find the unit rate for each job.
(b) Which job has a better pay rate?
(c) Complete the rate table below.

Number of Hours	1	10	20	30	40
Job 1 Pays ($)					
Job 2 Pays ($)					

22. Maria's car travels 650 miles on 20 gallons, and Ralph's car travels 426 miles on 12 gallons.
(a) Find the unit rate for each car.
(b) Which car gets better gas mileage?
(c) Complete the following rate table.

Gas Used (gallons)	1	5	10	15
Maria's Distance Driven (miles)				
Ralph's Distance Driven (miles)				

23. A 32-oz carton of orange juice at Fresh Mart costs $1.29. Write two different unit rates, and tell what each one means.

24. A 75-gram serving of ice milk has 120 calories. Write two different unit rates, and tell what each one means.

25. (a) An 18-ounce jar of peanut butter costs $1.98. What is the cost per ounce (called the **unit price**)?
(b) A 32-ounce jar of peanut butter costs $3.88. What is the cost per ounce?
(c) Which jar is a better buy (which has the lower unit price)?

26. A 12-oz box of corn flakes costs $2.37, and an 18-oz box costs $2.99. Which is a better buy?

27. You can use colored counters to show equivalent ratios.
(a) A child shows the ratio three red to four yellow counters. If you add two more sets of counters just like this, what new ratio do you obtain?
(b) Write the proportion formed by these two equivalent ratios.

28. You can use colored counters to show equivalent ratios.
(a) A child shows the ratio five red to two yellow counters. If you add three more sets of counters just like this, what new ratio do you obtain?
(b) Write the proportion formed by these two equivalent ratios.

29. A store gives you 4 comic books for every 5 you trade in. How many will the store give you for 35? What are different ways a child might solve this problem?

30. Most sunscreen lotions list the sun protection factor (SPF). Suppose an SPF of 10 will protect you from burning for 50 minutes. Assuming that the length of time is proportional to the SPF, for how long will an SPF of 15 protect you? What are different ways a child might solve this problem?

31. A child solves Exercise 29 as follows. Since $5 - 4 = 1$, they will give you $35 - 1 = 34$ comic books. What would you tell the child?

32. A child solves Exercise 30 as follows. The proportion is $\frac{50}{15} = \frac{x}{10}$. I solve that for x and get $33\frac{1}{3}$. What would you tell the child?

33. Solve for R in each proportion just by looking at the relationship between each pair of fractions.
(a) $\frac{5}{7} = \frac{R}{14}$ (b) $\frac{3}{R} = \frac{6}{10}$ (c) $\frac{R}{45} = \frac{2}{6}$

34. Which of the following proportions can be solved easily without computing cross products? (*Hint*: Look for a simple relationship within or between the fractions.)
(a) $\frac{3}{7} = \frac{T}{14}$ (b) $\frac{4}{5} = \frac{3}{N}$
(c) $\frac{5}{10} = \frac{13}{J}$ (d) $\frac{9}{7} = \frac{10}{K}$

35. Estimate the answer, and then solve for Q in each proportion.
(a) $\frac{Q}{5} = \frac{7}{3}$ (b) $\frac{Q}{12} = \frac{4}{Q}$ (c) $\frac{7}{Q} = \frac{5}{9}$

36. Solve for N in each proportion.
(a) $\frac{3}{8} = \frac{10}{N}$ (b) $\frac{N}{6} = \frac{12}{20}$ (c) $\frac{N}{2} = \frac{7}{N}$

37. Suppose $\frac{5}{7} = \frac{N}{2}$. Write three other proportions that have the same two cross products.

38. Suppose $\frac{2}{x} = \frac{3}{y}$. Write three other proportions that have the same two cross products.

39. Proportions with one unknown can be solved without using cross products by people who find cross products confusing. If the variable is in the numerator of one side, multiply both sides by the denominator of that fraction. If the variable is in the denominator, invert the fractions on each side and then use multiplication. Solve the following proportions with these methods.

(a) $\dfrac{x}{2} = \dfrac{3}{5}$ (b) $\dfrac{4}{y} = \dfrac{15}{2}$ (c) $\dfrac{3}{4} = \dfrac{10}{n}$

40. A seventh grade class is asked to solve the proportion $\dfrac{4}{n} = \dfrac{5}{9}$.

(a) See if you can find three different ways to solve it.
(b) If you didn't already, solve it with cross products.
(c) If you didn't already, solve it by multiplying both sides of the equation by the common denominator.
(d) If you didn't already, invert the fractions on both sides and then solve it by multiplying both sides by 4.
(e) Solve the proportion $\dfrac{7}{3} = \dfrac{5}{x}$ three different ways.

41. A couple drinks 3 quarts of orange juice and 2 quarts of skim milk each week. At this rate, how much orange juice and skim milk do they drink in 4 days?

(a) Estimate the answer.
(b) Write a proportion. Tell how you set up the proportion. Then solve it by using cross products.
(c) Find a unit rate for each drink in quarts/day, and use them to find the answer. (Make a unit rate table if you wish.)
(d) How could you find the answer by finding a multiplier (scale factor)?

42. If 15 pounds of fertilizer take care of 2000 ft² of lawn, how much fertilizer is needed for a rectangular lawn that is 80 ft by 100 ft?

(a) Write a proportion. Tell how you set up the proportion. Then solve it by using cross products.
(b) How could you find the answer by finding a multiplier (scale factor)?
(c) Find a unit rate in ft²/lb and use it to find the answer. (Make a unit rate table if you wish.)

43. Maija paid \$9 for $\dfrac{3}{4}$ lb of crabs. What is the price per pound?

(a) How can you solve it with a ratio or a proportion?
(b) How can you solve it with a part-of-a-whole diagram?

44. If a British pound is worth \$1.63, how many pounds can I buy for \$10?

(a) Estimate the answer.
(b) Solve the problem by using a proportion. (Make sure that the units in the proportion match up.)
(c) Solve the problem without writing a proportion.

45. Last year at the Laundromat Users convention, 3,216 people ate 1,011 chickens at the Saturday afternoon picnic. This year, 3,800 people are expected to attend. How many chickens should be ordered?

46. A teacher wants to plan a party for her school. She surveys her own class to find out their preference among chocolate cookies, ice cream sandwiches, and peanuts. Out of her 28 students, 12 prefer the cookies, 10 prefer the ice cream, and 6 prefer the peanuts. On the basis of this, how many of each dessert should she order for the 312 students at the school?

47. The ice cream cone in the photo is 6 feet 9 inches tall. About how tall is the woman?

Photo courtesy of Thomas Sonnabend.

48. A town has a property tax of $21.62 per $1000 of assessed value. Tell how to estimate the property tax on a house valued at $86,400.

49. A 20-ft-long pipe of uniform width and density is cut into two pieces. One piece is 12 ft long and weighs 140 pounds. How much does the other piece weigh?

50. Frozen orange juice concentrate is usually mixed with water in a ratio of 3 parts water to 1 part concentrate. How much orange juice can be made from a 12-oz can of concentrate?

51. Consider the following problem. "Three years ago, Francisco and Melissa invested $3,000 and $5,500, respectively, in a business. Today the business is worth $10,000. What is Francisco's share of the business worth?" Devise a plan and solve the problem.

52. Suppose that a representative survey of adults in a large city shows that 663 support a sales tax increase and 837 are against it. If the adult population of the city is 4,200,000, predict how many adults support the sales tax increase.

53. In a large city, 2 million cars are used to commute to work. The average number of people per car is 1.2. If the average number of people per car were increased to 1.5, how many cars would be used? (*Hint:* Should the answer be more or less than 2 million?)

54. A school has 1200 students and a student-teacher ratio of 30 to 1. How many additional teachers must be hired to reduce the student-teacher ratio to 24 to 1?

Extension Exercises

55. A ranger catches and tags 10 fish in Lake Leisure. Then she tosses them back in the lake and lets all the fish mingle. Then the ranger catches 20 fish and finds that 6 of them are tagged. So she guesses that about 6/20 of the fish in the lake are tagged. Estimate the number of fish in Lake Leisure.

56. The ranger in the preceding problem goes to Lake Boggy. She catches and tags 20 fish. Then she releases them back into the lake. Now she catches 50 fish and finds that 4 of them are tagged. Estimate the number of fish in Lake Boggy.

57. A 24-oz jar of Lipton iced tea mix costs $3.29. It contains about $\frac{1}{2}$ oz of tea and $23\frac{1}{2}$ oz of sugar. How much would it cost you to mix your own tea and sugar in the same way, using Lipton tea (8 oz for $3.79) and sugar (5 lb for $2.39)?

58. In one year, a fast-food restaurant pays about $190 million for 45 million gallons of cola syrup. (A gallon is 128 fluid oz.) A soda is one part syrup and five parts carbonated water. The carbonated water costs about $0.005 per ounce. About how much do the contents of a 21-oz "medium" cola cost the restaurant?

59. A car averages about 24 miles per gallon, a bus averages about 6 miles per gallon, and an electric train averages about 2 miles per gallon. Under what conditions would each vehicle be the most fuel-efficient?

60. It takes 15 minutes to cut a log into 3 pieces. How long would it take to cut a similar log into 4 pieces? (The answer is not 20 minutes.)

61. A car gets M miles from G gallons. What is the distance it can travel on H gallons?

62. A college has a male-female ratio of M to F. What fraction of all the students are female?

63. A function has exactly one output for each input or set of inputs. Suppose salmon costs $5/lb. Then 1 pound of salmon costs $5, 2 pounds costs $10, and so on. Describe a function that goes with this rate.

64. Describe a function that goes with the rate 34 miles/gallon.

Labs and Projects

65. Draw a floor plan of your home that is to scale.

66. Go to your local supermarket and determine whether larger sizes of products always have lower unit prices. Write a summary of your findings.

7.4 PERCENTS

NCTM Standards

- recognize and generate equivalent forms of commonly used fractions, decimals, and percents (3–5)
- develop meaning for percents greater than 100 and less than 1 (6–8)
- recognize and apply mathematics in contexts outside of mathematics (pre-K–12)

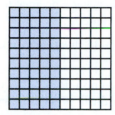

50% (50 per 100)

Figure 7–7

Do you think it is fair that a salesperson who sells twice as much should earn twice the commission, or that someone who owes three times as much money should pay three times as much interest? In situations such as these, many people consider it fair to charge the same rate on amounts of all different sizes. Percents were devised for just such situations!

First, it would be helpful to know what the word "percent" means. "Cent" in "per-cent" means *100*, just as it does in the words "century" and "centipede." "Percent" means *per 100*. For example, "50 percent" means 50 per 100, or $\frac{50}{100}$, or $\frac{1}{2}$. Figure 7–7 illustrates a percent as a part of a whole.

> **Definition: Percent**
>
> $n\% = \dfrac{n}{100}$ where % denotes **percent.**

A percent is a ratio that compares a number to 100.

PERCENTS, FRACTIONS, AND DECIMALS

The concepts of percents, fractions, and decimals are interrelated. The definition of a percent (%), $n\% = \dfrac{n}{100}$, is used to convert it to a fraction or decimal. Consider the following examples.

$$4\% = \frac{4}{100} = 0.04 \qquad 0.8\% = \frac{0.8}{100} = \frac{8}{1000} = 0.008$$

LE 1   Reasoning

(a) On the basis of the preceding examples, a shortcut for changing a percent to a decimal involves dropping the percent sign *and* moving the decimal point ＿＿＿ places to the ＿＿＿ to compensate.

(b) Use the fact that $n\% = n \div 100$ to explain your shortcut in part (a).

The process is reversed in changing decimals or elementary fractions to percents. The decimals 0.37 and 0.06 can be converted to percents as follows.

• EXAMPLE 1

Give the percent equivalents of 0.37 and 0.06.

SOLUTION

Write each number as a fraction with a denominator of 100. Then change it to a percent.

$$0.37 = \frac{37}{100} = 37\% \qquad \text{and} \qquad 0.06 = \frac{6}{100} = 6\%$$ ●

Use the results of the preceding example to find a shortcut for converting decimals to percents. Then convert a fraction to a percent.

LE 2   Skill

(a) The results of Example 1 suggest that a shortcut for changing a decimal to a percent involves moving the decimal point _____ places to the _____.

(b) Change $\frac{1}{8}$ to a decimal, and then use the method of part (a) to change it to a percent.

Why would someone use a percent rather than a fraction or a decimal? A percent gives a fixed rate per hundred. This is especially suitable for financial applications such as tax rates, sales commissions, and interest rates. Percents are also used to describe statistical data in sports, education, and science, because percents make it simple to compare rates in sets of different sizes.

Percents, fractions, decimals, and ratios are all used in comparisons.

LE 3   Connection

A survey of 80 students finds that 48 play soccer and 32 do not. On the basis of this survey, complete the following.

(a) _____ out of _____ students play soccer.
(b) _____ % of the students play soccer.
(c) The fraction of students who do not play soccer is _____.
(d) The ratio of students who play soccer to those who do not is _____ to _____.

BASIC PERCENT PROBLEMS

How much is 28% tax on $24,000? Which test score is better, 14 out of 20 right or 18 out of 25 right? A shirt is on sale for $13, which is 80% of its regular price. What is the regular price? These are examples of the three types of basic problems. How would you solve them?

 LE 4   Opener

(a) How much is 28% tax on $24,000?
(b) Use percents to determine which test score is better, 14 out of 20 right or 18 out of 25 right.
(c) A shirt is on sale for $13, which is 80% of its regular price. What is the regular price?

Most children study two approaches for solving basic percent problems: the percent proportion and the percent equation. The percent proportion states,

$$\frac{\text{part}}{\text{whole}} = \frac{n}{100} \quad \text{percent as fraction}$$

• EXAMPLE 2

Solve the problems in LE 4 with the percent proportion.

SOLUTION

(a) $24,000 is the whole, and the tax is a part of it.

$$\frac{\text{part}}{24{,}000} = \frac{28}{100}$$

Then find the cross products. $100(\text{part}) = 24{,}000(28)$ so part $= 6{,}720$. The tax is $6,720.

(b) Each test score gives the part and the whole.

$$\frac{14}{20} = \frac{n}{100} \text{ and } \frac{18}{25} = \frac{x}{100}$$

$$1400 = 20n \text{ and } 1800 = 25x$$

$$n = 70 \text{ and } x = 72$$

The scores are 70% and 72%, so the second score is higher.

(c) The sale price is the part, and the regular price is the whole.

$$\frac{13}{\text{whole}} = \frac{80}{100}$$

This can be simplified to $\dfrac{13}{\text{whole}} = \dfrac{4}{5}$. Then find the cross products. $65 = 4(\text{whole})$, so whole $= 16.25$. The regular price is $16.25. ●

LE 5   Skill

Solve the following problems with the percent proportion.

(a) 12 out of 60 is what percent?
(b) What is 20% of 60?
(c) 12 is 20% of what number?

You can also solve LE 5(b) and (c) with a percent model.

LE 6   Connection

(a) Complete the following percent model, and use it to find 20% of 60.

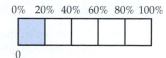

(b) Make a percent model for LE 5(c), and solve the problem.

You can derive the percent equation from the percent proportion.

LE 7   Reasoning

Replace $\dfrac{n}{100}$ by P (for percent) in the percent proportion, and solve the equation for "part."

The result of LE 7 is the percent equation, which states:

$$\text{part} = P \cdot \text{whole} \qquad \text{where } P \text{ is the percent}$$

● EXAMPLE 3

Solve the problems in LE 4 with the percent equation.

SOLUTION

(a) part $= 28\% \cdot 24{,}000$. Then part $= 6{,}720$. The tax is $6,720.

Most people do this by multiplying the percent times the whole without writing an equation.

(b) $14 = P \cdot 20$ or $\dfrac{14}{20} = P$. Then P $= 0.7 = 70\%$ right.

$18 = P \cdot 25$ or $\dfrac{18}{25} = P$. Then P $= 0.72 = 72\%$ right.

The second score is higher. Most people do this by dividing the part by the whole without writing an equation.

(c) $13 = 80\% \cdot$ whole. Then $\dfrac{13}{80\%} = $ whole.

Then whole $= 16.25$. The regular price is $16.25.   ●

LE 8   Skill

Solve the following problems with the percent equation (or multiplication or division).

(a) 24 out of 80 is what percent?
(b) 24 is 30% of what number?
(c) What is 30% of 80?

Children typically solve type 1 problems (such as LE 8(c)) with multiplication (the percent equation). Next, they learn to solve type 2 and 3 problems (such as LE 8(a) and (b)) with the percent proportion. In a later grade, they may learn to solve all three types of problems with the percent proportion and/or the percent equation.

SIMPLE AND COMPOUND INTEREST

Do you have money in the bank? What kind of interest do you earn? **Simple interest** pays interest only on the original money (the principal) you deposited. With **compound interest**, you also earn interest on your interest. Most banks pay compound interest on savings, usually more than once a year—either semiannually (twice a year), quarterly (four times a year), monthly, or daily (360 or 365 times a year).

LE 9   Reasoning

Suppose you invest $2000 in the bank, and it earns an annual rate of 3% simple interest.

(a) How much interest would you earn in one year?
(b) How much interest would you earn in three years?
(c) How much interest would you earn in half a year?
(d) The original amount ($2000) is called the **principal,** p. The interest rate (3%) is r. The time in years is t. Write a formula for I, the amount of interest earned in terms of p, r, and t.

$$I = \underline{\hspace{2cm}}$$

Did you obtain the following formula?

Formula for Simple Interest

$I = prt$, where I is the interest, p is the principal, r is the annual rate of interest, and t is the time in years.

Use this formula in the next exercise.

LE 10   Skill

I had $1000 in the bank earning 4% simple interest for 2 years. The 2 years are over. How much money do I have in that account now?

In LE 10, would it be better to receive interest that is compounded annually?

LE 11   Skill

You deposit $1000 earning 4% interest compounded annually.

(a) Find the interest and account balance after 1 year.
(b) In the second year, you receive 4% interest on the account balance from the first year. How much money will you have in the account after 2 years?
(c) How much more money do you have after 2 years than in LE 10?

You can study compound interest further in the Extension Exercises.

LE 12   Summary

Tell what you learned about percents in this section.

ANSWERS TO SELECTED LESSON EXERCISES

1. (a) 2; left
   (b) To divide 100, move the decimal point 2 places to the left.

2. (a) 2; right   (b) $\frac{1}{8} = 0.125 = 12.5\%$

3. (a) 48; 80   (b) 60   (c) $\frac{2}{5}$ (d) 3; 2

4. (a) $6,000   (b) 14 out of 20   (c) $15

5. (a) 20%   (b) 12   (c) 60

6. (b) 0% 20% 40% 60% 80% 100%

$12 = 20\%$ of 60

0   12   24   36   48   60

8. (a) $24 \div 80 = 0.3 = 30\%$
(b) $24 = 0.3x$ and $x = 80$
(c) $0.30 \cdot 80 = 24$

9. (a) \$60   (b) \$180   (c) \$30

10. $\$1000 + \$80 = \$1080$

11. (a) $\$1000 + \$40$ interest $= \$1040$
(b) \$1081.60   (c) \$1.60

7.4   HOMEWORK EXERCISES

Basic Exercises

1. What does the word "percent" mean?

2. (a) 14% means _____ per 100.
(b) Shade a decimal square (Activity Card 3) to represent 14%.
(c) 14% means _____ per 50.
(d) 14% means _____ per 10.

3. (a) 150% means _____ per 100.
(b) Shade decimal squares (Activity Card 3) to represent 150%.
(c) 150% means _____ per 20.
(d) 150% means _____ per 2.

4. (a) 125% means _____ per 100.
(b) Shade decimal squares to represent 125%.
(c) 125% means _____ per 20.
(d) 125% means _____ per 4.

5. Write each percent as a fraction and as a decimal.
(a) 34%   (b) 180%   (c) 0.06%

6. Write each percent as a fraction and as a decimal.
(a) 17%   (b) 0.01%   (c) 200%

7. Children have more difficulty with percents greater than 100%. A percent that is greater than 100% represents a number greater than _____.

8. Children have more difficulty with percents less than 1%. A percent that is less than 1% represents a number less than _____.

9. Write each decimal as a percent.
(a) 0.23   (b) 0.00041   (c) 24

10. Write each decimal as a percent.
(a) 0.79   (b) 5.24   (c) 0.00083

11. Write each fraction as a percent.
(a) $\dfrac{1}{25}$ (b) $\dfrac{3}{8}$ (c) $1\dfrac{3}{4}$

12. Write each fraction as a percent.
(a) $\dfrac{3}{4}$ (b) $3\dfrac{1}{2}$ (c) $\dfrac{2}{5}$

13. A survey of 200 adults finds that 68 read a book this month and 132 did not. Complete the following.
(a) _____ out of _____ adults didn't read a book.
(b) _____ % of the adults didn't read a book.
(c) The fraction of adults who did read a book is _____.
(d) The ratio of adults who read to those who did not is _____ to _____.
(e) Adults who did not read books outnumber those who did read a book by _____.

14. A survey of 40 cities finds that 14 met air pollution standards and 26 did not. Complete the following.
(a) _____ out of _____ cities met the standard.
(b) _____% of the cities met the standard.
(c) The fraction of cities that did not meet the standard is _____.
(d) The ratio of cities that met the standard to those that did not is _____ to _____.
(e) Cities that did not meet the standard outnumber cities that did by _____.

15. A survey of 500 adults finds that 495 watched television in the past week and 5 did not. Write four different statements that summarize the results with a difference, fraction, decimal, percent, or ratio.

16. A drink mix has 3 parts orange juice for every 2 parts of carbonated water.
(a) What fraction of the mix is carbonated water?
(b) What percent of the mix is orange juice?

17. Solve the following problems with the percent proportion.
  (a) What is 30% of 500?
  (b) 25 is 40% of what number?
  (c) 28 out of 40 is what percent?

18. Solve the following problems with the percent proportion.
  (a) 200 is 80% of what number?
  (b) What is 40% of 80?
  (c) 45 out of 300 is what percent?

19. Show how to work out Exercises 17(a) and (b) with a percent model.

20. Show how to work out Exercises 18(a) and (b) with a percent model.

21. Solve the following problems with the percent equation (or multiplication or division).
  (a) 120 is 60% of what number?
  (b) 180 out of 200 is what percent?
  (c) What is 70% of 20?

22. Solve the following problems with the percent equation (or multiplication or division).
  (a) 450 is 90% of what number?
  (b) What is 40% of 85?
  (c) 7 out of 350 is what percent?

23. A salesperson earns a 12% commission. If she sells a $2400 computer, what is her commission?
  (a) Estimate the answer.
  (b) Solve the problem using multiplication.
  (c) Solve the problem with a proportion, and draw a number-line model.
  (d) If she sells 3 times as many computers, her commission will be _____ times as much.

24. The sales tax on a $9.59 item is $7\frac{1}{2}$%.
  (a) Estimate the tax.
  (b) Exactly how much is the tax?

25. Each year in the United States, we produce about 110 million tons of air pollution from carbon monoxide. Motor vehicles produce about 82 million tons of this pollutant. What percent of the carbon monoxide do motor vehicles produce?
  (a) Solve the problem using division.
  (b) Solve the problem with a proportion, and draw a number-line model.

26. Ken made 54 out of 80 free throws. Don made 43 out of 61 free throws. Who is the better free-throw shooter? Give evidence to support your answer.

27. Consider the following problem. "A salesperson earns a commission of 8% of total sales. This week she earned $798.40. What were her total sales for the week?"
  (a) Write an equation (that is not a proportion) and solve it.
  (b) Write a proportion and solve it.

28. There are 184 comedy films at Fantasy Island Video. This represents about 18% of their collection. How many films do they have? Give all possible answers. (There are 57 of them!)

29. The drawing represents 40% of a figure. Draw 100% of the figure.

30. The drawing represents 75% of a figure. Draw 100% of the figure.

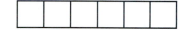

31. As of 2003, the United States and Canada have different kinds of health-care systems. In Canada, all citizens are insured; in the United States, about 41 million people (14%) are uninsured. Canada uses a single insurer in each of its ten provinces; the United States has about 1,500 different health insurers. The Organization for Economic Cooperation and Development collects data on health care.

	Population (2000)	Total Cost (2000)	Total Administrative Costs (2000)
U.S.	281 million	$1301 billion	$286 billion
Canada	31 million	$79 billion	$8.6 billion

  (a) Compute the per-capita (per-person) costs in the United States and Canada.
  (b) What percent of the total cost represents administrative costs in each country?
  (c) If the United States reduced its administrative costs from 22% to 11%, how much would be saved?

(Continued on next page)

	Infant Mortality Deaths per 1000 Births (2002)	Persons per Physician (2000)	Life Expectancy (2000)
U.S.	6.7	370	76.8 years
Canada	5.0	475	79.1 years

(d) What do the data suggest about the comparative performance of the two health-care systems?

32. Grapefruit Cocktail costs $3.15 for 64 oz. It contains 30% grapefruit juice and about 1.3 oz of sugar, plus water. How much would it cost you to make 64 oz of grapefruit cocktail yourself, using canned grapefruit juice ($2.19 for 46 oz), sugar ($2.39 for 5 lb), and water?

33. Consider the following problem. "A business manager must pay $35,000 per year in fixed expenses, and she pays 40% of her total sales to her workers. How much money must she make to earn a $40,000 profit after paying her expenses and her workers?" Devise a plan and solve the problem.

34. You are offered two real estate sales jobs. One pays 9% commission. The other job pays $50 per week and 6% commission. If sales tend to average about $2000 per week, which job would pay more?

35. You buy a $5000 savings certificate that pays 6% simple annual interest. How much interest will you earn in 6 months?

36. Your new credit card has an annual finance charge of 15% simple annual interest.
(a) Suppose you borrow $500 for a year. How much interest will you owe?
(b) Suppose you cannot afford to pay back the $500 for an additional 9 months. How much additional interest will you owe?

37. A credit card company charges 12% simple annual interest on debts.
(a) What percent interest does the company charge per month?
(b) What percent interest does the company charge per day?

38. Suppose you receive a credit card bill for $340. The minimum required payment is $40. You decide to pay the minimum and wait until next month to pay the balance. The annual simple interest rate on the balance is 18%. How much will your monthly finance charge be?

39. You deposit $8000 earning 6% interest compounded annually.
(a) Find the interest and account balance after one year.
(b) In the second year, you receive 6% interest on the account balance from the first year. How much money will you have in the account after two years?

40. You deposit $2500 earning 5% interest compounded annually.
(a) Find the interest and account balance after one year.
(b) In the second year, you receive 5% interest on the account balance from the first year. How much money will you have in the account after two years?

41. You deposit $6000 earning 4% interest compounded semiannually.
(a) What percent interest does the bank pay every 6 months?
(b) Find the interest and account balance after 6 months.
(c) In the second 6 months, you receive interest on the account balance from the first 6 months. How much money will you have in the account after a year?

42. You deposit $9000 earning 4% interest compounded semiannually.
(a) Find the interest and account balance after 6 months.
(b) In the second 6 months, you receive another 2% interest on the account balance from the first 6 months. How much money will you have in the account after a year?

Extension Exercise

43. Taxes are commonly classified as progressive, intermediate, or regressive. A **progressive tax** requires people with higher income to pay a higher percent of their income as tax. An **intermediate tax** requires people at every income level to pay the same percent income tax, and a **regressive tax** requires people with lower incomes to pay a higher percent of their income as tax. Classify each of the following tax structures as progressive, intermediate, or regressive.

(a)

Total Yearly Income	Percent Tax
$30,000	8%
$25,000	10%
$20,000	12%

(b)

Total Yearly Income	Tax
$30,000	$2000
$25,000	$2000
$20,000	$2000

(c) Maryland has a sales tax rate of 5%. The average person with $30,000 in yearly income buys about $4000 worth of sales-taxed items, and the average person with $10,000 in yearly income buys about $2000 worth of sales-taxed items. Classify this sales tax as progressive, intermediate, or regressive.

44. Fill in tax amounts so that the tax is *regressive* and those with higher incomes pay more money in tax.

Total Yearly Income	Tax Amount
$30,000	_____
$25,000	_____
$20,000	_____

 45. You deposit $1000 for two years. A bank pays 4% interest compounded annually.
(a) After one year, you have _____% of the original amount.
(b) By what single number could you multiply the principal to find the amount in the account after one year?
(c) To find the amount after two years, you could multiply the principal by 1.04 to what power?

(Continued in the next column)

(d) Suppose you left the money and interest in the bank for 8 years. What is a quick way to compute the account balance after 8 years with a scientific or graphing calculator?

 46. You deposit $4000 in an account. A bank pays 6% interest compounded annually. Use the method of the preceding exercise to find the account balance after
(a) 2 years.   (b) 5 years.

 47. You deposit $6000 in an account. A bank pays 6% interest compounded monthly.
(a) What percent interest would you receive each month?
(b) By what single number could you multiply the principal to find the amount in the account after one month?
(c) To find the amount after one year, you could multiply the principal by 1.005 to what power?
(d) Find the balance after one year.

 48. You deposit $8000 in an account. A bank pays 4% interest compounded quarterly.
(a) What percent interest would you receive each quarter?
(b) By what single number could you multiply the principal to find the amount in the account after one quarter?
(c) To find the amount after one year, you could multiply the principal by 1.01 to what power?
(d) Find the balance after one year.
(e) Find the balance after two years.

 49. The world population in 2003 was about 6.4 billion. It is expected to grow about 1.3% annually in the near future.
(a) Forecast the world population in 2004.
(b) Forecast the world population in 2015. (*Hint:* Use an exponent.)
(c) Forecast the world population in 2050.

50. Three percent annual inflation is the same mathematically as 3% interest compounded annually. If a car cost $20,000 in 2000, what is an equivalent price in 2010, assuming an annual inflation rate of 3%?

Enrichment Topic

The U.S. Department of Agriculture (USDA) has certain standards for names used to describe various meat products. Some of them follow.

USDA Standards for Meat
Hamburger: no more than 30% fat
Hot dog or **bologna:** no more than 30% fat and 10% added water
Smoke pork sausage: no more than 50% trimmable fat
Egg rolls with poultry: at least 12% poultry meat
Breaded poultry: no more than 30% breading
Poultry chop suey: at least 4% poultry meat
Poultry pies: at least 14% poultry meat
Poultry with gravy: at least 35% poultry meat
Poultry salad: at least 25% poultry meat
Poultry soup (condensed): at least 4% poultry meat
Poultry soup (ready-to-eat): at least 2% poultry meat

51. According to the USDA standards, which is required to have more meat: chicken salad, chicken pot pie, or chicken chop suey?

52. On the basis of USDA standards, give the government restrictions on each of the following products.
   (a) Chun King chicken egg rolls
   (b) Weaver chicken rondelets (breaded chicken patties)
   (c) Ball Park franks
   (d) Swift Premium turkey roast (boneless meat with gravy)
   (e) Campbell's chicken noodle soup (condensed)

7.5 PERCENTS: MENTAL COMPUTATION, ESTIMATION, AND CHANGE

NCTM Standards

• recognize and apply mathematics in contexts outside of mathematics (pre-K–12)

Could you quickly compute the value of a 50% discount? How about 10% tax on a $200 item? These same techniques could be used to estimate a 52% discount or a 9% tax on $211. First, consider mental computation.

COMPUTING 10%, 25%, OR 50% OF A NUMBER

Some can be done mentally. You should be able to compute 10%, 25% and 50% of many whole numbers mentally, using shortcuts. See if you can describe these shortcuts in LE 1.

LE 1   Skill
   (a) Write 50%, 25%, and 10% as fractions in simplest form.
   (b) Computing 50% of a number is the same as dividing the number by _____.
   (c) Computing 25% of a number is the same as dividing the number by _____ .
   (d) Computing 10% of a number is the same as dividing the number by _____.
   (e) Finding 10% of a number is the same as dividing by _____, which means you could move the decimal point in the number _____ place(s) to the _____.

As LE 1 suggests, you can use the following procedures.

Computing 50%, 25%, and 10% of a Number

- 50% of a number is half the number, so divide the number by 2.
- 25% of a number is $\frac{1}{4}$ of the number, so divide the number by 4.
- 10% of a number is $\frac{1}{10}$ of the number, so divide the number by 10, moving the decimal point one place to the left.

Apply these shortcuts in LE 2 and LE 3.

LE 2   Skill

Mentally compute
(a) 50% of 286.   **(b)** 25% of 4,000.

LE 3   Skill

A $25 shirt is selling at a 10% discount. Mentally compute the discount and the sale price.

COMPUTING OTHER PERCENTS

Now for some more amazing feats of mental computation! Certain other percents of a number, such as 40% and 75%, can sometimes be computed mentally, using our understanding of 10% and 25%, respectively.

• EXAMPLE 1

How could one mentally compute
(a) 40% of 620?   **(b)** 75% of 44?

SOLUTION

(a) To find 40% of 620, compute 4 times 10% of 620.

$$4 \times 62 = 248$$

(b) 75% of 44 = 3 × (25% of 44) = 3 × 11 = 33   ●

LE 4   Skill

How could you mentally compute the exact value of
(a) 75% of 48?   **(b)** 70% of 300?

ESTIMATING PERCENTS

It is helpful to know some percents and their fractional equivalents.

LE 5   Skill

Complete the second row of the table.

Fractions	$\frac{1}{20}$	$\frac{1}{10}$	$\frac{1}{5}$	$\frac{1}{4}$	$\frac{1}{3}$	$\frac{1}{2}$	$\frac{2}{3}$	$\frac{3}{4}$	$\frac{1}{1}$
Percents									

Simple fractions can be used to estimate percent problems.

LE 6   Skill

Two seventh graders are estimating 32% of 527. Maurice uses rounding and breaking apart. He says 32% of 527 is about 30% of 500. Then 10% of 500 is $\frac{1}{10}$ of 500 or 50. Then 30% is $3 \times 50 = 150$. Candace uses compatible numbers. She says 32% is about $\frac{1}{3}$. Then change 32% of 527 to $\frac{1}{3}$ of 540, which is 180. Use each child's method to estimate a 22% commission for selling a $648 washing machine.

ESTIMATING TIPS

You've just finished a feast. Chef Skippy fixed his special baked zucchini with peanut sauce, buttered apples, and licorice balls. The bill came to $12.25 plus $0.74 tax. What's a reasonable tip?

One common application of percent estimation is estimating the tip in a restaurant. People typically leave about 15% of the bill.

Discussion

LE 7   Skill

Describe every method you know for estimating a 15% tip on a bill of $12.25 plus $0.74 tax.

To estimate 15%, did you use breaking apart? Estimate 10%, and then add half of this estimate (5%) to obtain 15%. If the bill is $12.25, 10% is about $1.22, and half of $1.22 is about $0.60. A 15% tip is about $1.22 + $0.60 = $1.82. Leave $2.

A second method is to use the tax. If you realized the tax in this example was 6%, you could multiply the tax by 3 to obtain a reasonable tip (18%). This would be $0.74 × 3 = $2.22. Leave $2 or $2.25.

Sometimes people overestimate percents, as the following cartoon shows.

LE 8   Concept

What is wrong with saying "110%" in the cartoon?

PERCENT OF CHANGE

Suppose an experienced teacher who earns $43,870 and a fairly new teacher who earns $28,800 both received their annual raises. How can the two raises be compared? One way is to compare the percent increase (the change in the number of dollars *per each $100* earned) that each teacher received.

How is the percent change computed? The two-step method is as follows. First, compute the amount of the increase or decrease. Second, compute what percent of the *original* amount this amount is.

 • **EXAMPLE 2**

This year, Nancy Shaw's salary increased from $28,800 to $32,256. What percent increase is this?

SOLUTION

Equation Method

First, compute the amount of the increase: $32,256 − $28,800 = $3456. Then the amount of change = P · original amount. In this case, $3456 = P \cdot 28,800$. So $P = \dfrac{3456}{28,800}$ or 0.12. Nancy received a 12% increase.

Proportion Method

First, compute the amount of the increase: $32,256 − $28,800 = $3456. Then $\dfrac{\text{amount of change}}{\text{original amount}} = \dfrac{n}{100}$. In this case, $\dfrac{3456}{28,800} = \dfrac{n}{100}$. The equal cross products are $345,600 = 28,800n$ and $n = 12$. Nancy received a 12% increase.   •

The following exercise is similar to Example 2.

Discussion

LE 9 Skill

This year, Margaret Latimer's salary increased from $43,870 to $48,300. Find what percent increase this is, to the nearest tenth of a percent, by

(a) using the equation method.
(b) using the proportion method.
(c) How does this increase compare to Nancy Shaw's increase in Example 2?
(d) Describe another way of comparing their increases that would result in Margaret's increase being larger than Nancy's.
(e) Would you say Margaret and Nancy's raises are comparable, or did one come out better than the other?

Classroom Connection

LE 10 Reasoning

From 1990 to 2000, the population of Harmony increased from 10,000 to 12,000, and the population of Pine Valley increased from 20,000 to 23,000. Rico says that Pine Valley had a greater increase in population than Harmony. Ananya says that Harmony had a greater increase than Pine Valley. If both children did correct mathematics, tell how they reached their conclusions.

Some stores compute discount prices by decreasing the regular price by some percent. Here are two methods for finding a discounted price.

LE 11 Skill

A coat at the Outerwear House costs $140, but today it's 30% off! What is the sale price?

(a) Find the amount of the discount, and subtract it from the regular price.
(b) What percent of the regular price is the discounted price? Take this percent of the regular price, and give the discounted price.

LE 12 Summary

Tell what you learned about mentally computing percents in this section.

ANSWERS TO SELECTED LESSON EXERCISES

1. (a) $\frac{1}{2}, \frac{1}{4}, \frac{1}{10}$ (b) 2   (c) 4
(d) 10   (e) 10; 1; left

2. (a) 143   (b) 1000

3. $2.50 discount; $22.50 sale price

4. (a) $3 \times (25\% \text{ of } 48) = 3 \times 12 = 36$
(b) $7 \times (10\% \text{ of } 300) = 7 \times 30 = 210$

5.

Fractions	$\frac{1}{20}$	$\frac{1}{10}$	$\frac{1}{5}$	$\frac{1}{4}$	$\frac{1}{3}$
Percents	5%	10%	20%	25%	$33\frac{1}{3}\%$

Fractions	$\frac{1}{2}$	$\frac{2}{3}$	$\frac{3}{4}$	$\frac{1}{1}$
Percents	50%	$66\frac{2}{3}\%$	75%	100%

6. Maurice: 22% of $648 ≈ 20% × $650 =
$$2 × (10\% \text{ of } 650) = 2 × 65 = \$130$$

Candace: 22% of $648 ≈ $\frac{1}{5}$ of $650 = $130

8. A person cannot commit more than 100% to anything.

9. (a) $48,300 − $43,870 = $4430

$$\frac{\$4430}{\$43,870} ≈ 0.101 = 10.1\%$$

(b) $48,300 − $43,870 = $4430

$$\frac{\$4430}{\$43,870} = \frac{n}{100}$$

$n = 10.1$; a 10.1% increase

(c) It is lower.
(d) Compare the dollar amount of each increase.

10. Rico: Pine Valley increased by 3,000, and Harmony increased by 2,000.
Ananya: Harmony increased 20%, and Pine Valley increased 15%.

11. (a) $98   (b) 70%; $98

7.5 HOMEWORK EXERCISES

Basic Exercises

1. How can you mentally compute each of the following?
(a) 50% of 222
(b) 25% of 6
(c) 10% of 470

2. A $400 television is selling at a 25% discount. Mentally compute its sale price.

3. How could you mentally compute the exact value of each of the following?
(a) 75% of 12
(b) 80% of 400

4. How could you mentally compute the exact value of each of the following?
(a) 75% of 240
(b) 70% of 210

5. In 2002, the voting-age population in the United States was about 202 million, of which about 40% voted. Explain how to estimate how many people voted.

6. Many nutritionists recommend getting about 60% of our calories from carbohydrates. How could you mentally compute the exact number of calories that should come from carbohydrates in a 2000-calorie-per-day diet?

7. Compute mentally, and fill in each blank.
(a) 50% of _____ is 22.
(b) 25% of _____ is 80.
(c) 20% of _____ is 110.

8. (a) 50% of _____ is 4.
(b) 20% of _____ is 10.
(c) 25% of _____ is 20.

9. The rectangular region below represents 25% of a pizza. Draw

(a) 50% of the pizza.
(b) 100% of the pizza.
(c) 75% of the pizza.

10. The rectangular region below represents 150% of a pizza. Draw

```
┌─────────────────────────┐
│                         │
│                         │
└─────────────────────────┘
```

(a) 50% of the pizza.
(b) 100% of the pizza.
(c) 75% of the pizza.

11. Finding 1% of a number is the same as dividing by _____, which means you could move the decimal point in the number _____ places to the _____.

12. How could you mentally compute each of the following?
(a) 1% of 800   (b) 1% of 37

13. You can use 1% of a number to find other percents mentally. How would you mentally compute the following?
(a) 8% of 400   (b) 7% tax on $900

14. You can use 1% of a number to find other percents mentally. How would you mentally compute the following?
(a) 2% of 340   (b) 4% tax on $70

15. A fruit juice container shows the following list of ingredients: grape juice, apple juice, and passion fruit juice. Ingredients are listed in order from most abundant to least abundant. At least what percent of the drink is grape juice?

16. Estimate what percent of the square is shaded.

17. The Cereal Bowl seats 95,000. The stadium is 64% full for a clash between the Ballerinas and the Field-mice. Explain how to estimate the attendance mentally
(a) with rounding.
(b) with compatible numbers.

18. If you take an extra summer job that pays $4200, you will have 35% taken out of your paycheck in taxes. How can you estimate your take-home pay? (*Hint:* First estimate the amount taken out of your paycheck.)

19. Use estimation to determine whether each of the following calculator answers is reasonable. Explain why or why not.
(a) 15% of 724 = ☐ 10860
(b) 22% of 58,000 = ☐ 12760
(c) 86% of 94 = ☐ 109.3023256

20. Use estimation to order the following expressions from smallest to largest.

80% of 28	52% of 807
37% of 420	19% of 400

21. Zine answered 34 questions correctly out of 48. Show how to estimate what percent of the questions he answered correctly.

22. Four out of 17 children in a class like olive loaf. How could you estimate what percent of the children like olive loaf?

In Exercises 23 and 24, estimate a 15% tip on the bill in two different ways.

23.   Bill $21.07
     7% tax $ 1.48
      Total $22.55

24.   Bill $42.87
     5% tax $ 2.14
      Total $44.92

 25. My car was worth $16,600 when it was new. Now it is worth only $7000. By what percent has its value depreciated (decreased)?
(a) Use the equation method.
(b) Use the proportion method.

26. In 2003, our annual car insurance for one car rose from $463 to $574. What percent increase is that?
(a) Use the equation method. (Round to the nearest tenth of a percent.)
(b) Use the proportion method.
(c) Given that we had no accidents or traffic violations, was this a reasonable increase?

27. The boss gives you a raise in salary from $30,000 to $33,000.
   (a) Make an additive comparison of the old and new salaries.
   (b) Make a multiplicative comparison of the old and new salaries.

28. Sam invested $2,000 last year and ended up with $3,000. Mike invested $4,000 and ended up with $5,500.
   (a) Tell two ways in which we can compare the results of their investments.
   (b) What is the best way to compare their investments? Tell why.

 29. (a) A baseball glove sells for $15.98. How much will it cost at 20% off?
   (b) If the sales tax is 4%, what is the total price of the glove?

 30. Driving 40 mph instead of 60 mph on the highway increases gas mileage by 20%. A car that gets 28 mpg at 60 mph will get _____ mpg at 40 mph.

31. (a) The price of a hotel room increases by 6%. The new price is _____ times the old price.
   (b) The population of a town increases by 12%. The new population is _____ times the old population.

32. (a) The number of crimes in a town decreases by 4%. The new number of crimes is _____ times the old number.
   (b) The number of students in a school increases by n%. The new number of students is _____ times the old number.

33. A seventh grader says if 30 is 70% of an unknown number, you find it as follows. You need 30% more, so find 30% of 30, which is 9. Then $30 + 9 = 39$ is the unknown number.
   (a) Check this answer.
   (b) What doesn't this child understand about the problem?

 34. A copying machine can reduce an image to 50% or 75% of its original size. By using these buttons two or three times, what other size reductions can be made?

35. A store owner makes a 25% profit (on the wholesale price) by selling a dress for $80. How much is the profit in dollars?

 36. An airline ticket costs $218, including 8% tax. What was the base fare?

37. Write two multistep mathematical questions that can be answered using the following data.

	Regular Price	Sale Price
Gazelle Joggers	$78.50	$75.00
Hippo Running Shoes	$29.95	$19.95
Fatiguers Running Shoes	$35.00	$31.99
Bolt Running Shoes	$50.00	$40.00

38. Consider the following problem. "The number of boys in a class is 25% of the number of girls. What percent of the whole class is boys?"
   (a) Devise a plan and solve the problem.
   (b) Make up a similar problem and solve it.

 39. A store marks down an item 30%, to a price of $225.60.
   (a) Complete the chart.

Original price	$0	$100	$200		
30% off the price	$0				

   (b) Plot your points on graph paper, using axes like those shown.

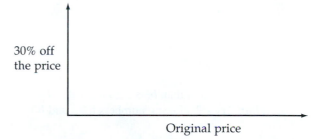

   (c) Use your graph to estimate the original price if the sale price is $225.60.

40. (a) Complete the chart.

Number	0	10	20	30	40
38% of the number	0	3.8			

(b) Plot your points on graph paper, using axes like those shown.

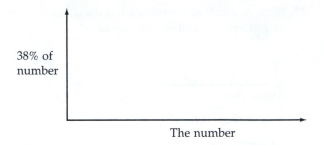

(c) Use your graph to estimate the following:
   10 is 38% of _____.

Extension Exercises

41. (a) The sign in the cartoon is making a prediction. What sort of prediction is it?

(b) If a is 15% less than b, is b 15% greater than a? (*Hint:* Try substituting numbers for a and b.)

42. If a is 25% more than b, then b is _____% of a.

43. (a) Place the digits 4, 5, 6, 8, and 9 in the blanks to obtain an answer as close as possible to 300.

   ___ ___% of ___ ___ ___ ≈ 300

(b) Use a calculator to see how close you came.
(c) Make a second guess and try to get closer to 300.
(d) Check your second guess on a calculator.

44. Use estimation to fill in each blank with a number that will make the statement true. Check your guess with a calculator, and revise it as needed.
(a) 42% of _____ is between 600 and 700.
(b) 64% of _____ is between 1000 and 1100.
(c) 7% of _____ is between 11 and 14.

45. A car is marked up 30% from its wholesale price. During a sale, the retail price is reduced by 20%. What percent higher is the sale price than the wholesale price?

46. A $80 dress is marked down 10% one week. The next week, an additional 25% is taken off the sale price. Find a single percent discount equal to these two successive discounts.

47. After two annual raises of 20%, Margaret receives an annual salary of $69,840. What was her salary 2 years ago?

48. The manager of a music store orders a compact stereo for $200. She wants to price it so she can offer a 10% discount off the posted price and make a profit of 40% of the price she paid. What will the posted price of the stereo be?

49. A car accelerates from a speed of N mph to M mph ($M > N$). What percent increase is this?

50. A student answers A out of B questions correctly on a test. If all the questions count the same, write the student's test score as a percent.

Technology Exercises

51. A woman is using a spreadsheet to keep a record of her investments. (Use a computer spreadsheet if you have one.)

	A	B	C	D	E
1	Fund	1-1-03	1-1-04	$ Change	% Change
2	Wings	$8342.61	$8680.10		
3	Rearguard	$12471.82	$11871.17		

(a) What does the number in cell C3 represent?
(b) What are the formulas for D3 and E3?
(c) If you have a computer, enter the formulas, and give the results for each cell. If not, use a calculator.

52. A store is using a spreadsheet to keep track of its discounted prices and the total price. (Use a computer spreadsheet if you have one.)

	A	B	C	D	E	F
1	Item	Retail	Discount	Sale Price	5% Tax	Total Price
2	Shoes	$60	20%			
3	Pants	$48	10%			

(a) What does the number in cell B3 represent?
(b) What are the formulas for D2, E2, and F2?
(c) If you have a computer, enter the formulas, and give the result for each cell. (Find out how to enter formulas in your spreadsheet program.) If not, use a calculator to find the values in the empty cells.

7.6   RATIONAL, IRRATIONAL, AND REAL NUMBERS

NCTM Standards

- use the associative and commutative properties of addition and multiplication and the distributive property of multiplication over addition to simplify computations with integers, fractions, and decimals (6–8)
- recognize and use connections among mathematical ideas (pre-K–12)
- work flexibly with fractions, decimals, and percents to solve problems (6–8)

What is the relationship between rational numbers and decimals? Can all rational numbers be written as decimals? Do all decimals represent rational numbers?

RATIONAL NUMBERS AS DECIMALS

The division model of fractions tells us that a rational number $\frac{p}{q}$ equals $p \div q$. Any rational fraction can be changed to a decimal by dividing the numerator by the denominator.

Discussion

LE 1   Opener

Change some rational fractions into decimals. Use long division rather than a calculator. What kind of decimals do you obtain?

Decimals come in three forms: terminating decimals, infinite repeating decimals, and infinite decimals that have no infinite repeating block. Examples are 0.3 (terminating), 0.313131 . . . (infinite repeating), and 0.31643847162 . . . (infinite with no repeating block).

LE 2   Concept

Given a decimal, how can you tell if it represents a rational fraction?

As you may have surmised, all rational fractions can be represented by terminating or (infinite) repeating decimals. Why is this the case? In converting a rational fraction $\frac{p}{q}$ to a decimal, we divide q into p. If at some point the division works out evenly, we have a terminating decimal, as is the case for $\frac{3}{8}$.

$$\frac{3}{8} = \frac{0.375}{8\overline{\smash{)}3.000}}$$

If each step in the division has a remainder, the remainders repeat at some point. For example, dividing 7 into 2 $\left(\text{to convert } \frac{2}{7} \text{ to a decimal} \right)$,

$$\frac{2}{7} = \frac{0.2\ 8\ 5\ 7\ 1\ 4\ 2\ 8\ 5\ldots}{7\overline{\smash{)}2.0^60^40^50^10^30^20^60^40\ldots}}$$

$$\uparrow$$

Remainders begin
to repeat

Why must the remainders begin to repeat? In this case, with 7 as the divisor, there are only six possible nonzero remainders—1, 2, 3, 4, 5, and 6—so the remainders must start to repeat after no more than six divisions.

A decimal such as 0.285714285714 . . . is a **repeating decimal,** with an infinite number of digits to the right of the decimal point and a repeating block of digits called the **repetend**. A bar indicates that the block of digits beneath it repeats an infinite number of times.

$$0.\overline{285714} = 0.285714285714285714\ldots$$

All rational fractions can be written as terminating or repeating decimals. Is it also true that all terminating or repeating decimals can be written as rational fractions? Do you know how to use place value to convert terminating decimals to fractions? Do so in LE 3.

LE 3   Skill

Write 0.072 as a fraction.

Infinite (repeating) decimals can be converted to fractions using patterns.

Discussion

LE 4   Reasoning

(a) Express $\frac{1}{9}$ and $\frac{2}{9}$ as decimals.

(b) On the basis of the pattern in part (a), write $\frac{7}{9}$ in decimal form.

(c) Express $\frac{1}{99}$ and $\frac{2}{99}$ as decimals.

(d) On the basis of the pattern in part (c), write $\frac{7}{99}$ in decimal form.

(e) On the basis of the pattern in part (c), write $\frac{13}{99}$ in decimal form. Use division to check your guess.

(f) On the basis of the patterns you have seen, conjecture how to write $\frac{278}{999}$ as a decimal.

The number n of repeating places equals the number of 9s in the denominator (which is $10^n - 1$). The numerator shows what digits are in the repetend. The general formula is

$$0.\overline{a_1a_2a_3 \ldots a_n} = \frac{a_1a_2a_3 \ldots a_n}{10^n - 1}$$

in which $a_1 \ldots a_n$ are the digits. Apply this formula in the following exercise.

LE 5 Skill

Write each of the following repeating decimals as a fraction.

(a) $0.\overline{8}$ **(b)** $0.\overline{37}$ **(c)** $0.\overline{02714}$

The preceding examples and exercises suggest that any repeating or terminating decimal represents a rational number. Some other types of repeating decimals appear in the homework exercises.

IRRATIONAL NUMBERS

A standard number line includes the location of every rational number. For example, the rational numbers $\frac{3}{4}$ and $\frac{15}{8}$ are shown in Figure 7–8. Are there points on the number line that are not named by rational numbers?

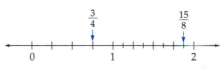

Figure 7–8

Early mathematicians believed that each point on the number line could be named by a rational number. Then, about 2400 years ago, Pythagoras (or one of his followers) wondered about the answer to a geometry problem like the following. What is the length of the diagonal of a square with sides of length 1? Using the Pythagorean Theorem (see Chapter 10),

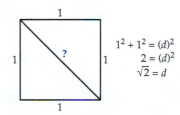

Where is this number $\sqrt{2}$ on a number line? It is between 1 and 2. A calculator might show $\sqrt{2}$ as 1.414213562. Is this the exact value of $\sqrt{2}$? No, because 1.414213562 × 1.414213562 would end in a 4 (from the last digit 2 × 2). All rational numbers can be written as terminating or repeating decimals. What about $\sqrt{2}$?

People have used computers to determine $\sqrt{2}$ to millions of decimal places. And still it has no repeating block of digits!

$$\sqrt{2} = 1.41421\ 35624\ 19339\ 16628\ 19759\ 88713\ 07959 \ldots \text{and so on.}$$

Since the Pythagoreans believed the world was organized around rational numbers, they were quite disturbed about discovering numbers like $\sqrt{2}$. The Pythagoreans vowed to kill anyone who revealed this discovery.

LE 6 Concept

What other kinds of decimals are there besides terminating and repeating decimals?

A **nonrepeating** (infinite) **decimal** has an infinite number of nonzero digits to the right of the decimal point, but it does not have a repeating block of digits that repeats an infinite number of times. Two more examples begin as follows.

$$3.141592653589793 \ldots$$

$$0.010010001 \ldots$$

Nonrepeating decimals are called **irrational** (not rational) **numbers** because they cannot be written as rational fractions. How did the ancient Greeks know they had discovered an irrational number? One of them wrote a deductive proof that $\sqrt{2}$ is irrational. Aristotle describes the proof in one of his books.

To prove that $\sqrt{2}$ is irrational, we shall use the fact that the square of any counting number greater than 1 has an even number of factors in its prime factorization. For example, $36 = 6 \cdot 6 = 2 \cdot 3 \cdot 2 \cdot 3$, and $400 = 20 \cdot 20 = 2 \cdot 2 \cdot 5 \cdot 2 \cdot 2 \cdot 5$. (The same set of prime factors will repeat twice.)

LE 7 Concept

Give another example of the square of a counting number, and show that it has an even number of factors in its prime factorization.

Now to show that $\sqrt{2}$ is irrational.

LE 8 Reasoning

We'll assume that $\sqrt{2}$ is a *rational* number $\frac{p}{q}$ and show that this is impossible.

(a) Suppose $\sqrt{2} = \frac{p}{q}$ in which p and q are counting numbers. Square both sides, and you obtain _____.

(b) Solve your equation from part (a) for p^2.

(c) So $2q^2 = p^2$. Imagine prime factoring $2q^2$ and p^2. Since p^2 is the square of a counting number, it has an _____ number of prime factors.
    (even, odd)

Since $2q^2$ is 2 times a perfect square, it has an _____ number of prime factors.
    (even, odd)

(d) Each step in the proof in parts (a)–(c) involves _____ reasoning.

In LE 8(c), you found that two equal numbers, $2q^2$ and p^2, appear to have different numbers of prime factors. This is impossible, according to the Fundamental Theorem of Arithmetic. Therefore, our assumption that $\sqrt{2} = \frac{p}{q}$ must be wrong, since all our deductive reasoning from that point on was valid. Thus, $\sqrt{2}$ is irrational!

LE 9 Concept

(a) Name two more square roots that you think are irrational numbers.

(b) Name a square root that *is* a rational number.

Soon after the Greeks proved that $\sqrt{2}$ is irrational, they proved that the square roots of 3, 5, 6, 7, 8, 10, 11, 12, 13, 14, 15, and 17 are also irrational.

The most famous irrational number of all is π. For thousands of years, people have calculated how the distance around a circle (the circumference) compared to the distance across the circle through the center (the diameter).

Discussion

LE 10 Reasoning

To measure in this problem, you will need a string and a ruler. (A compass would also be useful.)

(a) Draw three fairly large circles of different sizes, or find circular shapes in your classroom.

(b) Measure the circumference C and diameter d of each, and complete the following chart.

C	d	$C - d$	$C \div d$

(c) Is there a pattern in either of the last two columns?

(d) What generalization does part (c) suggest?

Did you find that $\dfrac{C}{d}$ is always a little more than 3? The exact quotient is represented by the symbol π. By definition, $\pi = \dfrac{C}{d}$, or $C = \pi d$.

At first, people assumed that π would be a simple number such as 3 or 3.1. The Rhind Papyrus (1650? B.C.) shows that the Egyptians used a value of $\dfrac{256}{81}$ or about 3.16 for π to solve problems involving circles.

Archimedes (240 B.C.) drew inscribed and circumscribed polygons around a circle and computed their perimeters (Figure 7–9). He knew that the circumference of the circle was between these two numbers. Archimedes showed that π was between $3\dfrac{10}{71}$ and $3\dfrac{1}{7}$. In the fifth century, the Chinese mathematician Tsu Ch'ung-Chih computed π to six decimal places $\left(\dfrac{355}{113}\right)$. In 1706, John Machin used the methods of calculus to calculate π to 100 decimal places, and William Jones introduced the symbol for π. Jones used π to

Figure 7–9

represent the circumference of a circle with a diameter of 1. Jones chose π, the Greek letter for p, since it was the first letter of the word "periphery." It wasn't until 1761 that Johann Lambert *proved* that π is irrational. In 2002, Yasumasa Kanada's team of computer scientists at the University of Tokyo used computers to find over 1.24 trillion digits of π.

$$\pi = 3.14159\ 26535\ 89793\ 23846\ 26434\ldots$$

Will you ever use π to hundreds of decimal places? Probably not. One can calculate the circumference of the earth to the nearest inch with only 10 decimal places of π. This is just the sort of approximation that a graphing calculator uses.

The only way to write the exact value of an irrational number such as $\sqrt{2}$ or π is to use the $\sqrt{}$ symbol or letter. The exact value cannot be written using a decimal number.

LE 11   Skill

A wheel has a diameter of 30 in. Approximate its circumference.

REAL NUMBERS

To name every point on a standard number line, we need both rational and irrational numbers. The union of the set of rational numbers and the set of the irrational numbers is the **real numbers.** There is one-to-one correspondence between real numbers and the points of the number line. Real numbers are all numbers that can be represented as decimal numbers.

LE 12   Concept

Give two examples of real numbers.

The following chart compares the two types of real numbers, rational and irrational.

Real Numbers

Rational numbers	Repeating or terminating decimals
Irrational numbers	Infinite nonrepeating decimals

All of the numbers in this book can be organized into an overall set picture.

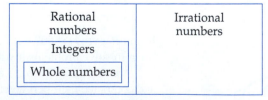

You may be wondering if there are any other kinds of numbers besides real numbers. To find solutions to $x^2 = a$, in which $a < 0$, we need imaginary numbers such as $\sqrt{-3}$ which are covered in some high-school and college mathematics courses.

• **EXAMPLE 1**

Which of the following describe 2.6?

(a) A whole number   **(d)** An irrational number
(b) An integer   **(e)** A real number
(c) A rational number

SOLUTION

2.6 is not a whole number or integer because of the .6. But 2.6 is rational because it can be written as $\frac{26}{10}$. Since 2.6 is rational, it cannot be irrational. Since 2.6 is rational, it is also a real number.   ●

LE 13 Concept

Complete the following chart by placing a $\sqrt{}$ in the appropriate columns.

	Whole Number	Integer	Rational Number	Irrational Number	Real Number
2.6			$\sqrt{}$		$\sqrt{}$
3					
$\sqrt{2}$					
−1					
$-2\frac{1}{2}$					

Because every real number corresponds to a point on the number line, solutions to inequalities for real numbers can be represented on a number line.

LE 14 Connection

(a) Write the inequality represented by the following number-line graph.

10

(b) Suppose Christine won't be a teacher for at least 5 years. Graph $n \geq 5$ on a number line, in which n represents the number of years.

MULTIPLE REPRESENTATIONS OF REAL NUMBERS

Why do we need both fractions and decimals to represent quantities? Each notation has its advantages. Sometimes it is easier to represent a number as a fraction than as a decimal, and sometimes it isn't.

LE 15 Reasoning

Name a rational number that is easier to represent with a fraction than with a decimal.

It is usually easier to compare the sizes of numbers in decimal form than in fraction form.

LE 16   Skill

If you have a calculator, what is an easy way to determine whether $\frac{3}{8}$ or $\frac{11}{29}$ is larger?

Some addition problems are easier to do with fractions; some are easier with decimals.

LE 17   Reasoning

In parts (a) and (b), state whether you think the addition problem is easier to do with fractions or with decimals.

(a) $\frac{2}{7} + \frac{3}{7}$ or $0.\overline{285714} + 0.\overline{428571}$ (b) $\frac{1}{4} + \frac{7}{100}$ or $0.25 + 0.07$

(c) Describe more generally when it is easier to do addition with fractions and when it is easier to use decimals. (Assume you do not have a calculator.)

In some problems, the preferred notation may depend on whether you want an exact or an approximate answer.

LE 18   Connection

Last year, your salary increased from \$21,000 to \$27,000, an increase of $28\frac{4}{7}\%$, or 28.6%.

(a) Classify each percent as exact or approximate.
(b) Describe an advantage of each answer.

LE 19   Connection

The numbers $\frac{1}{4}$, 0.25, and 25% all represent the same amount. Describe a situation in which each representation would be the best choice. For example, $\frac{1}{4}$ would be a good notation for a probability.

Why do we use special irrational number symbols (such as $\sqrt{2}$ and π) and decimal representations for the same numbers?

LE 20   Connection

(a) Why is it easier to write $\sqrt{273}$ than its decimal representation?
(b) A farmer wants to buy some fencing. You determine that the length of fence he needs is $\sqrt{273}$ ft. How much fence would you tell the farmer to buy?

RATIONAL NUMBER PROPERTIES RETAINED!

What properties do real numbers have? In computations with irrational numbers, we find that $(\pi + 2\pi) + 3\pi = \pi + (2\pi + 3\pi)$, $\sqrt{2} \cdot \sqrt{3} = \sqrt{3} \cdot \sqrt{2}$, and the additive inverse of $\sqrt{8}$ is $-\sqrt{8}$. Examples such as these suggest how real-number operations retain all the properties of rational numbers. The following list summarizes these properties.

> **Some Properties of Real-Number Operations**
>
> Addition, subtraction, and multiplication of real numbers are closed.
>
> Addition and multiplication of real numbers are commutative.
>
> Addition and multiplication of real numbers are associative.
>
> The unique additive identity for real numbers is 0, and the unique multiplicative identity for real numbers is 1.
>
> All real numbers have a unique additive inverse that is real, and all nonzero real numbers have a unique multiplicative inverse that is real.
>
> Multiplication is distributive over addition, and multiplication is distributive over subtraction for the set of real numbers.

The following exercises make use of these properties.

LE 21   Concept
Give an example showing that real-number (decimal) addition is commutative.

LE 22   Concept
$3.6 + (-3.6) = -3.6 + 3.6 = 0$ illustrates what property of real numbers?

LE 23   Concept
(a) Factor $2x + \sqrt{2}x$.
(b) What property is used to factor in part (a)?

What about properties that did not work for other sets of numbers in this course?

LE 24   Reasoning
The real numbers include all the other sets of numbers in Chapters 3, 5, and 6. If a property such as the associative property for subtraction does not work for whole numbers, explain why it could not work for real numbers.

Some of the properties of all the number systems you have studied are summarized in the following chart.

Property	Operation	Whole Numbers	Integers	Rational Numbers	Irrational Numbers	Real Numbers
COMMUTATIVE	+	√	√	√	√	√
	×	√	√	√	√	√
ASSOCIATIVE	+	√	√	√	√	√
	×	√	√	√	√	√
IDENTITY	+	√	√	√		√
	×	√	√	√		√
INVERSES	+		√	√		√
	×			√		√

LE 25   Concept

Tell why 0 cannot be the additive identity and 1 cannot be the multiplicative identity for irrational numbers.

LE 26   Summary

Tell what you learned about real numbers in this section.

ANSWERS TO SELECTED LESSON EXERCISES

1. Terminating and infinite repeating

2. Terminating and repeating

3. $\dfrac{72}{1000} = \dfrac{9}{125}$

4. (a) $0.\overline{1}$ and $0.\overline{2}$ (b) $0.\overline{01}$ 0.02
(c) $0.\overline{01}$, 0.02 (d) $0.\overline{07}$
(e) $0.\overline{13}$ (f) $0.\overline{278}$

5. (a) $\dfrac{8}{9}$ (b) $\dfrac{37}{99}$ (c) $\dfrac{2714}{99,999}$

6. Infinite nonrepeating

7. $81 = 3 \cdot 3 \cdot 3 \cdot 3$ (4 factors)

8. (a) $2 = \dfrac{p^2}{q^2}$ (b) $p^2 = 2q^2$
(c) even; odd

9. (a) $\sqrt{3}$ and $\sqrt{34}$ (b) $\sqrt{4}$

11. $30\pi \approx 94.2$ in.

12. 3 and 8.765

13. 3 is whole, integer, rational, and real; $\sqrt{2}$ is irrational and real; -1 is integer, rational, and real; $-2\dfrac{1}{2}$ is rational and real.

14. (a) $n < 10$
(b)

15. $\dfrac{3}{7}$

16. For each fraction, divide the numerator by the denominator and compare the decimal representations.

17. (a) Fractions   (b) Decimals
(c) Fractions are easier when there is a common denominator that has factors other than 2 and 5. Decimals are easier for fractions that are easily converted to decimals.

18. (a) The fraction is exact; the decimal is approximate.
(b) The fraction is more precise. The decimal is usually easier to work with.

19. The decimal 0.25 could be an amount of money and 25% could be a tax rate.

20. (a) The decimal representation is an infinite nonrepeating decimal.
(b) 16.5 ft (a decimal approximation)

21. $0.2 + 0.3 = 0.3 + 0.2$

22. Additive inverse property

23. (a) $(2 + \sqrt{2})x$
(b) Distributive property of multiplication over addition

24. All counterexamples for whole numbers are also counterexamples for real numbers.

25. They are not part of the set of irrational numbers.

7.6 HOMEWORK EXERCISES

Basic Exercises

1. Write the following fractions as decimals.

   (a) $\dfrac{3}{5}$ (b) $2\dfrac{2}{9}$

2. Write the following fractions as decimals.

   (a) $\dfrac{7}{20}$ (b) $\dfrac{7}{11}$

3. Write the following decimals as fractions.

   (a) 0.731   (b) -13.04

4. Write the following decimals as fractions.

   (a) -0.013 (b) 4.712

5. Which is greater, $0.3\overline{4}$ or $0.\overline{34}$?

6. Which is greater, $0.5\overline{2}$ or $0.\overline{52}$?

7. Write the following decimals as fractions.

   (a) $0.\overline{4}$ (b) $0.\overline{32}$ (c) $0.\overline{267}$

8. Write the following decimals as fractions.

   (a) $52.\overline{35}$ (b) $0.39\overline{17}$ (c) $0.\overline{9}$

9. One can use the results from the lesson to convert other types of repeating decimals to fractions. For example,

   $$0.43\overline{2} = \frac{43\frac{2}{9}}{100} = \frac{\frac{389}{9}}{100} = \frac{389}{900}$$

   Use this method to convert the following decimals to fractions.

   (a) $0.2\overline{1}$ (b) $0.34\overline{1}$

10. Use the method of the preceding exercise to convert the following decimals to fractions.

    (a) $0.5\overline{34}$ (b) $0.1\overline{23}$

11. The standard method for converting a repeating decimal to a fraction is as follows for $0.\overline{2}$.

    (a) Let $N = 0.\overline{2}$ What does $10N$ equal?

    (b) Complete the following and subtract.

    $$10N = \underline{\hspace{2cm}}$$
    $$- \ \underline{\ N} = \ \underline{\ 0.\overline{2}\ }$$

    (c) Solve the resulting equation from part (b) for N.

    (d) Use this same method to convert $0.\overline{31}$ to a fraction. (*Hint*: Use $100N$.)

12. Use the method of the preceding exercise to convert each repeating decimal to a fraction.

    (a) $0.\overline{72}$ (b) $0.8\overline{3}$

13. Make up a geometric problem whose answer is $\sqrt{5}$.

14. Make up an algebraic equation (other than $x = \sqrt{3}$) with a solution of $x = \sqrt{3}$.

15. (a) Find the value of $\sqrt{5}$ on a calculator.

    (b) Why can't this number be the exact value of $\sqrt{5}$?

16. (a) Enter any positive number. Press the square-root key repeatedly. What number do you reach eventually?

    (b) Repeat part (a), starting with a different number.

    (c) Generalize your results.

17. A wheel has a circumference of 76 in. What is its diameter?

18. The origin of π was in computing the ratio of the _____ to the _____.

19. Classify the following numbers as rational or irrational.

    (a) $\sqrt{11}$ (b) $\dfrac{3}{7}$ (c) π (d) $\sqrt{16}$

20. Classify the following numbers as rational or irrational.

    (a) -3 (b) $\sqrt{\dfrac{4}{9}}$ (c) $\sqrt{7}$ (d) $\dfrac{7}{3}$

21. Classify the following numbers as rational or irrational.

    (a) $0.34938661\ldots$ (b) $0.\overline{26}$

    (c) $0.565665666\ldots$ (d) 0.56

22. Classify the following numbers as rational or irrational.

    (a) $0.3333333\ldots$

    (b) 0.817

    (c) $0.42638275643\ldots$

    (d) $0.121121112\ldots$

23. An eighth grader thinks that $\dfrac{3}{5}$ and $-\dfrac{7}{2}$ are rational but 0.2 and $\sqrt{49}$ are not rational. What would you tell the child?

24. An eighth grader thinks the decimal $0.232232223\ldots$ is rational. What would you tell the child?

25. Police can estimate the speed s (in miles per hour) of a car by the length L (in feet) of its skid marks. On a dry highway, $s \approx \sqrt{24L}$. On a wet highway, $s \approx \sqrt{12L}$.
(a) Estimate the speed of a car that leaves 54-ft skid marks on a wet highway.
(b) Estimate the speed of a car that leaves 54-ft skid marks on a dry highway.
(c) Classify your answers to parts (a) and (b) as rational or irrational.

26. A pendulum is constructed by tying a weight on a string. The time to make one complete swing back and forth depends upon the length of the string. Suppose the following procedure can be used to find the time required for a complete swing.

Length in cm	→	Take square root	→	Divide by 5	→	Time in seconds

(a) How long would a 40-cm pendulum take to complete one swing?
(b) A pendulum takes 4 seconds to complete a swing. How long is the pendulum?
(c) A formula relating time T to length L is $T =$ _____.
(d) A formula relating length L to time T is $L =$ _____.

27. Match each word in column A to a word in column B.

A	B
Terminating	Rational
Repeating	Irrational
Infinite nonrepeating	

28. What set of numbers is used to represent every point on a number line?

29. (a) How many whole numbers are there between 3 and -3 (not including 3 and -3)?
(b) How many integers are there between 3 and -3?
(c) How many real numbers are there between 3 and -3?

30. Give an example of a number on a standard number line that is not a rational number.

31. Your calculator display shows $\boxed{0.3333333}$. Name two different exact values this might represent.

32. Compute $2 \div 3$ on your calculator. Does it round to the nearest or truncate?

33. Complete the chart.

	Whole	Integer	Rational	Irrational	Real
$\frac{1}{3}$			√		√
$\sqrt{13}$					
-6					
$\sqrt{9}$					
-0.317					

34. Complete the chart.

	Whole	Integer	Rational	Irrational	Real
$\sqrt{\frac{7}{2}}$				√	√
4.21					
$-\sqrt{16}$					
π					
51					

35. Draw a Venn diagram with these sets: real numbers, rational numbers, whole numbers.

36. True or false? All rational numbers are real numbers.

37. Graph the following inequalities on a number line.
(a) $x \leq 1$ (b) $x > -3$

38. The temperature T yesterday was between 54°F and 76°F. Graph this interval on a number line.

39. Name a real number that is not easily represented by either a decimal or a rational fraction.

40. Name a rational number that can be easily represented both as a fraction and as a decimal.

41. If you have a calculator, what is an easy way to determine which fraction is larger, $\frac{5}{7}$ or $\frac{12}{17}$?

42. A telephone company plans to raise monthly rates by 3%, which amounts to $28 million. Which number would probably be more upsetting to consumers?

43. 227.35682 ÷ 0.2379654 is about
(*Hint*: Change the divisor to a simple fraction.)
(a) 55   (b) 900   (c) 225   (d) 10

44. 82.35682 ÷ 0.1179654 is about
(a) 80   (b) 0.012   (c) 750   (d) 9

45. Make up an addition problem that is easier to do with fractions than with decimals.

46. Make up a subtraction problem that is easier to do with decimals than with fractions.

47. In school, 3.14 and $\frac{22}{7}$ are often used as approximations for $\pi = 3.141592654\ldots$.
(a) Which school approximation is closer to the actual value of π?
(b) If your calculator has a π key, write the decimal value it displays for π.

48. Different mathematicians have used fractional approximations for π. Which of the following is closest to $\pi = 3.141592654\ldots$?
(a) Babylonians (around 1700 B.C.): $\frac{25}{8}$
(b) Archimedes (around 200 B.C.): $\frac{223}{71}$
(c) Tsu Ch'ung-chih (around A.D. 480): $\frac{355}{113}$
(d) Bhaskara (around A.D. 1150): $\frac{3927}{1250}$

49. Why is it easier to use the symbol π than its decimal representation?

50. The distance around a running track is $50\pi + 360$ yd.
(a) What is a more useful way to express this distance?
(b) Is your answer to part (a) exact or approximate?

51. $0.37 + 0.516 = 0.516 + 0.37$ illustrates the _____ property of _____.

52. What properties guarantee that $-4x + 3 + 5.5 + 6x = (-4x + 6x) + (3 + 5.5)$ for any decimal x?

53. What is the additive inverse of -0.41798?

54. What is the multiplicative inverse of 0.1?

55. Give a counterexample that shows that real-number subtraction is not commutative.

56. According to the distributive property of multiplication over subtraction, $6.4x - 2.1x =$ _____.

57. How would you mentally compute 6.5×8?

58. What is the easiest way to multiply $0.6 \times 4.2 \times 5$?

59. Tell whether each equation is true for no real numbers, some pairs of real numbers, or all pairs of real numbers.
(a) $x + y = 3$
(b) $xy = yx$
(c) $x + y = x + y + 3$
(d) $\sqrt{x^2 + y^2} = x + y$

60. Tell whether each equation is true for no real number, some real number(s), or all real numbers.
(a) $x \cdot 1 = x$ (b) $2x - 4 = 10$
(c) $2 + x = x$ (d) $\sqrt{x + 4} = \sqrt{x} + \sqrt{4}$

61. $\frac{1}{5^1} = 0.2$ $\frac{1}{5^2} = 0.04$
(a) Write the next two examples that continue the pattern, and see if they are true.
(b) Write a generalization about $\frac{1}{5^N}$ based on your results.

62. $\frac{1}{2^1 \cdot 5^1} = 0.1$ $\frac{1}{2^2 \cdot 5^1} = 0.05$ $\frac{1}{2^1 \cdot 5^2} = 0.02$
Try some more examples, and hypothesize how many decimal places $2^N \cdot 5^M$ has.

Extension Exercises

63. (a) Which of the following fractions can be rewritten with denominators of 10, 100, 1000, and so on?
(1) $\frac{4}{5}$ (2) $\frac{3}{25}$ (3) $\frac{7}{9}$ (4) $\frac{3}{200}$ (5) $\frac{1}{7}$
(b) Which fractions in part (a) can be represented by terminating decimals?
(c) Prime-factor each denominator in part (a).
(d) Try to generalize your results by filling in the blanks. Fractions (in simplest form) that can be rewritten as terminating decimals have denominators that are divisible by no prime number other than _____ or _____.

64. As the preceding exercise suggests, a rational number $\frac{a}{b}$ in simplest form is a terminating decimal if and only if b is divisible by no prime numbers other than 2 or 5. Factor the denominator of each fraction, and determine whether it is a terminating or repeating decimal.

(a) $\frac{8}{9}$ (b) $\frac{7}{500}$ (c) $\frac{11}{30}$ (d) $\frac{7}{20}$

 65. A common algebraic error is to assume that $\sqrt{M + N} = \sqrt{M} + \sqrt{N}$ for all real numbers M and N. For what decimal values of M and N does $\sqrt{M + N} = \sqrt{M} + \sqrt{N}$? Make an educated guess after trying some examples.

66. Consider the following problem. "Name two irrational numbers whose product is 6."
(a) Devise a plan and solve the problem.
(b) Make up a similar problem.

67. (a) $0.010110111 \ldots + 0.101001000 \ldots =$
_____
(b) Part (a) shows that the sum of two
_____ numbers can be a
_____ number.

68. The sum of a rational number and an irrational number
(a) is rational   (b) is irrational
(c) could be either

69. Prove that $\sqrt[3]{2}$ is not rational.

70. Prove that $\sqrt{3}$ is not rational

Technology Exercise

71. Some calculators such as the TI-15 and TI-73 convert fractions to decimals and decimals to fractions. Use a calculator with an $\boxed{F \leftrightarrow D}$ key to
(a) convert $\frac{3}{16}$ to a decimal
(b) convert 0.316 to a fraction in simplest form.

Labs and Projects

72. Write a short report on the history of π.

CHAPTER 7 SUMMARY

The rules for decimal place value and arithmetic are extensions of the rules for whole-number place value and arithmetic. The primary question in decimal arithmetic is where to place the decimal point in the answer. The rules for placing decimal points can be explained using place value and rational-number properties. The models for whole-number operations apply to many decimal problems.

Decimal estimation relies primarily on the same strategies as fraction estimation—namely, rounding and the compatible-numbers strategy. One can mentally multiply or divide a decimal number by a power of 10 by moving the decimal point of the number. This same technique is used in converting a decimal number to scientific notation, an exponential shorthand used by scientists.

Many everyday applications of mathematics involve proportional relationships. A proportion states that two ratios are equal. Proportions with one missing quantity can be solved using cross products or, more intuitively, with a unit-rate or multiplier approach. Common ratios that occur in proportion problems include male-female ratios, student-teacher ratios, and miles per gallon. Proportions also occur in situations involving water bills, real estate taxes, map scales, surveys, and foreign currency rates.

Percents developed from the use of hundredths in fractions and decimals. People use percents to describe taxes, inflation, interest rates, finance charges, salary increases, discounts, and test scores. In all these cases, percents give the rate per 100. These applications of percents can usually be solved with equations or proportions and modeled by number-line pictures.

It is helpful to know the fractional equivalents of some basic percents in order to compute some percent problems mentally using shortcuts. Rounding or compatible numbers can be used to estimate solutions to many other percent problems.

The set of all decimal numbers is called the real numbers. Decimals are either terminating, infinite repeating, or infinite nonrepeating decimals. Terminating and repeating decimals represent rational numbers. With rational numbers, one has the option of using either decimal or fraction notation.

Nonrepeating decimals are called irrational numbers. Examples of irrational numbers include $\sqrt{2}$, $\sqrt{5}$, and π. The only way to write the exact value of an irrational square root or π is to use a special symbol.

Real-number operations possess all the properties that whole-number, integer, and rational-number operations possess. Each real number corresponds to a point on the number line.

STUDY GUIDE

To review Chapter 7, see what you know about each of the following ideas or terms that you have studied. You can also use this list to generate your own questions about the chapter.

DECIMALS AND PERCENTS IN GRADES 1–8

The following chart shows at what grade levels selected decimal topics typically appear in elementary- and middle-school mathematics textbooks.

Topic	Typical Grade Level in Current Textbooks
Decimal place value	3, 4, 5, 6
Decimal estimation	4, 5, 6, 7
Scientific notation	7, 8
Decimal addition and subtraction	3, 4, 5, 6
Decimal multiplication	5, 6, 7
Decimal division	5, 6, 7
Ratio	5, 6, 7
Proportion	6, 7, 8
Percents	5, 6, 7, 8

REVIEW EXERCISES

1. (a) Write 0.0037 in expanded notation.
(b) Write 0.0037 in scientific notation.

2. (a) Why do some children think that $0.3 < 0.25$?
Explain why $0.3 > 0.25$
(b) with a decimal-square picture.
(c) with a number line from 0 to 1.

(d) by putting the numbers in place-value columns of ones, tenths, and hundredths.

3. Suppose a child knows about positive whole-number exponents. Show how to find the value of 4^0 by either extending a pattern or by using the addition rule for exponents on an example such as $4^2 \cdot 4^0$.

(Continued in next column)

4. If 100 balloons cost $0.76, *explain* how you would mentally compute the charge per balloon.

5. Write 23.7 million in scientific notation.

6. A fourth grader computes $0.4 + 0.8 = 0.12$. What would you tell the child?

7. Draw a decimal-square picture that shows the result of $0.7 - 0.2$. State the result in an equation.

8. (a) *Explain* how to compute 0.2×0.4 using two decimal-square pictures.
   (b) Work out the same problem using fractions.
   (c) Work out the same problem using words for place value.

9. The quotient of $12.3 \div 0.2$ has the digits 615. Explain how to place the decimal point with estimation.

10. A child wants to know why we can move the decimal point to change $42 \div 0.06$ to $4,200 \div 6$. Write $42 \div 0.06$ as a fraction, and show how the process works.

11. Compute the following without a calculator.
   (a) $8 - 0.79$ (b) $0.3 \times 0.5 \times 0.8$
   (c) $0.076 \div 0.8$

12. What operation and category are illustrated in the following problem? "A man loses 8 lb in 10 days. What is his average weight loss per day?" (area, array, partition, repeated)

13. What operation and category are illustrated in the following problem? "Before driving to work, your odometer reads 26,872.6 miles. When you arrive at work, it reads 26,880.4. How long is the drive?" (compare, missing part, take away)

14. Complete the last example, repeating the same error pattern in the completed examples.
   $6 \div 0.3 = \underline{\ 0.2\ }$ $0.8 \div 0.4 = \underline{\ 0.02\ }$
   $1.2 \div 0.3 = \underline{\hspace{1cm}}$

15. Use a rectangle to represent a submarine sandwich.
   (a) Divide the sandwich into two sections A and B so that the ratio of A to B is $4:1$.
   (b) What fraction of the whole sandwich is section A?
   (c) Section B is _____ times as long as section A.

16. In a survey, 18 out of 24 third-graders said they liked recess. The rest did not like recess.
   (a) Make an additive comparison of the number of students who liked recess and the number that did not.
   (b) Make a multiplicative comparison of the number of students who liked recess and the number that did not.

 17. Consider the following problem. "The average 130-pound person needs 2,000 calories per day. How many calories does the average 150-pound person need?"
   (a) Write a proportion. Tell how you set up the proportion. Then solve it by using cross products.
   (b) Find a unit rate in calories/lb, and use it to find the answer. (Make a unit rate table if you wish.)
   (c) How could you find the answer by finding a multiplier (scale factor)?

18. Maria ate 5 times as many peanuts as Debbie. They ate a total of 96 peanuts. How many peanuts did each one eat?

19. It took G gallons of paint to paint 3 classrooms. How much paint would it take to paint C similar classrooms?

20. A survey of 800 adults found that 480 ate ice cream and 320 did not. On the basis of this survey, complete the following.
   (a) _____ % of the adults ate ice cream.
   (b) The fraction of adults who didn't eat ice cream is
   _____.
   (c) The ratio of adults who ate ice cream to those who didn't is _____ to _____.

 21. Solve the following problems with the percent proportion.
   (a) What is 34% of 40?
   (b) 3 out of 250 is what percent?
   (c) 43 is 20% of what number?

22. Show how to work out Exercise 21(c) with a percent model.

 23. Solve the following problems with the percent equation (or multiplication or division).
   (a) What is 34% of 40?
   (b) 3 out of 250 is what percent?
   (c) 43 is 20% of what number?

 24. You deposit $8000 earning 6% simple annual interest. Find the interest and account balance after 9 months.

25. The rectangular region below represents 40% of a regular pizza. Draw 100% of the pizza.

26. *Explain* how to compute 40% of $7,000 mentally.

27. Estimate 27% of 489
    (a) with rounding.   (b) with compatible numbers.

28. The price of a dress increases from $80 to $110. Find the percent increase in price with
    (a) the equation method.
    (b) the proportion method.

29. The price of a textbook increases by 12%. The new price is _____ times the old price.

30. While you are on a trip, you buy a $9.49 shirt. The cashier charges you $10.06. What percent is the sales tax?

31. Write the following decimals as fractions.
    (a) $0.\overline{78}$ (b) $0.9\overline{24}$

32. Classify the following numbers as rational or irrational.
    (a) $0.1\underline{4379}\ldots$ (b) $\sqrt{7}$
    (c) $0.\overline{23}$ (d) -3

33. Write a sentence or two about π.

34. $\sqrt{17}$ is
    (a) a whole number
    (b) an integer
    (c) a rational number
    (d) an irrational number
    (e) a real number

35. -5 is
    (a) a whole number
    (b) an integer
    (c) a rational number
    (d) an irrational number
    (e) a real number

36. Write a paragraph describing how the following are related: real numbers, rational numbers, irrational numbers, terminating decimals, repeating decimals, nonrepeating decimals.

37. $42.764764 \div 0.48917563645639$ is about
    (a) 9   (b) 40   (c) 1   (d) 90   (e) 20

38. Give an example of the distributive property of multiplication over subtraction for real numbers.

39. Name a property that real numbers have that the whole numbers do not.

ALTERNATE ASSESSMENT

Do one of the following assessment activities: add to your portfolio, add to your journal, write another unit test, do another self-assessment, or give a presentation.

SUGGESTED READINGS

Barnett, C. et al. *Fractions, Decimals, Ratios, and Percents*. Portsmouth, NH: Heinemann, 1994.

Curcio, F., N. Bezuk, and others. *Understanding Rational Numbers and Proportions*. Reston, VA: NCTM, 1994.

Davis, P. *The Lore of Large Numbers*. New York: Random House, 1961.

National Council of Teachers of Mathematics. 2002 Yearbook. *Making Sense of Fractions, Ratios, and Proportions*. Reston, VA: NCTM, 2002.

Navigating Through Number and Operations (4 vols). Reston, VA: NCTM, 2003

Use Infotrac to find articles in *Teaching Children Mathematics* on teaching children about decimals.

8 | INTRODUCTORY GEOMETRY

Geometry analyzes visual patterns and connects mathematics to the physical world. The oldest recorded examples of geometry are ancient cave drawings of circles, squares, and triangles. Ancient pottery and weaving bear examples of geometric designs. The root meaning of the word "geometry" is "earth measure," which derives from the use of geometry by the Egyptians and Babylonians in land surveying over 4000 years ago. The Egyptians and Babylonians also used geometry in agriculture, architecture, and astronomy. The Great Pyramid of Gizeh in Egypt, built over 4000 years ago, has a square base with sides that are all 756 ft long, with an error of less than 1 inch!

About 2500 years ago, the Greeks developed an important new approach to geometry, organizing geometric ideas into a logical sequence. They began with the most basic figures and terms and deduced all other concepts of plane geometry.

Today, school geometry follows this historical sequence. Children first study geometry in an informal, concrete manner. Then, in high school, many students study geometry using a more formal, deductive approach.

8.1 BEGINNING GEOMETRY

NCTM Standards
- understand how mathematical ideas interconnect and build upon one another to produce a coherent whole (pre-K–12)

We all begin exploring geometry when we notice shapes in our surroundings. Children come to elementary school with an awareness of objects and shapes. School geometry builds on this awareness by beginning with space figures (such as cubes and cylinders) and plane figures (such as rectangles and circles) suggested by objects in the environment. Then a more systematic study of geometry begins with the most basic figures—points, lines, and planes—and proceeds in a deductive sequence.

SHAPES IN OUR WORLD

In grades 1 and 2, children learn the names of common shapes as they study objects around them.

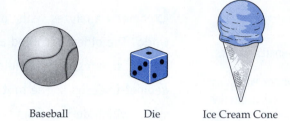

Baseball   Die   Ice Cream Cone

Geometric shapes are idealized versions of objects we see.

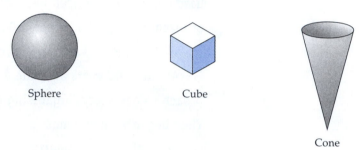

Sphere   Cube

Cone

According to Plato, the perfect geometric shapes in our minds are part of the ultimate reality, whereas objects in the world are imperfect.

The surfaces of objects also suggest geometric figures.

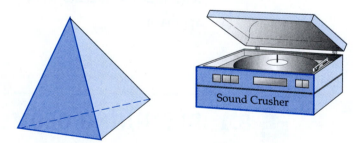

Sound Crusher

Discussion

LE 1   Opener

(a) Find three objects in your classroom that suggest common geometric figures.

(b) Find three objects in your classroom with surfaces that suggest common geometric figures.

(c) Compare your answers to others in your class.

Next, children study properties of figures. They count corners and sides of plane figures and faces (or "flat surfaces") of solid figures.

POINTS, LINES, AND PLANES

After an initial introduction to geometry through shapes suggested by objects, how should one begin a more systematic study of geometry? In his famous book *The Elements,* Euclid (c. 300 B.C.) begins **Euclidean geometry** by describing basic geometric terms and stating ten assumptions (postulates). Then he builds geometry deductively from those basic terms and assumptions. About 650 years later, Theon of Alexandria published the version of Euclid's *Elements* that we use in modern translations.

Theon's daughter, Hypatia (A.D. 370–415) (Figure 8–1), is known as the first woman mathematician. Thanks to her father's support and encouragement, Hypatia became one of Athens' finest mathematicians. Then Hypatia moved to Alexandria, where she gained popularity as a lecturer on mathematics and philosophy. Most of Hypatia's mathematical work has been lost, but she is known to have written a commentary on the work of Diophantus in number theory. Sadly, Hypatia was murdered by a Christian mob because of her outspoken defense of scientific methods and Greek paganism.

Today, a formal presentation of Euclidean geometry begins with a description of points, lines, and planes. "Point," "line," and "plane" are undefined terms that we can understand using everyday objects and our intuition. Points, lines, and planes are suggested by our surroundings, as shown in Figure 8–2.

© *Courtesy of Bettmann Archives.*

Figure 8–1 *Hypatia*

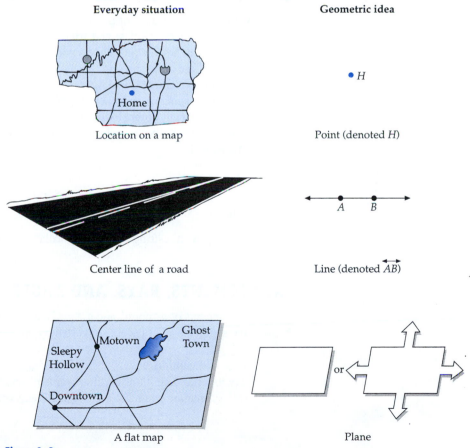

Figure 8–2

A location on a map and a grain of sand each suggest a point. A **point** represents an exact location. A taut wire and the center line on a road suggest a line. A **line** is straight; it extends indefinitely in two directions and has no thickness. The arrows on a line indicate that it extends without limit. The notation $\overleftrightarrow{AB}$ represents the line through points A and B.

A map and a piece of paper each suggest a section of a plane. A **plane** is flat; it has no thickness and extends endlessly in all directions (Figure 8–2 on page 391).

Mathematicians and philosophers have debated where geometric ideas such as a line come from. Why all the fuss, you ask? Because, in your whole life, you will never see a geometric line!

Discussion

LE 2   Concept

How does a geometric line differ from a drawing of a line?

Discussion

LE 3   Reasoning

(a) True or false? If A and B are points in a plane, then $\overleftrightarrow{AB}$ lies entirely in the plane.

(b) Draw a picture that supports your answer.

PLANE FIGURES, SPACE FIGURES, AND DIMENSION

A **plane figure** is a set of points in a plane. Examples of plane figures include lines, line segments, planes, circles, and triangles. Lines and line segments are one-dimensional figures. They have length but no width or height. A **two-dimensional figure** is a set of points that are all in the same plane but not all on the same line. Planes, circles, and triangles are two-dimensional figures.

Cubes, spheres, and everyday objects are examples of three-dimensional or space figures. A **three-dimensional** or **space figure** does not lie in a single plane, and it has volume. Informally, any shape that can be constructed from modeling clay is three-dimensional.

LE 4   Concept

What is the dimension (1, 2, or 3) of the geometric figures *suggested* by each of the following?

(a) A very thin string   (b) A gorilla   (c) The surface of a table

LINE SEGMENTS, RAYS, AND ANGLES

The concepts of a "point" and a "line" can now be used to define line segments and rays. Line segments and rays are suggested by our surroundings, as shown in Figure 8–3.

LE 5   Reasoning

Try to write definitions of "line segment" and "ray" using the terms "point" and "line." (*Hint:* Imagine you had a line. What would you do to it to obtain a line segment?)

Everyday situation   Geometric idea

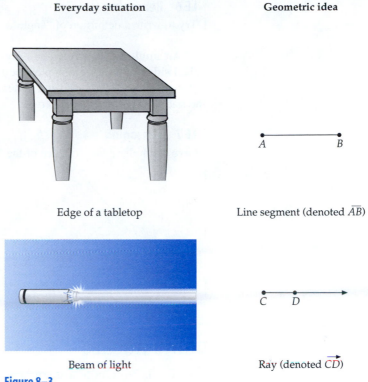

Edge of a tabletop   Line segment (denoted $\overline{AB}$)

Beam of light   Ray (denoted $\overrightarrow{CD}$)

Figure 8–3

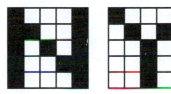

Figure 8–4

Line segments and rays are subsets of a line. A **line segment** consists of two points on a line and all the points between them. The idea of making a line segment out of "points" is used in dot-matrix printers and stadium scoreboards, as shown in Figure 8–4. The N and Y are each formed by three sets of squares in vertical or diagonal rows. Each row of squares is analogous to a line segment.

A **ray** (for example, $\overrightarrow{CD}$ in Figure 8–3) is a subset of a line that consists of a point (C) together with all points on the line ($\overleftrightarrow{CD}$) on one side of the point (the "D side" of C).

Next, consider the ways in which angles are also suggested by our surroundings (Figure 8–5).

Everyday situation   Geometric idea

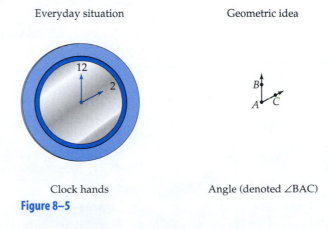

Clock hands   Angle (denoted $\angle BAC$)

Figure 8–5

LE 6   Reasoning

Try to write a definition of "angle" using the term "ray."

An **angle** is formed by two rays that have the same endpoint. The two rays are called **sides,** and the common endpoint is called the **vertex.** The angle $\angle BAC$ in Figure 8–5 has sides $\overrightarrow{AB}$ and $\overrightarrow{AC}$ and vertex A. When three letters are used to name an angle, the middle letter names the vertex.

LE 7   Reasoning

Give a simpler name for each of the following.

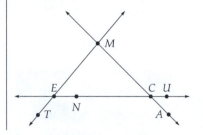

(a) $\overline{ME} \cup \overline{ET}$ (*Hint:* Trace over the two segments in the diagram.)

(b) $\overrightarrow{NU} \cap \overrightarrow{CE}$

(c) $\overleftrightarrow{ME} \cap \overline{CN}$

MEASURING ANGLES

North

30°

- - - - - - - - East

Storm sighted 30° East of North

Figure 8–6

To give the position of a storm in relation to a town (see Figure 8–6), a meteorologist might use an angle measure. Surveyors, navigators, and meteorologists all measure angles as a part of their work. Mathematicians use angle measures to classify angles. In Section 8.3, you will study a surprising pattern in the angle measures of polygons. The following background information will prepare you.

Measuring an angle is quite different from measuring a length. The measure of an angle tells the amount of rotation involved in moving from one ray to the other. Over 4000 years ago, the Babylonians chose a degree as the unit for angle measure. They made a complete rotation 360 degrees, so a degree is $\frac{1}{360}$ of a complete rotation.

You can measure angles in degrees using a protractor. Place the mark at the center of the protractor on the vertex of the angle (B in Figure 8–7) while lining up one side of the angle ($\overrightarrow{BA}$ in Figure 8–7) with the base of the protractor and the 0 degree mark. The other side of the angle ($\overrightarrow{BC}$) will pass through a number on the protractor representing $m\angle ABC$ in degrees.

As you can see, $\angle ABC$ measures 75° on the protractor pictured in Figure 8–7. This is written $m\angle ABC = 75°$.

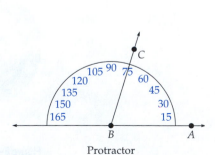

Figure 8–7   Protractor

Discussion

LE 8   Skill

A ship S is sighted from point U on the shoreline $\overleftrightarrow{UN}$.

(a) Estimate the measure of $\angle SUN$.
(b) Measure $\angle SUN$ with a protractor.

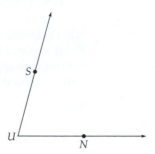

Classroom Connection

LE 9   Reasoning

A fourth grader is using a protractor that has two sets of numbers. She measures an angle and reads 60 and 120 off the protractor. How could she figure out which measurement is correct?

TYPES OF ANGLES

Mathematicians describe and categorize geometric shapes. Then they look for common properties within a category.

Throughout this chapter, geometric figures will be placed into categories. The first example of classifying shapes in this chapter is the categorization of angles according to their measure. Angle categories are suggested by our surroundings.

LE 10   Connection

What type of angle is most common in the room you are in now?

The most common angles in a classroom are usually **right angles** (90°) and **straight angles** (180°). Right angles appear in everything from books to boxes to corners of rooms (Figure 8–8).

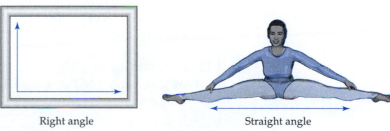

Right angle   Straight angle

Figure 8–8

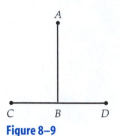

Figure 8–9

LE 11 Connection

Two right angles with a common side, ∠ABC and ∠ABD (Figure 8–9), form a straight angle (∠CBD).

(a) Give an example of an object that illustrates this property.
(b) Does having this property serve any purpose for the object in your example?

Right and straight angles are benchmarks for other, less common angle measures. Other angles are placed into three groups: **acute angles** (greater than 0° and less than 90°), **obtuse angles** (greater than 90° and less than 180°), and **reflex angles** (greater than 180° and less than 360°). Some examples are shown in Figure 8–10.

Acute angle   Obtuse angle   Reflex angle

Figure 8–10

LE 12 Connection

Name an object in your classroom that forms an acute angle.

Certain pairs of angles have special classifications.

LE 13 Reasoning

Use the following information to write definitions of "complementary" and "supplementary" pairs of angles.

Angle measures	42°, 48°	60°, 30°	78°, 20°	140°, 40°	82°, 98°
Complementary?	Yes	Yes	No	No	No
Supplementary?	No	No	No	Yes	Yes

Compare your definitions to the following. Two angles whose measures have a sum of 90° are **complementary angles.** Two angles whose measures have a sum of 180° are **supplementary angles.**

LE 14 Reasoning

(a) Are ∠ACE and ∠ACT complementary, supplementary, or neither?
(b) Using a definition to answer part (a) involves _____ reasoning.

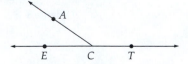

Congruent geometric figures are the same size and shape. How can we tell if two angles are congruent? We can measure them. Two angles with the same measure are **congruent angles.**

LE 15   Concept
(a) True or false? All acute angles are congruent.
(b) Give an example that supports your answer.

PARALLEL AND INTERSECTING LINES

Our surroundings suggest the common relationships between two distinct lines, as shown in Figure 8–11.

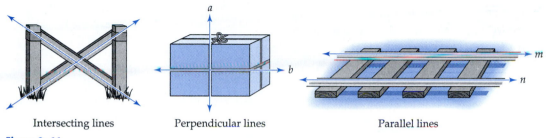

Intersecting lines   Perpendicular lines   Parallel lines

Figure 8–11

LE 16   Reasoning

Complete the following definitions.

(a) Two *intersecting lines* _____.
(b) Two lines are *perpendicular* if and only if _____.
(c) Two *parallel lines* lie in the same plane and _____.

Compare your definitions to the following. Two **intersecting lines** have exactly one point in common. Two lines that intersect at right angles are **perpendicular.** Two lines are **parallel** if and only if they lie in the same plane and do not intersect. In Figure 8–11, line *a* is perpendicular to line *b,* written $a \perp b$, and line *m* is parallel to line *n,* written $m \parallel n$.

Discussion

LE 17   Reasoning
(a) True or false? In a plane, two lines that are both parallel to a third line must be parallel to each other.
(b) In your classroom, identify a model of this situation that supports your answer.

LE 18   Summary
Tell what you learned about lines and angles in this section.

ANSWERS TO SELECTED LESSON EXERCISES

2. A geometric line is perfectly straight, extends forever in two directions, and has no width; it has no arrowheads.

3. (a) True
(b)

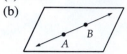

4. (a) 1
(b) 3
(c) 2

7. (a) $\overline{MT}$
(b) $\overline{NC}$ (what $\overline{NU}$ and $\overrightarrow{CE}$ have in common)
(c) {   }

8. (b) About 74°

9. By estimating

11. (a) Doors of a cabinet
(b) The property enables them to fit together and the bottom of the two doors to be level.

14. (a) Supplementary
(b) Deductive

15. (a) False
(b) A 30° angle and a 40° angle are acute, but they are not congruent.

17. (a) True

8.1   HOMEWORK EXERCISES

Basic Exercises

1. The word "geometry" can be split into "geo-" and "-metry." What do you think these two roots mean?

2. Describe a specific example of each of the following.
(a) Geometry in nature
(b) An everyday use of geometry
(c) Geometry in art
(d) Geometry in architecture

3. (a) Find three objects in your home that suggest common space figures.
(b) Find three objects in your home that suggest common plane figures.

4. (a) Find three objects in a grocery store that suggest common space figures.
(b) Find three objects in a grocery store with surfaces that suggest common plane figures.

5. Tell what geometric figure each of the following suggests.
(a)

Photo courtesy of the Department of the Army.

(b) A special piece of glass that disperses light into its color components
(c) The edge of a box

6. What shapes do you see in each of the structures pictured?
(a) Water tank

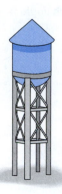

(b) Eskimo igloo (*Hint:* It's half of something.)

7. Euclid lived about _____ years ago.
(a) 50   (b) 200   (c) 500
(d) 2,000   (e) 10,000

8. Who is the first known woman mathematician?

9. How is a geometric point different from a dot?

10. How does the dictionary define the word "point"?

11. (a) True or false? If a line $\overleftrightarrow{AB}$ intersects a point in plane *c*, then $\overleftrightarrow{AB}$ lies in plane *c*.
(b) Make a drawing that supports your answer.

12. Suppose a point *P* lies in a plane. Are there any points in the plane that are exactly 100 miles from *P*? If so, how many?

13. What is the dimension (1, 2, or 3) of the geometric figure *suggested* by each of the following?
(a) A pillow   (b) A piece of very thin wire
(c) A sheet of paper

14. Give the number of dimensions in each figure.
(a) Line   (b) Cube   (c) Rectangle

15. How is a line segment different from a line?

16. Write a definition of "angle" using the term "ray."

17. (a) What is the vertex of $\angle PAT$?
(b) How is $\overrightarrow{AB}$ different from $\overrightarrow{BA}$?

18. Give three possible names for the line shown here.

19. Give a simpler name for each of the following.
(a) $\overrightarrow{AB} \cup \overrightarrow{AC}$
(b) $\overrightarrow{AB} \cup \overrightarrow{AD}$
(c) $\overrightarrow{BA} \cap \overrightarrow{AC}$

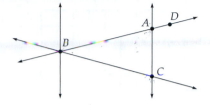

20. Give a simpler name for each of the following.
(a) $\overrightarrow{DA} \cap \overrightarrow{AC}$
(b) $\overrightarrow{AD} \cup \overrightarrow{AC}$
(c) $\overrightarrow{CB} \cup \overrightarrow{CD}$

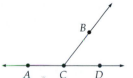

21. To tell other bees where flowers are, a bee measures the angle formed by the sun, the hive, and the flowers with the hive as the vertex. Estimate this angle in part (a) and the angle formed by the scissors in part (b). Then measure each angle with protractor.
(a)

(b)

22. Make an enlarged copy of the following drawing. Use a protractor to draw a ray with endpoint *C* in the given direction.
   (a) A storm is observed 30° east of south.
   (b) A ship is sighted 25° south of west.

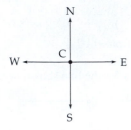

23.

 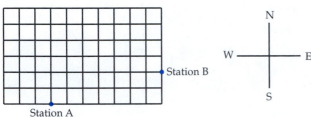

A ship reports a bearing of 50° north of east from station *A* and 30° west of north from station *B*. Locate the ship on the grid.

24. Pete and Sadie are on duty at observation towers *A* and *B* in a national forest. If there is a fire at *F,* how can Pete and Sadie measure angles and precisely locate the fire without leaving their towers? Assume that they can communicate by radio and that each has a map of the park and a protractor.

25. A fourth grader thinks that *m∠A* is larger than *m∠B.*

   (a) Why might the child think this?
   (b) How could you put the angles together to show that *m∠B > m∠A?*

26. You ask a third grader to draw a line, and he draws a line segment. What would you tell the child?

27. Name two capital letters that contain two or more acute angles and no obtuse or right angles.

28. Tell whether each angle is acute, right, or obtuse.
   (a)   (b)   (c)

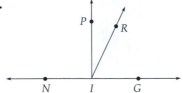

29.

Suppose *m∠PIN* = 90°.
   (a) Name two supplementary angles.
   (b) Name two complementary angles.
   (c) Name two more supplementary angles.

30. An angle measures 20°. What is the measure of
   (a) its supplement?   (b) its complement?
   (c) Do parts (a) and (b) involve induction or deduction?

31. True or false? Two complementary angles are congruent.

32. True or false? Two right angles are supplementary.

33. Which of the angles are congruent to ∠*A*?

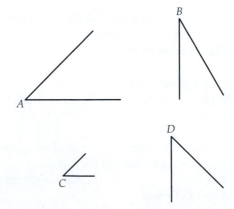

34. Which segments are congruent to $\overline{AB}$?

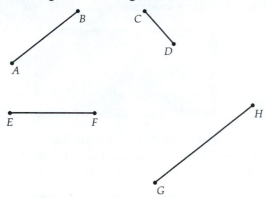

35. Which segments are congruent to $\overline{AB}$?

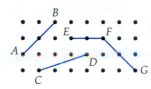

36. True or false? All right angles are congruent.

37. (a) True or false? In a plane, two lines that are perpendicular to a third line are parallel to each other.
(b) In your home, identify a model of this situation that supports your answer.
(c) How might the property in part (a) be useful to a carpenter?

38. True or false? In a plane, two lines that intersect a third line cannot be perpendicular to each other.

39. A fourth grader says that two parallel lines are an equal distance apart. Is this right? If not, what would you tell the child?

40. A fifth grader says that two parallel lines are tilted the same amount? Is this right? If not, what would you tell the child?

41. What does "perpendicular" mean in the sign?

> **PERPENDICULAR**
> **PARKING**
> **ONLY**

42. (a) Which of the following road intersection designs makes it easier to see cars coming from other directions?

(b) What term describes each pair of angles?

43. If possible, draw four lines in a plane that separate the plane into
(a) 10 sections. (b) 11 sections.

44. Into how many sections is a plane separated by the removal of each of the following?
(a) One line (b) Two parallel lines
(c) Two intersecting lines

45. Draw four points A, B, C, and D so that $\overleftrightarrow{AB}$, $\overleftrightarrow{AC}$, $\overleftrightarrow{AD}$, $\overleftrightarrow{BC}$, $\overleftrightarrow{BD}$, and $\overleftrightarrow{CD}$ are
(a) the same line.
(b) six different lines.
(c) four different lines.

46. If possible, draw two angles that intersect at exactly
(a) one point.
(b) two points.
(c) three points.

47.

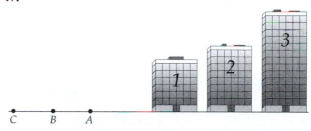

Which buildings can be seen from each of the following?
(a) A (b) B (c) C
(d) Shade the region of points from which only building 1 can be seen.

48.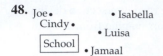

Joe •   • Isabella
Cindy •
             • Luisa
School • Jamaal

Assume that no one can see anyone else by looking
through the school building.
(a) Who can see Joe?
(b) Who can see Cindy?
(c) Who can see Luisa and Jamaal?

Extension Exercises

49. Name two morning times of day at which the minute
and hour hands form a
(a) 150° angle.   (b) 105° angle.

50. Give the angle formed at each of the following times
of day by the minute and hour hands of a clock.
(a) 5:00   (b) 3:30   (c) 2:06

51. Place eight dots in the following figure so that there
are exactly two dots on each circle and each line.

52. When a billiard ball bounces off the side of a billiard
table, it forms two congruent angles, as shown.

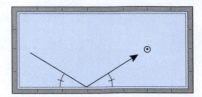

Use a protractor and trial and error to find how billiard
ball *A* should be aimed so that it hits the bottom and
right cushions (sides) and then strikes ball *B*.

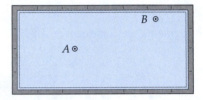

Labs and Projects

53. Read *The Dot and the Line* by Norton Juster, and describe how the author uses geometric shapes as symbols.

54. Measure the angles that different chair legs make with
the floor. Write a summary of your results.

8.2 POLYGONS

NCTM Standards

- recognize, name, build, draw, compare, and sort two- and three-dimensional
  shapes (K–2)
- classify two- and three-dimensional shapes according to their properties
  and develop definitions of classes and shapes such as triangles and pyramids
  (3–5)

After seeing how surfaces of objects suggest plane figures and studying basic terms such as
"line segment," "angle," "parallel," and "perpendicular," one is prepared for a more detailed
study of plane figures, beginning with polygons.

LE 1   Opener

What is a polygon?

The first part of the section will develop the definition of a polygon.

SIMPLE CLOSED CURVES

In elementary school, children study plane figures that are simple closed curves. See whether you can write a definition of a simple closed curve in the following exercise.

LE 2   Reasoning

Based on the diagrams in Figure 8–12, write a definition of a simple closed curve. (Note that "curves" can be composed of line segments.)

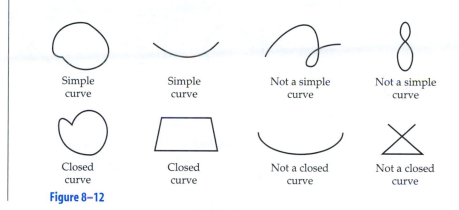

Figure 8–12

A **simple closed curve** does not cross itself and encloses a part of the plane. The **Jordan Curve Theorem** states that a simple closed curve divides the plane into three disjoint sets of points: the interior, the curve itself, and the exterior (Figure 8–13).

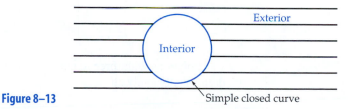

Figure 8–13

In elementary school, children study two main groups of simple closed plane figures: polygons and circles.

POLYGONS

Many surfaces in our environment approximate polygons (Figure 8–14). See if you can write a definition of a polygon in the following exercise.

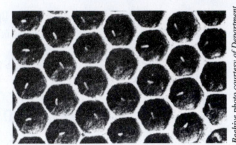

© Photo of school crossing sign by Tom Sonnabend.

Beehive photo courtesy of Department of Library Services, American Museum of Natural History.

Figure 8–14

LE 3   Reasoning
On the basis of the diagrams in Figure 8–15, write a definition of a polygon.

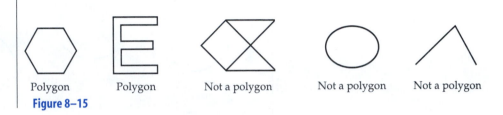

Figure 8–15

What is a polygon? After a while, geometry teachers tire of the well-known response "A dead parrot." A **polygon** is a simple closed plane curve formed by three or more line segments.

Classroom Connection

LE 4   Concept
A third grader draws the three figures below. She is not sure which ones are polygons. What would you tell the child?

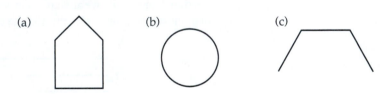

Polygons with three to ten sides are shown in Figure 8–16. Elementary-school children usually study all of them.

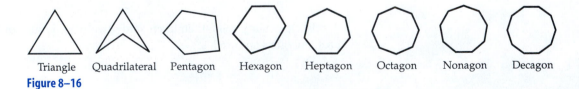

Figure 8–16

These polygons also have three to ten angles, respectively. Each vertex of an angle is also called a **vertex** (plural vertices) of the polygon.

A **geoboard** is a square board that has pegs in grid patterns. Usually there is a square grid pattern on one side, and sometimes there is a circular grid pattern on the other side. By placing rubber bands over the pegs, children can construct various shapes and examine their properties. In Chapters 8–10, you will find numerous exercises that make use of square grid patterns.

Give the name of each polygon shown on the following geoboard grid patterns.

(a) 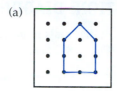   (b)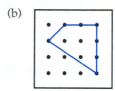

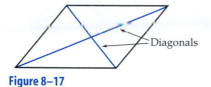

Diagonals

Figure 8–17

A **diagonal** is a line segment other than a side that joins two vertices of a polygon (Figure 8–17). In buildings and bridges, engineers sometimes use diagonals for reinforcement, as shown in Figure 8–18.

Diagonals are used to define convex polygons, the kind children study in elementary school. No portion of any diagonal of a **convex** polygon lies in its exterior (see Figure 8–19). (All triangles are convex polygons, since they do not have diagonals.)

© Photo courtesy of Library of Congress.

Figure 8–18

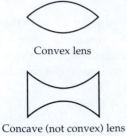

Convex lens

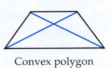

Convex polygon

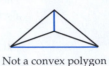

Convex polygon

Not a convex polygon

Not a convex polygon

Figure 8–19

Concave (not convex) lens

Figure 8–20

The term "convex" is also used for certain lenses (Figure 8–20).

AN INVESTIGATION: DIAGONALS OF POLYGONS

The numbers of diagonals in polygons follow a pattern. A triangle has no diagonals. A quadrilateral (Figure 8–21) has a total of two diagonals, one from each vertex.

 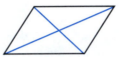 

Figure 8–21

These results are included in the following table.

Polygons

Number of vertices	3	4	5	6
Number of diagonals from each vertex	0	1		
Total number of diagonals	0	2		

Discussion

LE 6   Reasoning

(a) Devise a plan and complete the table just presented.
(b) On the basis of the pattern in your completed table, predict how many diagonals are in a heptagon (a seven-sided polygon).
(c) Draw a heptagon and its diagonals, and check your prediction.

What is the general formula for the number of diagonals in a polygon that has N sides? The following exercise will help you answer this intriguing question.

Discussion

LE 7   Reasoning

Now consider a polygon that has N sides.

Polygons

Number of vertices	3	4	5	6	7	N
Number of diagonals from each vertex	0	1	2	3	4	
Total number of diagonals	0	2	5	9	14	

(a) How many vertices does it have?
(b) How many diagonals can be drawn *from* a vertex?
(c) How is the total number of diagonals related to the number of vertices and the number of diagonals from each vertex?
(d) How many diagonals does the polygon have?

Were you able to complete LE 7? The result is as follows.

Diagonals of a Polygon

A polygon that has N sides has $\dfrac{N(N-3)}{2}$ diagonals.

LE 8   Reasoning

(a) Use the formula to find the number of diagonals that a decagon has.
(b) The process of assuming that the formula is true and applying it in part (a) involves _____ reasoning.

LE 9   Summary

Tell what you learned about polygons in this section.

ANSWERS TO SELECTED LESSON EXERCISES

4. Refer the child to a definition and check each figure.
(a) is a polygon; (b) is not bounded by line segments; (c) is not closed.

5. (a) Pentagon   (b) Quadrilateral

6. (a)

Number of vertices	3	4	5	6
Number of diagonals from each vertex	0	1	2	3
Number of diagonals	0	2	5	9

(b) 14

7. (a) N (b) $N - 3$
(c) Multiply the number of vertices times the number of diagonals from each vertex. But this counts each diagonal twice (at each end). So divide the result by 2.
(d) $\dfrac{N(N-3)}{2}$

8. (a) $\dfrac{10(7)}{2} = 35$ (b) deductive

8.2   HOMEWORK EXERCISES

Basic Exercises

1. Draw a curve that is simple but not closed.

2. Draw a curve that is closed but not simple.

3. Which of the following shapes are polygons? If a shape is not a polygon, explain why not.

(a)   (b)   (c)   (d)

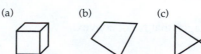
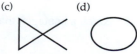

4. Which of the following shapes are polygons? If a shape is not a polygon, explain why not.

   (a)   (b)   (c)   (d)

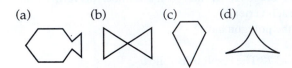

5. What polygon is suggested in each case?

   (a)   (b)

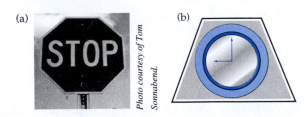

Photo courtesy of Tom Sonnabend.

6. Give the name of each polygon shown on the following geoboard grid patterns.

   (a)   (b)

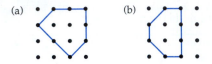

7. Sketch a hexagon that has exactly two acute angles.

8. Sketch a pentagon that has exactly three right angles.

9. Form the following shapes on a geoboard or on geoboard dot paper.
   (a) A pentagon with exactly two equal sides
   (b) A quadrilateral with exactly one pair of parallel sides

10. Form the following shapes on a geoboard or on geoboard dot paper.
    (a) A pentagon that has one pair of parallel sides
    (b) A quadrilateral that has no parallel sides but has two pairs of congruent, adjacent sides

11. Which of the following polygons are convex?

   (a)   (b)   (c)   (d)

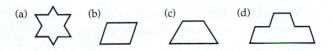

12. Sketch a heptagon that is not convex.

13. Draw all the diagonals of the hexagon.
    (a)

   (b) How many diagonals are there?

14. A polygon that has N sides has $\dfrac{N(N-3)}{2}$ diagonals.

   (a) How many diagonals does an octagon have?
   (b) How many diagonals does a nonagon have?
   (c) The process of using the formula to work out parts (a) and (b) involves _____ reasoning.

Extension Exercises

15. Assuming the two figures share no common sides, find the maximum number of intersection points for
    (a) a triangle and a square.
    (b) a triangle and a convex pentagon.
    (c) a square and a convex pentagon.
    (d) a regular polygon that has n sides and a regular polygon that has m sides, $n < m$.

16. Draw on the grid a polygon that meets all of the following conditions (Activity Card 4).
    (a) It is a hexagon.
    (b) Not all of its sides are congruent.
    (c) It has exactly two acute angles.
    (d) It has exactly three dots in its interior.

17. In 1884, Ezra Gilliland designed a phone system that allowed 15 people to speak to one another. How many connections were needed? (*Hint:* How many segments are needed to connect 15 points in pairs?)

18. At a party, all 15 friendly people want to shake hands with every other person there. How many handshakes would this require?

19. What kinds of polygons do you see in the shapes of traffic signs?

8.3 TRIANGLES, QUADRILATERALS, AND CIRCLES

NCTM Standards

- describe attributes and parts of two- and three-dimensional shapes (K–2)
- classify two- and three-dimensional shapes according to their properties and develop definitions of classes of shapes such as triangles and pyramids (3–5)
- precisely describe, classify, and understand relationships among types of two- and three-dimensional objects using their defining properties (6–8)

The most important two-dimensional figures are triangles, quadrilaterals, and circles. They are the most common shape in our environment. Mathematicians have classified triangles and quadrilaterals and analyzed their properties. The section begins with further study of triangles.

TRIANGLES

Triangular shapes appear occasionally in nature. More often, one sees them in support structures for buildings and furniture (Figure 8–22).

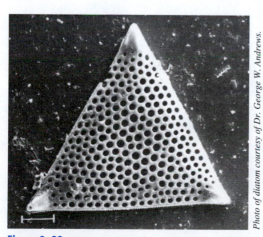

Photo of diatom courtesy of Dr. George W. Andrews.

Photo of bridge courtesy of Library of Congress.

Figure 8–22

LE 1   Connection

Why do builders use triangles rather than rectangles for support structures?

Discussion

LE 2   Opener

Mathematicians like to classify shapes.

(a) What are some ways to group the triangles below by their characteristics?

(b) Give the name for each group of triangles you formed in part (a) if you know it.

In LE 2, did you group triangles by properties of their sides or angles? One way to classify triangles is by how many congruent sides they have. **Scalene triangles** have no congruent sides; **isosceles triangles** have at least two congruent sides; and **equilateral triangles** have three congruent sides (Figure 8–23).

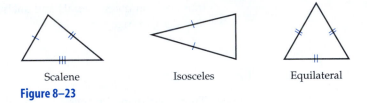

Scalene   Isosceles   Equilateral

Figure 8–23

A second way to classify triangles is by their largest angle. All triangles have at least two acute angles. A **right triangle** has one right angle; an **obtuse triangle** has one obtuse angle; and all the angles are acute in an **acute triangle.**

LE 3   Concept

Classify each of the following triangles by its sides and by its angles.

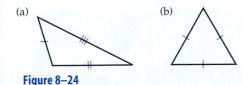

(a)   (b)

Figure 8–24

Children tend to prefer shapes that are equilateral, are symmetric, and have a horizontal base. Children first learning about triangles usually recognize an equilateral triangle with a horizontal base. However, they may not recognize other triangles that look much different, such as long skinny triangles or triangles that are rotated.

QUADRILATERALS

How should children learn about geometric shapes? While teaching high-school mathematics in Holland, Dina van Hiele-Geldof and Pierre van Hiele developed a theory about stages of learning in geometry. According to the van Hieles, school geometry should begin with recognition, followed by analysis and then informal deduction.

Students at the first level, **recognition,** recognize whole shapes by their appearance. They might think that a figure with three curved sides is a triangle. Students at this level do not analyze the components (such as sides and angles) and properties. At the second level, **description,** students are able to describe the component parts and properties of a shape, such as how many sides it has and whether it has some congruent sides or angles. Students use the parts of a figure to describe and define the figure. At the third level, **relationships,** students become aware of relationships between different shapes (for example, they see that all squares are parallelograms). At this level, students also deduce relationships among the properties of a figure; for example, they determine whether a quadrilateral that has four congruent sides must also have four congruent angles.

The two highest levels are usually developed in high-school and college geometry. At the **formal deduction** level, students employ deduction to develop new ideas in a logical system, such as Euclidean geometry. They study logical systems consisting of axioms, undefined terms, theorems, and definitions and learn to write proofs of new theorems. At the highest level, **rigor,** students work at an abstract level in a variety of axiomatic systems and compare them.

Children in elementary-school mathematics generally work at the first three van Hiele levels: recognition, description, and relationships. How would a teacher present a unit on quadrilaterals using the first three van Hiele levels?

At the first level, we *recognize* quadrilaterals in our environment (Figure 8–25). At the recognition level, the student compares shapes that are quadrilaterals with shapes that are not quadrilaterals. This same comparison can be made with different types of quadrilaterals. Figure 8–26 on the next page shows some examples.

Photos of door and street lamp by Tom Sonnabend.

Photo of halite crystals by A. Singer, Courtesy of Department of Library Services, American Museum of Natural History.

Figure 8–25

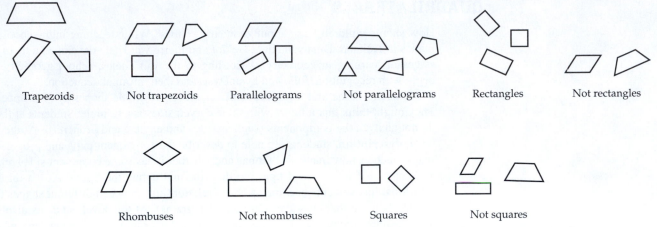

Trapezoids   Not trapezoids   Parallelograms   Not parallelograms   Rectangles   Not rectangles

Rhombuses   Not rhombuses   Squares   Not squares

Figure 8–26

At the next level, description, the student examines properties of quadrilaterals while still being unaware of any formal definition.

LE 4   Concept

Which of the five types of quadrilaterals in Figure 8–26 have both pairs of opposite sides congruent?

The next phase of the description level is to compare and contrast properties of different quadrilaterals.

LE 5   Reasoning

(a) Name two properties that all parallelograms and all rectangles share.
(b) Name a property that all rectangles have that some parallelograms do not have.

LE 6   Reasoning

(a) Name two properties of sides or angles that all squares and all rhombuses share.
(b) Name a property of sides or angles that all squares have that some rhombuses do not have.

At the third level, relationships, students use properties to define shapes and examine alternative definitions.

LE 7   Concept

On the basis of the preceding exercises and diagrams, write definitions of a trapezoid, a parallelogram, a rectangle, a rhombus, and a square.

Here are some of the most commonly used definitions in elementary- and middle-school textbooks.

Definitions: Common Quadrilaterals

- A **trapezoid** is a quadrilateral that has exactly one pair of parallel sides.

- A **parallelogram** is a quadrilateral in which each pair of opposite (nonintersecting) sides is parallel.

- A **rectangle** is a parallelogram that has four right angles.

- A **rhombus** is a parallelogram that has four congruent sides.

- A **square** is a rectangle that has four congruent sides.

Following are illustrations of these shapes. The markings on the angles of the rectangle and square denote right angles.

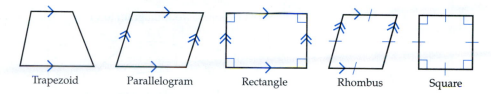

Trapezoid   Parallelogram   Rectangle   Rhombus   Square

Definitions such as these enable mathematicians to communicate using a shared terminology. A definition *may* give a minimum set of properties that defines a shape; it may also list additional properties that clarify what the shape is like.

For example, the preceding definition of a rectangle does not give a minimum set of properties. The next exercise addresses this issue.

LE 8   Reasoning

A rectangle can be defined as a parallelogram that has four right angles. What is the *minimum* number of right angles that makes a parallelogram a rectangle? (*Hint:* Try to draw parallelograms with one, two, or three right angles that are *not* rectangles.)

As LE 8 suggests, a rectangle *can* be defined as "a parallelogram that has one right angle" (the minimum), but elementary- and secondary-school texts usually state additional properties that clarify what a rectangle is.

Classroom Connection

LE 9   Reasoning

A fifth grader says a square is not a rectangle because a square has four congruent sides and rectangles don't have that. A second fifth grader says a square is a type of rectangle because it is a parallelogram and it has four right angles.

(a) Which child is right?

(b) How can you use the definitions to help the other child understand?

The five classes of quadrilaterals overlap. At the level of relationships, definitions are used to find relationships among classes of figures. See if you can answer the following questions by referring to the definitions of the five common quadrilaterals.

Discussion

LE 10 Reasoning

(a) According to the preceding definitions, would every square be a type of rectangle?
(b) Is every trapezoid also a parallelogram?
(c) Is every square also a rhombus?

Discussion

LE 11 Reasoning

Suppose $P = \{parallelograms\}$, $Rh = \{rhombuses\}$, $S = \{squares\}$, $Re = \{rectangles\}$, $T = \{trapezoids\}$, and $Q = \{quadrilaterals\}$.

(a) $Rh \cap Re = $ _____   (b) $T \cap P = $ _____

Discussion

LE 12 Reasoning

(a) Use the information from the preceding three exercises to organize the following into a Venn diagram: trapezoids, parallelograms, quadrilaterals, rhombuses, squares, rectangles.
(b) Compare your diagram to the following ones. Which one of them is correct?

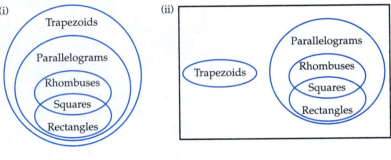

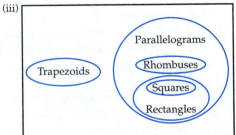

AN INVESTIGATION: GEOBOARD TRIANGLES AND QUADRILATERALS

LE 13 Concept

See if you can construct each of the following figures on a geoboard or on geoboard dot paper (Activity Card 4). Some of them may be impossible.

(a) A quadrilateral with exactly one right angle
(b) A triangle with exactly two right angles
(c) A square with no sides parallel to the edges of the geoboard
(d) A triangle that touches six pegs and has one peg inside

CIRCLES

Another shape has proven to be quite useful. Without it, it would be a lot harder to transport people and goods, play music, or can foods. You know this plane figure well: It is the circle (Figure 8–27).

Photo of clock courtesy of Library of Congress.

Photo of diatom courtesy of Dr. George Andrews.

Figure 8–27

LE 14   Concept

(a) Suppose that a radio program can be received anywhere within 20 km of the radio station. If the land is flat and has few obstacles, what shape is the border of the region in which the station can be received?

(b) On the basis of part (a), complete the following definition of a circle. A circle is the set of points in a plane that are _____.

The preceding exercise suggests the following definition of a circle.

Definitions: Circle, Radius, and Center

A **circle** is the set of points in a plane that are the same distance (the **radius**) from a given point (the **center**).

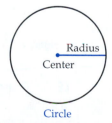

Radius

Center

Circle

Note that the word "radius" is used to refer to both a *segment* joining the center to the circle and the *length* of that segment.

Can you write a definition of a chord in the following exercise?

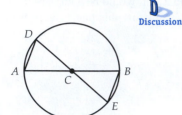

Figure 8–28

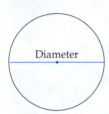

Diameter

Figure 8–29

Discussion

LE 15 Concept
In Figure 8–28, $\overline{AB}$, $\overline{AD}$, $\overline{BE}$, and $\overline{AE}$ are chords of a circle. $\overline{AC}$ and $\overline{BC}$ are not chords. Define "chord."

A **chord** is a line segment with endpoints on a circle. What about a "diameter"?

LE 16 Concept
(a) In Figure 8–28, $\overline{AB}$ and $\overline{DE}$ are diameters of a circle with center C, and $\overline{AD}$ and $\overline{BC}$ are not diameters. Define "diameter."
(b) How could you describe the size of a circular object such as a bicycle wheel?

A **diameter** is a line segment that (1) has two points of the circle as endpoints and (2) passes through the center of the circle (Figure 8–29). The diameter is a commonly used measurement of circular objects. (Two other common measures, area and circumference, will be examined in Chapter 10.)

Just as with the radius, the term "diameter" can be used to describe the segment or its length. The intersecting diameters of some wheels suggest central angles. So do the hands of a clock (Figure 8–30). A **central angle** is an angle whose vertex is the center of a circle.

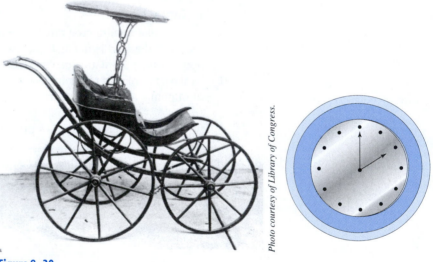

Figure 8–30

Photo courtesy of Library of Congress.

LE 17 Reasoning
What are some properties of a circle? They might involve radii, diameters, arcs, chords, or central angles.

Discussion

LE 18 Reasoning
Consider the following problem. "Name a time of day when the minute and hour hands of a clock form a 75° angle." Devise a plan and solve the problem.

DYNAMIC GEOMETRY SOFTWARE

Dynamic geometry software allows the user to construct and measure a variety of plane figures, including polygons and circles. A figure can then be manipulated so that it changes position or shape. When the position or shape changes, what properties of the figure stay the same? Answering this question can lead to useful generalizations.

Examples of dynamic geometry software include *The Geometer's Sketchpad* (Key Curriculum Press), *Cabri* (Thomson Publishing), and *The Geometric superSupposer* (Sunburst Communications). Geometry software is also available for graphing calculators. The exercises in this book are based on the capabilities of *The Geometer's Sketchpad*.

In the following exercise, follow along as your instructor uses dynamic geometry software to explore the properties of a parallelogram. (If you already know how to use the software, you can do this exercise yourself.)

LE 19   Reasoning

(a) Construct a line segment and a point not on the line.
(b) Select (highlight) the line segment and point. Construct a new line parallel to the line segment through the point.
(c) Construct another side of the parallelogram. Now select the new side and the vertex that is not on that side. Construct the other parallel side. Finally, hide the lines.
(d) Use the text tool to label the four vertices of the parallelogram.
(e) Now select the four sides in turn and measure them.
(f) Drag to create two more parallelograms, and record the lengths of the four sides.
(g) Make a generalization (conjecture) about the lengths of the sides of a parallelogram.
(h) Now select the four angles in turn and measure them.
(i) Drag to create two more parallelograms, and record the measures of the four angles.
(j) Make a generalization about the angle measures of a parallelogram.

LE 20   Summary

Tell what you learned about classifying triangles and quadrilaterals in this section.

ANSWERS TO SELECTED LESSON EXERCISES

1. Triangles are rigid. Rectangles are not.

3. (a) Scalene; obtuse
(b) Equilateral; acute

4. Parallelogram, rectangle, rhombus, square

5. (a) Both pairs of opposite sides are parallel and equal.
(b) Four right angles

6. (a) Both pairs of opposite sides are parallel and all four sides are equal.
(b) Four right angles

8. One

9. (a) The second child
(b) Show the definition. Ask if the square satisfies it.

10. (a) Yes (every square is a quadrilateral with four right angles)
(b) No   (c) Yes

11. (a) S (b) {   }

12. (b) (ii)

13. (b) is impossible

14. (a) Circle

16. (b) You could measure the diameter, circumference, or area.

18. Possible answer: 3:30 (*Hint:* Each minute has a measure of $\frac{1}{60}$ (360°) = 6°.)

8.3   HOMEWORK EXERCISES

Basic Exercises

1. Some painters utilize the triangular form. Leonardo DaVinci used it in the *Mona Lisa,* and Hans Memling used it in his *Madonna and Child.* What mood or effect might the triangular form create?

2. What is the advantage of a triangular rooftop over a flat roof?

3. Classify each triangle by its sides and by its angles.

(a)   (b)

(c)

4. True or false? No scalene triangle is isosceles.

5. Draw each type of triangle on the dot paper (Activity Card 4).

(a)   Isosceles right   (b)   Acute scalene

6. Draw each type of triangle on the dot paper.

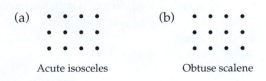

(a)   Acute isosceles   (b)   Obtuse scalene

7. Complete a square that has the segment shown as one of its sides.

8. Which of the following figures are rhombuses?

(a)    (b)

(c)

9. A second grader is having trouble recognizing a rectangle when it has been rotated less than 90 degrees. What would you tell the child?

10. What shape is the diamond in a deck of cards?

11. Consider different ways to divide a square into smaller squares. For example, the figure below shows how to divide a square into 7 smaller squares. Find out all the other possible numbers of smaller squares between 4 and 10 (inclusive) that you can obtain.

12. How many squares are in the following design?

13. How could you use some of the segments shown to form each of the following?
(a) A parallelogram
(b) A trapezoid
(c) A rhombus

Group 1   Group 2   Group 3

_____
                _____
_____   _____   _____
_____

14. Give all names that apply to each figure.

(a) 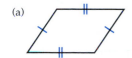   (b)

15.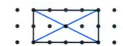

Refer to the drawing. Which of the following appear to be true?
(1) The diagonals have the same length.
(2) The diagonals are perpendicular.
(3) The diagonals bisect each other.

16.

(a) Refer to the drawing. Which of the following appear to be true?
(1) The diagonals have the same length.
(2) The diagonals are perpendicular.
(3) The diagonals bisect each other.
(b) What van Hiele level is represented in part(a)?

17. (a) Draw a parallelogram with two diagonals and a rhombus with two diagonals.
(b) What properties do the diagonals of both figures have in common?
(c) What property do the diagonals of the rhombus have that the diagonals of the parallelogram may not have?

18. (a) Draw a parallelogram and a rectangle.
(b) What properties do the angles of both figures have in common?
(c) What property do the angles of the rectangle have that the angles of the parallelogram may not have?

19. Tell whether each of the following shapes must, can, or cannot have at least one right angle.
(a) Rhombus   (b) Square
(c) Trapezoid   (d) Rectangle
(e) Parallelogram

20. In which of the following shapes are both pairs of opposite sides parallel?
(a) Rhombus   (b) Square
(c) Trapezoid   (d) Rectangle
(e) Parallelogram

21. Name properties that a square, parallelogram, and rhombus have in common.

22. Name properties that a trapezoid and a parallelogram have in common.

23. Cut a rectangular shape along its diagonal to form two congruent pieces. Sketch and name all the shapes you can form by putting the two pieces together in different ways without overlapping them.

24. A carpenter has made a rectangular door. How can he use a tape measure to be sure that the door has right angles?

25. Tell whether each definition has sufficient information. If it is not sufficient, tell what information is missing.
(a) A square is a polygon with four right angles and four congruent sides.
(b) A rectangle is a parallelogram that has opposite sides congruent and one right angle.
(c) A rhombus is a parallelogram with three congruent sides.

26. Tell whether each definition has sufficient information. If it is not sufficient, tell what information is missing.
    (a) A rhombus is a quadrilateral with both pairs of opposite sides parallel.
    (b) A square is a quadrilateral with four congruent sides.
    (c) A rhombus is a quadrilateral that has four congruent sides.

27. A rectangle can be defined as a quadrilateral that has right angles. What is the *minimum* number of right angles that makes a quadrilateral a rectangle? (*Hint:* Try to draw quadrilaterals with one, two, and three right angles that are *not* rectangles.)

28. A square is defined in the text as "a quadrilateral that has four congruent sides and four right angles," but this is not a *minimum* set of conditions. Make drawings to determine which of the following also define a square. If a definition is incorrect, tell why.
    (a) A rectangle that has two congruent sides
    (b) A rectangle that has three congruent sides
    (c) A rectangle that has two *adjacent* congruent sides
    (d) A rectangle that has congruent diagonals

29. Draw each of the following shapes.
    (a) A parallelogram   (b) A rectangle that
        that is not a rhombus   is also a square

30. Is every rhombus a parallelogram? If so, give an example. If not, draw a counterexample.

31. A square is also which of the following?
    (a) Quadrilateral   (b) Parallelogram
    (c) Rhombus   (d) Rectangle

32. Fill in the blank with "All," "Some," or "No."
    (a) _____ rectangles are squares.
    (b) _____ parallelograms are trapezoids.
    (c) _____ rhombuses are quadrilaterals.

33. A fourth grader does not think a square is a type of rectangle. Tell why it is.

34. A fifth grader does not think a rectangle is a type of parallelogram. Tell why it is.

35. "The diagonals of a rectangle are congruent." Why does this statement imply that the diagonals of a square must also be congruent?

36. You learn the theorem that the diagonals of a parallelogram bisect each other.
    (a) What other quadrilaterals must also have this property?
    (b) Are you using inductive or deductive reasoning in part (a)?

37. Which set picture best represents the relationship between rhombuses and rectangles? Label the correct circles appropriately.

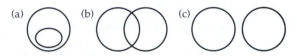

38. Which set picture best represents the relationship between parallelograms and trapezoids? Label the correct circles appropriately.

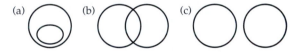

39. Suppose that P = {parallelograms}, S = {squares}, T = {trapezoids}, and Q = {quadrilaterals}.
    (a) $P \cap S$ = _____
    (b) Is $T \subseteq Q$?
    (c) $P \cup Q$ = _____

40. Draw a Venn diagram showing the relationship among parallelograms (P), rectangles (RE), and rhombuses (RH).

41. Fill in the blanks to describe the following circle with center N. Circle N is the set of _____ in a plane that are _____ from _____.

42. True or false? Every chord is also a diameter of the circle.

43. The diagram of a bicycle wheel has C as its center.

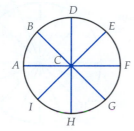

(a) Name a central angle and give its measure.
(b) What kind of triangle is $\triangle CDE$?
(c) Name a chord of the circle.

44. Draw a circle with center C, a central angle $\angle ACB$, and chords AB and AD.

45. What property of a circle makes it work as a wheel?

46. The diameter of a circle divides its interior into two congruent regions. How can someone use this property in dividing a circular pizza or pie in half?

Extension Exercises

47. The midpoint of a side divides it into two equal lengths.
(a) Draw a large quadrilateral on a sheet of paper.
(b) Use a ruler to locate the midpoint of each side.
(c) Connect the midpoints to form a quadrilateral.
(d) What shape appears to be the result?
(e) Repeat steps (a)–(d) for a second quadrilateral.
(f) Make a generalization based on your answers to part (d).
(g) Does part (f) involve induction or deduction?

48. (a) Draw a large rectangle on a sheet of paper.
(b) Use a ruler to locate the midpoint of each side.
(c) Connect the midpoints to form a quadrilateral.
(d) What shape appears to be the result?
(e) Repeat steps (a)–(d) for a second rectangle.
(f) Make a generalization based on your answers to part (d).

49. Some mathematics books define a trapezoid as "a quadrilateral that has at least one pair of parallel sides." If this definition is used, which of the following Venn diagrams is correct?

(a)
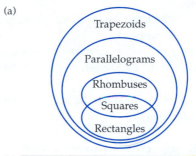

(b)
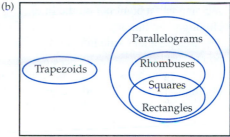

(c)
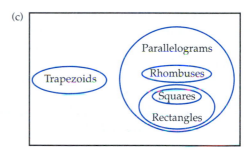

50. If possible, draw a triangle and a circle that intersect at exactly
(a) one point.   (b) two points.
(c) three points.   (d) four points.

51. If possible, sketch two parallelograms that intersect at exactly
(a) one point.   (b) two points.
(c) three points.   (d) four points.

52. If possible, sketch two parallelograms that intersect at exactly
(a) five points.   (b) six points.
(c) seven points.   (d) eight points.

Technology Exercises

53. (a) Use dynamic geometry software to construct a triangle from three line segments.
  (b) Use the text tool to label the three vertices of the triangle.
  (c) Highlight the three angles in turn, measure them, and find the sum.
  (d) Drag to create two more triangles. Record the measures of the three angles, and find the sum for each triangle.
  (e) Make a conjecture about the sum of the angle measures of a triangle.

54. (a) Use dynamic geometry software to draw a line segment and a point not on the line.
  (b) Highlight the line segment and point. Construct a new line parallel to the line segment through the point.
  (c) Construct another side of the parallelogram. Now highlight the new side and the vertex that is not on that side. Construct the other parallel side. Finally, hide the lines.
  (d) Use the text tool to label the four vertices of the parallelogram.
  (e) Now draw the two diagonals of the parallelogram, and use the text tool to label the point where they intersect.
  (f) Highlight the two parts of each diagonal in turn and measure them.
  (g) Drag to create two more parallelograms, and record the measures of the two parts of each diagonal.
  (h) Make a conjecture about the diagonals.

Enrichment Topics

55.

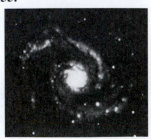

Photo courtesy of NASA.

Photo by A. Singer. Courtesy of Department of Library Services, American Museum of Natural History.

The beautiful shape of the equiangular spiral appears in the stars and in the shell of the chambered nautilus. You can graph one using a circular grid.

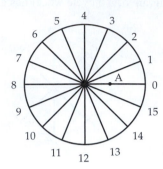

(a) Draw a segment from radius 1 that is perpendicular to radius 0 at point A. Label the intersection point with radius 1 as B.
(b) Draw a segment from radius 2 that is perpendicular to radius 1 at B. Label the intersection point with radius 2 as C.
(c) Continue this pattern for all 15 radii.
(d) Draw a smooth curve through points A, B, C, and so on that resembles an equiangular spiral.

56. A series of straight line segments can be used to define a curved figure. Strange, but true! Trace the following circle on a piece of paper.

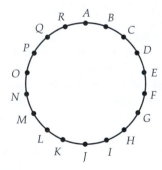

(a) Draw the segments $\overline{AI}$, $\overline{BJ}$, $\overline{CK}$, $\overline{DL}$, . . . , $\overline{RH}$.
(b) Using a different color, draw $\overline{AH}$, $\overline{BI}$, $\overline{CJ}$, $\overline{DK}$, . . . , $\overline{RG}$.
(c) Using another color, draw the next set of segments that continues the pattern.

8.4   ANGLE MEASURES OF POLYGONS

NCTM Standards

* make and test conjectures about geometric properties and relationships and develop logical arguments to justify conclusions (3–5)

Although most floor tiles are square, many other designs can be used. Figure 8–31 shows some examples. Some quilts also utilize interlocking, repeating patterns. What shapes can be used for floor tiles or quilt patterns?

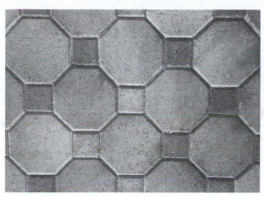

Photos by Tom Sonnabend.

Figure 8–31

We can answer this question by studying the angle measures of polygons, beginning with the simplest polygon: the triangle.

THE ANGLES OF A TRIANGLE

LE 1   Opener

What do you know about the sum of the angle measures in a
(a) triangle?   (b) quadrilateral?

The sum, in degrees, of the angle measures of any triangle is always the same. Do you know what the sum is? Verify it in the following exercise.

Discussion

LE 2   Reasoning

(a) If you have scissors and paper, do the following. If not, skip to part (c). Stack three small sheets of paper, and draw a triangle with sides and angles of your choice on the top sheet. Cut out the triangle, creating three copies. Number the angles in each triangle as shown in Figure 8–32.

Figure 8–32

(**b**) Cut out the triangles.

(**c**) Now place the three triangles together in a tiling pattern, as shown in Figure 8–33.

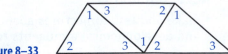

Figure 8–33

(**d**) What does the sum of the three angle measures (∠1, ∠2, and ∠3) appear to be?

LE 2(d) suggests the following property.

Angle Measures of a Triangle

The sum of the three angle measures of a triangle is 180°.

This property is helpful in finding angle measures of triangles.

LE 3   Reasoning

An equilateral triangle has three congruent sides and three congruent angles. What is the measure of each angle?

LE 4   Reasoning

Why is it impossible for a triangle to have two obtuse angles?

THE ANGLES OF A POLYGON

It is possible to find the sum of the angle measures of any polygon without measuring its angles! One can use the sum of the angle measures of a triangle to deduce the sum of the angle measures in any convex polygon. First, consider quadrilaterals.

LE 5   Concept

What is the sum of the angle measures of any square or rectangle?

The angle measures of all convex quadrilaterals have the same sum! This fact can be proved by drawing a diagonal of any convex quadrilateral to form two triangles, as shown in Figure 8–34.

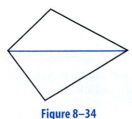

Figure 8–34

LE 6   Reasoning

(**a**) What is the sum of the three angle measures of each triangle in the quadrilateral shown in Figure 8–34?

(**b**) Mark each angle of the triangle with an arc.

(**c**) What is the sum of the four angle measures of the quadrilateral?

LE 6 verifies the following property.

> **Angle Measures of a Convex Quadrilateral**
>
> The sum of the four angle measures of any convex quadrilateral is 360°.

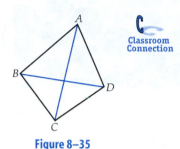

Figure 8–35

Classroom Connection

LE 7   Reasoning

A seventh grader says that the sum of the angle measures of a quadrilateral is 720°, not 360°. She draws a quadrilateral *ABCD* with two diagonals (Figure 8–35). The interior angle measures of each triangle add up to 180°, so the interior angle measures of *ABCD* add up to 4 · 180° = 720°.

(a) What is wrong with the child's reasoning?
(b) How could you compute the sum of the interior angle measures of the quadrilateral using the four triangles?

Discussion

LE 8   Reasoning

Draw a convex pentagon. Now draw all the diagonals from one vertex. Explain how to use the angle sum property of a triangle and the drawing to find the sum of the interior angle measures of the pentagon.

Discussion

LE 9   Reasoning

(a) Use inductive reasoning or more drawings to complete the following table.

Polygons (convex)

Number of sides	3	4	5	6	7	N
Number of triangles formed	1	2				
Sum of interior angle measures	180°	360°				

(b) Write a generalization of your results.

As the pattern in LE 9 suggests, you can divide any convex polygon that has *N* sides into *N* − 2 triangles by drawing all the possible diagonals from any single vertex. This result leads to the following conclusion.

> **Angle Measures of a Convex Polygon**
>
> The sum of the angle measures of any *N*-sided convex polygon is $(N-2) \cdot 180°$.

This property of angle measures also applies to polygons that are *not* convex; however, children rarely study the angle measures of concave polygons in elementary or middle school.

LE 10   Skill

What is the sum of the angle measures of a convex octagon?

Many shapes in the world suggest regular polygons (Figure 8–36).

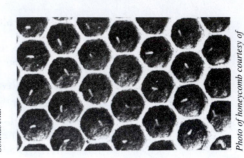

Stop sign photo courtesy of Tom Sonnabend.

Photo of honeycomb courtesy of Department of Library Services, American Museum of Natural History.

Figure 8–36

A **regular polygon** has sides that are all congruent and angles that are all congruent.

LE 11   Skill

You want to construct a STOP sign in the shape of a regular octagon. How many degrees are there in each interior angle of a stop sign?

Logo is geometry software that allows you to use angle measures to draw plane figures. See the textbook Web site for further information.

TESSELLATIONS

To return to the opening question, what shapes can be used as tiles?

Figure 8–37

Figure 8–38

All of the patterns in Figure 8–37 are called tessellations. A plane **tessellation** is a complete covering of a plane by shapes in a repeating pattern, without gaps or overlapping. Tessellations are also used to make art (Section 9.2).

Archimedes (287–212 B.C.) figured out which regular polygons tessellate the plane. What were his results? First, consider squares. If you've seen a few tile floors, you know that squares can be used. But why?

Pick a vertex of a square (Figure 8–38). Can we cover the plane region around the chosen point with additional squares, with no gaps or overlaps? Yes! Squares tessellate, as shown in Figure 8–39. Since the angle measures of the four squares around the point add up to 360°, the squares fit together with no gaps.

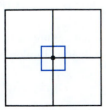

Figure 8–39

LE 12   Skill

In Figure 8–39 the angle measures around the marked point add up to _____ degrees.

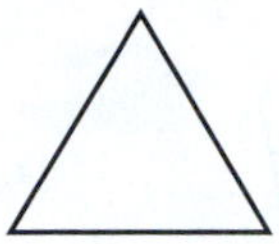

Figure 8–40

Discussion

LE 13   Reasoning

Do equilateral triangles tessellate?

(a) Make a drawing or cut out six copies of the equilateral triangle in Figure 8–40 (or use pattern blocks or Activity Card 5 or 6), and see if the triangles tessellate.

(b) What is the measure of each angle of an equilateral triangle? How can you use this measurement to explain your result in part (a)?

Discussion

LE 14   Reasoning

Do regular pentagons tessellate?

(a) Make a drawing or cut out four copies of the regular pentagon (Activity Card 5) in Figure 8–41, and see if the pentagons tessellate.

(b) What is the measure of each angle of a regular pentagon? How can you use this measurement to explain your result in part (a)?

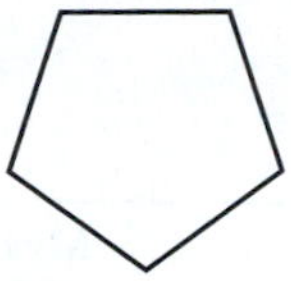

Figure 8–41

Discussion

LE 15   Reasoning

(a) Complete the table.

Regular Polygons	Total Number of Degrees	Measure of Each Interior Angle
Triangle		
Quadrilateral (square)	360°	90°
Pentagon		
Hexagon		
Heptagon		
Octagon		
Decagon		
Dodecagon (12 sides)		

(b) Without drawing the shapes or making cutouts, tell which figures in the table tessellate the plane.

LE 16   Summary

Tell what you learned about the angles of a polygon in this section.

ANSWERS TO SELECTED LESSON EXERCISES

3. 60°

4. Two obtuse angle measures have a sum greater than 180°.

5. 360°

6. (a) 180°   (c) 360°

7. (a) Together, the angles of the four triangles do not form the four angles of the quadrilateral.
(b) $(4 \cdot 180°) - 360° = 360°$

8.

The diagonals divide the pentagon into 3 triangles. The sum of the angle measures of the 3 triangles is $3 \times 180° = 540°$. Therefore, the sum of the angle measures of the pentagon is also 540°.

9. (a)

Polygons (convex)

Number of sides	3	4	5	6	7	N
Number of triangles formed	1	2	3	4	5	$N - 2$
Sum of interior angle measures (in degrees)	180°	360°	540°	720°	900°	$(N - 2)180°$

10. $(8 - 2) \cdot 180° = 1080°$

11. $1080°/8 = 135°$

12. 360

13. Yes

14. (a) No

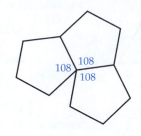

15. (a)

Regular Polygons	Total Number of Degrees	Measure of Each Interior Angle
Triangle	180°	60°
Quadrilateral	360°	90°
Pentagon	540°	108°
Hexagon	720°	120°
Heptagon	900°	$128\frac{4}{7}°$
Octagon	1080°	135°
Decagon	1440°	144°
Dodecagon	1800°	150°

(b) Regular triangles, quadrilaterals, and hexagons tessellate the plane.

8.4   HOMEWORK EXERCISES

Basic Exercises

1. (a) Draw a large triangle on a sheet of paper
   (b) Measure each interior angle.
   (c) Do the three angle measures add up to 180°? If not, why not?

2. How does the series of drawings below show that the angle sum of the triangle is 180°?

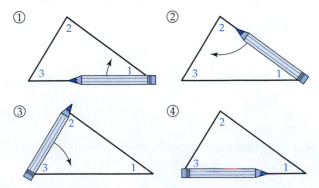

3. Explain why a triangle cannot have two right angles.

4. Explain why a triangle cannot have an obtuse angle and a right angle.

5. $\triangle KAR$ is regular and $m\angle PKR = 108°$. Fill in the missing angle measures.

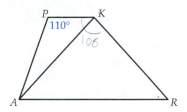

6. Is a triangle acute, right, or obtuse if two of its angle measures are
(a) 30° and 50°? (b) 25° and 90°?
(c) 70° and 40°?

7. A quadrilateral has two right angles. What can you deduce about the measures of the other two angles?

8. (a) Draw a quadrilateral that is not a rectangle.
(b) Use the method of Exercise 2 to show that the sum of the angles of the quadrilateral is 360°.

9. Use the angle sum property of a triangle and a drawing to explain why the interior angle measures of a convex hexagon add up to 720°.

10. Use the angle sum property of a triangle and a drawing to explain why the interior angle measures of a convex octagon add up to 1080°.

11. What is wrong with the following explanation?

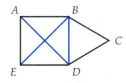

Pentagon *ABCDE* can be divided into five triangles, as shown. Therefore, the sum of the interior angles measures of the pentagon is 5 · 180° = 900°.

12. Draw a quadrilateral that is not convex. Show whether or not the sum of its interior angle measures is 360°.

13. What is the sum of the interior angle measures of a 40-sided convex polygon?

14. A Canadian nickel has the shape of a regular dodecagon (12 sides). How many degrees are in each angle?

15. Beehive cells approximate regular hexagons.
(a) When bees make the cells, what size interior angles do they make?
(b) Use a protractor to draw a regular hexagon. Then divide it into three congruent rhombuses.

16. (a) Is a rhombus a regular polygon? Why or why not?
(b) Is a rectangle a regular polygon? Why or why not?

17. In this lesson, you saw that regular triangles, squares, and regular hexagons tessellate the plane. There are also eight possible semiregular tessellations that use two or more different regular polygons to tessellate the plane. The following figure shows a semiregular tessellation that uses two regular triangles and two regular hexagons.

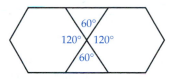

(a) Draw a repeating tile pattern that uses equilateral triangles, squares, and regular hexagons. (*Hint:* Use the angle measures or Activity Card 5.)
(b) Using the completed table from LE 15, try to sketch the other six possible semiregular tessellations.

18. (a) Determine what types of triangles tessellate and what types do not. You may cut out triangles as part of your investigation.
(b) How can this diagram be used to find the sum of the three angles of a triangle?

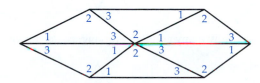

19. Draw or cut out four congruent copies of each type of figure, and determine if they tessellate. (Use pattern blocks if you have them.)
(a) A parallelogram (b) A trapezoid

20. Make copies of the following figure on the square grid (Activity Card 4) to show that it tessellates.

Extension Exercises

21. (a) How many rectangles are in the following diagram?

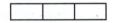

(b) How many rectangles are in the following diagram?

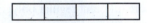

(c) How many rectangles are in the following diagram?

(d) Describe a general rule relating the total number of rectangles to the number of sections in a diagram of this type.

(e) Does part (d) involve induction or deduction?

(f) Use your rule from part (d) to determine how many rectangles are in the following diagram.

22. (a) How many equilateral triangles are shown?

(b) How many equilateral triangles are shown?

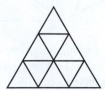

(Continued in the next column)

(c) How many equilateral triangles are shown?

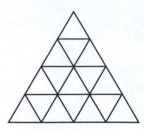

23. A certain regular polygon has *n* sides.
(a) What is the measure of each interior angle in terms of *n*?
(b) How do you know that the expression in part (a) is less than 180°?

24. Each angle of a certain regular polygon measures 174°. How many sides does it have?

25. Consider the following problem. "In a basketball tournament, each of eight teams plays every other team once. How many matches are there?" Devise a plan and solve the problem. (*Hint:* Draw a regular octagon.)

26. What is the measure of ∠*A* in the pentagram? (Assume that *FGHIJ* is regular.)

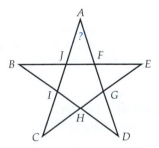

Puzzle Time

27. (a) Trace the square and cut it along the lines into seven pieces (Activity Card 7).

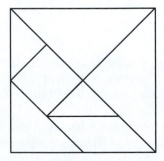

You have just created a Chinese tangram puzzle set. This type of geometric puzzle is at least 4000 years old! How does it work? You can rearrange the seven pieces into a variety of shapes. For example, after taking it apart, you can try to reconstruct a square with all seven pieces.

(b) Try an introductory problem. Cover the following picture with two tangram pieces.

(c) What fraction of the original square is covered by the parallelogram piece?

(d) Cover the following figure with all seven pieces.

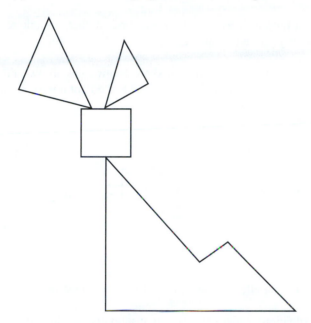

(e) Use all seven pieces to cover a whale.

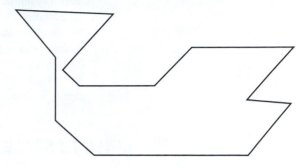

(f) Use all seven pieces to make a triangle.

28. If possible, form each of the following shapes with exactly four tangram pieces.
    (a) Triangle
    (b) Square
    (c) Parallelogram
    (d) Trapezoid

8.5 THREE-DIMENSIONAL GEOMETRY

NCTM Standards

- describe attributes and parts of two- and three-dimensional shapes (K–2)
- classify two- and three-dimensional shapes according to their properties and develop definitions of classes of shapes such as triangles and pyramids (3–5)
- precisely describe, classify, and understand relationships among types of two- and three-dimensional objects using their defining properties (6–8)

Space figures are the first shapes children perceive in their environment. But children cannot systematically study and categorize three-dimensional shapes in school until they have studied one- and two-dimensional shapes.

In elementary school, children learn about the most basic classes of space figures: prisms, pyramids, cylinders, cones, and spheres. This lesson concerns the definitions of these space figures and the properties of their faces, vertices, and edges.

As a basis for studying space figures, first consider relationships among lines and planes in space. These relationships are helpful in analyzing and defining space figures.

LINES IN SPACE

What are possible relationships between two lines in space? As in two dimensions, two distinct lines can be either parallel or intersecting. But there is another possibility in three dimensions!

LE 1   Opener

Discussion

Using two pencils to represent lines in space, see if you can find another possible relationship in addition to their being parallel or intersecting.

In space, two lines can be parallel, intersecting, or skew. **Skew** lines are two lines that cannot be contained in a plane. Skew lines are not parallel and do not intersect. Figure 8–42 shows two examples of skew lines.

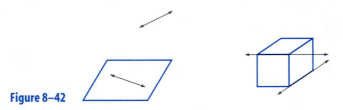

Figure 8–42

LE 2   Reasoning

Discussion

(a) True or false? A line that is parallel to one of two skew lines must intersect the other skew line.

(b) Identify a model of this situation in your classroom that supports your answer.

A LINE AND A PLANE IN SPACE

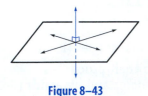

Figure 8–43

Some solids are defined in terms of a line that is perpendicular to a plane. A **line and a plane are perpendicular** if and only if they intersect at one point and the line is perpendicular to every line in the plane that passes through that point (Figure 8–43).

LE 3   Connection

In everyday life, legs of a table are approximately perpendicular to the plane of the floor. Give another example from everyday life of something that approximates a line that is perpendicular to a plane.

LE 4   Concept

(a) True or false? If two distinct lines are both perpendicular to the same plane, then the two lines are parallel to each other.
(b) In your classroom, identify a model of this situation that supports your answer.

PLANES IN SPACE

What are the possible relationships between two planes in three dimensions? One possibility is that the two planes are parallel. Two planes are **parallel** if and only if they do not intersect.

LE 5   Opener

Using two pieces of paper to represent planes, find the possible relationships between two planes.

Did you find the following possibilities in LE 5?

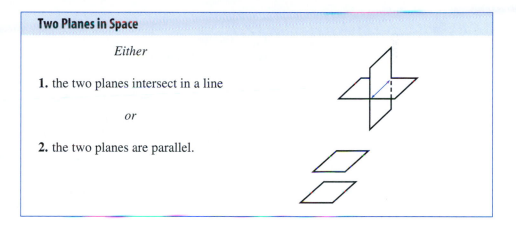

Two Planes in Space

Either

1. the two planes intersect in a line

or

2. the two planes are parallel.

LE 6   Reasoning

(a) True or false? Two planes that are parallel to a third plane must be parallel to each other.
(b) In your classroom, identify a model that supports your answer.

Parallel planes are used in defining a prism, the most common type of polyhedron (plural, polyhedra or polyhedrons) in school geometry.

POLYHEDRA

Many everyday objects, such as boxes and crystals (Figure 8–44 on the next page), resemble polyhedra. Do you recognize a polyhedron when you see one?

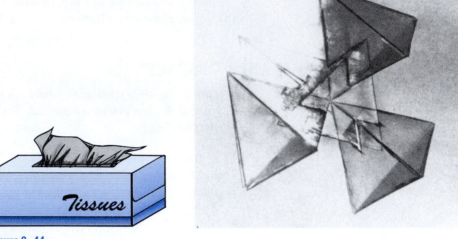

Photo (neg. no. 116114) by Julius Kirshner. Courtesy of Library Services, American Museum of Natural History.

Figure 8–44

LE 7 Opener

Which of the following are polyhedra?

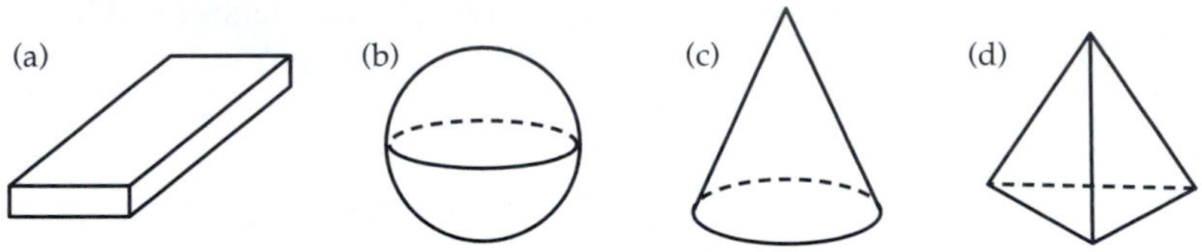

(a) 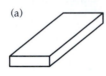 (b)   (c)   (d)

The shapes in Figure 8–45 are polyhedra.

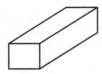

Figure 8–45 Rectangular prism Triangular pyramid

The shapes in Figure 8–46 are *not* polyhedra.

Figure 8–46 Sphere Cone

LE 8   Concept

Which of the shapes in the preceding lesson exercise have the following properties?

(a) All faces are triangles?
(b) Two faces are parallel and congruent.

To define a polyhedron, we will use the terms "simple closed surface" and "polygonal region." A simple closed surface in space is analogous to a simple closed curve in a plane. A **simple closed surface** separates space into three disjoint sets: points inside the surface, points on the surface, and points outside the surface. A **polygonal region** is a polygon together with its interior (Figure 8–47).

Figure 8–47   Polygonal region   Polygon

LE 9   Concept

Use the preceding information to write a definition of a polyhedron.

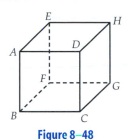

Figure 8–48

A **polyhedron** is a simple, closed space figure bounded by polygonal regions. All of the surfaces of a polyhedron are flat, not curved. The polygonal surfaces of polyhedra are called **faces.** The sides of each face (polygonal region) are called **edges,** and the vertices of the polygons are also **vertices** (corners) of the polyhedron.

For example, in the cube shown in Figure 8–48, square *ADHE* and its interior make up a *face, AD* is an *edge,* and *A* is a *vertex.*

LE 10   Concept

(a) How many faces does a cube have?
(b) How many vertices (corners) does a cube have?
(c) How many edges does a cube have? Count them in groups, beginning on the top of the cube.

A cube and its interior comprise a solid. A **solid** is the union of a simple closed surface and its interior.

Classroom Connection

LE 11   Communication

You ask a second-grade class how many edges a cube has. A child responds, "Each face has 4 edges. There are 6 faces. The total would be 4 + 4 + 4 + 4 + 4 + 4 = 24 edges." How would you explain the correct solution to the child?

PRISMS

A cube and a rectangular solid (Figure 8–49) are both a special type of polyhedron called a prism. A triangular pyramid, a sphere, and a cone, on the other hand, are not prisms.

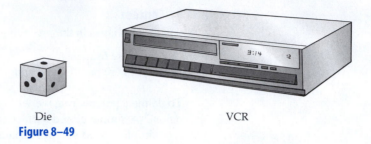

Die   VCR
Figure 8–49

LE 12   Reasoning
Describe a property prisms have that pyramids, spheres, and cones do not have.

A **prism** has two congruent polygonal regions that are opposite faces (usually the top and bottom), and the corresponding vertices are connected by parallel line segments. The opposite faces, called **bases,** lie in parallel planes. The faces that are not bases are called **lateral faces.**
   A prism is named by the kind of base it has (Figure 8–50).

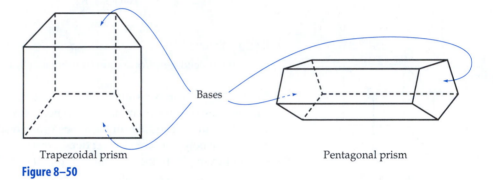

Trapezoidal prism   Pentagonal prism
Figure 8–50

In more formal terms, a **prism** is a polyhedron formed by two congruent polygonal bases in parallel planes connected by three or more parallelogram-shaped regions. A prism with lateral faces that are rectangular regions is a **right prism** (Figure 8–51).

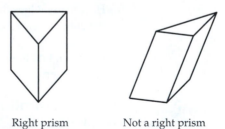

Figure 8–51   Right prism   Not a right prism

Note that any kind of prism can be split into congruent layers, as shown in Figure 8–52.

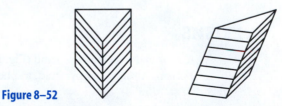

Figure 8–52

LE 13   Concept

(a) How many faces does a triangular prism have?

(b) How many vertices does it have?

(c) How many edges does it have?

Your results from LE 10 and LE 13 should have been as follows.

Figure	Faces	Vertices	Edges
Cube	6	8	12
Triangular prism	5	6	9

Figure 8–53 shows a chart that second-graders complete in *Harcourt Math.*

Classroom Connection

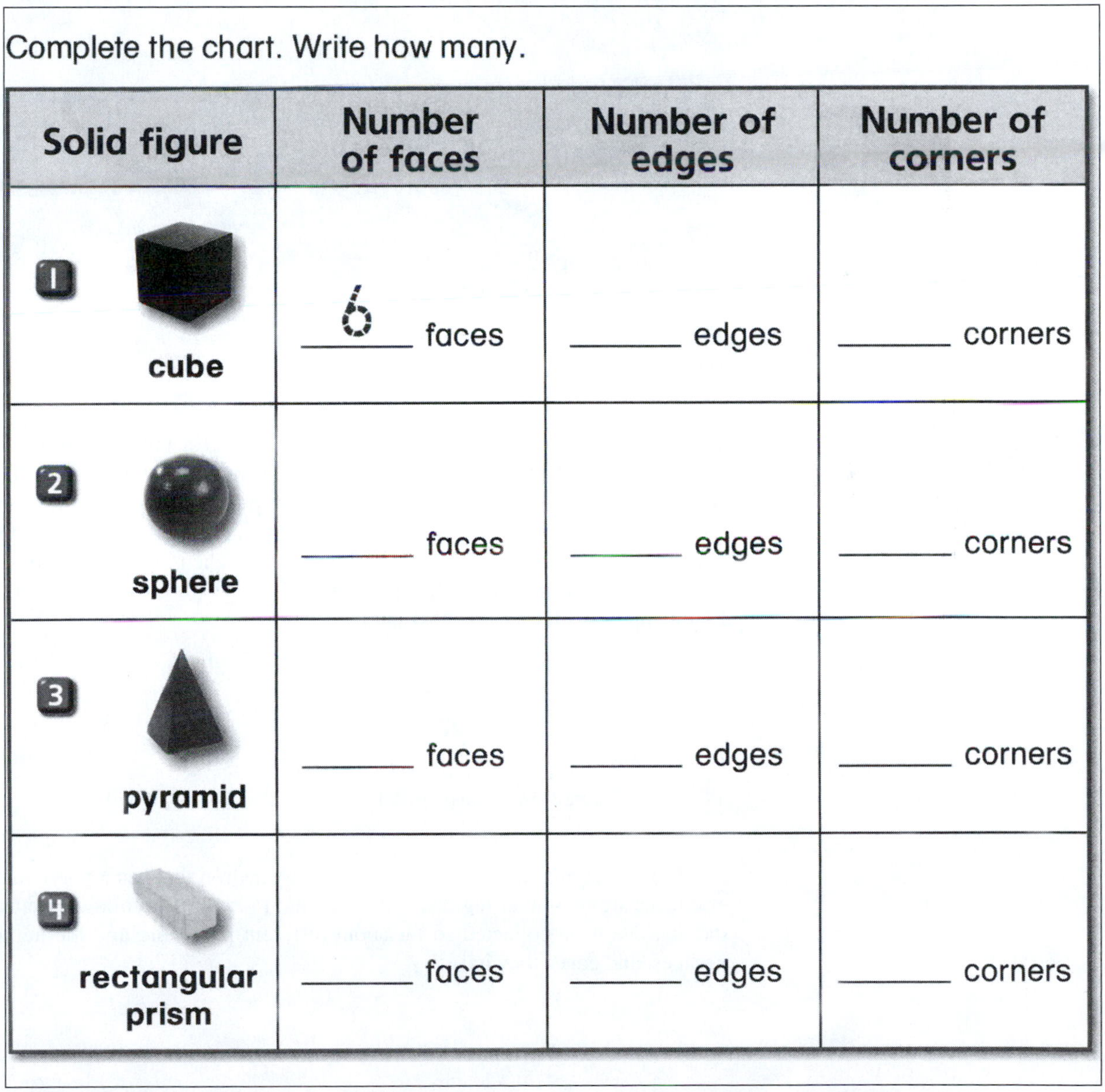

Complete the chart. Write how many.

Solid figure	Number of faces	Number of edges	Number of corners
1 cube	6 faces	_____ edges	_____ corners
2 sphere	_____ faces	_____ edges	_____ corners
3 pyramid	_____ faces	_____ edges	_____ corners
4 rectangular prism	_____ faces	_____ edges	_____ corners

Figure 8–53

From Harcourt Math, Grade 2 (Orlando, FL: Harcourt School Publishers, 2002), p. 261.

Discussion

LE 14   Reasoning

(a) Add two more rows to the preceding table: one for a rectangular prism and one for a pentagonal prism. Then fill in the numbers of faces, vertices, and edges each has.

(b) What pattern do you see in each row of the table?

PYRAMIDS

Figure 8–54

Photo: Lowenfish. Courtesy of Department of Library Services, American Museum of Natural History.

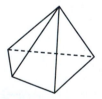

Figure 8–55

The ancient Egyptians constructed some of the largest buildings in the world (Figure 8–54). Some contain over 2 million stones, each weighing at least a ton! These buildings suggest another important group of polyhedra: pyramids. A pyramid can be formed by taking any polygon and a point above the plane of the polygon and connecting all the vertices of the polygon to that point (Figure 8–55). The polygon is the base.

A pyramid is named by the kind of base it has (Figure 8–56). The Great Pyramid of Egypt approximates a square pyramid.

Figure 8–56   Square pyramid   Triangular pyramid

In more formal terms, a **pyramid** is a polyhedron that has a polygonal base. Its lateral faces are triangular regions with a common vertex. The following table summarizes the data we have collected so far about different polyhedra and the numbers of faces, vertices, and edges they have.

Figure	Faces	Vertices	Edges
Cube	6	8	12
Triangular prism	5	6	9
Rectangular prism	6	8	12
Pentagonal prism	7	10	15
Triangular pyramid			
Square pyramid			

Discussion

LE 15   Reasoning

(a) Fill in the last two rows of the preceding table.

(b) Use inductive reasoning to hypothesize how the number of edges of each polyhedron is related to the numbers of faces and vertices it has.

The pattern in LE 14 and LE 15 holds for all polyhedra. Euler's formula describes this pattern algebraically.

> **Euler's Formula**
>
> For all polyhedra, $F + V - E = 2$, in which F is the number of faces, V is the number of vertices, and E is the number of edges.

Leonhard Euler (pronounced "oiler") (1707–1783) was the most prolific mathematician who ever lived (Figure 8–57). Even during the last 17 years of his life, when he was blind, Euler developed many new mathematical ideas.

Figure 8–57 *Leonard Euler*

Photo Courtesy of Library of Congress.

REGULAR POLYHEDRA AND EULER'S FORMULA

Although there is an infinite number of regular polygons, there are only five regular polyhedra! A **regular polyhedron** is a polyhedron whose faces are congruent regular polygonal regions and in which the number of edges that meet at each vertex is the same. The ancient Greeks proved that the only regular polyhedra are the five shown in Figure 8–58.

Tetrahedron
(triangular pyramid with
4 equilateral triangles)

Octahedron
(8 equilateral triangles)

Icosahedron
(20 equilateral triangles)

Cube
(6 squares)

Dodecahedron
(12 regular pentagons)

Figure 8–58

Natural crystals occur in the shapes of the tetrahedron (for example, chrome alum), the cube (salt), and the octahedron (sodium sulphantimoniate).

In the following exercise, verify that Euler's formula works for regular polyhedra.

LE 16   Concept

(a) Complete the table.

Regular Polyhedron	Faces	Vertices	Edges
Cube	6	8	12
Tetrahedron			
Octahedron			
Icosahedron		12	30
Dodecahedron		20	30

(b) Does Euler's formula work for all regular polyhedra?

CYLINDERS, CONES, AND SPHERES

Some everyday objects suggest space figures that are not polyhedra: cylinders, cones, and spheres (Figure 8–59).

Photos by Tom Sonnabend

Cylinder

Cone

Sphere

Figure 8–59

LE 17 Concept
Why aren't cylinders, cones, and spheres polyhedra?

The most common types of cylindrical shapes in our environment are right circular cylinders. Circular cylinders have parallel bases that are circular regions. In a **right circular cylinder** (Figure 8-60), the line segment connecting the centers of the circular bases is perpendicular to the bases.

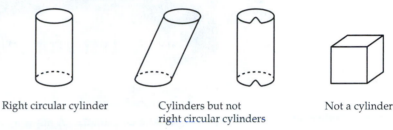

Right circular cylinder   Cylinders but not   Not a cylinder
right circular cylinders

Figure 8–60

LE 18 Concept
You can construct the lateral surface of a right circular cylinder from a certain polygon. Which polygon is it?

The familiar type of cone is called a right circular cone.

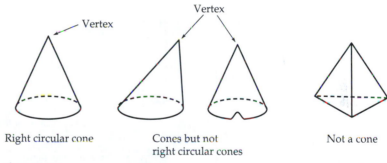

Right circular cone   Cones but not   Not a cone
right circular cones

Figure 8–61

A circular cone has a circular region for a base. In a **right circular cone,** the line segment connecting the center of the circular base to the vertex (Figure 8–61) is perpendicular to the base.

LE 19 Concept
(a) Name a property that all cones and pyramids have in common.
(b) At what van Hiele level is the question in part (a)?

NETS

By constructing solids from two-dimensional shapes called nets, students learn more about the properties of solids. A **net** is a two-dimensional pattern that can be folded into a three-dimensional shape such as a prism or pyramid. Figure 8–62 shows a net for a cube in the fourth-grade textbook for *Harcourt Mathematics*.

C
Classroom Connection

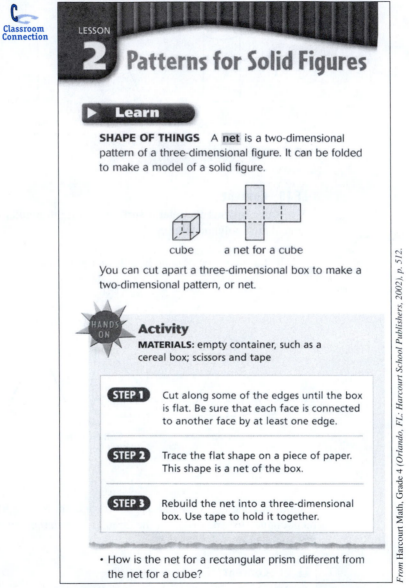

LESSON

2 Patterns for Solid Figures

▶ **Learn**

SHAPE OF THINGS   A **net** is a two-dimensional pattern of a three-dimensional figure. It can be folded to make a model of a solid figure.

cube   a net for a cube

You can cut apart a three-dimensional box to make a two-dimensional pattern, or net.

HANDS ON   **Activity**

MATERIALS: empty container, such as a cereal box; scissors and tape

STEP 1   Cut along some of the edges until the box is flat. Be sure that each face is connected to another face by at least one edge.

STEP 2   Trace the flat shape on a piece of paper. This shape is a net of the box.

STEP 3   Rebuild the net into a three-dimensional box. Use tape to hold it together.

• How is the net for a rectangular prism different from the net for a cube?

From Harcourt Math, Grade 4 (Orlando, FL: Harcourt School Publishers, 2002), p. 512.

Figure 8–62

LE 20   Connection

Name the solid figure that can be made with each net.

(a)   (b)

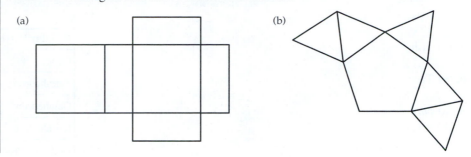

LE 21   Concept

Draw a net for each of the following shapes. (*Hint:* Draw the bottom and then visualize folding the other faces to complete the solid.)

(a) A hexagonal prism   **(b)** A square pyramid

 ## AN INVESTIGATION: NETS FOR A CUBE

LE 22   Reasoning

Figure 8–62 shows a net for a cube. There are 10 other ways to make a net for a cube. Which of the following are also nets for a cube? (If necessary, copy the nets on graph or grid paper, cut them out, and fold them.)

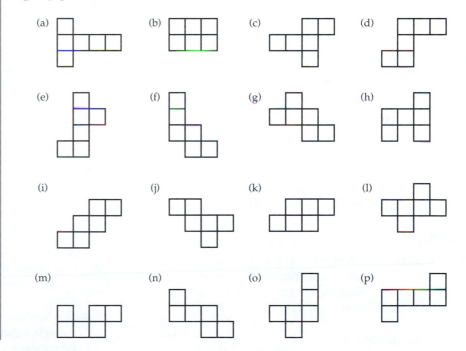

LE 23   Summary

Tell what you learned about polyhedra in this lesson.

ANSWERS TO SELECTED LESSON EXERCISES

2. (a) False

4. (a) True

6. (a) True

8. (a) part (d)
(b) part (a)

10. (a) 6
(b) 8
(c) 12 (4 on top, 4 connecting top to bottom, 4 on the bottom)

11. Show the class a model of a cube. Have the child count the edges. If the child doesn't notice he is counting the same edges more than once, ask about that.

12. Two congruent bases

13. (a) 5
(b) 6
(c) 9

14. (a) Rectangular prism: 6, 8, 12; pentagonal prism: 7, 10, 15

15. (a) 4, 4, 6 and 5, 5, 8

16. (a)

Regular Polyhedron	Faces	Vertices	Edges
Cube	6	8	12
Tetrahedron	4	4	6
Octahedron	8	6	12
Icosahedron	20	12	30
Dodecahedron	12	20	30

(b) Yes

17. Their faces are not polygonal regions.

18. A rectangle

19. (a) They have exactly one base.
(b) Description

20. (a) A rectangular prism   (b) A pentagonal pyramid

21. (a) 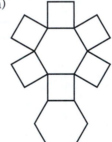   (b)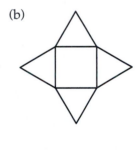

8.5   HOMEWORK EXERCISES

Basic Exercises

1. (a) True or false? In space, two lines that are perpendicular to a third line are parallel to each other.
(b) In your surroundings, identify a model of this situation that supports your answer.

2. (a) What are skew lines?
(b) Identify a model of skew lines in your home or classroom.

3. In space, point A is not on a line b. How many lines pass through A that
(a) intersect b?   (b) are perpendicular to b?
(c) are parallel to b?   (d) are skew to b?

4. In space, point A is on line b. How many lines pass through A that
(a) intersect b?   (b) are perpendicular to b?
(c) are parallel to b?   (d) are skew to b?

5. True or false? Suppose line *a* lies in a plane *A* and line *b* does not. If line *b* is perpendicular to line *a*, then line *b* is perpendicular to plane *A*.

6. True or false? Suppose two distinct lines *a* and *b* lie in plane *A* and line *c* does not. If line *c* is perpendicular to lines *a* and *b*, then line *c* is perpendicular to plane *A*.

7. True or false? If a line is perpendicular to one of two parallel planes, then it is perpendicular to the other.

8. True or false? Two planes, *a* and *b*, intersect a third plane in parallel lines. Planes *a* and *b* are parallel.

9. (a) Draw a point *A*. How many different planes pass through *A*?
(b) Draw a second point *B*. How many planes pass through both *A* and *B*?
(c) Draw a third point *C*. How many planes pass through *A*, *B*, and *C*?
(d) Draw a fourth point *D* so that you can find a different answer for *A*, *B*, and *D* than you did for *A*, *B*, and *C*.

10. Point *A* is not on line *b*. How many different planes contain point *A* and line *b*?

11. *ABCDEFGH* is a cube.

(a) Name two skew lines in the drawing.
(b) The plane that contains *ABEF* is parallel to the plane that contains _____.
(c) The plane that contains *ABCD* is _____ to the plane that contains *BCGF*.
(d) What is the intersection of the plane containing *E*, *F*, *G*, and *H* with the plane containing *B*, *C*, *F*, and *G*?

12. (a) The ceiling and floor of a room suggest _____ planes.
(b) The ceiling and side wall of a room suggest _____ planes.

13. (a) True or false? If a line intersects one of two parallel lines, then it intersects the other also.
(b) In your home, identify a model of this situation that supports your answer.

14. (a) True or false? Every set of four points is contained in one plane.
(b) In your home, identify a model of this situation that supports your answer.

15. True or false? Through a given point not on plane *P*, there is exactly one plane parallel to *P*.

16. True or false? Through a given point not on plane *P*, there is exactly one line parallel to *P*.

17. Tell whether each of the following suggests a polygon or a polygonal region.
(a) A picture frame
(b) A page in this book
(c) A face of a triangular prism

18. Tell whether each of the following suggests a polygon or a polygonal region.
(a) A stop sign
(b) A face of a cube
(c) A mold for a star-shaped cookie

19. Which of the following are polyhedra?
(a)   (b)

(c) Cylinder   (d) Hexagonal prism

20. Give an example of a simple closed solid figure that is not bounded by polygonal regions.

21. Name a solid figure that approximates the shape of each of the following.
(a) A new piece of chalk
(b) Your refrigerator at home

22. Name each figure.

(a)   (b)

23. The figure shown is a cube.

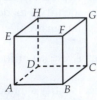

   (a) What kind of quadrilateral is *EGCA*?
   (b) What kind of triangle is △*EGH*?

24. Name a solid figure that has 2 bases and 6 other faces.

25. Name a property that prisms have that pyramids do not.

26. (a) Name two properties that prisms and pyramids have in common.
   (b) At what van Hiele level is the question in part (a)?

27. (a) How many faces does a hexagonal prism have?
   (b) How many edges?
   (c) How many vertices?

28. A pyramid has a heptagonal base. How many faces, vertices, and edges does it have?

29. A certain pyramid has 10 edges. How many vertices and faces does it have?

30. A certain prism has 20 vertices. Has many faces and edges does it have?

31. Determine if Euler's formula is true for the following figures.

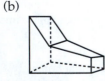

  (a)   (b)   (c)

32. (a) Sketch a pyramid that has a hexagonal base.
   (b) How many faces does it have?
   (c) How many vertices?
   (d) How many edges?
   (e) Show that Euler's formula works for this pyramid.

33. Name the space figures that make up each object.

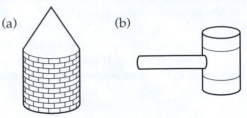

(a)   (b)

34. Name the space figures that make up each object.

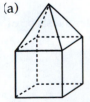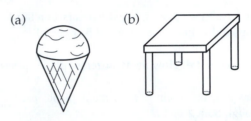

(a)   (b)

35. You show a group of first graders a cylinder and ask them what they see. What different responses might you get?

36. You show a group of first graders a rectangular prism and ask them what they see. What different responses might you get?

37. What shape has no parallel faces, one face that has five edges, and five faces that are triangles?

38. What shape has all faces that are rectangles and two faces that are parallel and congruent?

39. What property of (right circular) cones and cylinders makes them more suitable than prisms or pyramids for the shape of an ice cream cone?

40. What property of (right) prisms and cylinders makes them more suitable than pyramids or cones for the shape of a garbage can?

41. How is a basketball different than a geometric sphere?

42. (a) What simple change in the definition of a circle would make it the definition of a sphere?
   (b) What are the possible shapes of the intersection of a plane and a sphere?

43. Name the solid figure that can be made with each net.

(a)   (b)   (c)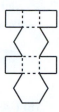

44. Name the solid figure that can be made with each net.

(a)   (b) 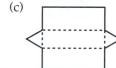   (c)

45. Draw a net for each of the following shapes.
   (a) A triangular pyramid   (b) A pentagonal prism

46. Draw a net for each of the following shapes.
   (a) A rectangular prism   (b) A cylinder

47. Which of the following patterns fold into a cube? (Use paper cutouts as needed.)

(a)   (b)

(c)   (d)

48. Which of the following nets fold into a triangular pyramid?

(a)   (b)   (c)

49. After you have finished a roll of toilet paper at home, save the cardboard roll.
   (a) Guess what shape you will obtain if you cut it along the curved line on its surface and unroll it. Then cut it and see what happens.
   (b) Why does the manufacturer use the shape in part (a) rather than a rectangle?

50. After you have finished using a box at home (cereal, crackers, etc.), disassemble it and sketch the shape of the net.

Extension Exercises

51. Decide whether each of the following describes a point, line, plane, segment, or ray.
   (a) All points that are equidistant from two parallel planes
   (b) In space, all points that are equidistant from both endpoints of a line segment
   (c) All points that are 3 inches below a line

52. Why are there no such things as skew planes?

53. A prism has a base with n sides.
   (a) How many faces does it have?
   (b) How many vertices does it have?
   (c) How many edges does it have?
   (d) Does Euler's formula work for prisms?

54. A pyramid has a base with n sides.
   (a) How many faces does it have?
   (b) How many vertices does it have?
   (c) How many edges does it have?
   (d) Does Euler's formula work for pyramids?

55. Take a regular polyhedron. Connect the centers of adjacent faces and you get another polyhedron (called the **dual**)!
   What shape is obtained by connecting the centers of adjacent faces of

(a)   (b)   (c)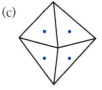

   A cube?   A tetrahedron?   An octahedron?

(*Note:* The back of the octahedron is not shown.)
   (d) A regular polyhedron has x vertices, y faces, and z edges. Conjecture how many vertices, faces, and edges its dual has.

56. Construct or obtain models of a square pyramid and a triangular pyramid with congruent equilateral triangles for all their triangular faces. (You could use cardboard.) Place the two shapes together to form a polyhedron with the smallest number of faces. How many faces does it have?

Labs and Projects

57. A plane surface suggested by a sheet of paper has two sides. In fact, a piece of paper always has two sides, right? Not always. The remarkable Möbius strip has only one side! You can make one yourself using scissors, paper, and tape.

(a) Cut out two strips of paper. Tape the ends of one strip to make a regular loop. Before taping together the ends of the second strip, give it a half-twist to create a Möbius strip.

(b) Use a pencil to draw a line down the middle of one side of each paper loop. Continue each line until you reach your starting point.

(c) What is the difference in your results for the two loops in part (b)?

8.6 VIEWING AND DRAWING SOLID FIGURES

NCTM Standards

- recognize and represent shapes from different perspectives (K–2)
- build and draw geometric objects (3–5)
- identify and build a three-dimensional object from two-dimensional representations of that object (3–5)

It's amazing that a figure sketched on a flat surface can appear to have a third dimension, as in Figure 8–63. Representing three-dimensional figures on a two-dimensional surface posed a serious challenge for artists until the Renaissance. Today, artists have no problem making flat drawings that create the illusion of three dimensions. This knowledge resulted from many years of experimentation.

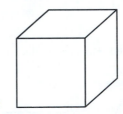

Figure 8–63

VISUAL PERCEPTION

Although the M. C. Escher drawing in Figure 8–64 is two-dimensional, it appears to be three-dimensional. Our brains tend to look for depth, the third dimension, even in a flat picture. Some optical illusions are based on the tendency to perceive depth in a flat drawing.

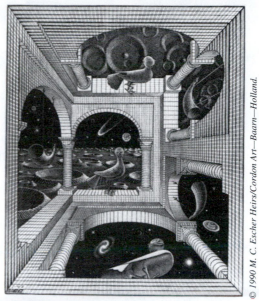

Figure 8–64

Figure 8–65

Figure 8–66

In LE 1 and LE 2, use your eyes. *No measuring allowed!*

LE 1   Concept

Which one of the following is true?

(a) The circles in Figure 8–65 are the same size.
(b) The top circle is larger.
(c) The bottom circle is larger.

LE 2   Concept

Will a dime fit inside the parallelogram drawing in Figure 8–66?

In LE 1 and LE 2, we perceive depth. In LE 1, the top circle appears farther back. Our brains expect objects that are farther away to look smaller. Since the circle in back is actually the same size as the "closer" black circle, we see the circle in back as the larger one. In LE 2, we assume that the top of the parallelogram is farther back than the bottom, making it seem larger.

Although our eyes help us make sense of our surroundings, these examples illustrate that our visual perception is not always reliable for evaluating spatial relationships. On the other hand, once we understand visual perception, we can make sketches of solids on paper that are helpful in analyzing three-dimensional relationships.

DIFFERENT VIEWS OF SOLID FIGURES

We never see a solid in its entirety; we see it from a viewpoint. Consider the solid figure and its three viewpoints (Figure 8–67 on the next page). The top view is sometimes called the **base outline.** Thinking of the top view as a base outline may help you in drawing it.

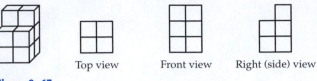

Top view   Front view   Right (side) view

Figure 8–67

LE 3   Concept

Draw the top view (base outline), front view, and right (side) view for the following solids. Use the height of the building to draw the front and right views.

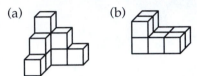

(a)   (b)

LE 4   Concept

A **base plan** shows how high each stack of cubes is. Below is the base plan for the cube building in part (a) of LE 3. The bottom is the front of the building.

Draw the base plan for the cube building from LE 3(b).

Now try to draw the views from a base plan.

LE 5   Skill

Draw the top view, front view, and right (side) view for each base plan.

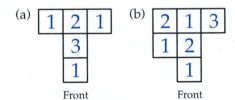

(a)   (b)

Front   Front

Now try making a base plan from the top view, front view, and right view.

LE 6   Skill

Draw a possible base plan for the following.

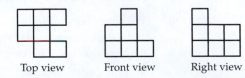

Top view   Front view   Right view

Artists, architects, and photographers study how objects appear from different viewpoints. Such knowledge enriches our understanding and our experience in a world replete with geometric shapes and spatial relationships. Architects draw plans using views (called elevations). A view accurately shows the proportions, whereas a perspective drawing does not.

Figure 8–68 shows different views of a cylinder. A **cross section** of a solid is the intersection of the solid and a plane.

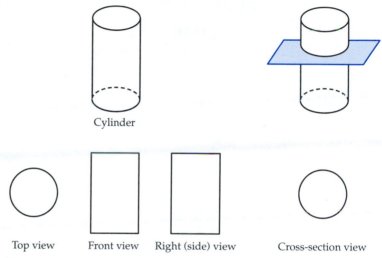

Cylinder

Top view   Front view   Right (side) view   Cross-section view

Figure 8–68

LE 7   Reasoning

Name the solid figure that has the following front view, top view, and side view.

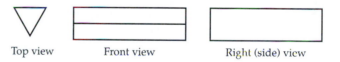

Top view   Front view   Right (side) view

Triangular pyramid

Figure 8–69

LE 8   Skill

Draw a cross-sectional view of the solid figure in Figure 8–69.

DRAWING PRISMS

How do you draw a picture of a cube? Isometric dot paper helps.

LE 9   Opener

(a) Draw a regular hexagon on the isometric dot grid (Activity Card 4) below.
(b) What is the measure of the angles that are formed?

The preceding exercise clarifies the design of isometric dot paper. Now, how do you draw a cube or a stack of cubes on the dot paper?

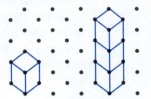

LE 10   Skill

Make two different drawings that show four cubes.

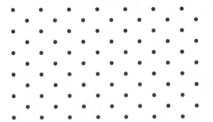

LE 11   Concept

A view of a cube building is shown from the front right corner. Draw the top view (base outline), front view, and right view.

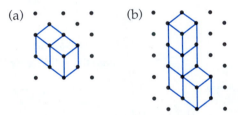

(a)   (b)

PERSPECTIVE DRAWING

In the Middle Ages in Europe, many artists painted scenes that had religious themes. They were not concerned about accurately representing people and objects in these symbolic works. In the late thirteenth, fourteenth, and fifteenth centuries, however, European painters became more interested in accurately drawing their surroundings, but they were not sure exactly how to do it.

LE 12   Concept

What is unrealistic about the fifteenth century painting shown in Figure 8–70?

Figure 8–70

The painting in Figure 8–70 does not realistically depict how chairs and desks look.

Renaissance painters learned how to portray three dimensions realistically in paintings. They realized that they needed geometry to solve the problem of depicting three dimensions on a flat canvas. Piero della Francesca, the great fifteenth century painter and mathematician, used the idea of a vanishing point to create more realistic paintings, such as the one in Figure 8–71.

Figure 8–71

Figure 8–72

When the lines running *from the front to the back* of the picture are extended, they intersect at one point (the vanishing point) just below the middle of the picture.

LE 13   Concept

Find two parallel lines in the painting in Figure 8–71 that do not pass through the vanishing point. Explain why not.

How is the illusion of depth created in a drawing? In a drawing of railroad tracks that go off into the horizon, like those in Figure 8–72, the parallel rails appear to meet at a point on the horizon called the **vanishing point.** Drawings based on a vanishing point are called **one-point perspective drawings.**

A second method of drawing a cube utilizes one-point perspective. This method is more difficult than the method of LE 10, but the result is more realistic.

LE 14   Skill

To draw a cube with one-point perspective, follow these steps.

Step 1: Draw the front of the cube and choose a vanishing point (Figure 8–73).
Step 2: Use a ruler to connect the vertices to the vanishing point (Figure 8–74).

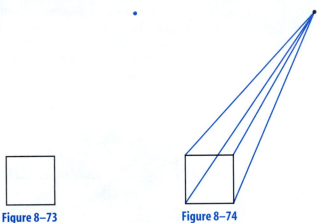

Figure 8–73   **Figure 8–74**

Step 3: Draw the back of the cube as shown in Figure 8–75.
Step 4: Darken the edges that would be seen from the front (Figure 8–76).

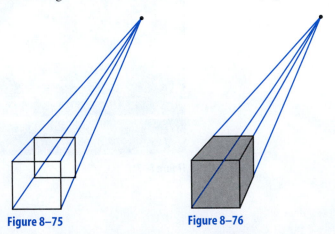

Figure 8–75   **Figure 8–76**

Figure 8–77

LE 15   Skill

Where is the vanishing point for the drawing in Figure 8–77?

LE 16   Summary

Tell what you learned about drawing solids and different views in this section.

ANSWERS TO SELECTED LESSON EXERCISES

1. (a)

2. No

3. (a)

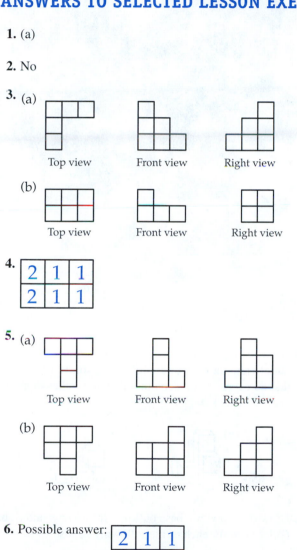

4.

2	1	1
2	1	1

5. (a)

(b)

6. Possible answer:

2	1	1	
	1	2	
		3	1

7. Triangular prism

8.

9. (b) 120°

11. (a)

     Top view   Front view   Right view

(b)

     Top view   Front view   Right view

12. All of the furniture looks crooked.

13. Lines going from left to right across the floor or ceiling are drawn parallel and do not pass through the vanishing point. Only parallel lines from the front to back are drawn so that they meet at the vanishing point.

15.

8.6 HOMEWORK EXERCISES

Basic Exercises

1.

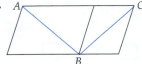

(a) Which appears to be longer, $\overline{AB}$ or $\overline{CD}$?
(b) Measure and find the correct answer.
(c) Why does our visual perception tend to elongate $\overline{AB}$?

2.

(a) Which one of the following statements appears to be true?
(1) $\overline{AB}$ and $\overline{BC}$ are the same length.
(2) $\overline{AB}$ is longer.
(3) $\overline{BC}$ is longer.
(b) Measure and find the correct answer.

3.

These two photos show the same person on opposite sides of the same room! Explain how this optical illusion is created.

4. What is odd about the people on the staircase in the following Escher drawing?

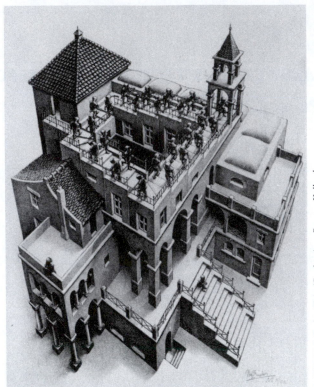

© 1990 M.C. Escher Heirs/Cordon Art—Baarn—Holland.

5. Draw the top view (base outline), front view, and right (side) view for each of the following solids. (Assume that no cubes are hidden other than those that support visible cubes.)

(a)   (b)

6. Draw the top view (base outline), front view, and right (side) view for each of the following solids.

(a) 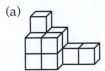   (b)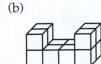

7. Draw the base plan for each solid in Exercise 5.

8. Draw the base plan for each solid in Exercise 6.

9. Draw the front and top views of the buildings.

10. Sketch the front and top views of the building in which you live.

11. Draw the top view, front view, and right (side) view for each base plan.

(a)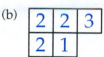

(b)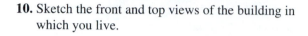

12. Draw the top view, front view, and right (side) view for each base plan.

(a)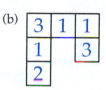

(b)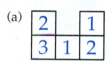

13. Draw a possible base plan for the following.

Top view   Front view   Right view

14. Draw a possible base plan for the following.

  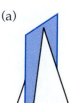

Top view   Front view   Right view

15. Tell why the following top view, front view, and right view do not match.

Top view   Front view   Right view

16. Tell why the following top view, front view, and right view do not match.

Top view   Front view   Right view

17. Identify the solid figure that has the top view, front view, and right view shown.

Top view   Front view   Side view

18. Identify the solid figure that has the top view, front view, and right view shown.

Top view   Front view   Right view

19. Each solid is cut by a plane as shown. What is the resulting cross section?

(a) (b)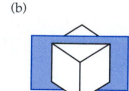

20. In each case, identify the cross section cut by the plane.

(a) 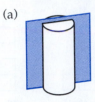 (b)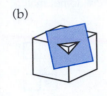

21. Name two figures other than a cube that have a square cross section. Use drawings to support your answers.

22. Why do drinking glasses usually have circular cross sections?

23. Make two different drawings that show three cubes.

24. Make two different drawings that show two cubes.

25. A view of a cube building is shown from the front.

(a) Draw a copy of the building on an isometric dot grid.
(b) Draw the top view (base outline), front view, and right view.

26. A view of a cube building is shown from the front.

(a) Draw a copy of the building on an isometric dot grid.
(b) Draw the top view (base outline), front view, and right view.

27. Find the vanishing point in the drawing.

28. (a) Locate the vanishing point in Albrecht Dürer's *St. Jerome in His Study.*

Dürer, St. Jerome in His Study, engraving. © Marburg/Art Resource, N.Y.

(b) Find two parallel lines that do not pass through the vanishing point. Explain why not.

29. You can draw block letters using a vanishing point, as shown. Draw your first or last name in block letters, using a vanishing point.

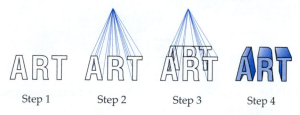

Step 1   Step 2   Step 3   Step 4

30. Draw a cube with one-point perspective.

31. In drawing railroad tracks in perspective, how much space should there be between parallel tracks? Look at drawings 1 and 2, and answer the questions that follow.

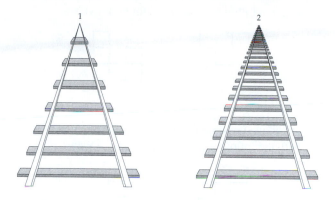

(a) Which picture of the railroad tracks, 1 or 2, looks more realistic?

(b) Measure the distances between successive horizontal ties in each drawing. Which drawing has evenly spaced ties?

(c) The vanishing point can be used to draw tracks correctly. Following these four steps, make your own drawing with a ruler and pencil.

Step 1: Use a vanishing point to draw the rails. Then draw the front and back ties.

Step 2: Label the front $\overline{AB}$ and the back $\overline{CD}$.

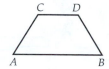

Step 3: To locate the tie between $\overline{AB}$ and $\overline{CD}$, draw $\overline{AD}$ and $\overline{BC}$. Then draw a new tie parallel to $\overline{AB}$, through the point where $\overline{AD}$ and $\overline{BC}$ intersect.

Step 4: Erase your guidelines.

32. How would you locate the first tie in the following drawing? (*Hint:* Use the midpoint of the second tie, and see the preceding exercise.)

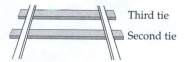

Third tie

Second tie

33. In the game pick-up sticks, a player picks up the sticks one at a time, always taking the stick on top. In what order should the sticks in the diagram be picked up?

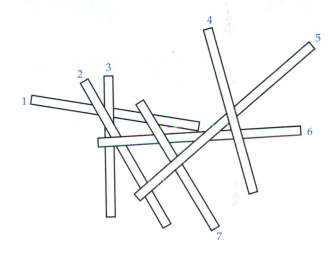

34. (a) Copy the figure.

(b) What is odd about the figure?

35. This drawing by William Hogarth is called *False Perspective*. Make a list of all the perspective errors you can find.

The Bettmann Archives.

36. Explain why the man reading the book on perspective in the cartoon thinks it's hogwash.

Extension Exercises

37. Make an isometric drawing from the front right corner for the base plan below.

38. Make an isometric drawing from the front right corner for the base plan below.

39. Draw one possible front view and a possible base plan with the following top and right (side) views.

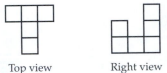

Top view   Right view

40. Draw one possible front view and a possible base plan with the following top and right (side) views.

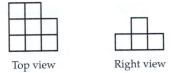

Top view   Right view

By permission of Johnny Hart and Creators Syndicate, Inc.

41. Draw the front, top, and side views of
   (a) a cup.   (b) a table.

42. Create your own drawing using one-point perspective.

Puzzle Time

43. Suppose that a wooden cube is painted blue on the outside, as shown, and then cut into smaller cubes.

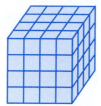

   (a) How many small cubes would there be?
   (b) How many cubes would have exactly one surface painted blue? (*Hint:* There is a pattern in the locations of these cubes.)
   (c) How many cubes would have exactly two surfaces painted blue?
   (d) How many cubes would have exactly three surfaces painted blue?
   (e) How many cubes would have exactly four surfaces painted blue?
   (f) How many cubes would have no surfaces painted blue?

44. Someone builds a 5-by-5-by-5 model of cubes, as shown, and paints the outside of the model.

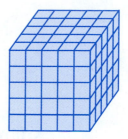

How many of the smallest cubes have each of the following?
   (a) Exactly one face painted
   (b) Exactly two faces painted
   (c) Exactly three faces painted
   (d) Exactly four faces painted
   (e) Zero faces painted

45. Repeat the preceding question for the following models.
   (a) 6-by-6-by-6 model
   (b) *n*-by-*n*-by-*n* model

CHAPTER 8 SUMMARY

Geometry began when people first observed three-dimensional shapes in their surroundings. The surfaces of three-dimensional objects suggested two-dimensional shapes. Young children today discover geometry in the same way.

Ideal geometric shapes do not exist in the world. We abstract them from our surroundings. Geometric shapes then take on a life of their own. It was the work of the ancient Greeks, culminating with Euclid's *Elements* around 300 B.C., that gave us the logically organized Euclidean geometry. Euclidean geometry begins with the undefined terms "point," "line," "plane," and "space" and builds geometry from there. Formal geometry proceeds in a certain order because one cannot, for example, define a prism without first being familiar with points, line segments, and polygons.

In geometry, shapes are classified. Within each classification, mathematicians look for common properties. Once these properties are established, they can be applied to all kinds of objects in the world that approximate the shape being studied. This chapter contains classifications of angles, lines, polygons, and space figures.

In learning geometry, the van Hieles suggest first working on recognizing whole figures, then analyzing their properties, and then writing precise definitions and identifying relationships between different classes of figures. The van Hiele levels work especially well in the study of quadrilaterals.

Geometry offers interesting opportunities for investigations. For instance, the total number of diagonals of various polygons is related to the number of sides. The sums of the angle measures in convex polygons follow a pattern. Euler's formula describes another pattern in the numbers of faces, vertices, and edges in any polyhedron.

In studying solid figures, we confront the difficulty of visualizing spatial relationships represented by diagrams and the challenge of drawing different views of solid figures. Renaissance mathematician-artists were the first to develop a system of perspective drawing based on their understanding of the way we see the world.

STUDY GUIDE

To review Chapter 8, see what you know about each of the following ideas or terms that you have studied. You can also use this list to generate your own questions about the chapter.

GEOMETRY IN GRADES 1–8

The following chart shows at what grade levels selected geometry topics typically appear in elementary- and middle-school mathematics textbooks.

Topic	Typical Grade Level in Current Textbooks
Points, lines, and planes	4, 5, 6
Measuring angles	5, 6, 7
Two- and three-dimensional figures	1, 2, 3, 4, 5, 6, 7, 8
Polygons (five or more sides)	3, 4, 5, 6, 7
Radius, diameter, chord	4, 5, 6
Tessellations	6, 7
Faces, edges, and vertices	4, 5, 6, 7, 8
Nets	4, 5, 6
Drawing prisms	6, 7

REVIEW EXERCISES

1. Why does the *formal* study of geometry begin with points and lines rather than with three-dimensional figures?

2. Name two ways in which a sheet of paper differs from a plane.

3. Give a simpler name for each of the following.
   (a) $\overrightarrow{BA} \cup \overline{BC}$
   (b) $\overleftrightarrow{AD} \cap \overline{BE}$

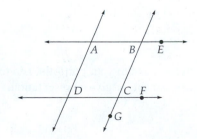

4. Suppose $\angle A$ is supplementary to $\angle B$, and $\angle C$ is supplementary to $\angle B$. What can you deduce about $\angle A$ and $\angle C$?

5. (a) Measure $\angle ABC$ with a protractor.

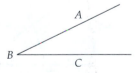

   (b) Is the angle acute, obtuse, or right?
   (c) Based on part (a), what is the measure of the complement of $\angle ABC$?

6. Draw a Venn diagram showing the relationship among rectangles, squares, and trapezoids.

7. Is every rhombus a type of square? If not, give a counterexample.

8. Fill in each blank with "All," "Some," or "No."
  (a) _____ parallelograms are rhombuses.
  (b) _____ squares are rectangles.

9. (a) Draw all the diagonals of a convex pentagon.
  (b) How many diagonals are there?

10. A rhombus is defined in the text as "a parallelogram that has four congruent sides." Make drawings to determine which of the following also defines a rhombus. If a definition is incorrect, tell why or make a drawing.
  (a) A parallelogram that has two congruent sides
  (b) A parallelogram that has two congruent adjacent sides
  (c) A parallelogram that has three congruent sides
  (d) A quadrilateral that has three congruent sides

11. Sketch two parallelograms that intersect at exactly five points.

12. A fifth grader defines a circle as "the set of all points that are the same distance from a point which is the center." Tell what is wrong with this definition.

13. Use the angle sum property of a triangle and a drawing to explain why the interior angle measures of a pentagon add up to 540°.

14. How many degrees are in each angle of a nonagon?

15. Assume *ABCDE* is a regular pentagon. Fill in all the missing interior angle measures in the diagram.

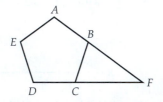

16. Is a rhombus a regular quadrilateral? Tell why or why not.

17. Do regular hexagons tessellate? Give evidence to support your answer.

18. Make a sketch of a repeating tile pattern that uses squares and regular octagons.

19. Point *A* is on line *b*. How many different planes contain point *A* and line *b*?

20. True or false? Through a point not on a given line, there is exactly one line that is skew to the given line.

21. True or false? Line *m* intersects plane *P* and is not perpendicular to it. Then there is no line in plane *P* that is perpendicular to *m*.

22. How many faces, vertices, and edges does a hexagonal pyramid have?

23. A certain prism has 21 edges. How many faces does it have?

24. Draw the front view and right (side) view for the base plan.

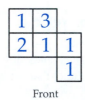

Front

25. Locate the next railroad tie in correct perspective.

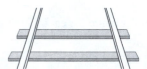

ALTERNATE ASSESSMENT

Do one of the following assessment activities: add to your portfolio, add to your journal, write another unit test, do another self-assessment, or give a presentation.

SUGGESTED READINGS

Battista, M. T. *Shape Makers: Developing Geometric Reasoning with Geometer's Sketchpad.* Emeryville, CA: Key Curriculum, 1998.

Del Grande, J., and others. *Geometry and Spatial Sense.* Reston, VA: NCTM, 1993.

Erickson, Tim. *Get It Together.* Berkeley, CA: Equals, 1989.

Fuys, D., D. Geddes, and R. Tischler. *The Van Hiele Model of Thinking in Geometry Among Adolescents.* Reston, VA: NCTM, 1988.

Gardner, M. *Aha!* New York: W. H. Freeman, 1978.

Juster, Norton. *The Dot and the Line.* New York: Seastar, 2001

National Council of Teachers of Mathematics. *Geometry in the Mathematics Classroom.* 1973 Yearbook. Reston, VA: NCTM, 1973.

National Council of Teachers of Mathematics. *Computers in Mathematics Education.* 1984 Yearbook. Reston, VA: NCTM, 1984.

National Council of Teachers of Mathematics. *Learning and Teaching Geometry, K–12.* 1987 Yearbook. Reston, VA: NCTM, 1987.

O'Daffer, P., and S. Clemens. *Geometry: An Investigative Approach.* 2nd ed. Reading, MA: Addison-Wesley, 1992.

Schifter, D. et al. *Examining Features of Shape: Casebook.* Parsippany, NJ: Dale Seymour, 2002.

Navigating Through Geometry (4 vols.). Reston, VA: NCTM, 2001.

Use Infotrac to find articles in *Teaching Children Mathematics* about how to teach children about geometry.

9 | CONGRUENCE, SYMMETRY, AND SIMILARITY

Many artistic and architectural designs utilize patterns of congruent shapes. Sometimes each individual shape is symmetric; that is, all of its parts match up to one another in a certain way.

Sometimes, people use similarity relationships in planning designs. In technical terms, similar figures are the same shape, although they may be of different sizes. For example, a scale model of a building is similar to the planned building.

Congruence, symmetry, and similarity are all related to transformation geometry. By moving a geometric figure to different positions or changing its size, one can tell if it is symmetric or if it is congruent or similar to another geometric figure.

Over 2500 years ago, the Greeks used straightedge-and-compass constructions to construct and copy geometric shapes. Many of these constructions can be analyzed by using congruence properties.

9.1 TRANSFORMATIONS AND CONGRUENCE

NCTM Standards

- predict and describe the results of sliding, flipping, and turning two-dimensional shapes (3–5)
- describe sizes, positions, and orientations of shapes under informal transformations such as flips, turns, slides, and scaling (6–8)
- examine the congruence, similarity, and line or rotational symmetry of objects using transformations (6–8)

In Chapter 8, you learned about congruent line segments and angles. Congruence is one of the most important ideas of geometry. A set of photocopies or a set of mass-produced items would be an example of approximately congruent objects (Figure 9–1). The Dutch artist M. C. Escher drew some interesting interlocking congruent shapes (Figure 9–2).

Figure 9–1

Figure 9–2

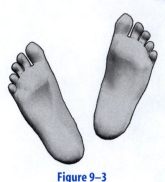

Figure 9–3

RIGID TRANSFORMATIONS

Ask a child if two flat shapes are the same (that is, congruent), and the child may place one shape on top of the other to be sure (Figure 9–3). In general, it is true that two plane figures are congruent if and only if one can fit exactly over the other.

LE 1   Opener
Trace triangle *A*, and describe how you would move it to the position of triangle *B*.

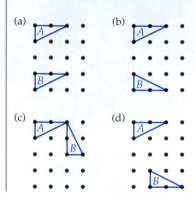

The set of **rigid transformations** (or **isometries**) consists of the various ways to move a geometric figure around while preserving the distances between points in the figure. The three basic rigid transformations are rotations (turns), translations (slides), and reflections (flips). The fifth-grade *Harcourt Math* textbook gives an example of each motion (Figure 9–4 on page 469).

How do these three basic transformations work? Begin with translation (slide).

There's a new geometric dance called the translation. You stand on one foot and slide 2 inches to the right (Figure 9–5). (It's easier on a recently waxed floor.)

How do you do a translation (slide)? A **translation** slides each point of the plane the same distance in the same direction along a line. If footprint *ABCD* in Figure 9–5 slides 4 centimeters to the right, it will coincide with footprint *EFGH*. So footprint *EFGH* is the **image** of footprint *ABCD* under a translation of 4 centimeters to the right. To describe a translation, you need a distance and a direction.

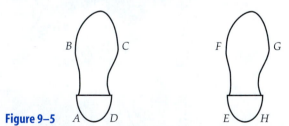

Figure 9–5

LE 2   Skill
(a) Copy the square grid and picture from Figure 9–6 onto a piece of paper (Activity Card 4).

(b) Show the image of △*ABC* after a translation of 2 units down and 1 unit to the right. (Adjacent dots in each column or row are 1 unit apart.)

Figure 9–6

Classroom
Connection

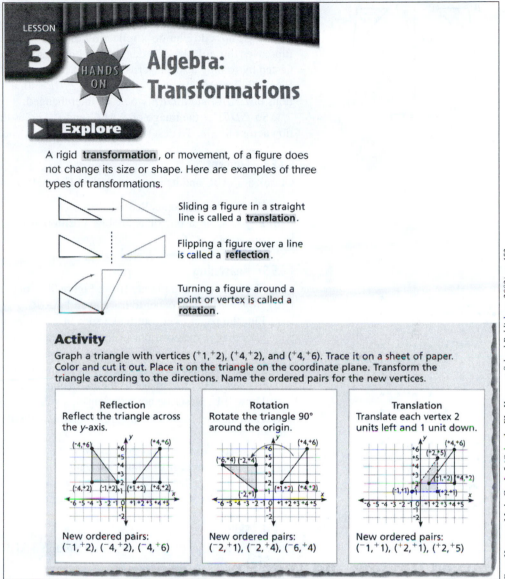

Figure 9–4

From Harcourt Math, Grade 5 (Orlando, FL: Harcourt School Publishers, 2002), p. 452.

Next, perform a translation on a coordinate graph. How will the coordinates of each point change?

LE 3 Skill

(a) Plot the points (2, 1) and (4, 2) on a graph, and draw the line segment that connects them.

(b) Now translate the line segment 3 units to the left and 2 units up. Draw the image, and give the coordinates of the endpoints.

(c) If you translate the point (a, b) left 3 units and up 2 units, what would be the coordinates of the image point?

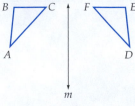

Figure 9–7

LE 3 is an example of how a translation of c units to the right or left changes the x-coordinate by c and a translation of d units up or down changes the y-coordinate by d.

Next, how do you do a reflection (a flip)? Wait a second. Don't start doing body flips over those desks. It's dangerous. Let's use a triangle instead. A reflection is suggested by folding a shape across a line. If this page of the book were folded on line m in Figure 9–7, △ABC would coincide with △DEF. In a reflection through line m, the positions of △ABC and △DEF would be interchanged.

So △DEF is the image of △ABC, and △ABC is the image of △DEF under a reflection across line m. To describe a reflection (a flip), you need a line of reflection.

LE 4 Concept

Look at △ABC and its image △DEF in Figure 9–7.

(a) Why is this motion called a reflection?
(b) If you have a mirror available, position it to verify that △DEF is the reflected image of △ABC.

Discussion

LE 5 Reasoning

(a) Copy the grid and picture (from Figure 9–8) onto a piece of paper (Activity Card 4).
(b) Now use tracing paper to trace m and △ABC.
(c) Flip the paper over, but make sure that line m stays in the same position. (Do *not* interchange the positions of the arrowheads of line m.)
(d) Trace over the triangle in its new position so that it shows up on your original paper. Then remove the tracing paper and draw the image in pencil or pen. Label the vertices of the image triangle A' (image of A), B', and C'.
(e) Draw $\overline{AA'}$. Describe any relationships you observe between $\overline{AA'}$ and line m.
(f) Explain how to find the image of △ABC without using tracing paper.

Figure 9–8

In LE 5(e), did you find that line m is the perpendicular bisector of $\overline{AA'}$? (*Note:* The **perpendicular bisector** of a line segment is a line that is perpendicular to the line segment and bisects it.) A **reflection** across line m transforms each point A (not on m) in a plane to a point A' so that m is the perpendicular bisector of $\overline{AA'}$. If A is on m, then $A' = A$.

LE 6 Skill

(a) Plot the points (2, 1) and (4, 2) on a graph, and draw the line segment that connects them.
(b) Now reflect the line segment across the y-axis. Draw the image, and give the coordinates of the endpoints.
(c) If you reflect the point (a, b) across the y-axis, what would be the coordinates of the image point?

As LE 6 indicates, the image of (a, b) after a reflection across the y-axis is $(-a, b)$. Next, consider a rotation (turn). A **rotation** holds one center point fixed and turns the points of a plane around this point by a certain number of degrees in a certain direction.

Figure 9–9

LE 7 Skill

(a) Place a piece of thin paper on this page, and trace the shape ABC and the point O from Figure 9–9.
(b) Now place your pencil point on O and turn the tracing paper a little. You have just done a rotation around the point O!

To describe any clockwise rotation, we specify a fixed center point, the measure of the angle through which we turn the plane around the center point, and the direction of the turn (clockwise).

LE 8   Concept

Suppose that you turn ⌐ a full turn clockwise around point C (Figure 9–10) so that it ends up back in the same place. How many degrees have you rotated the shape?

Figure 9–10

LE 9   Concept

Suppose that you rotate the shape clockwise halfway around point C, as shown in Figure 9–11. How many degrees have you rotated the shape?

Figure 9–11

LE 8 and LE 9 illustrate two important rotations: the 360° full turn and the 180° half-turn. Now try another rotation.

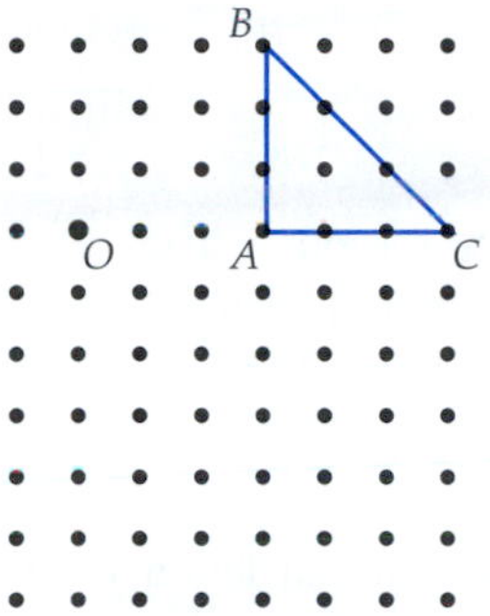

LE 10   Skill

(a) Copy the grid and picture from Figure 9–12 onto a piece of paper (Activity Card 4).
(b) Now use tracing paper to trace O and $\triangle ABC$, and then rotate $\triangle ABC$ clockwise 90° around point O.
(c) After completing the rotation, trace over the triangle in its new position (called the image) so that it shows up on your original paper drawing. Then remove the tracing paper, and draw the image in pencil or pen.

Figure 9–12

Classroom Connection

LE 11   Concept

A fifth grader says that she thinks a 180° clockwise rotation around a point is the same as a 180° counterclockwise rotation.

(a) Is that right?
(b) Can you name another rotation that has this property?

Now try a rotation on a coordinate graph.

LE 12   Skill

(a) Plot the points (2, 3) and (5, 4) on a graph, and draw the line segment that connects them.
(b) Now rotate the line segment 90° counterclockwise around (0, 0). Draw the image, and give the coordinates of the endpoints.
(c) If you rotate the point (a, b) by 90° counterclockwise around (0, 0), what would be the coordinates of the image point? (You may also use the results from the grade 5 textbook page (Figure 9–4)).

Next, consider the result of two consecutive reflections across parallel lines.

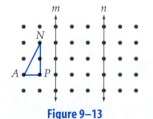

Figure 9–13

Discussion

LE 13   Reasoning

(a) Find the image of $\triangle NAP$ after a reflection across m (Figure 9–13). Label the image $\triangle N'A'P'$.
(b) Find the image of $\triangle N'A'P'$ after a reflection across n. Label the image $\triangle N''A''P''$.
(c) What single motion maps $\triangle NAP$ to $\triangle N''A''P''$?
(d) How is the distance from N to N'' related to the distance between m and n?

Two reflections across parallel lines are equivalent to a single translation.

RIGID TRANSFORMATIONS AND CONGRUENCE

When you perform a rigid transformation on a figure, what changes and what stays the same?

LE 14   Reasoning

Refer back to LE 2, LE 5, and LE 10, and compare the image to the original figure.

(a) What is the same about the image and the original figure in each exercise?
(b) How does the image differ from the original figure in each exercise?

LE 15   Concept

Suppose you have two congruent shapes. Do you think it is always possible to move one shape onto the other using a single rigid transformation or some combination of rigid transformations?

You can use rigid transformations to test for congruence.

> **A Test for Congruence**
>
> Two plane drawings are congruent if one can be moved onto the other using a rotation, a translation, a reflection, or some combination of these transformations.

LE 16   Concept

Trace one of each pair of geometric figures below, and use a rotation, translation, or reflection to determine if the two shapes are congruent. If the shapes are congruent, state which transformation you used to show this.

(a)   (b)   (c)

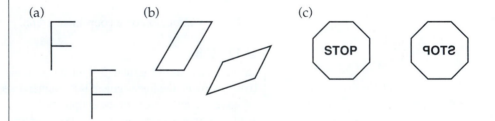

While it is probably obvious to you that rotating a shape does not change it, many young children have difficulty grasping this idea. A child may consider the orientation of a figure to be an essential property. Consequently, the child would not consider two figures with different orientations to be the same (congruent).

GLIDE REFLECTIONS

LE 17   Opener

Refer back to LE 1. In which parts is it impossible to move triangle *A* onto triangle *B* with a *single* translation, reflection, or rotation?

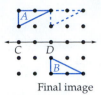

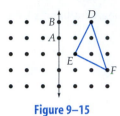

Final image

Figure 9–14

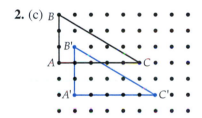

Figure 9–15

It is not always possible to move one of two congruent figures onto the other using a *single* rotation, translation, or reflection. This observation led to the introduction of a fourth rigid motion, the **glide reflection,** which combines a translation (glide) and a reflection.

In LE 1(d), the glide reflection $\overrightarrow{CD}$ (where → is a vector or arrow, not a ray) moves triangle A onto triangle B. First, translate triangle A the distance and direction of $\overrightarrow{CD}$ (two units to the right, as in Figure 9–14). Then reflect the image of triangle A across $\overleftrightarrow{CD}$.

LE 18   Skill

Find the image of △DEF after a glide reflection of vector $\overrightarrow{BA}$ (Figure 9–15). Translate △DEF the distance and direction of $\overrightarrow{BA}$. Then reflect the image across $\overleftrightarrow{AB}$ (Activity Card 4.).

Using the set of four rigid transformations, we can say: Two plane figures are congruent if and only if one figure can be moved onto the other using a *single* rotation, translation, reflection, or glide reflection.

LE 19   Summary

Tell what you learned about translations, reflections, and rotations in this section.

ANSWERS TO SELECTED LESSON EXERCISES

2. (c)

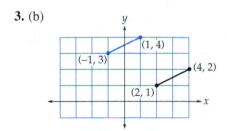

3. (b)

(1, 4) **(−1, 3)** **(4, 2)** **(2, 1)**

(c) $(a - 3, b + 2)$

4. (a) One triangle looks like a mirror reflection of the other when the line is used as the mirror.

5. (d)

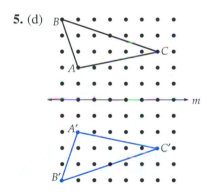

(f) A′, B′, and C′ are on the opposite side of m, the same distances from m as A, B, and C, respectively.

6. (b)

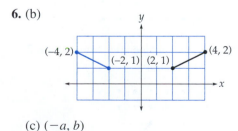

(−4, 2)   (−2, 1) (2, 1)   (4, 2)

(c) $(-a, b)$

8. 360°

9. 180°

10. (c)

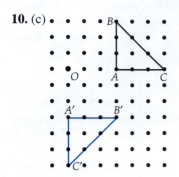

11. (a) Yes   (b) 360°

12. (b)

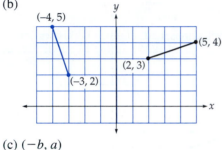

(c) $(-b, a)$

13. (c) Translation
(d) It is twice as much.

14. (a) They have the same size and shape.
(b) Its position is different.

15. Yes

16. Parts (a) and (c) contain congruent figures by a translation and a reflection, respectively.

18.

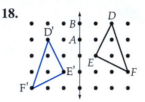

9.1   HOMEWORK EXERCISES

Basic Exercises

1. Show the image of the triangle after a translation 3 units to the right and 2 units down (Activity Card 4).

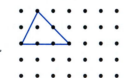

2. Show the image of the triangle after a translation 2 units to the left and 1 unit up.

3. (a) Plot the points $(-1, 3)$ and $(1, -2)$ on a graph, and draw the line segment that connects them.
(b) Now translate the line segment 2 units to the right and 4 units down. Draw the image, and give the coordinates of the endpoints.
(c) If you translate the point (a, b) right 2 units and down 4 units, what would be the coordinates of the image point?

4. (a) Plot the points $(-1, -2)$ and $(0, -4)$ on a graph, and draw the line segment that connects them.
(b) Now translate the line segment 2 units to the left and 1 unit down. Draw the image, and give the coordinates of the endpoints.
(c) If you translate the point (a, b) left 2 units and down 1 unit, what would be the coordinates of the image point?

5. (a) Plot the points $(1, 2)$, $(3, -1)$, and $(1, -2)$, and label them A, B, and C, respectively.
   (b) Suppose that a mapping changes the coordinates of each point as follows.

   $$(x, y) \rightarrow (x - 4, y + 1)$$

   Find image points A', B', and C' with this rule.
   (c) Plot A', B', and C'.
   (d) What motion maps $\triangle ABC$ to $\triangle A'B'C'$?

6. (a) Plot the points $(2, -1)$, $(4, -1)$, and $(2, 3)$, and label them A, B, and C, respectively.
   (b) Suppose that a mapping changes the coordinates of each point as follows.

   $$(x, y) \rightarrow (x + 3, y - 5)$$

   Find image points A', B', and C' with this rule.
   (c) Plot A', B', and C'.
   (d) What transformation maps $\triangle ABC$ to $\triangle A'B'C'$?

7. Reflect the figure across line m, and draw the image.

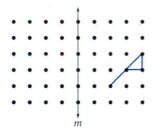

8. Show the image of the ▷ after a reflection through line m.

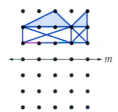

9. (a) Plot the points $(-1, 2)$ and $(2, 3)$ on a graph, and draw the line segment that connects them.
   (b) Now reflect the line segment across the x-axis. Draw the image, and give the coordinates of the endpoints.
   (c) If you reflect the point (a, b) across the y-axis, what would be the coordinates of the image point?

10. (a) Plot the points $(-2, -3)$ and $(-1, 1)$ on a graph, and draw the line segment that connects them.
   (b) Now reflect the line segment across the y-axis. Draw the image, and give the coordinates of the endpoints.
   (c) If you reflect the point (a, b) across the y-axis, what would be the coordinates of the image point?

11. Using tracing paper, find the line of reflection that flips one figure onto the other. (Guess and check.)

12. Using tracing paper, find the line of reflection that flips one figure onto the other. (Guess and check.)

13. Suppose you have a shape and its image from a reflection. Describe a procedure for locating the line of reflection.

14. If you reflect an isosceles triangle across a line that contains its nonequal side, what shape is formed by the triangle and its image?

15. Some ambulances have ƎƆИAⅬUᗺMA written on the front. Why is the word printed this way?

16. Different sets of golf clubs are made for right-handed and left-handed golfers. The two sets of clubs are congruent but have different **orientations.** Name two other kinds of objects that are sometimes made congruent but with different orientations.

17. (a) Copy the grid and picture onto a piece of paper.

   (b) Now use tracing paper to trace point O and the ⌐.
   (c) Rotate the ⌐ 180° clockwise around point O.
   (d) Show the image of the ⌐ on your original grid.

18. Find the image of the flag after a 270° clockwise rotation around point *C*.

19. (a) Plot the points (−2, 1) and (−3, 3) on a graph, and draw the line segment that connects them.
(b) Now rotate the line segment 90° counterclockwise around (0, 0). Draw the image, and give the coordinates of the endpoints.
(c) If you rotate the point (*a*, *b*) by 90° counterclockwise around (0, 0), what will be the coordinates of the image point?

20. (a) Plot the points (−2, −1) and (−3, −4) on a graph, and draw the line segment that connects them.
(b) Now rotate the line segment 180° counterclockwise around (0, 0). Draw the image, and give the coordinates of the endpoints.
(c) If you rotate the point (*a*, *b*) by 180° counterclockwise around (0, 0), what would be the coordinates of the image point?

21. *ABCDEF* is a regular hexagon, and *G* is the center point.

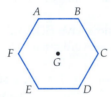

What is the image of *A* under a
(a) 120° clockwise rotation around *G*?
(b) 60° counterclockwise rotation around *G*?
(c) 240° clockwise rotation around *G*?
(d) reflection through $\overleftrightarrow{CF}$?

22. △*ABC* is equilateral, and *D* is the center point.

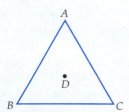

What is the image of *B* under a
(a) 120° clockwise rotation around *D*?
(b) 120° counterclockwise rotation around *D*?
(c) 240° clockwise rotation around *D*?
(d) reflection through $\overleftrightarrow{CD}$?

23. Without tracing paper, draw the image of each figure after a 180° rotation around *C*.

(a)   (b)

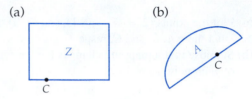

24. Without tracing paper, draw the image of each figure after a 180° rotation around *C*.

(a)   (b)

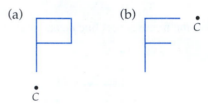

25. (a) Find the image of $\overline{AB}$ after a rotation of 90° counterclockwise around *B*. Label the image $\overline{A'B'}$.

(b) Find the image of $\overline{A'B'}$ after a translation two units to the right. Label the image $\overline{A''B''}$.
(c) What specific single motion maps $\overline{AB}$ to $\overline{A''B''}$?

26.

(a) Find the image of △*NAP* after a reflection across the *y*-axis. Label the image △*N'A'P'*.
(b) Find the image of △*N'A'P'* after a reflection across the *x*-axis. Label the image △*N"A"P"*.
(c) What specific single motion maps △*NAP* to △*N"A"P"*?

27. (a) Place a mirror on the line. What is the reflection of the name MATT?

(b) Make up another name that matches its own vertical reflection.

28. Rotate the following sign (from *Unexpected Hanging* by Martin Gardner) 180°. What does it say?

29. What rigid motion is suggested by each of the following?
(a) A tree falling to the ground
(b) A pair of shoes
(c) A train traveling straight down a railroad track

30. A dog is tied in front of a house. Draw the limits of the dog's range.

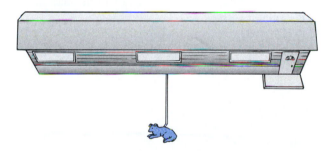

31. You can make a pattern using a rigid motion. Starting at the left, the following pattern was made using a reflection through the vertical line into the next frame.

Draw a figure in each left-hand box, and use a reflection or translation three times to create a pattern.

(a)

(b)

32. A heavy container must be moved as shown. The easiest way is to move it to an adjacent square by rotating the container around one of its corners.

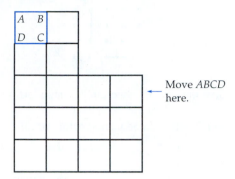

← Move *ABCD* here.

(a) Fill in the squares, including the positions of the corners of the container as it moves from start to finish.
(b) Describe the specific motion used in each step.

33. Trace one of each pair of figures, and use a rotation, translation, or reflection to determine if the two figures are congruent. If the figures are congruent, state which transformation you used to show their congruence. Describe the transformation as specifically as possible.

(a)   (b)   (c)

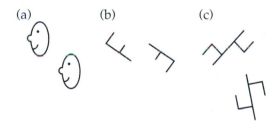

34. Draw two congruent figures such that one figure cannot be mapped onto the other using a *single* rotation, translation, or reflection.

35. (a) Give the center point and the degree of rotation that maps $\overline{AB}$ to $\overline{DC}$.

```
C     D
•     •

•  •  •

•     •
A     B
```

(b) Give the line of reflection that maps $\overline{AB}$ to $\overline{CD}$.

36. (a) Give the center point and the degree of rotation that maps $\overline{AB}$ to $\overline{A'B'}$.

(b) Give the line of reflection that maps $\overline{AB}$ to $\overline{A'B'}$.

37. A point P is translated using vector $\overrightarrow{AB}$. If P is not on $\overleftrightarrow{AB}$, then $\overleftrightarrow{PP'} \parallel$ _____, and $\overleftrightarrow{PA} \parallel$ _____.

38. What sets of points does a reflection map to itself?

39. (a) Do the segments in each picture appear to be congruent?

(1) $\overline{AB} \cong \overline{BC}$?   (2) $\overline{EF} \cong \overline{GH}$?

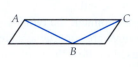

(b) Check their congruence with tracing paper.

40. The relationship between rigid transformations and congruence also applies to space figures. Which of the following pairs of figures are congruent?

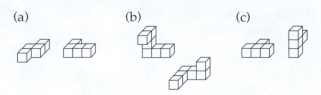

(a)   (b)   (c)

41. Find the image of the flag after a glide reflection of vector $\overrightarrow{AB}$.

42. Find the image of the flag after a glide reflection of vector $\overrightarrow{AB}$.

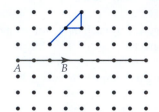

43. In the following cartoon, the motion of the man and the motion of the Z's do not correspond.
(a) Does the man's motion approximate a rotation, a translation, or a reflection?
(b) Which of the three transformations is applied to the Z's?

44. The following cartoon shows that two successive reflections through parallel lines are the same as what single motion?

Extension Exercises

45. How was this photograph taken?

46. Suppose that you want to be able to stand in front of a wall mirror and see yourself from head to toe. Find the minimum possible height of the mirror, and tell how it would be positioned.

In Exercises 47 and 48, fill in the blanks with any of the following words that make the statement true:

rotation   translation   reflection

47. Under a _____ in a plane, no points remain fixed.

48. After a _____ in a plane, two lines that are parallel have images that are parallel to each other.

49. A function has exactly one output for each input or set of inputs. Explain how a rigid transformation is a function.

Technology Exercises

50. Try translating a triangle as follows.
  (a) Construct a triangle with dynamic geometry software. ___
  (b) Construct $\overline{AB}$ (not too far from the triangle). Select point A and then B to slide in the direction of A to B, and select Mark Vector on the Transform menu.
  (c) Select the whole triangle. Then choose Mark Vector in the Transform Menu. Then choose Translate . . . By Marked Vector.

51. Try reflecting a triangle as follows.
  (a) Construct a triangle with dynamic geometry software.
  (b) Construct a segment near the triangle and select it with Mark Mirror from the Transform menu.
  (c) Select the whole triangle. Then choose Reflect from the Transform menu.

52. Try rotating a triangle as follows.
  (a) Construct a triangle with dynamic geometry software.
  (b) Construct a point and select it with Mark Center from the Transform menu.
  (c) Select the whole triangle. Then choose Rotate from the Transform menu. Choose an angle measure for the rotation.

Labs and Projects

53. Pentominos are made up of five connected squares. Each square must be connected to some other square on at least one *complete* side.

Pentomino   Pentomino   Not a pentomino

(a) Which two of the following pentominos are congruent? What motion maps one congruent pentomino onto the other?

(1)   (2)   (3)

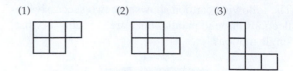

(b) Draw 12 different (noncongruent) pentomino shapes.

(c) Cut out the 12 different pentomino shapes. See whether you can put them together to form a single rectangle.

9.2 APPLICATIONS OF TRANSFORMATIONS

NCTM Standards

- explore congruency and similarity (3–5)

- create and critique inductive and deductive arguments concerning geometric ideas and relationships such as congruence, similarity, and the Pythagorean relationship (6–8)

You can study the properties of intersecting lines, parallel lines, and congruent figures with rigid transformations. Transformation can also be used to create interesting tessellation patterns.

VERTICAL, CORRESPONDING, AND ALTERNATE INTERIOR ANGLES

Rigid transformations indicate congruent pairs of angles or line segments. The symbol for congruence is $\cong$.

Figure 9–16

Discussion

LE 1   Opener

(a) Find a specific transformation that maps $\angle ATC$ in Figure 9–16 onto another angle.

(b) On the basis of part (a), it appears that $\angle ATC \cong$ _____.

(c) Using transformation geometry, find another pair of angles in Figure 9–16 that might be congruent.

Angles such as $\angle ATC$ and $\angle ETO$ in LE 1 that are formed by two intersecting lines are called **vertical angles.** Vertical angles have a common vertex but no sides in common. $\angle ETC$ and $\angle ATO$ are also vertical angles. Using rotations to *prove* that vertical angles are congruent requires a more rigorous study of rotations.

Next, consider two parallel lines and a line (called a **transversal**) that intersects them both.

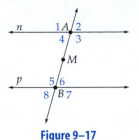

Figure 9–17

LE 2   Reasoning

In Figure 9–17, $n \parallel p$, and M is the midpoint of $\overline{AB}$.

(a) What specific transformation maps $\angle 1$ onto another angle in the diagram? Make a plan, and solve the problem.
(b) On the basis of part (a), it appears that $\angle 1 \cong$ _____.
(c) Repeat parts (a) and (b) until you have listed all possible sets of congruent angles.

Notice that each of the two parallel lines in Figure 9–17 forms four numbered angles with the transversal. Each angle (such as $\angle 1$) in one group corresponds in position to an angle in the other group (such as $\angle 5$). Thus, $\angle 1$ and $\angle 5$ are called **corresponding angles**. Other pairs of corresponding angles are $\angle 2$ and $\angle 6$, $\angle 3$ and $\angle 7$, and $\angle 4$ and $\angle 8$.

In Figure 9–17, **alternate interior angles** (such as $\angle 3$ and $\angle 5$) are nonadjacent angles on opposite (alternate) sides of the transversal ($\overleftrightarrow{AB}$) and between (interior to) the other two lines (n and p). The other pair of alternate interior angles is $\angle 4$ and $\angle 6$. A pair of alternate interior angles is like the angles in the letter **Z**.

In LE 2, you probably made conjectures about some of the corresponding or alternate interior angles. Propose more general conjectures in the following exercise.

LE 3   Reasoning

Look back at Figure 9–17, and make a conjecture about

(a) alternate interior angles.
(b) corresponding angles.

LE 2 and LE 3 should suggest the following theorem.

Parallel Lines and Corresponding and Alternate Interior Angles

If two parallel lines are cut by a transversal, then each pair of corresponding angles is congruent, and each pair of alternate interior angles is congruent.

Recall supplementary angles from Section 8.1. Figures 9–16 and 9–17 contain quite a few pairs of supplementary angles.

LE 4   Concept

Name every pair of supplementary angles in Figure 9–17 that includes $\angle 4$.

Two of the answers to LE 4, $\angle 1$ and $\angle 3$, are angles that are adjacent to $\angle 4$. **Adjacent angles** are two angles in a plane that have a common vertex and side but whose interiors do not intersect.

CONGRUENT FIGURES

Rigid transformations show that one whole figure is congruent to another. We can also show that two figures are congruent by matching up their corresponding parts.

LE 5   Concept

What set of facts about the sides and angles of △*ABC* and △*DEF* (Figure 9–18) would guarantee that the two triangles are congruent?

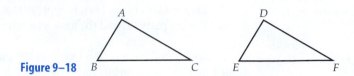

Figure 9–18

You can determine if two polygons are congruent from information about their corresponding sides and angles. Two **congruent polygons** have all their corresponding sides congruent and all their corresponding angles congruent. Consider △*ABC* and △*DEF* in Figure 9–18. If △*ABC* were translated onto △*DEF* in Figure 9–18, vertex *A* would move to *D*, vertex *B* would move to *E*, and vertex *C* would move to *F*. To indicate these corresponding vertices, we would write

$$A \leftrightarrow D \qquad B \leftrightarrow E \qquad C \leftrightarrow F$$

We could also say △*ABC* ≅ △*DEF*. In naming congruent triangles such as △*ABC* ≅ △*DEF*, the order of the letters indicates the corresponding vertices. The correspondences can also be used to match up congruent angles or congruent sides. For example, ∠*A* ≅ ∠*D*, and $\overline{BC} \cong \overline{EF}$.

You may be wondering about the difference between the congruence sign (≅) and the equal sign (=) in geometry. Congruent figures or parts of figures, such as $\overline{BC}$ and $\overline{EF}$, have the same shape and size, and this congruence is expressed as $\overline{BC} \cong \overline{EF}$. An equal sign is used most commonly to indicate that some *measure* of two geometric figures is the same. The expression "*BC* = *EF*" means that the length of $\overline{BC}$ equals the length of $\overline{EF}$. The expressions *BC* and *EF* represent *numbers* (lengths), whereas the terms $\overline{BC}$ and $\overline{EF}$ represent line segments.

The following chart shows the corresponding notations for equality and congruence of line segments and angles.

	Equality	Congruence
Line segments	$BC = EF$	$\overline{BC} \cong \overline{EF}$
Angles	$m\angle A = m\angle D$	$\angle A \cong \angle D$

LE 6   Concept

RANT ≅ *BCDA*. Complete the following expressions.

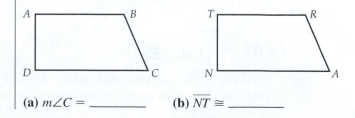

(a) $m\angle C =$ _____   **(b)** $\overline{NT} \cong$ _____

TESSELLATIONS AS ART

Artistic tessellations utilize congruent figures and transformations as a basis for creating designs. Recall that all triangles, quadrilaterals, and regular hexagons tessellate the plane.

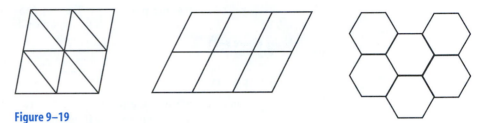

Figure 9–19

The Dutch artist M. C. Escher (1898–1972) used the shapes shown in Figure 9–19 as a basis for many ingenious tessellations (Figure 9–20). Escher was inspired by Moorish art he saw during his travels to Spain in the 1930s.

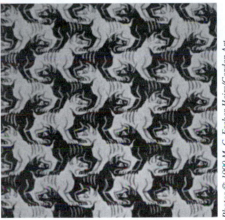

Photos © 1990 M. C. Escher Heirs/Cordon Art—Baarn–Holland.

Figure 9–20

LE 7   Skill

If each design in Figure 9–20 covered an entire plane, name a transformation that would map the design onto itself.

How did Escher create his tessellations? By altering polygons that tessellate with translations, rotations, or reflections. It is easiest to use translations. Start with a rectangle.

Now take out a piece from one side, and translate it to create a new figure that tessellates (Figure 9–21 on next page). (If you have scissors and tape, you could cut out the piece and tape it on the opposite side.)

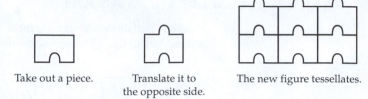

Take out a piece.   Translate it to   The new figure tessellates.
                           the opposite side.

Figure 9–21

LE 8   Skill

Translate each missing piece of the square to the opposite side to create a figure that tessellates. (Use tracing paper if you have it.)

(a)   (b)

(c) Make a tessellating pattern using 4 copies of the figure you drew in part (a). If you like, add color or a design to the interior of the figures.
(d) Make a tessellating pattern using 4 copies of the figure you drew in part (b).

You can make a more complicated pattern using two translations.

LE 9   Concept

The square has a piece removed from the left and a piece removed from the bottom.

(a) Translate the piece from the left to the right and the piece from the bottom to the top, and draw the result. (Use tracing paper if you have it.)
(b) Make a tessellating pattern using 4 copies of the figure you drew in part (a).

Now create one yourself.

LE 10   Concept

(a) Alter a square, rhombus, or regular hexagon by making two translations to create a new tessellating shape.
(b) Make a tessellating pattern with 4 copies of your shape. Color it or add drawings or designs inside each shape.

Tessellations can also be created with computer software.

LE 11   Summary

Tell what you learned about parallel lines and congruent angles in this section.

ANSWERS TO SELECTED LESSON EXERCISES

1. (a) 180° rotation around T (b) $\angle ETO$
  (c) A half-turn around T appears to map $\angle ATO$ onto
    $\angle ETC$.

2. (c) $\angle 1 \cong \angle 3 \cong \angle 5 \cong \angle 7$ and
    $\angle 2 \cong \angle 4 \cong \angle 6 \cong \angle 8$

4. $\angle 4$ and $\angle 1$, $\angle 4$ and $\angle 3$, $\angle 4$ and $\angle 5$, $\angle 4$ and $\angle 7$

5. $\angle A \cong \angle D$, $\angle B \cong \angle E$, $\angle C \cong \angle F$, $\overline{AB} \cong \overline{DE}$, $\overline{AC} \cong \overline{DF}$,
  $\overline{BC} \cong \overline{EF}$

6. (a) $m\angle A$ (b) $\overline{DA}$

7. Beetles: a reflection through a line that splits any beetle in half down the middle or a translation that moves
  one white beetle's left eye to the left eye of the white
  beetle above it

  Dogs: a translation up or to the right that moves the
  right ear of one white dog to the right ear of another
  white dog

9.2   HOMEWORK EXERCISES

Basic Exercises

1. Suppose $n \parallel p$

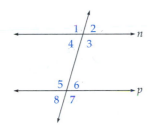

(a) Name two pairs of vertical angles.
(b) Name two pairs of alternate interior angles.
(c) Name two pairs of corresponding angles.
(d) Name a pair of adjacent angles.
(e) If $m\angle 3 = 110°$, give the measures of the other
    seven angles.
(f) Name a pair of supplementary angles.

2. Suppose $n \parallel p$.

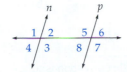

(a) Name two pairs of vertical angles.
(b) Name two pairs of alternate interior angles.
(c) Name two pairs of corresponding angles.
(d) Name a pair of adjacent angles.
(e) If $m\angle 4 = 62°$, give the measures of the other
    seven angles.
(f) Name a pair of supplementary angles.

3. *BART* is a parallelogram.

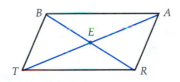

(a) What single transformation maps $\overline{BA} \to \overline{RT}$ and
    $\overline{BT} \to \overline{RA}$?
(b) On the basis of part (a), it appears that
    $\overline{BA}$ _____ $\overline{RT}$ and $\overline{BT}$ _____ $\overline{RA}$.
(c) Find a specific transformation that maps $\angle ABT$
    to another angle in the diagram.
(d) On the basis of part (c), it appears that
    $\angle ABT \cong$ _____.
(e) Repeat parts (c) and (d) for $\angle BTR$.

4. Assume that $\triangle ACT$ is an isosceles triangle, with $\overline{AC} \cong \overline{AT}$ and M the midpoint of $\overline{CT}$.

(a) Name a specific transformation that maps $\overline{AC} \rightarrow \overline{AT}$.
(b) Name a second specific transformation that maps $\overline{AC} \rightarrow \overline{AT}$.
(c) Find a transformation that suggests which two angles in the triangle are congruent, and name them.

5. In the diagram, $\overleftrightarrow{CD} \parallel \overleftrightarrow{GH}$.

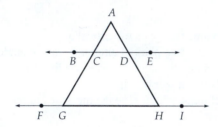

(a) Name four pairs of congruent corresponding angles.
(b) Name four pairs of congruent alternate interior angles.

6. The Parallel Line and Alternate Interior Angles Theorem can be used to prove that the sum of the interior angles of a triangle is $180°$.
(a) Draw $\triangle ABC$ and $\overleftrightarrow{AD}$ parallel to $\overleftrightarrow{BC}$, and label the angles as shown.

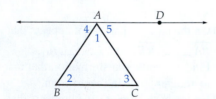

(b) What do we want to prove about $\angle 1$, $\angle 2$, and $\angle 3$?
(c) Name two congruent pairs of alternate interior angles.
(d) Explain why $m\angle 1 + m\angle 2 + m\angle 3 = 180°$. (*Hint:* Start with an equation that includes $m\angle 1$, $m\angle 4$, and $m\angle 5$.)
(e) Does part (d) involve induction or deduction?

7. Suppose $n \parallel p$. The **alternate exterior angles** (such as $\angle 1$ and $\angle 7$) are nonadjacent angles on opposite sides of the transversal.

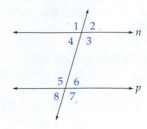

(a) Name another pair of alternate exterior angles.
(b) Make a conjecture about the alternate exterior angles.
(c) How could you prove $\angle 1 \cong \angle 7$ if you know that vertical angles are congruent and alternate interior angles are congruent?

8. Suppose $n \parallel p$.

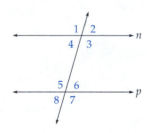

(a) Name two interior angles that are on the same side of the transversal.
(b) Are these two angles congruent, supplementary, or neither?

9. Under what conditions would each of the following be congruent?
(a) Two squares   (b) Two rays

10. Under what conditions would each of the following be congruent?
(a) Two line segments   (b) Two rectangles

11. $\triangle SUE \cong \triangle PAD$
(a) $\overline{ES} \cong$ _____
(b) What angle corresponds to $\angle DPA$?

12. $\triangle MAY \cong \triangle HUT$.
(a) Which angle of $\triangle HUT$ corresponds to $\angle M$?
(b) $TH =$ _____

13. (a) Select one of the repeated figures in the following drawing by Escher. Find a center point and the number of degrees of a clockwise rotation that maps it onto another figure.

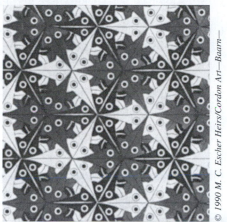

© 1990 M. C. Escher Heirs/Cordon Art—Baarn—Holland.

(b) Select one of the repeated figures. Find a line of reflection that maps it onto another figure of the same size and shape.

14.

© 1990 M. C. Escher Heirs/Cordon Art—Baarn—Holland.

Select one of the repeated figures. Describe a specific transformation that maps it onto another figure of the same size and shape.

15. (a) Alter a square or rhombus by making two translations to create a new tessellating shape.
(b) Make a tessellating pattern with 4 copies of your shape. Color it or add drawings or designs inside each shape.

16. (a) Alter a regular hexagon by making two translations to create a new tessellating shape.
(b) Make a tessellating pattern with 4 copies of your shape. Color it or add drawings or designs inside each shape.

17. You can also create a tessellating pattern with a rotation around a vertex. Take out a piece from one side, and rotate it to create a new figure that tessellates.

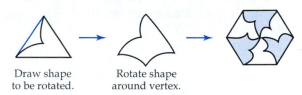

Draw shape   Rotate shape
to be rotated.   around vertex.

Rotate each missing piece of the square around a vertex that it touches to create a figure that tessellates (Use tracing paper if you have it.)

(a)   (b)

(c) Make a tessellating pattern using 4 copies of the figure you drew in part (a). If you like, add color or a design to the interior of the figures.
(d) Make a tessellating pattern using 4 copies of the figure you drew in part (b).

18. Use the method of the preceding exercise to create your own tessellation based on a rotation.

Extension Exercises

19. Suppose $\overleftrightarrow{AB} \parallel \overleftrightarrow{DE}$. Find all the missing angle measures.

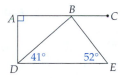

20. Suppose $\overleftrightarrow{AB} \parallel \overleftrightarrow{DE}$. Find all the missing angle measures.

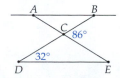

21. Suppose *ABCD* is a parallelogram. Explain why ∠1 ≅ ∠3.

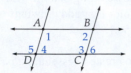

22. Use the following diagram to explain why vertical angles ∠1 and ∠3 are congruent. (*Hint:* How are they related to ∠2?)

Technology Exercises

23. Use dynamic geometry software to investigate the angles formed by two intersecting lines.

(a) Construct $\overleftrightarrow{AB}$ and $\overleftrightarrow{AC}$, and drag to make sure that *A* stays as the intersection point.

(b) Construct *D* on $\overleftrightarrow{AB}$ so that *A* is between *B* and *D*. Construct *E* on $\overleftrightarrow{AC}$ so that *A* is between *C* and *E*.

(c) Measure the angles formed by the intersecting lines.

(d) Drag points *B* and *C*, and check the angle measures again.

(e) Write your conclusions about the angle measures formed by two intersecting lines.

24. Investigate the angles formed by two parallel lines and a transversal.

(a) Construct $\overleftrightarrow{AB}$ and a point *C* not on the line.

(b) Construct $\overleftrightarrow{AC}$.

(c) Now construct a line through *C* that is parallel to $\overleftrightarrow{AB}$.

(d) Construct points on two more points on each of the three lines so that all the angles can be named with three letters.

(e) Measure the different angles formed by the intersecting lines.

(f) Drag point *A*, and check the measures of the angles.

(g) Drag $\overleftrightarrow{AC}$, and check the measures of the angles.

(h) Write your conclusions about the angle measures formed by two parallel lines and a transversal.

9.3   CONSTRUCTIONS AND CONGRUENCE

NCTM Standards

- explore congruence and similarity (3–5)
- create and critique inductive and deductive arguments concerning geometric ideas and relationships such as congruence, similarity, and the Pythagorean relationship (6–8)
- recognize and use connections among mathematical ideas (pre-K–12)

Ancient Greek mathematicians enjoyed the challenge of copying plane figures by drawing a series of line segments (using a straightedge) and circles (using a compass), since they considered lines and circles to be the basic units of geometry. Many plane figures can be copied or divided into two equal parts using only lines and circles.

The ancient Greeks succeeded in constructing nearly everything they attempted, but they were unable to trisect an angle or to construct a square equal in area to a given circle. More recently, it has been proven that these two constructions are impossible.

Today, we do constructions as historical rituals and as thought-provoking puzzles. As you do your constructions, think of the ancient Greeks who puzzled over the same problems, wondering what they could and could not construct with limited tools.

LE 1   Opener

Before doing anything fancy, draw the basic figures—a circle and a line segment—with your compass and straightedge, respectively.

CONSTRUCTING CONGRUENT FIGURES

What kind of congruent figures can be constructed with a compass and a straightedge? For starters, you should be able to draw a circle and a line segment congruent to those that you drew in LE 1.

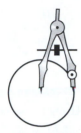

LE 2   Skill

Construct Congruent Circles

(a) Do two circles with radii of equal length have to be congruent?
(b) Part (a) is the basis for copying a circle. Place the point of your compass at the center of the circle (from LE 1) and the pencil point on any point of the circle (Figure 9–22).
(c) Hold this distance (the radius), and use it to draw a congruent circle elsewhere on the paper.

Figure 9–22

LE 3   Skill

Construct Congruent Line Segments

(a) What makes two line segments congruent?
(b) Draw a ray with endpoint *A* that is longer than your line segment from LE 1.

A

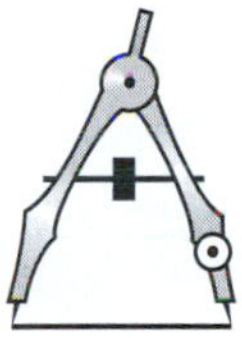

(c) Use your compass to span the length of the segment (from LE 1) by placing the compass point on one endpoint and the pencil point on the other (Figure 9–23).
(d) Using the compass width from part (c), move the compass point to *A*, and draw an arc that intersects the ray. Label the intersection point *B*. Segment *AB* is a copy of your original segment.

Figure 9–23

The first two constructions of congruent figures are not all that impressive. Let's move on to some more interesting constructions. (*Note:* If you prefer to use drawing software to investigate the conditions that make two triangles congruent, see the last computer exercise in the homework exercises.)

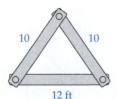

10   10

12 ft

Figure 9–24

LE 4   Connection

Suppose you are designing a bridge with triangular supports from steel beams of length 10 ft, 10 ft, and 12 ft (Figure 9–24).

(a) Will these triangles be rigid?
(b) If each person in the work crew makes one of these triangles correctly, will the triangles all be approximately congruent?

The next construction shows how the shape and size of a triangle are determined by specifying the lengths of its three sides.

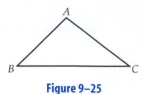

Figure 9–25

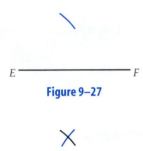

Figure 9–27

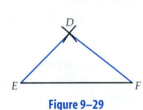

Figure 9–28

Figure 9–29

LE 5 Skill

Construct Congruent Triangles (SSS Method)

(a) Can you figure out a way to construct a triangle congruent to △*ABC* (Figure 9–25) by copying each of its sides? If not, go on to part (b).

If your method worked, compare it to the method outlined in parts (b)–(e).

(b) Copy $\overline{BC}$ using the method of LE 3. Label the copy $\overline{EF}$ (Figure 9–26).

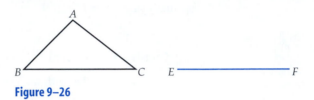

Figure 9–26

(c) Place the compass point on *B* and the pencil point on *A*. Using the same compass opening, place the compass point on *E* and draw an arc above *EF* (Figure 9–27).

(d) Place the point of your compass on *C* and the pencil point on *A*. Now use the same compass opening to draw an arc with center *F* and radius *CA* (Figure 9–28) that intersects the last arc you drew.

(e) Label the intersection point *D*. Draw $\overline{DE}$ and $\overline{DF}$. Voila! You have constructed △*DEF* (Figure 9–29).

(f) Use tracing paper to check if △*DEF* ≅ △*ABC*.

The construction in LE 5 suggests that making *EF* = *BC*, *DE* = *AB*, and *DF* = *AC* will make △*DEF* ≅ △*ABC*. This is an example of a property called the SSS (three-sides) property of triangle congruence.

> ### SSS Triangle Congruence
>
> If three sides of one triangle are congruent to the three corresponding sides of another triangle, then the two triangles are congruent.

LE 6 Concept

Which pairs of triangles must be congruent by SSS?

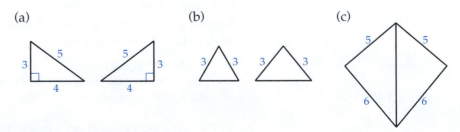

You've copied circles, line segments, and triangles. Now try copying an angle.

Figure 9–30

Figure 9–31

Figure 9–32

Figure 9–33

LE 7 Skill

Construct Congruent Angles

To construct an angle congruent to ∠A (Figure 9–30):

(a) Draw a ray with endpoint *D* (Figure 9–31).
(b) Place the compass point on *A*, and draw an arc through ∠A. Label the points of intersection *B* and *C* (Figure 9–32).
(c) Using *the same compass opening*, place the compass point on *D* and draw an arc that intersects the ray as shown. Label the point of intersection *E* (Figure 9–33).
(d) See if you can complete the construction on your own. If not, go on to parts (e) and (f).
(e) Now measure *BC* with the span of your compass, and use the same compass opening as you place the compass point on *E* and draw an arc intersecting the arc you drew before (Figure 9–34).

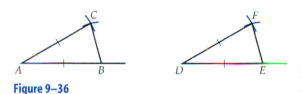

Figure 9–34   **Figure 9–35**

(f) Label the intersection of your two arcs *F*, and draw $\overrightarrow{DF}$ (Figure 9–35). ∠*FDE* ≅ ∠*CAB*!

Why does the procedure for copying an angle work? You begin with ∠A and mark an equal distance on each side (Figure 9–36). That distance is copied on each side of ∠D.

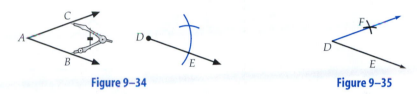

Figure 9–36

You measure $\overline{BC}$ and make $\overline{EF}$ the same length. So you have constructed two congruent triangles, △*ABC* and △*DEF* (Figure 9–37).

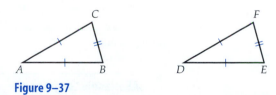

Figure 9–37

Discussion

LE 8 Reasoning

The following questions concern how the preceding construction works.

(a) Why is △*ABC* ≅ △*DEF*?
(b) From part (a), we know that △*ABC* ≅ △*DEF*. We can now deduce that
    ∠*A* ≅ _____.
(c) Explain why ∠*A* ≅ ∠*D* in part (b).

So the construction for copying an angle is based on the SSS triangle property! Now try copying a triangle, given two sides and the angle between them.

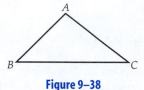

Figure 9–38

LE 9 Skill
Construct Congruent Triangles (SAS Method)

(a) Can you figure out a way to construct a triangle congruent to △*ABC* (Figure 9–38) by copying two of its sides and the angle formed by those sides (such as $\overline{AB}$, $\overline{BC}$, and ∠*B*)? If not, go on to part (b).

If your method worked, compare it to the method outlined in parts (b)–(e).

(b) Copy angle *B*, employing the method of LE 7. Label the vertex of the new angle *E* (Figure 9–39).

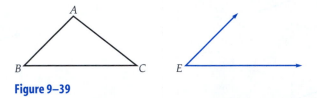

Figure 9–39

(c) See if you can complete the construction on your own. If not, go on to parts (d) and (e).

(d) Use the compass opening to mark the lengths *BC* and *BA* from point *E* on the two rays, and label the points *F* and *D*, respectively (Figure 9–40).

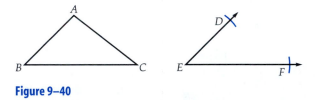

Figure 9–40

(e) Draw $\overline{DF}$ to complete △*DEF*.

(f) Use tracing paper to check if △*DEF* ≅ △*ABC*.

The construction in LE 9 suggests that making $m\angle B = m\angle E$, $BC = EF$, and $AB = DE$ will make △*DEF* ≅ △*ABC*. This is an example of the SAS property of triangle congruence.

> **SAS Triangle Congruence**
>
> If two sides and the included angle of one triangle are congruent to two corresponding sides and the included angle of another triangle, respectively, then the two triangles are congruent.

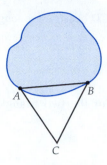

LE 10   Reasoning

You want to measure indirectly the distance *AB* across a lake (Figure 9–41).
After selecting a point *C*, measure *AC* and *BC*.
Next, construct ∠*BCD* so that ∠*BCD* ≅ ∠*BCA* and *CD* = *CA* (Figure 9–41).
Explain why *BD* must be the same as the length across the lake (*BA*). (*Hint:* First explain why △*ACB* ≅ △*DCB*. Then tell why *BD* = *BA*.)

CONSTRUCTING BISECTORS

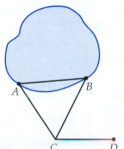

A compass and a straightedge can also be used to divide an angle or a line segment into two congruent parts. This procedure is called **bisection.** An **angle bisector** is a ray that divides an angle into two congruent angles. In Figure 9–42, $\overrightarrow{BD}$ bisects ∠*ABC*.

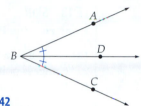

Figure 9–41

Figure 9–42

A **midpoint** (bisector) is a point that divides a line segment into two congruent segments. In the following diagram, *M* is the midpoint of $\overline{AY}$.

First, try bisecting an angle.

Figure 9–43

LE 11   Skill

Construct an Angle Bisector (Figure 9–43)

(a) Place the compass point on *A*. Draw an arc that intersects both sides ∠*A*. Label the intersection points *B* and *C* (Figure 9–44).

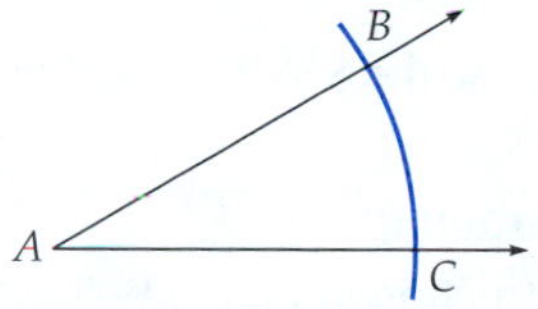

Figure 9–44 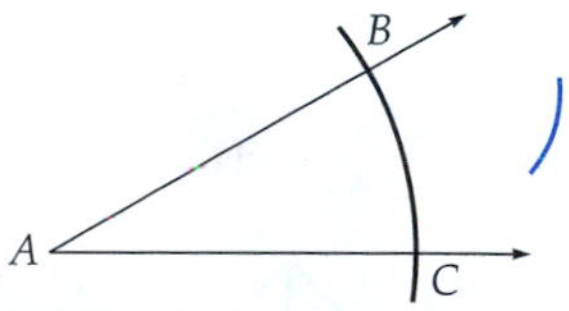   **Figure 9–45**

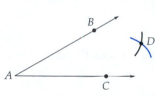

Figure 9–46

(b) Place the compass point on *B* and draw an arc across the interior of the angle (on the opposite side of $\overline{BC}$ from *A*) (Figure 9–45).

(c) Place the compass point on *C*, and use the same compass opening to draw an arc that intersects the arc you just drew (Figure 9–46). Label the intersection point *D*.

(d) Draw $\overrightarrow{AD}$ which is the angle bisector of ∠*BAC* (Figure 9–47).

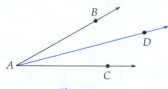

Figure 9–47

This construction also works because of the SSS property. (This SSS property is very handy for explaining constructions!)

In the construction in Figure 9–48, $AB = AC$ and $BD = CD$ because the lengths in each pair were drawn with the same radius.

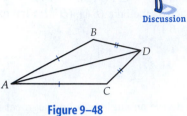

Figure 9–48

Discussion

LE 12   Reasoning

(a) In Figure 9–48, why is $\triangle ABD \cong \triangle ACD$ by SSS?
(b) If $\triangle ABD \cong \triangle ACD$, what other corresponding parts of the two triangles are congruent?
(c) Explain why $\overrightarrow{AD}$ must be an angle bisector of $\angle BAC$.

In the next construction, you will locate the midpoint of a line segment using a line perpendicular to the line segment. In other words, the construction that bisects the line segment also creates the perpendicular bisector.

LE 13   Skill

Construct the Perpendicular Bisector of a Line Segment

(a) Draw a line segment $\overline{AB}$. Select a compass radius slightly less than AB. Place the compass point on A, and draw an arc through $\overline{AB}$ as shown in Figure 9–49.
(b) Use the same compass opening. Place the compass point on B, and draw an arc through $\overline{AB}$ that intersects the other arc at two points (Figure 9–50). Label the intersection points C and D.

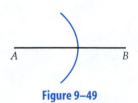

Figure 9–49

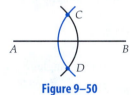

Figure 9–50   **Figure 9–51**

(c) Draw $\overline{CD}$. Label as M the point where $\overline{CD}$ intersects $\overline{AB}$. Not only is M the midpoint of $\overline{AB}$, but $\overleftrightarrow{CD}$ is the perpendicular bisector of $\overline{AB}$ (Figure 9–51). So this procedure can also be used to construct a 90° angle.

You can use the SSS and SAS properties to prove why this construction works.

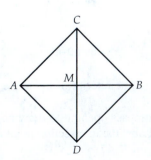

Figure 9–52

LE 14   Reasoning

In the construction in Figure 9–52, $AC = CB = AD = DB$ because the lengths were all drawn with the same radius.

(a) Why is $\triangle ACD \cong \triangle BCD$?
(b) Why is $\angle ACM \cong \angle BCM$?
(c) Why is $\triangle ACM \cong \triangle BCM$?
(d) Why is M the midpoint of $\overline{AB}$?

THE ISOSCELES TRIANGLE THEOREM

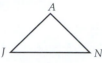

Figure 9–53

All isosceles triangles have at least two congruent sides. Must an isosceles triangle also have two congruent angles? The SSS triangle congruence property can be used to deduce the answer to this question.

Suppose that △*JAN* (Figure 9–53) is an isosceles triangle with $JA = AN$. Is $\angle J \cong \angle N$?

LE 15   Reasoning

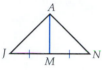

Figure 9–54

(a) In △*JAN*, draw a line segment from *A* to the midpoint *M* of $\overline{JN}$ (Figure 9–54).
(b) Explain why △*JAM* ≅ △*NAM*.
(c) Explain why $\angle J \cong \angle N$.
(d) Do parts (b) and (c) involve induction or deduction?

The angle property you proved in LE 15 is called the Isosceles Triangle Theorem.

> **The Isosceles Triangle Theorem**
>
> If two sides of a triangle are congruent, then the angles opposite them are congruent.

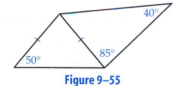

Figure 9–55

The Isosceles Triangle Theorem can sometimes be used to find missing angle measures in polygons.

LE 16   Reasoning

Fill in the missing angle measures in Figure 9–55.

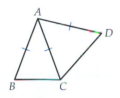

Figure 9–56

Classroom Connection

LE 17   Reasoning

An eighth grader says that since $AB = AC = AD$ (Figure 9–56), then $\angle B \cong \angle C \cong \angle D$. Is this right? If not, what would you tell the child?

CONSTRUCTING PERPENDICULAR AND PARALLEL LINES

Suppose you have a point *B* not on a line $\overleftrightarrow{AC}$.

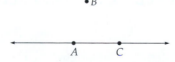

It is possible to construct a line through *B* that is perpendicular to $\overleftrightarrow{AC}$ or a line through *B* that is parallel to $\overleftrightarrow{AC}$. Consider the following application.

Discussion

LE 18   Reasoning

You are camping and must return home for an emergency.

Your location
•

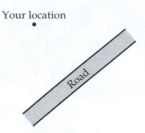

Road

(a) What is the shortest route to the road?

Construct a Line Perpendicular to a Given Line Through a Point Not on the Line

(b) Construct the shortest route to the road. (*Hint:* Place the point of the compass on "your location," and draw an arc that intersects the road at two points. Label the intersection points A and B. Then construct the perpendicular bisector of $\overline{AB}$.

   To construct parallel lines, make use of the preceding construction of a perpendicular line.

Discussion

LE 19   Skill

Construct Parallel Lines

Suppose B is a point not on $\overleftrightarrow{AC}$. Construct a line through B that is parallel to $\overleftrightarrow{AC}$. (*Hint:* Draw $\overleftrightarrow{AB}$. Then construct an angle with vertex B that is congruent $\angle BAC$.)

•B

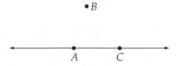

A   C

LE 20   Summary

Tell what you learned about constructions in this section.

ANSWERS TO SELECTED LESSON EXERCISES

2. (a) Yes

3. (a) They are the same length.

6. (a), (c)

8. (a) SSS triangle congruence
   (b) $\angle D$
   (c) The triangles are congruent, so the corresponding parts are congruent.

10. Since $AC = CD$, $BC = BC$, and $\angle ACB \cong \angle DCB$, $\triangle ACB \cong \triangle DCB$ by SAS. If the triangles are congruent, then $BD = BA$.

12. (a) $AB = AC$, $BD = CD$, and $AD = AD$.
   (b) $\angle B \cong \angle C$, $\angle BAD \cong \angle CAD$, $\angle BDA \cong \angle CDA$
   (c) Since $\triangle ABD \cong \triangle ACD$, $\angle BAD \cong \angle CAD$. This means that $\overrightarrow{AD}$ is the angle bisector of $\angle BAC$.

14. (a) By SSS (*AC* = *BC*, *AD* = *BD*, *CD* = *CD*)
  (b) Because △*ACD* ≅ △*BCD*
  (c) By SAS (*AC* = *BC*, ∠*ACM* ≅ ∠*BCM*, *CM*=*CM*)
  (d) Since △*ACM* ≅ △*BCM*, *AM* = *MB*. So *M* is the midpoint of $\overline{AB}$.

15. (b) By SSS (*JA* = *NA*, *JM* = *NM*, *AM* = *AM*)
  (c) Since △*JAM* ≅ △*NAM*, ∠*J* ≅ ∠*N*.

16.

17. No. Do the angles look congruent? How many triangles are involved? How would you apply the Isosceles Triangle Theorem to the two triangles?

9.3   HOMEWORK EXERCISES

Basic Exercises

1. (a) Construct a copy of △*MAN*, called △*BCD*, so that △*MAN* ≅ △*BCD*.

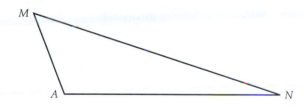

  (b) Explain why △*MAN* ≅ △*BCD*.

2. Construct a copy of △*CDE*.

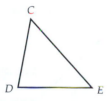

3. (a) Draw two noncongruent quadrilaterals, showing that an SSSS congruence property would not work.
  (b) How does part (a) relate to why quadrilaterals (without diagonals) are not used in structures such as bridges?

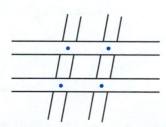

4. An architect constructs triangular roof supports, each with sides of length 1.4 m, 1.4 m, and 2.3 m. Are all the triangular supports congruent? Why or why not?

5.

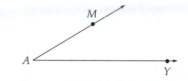

  (a) Construct a copy of ∠*MAY* called ∠*BCD*.
  (b) Explain why ∠*MAY* ≅ ∠*BCD*.
  (c) Does part (b) involve induction or deduction?

6. Construct a copy of ∠*TED*.

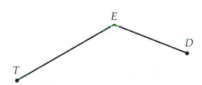

7.

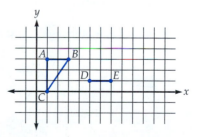

Give all possible coordinates of *F* that would determine a △*DEF* that is congruent to the triangle that has vertices *A*, *B*, and *C*.

8.

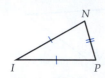

(a) Why is △*NIP* ≅ △*ACR*?
(b) Why is ∠*I* ≅ ∠*C*?

9. Copying a Triangle (ASA Method)

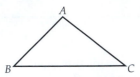

Show how to construct a copy of △*ABC* by copying two angles and the side between them (such as ∠*B*, ∠*C*, and $\overline{BC}$). This suggests the validity of the ASA triangle congruence property.

10. Would an AAS triangle congruence property work? Use the ASA property to show why. Assume that ∠*A* ≅ ∠*D*, ∠*B* ≅ ∠*E*, and $\overline{BC}$ ≅ $\overline{EF}$.

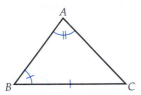

(a) Why is ∠*C* ≅ ∠*F*?
(b) Why is △*ABC* ≅ △*DEF* (showing that AAS leads to congruent triangles)?

11. Decide whether each pair of triangles must be congruent.

(a)    (b)

(c) 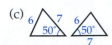   (d)

12. Decide whether each pair of triangles must be congruent.

(a)   (b)

(c)   (d)

13. Create a counterexample to an SSA congruence property by completing the following figure. Draw △*DEF* with *DE* = *AB*, *DF* = *AC*, and *m*∠*B* = *m*∠*E* so that △*DEF* is not congruent to △*ABC*.

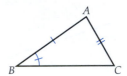

14. Show that an AAA Congruence Property would not work. Draw two noncongruent triangles with three congruent corresponding angles. (*Hint:* You could use all 60° angles.)

15. (a) Construct the angle bisctor of ∠*CAT* and call it $\overrightarrow{AB}$.

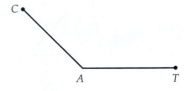

(b) Explain why your procedure works.
(c) Use congruent triangles to explain why $\overrightarrow{AB}$ bisects ∠*CAT*.

16.

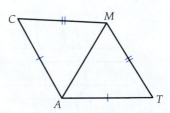

Explain why $\overrightarrow{AM}$ must be the angle bisector of ∠*CAT*.

17. A **median** of a triangle connects a vertex to the midpoint of the opposite side.
(a) Use a ruler to draw the three medians of the triangle.

(b) Construct a median from A to the midpoint of $\overline{BC}$.

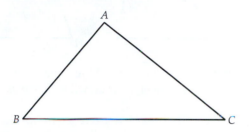

(c) Cut out a paper triangle, and fold the three medians of the triangle. What do you observe?

18. A plane is flying over water from A to B on a clear day with little wind. If there is trouble along the way, construct the point C that can be used to determine whether it would be easier for a plane with engine trouble to go on to B or to return to A.

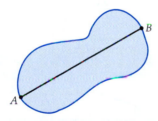

19. Construct a line segment that is three times the length of $\overline{AB}$.

20. Use a straightedge and a compass to divide a segment into four equal parts.

21. (a) Draw an angle on a piece of paper.
(b) Locate its angle bisector by paper folding.

22. (a) Draw a line segment on a piece of paper.
(b) Locate its midpoint by paper folding.

23. A MIRA is a red Plexiglas device that acts like a mirror. Use a MIRA to do the following constructions.
(a) Bisect an angle.
(b) Find the midpoint of a line segment.

24. Use a MIRA to do the following constructions.
(a) Copy an angle.
(b) Copy a triangle.

25. Show that a triangle $\triangle ABC$, with $\angle B \cong \angle C$, has two congruent sides. (*Hint:* Draw the angle bisector of $\angle BAC$.)

26. Use the Isosceles Triangle Theorem to explain why an equilateral triangle must have three congruent angles.

27. Find all missing angle measures in each diagram.

(a)   (b)   (c)

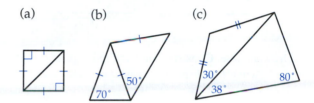

28. Find all the angle measures in the regular octagon.

29. Construct an equilateral $\triangle ABC$ with a side $\overline{AB}$. (*Hint:* C must be a distance AB away from both A and B.)

30. $m\angle PUN = 60°$. Construct an equilateral triangle $\triangle NUT$ with T on $\overrightarrow{UP}$.

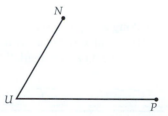

31. Construct a line perpendicular to m through P.

32. Construct a line parallel to *m* through *P*.

Extension Exercises

33. Suppose *B* is on a point $\overleftrightarrow{AC}$. Construct a line through *B* that is perpendicular to $\overleftrightarrow{AC}$. (*Hint:* Construct an angle bisector for ∠*ABC*.)

34. Draw a circle. Construct an inscribed regular hexagon as shown. (*Hint:* Find the measures of some central angles.)

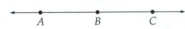

35. Imagine that you (*A*) see a boat at point *B* and want to know how far it is from the shore. Over 2000 years ago, Thales described the following method for finding the distance. Walk along the shore to a point *C*, and mark it.

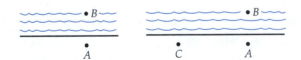

Continue walking an equal distance to point *D* so that *AC* = *CD*. Then walk directly away from the water until *C* is lined up with *B*, and mark this point *E*.

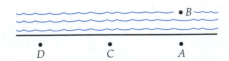

DE equals the distance from the shore to the boat! Explain why *DE* = *AB*. (*Hint:* Draw △*ABC* and △*DEC*.)

36. Prove that the diagonals of a rectangle are equal.

37. Suppose that two quadrilaterals have three congruent corresponding sides and two congruent corresponding included angles, creating SASAS. Show that the two quadrilaterals must be congruent.

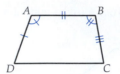 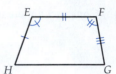

(*Hint:* Show that the remaining three corresponding parts are congruent, after drawing diagonals $\overline{AC}$ and $\overline{EG}$.)

38. Show that an ASASA quadrilateral property works for *ABCD* and *EFGH*. (*Hint:* Draw $\overline{AC}$ and $\overline{EG}$.)

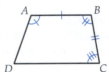 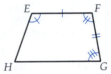

39. *AB* = *AC*. Find *m*∠*A*.

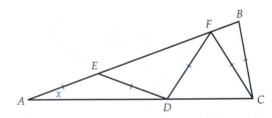

40. *AB* = *BC* = *BD*. Find *m* ∠*ADC*.

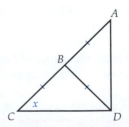

Technology Exercises

41. Will two triangles with three congruent corresponding sides (SSS) always be congruent?
   (a) Construct $\triangle ABC$, and measure each of its sides.
   (b) Draw $\overline{DE}$ congruent to $\overline{AB}$.
   (c) Construct a circle with center D and radius $\overline{AC}$.
   (d) Construct a circle with center E and radius $\overline{BC}$.
   (e) Label the intersection of the two circles F, and draw $\overline{DF}$ and $\overline{EF}$.
   (f) Now compare the measures of the two triangles and decide if they are congruent.

42. Use dynamic geometry software to test the following triangle congruence properties.
   (a) SS (*Hint:* Is it possible to draw two triangles with two congruent corresponding sides that are not congruent? If so, then SS does not show congruence.)
   (b) SAS   (c) AS   (d) SSA
   (e) ASA   (f) AAS

43. (a) Construct $\overline{AB}$.
   (b) Construct the perpendiular bisector of $\overline{AB}$.
   (c) Mark and label three different points on the perpendicular bisector.
   (d) Find the distance of each point in part (c) from A and from B.
   (e) Make a conjecture about perpendicular bisectors.

44. Follow the directions to construct an isosceles triangle with dynamic geometry software, and then check its angles.
   (a) Construct a circle with center A and a point B on the circle.
   (b) Construct radius $\overline{AB}$. Construct radius $\overline{AC}$. Drag C to verify that $\overline{AC}$ remains a radius.
   (c) Construct $\overline{BC}$. Hide the circle.
   (d) Highlight the three angles in turn, and measure them.
   (e) Drag the vertices of the triangle, and note the angle measures.
   (f) Make a conjecture about the angles of an isosceles triangle.

Labs and Projects

45. The lateral surface of a cone can be built from a region like the following one. Try it.

(a) Draw a circle with a compass, and cut it out.
(b) Draw two radii, and cut out the smaller closed region.

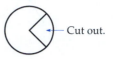

Cut out.

(c) Put the two cut edges together to form the lateral surface of a cone.

46. Using only a compass, a straightedge, and a pencil or pen, create one of the following designs or make one of your own.

(a)   (b)   (c)

47. A drawing may appear to represent a three-dimensional object that is impossible to make. The "Penrose triangle" is such a drawing.

You will need a compass and a ruler to make a Penrose triangle.
(a) Draw an equilateral triangle with sides 12 cm long. Draw the base with a ruler. Locate the top with a compass by measuring 12 cm from each end of the base.

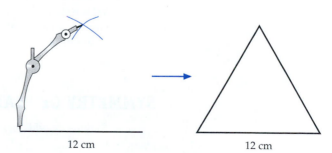

12 cm   12 cm

(b) Mark points 1 cm and 2 cm from each corner.

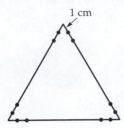

1 cm

(c) Draw line segments lightly, connecting the points as shown.

(d) Go over the lines shown. Erase the others.

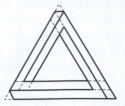

(e) Shade them as shown in the original Penrose triangle.

9.4 SYMMETRY

NCTM Standards

- identify and describe line and rotational symmetry in two- and three-dimensional shapes and designs (3–5)
- examine the congruence, similarity, and line or rotational symmetry of objects using transformations (6–8)
- recognize and use connections among mathematical ideas (pre-K–12)

The term "congruence" describes a relationship between two figures that are the same size and shape. The term "symmetry" describes a way in which a single figure can be divided into parts that are the same size and shape and possess a particular orientation.

SYMMETRY OF PLANE FIGURES

Symmetry adds beauty and balance to natural forms and architectural designs (Figure 9–57).

Figure 9–57

Can you recognize plane figures that have symmetry?

Discussion

LE 1   Opener

Which designs in Figure 9–58 possess some kind of symmetry?

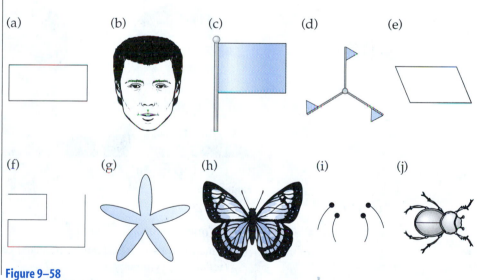

(a)   (b)   (c)   (d)   (e)

(f)   (g)   (h)   (i)   (j)

Figure 9–58

In LE 1, all but two of the figures are symmetric. You can check your answers with the following test for symmetry.

> **A Test for Symmetry**
>
> A plane figure is symmetric if you can copy it onto tracing paper and find a different position for the tracing paper in which the traced figure coincides with the original figure.

LE 2 Concept

Use tracing paper to determine which figures in LE 1 are symmetric.

For each symmetric figure, you moved the tracing paper to a new position, and the traced figure fit on top of the original figure. The motion you used was either a rotation (other than a full turn) or a reflection.

LINE (REFLECTION) SYMMETRY

A human face is not quite symmetric. Compare the real photograph of my face in Figure 9–59(a) to the two symmetric versions of me. Each of the symmetric versions is constructed from one half of my face. This example illustrates the most well-known type of symmetry: line (reflection) symmetry.

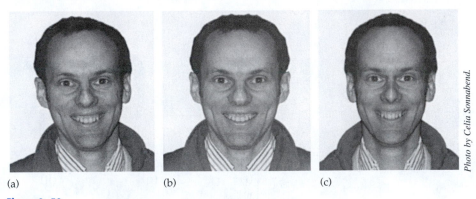

(a)   (b)   (c)

Photo by Celia Sonnabend.

Figure 9–59

LE 3 Connection

(a) Which shapes in LE 1 can be reflected onto themselves? (These figures have line (reflection) symmetry.)

(b) Select one of the shapes that has line symmetry. Draw a line that divides it into two equal parts such that one part is the reflection image of the other through the given line.

(c) Select one of the shapes that has line symmetry. How could you use paper folding to show the symmetry?

(d) Select one of the shapes that has line symmetry. How could you use a pocket mirror or a MIRA (if you have one) to show the symmetry?

A plane figure has **line (reflection) symmetry** if and only if it can be reflected across a line *m* so that the figure is its own image. This means that its image coincides with its original position. The reflection line *m* is called the **line of symmetry.** Each point in the figure is the same distance from the line of symmetry as its image point is.

Most elementary- and middle-school textbooks describe line symmetry in terms of paper folding. According to them, a plane figure has line symmetry if it can be folded along a line so that one half of the figure matches the other half. The tracing-paper test in this section is a more general test that also works for other kinds of symmetry. You may either fold paper or reflect tracing paper to check line symmetry.

You will need scissors and paper for the following exercise.

LE 4 Reasoning

Fold a piece of paper in half, and cut out the shape along the fold. Predict the shape that will result when you unfold the paper. Then see if your prediction was right.

Figure 9–60

Line symmetry is also called mirror symmetry, since a line of symmetry acts like a double-sided mirror. Points on each side are reflected to the opposite side. Many figures have several line (reflection) symmetries. The equilateral triangle in Figure 9–60 has three lines of symmetry.

LE 5 Concept

Draw all lines of symmetry for the regular pentagon in Figure 9–61.

Figure 9–61

Discussion

LE 6 Concept

Why isn't the line shown in Figure 9–62 a line of symmetry for the shape in the drawing?

LE 7 Skill

Complete each figure so that the line acts as a line of symmetry.

(a)   (b)   (c)

 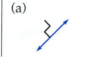

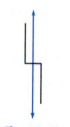

Figure 9–62

ROTATIONAL SYMMETRY

The surfaces of snowflakes (Figure 9–63 on page 506) suggest a second major type of plane symmetry: rotational symmetry. Natural and manufactured objects that have rotational symmetry possess an order that enhances their beauty and function.

Courtesy of National Oceanic and Atmospheric Administration.

Figure 9–63

LE 8   Opener

Which of the shapes in LE 1 can be rotated *less than 360°* (a full turn) onto themselves?

The figures you selected in LE 8 should have rotational symmetry. A plane figure has **rotational (turn) symmetry** if and only if it can be rotated more than 0° and less than 360° so that its image coincides with its original position.

Many shapes have several rotational symmetries. The equilateral triangle in Figure 9–64 can be turned 120° or 240° around its center to coincide with its original position. Therefore, it has 120° and 240° rotational symmetry.

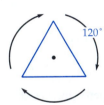

Figure 9–64

LE 9   Concept

What rotational (turn) symmetries does a regular pentagon have (greater than 0° and less than 360°)?

Can a plane figure have both rotational and reflection symmetry? Can it have one kind of symmetry without the other?

Classroom Connection

LE 10   Concept

A sixth grader says that any figure that has rotational symmetry must also have reflection symmetry. Is that correct? If not, find a counterexample among the figures in LE 1.

Discussion

LE 11   Connection

(a) Which shapes in LE 1 have rotational *and* line (reflection) symmetry?
(b) Which shapes in LE 1 have rotational symmetry but not line (reflection) symmetry?
(c) Which shapes in LE 1 have line (reflection) symmetry but not rotational symmetry?

Discussion

Figure 9–65

LE 12 Reasoning

Devise a plan and solve the following problem. Shade the smallest number of squares in Figure 9–65 so that the pattern has line *and* rotational symmetry. (*Hint:* First shade squares for line symmetry. Then shade additional squares for rotational symmetry.)

SYMMETRY OF SPACE FIGURES

Some three-dimensional figures have the same kinds of symmetry as two-dimensional shapes. The beetle and the chair in Figure 9–66 each have two halves that are virtually identical. These objects have approximate reflection (plane) symmetry.

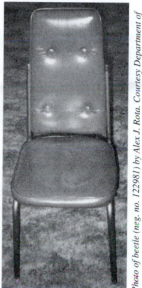

Photo of beetle (neg. no. 122981) by Alex J. Rota. Courtesy Department of Library Services, American Museum of Natural History. Photo of chair by Tom Sonnabend.

Figure 9–66

Reflection (plane) symmetry of space figures exists if a plane divides the figure in half so that one half of the figure is a mirror image of the other. To put it more precisely, each point on one side of the plane must have a corresponding point the same (perpendicular) distance away on the opposite side. This relationship is analogous to the distance relationship for line symmetry.

LE 13 Concept

The beetle in Figure 9–66 has approximate reflection symmetry. Where is the plane of symmetry?

Reflection symmetry gives animals balance. Many animals and many machines have reflection symmetry.

LE 14   Connection

Find an object in your classroom that has approximate reflection symmetry that gives it balance.

Figure 9–67

Figure 9–68   *Sonya Kovalevsky*

Courtesy of Library of Congress

Saturn has approximate rotational symmetry. After being rotated any amount around the line shown in Figure 9–67, its image will coincide with its original position.

Rotational symmetry of space figures exists if there is an axis of rotation (a line) around which the figure can be turned (less than a full turn) so that it coincides with itself. The rotational symmetry of an object gives the object "directional flexibility." For example, a square card table can be placed in four different positions that function in exactly the same way.

In 1888, the Russian mathematician Sonya Kovalevsky (1850–1891; Figure 9–68) gained prominence with her research paper *On the Rotation of a Solid About a Fixed Point*, which won the French Academy of Sciences prestigious Prix (Prize) Bordin.

As a child, Sonya Kovalevsky slept in a room that was wallpapered with her father's calculus notes from college, since the family couldn't afford to buy wallpaper. Perhaps that helped to stimulate her interest in mathematics.

In the nineteenth century, women were not allowed to attend universities in Russia. Kovalevsky married so she could emigrate to Germany, where she studied mathematics. Karl Weierstrass, one of Germany's greatest mathematicians, tutored her privately, since the University of Berlin would not admit her to his lectures. Thanks to Weierstrass's strong recommendation, the University of Göttingen awarded Kovalevsky a doctorate even though she never officially attended school there! Kovalevsky was clearly qualified for the degree on the basis of three outstanding research papers.

In 1883, Kovalevsky was invited to teach at the University of Stockholm. From 1889 until her death in 1891, she worked there as a full professor of mathematics.

Discussion

LE 15   Connection

(a) Find an object in your classroom that has approximate rotational symmetry that gives it directional flexibility.

(b) Describe the location of the axis of rotation.

(c) What rotational symmetries between 0° and 360° does the object have?

LE 16   Summary

Tell what you learned about symmetry is this section.

ANSWERS TO SELECTED LESSON EXERCISES

3. (a) a, b, g, h, i, j

5.

6. A reflection through the line does not leave the figure unchanged.

7.

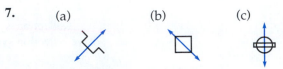

     (a)   (b)   (c)

8. a, d, e, g

9. 72°, 144°, 216°, 288° clockwise (or counterclockwise)

10. Part (d) is a counterexample.

11. (a) a, g   (b) d, e   (c) b, h, i, j

12.

13. Perpendicular to the plane of the wings and dividing its body in half

14. A garbage can, chair, or table

15. (a) A garbage can or table

9.4   HOMEWORK EXERCISES

Basic Exercises

1. Does symmetry enhance the beauty of a shape? Cover up the right half of the butterfly. Is it still as pleasing to the eye?

2. Describe a test for symmetry.

3. Which of the following figures have line (reflection) symmetry?

   (a)   (b)   (c)

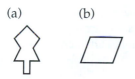

4. Which of the following figures have line (reflection) symmetry?

   (a)   (b)   (c)

5. (a) Draw two intersecting lines.

(b) Draw the two lines of symmetry for your lines in part (a).

(c) Repeat parts (a) and (b), starting with lines intersecting at a different angle.

(d) Propose a generalization of your results.

(e) Does part (d) involve induction or deduction?

6. Draw a line of symmetry for each figure.

   (a)   (b)

7. How many lines of symmetry does each of the following quadrilaterals have?

   (a)   (b)

     Square   Rectangle

   (c)   (d)

     Rhombus   Parallelogram

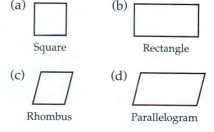

8. Draw all lines of symmetry for each figure. Verify your results with paper folding.

   (a)   (b)   (c)   (d)

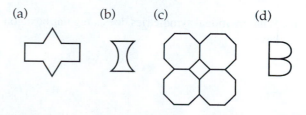

9. Complete the figure so that it is symmetric to the line.

10. (a) Fold a sheet of paper in half. Then make a cut so that the unfolded paper shows a heart.
(b) Repeat part (a), but create a pumpkin.
(c) Repeat part (a) and make your own design.

11. A third grader draws a line of symmetry for a rectangle as shown. Is this right? If not, what would you tell the child?

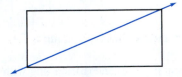

12. Suppose △*ABC* has a line of symmetry $\overleftrightarrow{AM}$. Name any sides or angles that must be congruent.

13. Sketch any of the following seven figures that are possible: hexagons that have exactly zero, one, two, three, four, five, or six lines of symmetry.

14. Draw an octagon that has
(a) no lines of symmetry.
(b) exactly four lines of symmetry.
(c) exactly two lines of symmetry.

15. How many rotational (turn) symmetries (greater than 0° and less than 360°) does each of the following quadrilaterals have?

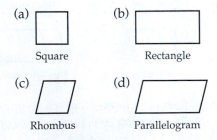

16. What rotational symmetries does a regular hexagon have?

17. What kind of symmetry does each flag have?

18. Three snowflakes follow. Which ones appear to possess (approximate) symmetry?

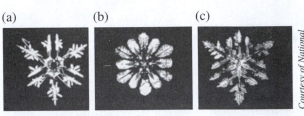

Courtesy of National Oceanic and Atmospheric Administration.

19. (a) Which letters in the word shown below have reflection symmetry?
(b) Which letters have rotational symmetry?

S Q U I D

20. The word HIDE has a horizontal line of symmetry. Write three more words that have a horizontal line of symmetry.

21. Examine a deck of playing cards.
(a) Which cards have rotational symmetry?
(b) Which cards have reflection (line) symmetry?

22. You are "O" in tic-tac-toe. For your next move, you have 8 choices, but some are equivalent. How many different kinds of moves are there?

23. Draw a plane figure that has rotational symmetry but no line (reflection) symmetry.

24. Draw a plane figure that has reflection (line) symmetry but no rotational symmetry.

25. For all plane figures having two or more reflection (line) symmetries, there is a relationship between the number of reflection symmetries and the number of rotational symmetries. Devise a plan and find the relationship.

26. Tell the number of rotational and reflection symmetries each hubcap has. (Disregard the valve stems and the logos in the center.)

(a) (b)

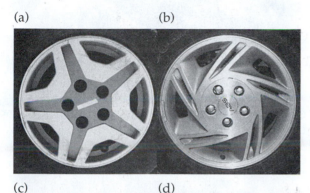

(c) (d)

Photos by Tom Sonnabend.

27. Shade the smallest number of squares so that each of the following diagrams has rotational symmetry.

(a) (b)

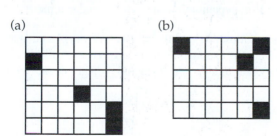

28. Shade the smallest number of squares to that each of the following figures has rotational and line symmetry.

(a) (b)

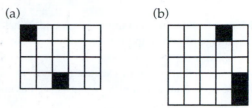

29. If an isosceles triangle, $\triangle ABC$, has a line of symmetry $\overleftrightarrow{CD}$, what can you say about each of the following?
(a) AD and DB
(b) $\angle A$ and $\angle B$

30. A plane figure that looks exactly the same upside down as right side up always has what specific symmetry?

31. (a) How many planes of symmetry does a right square pyramid have?
(b) How many axes (lines) of rotational symmetry does a right square pyramid have?

32. (a) How many planes of symmetry does a cube have?
(b) How many axes (lines) of rotational symmetry does a cube have?

33. If you rotate an isosceles triangle in space about its line of symmetry, what solid will you obtain?

34. If you rotate a rectangle in space around one of its lines of symmetry, what solid will you obtain?

35. In how many ways can each key be inserted into the corresponding hole from one side?

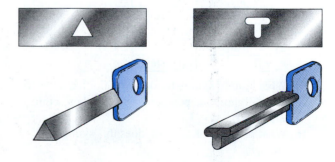

36. The driver's door of a two-door car is damaged and has to be refitted. The only available replacement is a passenger door. Will it fit?

37. Many windows are rectangular in shape. What is the advantage of this shape for a window?

38. A nut ⬡ that is used to secure a screw has approximate rotational symmetry. What is the advantage of this?

39. Find a tile floor in your home or school.
(a) What kind of symmetry do the tiles have?
(b) What is the practical benefit of this symmetry?

40. (a) What kind of symmetry do birds have?
(b) What is the practical benefit of this symmetry?

41. (a) Describe the symmetries of a tennis ball, a tennis racket, and a tennis court.
(b) Explain how the symmetries of each object relate to the function of the object.

42. (a) List five objects in your home that have reflection symmetry.
(b) List five objects in your home that have rotational symmetry.

Extension Exercises

43.

+	0	1	2	3	4	5	6	7	8	9
0	0	1	2	3	4	5	6	7	8	9
1	1	2	3	4	5	6	7	8	9	10
2	2	3	4	5	6	7	8	9	10	11
3	3	4	5	6	7	8	9	10	11	12
4	4	5	6	7	8	9	10	11	12	13
5	5	6	7	8	9	10	11	12	13	14
6	6	7	8	9	10	11	12	13	14	15
7	7	8	9	10	11	12	13	14	15	16
8	8	9	10	11	12	13	14	15	16	17
9	9	10	11	12	13	14	15	16	17	18

(a) What kind of symmetry does an addition table have? (Ignore the position of the number within each box.)
(b) What property of addition does this symmetry indicate?
(c) Does a multiplication table for 0 to 9 have the same symmetry?

44.

© 1990 M. C. Escher Heirs/Cordon Art—Baarn—Holland.

Describe all the symmetries of this pattern.

45. Trace the circle shown here on a sheet of paper. How can you locate its center with paper folding?

46. Mirror cards (developed by Marion Walters) are used to teach symmetry in elementary school. Using the mirror card at the left, where would you place a mirror on the card to obtain the figures shown in parts (a), (b), and (c)?

    (a)   (b)   (c)

47. (a) Which pentominos have reflection symmetry?
(b) Which pentominos have rotational symmetry?

48. A kaleidoscope usually has two mirrors placed at a 60° angle, as shown. One image is drawn. Fill in all the other images. (*Hint:* Include images of images.)

Labs and Projects

49. Use paper folding and cutting to create your own symmetric design.

9.5 SIMILARITY AND DILATIONS

NCTM Standards

- explore congruency and similarity (3–5)

- describe sizes, positions and orientations of shapes under informal transformations such as flips, turns, slides, and scaling (6–8)

- solve problems involving scale factors, using ratio and proportion (6–8)

Scale models of objects are the same shape but not the same size as the object. When you enlarge a photograph, you obtain a new picture in which everything is the same shape but a different size (Figure 9–69).

 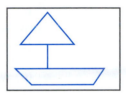

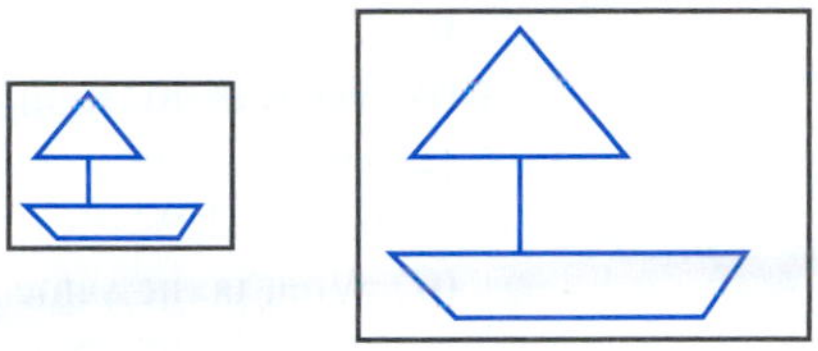

Figure 9–69

Beyond a certain age, dolphins grow larger but retain roughly the same shape (Figure 9–70).

Photo: Rob Mathewson. Courtesy Department of Library Services. American Museum of Natural History.

Figure 9–70

SIMILAR POLYGONS

Similar polygons have the same shape. What relationships exist between the angles and sides of similar polygons?

LE 1   Opener

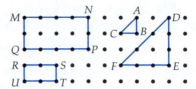

Rectangles *MNPQ* and *RSTU* are similar.

(a) What is the same about the two rectangles?
(b) What is different about the two rectangles?
(c) How are the lengths of the corresponding sides of the two rectangles related?
△ *ABC* and △*DEF* are similar.
(d) What is the same about the two triangles?
(e) What is different about the two triangles?
(f) How are the lengths of the corresponding sides of the two triangles related?

As the preceding exercise suggests, similar polygons can be defined by the following properties of their corresponding angles and sides.

Definition: **Similar Polygons**

Similar polygons have

1. congruent corresponding angles and

2. proportional corresponding lengths.

The two triangles in Figure 9–71 are similar. They have congruent corresponding angles and proportional corresponding lengths.

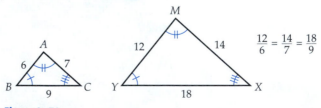

Figure 9–71

The corresponding vertices are $A \leftrightarrow M$, $B \leftrightarrow Y$, and $C \leftrightarrow X$. The relationship can be written symbolically as $\triangle ABC \sim \triangle MYX$. The symbol $\sim$ means "similar to." Note how corresponding vertices are written in corresponding positions in the names of the similar triangles.

LE 2   Concept

In each case, use the definition of similar polygons to decide if the two figures are similar.

(a)

(b)

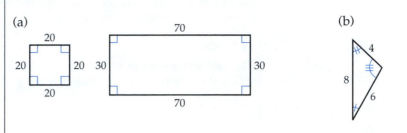

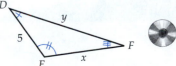

Classroom Connection

LE 3   Concept

A fifth grader asks if all squares are similar to each other. How would you respond?

When you enlarge a photograph, you create a similar photograph. Suppose you know one dimension you want for the larger photograph but not the other. You can find the missing dimension with a proportion.

• EXAMPLE 1

You want to enlarge a photograph that is 4 in. by 6 in. so that its longer dimension is 9 in. How long is the shorter dimension?

SOLUTION

An enlarged photograph should retain the same *shape*. Draw two similar rectangles to represent the photograph before and after enlargement.

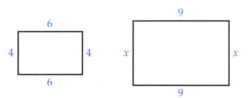

Since the rectangles are similar, corresponding lengths are proportional. So

$$\frac{6}{9} = \frac{4}{x}$$

This is the same as

$$6 \cdot x = 4 \cdot 9$$
$$6x = 36$$
$$x = \textbf{6 in.}$$

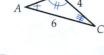

Figure 9–72

LE 4   Skill

The two triangles in Figure 9–72 are similar.

(a) $\triangle ABC \sim$ _____.

(b) Find x and y.

SIMILAR SOLIDS

Like similar plane figures, similar solids are the same shape but not necessarily the same size. In similar polyhedra, the lengths of corresponding edges are proportional, and corresponding angles are congruent.

LE 5   Skill

The two rectangular prisms in Figure 9–73 are similar. *AB* corresponds to *EF*, *BC* to *FG*, and *CD* to *GH*. Find the lengths x and y.

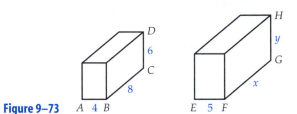

Figure 9–73

APPLICATIONS OF SIMILARITY

On a sunny day, you can use the length of your shadow to compute the height of a building. It's a good way to impress your friends. How does it work?

You and your shadow determine a triangle. This triangle is similar to the triangle determined by nearby objects and their shadows at the same time of day!

In Figure 9–74, $\triangle ABC \sim \triangle DEF$. Now suppose that you want to estimate how tall the building is. You can use your height, the length of your shadow, and the length of the building's shadow to compute the height of the building. Example 2 illustrates how this indirect measurement is done.

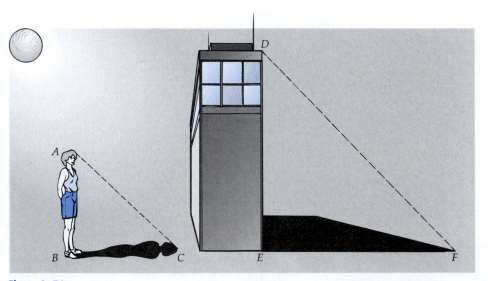

Figure 9–74

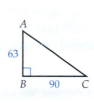

EXAMPLE 2

Jane is 5 ft 3 in. tall. At 4 P.M., her shadow is 7 ft 6 in. long. She measures the shadow of a nearby building. It is 500 in. long. About how tall is the building?

SOLUTION

Understanding the Problem   The length of each object and its corresponding shadow have the same ratio.

Devising a Plan   Draw a picture, and set up a proportion with the appropriate units.

Carrying Out the Plan   △*ABC* and △*DEF* represent cross sections in Figure 9–75.

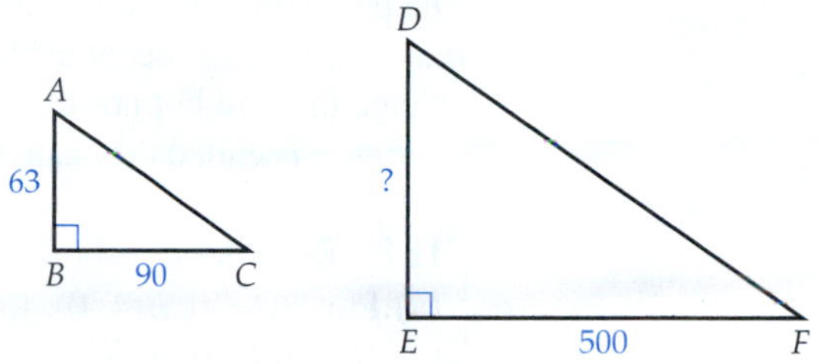

Figure 9–75

Convert the measurements to inches.

$$5 \text{ ft } 3 \text{ in.} = 63 \text{ in.} \quad \text{and} \quad 7 \text{ ft } 6 \text{ in.} = 90 \text{ in.}$$

$$\triangle ABC \sim \triangle DEF \quad \text{so} \quad \frac{AB}{DE} = \frac{BC}{EF} = \frac{AC}{DF}$$

Fill in the given information,

$$\frac{63}{?} = \frac{90}{500} = \frac{AC}{DF}$$

Use the first two ratios and solve.

$$\frac{63}{?} = \frac{90}{500} \quad \text{means} \quad 63 \cdot 500 = 90 \cdot (?)$$

Using a calculator, $63 \times 500 \div 90 = 350$. So the building is about 350 in. or 29 ft 2 in. tall. (On a calculator, $350 \div 12 = 29.166666$ ft or something similar. The .166666 ft = (.16666)(12 in.) = 2 in.)

Looking Back   The building's shadow is about 5 times the length of Jane's shadow. So the height of the building should be about 5 times Jane's height. 29 ft 2 in. is about 5 times 5 ft 3 in.   ●

LE 6   Skill

One sunny morning, Arturo, who is 5 ft tall, casts a shadow 7 ft 10 in. long. At the same time, a nearby tree casts a shadow 10 ft 10 in. long. About how tall is the tree?

Computing the distance between two cities on a map involves similar figures. A map shape is approximately similar to the region it represents (although maps are flat and the earth is not). Therefore, map distances are approximately proportional to actual distances.

LE 7   Skill

A map has a scale ratio of 1 in. $\approx$ 120 miles. On the map, New York is $1\frac{5}{8}$ in. from Boston. Approximately how far is New York from Boston?

DILATIONS

The pupils in your eyes dilate in response to light, enlarging when it becomes darker and reducing when it becomes lighter. In geometry, a dilation is a transformation that may change the size but not the shape of a figure. The ratio of the dilated image to the original figure is called the scale factor.

LE 8   Reasoning

(a) Plot $A(1, 0)$, $B(4, 0)$, $C(3, 2)$, and $D(1, 2)$ on a graph. Connect A to B, B to C, C to D, and D to A. What shape do you obtain?

(b) To perform a dilation, find image points that are twice as far from the origin and in the same direction. For example, A is 1 right and 0 up from $(0, 0)$, so A' is 2 to the right and 0 up from $(0, 0)$. This means that $A' = (2, 0)$. Find B', C', and D', and plot all the image points.

(c) Connect A' to B', B' to C', C' to D', and D' to A'. What shape do you obtain?

(d) How do $ABCD$ and $A'B'C'D'$ compare?

(e) How does $A'B'$ compare to AB? How does $C'D'$ compare to CD?

(f) How do the corresponding angle measures of the two figures compare?

(g) Now find E, F, G, and H that are three times as far from the origin as A, B, C, and D in corresponding directions. Plot E, F, G, and H.

(h) Connect E to F, F to G, G to H, and H to E. How does $EFGH$ compare to the other two figures?

(i) How does EF compare to AB? How does GH compare to CD?

(j) How do the corresponding angles of $EFGH$ compare to $ABCD$?

(k) Propose generalizations of your results.

LE 8(b) and LE 8(g) are examples of dilations. A dilation has a **center** and a positive **scale factor.** A **dilation** in a plane multiplies the distance of each point in the plane from the center by the scale factor. In LE 8(b), the dilation assigns an image to each point of $ABCD$ that is twice as far from the center point $(0, 0)$. The dilation produces an image that has the same shape as the original figure. The dilation multiplies the lengths of the original figure by the scale factor but does not change the measures of the angles.

A dilation can also be performed on a figure that is not in a coordinate plane.

Discussion

LE 9   Reasoning

Dilate $\triangle ABC$ by a scale factor of 0.5 from center A, by following these instructions. Use a ruler.

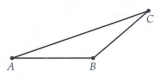

(a) A is the center of dilation, so $A' = A$. Label A'.

(b) B' is on $\overline{AB}$, and it is 0.5 times as far from A as B is. Label B'. In the same way, find image point C' on $\overline{AC}$.

(c) Describe any relationships that exist between the corresponding sides, angles, and vertices of $\triangle ABC$ and $\triangle A'B'C'$.

Next, try to find some general properties of dilation.

LE 10   Reasoning

(a) In LE 8 and LE 9, how are the corresponding angles and sides of the image related to the original figure?

(b) Part (a) shows what relationship between a figure and its dilated image?

(c) In the dilation from LE 8(b), draw a line through each point and its corresponding image point. Where do the lines intersect?

(d) Repeat part (c) for LE 9.

(e) Generalize the result of parts (c) and (d).

(f) Does part (e) involve induction or deduction?

As LE 10 suggests, the image of a dilation is similar to the original figure. Furthermore, every line containing a point and its image intersects the center of dilation.

LE 11   Summary

Tell what you learned about similarity and dilations in this section.

ANSWERS TO SELECTED LESSON EXERCISES

2. (a) No   (b) Yes

3. Would two squares satisfy the two conditions of the definition of similar polygons? (Yes.)

4. (b) $\dfrac{3}{5} = \dfrac{4}{x} = \dfrac{6}{y}$, so $x = 6\dfrac{2}{3}$ and $y = 10$.

5. $\dfrac{4}{5} = \dfrac{8}{x} = \dfrac{6}{y}$, so $x = 10$ and $y = 7\dfrac{1}{2}$.

6. About 6 ft 11 in.

7. About 195 miles

8. (a) Trapezoid
   (d) They are the same shape.
   (e) $A'B' = 2(AB)$; $C'D' = 2(CD)$
   (f) They are equal.
   (h) It is the same shape but larger.
   (i) $EF = 3(AB)$; $GH = 3(CD)$
   (j) They are equal.

9. (b)

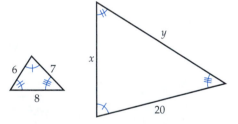

   (c) $\triangle ABC$ and $\triangle A'B'C'$ are the same shape. They have congruent corresponding angles. The corresponding sides of $\triangle A'B'C'$ are 0.5 times as long.

9.5   HOMEWORK EXERCISES

Basic Exercises

1. Tell why the following two figures are not similar.

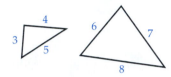

2. In each case, decide if the two figures are similar.

(a)   (b)

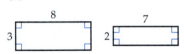

3. A rectangle has length L and width W. A second rectangle has length KL and width KW, with $K > 0$. Are the rectangles similar?

4. A rectangle has dimensions of b and h. A second rectangle has dimensions of $b + 6$ and $h + 6$. Are the rectangles similar?

5. $\triangle ANT \sim \triangle REP$. Fill in the blanks.

   (a) $\angle T \cong$ _____   (b) $\dfrac{ER}{\square} = \dfrac{\square}{TN}$

6. Fill in the blank, putting the vertices for $\triangle NMR$ in the corresponding order. $\triangle ABC \sim$ _____.

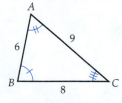

7. Are any two rhombuses similar? Why or why not?

8. Recall that similar figures have the same shape but not necessarily the same size. Are any two circles similar?

9. A quadrilateral has sides of length a, b, c, and d. A second quadrilateral has sides of length $2a$, $2b$, $2c$, and $2d$. Must the quadrilaterals be similar?

10. Are two congruent figures also similar?

11. The two triangles shown are similar. Find x and y.

12. Suppose you want to enlarge a photograph that is 3 in. by 5 in. so that its longer dimension is 8 in. What will the shorter dimension be?

13. Assume that all three triangles are similar. Find x.

14. Assume that the two triangles in each part are similar, and find x.

(a)
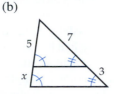

(b)
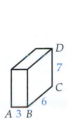

15. The two rectangular prisms shown here are similar; $\overline{AB}$ corresponds to $\overline{EF}$, $\overline{BC}$ corresponds to $\overline{FG}$, and $\overline{CD}$ corresponds to $\overline{GH}$. Find the lengths x and y.

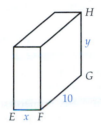

16. Would any two cylinders be similar? Why or why not?

17. Bill is 4 ft 4 in. tall. At 3 P.M., his shadow is 5 ft 10 in. long. The shadow of a nearby building is 33 ft 4 in. long. How tall is the building?

18. Kelly is 5 ft 3 in. tall. At 11 A.M., her shadow is 3 ft 8 in. tall. The shadow of a nearby flagpole is 18 ft 10 in. long. How high is the flagpole?

19. A drawing is 10 cm by 15 cm. A copying machine reduces the length and width to 68% of their original sizes.
(a) What are the dimensions of the reduced drawing?
(b) Is the reduced drawing similar to the original?
(c) The area of the reduced image is _____% of the area of the original drawing.

20. Using a particular magnifying glass, all length measurements appear to be three times their original sizes. What percent *increase* in length measurements is this?

21. Make a scale drawing of the house shown here. In your drawing, let 1 cm represent 4 m.

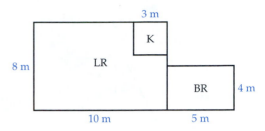

22. A map has a scale ratio of 1 in. ≈ 60 miles. Sleepytown is $3\frac{1}{4}$ in. from Swingtown on the map. Based on this, the actual distance is _____.

23. You will need a protractor and a ruler. Make a scale drawing to answer the following. An 18-foot ladder makes an angle of 70° with the ground. How high up a wall will it go?

24. An engineer wants to build a tunnel from A to B through a mountain, as shown. To find the length AB, the engineer sights A and B from C. $AC = 60$ ft, $m\angle A = 90°$, $m\angle B = 32°$, and $m\angle C = 58°$. Make a scale drawing and approximate AB.

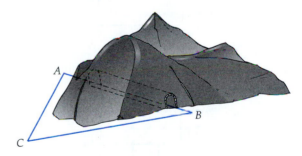

25. Dilate $\triangle ABC$ by a scale factor of 1.5 with $(0, 0)$ as the center.

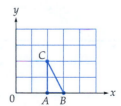

26. Dilate $\triangle ABC$ by a scale factor of $\frac{1}{3}$ with $(0, 0)$ as the center.

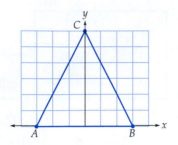

27. Dilate rectangle $ABCD$ by a scale factor of 2 with center A.

28. Dilate rectangle $ABCD$ by a scale factor of $\frac{1}{2}$ with center B.

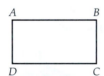

29. Find the center and scale factor for the dilation shown. The original figure is black.

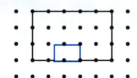

30. Find the center and the scale factor for the dilation shown. The original figure is black.

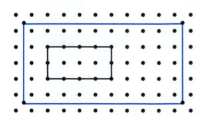

Extension Exercises

31. If two triangles have equal corresponding angle measures, do they have to be similar? Try the following. You will need these materials: one ruler, one protractor, and one live brain.
- (a) Draw a triangle, and measure its three angles. Now draw a second triangle that has the same three angle measures as the first triangle but has longer or shorter sides.
- (b) Now look at your two triangles. What strikes you about their appearance (one compared to the other)?
- (c) Measure the sides of each triangle. How do they compare?
- (d) Complete the following generalization suggested by this exercise. If three angles of one triangle are congruent to three angles of a second triangle, then the triangles are _____ (called the AAA similarity property).
- (e) Does part (d) involve induction or deduction?

32. The preceding exercise concerned the AAA similarity property. Now suppose you have two triangles and two angles in one triangle are congruent to two corresponding angles in the other triangle.
- (a) What do you know about the third angle in each triangle?
- (b) What can you then conclude about the two triangles?
- (c) Parts (a) and (b) illustrate that the AAA similarity property can be simplified as the _____ property.

33. (a) Draw a triangle with sides of 2 in., 3 in., and 4 in.
- (b) Draw a second triangle with sides of 1 in., $1\frac{1}{2}$ in., and 2 in.
- (c) Do the triangles appear to be similar?
- (d) Repeat parts (a)–(c) for two other triangles with proportional sides.
- (e) Based upon your two examples, two triangles with proportional sides appear _____ similar (called the SSS similarity property).
  (to be, not to be)

34. (a) Plot $A(0, 0)$, $B(4, 0)$, $C(2, 2)$, and $D(0, 2)$ on a graph, and connect A to B, B to C, C to D, and D to A.
- (b) What happens if you multiply all the coordinates by $k < 0$, plot the resulting points, and connect them in the same order?

35. If every teacher's salary increases by 10%, that is like a dilation with a scale factor of _____. (*Hint:* The answer is not 0.10.)

36. When a population increases by 8%, this is like a dilation with a scale factor of _____.

37. In 2003, most movie screens and television sets have different rectangular shapes. Most movies are filmed with a width-to-height ratio of 16 to 9. Most television screens have a width-to-height ratio of 4 to 3. Suppose you have a television set with a width of 24 inches and a height of 18 inches. If a movie picture is reduced to fit the width of the screen ("letter box"), what will the height of the picture be?

38. A function has exactly one output for each input or set of inputs. Explain how a dilation is a function.

Technology Exercises

39. (a) Use a graphing calculator to graph a triangle with vertices $(0, 0)$, $(3, 0)$, and $(2, 1)$. Enter the points as data in columns (such as L_1 and L_2 in the TI-83). Put $(0, 0)$ again as the fourth entry. Then plot a connected graph.
Graph the following points in L_3 and L_4, and tell how each image is related to the original triangle.
(b) $3L_1 \rightarrow L_3$ and $3L_2 \rightarrow L_4$
(c) $2L_1 \rightarrow L_3$ and $4L_2 \rightarrow L_4$

40. Use dynamic geometry software to see if two triangles with congruent corresponding angles must be similar.

(a) Construct $\triangle ABC$
(b) Now construct $\angle EDF \cong \angle BAC$.
(c) Construct $\angle DEF \cong \angle ABC$.
(d) Measure the corresponding angles and sides of $\triangle ABC$ and $\triangle DEF$, and decide if the triangles are similar.

41. How does a quadrilateral compare to its image after a dilation?
(a) Construct two unequal line segments $\overline{AB}$ and $\overline{CD}$. Make $\overline{CD}$ longer than $\overline{AB}$.
(b) Select $\overline{AB}$ and $\overline{CD}$, and mark CD/AB as a ratio with the Transform menu. Measure the ratio CD/AB. This will be the scale factor for a dilation.
(c) Construct a quadrilateral and its interior.
(d) Construct a point outside the quadrilateral. Double-click the point to designate it as the center of your dilation.
(e) Now select the sides, vertices, and interior of your quadrilateral. Dilate the quadrilateral by the ratio CD/AB.
(f) Measure the sides and corresponding angles of each quadrilateral.
(g) What can you conclude about the quadrilaterals?

Labs and Projects

42. Go outdoors on a sunny day, and use indirect measurement and shadows to find the height of a tree or pole in your neighborhood.

43. Find a doll, and measure its height and waist size. Explain whether or not the body proportions are realistic.

CHAPTER 9 SUMMARY

Anyone who spends time studying shapes is likely to notice congruence, symmetry, and similarity. Congruence and similarity are two important relationships between pairs of geometric figures, whereas symmetry is a property of one figure. Congruent figures are identical in shape and size, whereas similar figures are the same shape but not necessarily the same size. Symmetry describes a certain kind of balance possessed by a figure.

Congruence and symmetry of plane figures are related to translations, reflections, and rotations. Ask a young child whether two shapes are identical, and the child will attempt to fit one shape on top of the other. This example illustrates the fact that two plane figures are congruent if and only if one can be mapped onto the other using a translation, reflection, rotation, or some combination of these motions. A plane figure is symmetric if you can find a different position in the plane in which the newly positioned figure coincides with the original figure.

Similarity is related to dilations. A dilation uniformly multiplies the dimensions of a figure without changing the angle measures or shape. The resulting image is similar to the original figure.

Symmetry contributes to the beauty of nature. Line symmetry gives balance to birds and kites. Rotational symmetry gives directional flexibility to starfish and bolts. Similarity is useful in measuring heights indirectly and in finding distances represented on maps.

Constructions are a geometric ritual that has been passed on for thousands of years. Using only a compass and a straightedge, it is possible to copy or bisect all sorts of plane figures.

STUDY GUIDE

To review Chapter 9, see what you know about each of the following ideas or terms that you have studied. You can also use this list to generate your own questions about the chapter.

CONGRUENCE, SYMMETRY, AND SIMILARITY IN GRADES 1–8

The following chart shows at what grade levels selected geometry topics typically appear in elementary- and middle-school mathematics textbooks.

Topic	Typical Grade Level in Current Textbooks
Slides, flips, turns	4, 5, 6, 7, 8
Congruent figures	1, 2, 3, 4, 5, 6, 7, 8
Constructions	6, 7, 8
Line symmetry	2, 3, 4, 5, 6
Rotational symmetry	5, 6, 7
Similar figures	4, 5, 6, 7, 8
Dilations	8

REVIEW EXERCISES

1. Show the image of $\triangle ABC$ after a reflection across line m followed by translation 2 units up and 1 unit to the right (Activity Card 4).

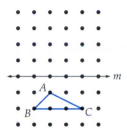

2. Show the image of $\triangle ABC$ after a 90° rotation counter-clockwise around A followed by a reflection across line m.

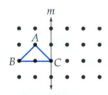

3. Plot the points $(2, 1)$, $(-3, 2)$, and $(1, -4)$, and label them A, B, and C, respectively. Suppose that a mapping changes the coordinates of each point as follows.

$$(x, y) \rightarrow (x - 3, y + 1)$$

(a) Find the image points of A, B, and C with this rule and plot them.
(b) What specific transformation maps $\triangle ABC$ to $\triangle A'B'C'$?

4. A point P is reflected through line m. If P is not on m, then line m _____ $\overline{PP'}$.

5. What is the relationship between *congruence* and rotations, translations, and reflections?

6. Find the image of $\triangle DEF$ after a glide reflection of vector $\overrightarrow{BA}$.

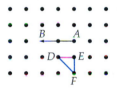

7.

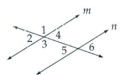

Lines m and n are parallel.
(a) Name a pair of vertical angles.
(b) Name a pair of alternate interior angles.
(c) Name a pair of corresponding angles.
(d) $m\angle 2 = 64°$. Find the measures of all the other numbered angles.

8. *ABCD* is a rectangle.

(a) What is the image of *B* under a 180° counter-clockwise rotation around *E*?

(b) Select a pair of angles that appear to be congruent, and name a specific transformation that maps one angle onto the other.

(c) Repeat part (a) for another pair of angles.

9. (a) Alter a square by making a translation to create a new tessellating shape.

(b) Make a tessellating pattern with four copies of your shape.

10. Construct a triangle that is congruent to △*ABC*.

11. (a) Construct the angle bisector of ∠*AIM* and call it $\overrightarrow{IT}$. (Assume *AI* = *MI*.)

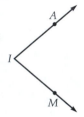

(b) Use congruent triangles to explain why $\overrightarrow{IT}$ bisects ∠*AIM*.

(c) Does part (b) involve induction or deduction?

12. Shown here is a regular hexagon. Find all the angle measures in the diagram.

13. Draw all lines of symmetry for the rectangle shown.

14. Draw a figure that has four lines of symmetry and three rotational symmetries less than 360°.

15. Shade the smallest number of squares so that the following diagram has rotational symmetry.

16. Sketch any of the following six figures that are possible: pentagons that have exactly zero, one, two, three, four, or five lines of symmetry.

17. Are any two regular pentagons similar? Why or why not?

18. If △*RUN* ∼ △*RAC*, find *AC*.

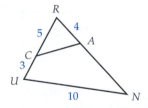

19. Sandy is 4 ft 10 in. tall. At 1 P.M., her shadow is 5 ft 5 in. long. The shadow of a nearby building is 30 ft 2 in. long. How tall is the building?

20. A rectangle has length *L* and width *L* + 6. A similar rectangle has length 10. What is its width?

21. Dilate △*ABC* by a scale factor of 2 from center *B*.

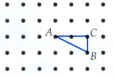

ALTERNATE ASSESSMENT

Do one of the following assessment activities: Add to your portfolio, add to your journal, write another unit test, do another self-assessment, or give a presentation.

 ## SUGGESTED READINGS

Britton, J. *Investigating Patterns: Symmetry and Tessellations*. White Plains, NY: Dale Seymour, 1999.

Geddes, D., and others. *Geometry in the Middle Grades*. Reston, VA.: NCTM, 1992.

Holden, A. *Shapes, Space, and Symmetry*. Mineola, NY: Dover, 1991.

Kroner, L. *Slides, Flips, and Turns*. Palo Alto, CA: Creative Publications, 1995.

National Council of Teachers of Mathematics. *Connecting Mathematics Across the Curriculum*. 1995 Yearbook. Reston, VA: NCTM, 1995.

Navigating Through Geometry (4 vols.). Reston, VA: NCTM, 2001.

O'Daffer, P., and S. Clemens. *Geometry: An Investigative Approach*. 2nd ed. Reading, MA: Addison-Wesley, 1992.

Olson, T. A. *Mathematics Through Paper Folding*. Reston, VA: NCTM, 1975.

Weyl, H. *Symmetry*. Princeton, NJ: Princeton University Press, 1983.

Use Infotrac to find articles in *Teaching Children Mathematics* on teaching children about transformation geometry, congruence, symmetry, and similarity.

10 | MEASUREMENT

People created measurement systems to satisfy practical needs such as finding the distances between places, the heights of horses, and the weights of grains. They developed calendars to keep track of days, seasons, and years.

Every measurement requires a unit of measure. Ancient people used body measurements as units of length and seeds and stones as units of weight. The first units of time were the cycles of the sun and moon. In recent times, most countries have developed uniform units of measure. Today, most countries in the world use the metric system. The United States uses both the customary (old English) system and the metric system.

About 3,500 years ago, the Babylonians developed methods for finding the area of some triangles and trapezoids. They also solved problems involving right triangles. Around the same time, the ancient Egyptians knew how to approximate the areas of circles. Since that time, people have developed ingenious formulas and methods for computing the areas and volumes of geometric shapes, and they use these formulas to approximate the measures of many everyday objects.

10.1 SYSTEMS OF MEASUREMENT

NCTM Standards
- understand how to measure using nonstandard and standard units (pre-K–2)
- carry out simple unit conversions such as from centimeters to meters, within a system of measurement (3–5)
- understand that measurements are approximations and understand how differences in units affect precision (3–5)
- select and use benchmarks to estimate measurements (3–5)

Any object has many attributes that can be measured. When you measure something, you focus on one attribute and ignore all other attributes and qualities. Measurement involves (1) identifying an attribute, (2) selecting a unit of measure, and (3) comparing the attribute of the object to the unit of measure. The first units of measure were nonstandard units.

NONSTANDARD MEASURING UNITS

Historical records indicate that the first units of length were based on people's hands, feet, and arms (Figure 10–1).

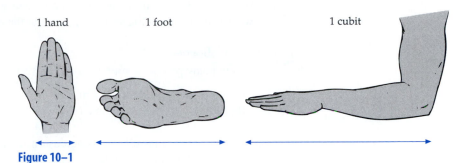

1 hand   1 foot   1 cubit

Figure 10–1

LE 1   Concept

(a) How many hand widths high is this page of your book?

(b) Why might others in the class obtain different answers?

(c) If someone else in the class has a larger number than yours for an answer, what do you know about that person's hand?

Hands, feet, and arms are convenient for measuring, but different people have different-sized hands, feet, and arms. That is why we need standard measuring units such as those of the metric system. By using standard units, scientists and manufacturers around the world communicate easily with one another about measurements. Children develop a better understanding of measuring units when they have experiences using both nonstandard and standard units.

THE CUSTOMARY AND METRIC SYSTEMS

The customary (English) system developed from nonstandard units. The inch was the length of three barleycorns placed end to end. The foot was about the length of an adult's foot, and the yard was the distance from the tip of a person's nose to the end of either outstretched arm. The yard was used to measure fabric and other "yard goods."

Once people realized the need for standardized measurements, the foot was defined as the length of a special metal bar. Then the inch, yard, and mile were defined in terms of a foot.

LE 2   Concept

Define 1 inch, 1 yard, and 1 mile in terms of 1 foot.

Since customary units developed independently, there are no simple relationships among them. This makes converting from one unit to another fairly difficult.

The metric system was developed in France in the late 1700s. The word "metric" comes from the Latin word "metricus," which means measure. A meter was originally defined as $\dfrac{1}{10,000,000}$ of the distance from the North Pole to the equator. The meter is now defined as the distance light travels in a vacuum in $\dfrac{1}{299,792,458}$ second.

Because of the historical connections between the United States and England, the United States adopted the English system rather than the metric system of England's former enemy, France. Today, nearly every country in the world except the United States and Australia uses the metric system. In the United States, the metric system is widely used in many areas, including science and photography but not carpentry. In what ways is the metric system superior to the customary (U.S.) system?

Discussion

LE 3   Opener

How many people in your class can correctly answer each of the following questions?

(a) How many yards are in a mile?
(b) How many meters are in a kilometer?

Most of you have been using customary measure all your life and metric measure for only a short time. Despite this fact, more adults in the United States can correctly answer LE 3(b) than LE 3(a)!

Two advantages of the metric system are that (1) metric prefixes are the same for length, weight, and liquid volume and (2) all conversions within the metric system involve powers of 10.

Each pair of students will need a meter stick and a 30-cm ruler to do LE 4, LE 5, LE 8, and LE 10.

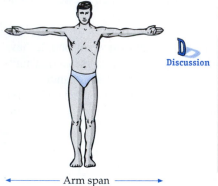

Discussion

LE 4   Skill

Examine your meter stick, and complete the following.

(a) Find a centimeter on your meter stick.
(b) How many centimeters are in a meter?
(c) The millimeter is the smallest unit of measure on your meter stick. Find a millimeter on your meter stick.
(d) How many millimeters are in a centimeter?
(e) How many millimeters are in a meter?

— Arm span —

Figure 10–2

How does your arm span compare to your height (Figure 10–2)?

Discussion

LE 5   Skill

(a) What is your height in centimeters? (Stand against a blackboard, and have another student mark your height.)
(b) Give your height, using meters and centimeters (such as 1 m 37 cm).
(c) What is your height in meters?
(d) How long is your arm span?
(e) How does your arm span compare to your height?

In the metric system, people's heights are most commonly recorded in centimeters, as in LE 5(a).

BENCHMARKS

A good way to learn the approximate sizes of unfamiliar units of measure is to relate them to measurements of your body or your environment.

LE 6   Connection

(a) Find an object in your classroom that is about 1 m high or 1 m above the floor.
(b) Find an object in your classroom that is about 1 m long.
(c) Find some part of your hand that is about 1 cm long.
(d) Find an object that is about 1 mm long.

Your answers to LE 6 are called benchmarks. **Benchmarks** are common environmental or body measurements that approximate basic measuring units such as a meter. Benchmarks help to develop an intuitive understanding (number sense) of metric measure.

You can use benchmarks to estimate other measurements. In LE 7, use a benchmark from LE 6 to select the best estimate.

LE 7   Concept

Using your benchmark, select the best estimate for the length of your professor's shoe.

(a) 25 mm   **(b)** 25 cm   **(c)** 2.5 m   **(d)** 25 m

LE 8   Skill

(a) Use your benchmark for 1 cm to draw a line segment that is about 8 cm long.
(b) Measure your segment to see how close you came.

LE 9   Reasoning

Use your benchmark for 1 m to obtain an estimate of the area of your classroom floor, in square meters. (*Note:* Area will be discussed further in the next section.)

How do metric units compare in size to customary units? Examine your rulers to answer the following questions.

LE 10   Connection

(a) What customary unit of length is about the same length as a meter?
(b) Which is longer, an inch or a centimeter?

Why do we need units of different sizes, such as centimeters, meters, and kilometers? If we measured traveling distances in centimeters, the numbers would be so great that they would be hard to grasp. For example, the distance from Washington, D.C., to Boston is about 66,000,000 cm. Conversely, measuring the length of your foot in kilometers, you might get a result such as 0.00021 km, which would not be very meaningful.

CONVERSIONS WITHIN THE METRIC SYSTEM

You have already seen how meters, centimeters, millimeters, and kilometers are related by powers of 10. This same relationship holds for all metric measures, and it makes conversions within the metric system easier than those in the customary system.

The metric prefixes are presented in the following table.

Prefix	kilo-	hecto-[†]	deka-[†]	(none)	deci-	centi-	milli-
Symbol	k	h	dk		d	c	m
Meaning	1000	100	10	1	$\frac{1}{10}$	$\frac{1}{100}$	$\frac{1}{1000}$

[†]Not commonly used.

Every metric length measurement includes one of these prefixes in front of the basic length unit, the meter. For most everyday measures, kilometers, meters, centimeters, and millimeters are sufficient.

Kilometer	Hectometer[†]	Dekameter[†]	Meter	Decimeter	Centimeter	Millimeter
km	hm	dkm	m	dm	cm	mm

[†]Not commonly used in U.S. elementary and secondary schools.

Example 1 shows how to convert within the metric system.

• EXAMPLE 1

350 cm = _____ m.

SOLUTION

Compare centimeters to meters. Since 100 centimeters make a meter, divide 350 cm by 100 to convert it to meters: $350 \div 100 = 3.50$. (Move the decimal point two places to the left.) You can also use a prefix chart and see that to go from centimeters to meters, one moves two places to the left. This corresponds to the movement of the decimal point in converting from centimeters to meters.

More formally, one can include the units in the computation and multiply 350 cm by 1 in the form $\dfrac{1 \text{ m}}{100 \text{ cm}}$.

$$350 \text{ cm} \times \frac{1 \text{ m}}{100 \text{ cm}} = \frac{350}{100} \text{ m} = 3.5 \text{ m}$$

This more formal approach with units is part of **dimensional** (unit) **analysis.**   ●

As suggested by converting your height from centimeters to meters and by the preceding example (350 cm = 3.5 m), metric conversions require nothing more than moving the decimal point, since they all involve multiplying or dividing by a power of 10. When you change larger units to smaller units, you multiply. When you change smaller units to larger units, you divide.

Try the following conversions.

Discussion

LE 11   Skill

(a) 0.5 m = _____ mm   **(b)** 80 mm = _____ cm
(c) A kilometer is _____ meters (about $\frac{3}{5}$ of a mile).

Classroom Connection

LE 12   Concept

A fourth grader says that 80 mm = 800 cm. What would you tell the child?

MASS AND WEIGHT

> **Miracle Diet Program!**
> **Lose 5/6 of Your Weight:**
> **Take a Trip to the Moon**

The gram is the basic unit of *mass* in the metric system. People sometimes confuse mass and weight. **Mass** measures the quantity of matter a body contains. **Weight** measures the force exerted by gravity on a body.

If you went to the moon, your mass would remain the same, but your weight would be much less (about $\frac{1}{6}$ of your weight on Earth) because the force of gravity is much weaker on the moon. In everyday usage, kilograms (the metric unit of mass) and pounds (the customary unit of weight) are used interchangeably. The conversion 1 kg = 2.2 lb works on the surface of Earth, but on the moon, a mass of 1 kg weighs about 0.4 lb.

Discussion

LE 13   Concept

(a) If you flew to Jupiter, the largest planet, would your *mass* be more than, the same as, or less than your mass on Earth?

(b) Would your *weight* be more than, the same as, or less than your weight on Earth?

The most commonly used units of metric mass are milligrams (very small), grams, and kilograms. The prefixes *milli-* $\left(\frac{1}{1000}\right)$ and *kilo-* (1000) have the same meaning with grams as with meters. Do you know any benchmarks for grams and kilograms?

LE 14   Connection

Name something that has a mass of about 1 g (gram).

LE 15   Connection

Name something that has a mass of about 1 kg (kilogram).

A raisin or a solid, well-made paper clip has a mass of 1 g. A textbook you can finish in one semester or two loaves of bread may have a mass of about 1kg. A milligram is very small, like the mass of a feather or a few grains of salt. The smallest automobiles have a mass of about one metric ton (t). A metric ton is one million grams and about 10% more than a ton.

Customary units of weight include ounces (oz), pounds (lb), and tons. A pound is 16 ounces, and a ton is 2000 pounds.

CAPACITY

The liter (L) is the metric unit for capacity. Do you know any benchmarks for a liter?

LE 16   Connection

Name an everyday object that holds about 1 L (liter).

A drop of water from an eyedropper is about 1 mL. The metric system also established the following simple relationships among capacity, mass, and volume that are used in Section 10.6: A capacity of 1 L of pure water has a mass of 1 kg and a volume of 1000 cm^3.

LE 17   Skill

Grams and liters use the same prefixes as meters.

(a) 80 g = _____ kg
(b) N mL = _____ L (N is a positive number.)

Customary units of capacity include cups (c), pints (pt), quarts (qt), and gallons (gal).

LE 18   Connection

Define a pint, quart, and a gallon in terms of cups.

TEMPERATURE

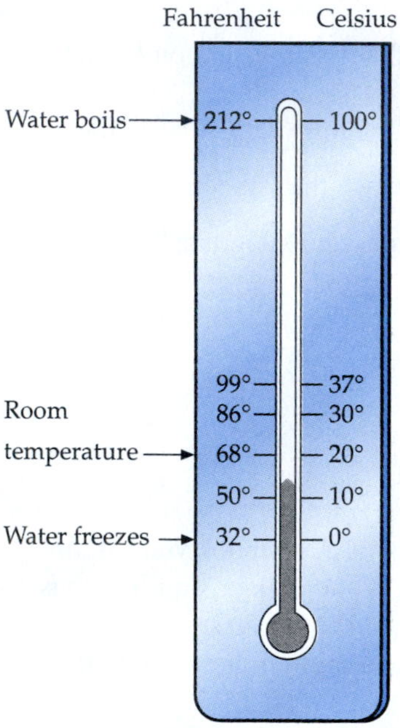

Do you know what a 35° Celsius day would feel like? Hot! Metric temperatures are often measured in degrees Celsius (C). In this system, 0°C is the freezing point of water and 100°C is the boiling point of water. These temperatures are easier to remember than those in the Fahrenheit scale, in which water has a freezing point of 32°F and a boiling point of 212°F. The thermometer shown in Figure 10–3 gives some other common temperatures in Fahrenheit and Celsius.

LE 19   Connection

The temperature inside a working refrigerator is about

(a) 5°C   **(b)** 15°C   **(c)** 25°C   **(d)** 35°C

The German scientist Gabriel Fahrenheit made the first mercury thermometer in 1714. On his scale, he defined 0° as the temperature of a mixture of equal parts of salt and ice. On this basis, Fahrenheit assigned 32° to the freezing point of water, 96° to normal body temperature, and 212° to the boiling point of water. After his death, an error was discovered in his measurements, and normal body temperature was changed to 98.6°.

In 1742, Anders Celsius, a Swedish astronomer developed the Celsius (or centigrade) system, in which the boiling point of water was 0° and the freezing point was 100°. Later on, the freezing point and boiling point were interchanged.

Fahrenheit   Celsius

Water boils → 212° — 100°

99° — 37°
Room   86° — 30°
temperature → 68° — 20°
50° — 10°
Water freezes → 32° — 0°

Figure 10–3

PRECISION AND SIGNIFICANT DIGITS

Whereas exact measurements can be given for perfect geometric shapes, no measurement of an actual object (other than counting a set of objects) is exact. In theory, perfect geometric figures have exact measures, whereas actual objects have approximate measures. In Chapter 3, this textbook distinguished between sets and measures. To be consistent, counting a set of objects will not be referred to as a measurement in this chapter.

Geometric shape → Exact measurement

Everyday object → Approximate measurement

A reported numerical measurement should reflect the precision with which it was made. The precision of an approximate measurement is determined by the unit of measure.

LE 20   Concept

(a) Measure the height of this page to the nearest centimeter.
(b) Measure the height of this page to the nearest millimeter.
(c) Which measurement is more precise?

The **precision** of a measurement is determined by the smallest unit of measurement used. In LE 20, neither measurement of page height was exact, but using the smaller unit (millimeters) resulted in a more *precise* measure. Similarly, a measurement of height to the nearest foot (such as 5 ft) is less precise than a measurement to the nearest inch (such as 5 ft 4 in.).

The author's height to the nearest centimeter is 171 cm. Using a more precise electronic ruler, the author's height is 171.48 cm. The first measurement has 3 significant digits. The second has 5 significant digits. The **significant digits** are all the digits in a measurement that are known to be exact.

All nonzero digits are significant. However, zeroes at the beginning or end of a number may indicate rounding rather than an exact measurement. For this reason, two types of zeroes are not counted as significant: zeroes at the end of a whole number (3000 has 1 significant digit) and zeroes at the beginning of a number between 0 and 1 (0.002 has 1 significant digit).

LE 21   Concept

How many significant digits are in each measurement?

(a) 38.7 m   (b) 52,000 years   (c) 0.0310 g

In science, people use the following rules for arithmetic of measurements. When adding or subtracting two measurements with different levels of precision, the answer should have the same level of precision as the least precise measurement. When multiplying or dividing two measurements with different numbers of significant digits, the answer should have the same number of significant digits as the measurement with the lower number of significant digits.

LE 22   Concept

People usually give their weight to the nearest pound or kilogram. Suppose that John's correct mass, rounded to the nearest kilogram, is 53.

(a) What is the smallest decimal number that would be rounded to 53 as the nearest whole number?
(b) Any decimal number greater than or equal to 52.5 and less than _____ would be rounded to 53 as the nearest whole number.
(c) How far from 53 kg could John's exact mass be?

In LE 22, John's exact mass is between 0 and 0.5 kg away from 53 kg. The greatest possible error is 0.5 kg. This could be obtained by taking half of the smallest unit (kg) that was used in measuring 53 kg. The **greatest possible error** (GPE) is half of the smallest unit used in the measurement.

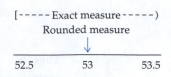

| 52.5 | 53 | 53.5 |

John's actual mass is somewhere between 52.5 and 53.5 kg (including 52.5). In practice, scientists say "between 52.5 and 53.5 kg" without mentioning whether the end-point measures are included, since no object is likely to weigh *exactly* 52.5 or 53.5 kg.

LE 23 Concept
Janie's shoe is 27.6 cm long.

(a) What is the smallest unit of measure used?
(b) Her actual shoe length is between _____ cm and _____ cm.
(c) What is the GPE?

Which would be more significant, an error of 0.5 kg in a measurement of John's mass as 53 kg or an error of 0.05 cm in a measurement of Janie's shoe as 27.6 cm? Relative error answers this by comparing the size of the greatest possible error to the measurement. The **relative error** of a measurement is the GPE divided by the measurement.

LE 24 Skill
(a) Find the relative error of the 53 kg from LE 22.
(b) Find the relative error of the 27.6 cm from LE 23.
(c) Which measurement has the smaller relative error?

LE 25 Summary
Tell what you learned about systems of measurement in this section.

ANSWERS TO SELECTED LESSON EXERCISES

1. (a) About 3 or 4
(b) Their hands are different sizes.

2. 1 inch = $\frac{1}{12}$ foot, 1 yard = 3 feet, 1 mile = 5,280 feet

4. (b) 100   (d) 10   (e) 1000

6. (a) *Hint:* Try a doorknob.

7. 25 cm

10. (a) A yard   (b) An inch

11. (a) 500   (b) 8   (c) 1000

12. Ask which units are larger, mm or cm? When you use larger units, what will happen to the *number* of units?

13. (a) The same as   (b) More than

16. A quart milk carton

17. (a) 0.08   (b) $\frac{N}{1000}$

18. 1 pint = 2 cups, 1 quart = 4 cups, 1 gallon = 16 cups

19. 5°C

20. (c) Part (b)

21. (a) 3   (b) 2   (c) 3 (includes the last zero)

23. (a) One tenth of a centimeter
(b) 27.55; 27.65
(c) 0.05 cm

24. (a) $\dfrac{0.5}{53} = 0.0094$ or 0.94%

(b) $\dfrac{0.05}{27.6} = 0.0018$ or 0.18%

(c) Part (b)

10.1   HOMEWORK EXERCISES

Basic Exercises

1. Make up a measuring unit for length. Use it to measure the height of a page in this book.

2. How many paper clips long is a pen?

3. (a) How long is your foot in centimeters?
(b) Measure the width of your front door using your foot.
(c) Convert the width of your door in part (b) to centimeters.

4. (a) How long is your pace (one step) in centimeters?
(b) Measure the length of a room in paces.
(c) Convert the length of the room in part (b) to centimeters.

5. Research suggests that children need experiences comparing lengths. Try the following. Measure a pen and pencil with paper clips.
(a) Which is longer?
(b) How much longer is it?

6. Measure the length of a fork and the length of a sponge in key lengths. Tell which one is longer and by how much.

7. A second grader measures a line segment as shown. She says that it is 7 cm. long.

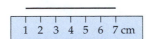

(a) How did the child obtain this answer?
(b) What is she confused about?

8. You ask two second graders to measure a pen that is 5 paper clip lengths. The first child says that it is 4 paper clips long. The second child says that it is 7 paper clips long. What might each child have done?

9. Why did people invent standard measuring units?

10. In what ways is the metric system easier to work with than the customary system?

11. Complete the following conversions.
(a) 3 feet = _____ inches
(b) 2 miles = _____ feet
(c) 5 feet = _____ yards

12. Complete the following conversions.
(a) 7 yards = _____ feet
(b) 9 inches = _____ feet
(c) 500 yards = _____ miles

13. Give a reference measure for
(a) 1 cm.   (b) 1 m.

14. The width of an adult's hand is a benchmark for what metric measure?

15. Approximate the floor area of your bedroom in metric units.

16. (a) Estimate the length of your arm in centimeters.
(b) Measure your arm length in centimeters.

17. Give a benchmark for
(a) a inch.   (b) a foot.

18. Give a benchmark for
(a) a yard.   (b) a mile.

19. The distance from New York to Los Angeles is 2,800 miles. Use a benchmark for a smaller distance that is familiar to you, and describe what a distance of 2,800 miles is like.

20. Learn More U. has 5,000 students. Use a benchmark for a smaller group of people that is familiar to you, and describe what a group of 5,000 is like.

21. You just measured the length of an object. Now you decide to use a smaller unit of measurement. What will happen to the number of units in the measurement?

22. A third grader measures her height in both centimeters and meters. Which measurement will have more units?

23. Complete the following conversions.
    (a) 0.02 m = _____ mm
    (b) 16 cm = _____ mm
    (c) 82 m = _____ km

24. A film is 35 mm wide. How many centimeters wide is it?

25. Lynn Matthews wants to swim 1 km in a 50-m-long pool. How many pool lengths must she swim?

26. A town is T km away. How many meters away is it?

27. List the following measurements in increasing order: 37 m, 46 cm, 871 mm, 3 km, 137 cm.

 28. 54 cm/s = _____ km/h

29. The average adult man has a mass of about
    (a) 7 kg   (b) 700 kg
    (c) 700 g   (d) 70 kg

30. A pound (on Earth) is about
    (a) 5 g   (b) 5 kg   (c) 50 g   (d) 500 g

31. A box contains 4 kg of paper clips. Each paper clip has a mass of about 0.8 g.
    (a) About how many paper clips are in the box?
    (b) Does this problem illustrate repeated measures or partition of a measure?

32. A metric ton is 1,000 kilograms, How many metric tons is a car that weights 2,480 kg?

33. Complete the following conversions.
    (a) 7 pounds = _____ ounces
    (b) 300 pounds = _____ tons
    (c) 12 ounces = _____ pounds

34. How many pounds in an ounce?

35. A container holds 3.24 liters of milk. How many milliliters is that?

 36. Your heart pumps about 60 mL of blood per heartbeat. About how many liters of blood will it pump in a day?

37. (a) 9.4 L = _____ mL
    (b) 37 mg = _____ g
    (c) 0.082 kg = _____ g

38. (a) 65 g = _____ mg
    (b) 47 g = _____ kg
    (c) 346 mL = _____ L

39. A coffee cup holds about
    (a) 25 mL   (b) 250 mL
    (c) 2.5 L   (d) 25 L

40. Give a benchmark for a milliliter.

41. A nurse wants to give a patient 0.3 mg IV. The drug comes in a solution containing 0.5 mg per 2 mL. How many milliliters should be used?

42. A nurse wants to give a patient 3 gm of sulfisoxazole. It comes in 500 mg tablets. How many tablets should be used?

43. Complete the following conversions.
    (a) 3 gallons = _____ quarts
    (b) 5 cups = _____ pints

44. Complete the following conversions.
    (a) 7 pints = _____ quarts
    (b) 12 cups = _____ gallons

45. Give a benchmark for
    (a) an ounce   (b) a quart

46. Give a benchmark for
(a) a pound   (b) a cup

47. It is 2050, and the United States has finally adopted the metric system. You have been assigned to determine the new standard metric size for each of the following. What will you use?
(a) a yardstick   (b) a half-gallon of milk
(c) a sheet of notebook paper
(d) a quarter-pound hamburger

48. True or false?
(a) 1 mm is longer than 1 in.
(b) 1 m is longer than 1 km.
(c) 1 g is heavier than 1 lb.
(d) 1 gallon is more than 1 L.

49. For each item, select the most appropriate measuring unit from the following list: mm, cm, m, km, g, kg.
(a) The weight of a penny
(b) Your waist size
(c) The thickness of a page

50. For each item, select the most appropriate measuring unit from the following list: mm, cm, m, km, g, kg.
(a) Your weight
(b) The distance from Miami to Atlanta
(c) The height of a room

51. Someone who tells you that the temperature is 80° outside
(a) is using Fahrenheit   (b) is using Celsius
(c) could be using either Fahrenheit or Celsius

52. A temperature of $-10°C$ is about
(a) $-20°F$ (b) $10°F$
(c) $40°F$ (d) $70°F$

53. The weather outside is sunny and 15°C. Which of the following would you wear?
(a) A heavy coat   (b) A light sweater
(c) A swimsuit

54. It is a cold day. Which would be a greater increase, an increase of 10°F or an increase of 10°C?

55. Tell whether each number is probably exact or approximate.
(a) Joe bought 1 kg of apples.
(b) Bill bought 2 ears of corn.

56. Tell whether each number is probably exact or approximate.
(a) The room is 5 m long.
(b) Sally has 3 brothers.

57. Measure a pen
(a) to the nearest inch.   (b) to the nearest half-inch.
(c) to the nearest eighth-inch.
(d) Which measurement is closest to the actual length of the pen?

58. Which measurement in each pair is more precise?
(a) 8 cm or 78 mm
(b) 6 kg or 5,820 g

59. How many significant digits are in each measurement?
(a) 380 miles   (b) 3.056 cm   (c) 0.0030 mg

60. How many significant digits are in each measurement?
(a) 5.003 kg   (b) 0.0012 L   (c) 200 days

61. (a) Measure the following line segment to the nearest millimeter. _____
(b) The actual length of the line segment is between _____ and _____.
(c) What is the GPE?

62. I measure the length of my shoe as 28 cm.
(a) The actual length of the shoe is between what two measures?
(b) What is the GPE?

63. Complete the table.

Measurement	Precision	Actual Length Is Between	GPE
32.1 m	Nearest tenth of a meter (or nearest dm)	32.05 m, 32.15 m	0.05 m
(a) 62 g			
(b) 5.12 cm			
(c) 4.6 km			

64. Give the GPE of each measurement.
(a) 37 m   (b) 5.21 g

65. Find the relative error for the measurements in parts (a)–(c) of Exercise 63.

66. (a) Measure the length and width of a dollar bill to the nearest centimeter.
    (b) Give the relative error of each measurement.

Extension Exercises

67. A passenger is riding in a train that is going 150 km/h. The passenger sees a second train pass by in 2 seconds. If the second train is moving at a speed of 150 km/h in the opposite direction, how long is the second train?

68. You have a large set of black and white centimeter cubes. You could make two different towers that are 1 cm high: a black tower and a white tower.

(a) How many different towers could you make that are 2 cm high?
(b) How many different towers could you make that are 3 cm high? (Draw pictures.)
(c) How many different towers could you make that are 4 cm high?
(d) How many different towers could you make that are N cm high? (N is a counting number.)
(e) Does part (d) involve induction or deduction?

69. (a) The prefix *micro*- represents 10^{-6}, or $\frac{1}{1,000,000}$. A beam of light travels 1 meter in about 3 microseconds. How many seconds it this?
    (b) The prefix *nano*- represents 10^{-9}, or $\frac{1}{1,000,000,000}$. The radius of a chlorine atom is 0.1 nanometer. How many meters is this?
    (c) A computer might add ten numbers in about $1.6 \cdot 10^{-8}$ sec. How many nanoseconds is this?

70. (a) The prefix *giga*- means 1 billion. Suppose a computer has a 160-gigabyte hard drive. How many bytes is this?
    (b) The prefix *mega*- means 1 million. Suppose a computer has an 80-megabyte hard drive. How many gigabytes is this?
    (c) A human being might live for 2,400 megaseconds. How many seconds is this?

Labs and Projects

71. Make a scale drawing of the solar system in which the distance from the sun to Pluto is somewhere between 30 cm and 1 m. The actual distances of planets from the sun, in meters, are Mercury, 5.8×10^{10}; Venus, 1.1×10^{11}; Earth, 1.5×10^{11}; Mars, 2.3×10^{11}; Jupiter, 7.8×10^{11}; Saturn, 1.4×10^{12}; Neptune, 2.9×10^{12}; Uranus, 4.5×10^{12}; Pluto, 5.9×10^{12}.

72. Make a scale drawing of your classroom on centimeter graph paper.

10.2 PERIMETER AND AREA

NCTM Standards

- explore what happens to measurements of a two-dimensional shape such as its perimeter and area when the shape is changed in some way (3–5)

- develop strategies for estimating the perimeters, areas, and volumes of irregular shapes (3–5)

- develop and use formulas to determine the circumference of circles and the area of triangles, parallelograms, trapezoids, and circles and develop strategies to find the area of more complex shapes (6–8)

LE 1   Opener

Discussion

In Figure 10–4, which is larger, the site or the floor plan?

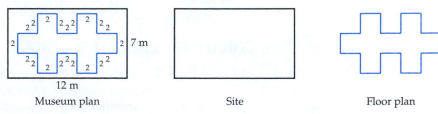

Figure 10–4

Children often have difficulty grasping the fact that more than one aspect of a figure can be measured. To answer the question in LE 1, one must first decide whether to compare perimeter or area.

PERIMETER

What length of fence will enclose a field? What length of a decorative border will you need for a classroom bulletin board? These situations both call for measuring the perimeter.

In elementary and secondary school, children study the perimeters of polygons and circles. The **perimeter** of a simple, closed plane figure is the length of its boundary. The perimeter is always measured in units of length, such as feet or centimeters.

LE 2   Reasoning

Discussion

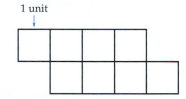

(a) The perimeter is _____.

(b) Suppose you rearrange the squares in other configurations in which each square shares at least one side with another square. What is the largest perimeter you can obtain? What is the smallest perimeter you can obtain? Make a plan, and solve the problem. (You could use graph paper, a square dot grid, or square tiles.)

The perimeter of a polygon is the sum of the lengths of its sides. Some polygons have fairly simple perimeter formulas that can be used instead of adding up the lengths of all the sides. Find the appropriate formula for a rectangle in the following exercise.

LE 3   Concept

A rectangle has a length l and a width w. What is a formula for the perimeter P?

The perimeter formula from LE 3 can be used in the following exercise.

LE 4   Skill

Refer back to the diagram at the beginning of the section. Which has a larger perimeter, the site or the floor plan?

THE CIRCUMFERENCE OF A CIRCLE

How far is one lap around a track with semicircular arcs? What is the perimeter of a bicycle tire with a diameter of 26 in. (Figure 10–5)? To answer these questions, one computes the **circumference** (perimeter) of a circle.

Figure 10–5

In Section 7.6, you saw how π originated from studies of the relationship between the circumference and the diameter. The circumference C equals π times the diameter d, or $C = \pi d$.

The value of π is an irrational number, an infinite, nonrepeating decimal that begins as 3.141592654. . . . In 1995, Hiroyuki Goto memorized and recited the first 42,000 places of π. Most of us can get by with 3.14 or 22/7 as an approximation of π.

The circumference formula, $C = \pi d$, can be rewritten using $2r$ in place of d as $C = 2\pi r$.

The Circumference Formula

A circle with diameter d and radius r has circumference $C = \pi d$, or $C = 2\pi r$.

You can use this formula to find the circumference when you know the length of the radius or diameter. In computations involving π, use π in situations involving perfect geometric figures, and use an approximate value of π such as 3.14 (or 22/7) in situations involving everyday objects.

Perfect geometric shapes $\rightarrow$ exact answers $\rightarrow$ Use π.

Everyday objects $\rightarrow$ approximate answers $\rightarrow$ Use a decimal approximation.

 LE 5   Skill

A bicycle wheel has a diameter of 26 in. How far would a rider travel in one full revolution of the tire? (Use 3.14 for π.)

AREA

If you want to know the size of the interior of a field, or which package of gift wrap is a better buy, you measure the area. **Area** is the measure of a closed, two-dimensional region.

LE 6   Opener

(a) How are squares used to measure area?

(b) What is another shape that could be used as a unit for area?

LE 7   Concept

Determine how many of each shape shown are needed to fill in the design in Figure 10–6.

(a) Trapezoids

(b) Triangles

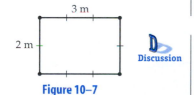

Figure 10–6

Although area can be measured with trapezoids or triangles as units, it is usually measured in square units. An area of "10 square units" means that 10 unit squares are needed to cover a flat surface. A square centimeter (cm^2) is an example of a square unit.

What is the area of your hand? To find out, you will need your hand and some centimeter graph paper.

LE 8   Reasoning

(a) Trace an outline of your hand on the centimeter graph paper. (Keep your fingers together.)

(b) How long is your hand?

(c) Estimate the area of your hand in square centimeters.

(d) Explain how you obtained your answer in part (c).

(e) If your teacher records the class's results, describe how they came out.

LE 9   Concept

(a) Divide up the interior of the rectangle in Figure 10–7 to show how many square meters would cover it.

(b) What is the area of the rectangle?

Figure 10–7

LE 10   Concept

A child says that there are 10 mm in a cm, so there must be 10 mm^2 in a cm^2. Is that right? If not, make a drawing that would guide the child to the correct result.

LE 11   Concept

Each square inside the rectangle in Figure 10–8 is 1 cm by 1 cm.

(a) How many rows and how many columns of squares are there?

(b) What is the area of the rectangle?

(c) Find the length and width of the rectangle.

(d) How would you compute the area using the length and width?

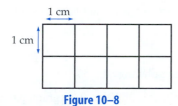

Figure 10–8

The rectangle in LE 11 has 2 rows and 4 columns of unit squares. While it is clear to you how many rows and columns are in Figure 10–8, some children have difficulty seeing this. They may need practice building arrays with square tiles.

The total number of squares is the number of rows times the number of columns (the array model of multiplication). Since the number of columns is determined by the length of the rectangle and the number of rows is determined by the width, the area can be obtained more quickly by multiplying the length (the number of columns) by the width (number of rows).

Area of a Rectangle

The area A of a rectangle that has length l and width w is

$$A = lw$$

Area formulas are especially useful in problems involving larger numbers, when counting squares might take a very long time.

LE 12   Concept

Refer back to the diagram at the beginning of the section. Which has a larger area, the site or the floor plan?

Consumers use the area formula for a rectangle to decide which package of wrapping is a better buy. Consider the following exercise.

LE 13   Connection

A store sells two kinds of wrapping paper. Package A costs $4 and has 3 rolls, each $2\frac{1}{2}$ ft by 6 ft. Package B costs $3.25 and has 4 rolls, each 2 ft by 5 ft. Which is the better buy? (*Hint:* Compare the cost per square foot for each package.)

AREA, PERIMETER, AND CONGRUENCE

Do you wonder how area, perimeter, and congruence are related to one another? The following exercises address this question.

LE 14   Concept

Do two congruent figures have equal area and equal perimeter?

Next, consider the relationship between area and perimeter. Children in both elementary and secondary school have trouble distinguishing between perimeter and area.

On farms, some animals live outdoors in corrals. Suppose you want to design a rectangular corral.

Discussion

LE 15   Reasoning

Suppose you have 20 m of fencing to construct a rectangular corral.

(a) Draw four different rectangles that each have a perimeter of 20 m. (Use graph paper if you have it.)
(b) Do they all have the same area?
(c) On the basis of your results, what kind of rectangle tends to have a smaller area?

LE 16   Reasoning

Discussion

You want to build a corral that will have an area of 16 m².

(a) Draw three different rectangles that each have an area of 16 m².
(b) Do they all have the same perimeter?
(c) On the basis of your results, what kind of shape tends to have a smaller perimeter?

LE 17   Reasoning

Discussion

On the basis of LE 14–LE 16, which of the following statements about plane figures *A* and *B* appear to be true? If you think a statement is false, give a counterexample.

(a) If the area of *A* equals the area of *B*, then $A \cong B$.
(b) If $A \cong B$, then the area of *A* equals the area of *B*.
(c) If *A* has a larger perimeter than *B*, then *A* has a larger area than *B*.
(d) If *A* has a larger area than *B*, then *A* has a larger perimeter than *B*.

As you may have guessed in LE 14, if two figures are congruent, then they have equal areas and perimeters. As you saw in LE 15 and LE 16, knowing which of two figures has a larger area does not determine which has a larger perimeter.

LE 18   Summary

Tell what you learned about perimeter and area in this section.

ANSWERS TO SELECTED LESSON EXERCISES

2. (a) 14   (b) 18; 12

3. $P = 2l + 2w$

4. The floor plan (40 m)

5. 81.64 in.

7. (a) 4   (b) 12

9. (a)

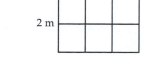

(b) 6 m²

10. No

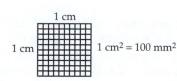

1 cm² = 100 mm²

11. (b) 8 cm²

12. The site

13. B ($3.25 for 40 ft² or $0.08125/ft²)

14. Yes

15. (b) No   (c) A longer, thinner shape

16. (b) No   (c) A square shape

17. Only part (b) is true.

10.2 HOMEWORK EXERCISES

Basic Exercises

1. (Adapted from *Measuring Space in One, Two, and Three Dimensions* by D. Schifter et al.)
   (a) When third graders are asked which of the following rectangles is the biggest, which one do you think most of them would choose?

   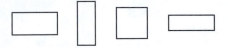

   (b) What measurement are these children comparing?
   (c) If you told them that each rectangle showed a chocolate bar, which one would they pick as the biggest?
   (d) What measurement are the children comparing in part (c)?

2. A fourth grader says that one rectangle is twice as big as another. What is unclear about this statement?

3. Measure the perimeter of each figure to the nearest tenth of a centimeter.

4. Measure the perimeter of the figure to the nearest tenth of a centimeter.

5.   1 unit

   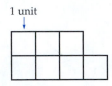

   (a) The perimeter is _____.
   (b) Suppose you rearrange the squares in other configurations in which each square shares at least one side with another square. What is the largest

perimeter you can obtain? What is the smallest perimeter you can obtain? (Devise a plan, and solve the problem.)

6.   1 unit

   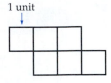

   (a) The perimeter is _____.
   (b) Rearrange the squares in other configurations, and compute the perimeter. (Each square must share at least one side with another square.) What is the largest possible perimeter? What is the smallest possible perimeter?

7. Draw a square. Draw a second square whose sides are twice as long.
   (a) How do their perimeters compare?
   (b) Start with a new square, and repeat the exercise.
   (c) Make a generalization based on parts (a) and (b).
   (d) Does part (c) involve induction or deduction?

8. A lot is 21 ft by 30 ft. To support a fence, an architect wants an upright post at each corner and an upright post every 3 ft in between. How many of these posts are needed?

9. Why is $C = 2\pi r$ the same as $C = \pi d$?

10. In a circle, how many times longer is the circumference C than the diameter d?

11. The circumference (equator) of the Earth is about 25,000 miles. What is the Earth's approximate diameter? (Use a decimal approximation for π.)

12. The dome of the New Orleans Superdome has a diameter of 680 ft. What is the circumference? (Use a decimal approximation for π.)

13. The Earth travels around the sun in an approximately circular path with a radius of 9.3×10^7 miles.
    (a) How far does the Earth travel in one trip around the sun? (Use a decimal approximation for π.)
    (b) If the Earth takes 365 days to make one revolution, how far does it travel in 1 hour?
    (c) If the Earth is moving so fast, why don't we fall off?

14. A car has wheels with radii of 40 cm. How many revolutions per minute must a wheel turn so that the car travels 50 km/h?

 15. A track consists of straightaways and circular arcs.

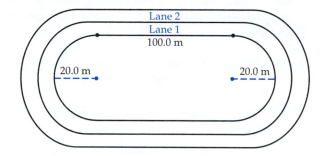

(a) Find the perimeter of the inside edge of lane 1.
(b) If lane 1 is 1.0 m wide, how much farther is a lap around the inside edge of lane 2 than a lap around the inside edge of lane 1?

 16. An engineer is designing a track with a semicircular arc on each end, as shown. One lap around the track should be 400 m.

(a) If $CD = 40$ m, how long is AB?
(b) If $AB = 150$ m, how long is CD?

17. A tile design for a pool that is 4 m by 2 m has a border with 16 one-meter-square tiles.

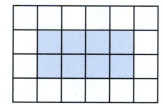

How many tiles would an x meter by y meter pool with the same design have? (Assume x and y are counting numbers.) Make a plan and solve the problem.

18. When the radius of a circle increases by 1, the circumference increases by _____.

19. What is the area of the following figure in unit hexagons?

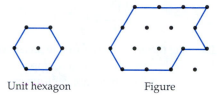

Unit hexagon   Figure

20. How many of each shape would it take to cover the following hexagon?

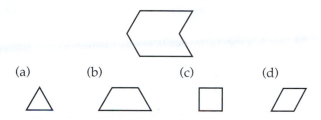

(a)   (b)   (c)   (d)

21. Find the area, in square units, of the shaded region (Activity Card 4).

1 unit

22. Find the area of the figure in square units.

23.

(a) Estimate the area of the footprint on each grid, in the square units shown.
(b) Which estimate is more precise?

24. Approximate the area of the circle by counting squares.

25. (a) Estimate the area of your foot in square centimeters.
    (b) Use centimeter graph paper to approximate the area of your foot in square centimeters.

26. (a) Estimate the seat area of a chair in your home.
    (b) Measure the seat and approximate its area.

27. What aspect of a polygon or circle does area measure?

28. What is the difference between a perimeter of 6 inches and an area of 6 square inches?

29. A third grader cannot decide which would have more space for planting, a rectangle that is 2 feet by 4 feet or one that is 3 feet by 3 feet. Show how to compare the two gardens by using a drawing or square tiles.

30.

5 m

2 m

(a) Why can't you measure the area of a rectangle in meters?
(b) Show how many 1-m squares cover the interior of the rectangle.
(c) What is the area of the rectangle?

31. A pool has two different sections. Show two different ways to compute the surface area of the swimming pool.

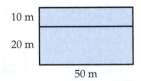

10 m
20 m
50 m

32. The Federal Housing Administration (FHA) requires that the window area of a house be at least 10% of the floor space. How many 15-ft² windows are needed to meet FHA requirements for a house with a floor that is 25 ft by 40 ft?

33. In November 2002, the Galicia oil spill off the coast of Spain covered 800 square kilometers. Find the land area of some place where you live, and compute how many of those would be needed to cover 800 square kilometers.

34. A major-league baseball field has an area of about
    (a) 6 m²   (b) 6000 cm²   (c) 6000 m²

35. What decimal multiplication equation is represented by the picture?

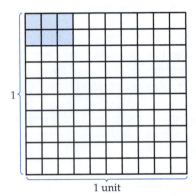

1

1 unit

36. What fraction multiplication is represented by the picture?

1

1 unit

37. (a) How many square feet are in a square yard? (*Hint:* Draw 1-ft² squares in the interior.)

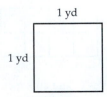

1 yd

1 yd

(b) As you know, 3 ft = 1 yd. So (3 ft)(3 ft) = (1 yd) (1 yd). Simplify both sides, and write the resulting equation.

38. A store sells carpet for $8.99 per square yard. How much does a piece of carpet 7 ft by 9 ft cost? (*Hint:* See the preceding exercise.)

39. (a) 1 cm^2 = _____ mm^2 (*Hint:* First solve
     1 cm = _____ mm, and square both sides.)
   (b) 1 cm^2 = _____ m^2

40. (a) 8 cm^2 = _____ mm^2
   (b) 500 m^2 = _____ cm^2

 41. (a) How many tiles, each 25 cm by 25 cm, would be needed to cover a floor that is 2 m by 4 m?
   (b) How many tiles, each 25 cm by 25 cm, would be needed to cover a floor that is 2.2 m by 4.6 m?

 42. A mapmaker wants to represent an area of roughly 20,000 ft^2 using the scale 1 in. = 20 ft. What area will the map have?

 43. A store sells two kinds of wrapping paper. Package A costs $5.25 and has 6 rolls, each 2 ft by 5 ft. Package B costs $7 and has 4 rolls, each 4 ft by 6 ft. Which package is the better buy?

 44. A photo enlargement that is 6 in. by 9 in. costs $1.75, and an enlargement that is 8 in. by 10 in. costs $2.75. Which enlargement costs less per square inch?

45. A rectangular garden is 10 ft by 12 ft. A rectangular sidewalk 1 ft wide is built around the outside. What is the area of the sidewalk? (Draw a picture.)

46. A rectangle and a square have equal areas. If the rectangle is 6 ft by 8 ft, how long is a side of the square?

 47. Suppose that the housing authority has valued the house shown here at $220 per ft^2.

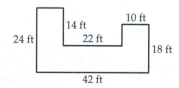

(a) Show two different ways of subdividing the region into rectangles.
(b) Find the assessed value of the house.

 48. Shown here is a scale drawing $\left(\frac{1}{4} \text{ in.} \approx 12 \text{ ft}\right)$ of the floor plan for the bedroom, hall, and living room of your new apartment.

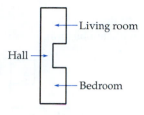

(a) Fill in the missing lengths.
(b) How many square feet of carpet will be needed to cover the floors?
(c) The carpet you want costs $8 per square *yard*. What will the carpet for your apartment cost?
(d) How many feet of molding would you need around the edges of the floor to hold the carpet in place?
(e) If the molding costs $0.59/ft, find the total cost of the molding.

49. (a) Complete the third example, repeating the error pattern from the completed examples.
   (b) Describe the error pattern.

$$7 \text{ cm}^2 = \underline{\quad 70 \quad} \text{ mm}^2$$
$$80 \text{ m}^2 = \underline{\quad 8000 \quad} \text{ cm}^2$$
$$26 \text{ mm}^2 = \underline{\quad\quad\quad} \text{ cm}^2$$

50. (a) Complete the third example, repeating the error pattern from the completed examples.
   (b) Describe the error pattern.

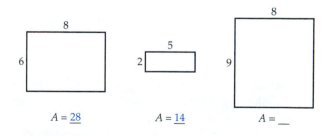

51. Using centimeter graph paper, draw seven different shapes, composed of centimeter squares, that each have a perimeter of 16 cm and an area of 12 cm^2. Each centimeter square should have a common side with at least one other square unit.

52. Use a geoboard or a geoboard dot grid, and show a shape that has
   (a) a perimeter of 8 units and an area of 3 square units.
   (b) a perimeter of 10 units and an area of 6 square units.

53. Add a square to the figure that will increase the area of the figure without changing its perimeter.

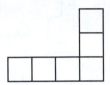

54. A 3-by-5-in. photo print costs $0.30. I want an enlargement with twice the length and twice the width. I figure that it should cost about twice as much. Am I right?

55. A fourth grader tells you that as the perimeter of a rectangle increases, the area will also increase. Her examples are rectangles that are 10 by 10, 10 by 15, and 10 by 20. How would you respond?

56. Which of the following statements about plane figures A and B are true?
   (a) If $A \cong B$, then the area of A equals the area of B.
   (b) If the area of A equals the area of B, then $A \cong B$.
   (c) If the area of A equals the area of B, then the perimeter of A equals the perimeter of B.

 57. (a) Consider the following problem. "A rectangle has an area of 4 m². What is the smallest possible perimeter it could have?" Devise a plan, and solve the problem.
   (b) Repeat part (a) for a rectangle with an area of 9 m².
   (c) Generalize your results for a rectangle with an area of N m².

58. Try the following if you have a set of 36 square tiles. (Graph paper could also be used.) Suppose a rectangle has an area of 36 square units.
   (a) Use the tiles to construct all possible rectangles with a length and a width that are whole numbers. Find the perimeter of each rectangle, and record the results.
   (b) What are the length and width of the rectangle with the greatest perimeter?

Extension Exercises

59. (a) A rectangle has a whole-number length and a whole-number width, and its area and perimeter are the same number. What are its dimensions?
   (b) Find a second solution to part (a).

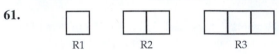 **60.** A rectangle has a length that is 16 m more than its width, and its area is 2961 m². What are the length and width? (Guess and check.)

61.

   (a) If each side has a length of 1 unit, find the perimeters of R1, R2, and R3.
   (b) Draw R4 and find its perimeter.
   (c) What is a formula for the perimeter of RN, in which N is a whole number?
   (d) Find the perimeter of R10.

62.

   (a) If each side has a length of 1 unit, find the perimeters of H1, H2, and H3.
   (b) Draw H4 and find its perimeter.
   (c) What is a formula for the perimeter of HN, in which N is a whole number?
   (d) Find the perimeter of H10.

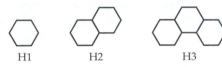 **63.** A cowhand has 350 yards of fence to enclose equal adjacent rectangular areas for horses and cattle. Find an arrangement that will enclose more area than either one large square or two adjacent squares. (Guess and check.)

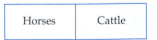

| Horses | Cattle |

64. Mrs. Cunningham wants to build a pen for her prize-winning hogs next to the barn with a three-sided, rectangular 24-yard fence. What dimensions will enclose the largest area? If you don't have a computer spreadsheet program, then just guess and check. If you do have a spreadsheet program, make three columns labeled "Base," "Height," and "Area." Fill in all possible whole-number values for the base, and use formulas to generate a table of values in the other two columns.

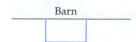

Barn

65. Make a graph showing the relationship between length and width for all rectangles that have perimeters of 20.

Width

Length

66. Make a graph showing the relationship between length and width for all rectangles that have areas of 20.

Width

Length

67. Many algebraic formulas can be represented geometrically. Refer to the figure, and write an equation related to the area of the entire region.

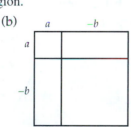

(a)

	a	2
a	a^2	$2a$
2	$2a$	4

(b)

	a	$-b$
a		
$-b$		

68. What algebraic equation is suggested by each diagram?

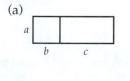

(a)

a

b c

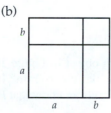

(b)

b

a

a b

69. About 2,200 years ago, Eratosthenes measured the circumference of the Earth. (That's surprising, considering that many people in more recent civilizations did not even know the Earth was approximately spherical.) Eratosthenes measured the angle of a shadow in Alexandria to be 7.5° at the same time the sun was directly overhead in Aswan, 500 miles away.

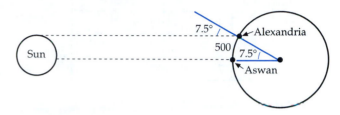

Sun   7.5°   Alexandria   500   7.5°   Aswan

From this, he knew that the central angle shown was 7.5°. Explain how he would compute the circumference of the Earth. (Its actual circumference is about 24,902 miles.)

70. A function has exactly one output for each input or set of inputs. Explain how measuring the area of a plane figure creates a function.

Technology Exercises

71. A store records the price of its rugs on a spreadsheet. Length and width are given in feet. (Use a computer spreadsheet if you have one.)

	A	B	C	D	E
1	Rugs				
2	Length	Width	Area	Price/ft^2	Total Price
3	9	12		$6.99	
4	10	20		$8.99	

(a) What does the number in cell D3 represent?

(b) What are the formulas to determine the numbers in cells C3 and E3?

(c) Complete column E. Use a spreadsheet program if you have it.

72. A student wants to study the areas of various rectangles with perimeters of 16, using a spreadsheet.

	A	B	C
1	Rectangles	$P = 16$	
2	Length	Width	Area
3	4	4	
4	5	3	
5	6	2	
6	7	1	

(a) What is the formula to determine the number in cell C3?
(b) Complete column C. Use a spreadsheet program if you have it.
(c) How is the size of the area related to the length and width?

Labs and Projects

73. Draw a floor plan for a two-bedroom apartment that will have 1,300 ft² of floor space.

10.3 AREAS OF QUADRILATERALS, TRIANGLES, AND CIRCLES

NCTM Standards

- develop, understand, and use formulas to find the areas of rectangles and related triangles and parallelograms (3–5)

- develop and use formulas to determine the circumference of circles and the area of triangles, parallelograms, trapezoids, and circles and develop strategies to find the area of more complex shapes (6–8)

- understand how mathematical ideas interconnect and build upon one another to produce a coherent whole (pre-K–12)

Tennessee

Sail

Dinner plate

Figure 10–9

To approximate the surface area of each object in Figure 10–9, you could use an area formula. Do you know where such formulas come from? In Section 10.2, you studied the concept of measuring area in square units and how that leads to a more efficient area formula for the rectangle (and square).

Concept of area $\longrightarrow$ Area of a rectangle

We'll pick things up from there. In this section, we will derive in a deductive sequence the area formulas for a triangle, a parallelogram, and a circle.

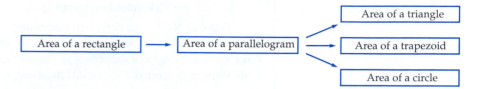

 LE 1   Opener

(a) What are the area formulas for a parallelogram, triangle, and trapezoid?
(b) How do you derive the area formula for a parallelogram from the area formula for a rectangle?

The activities in this section require scissors, a 30-cm ruler, and some paper.

THE AREA OF A PARALLELOGRAM

Before you begin cutting, here is some background information about parallelograms. Any side of a parallelogram can be designated as the **base** (usually it is the bottom side). Then the **height** is the distance from the base to the opposite side. The height is always measured with a segment that is *perpendicular* to the base (or an extension of the base) and the opposite side (Figure 10–10).

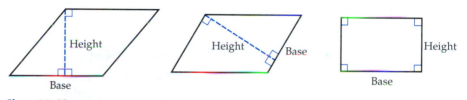

Figure 10–10

The terms "base" and "height" are also used for triangles and trapezoids (more on that later). The following exercise will show you how the area formula for a parallelogram is related to the area formula for a rectangle.

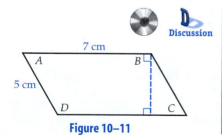

Figure 10–11

LE 2   Skill

(a) Cut out a rectangle that is 7 cm by 5 cm. (Use centimeter graph paper if you have it.)
(b) Cut out a parallelogram with the dimensions and shape shown in Figure 10–11, and label its vertices A, B, C, and D inside the parallelogram, as shown.
(c) What is the perimeter of the rectangle?
(d) What is the perimeter of the parallelogram?
(e) Compare the areas of the two figures by placing one paper over the other. Is one larger in area, or are they equal in area?

LE 3   Reasoning

To find the exact area of the parallelogram, do the following.

(a) Cut the parallelogram into pieces by cutting along the perpendicular segment from B to $\overline{CD}$. Rearrange these two pieces into a rectangle.

(b) What are the dimensions of the rectangle you formed?

(c) What is the area of the rectangle you formed from the parallelogram?

(d) What is the area of the original parallelogram?

LE 4   Reasoning

(a) On the basis of LE 2 and LE 3, describe a general method for finding the area of a parallelogram such as the one in Figure 10–12.

(b) What is the area of any parallelogram in terms of its base b and its height h?

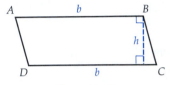

Figure 10–12

The answer to LE 4(b) is the parallelogram area formula.

Area of a Parallelogram

The area A of a parallelogram that has a base of length b and height h is

$$A = bh$$

Figure 10–13

LE 5   Skill

(a) Use the area formula for a parallelogram to find the area of *JANE* in Figure 10–13.

(b) Does part (a) involve induction or deduction?

THE AREA OF A TRIANGLE

How big are the sails on your new sailboat? The formula for the area of a triangle might tell you. This formula can be derived from the area formula for a parallelogram. Do you have scissors ready?

LE 6   Reasoning

(a) Cut out two congruent triangles. Label one side as the base b, and draw the perpendicular height from b to the opposite vertex (Figure 10–14).

(b) Figure out how to put the two triangles together to form a parallelogram.

(c) What is the area of the parallelogram in terms of b and h?

(d) What is the area of each triangle in terms of b and h?

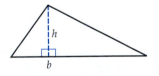

Figure 10–14

As you saw in LE 6, two congruent triangles can be put together to form a parallelogram. Since the two triangles have the same area, each has half the area of the parallelogram.

Area of a Triangle

The area A of a triangle that has a base of length b and height h is

$$A = \frac{1}{2} bh$$

The ancient Egyptians, Babylonians, and Chinese were all familiar with the standard area formulas for rectangles, parallelograms, and triangles.

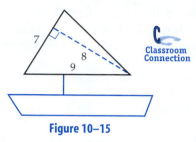

Figure 10–15

Classroom Connection

LE 7   Skill

A sixth grader says the area of the triangle in Figure 10–15 is $\frac{1}{2} \cdot 9 \cdot 8 = 36$ square units. Is this right? If not, what would you tell the child?

The area of any polygon can be found by subdividing it into rectangles, triangles, or both! After finding the area of each rectangle and triangle, add up the areas to obtain the area of the polygon.

THE AREA OF A TRAPEZOID

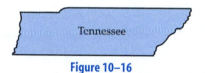

Figure 10–16

The state of Tennessee is shaped like a trapezoid (Figure 10–16). You could approximate its land area if you had an area formula for a trapezoid.

The area formula for a parallelogram can be used to deduce the formula for the area of a trapezoid. Two congruent triangles can be put together to form a parallelogram, and so can two congruent trapezoids. You don't believe me? Try LE 8.

Discussion

LE 8   Reasoning

(a) Cut out two congruent trapezoids, and label them as shown in Figure 10–17. (The parallel sides of the trapezoids are its bases, b_1 and b_2.)
(b) Put them together to form a parallelogram.
(c) What is the area of the parallelogram in terms of b_1, b_2, and h?
(d) What is the area of the trapezoid?

Figure 10–17

You have discovered another area formula!

Area of a Trapezoid

The area A of a trapezoid that has parallel sides of lengths b_1 and b_2 and height h is

$$A = \frac{1}{2}(b_1 + b_2)h$$

LE 9 Skill

Tennessee is approximately a trapezoid (Figure 10–18).

Figure 10–18

(a) What is the approximate surface area of Tennessee?
(b) Land is not perfectly flat. Would this fact make the surface area greater or less than the area of a perfectly flat surface?

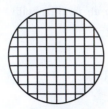

Figure 10–19

Figure 10–20

Figure 10–21

THE AREA OF A CIRCLE

What is the surface area of a dinner plate that has a 12-in. diameter? To answer this question, one would compute the area of a circle. It is difficult to count the number of squares inside a circle. Figure 10–19 indicates why circle areas rarely come out to whole numbers of unit squares.

As with the triangle and the trapezoid, the formula for the area of a circle is suggested by its relationship to the area formula for a parallelogram! However, in this case, the parallelogram representation is approximate. Calculus is needed for a precise derivation of the area formula for a circle.

Again, the idea is to make the new figure (a circle) look like a parallelogram. Suppose a circle is cut into six equal parts, as shown in Figure 10–20. The parts can be arranged as shown in Figure 10–21.

Discussion

LE 10 Reasoning

(a) What shape does the rearranged circle in Figure 10–21 approximate?

(b) The "base" of the new figure is about $\frac{1}{2}$ the _____ of the original

circle, and the "height" of the figure is about the same as the _____ of the original circle.

(c) Using the parallelogram formula $A = bh$, the area of the circle is about _____.

LE 10 suggests that the area of the circle is $\frac{1}{2} \cdot C \cdot r$. Using $C = 2\pi r$, we can obtain the familiar area formula from $A = \frac{1}{2} \cdot C \cdot r$. It goes like this (see Figure 10–22):

$$A = \frac{1}{2} \cdot C \cdot r = \frac{1}{2} \cdot 2\pi r \cdot r = \pi r^2$$

is approximately

$A = \pi r^2$

Figure 10–22

Figure 10–23 shows a lesson in the sixth-grade *Harcourt Math* textbook, which derives the area formula of a circle in a similar way. The more pieces the circle is cut into, the closer the shape is to a parallelogram with base πr and height r. This connection suggests the formula for the area of a circle.

Area of a Circle

The area A of a circle that has radius r is

$$A = \pi r^2$$

You can use this formula to find the area of a circle when you know the radius or diameter.

Classroom Connection

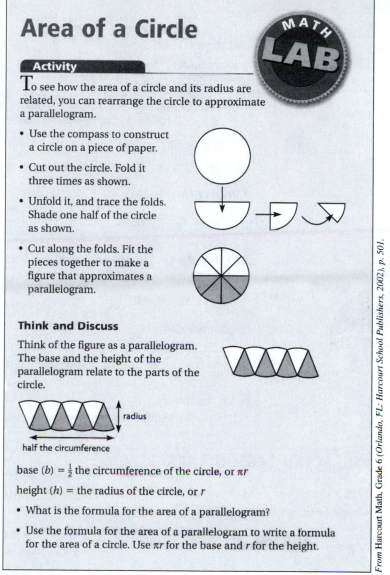

Area of a Circle

Activity

To see how the area of a circle and its radius are related, you can rearrange the circle to approximate a parallelogram.

- Use the compass to construct a circle on a piece of paper.

- Cut out the circle. Fold it three times as shown.

- Unfold it, and trace the folds. Shade one half of the circle as shown.

- Cut along the folds. Fit the pieces together to make a figure that approximates a parallelogram.

Think and Discuss

Think of the figure as a parallelogram. The base and the height of the parallelogram relate to the parts of the circle.

radius

half the circumference

base (b) = $\frac{1}{2}$ the circumference of the circle, or πr

height (h) = the radius of the circle, or r

- What is the formula for the area of a parallelogram?

- Use the formula for the area of a parallelogram to write a formula for the area of a circle. Use πr for the base and r for the height.

From Harcourt Math, Grade 6 (Orlando, FL: Harcourt School Publishers, 2002), p. 501.

Figure 10–23

LE 11   Skill

You are at a pizza parlor with a friend. You suggest ordering two small (8-in.-diameter) pizzas. She suggests getting one 16-in.-diameter pizza, since you both like the same toppings. Is the one large pizza the same size as the two small ones?

(a) Guess the answer.

(b) Work out the areas, and find out if you're right.

Some area problems require using more than one area formula.

8

Figure 10–24

- **EXAMPLE 1**

Figure 10–24 shows a square and a semicircle. Find the exact shaded area.

SOLUTION

Understanding the Problem   The shaded area is inside the square but outside the semicircle.

Devising a Plan   In this type of problem, (1) identify familiar shapes and their area formulas, and (2) determine how the shaded area is related to the familiar shapes.

The familiar shapes are a square $(A = s^2)$ and half of a circle $\left(A = \frac{1}{2}\pi r^2\right)$. The shaded area is the area of the square minus the area of the half circle.

Carrying Out the Plan

$$A_{\text{shaded}} = A_\square - \frac{1}{2}A_\bigcirc = (8^2) - \frac{1}{2}\pi \cdot 4^2 = (64 - 8\pi) \text{ square units}$$

Looking Back   The same method would work if the square and circle were replaced by other shapes, such as triangles and rectangles.

Try one yourself.

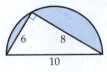
6   8
10

Figure 10–25

LE 12   Reasoning

Consider the following problem. "Figure 10–25 shows a semicircle. Find the exact shaded area." Devise a plan, and solve the problem.

LE 13   Summary

Tell what you learned about area in this section.

ANSWERS TO SELECTED LESSON EXERCISES

2. (c) 24 cm   (d) 24 cm
(e) The rectangle has a larger area.

4. (a) Translate the right triangle as shown to form a rectangle.

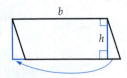
b

h

The area of the rectangle is bh.
Parallelogram area = rectangle area = bh
(b) $A = bh$

5. (a) 50 square units
(b) Deduction

6. (c) bh (d) $\frac{1}{2}bh$

7. What base goes with the height of 8? (Rotate the triangle if necessary.)

$$A = \frac{1}{2}(7)(8) = 28 \text{ square units}$$

8. (c) $(b_1 + b_2) \cdot h$

9. (a) About $\frac{1}{2}(439 + 350) \cdot 110 = 43{,}395$ square miles

(b) Greater

10. (a) A parallelogram
(b) circumference; radius   (c) $\frac{1}{2} \cdot C \cdot r$

11. (b) One 16-in. pizza has twice as much pizza as two 8-in. pizzas!

12. $12.5\pi - 24$ square units

10.3   HOMEWORK EXERCISES

Basic Exercises

1. Find the area of each polygon. (*Hint:* You could count squares and parts of squares, or you could subdivide each figure.)

(a)    (b)

2. Find the area of each polygon (Activity Card 4).

(a)   (b)

3. Which figure shown has the larger area?

(1)

8

10

8

10

(2)

10

8   8

10

(a) Figure 1   (b) Figure 2
(c) Their areas are equal.

4. A parallelogram has adjacent sides of length 4 cm and 6 cm.
   (a) Find the largest possible area the parallelogram could have.
   (b) Could the parallelogram have an area of 1cm²?

5. Find the area and perimeter of each parallelogram.

(a)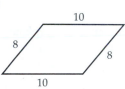

6 in. 5 in.

10 in.

(b)

2 m   3 m

4 m

6. Find the area and perimeter of each parallelogram.

(a)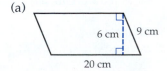

6 cm 9 cm

20 cm

(b)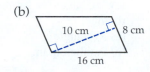

10 cm   8 cm

16 cm

7. Use the area formula for a rectangle ($A = bh$) to explain why the area of the parallelogram shown is $A = bh$.

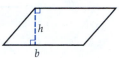

h

b

8.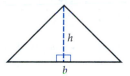

D   C

A   B

Suppose a fifth grader has not yet learned the area formula for a parallelogram. How could you develop the formula from the geoboard diagram?

9. Use the area formula for a parallelogram ($A = bh$) to explain why the area of the triangle shown is $A = \frac{1}{2}bh$.

h

b

10. (a) Suppose you have two congruent rectangles. You divide one into two smaller rectangles and divide the other into two triangles as shown. Tell how the area of one of the smaller rectangles compares to the area of one of the triangles. Show why.

   (b) How does part (a) suggest the area formula for the triangle?

11. Find the area of each triangle.

(a)   30 in.

YIELD

28.3 in.   28.3 in.   24 in.

(b)

6

10

4

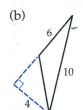

(c) If the metal for the sign in part (a) costs $0.79 per 100 in.², find the total cost of metal for the sign.

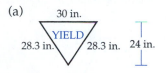

12. Find the area of the kite.

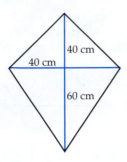

40 cm
40 cm
60 cm

13. Find the area of the triangle.

14. Use a geoboard or a geoboard drawing, and show a triangle with each of the following areas. (Make a drawing of your answers if you use a geoboard.)
(a) 2 square units   (b) 4.5 square units

15. Use the area formula for a parallelogram ($A = bh$) to explain why the area of the trapezoid shown is

$$A = \frac{1}{2}(b_1 + b_2) \cdot h.$$

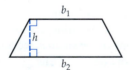

b_1
h
b_2

16. Find the area of the trapezoid.

 17. Use a ruler to find the area of the trapezoid in square millimeters.

 18. Find the approximate land area of Nevada.

310 miles
210 miles
Nevada
480 miles
410 miles

 19. Two adjacent lots are for sale. Lot A costs $20,000, and lot B costs $27,000. Which lot has the lower cost per square meter?

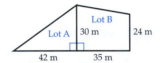

Lot B
Lot A   30 m   24 m
42 m   35 m

20. Following is a drawing of a Boeing 747 airplane wing. What is the area of the top surface of the wing? (Assume that the wing is flat.)

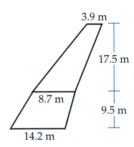

3.9 m
17.5 m
8.7 m
9.5 m
14.2 m

21. About 3600 years ago, Egyptians used the area formula

$$A = \frac{(a + c)(b + d)}{4}$$

for quadrilaterals with lengths of successive sides a, b, c, and d. For what types of quadrilaterals does this formula give the correct area?

22. Since a square is a type of rectangle, the area formula for a rectangle works for squares. Which of the following have area formulas that would work for any rhombus?
(a) Rectangle   (b) Triangle
(c) Parallelogram   (d) Square

23. (a) Show why $\frac{1}{2}Cr = \pi r^2$ for any circle.

(b) Does part (a) involve induction or deduction?

24. A circle is rearranged as shown.

Dashed line has length b

Show how to derive an approximate area formula for a circle using the area formula for a parallelogram.

 25. An ancient Egyptian document called the Rhind Papyrus (1650 B.C.) shows how to find the area of a circular field with a diameter of 9. It says to take $\frac{1}{9}$ of the diameter, which is 1. Square the remainder of the diameter $(9 - 1 = 8)$ to obtain 64, which is the area.

(a) What is the exact answer to the problem?
(b) Approximate the result of part (a), using 3.14 for π.
(c) What area formula did the ancient Egyptians use?
(d) What value of π would make the Egyptians' formula work?

26. Suppose that you know how to find the area of a square but not the area of a circle. How could you approximate the area of the circle shown?

 27. (a) Exactly how many times larger is the area of a pizza that has a 14-in. diameter than a pizza that has a 10-in. diameter?

(b) If a 10-in. pizza costs \$6, what is a fair price for a 14-in. pizza?

28. A woman wants to water a square plot.

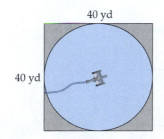

40 yd

40 yd

40 yd

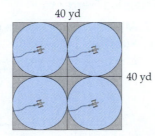

40 yd

40 yd

40 yd

Would 4 small sprinklers reach more area than 1 giant sprinkler? (Assume that the sprinklers spray the circular areas shown.)

29. (a) When the radius of a circle is doubled, the circumference is multiplied by _____.

(b) When the radius of a circle is doubled, the area is multiplied by _____.

30. If the radius of a circle increases by 30%, then the area of the circle increases by
(a) 9%   (b) 30%   (c) 69%
(d) 90%   (e) 130%

31. Approximate the area of the eating surface of a plate in your home.

32. Approximate the area of the surface of the lens from a pair of glasses you have at home.

33. A sixth grader works out the area of a triangle with base 4 and height 10 as follows:

$$\frac{1}{2}(4 \cdot 10) = \frac{1}{2} \cdot 4 \cdot \frac{1}{2} \cdot 10 = 2 \cdot 5 = 10 \text{ square units.}$$

Is this right? If not, what would you tell the child?

34. (a) Complete the third example, repeating the error pattern from the completed examples.

$A = 8\pi$ $A = 6\pi$ $A = \_\_$

(b) Describe the error pattern.

35. C is the center of the circle.

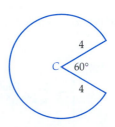

(a) Perimeter = _____
(b) Area = _____

36. What are the area and perimeter of the sector of the circle shown? C is the center.

3 ft

C 3 ft

37. The circumference of a circle is 20π ft. What is the exact area?

38. The area of a circle is 25π m². What is the exact circumference?

39. A circular garden has a diameter of 6.0 m. It has a circular path 1.0 m wide around its border. What is the area of the path?

40.

(a) Which do you think is larger, the area of the inner circle or the area of the shaded region?
(b) Using a ruler, compute each area.

41. C is the center.

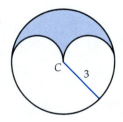

C 3

Shaded area = _____

 42. A 3.0-m fence is used to enclose a garden with a circular arc.

Fence

L
Garden

L

(a) How long is L?
(b) What is the area of the garden?

43. Find the area of the region on the square lattice. All curves are circular arcs.

44. The figure shows a square and a semicircle.

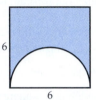

6

6

Shaded area = _____

45. C is the center. Find the shaded area.

6

C 6

46. Find the exact area of the geometric shape (used for Norman windows). Assume that the top (BC) is a circular arc.

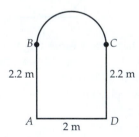

B C

2.2 m   2.2 m

A 2 m D

Extension Exercises

47. All of the regions shown have perimeters of 12π. Which has the largest area?

(a)    (b)   (c)

48. A lake has a surface area of 800 km².
   (a) What is the minimum length the shoreline could be?
   (b) What is the maximum length the shoreline could be?

49. Area *ARNO* = _____

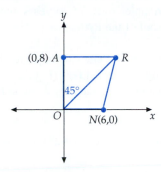

50. Make a graph showing the relationship between the radius and circumference of a circle.

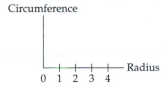

51. Express the area *A* of a circle in terms of its circumference *C*.

Labs and Projects

52. Express the perimeter *P* of a square in terms of its area *A*.

53. (a) Draw a large quadrilateral *ABCD* on a sheet of paper. Mark the midpoints of each side *E*, *F*, *G*, and *H*, and connect *E* to *F* to *G* to *H* to *F*.

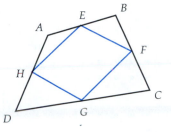

   (b) Cut out the four triangles and place them so that they cover *EFGH*.
   (c) What relationship does part (b) suggest?

54. Write a report on the history of π.

10.4   THE PYTHAGOREAN THEOREM

NCTM Standards

- create and critique inductive and deductive arguments concerning geometric ideas and relationships such as congruence, similarity, and the Pythagorean relationship (6–8)

The lengths in Figure 10–26 can be found using one of the most famous theorems in mathematics: the Pythagorean Theorem.

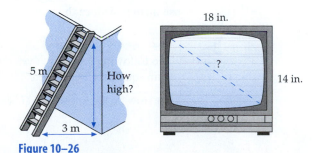

Figure 10–26

PYTHAGOREAN THEOREM

As you know from Chapter 4, Pythagoras and his followers were interested in numbers and their characteristics. They also studied geometry.

The famous Pythagorean Theorem describes the relationships between the lengths of the three sides of a right triangle (a triangle that has a right angle). Although the Babylonians knew this property of right triangles over 3,500 years ago, Pythagoras or one of his followers gave the first known proof of the Pythagorean Theorem about 2,500 years ago.

Figure 10–27

The sides of a right triangle have special names. The side opposite the right angle is called the **hypotenuse,** and the other two sides are called **legs** (Figure 10–27).

The lengths of the sides of any right triangle are related by a single formula.

Figure 10–28

LE 1   Reasoning

Suppose a right triangle has legs of length 1 (see Figure 10-28). Below it, the same triangle is shown with a square drawn on each side.

(a) Find the areas of the three squares.
(b) Now draw a right triangle with legs of length 2. Draw a square on each side.
(c) Find the areas of the three squares in part (b).
(d) Repeat parts (b) and (c) for a right triangle with legs of length 1 and 2.
(e) Generalize your results.

Your results from LE 1 should suggest the Pythagorean Theorem. And now, without further ado, here is one of the most famous and useful theorems in all of mathematics.

The Pythagorean Theorem

If a right triangle has legs of lengths a and b and a hypotenuse of length $c,$ then $c^2 = a^2 + b^2.$

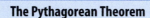

How do we know that this theorem is true? Not only has the theorem been proved; it has been proved in about 370 different ways! Even James Garfield (later President Garfield) thought up a new proof while he was in Congress.

One proof uses the area formulas for a triangle and a square and the sum of the angle measures in a triangle. These are all ideas you have studied in this course. Now use them to deduce the Pythagorean Theorem.

Discussion

LE 2   Reasoning

The following proof involves putting together four copies of a right triangle $\triangle FGH$ to form square $ACEG$ (Figure 10–29).

(a) How do we know $ACEG$ is a square?

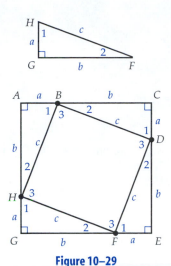

Figure 10–29

Figure 10–30

Figure 10–31

In parts (b)–(d), show that *BDFH* is a square.

(b) What is the sum of measures of $\angle 1$ and $\angle 2$?
(c) What is the measure of $\angle 3$?
(d) Why is *BDFH* a square?

In parts (e) and (f), find the area of the figure in two ways.

(e) Find the area of square *ACEG* by using the lengths of its sides. Your answer will be in terms of *a* and *b* only.
(f) Find the area of square *ACEG* by adding the area of the four right triangles to the area of square *BDFH*.

Answers to parts (e) and (f) both represent the area of the big square, so they must be equal.

(g) Set your answers to parts (e) and (f) equal to each other, and see if you can derive the Pythagorean Theorem from your equation.

At first, the ancient Greeks used only whole numbers and rational numbers. They may have discovered a geometric representation of a new type of number by solving the following problem.

LE 3   Connection

(a) Find *N* in Figure 10–30.
(b) What new set of numbers is suggested by part (a)?

The Pythagorean Theorem is used in a variety of applications. In any problem in which you know the lengths of two sides of a right triangle, you can use the Pythagorean Theorem to find the length of the remaining side.

Classroom Connection

LE 4   Skill

(a) A rectangular TV screen measures 14.0 in. by 18.0 in. How long is the diagonal, to the nearest tenth of an inch?
(b) A seventh grader solves this problem as follows. First, $c^2 = 14^2 + 18^2$. So $c = \sqrt{14^2 + 18^2} = 14 + 18 = 32$ in. Is this right? If not, what would you tell the child?

The Pythagorean Theorem is sometimes used to find a missing length in an area problem.

• EXAMPLE 1

Figure 10–31 shows an equilateral triangle inside of a square. Find the shaded area.

SOLUTION

Understanding the Problem   We want to find out the shaded area. This area is outside the triangle and inside the square.

Devising a Plan   Find the area of the square, and subtract the area of the triangle. This will require two area formulas.

Carrying Out the Plan

$$\text{Shaded area} = A_\square - A_\triangle$$

This requires two area formulas.

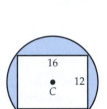

Figure 10–32

$$A_\square = s^2 \quad \text{and} \quad A_\triangle = \frac{1}{2}bh$$

$$A_\square = 6^2 = 36$$

To find $A_\triangle$, we need b and h. (See Figure 10–32.) Now $b = 6$, but what is h? Use the Pythagorean Theorem (see Figure 10–33.)

Figure 10–33

$$h^2 + 3^2 = 6^2$$

$$h^2 + 9 = 36$$

$$h^2 = 27$$

$$h = \sqrt{27} \ (\text{or } 3\sqrt{3})$$

$$A_\triangle = \frac{1}{2}(\sqrt{27}) \cdot 6 = 3\sqrt{27}$$

Therefore, $A_{\text{shaded}} = A_\square - A_\triangle = 36 - 3\sqrt{27}$ square units.

Looking Back   The answer is less than the area of the square (36) by the amount of the area of the triangle. This is reasonable.   ●

Now you try one.

LE 5   Reasoning

Find the area of the shaded region (Figure 10–34); C is the center. (*Hint:* Form a right triangle in the rectangle.)

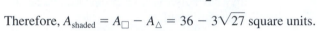

Figure 10–34

Discussion

THE RIGHT-TRIANGLE TEST

If you know the lengths of the three sides of a triangle, is there a way to tell if it is a right triangle? Yes. This is the converse of the Pythagorean Theorem. You can complete the proof in the homework exercises. (No need to thank me for putting it there.)

LE 6   Reasoning

Look at the Pythagorean Theorem, and see if you can write its converse.

According to the converse, any triangle in which the lengths have the relationship $a^2 + b^2 = c^2$ must be a right triangle. The converse, the Right-Triangle Test, can be used to determine if a triangle with three given lengths is a right triangle.

> **The Right-Triangle Test**
> **(Converse of the Pythagorean Theorem)**
>
> If a, b, and c are the lengths of the sides of a triangle and $a^2 + b^2 = c^2$, then the triangle is a right triangle.

In the Right-Triangle Test, the side of length c would have to be the longest side of the triangle. The Right-Triangle Test is based on the fact that the Pythagorean Theorem works only on right triangles. Use the test in the following exercise.

LE 7   Skill

Determine whether each of the following is a right triangle.

(a)

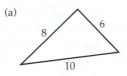

(b)

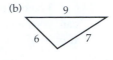

(c) Do parts (a) and (b) involve induction or deduction?

THE SIDES OF A TRIANGLE

The Right-Triangle Test enables us to recognize three lengths that could form a right triangle. Is it possible to tell whether three lengths could form an acute triangle (in which all three angles are acute), an obtuse triangle (in which one angle is obtuse), or no triangle at all?

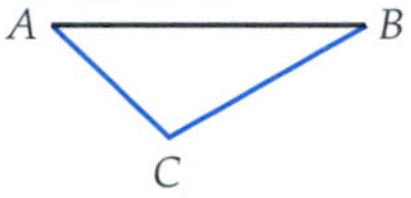

LE 8   Reasoning

Suppose that you are traveling from point A to point B (Figure 10–35).

(a) What is the shortest route?

(b) What does part (a) imply about the distance from A to C to B?

(c) Parts (a) and (b) suggest that if C is not on $\overline{AB}$, then $AC + CB$ _____ AB.

A — *B*
C
Figure 10–35

The result of LE 8(c) is called the Triangle Inequality.

The Triangle Inequality

The sum of the measures of any two sides of a triangle is greater than the measure of the third side.

The following exercises uses the Triangle Inequality and the Right-Triangle Test to develop a more general result.

Discussion

LE 9   Reasoning

Consider the lengths 3, 4, and c, where $c > 4$.

(a) For what values of c could these lengths form a triangle?

(b) For what values of c could these lengths form a right triangle, an acute triangle, or an obtuse triangle?

LE 10 Reasoning

(a) Three segments have lengths a, b, and c, with $a < b < c$. Under what conditions could segments of lengths a, b, and c be used to form a triangle? (*Hint:* Find lengths a, b, and c that could not form a triangle.)

(b) A triangle has sides of lengths a, b, and c, with $a < b < c$. What conditions for a, b, and c would make the triangle a right triangle, an acute triangle, or an obtuse triangle?

You can use the Triangle Inequality to test if three lengths can form a triangle. If they can, you can determine if the triangle is acute, right, or obtuse by comparing $a^2 + b^2$ to c^2. If $a^2 + b^2 > c^2$, the triangle is acute. If $a^2 + b^2 < c^2$, the triangle is obtuse.

LE 11 Reasoning

Check whether each set of lengths could form a triangle. If it does, determine whether the triangle is right, obtuse, or acute.

(a) 6, 7, 9   **(b)** 3, 7, 12   **(c)** 4, 5, 8

LE 12 Summary

Tell what you learned about triangles in this section.

ANSWERS TO SELECTED LESSON EXERCISES

2. (a) Each side has length $a + b$, and it has four right angles.
(b) 90°   (c) 90°
(d) It has four 90° angles and four congruent sides
(e) $(a + b)^2$ (f) $4\left(\frac{1}{2}ab\right) + c^2$
(g) *Hint:* $(a + b)^2 = a^2 + 2ab + b^2$

3. (a) $\sqrt{2}$ (b) Irrational numbers

4. (a) 22.8 in.
(b) No. Work out the problem correctly, and compare the answer to 32 in.

5. $100\pi - 192$ sq. units

7. (a) Yes, since $6^2 + 8^2 = 10^2$
(b) No, since $6^2 + 7^2 \neq 9^2$
(c) Deduction

8. (a) AB
(b) It is longer than AB.
(c) >

9. (a) $4 < c < 7$
(b) $4 < c < 5$; acute triangle
$c = 5$; right triangle
$5 < c < 7$; obtuse triangle

10. (a) $a + b > c$
(b) $a^2 + b^2 = c^2$; $a^2 + b^2 > c^2$; $a^2 + b^2 < c^2$

11. (a) Acute triangle, $6^2 + 7^2 > 9^2$
(b) Not a triangle, $3 + 7 < 12$
(c) Obtuse triangle, $4^2 + 5^2 < 8^2$

10.4   HOMEWORK EXERCISES

Basic Exercises

1. Many students know the Pythagorean Theorem as $a^2 + b^2 = c^2$ but do not know anything else about a, b, and c.
  (a) To what geometric shape does the Pythagorean Theorem apply?
  (b) What do a, b, and c represent?

2. Consider the following problem. "Find the area A in the diagram. The areas of the two smaller squares are 144 and 25 square units."

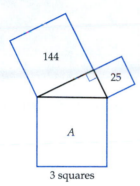

3 squares

Devise a plan, and solve the problem.

3. Before becoming president, James Garfield proved the Pythagorean Theorem using the following trapezoid.

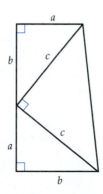

(a) Write formulas for the areas of the three triangles in terms of a, b, and c.
(b) Find the area of the trapezoid.
(c) Set your two areas equal to each other, and see whether you can derive the Pythagorean Theorem.
(d) Does part (c) involve induction or deduction?

4. $ACEG$ is a square. $\triangle ABH \cong \triangle CDB \cong \triangle EFD \cong \triangle GHF$. Explain why $\angle 3$ must be a right angle.

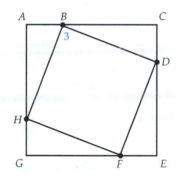

5. Draw a line segment of length $\sqrt{13}$ units on the square lattice (Activity Card 4).

6. Draw a line segment of length $\sqrt{17}$ units on a square dot grid.

7. Find the perimeter of the following parallelogram on a square lattice.

8. Find the area and perimeter of the following right triangle on a square lattice.

 9. A ladder is 4.0 m long. The bottom of the ladder is placed 0.8 m from the wall. How high up on the wall will the ladder reach?

 10.

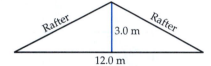

How long are the rafters on the roof?

11. Flo and Ken want to carry a tall piece of glass that is 9 ft square through a rectangular doorway that is 3 ft by 8 ft. Will it fit?

12. A rectangle has sides of length b and h. Compute the length of each diagonal, and show that they are equal in length.

13. Two forces act on an object, one pushing east at 4 mph and one pushing north at 4 mph. Where will the object go, and at what speed?

14. You want to hang a microphone 3 m above a stage, using two wires. About how much wire do you need?

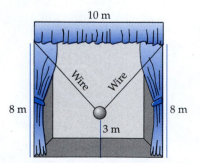

15. After a person in Kansas travels 7 miles north, then 3 miles east, and finally 3 miles south, how far is the person from the starting point?

 16. In walking from A to B on a square city block, cutting straight across the grass is about _____% the distance of using the sidewalk.

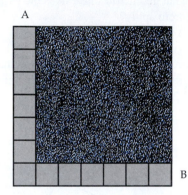

17. Find the area of the rhombus if $AC = 8$ and $AB = 6$.

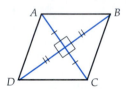

18. Find the area of the triangle.

19. C is the center of both circles. $CB = 5$ and $AB = 8$. Find the shaded area.

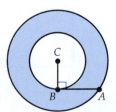

20. Find the area of the shaded region. C is the center of the circle.

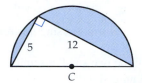

21. Find the perimeter and the area of the trapezoid.

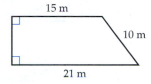

22. Find the perimeter of the square.

23. (a) Complete the third example, repeating the error pattern from the completed examples.

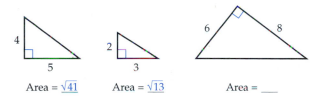

(b) Describe the error pattern.

24. A seventh grader thinks that $\sqrt{a^2 + b^2} = a + b$. Is this correct? If not, give a counterexample.

25. Use the Right-Triangle Test to determine if each of the following is a right triangle.

(a)   (b)

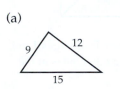 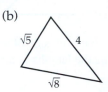

26. Use the Right-Triangle Test to determine if each of the following is a right triangle.

(a)   (b)

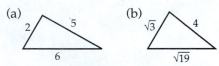

 27. (a) Select three lengths that could form a right triangle, and multiply them all by the same positive number. Do the resulting lengths determine a right triangle?
(b) Repeat part (a) using three new lengths.
(c) Propose a generalization of your results.

28. Select three lengths that could form a right triangle, and add the same whole number to all of them. Do the resulting lengths determine a right triangle?

29. In a large city, suppose your hotel is 20 miles from one airport and 30 miles from another airport. How far apart can the airports be?

30. A triangle has sides of lengths 5, 6, x. If the triangle is acute, give all possible values of x. (Use $<$ or $>$ in your answer.)

31. Tell whether each set of lengths could form a right triangle, an acute triangle, an obtuse triangle, or no triangle.
(a) 3, 5, 6   (b) 9, 12, 15
(c) 3, 5, 10   (d) 2, 3, 4

32. Tell whether each set of lengths could form a right triangle, an acute triangle, an obtuse triangle, or no triangle.
(a) 8, 9, 20   (b) 5, 6, 8
(c) 3, $\sqrt{7}$, 4   (d) 7, 8, 10

Extension Exercises

33. Pythagorean triples are three counting numbers that satisfy the relationship $a^2 + b^2 = c^2$. Over 2000 years ago, the Chinese knew the smallest triple: 3, 4, 5 (since $3^2 + 4^2 = 5^2$). Another triple is 5, 12, 13 (since $5^2 + 12^2 = 13^2$).
(a) Double 3, 4, and 5. Does this create a new Pythagorean triple?
(b) Name three more Pythagorean triples.
(c) Suppose you know that a, b, c is a Pythagorean triple with $a^2 + b^2 = c^2$. Show that ka, kb, kc, is also a Pythagorean triple for any counting number k.

34. One of the most amazing Babylonian tablets, dating from around 1700 B.C., is called Plimpton 322 (see the photo).

George A. Plimpton Collection, Rare Book and Manuscript Library, Columbia University.

It contains a list of Pythagorean triples! All the triples in the chart are listed in the first three columns (a, b, c). The last two columns list u and v such that $a = 2uv$, $b = u^2 - v^2$, and $c = u^2 + v^2$. Fill in the missing values in the following table, which shows the first five rows of Plimpton 322.

a	b	c	u	v
120	119	169	12	
3,456	3,367	4,825		27
4,800	4,601		75	
13,500		18,541		54
	65	97		

35. One of the most famous problems in mathematics is Fermat's Last Theorem. In 1631, Fermat considered equations of the form $x^2 + y^2 = z^2$, $x^3 + y^3 = z^3$, $x^4 + y^4 = z^4$, and so on. He said that only $x^2 + y^2 = z^2$ has a solution set made up of counting numbers. None of the other equations have any such solutions! After writing this idea down, Fermat wrote in the margin of the paper (in Latin): "I have discovered a truly wonderful proof of this, but the margin is too small to contain it." The theorem remained unproven for over 360 years! In May 1995, Princeton University mathematician Andrew Wiles published a proof of Fermat's Last Theorem.
(a) Give three sets of counting-number solutions to $x^2 + y^2 = z^2$.
(b) Consider $x^3 + y^3 = z^3$. Is $x = 5$, $y = 6$, and $z = 7$ a solution set?

36. People have found formulas for some Pythagorean triples. One is the following. For any odd number $N > 1$, N, $\dfrac{N^2 - 1}{2}$, and $\dfrac{N^2 + 1}{2}$ will be a Pythagorean triple. Write four triples using this formula.

37. The areas of the two smaller semicircles are 18π and 32π square units. Find A.

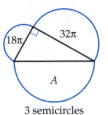

3 semicircles

38. The areas of the two smaller triangles are $\dfrac{9\sqrt{3}}{4}$ and $4\sqrt{3}$ square units. Find A.

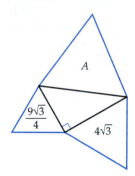

39. Tangrams (Activity Card 7) are formed by cutting a square into 7 pieces as shown.

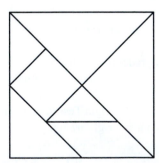

Suppose that each side of the small square is 1 cm long. Find the lengths of all the other sides in the figure.

40. Consider the following problem. "A standard stop sign measures 25.0 cm on each side. What is the area of the surface of a stop sign?" Devise a plan and solve the problem.

41. A cube has edges 2 cm long. How long is a diagonal of the cube? (*Hint:* Draw the diagonal of the base.)

42. A closet is 8 ft by 4 ft by 6 ft. Will a 10-ft pole fit inside?

43. Read the following proof of the Right-Triangle Test, and fill in the missing reasons.

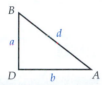

A triangle has lengths a, b, and d, with $a^2 + b^2 = d^2$. How do we show that it must be a right triangle?

   Draw another triangle with sides of lengths a and b and a right angle between them.

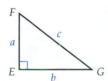

(a) Since $\triangle EFG$ is a right triangle, the Pythagorean Theorem says _____.
(b) Why must $c^2 = d^2$?
(c) Now, if $c^2 = d^2$, then $c = d$. Therefore, $\triangle ABD \cong \triangle$ _____ because of _____.
(d) If the two triangles are congruent, then $m\angle E =$ _____ $= 90°$. Therefore, $\triangle ABD$ is a right triangle!

44.

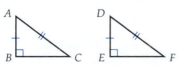

(a) Why is $\overline{BC} \cong \overline{EF}$?
(b) Why is $\triangle ABC \cong \triangle DEF$?
This verifies the *HL* (hypotenuse-leg) congruence property for right triangles.

Technology Exercises

45. Use dynamic geometry software to examine the Pythagorean Theorem.
(a) Construct $\overline{AB}$.
(b) Construct the midpoint M of $\overline{AB}$.
(c) Construct a circle with center M and radius MA.
(d) Construct $\overline{AC}$ and $\overline{BC}$, where C is on the circle.
(e) Hide the circle and midpoint.
(f) Write a script that constructs a square, and put it in the Script Tools Folder.
(g) Use your script to construct squares on $\overline{AB}$, $\overline{AC}$, and $\overline{BC}$.
(h) Measure the area of each square. How are the three areas related?
(i) Drag a vertex of the triangle. Make sure it stays a right triangle, and see if the relationship among the three areas still holds.

46. (a) Use dynamic geometry software to construct a triangle (from three line segments). Label the vertices *A, B,* and *C.* Then select the three line segments and measure them.
(b) If $m\angle ACB = 90°$, what do you think $a^2 + b^2 - c^2$ would equal?
(c) Mark and measure $\angle ACB$.
(d) Have the software calculate $a^2 + b^2 - c^2$.
(e) Drag to create more triangles in which $\angle ACB$ is either acute or obtuse, and repeat parts (c) and (d).
(f) How is the value of $a^2 + b^2 - c^2$ related to $m\angle ACB$?

Enrichment Topics

47. Read "Socrates and the Slave" by Plato (from the *Meno*) in *Fantasia Mathematica*, edited by Clifton Fadiman. Write a report that includes a summary of the main ideas and your reactions to them.

48. Write about the connections the Pythagoreans discovered between mathematics and music.

10.5 SURFACE AREA

NCTM Standards

- develop strategies to determine the surface areas and volumes of rectangular solids (3–5)

- use two-dimensional representations of three-dimensional objects to visualize and solve problems such as those involving surface area and volume (6–8)

- develop strategies to determine the surface area and volume of selected prisms, pyramids, and cylinders (6–8)

Photo by Tom Sonnabend.

Figure 10–36

LE 1   Opener

Name an application in which it is useful to measure surface area (Figure 10–36).

SURFACE AREA OF A RECTANGULAR PRISM

How much will a container cost to manufacture? How much paint do you need to paint your bedroom? To answer these questions, you must compute the surface area (Figure 10–37).

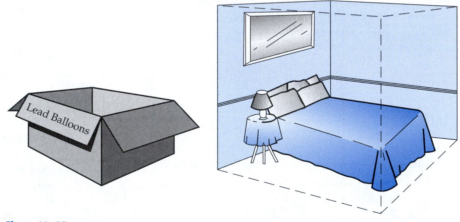

Lead Balloons

Figure 10–37

The total **surface area** of a closed space figure is the sum of the areas of all its surfaces. A surface area in square units indicates how many squares it would take to cover the outside of a space figure.

Classroom
Connection

Figure 10–38

LE 2   Skill

Consider the rectangular prism in Figure 10–38.

(a) Draw a net.

(b) Compute the total surface area. (In other words, determine how many squares are needed to cover all the faces.)

(c) A child says the surface area is 20 square units. What would you tell the child?

Computing the surface area of a rectangular prism can help you decide how much paint you need to paint a room.

LE 3   Skill

A box-shaped room is 18 ft long, 12 ft wide, and 8 ft high. There are two windows that are 3 ft by 5 ft and a doorway that is 4 ft by 7 ft.

(a) Sketch the room.
(b) What is the area of the ceiling?
(c) Suppose you want to paint the walls and ceiling. Excluding the windows and the door, how much area is there to paint?
(d) If 1 gallon of paint covers 400 ft^2, how many gallon cans of paint will be needed to paint the four walls and the ceiling?

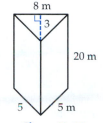

8 m

3

20 m

5   5 m

Figure 10–39

Figure 10–40

SURFACE AREAS OF PRISMS AND CYLINDERS

Computing the total surface areas of other prisms and cylinders also requires finding the area of each surface and adding the areas together.

• EXAMPLE 1

Find the total surface area of the triangular right prism in Figure 10–39.

SOLUTION

Understanding the Problem   Find the sum of the areas of the five faces of the triangular prism.

Devising a Plan   The triangular prism has five faces. There are two isosceles triangles (the bases) and three rectangles (the lateral faces), as shown in the net in Figure 10–40. (A **net** is a pattern that can be used to construct a polyhedron.)

$$A_{\text{surface}} = A_{\triangle 1} + A_{\triangle 2} + A_{\square 1} + A_{\square 2} + A_{\square 3}$$

Carrying Out the Plan   Each of the triangles has an area of $\frac{1}{2} \cdot 3 \cdot 8 = 12$. The rectangles are 5 by 20, 5 by 20, and 8 by 20. So their areas are 100, 100, and 160.

$$A_{\text{surface}} = 12 + 12 + 100 + 100 + 160 = 384 \text{ m}^2$$

Looking Back   The method for finding the total surface area can be used whenever you are able to find the area of each surface.   ●

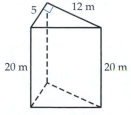

5   12 m

20 m   20 m

Figure 10–41

Now it's your turn.

LE 4   Skill

Consider the following problem. "Find the total surface area of the triangular prism in Figure 10–41." Devise a plan, and solve the problem.

How do you find the surface area of a cylinder?

Figure 10–42

Discussion

LE 5   Reasoning

(a) How many surfaces does the cylinder in Figure 10–42 have?
(b) What shape are its bases?
(c) What shape is the lateral surface? (*Hint:* Roll up a regular sheet of paper.)

A cylinder has two circles for bases. Figure 10–43 shows that the lateral surface of a cylinder is a rolled-up rectangle.

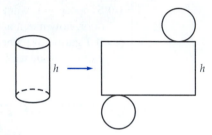

Figure 10–43

Discussion

LE 6   Reasoning

(a) If a cylinder has radius r and height h, what are the dimensions of the rolled-up rectangle (the lateral surface)? It might represent the label of a can.
(b) Compute the area of the two circles and the rectangle, and find a formula for the total surface area of a cylinder.

LE 7   Skill

A cylinder has a radius of 3 cm and a height of 8 cm.

(a) Sketch the cylinder.
(b) Find the total area of the bases.
(c) What is the area of the lateral surface?
(d) What is the surface area of the cylinder?

The surface area of any prism or cylinder can be computed using a more general formula.

LE 8   Reasoning

(a) How can the lateral surface area of a prism or cylinder be computed from the perimeter of the base?
(b) If B is the area of the base, p is the perimeter of the base, and h is the height, what is a formula for the surface area of a prism or cylinder?

The general surface-area formula for a prism or cylinder, $A = ph + 2B$, is not commonly used in elementary or middle-school mathematics.

SURFACE AREAS OF PYRAMIDS AND CONES

A square pyramid is a type of **right regular pyramid,** since its base is a regular polygon. How do you find the surface area of this type of figure?

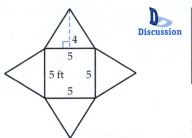

Figure 10–44

Figure 10–45

Figure 10–46

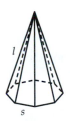

Figure 10–47

LE 9   Reasoning

(a) The net for the square pyramid (Figure 10–44) shows that its base has edges of length 5 ft. The height of its triangular faces is 4 ft. Find the surface area.
(b) Consider a more general square pyramid whose base has edges of length *s*. The height of each triangular face is *l*. What is the surface area?

The height *l* of each triangular (lateral) face of the pyramid is called the **slant height** of the pyramid. The base of the pyramid can be any polygon. How would you find its surface area?

LE 10   Reasoning

(a) Suppose the base of a pyramid is a regular *n*-sided polygon with edges of length *s* (Figure 10–45). The slant height is *l*, and the area of the base is *B*. What is the surface area of the pyramid?
(b) What is the perimeter *P* of the base of the pyramid?
(c) The surface area of the right regular pyramid is $A = B + n\left(\frac{1}{2} sl\right)$. Use algebra properties, and substitute *P* into the surface area. What is the result?

LE 10 leads to the result that the surface area *A* of a right regular pyramid whose base has area *B* and perimeter *P* and whose slant height is *l* is $A = B + \frac{1}{2} Pl$.

LE 11   Skill

A square pyramid has a base with edges of 8 cm (Figure 10–46). The edges connecting the base to the apex each have a length of 5 cm.

(a) What is the slant height?
(b) What is the surface area?

You can use the pyramid formula to find the formula for the surface area of a cone.

LE 12   Reasoning

The right regular pyramid and the cone (Figure 10–47) both have the surface area formula $A = B + \frac{1}{2} Pl$. Suppose the cone has a radius *r* and slant height *l*. Write the surface area of the cone in terms of *r* and *l* (while eliminating *B* and *P* from the equation).

Did you find that the cone has $A = \pi r^2 + \pi rl$? Use this formula in the following exercise.

LE 13   Skill

An ice cream cone has a diameter of 6 cm and a slant height of 13 cm. What is its lateral surface area?

LE 14   Summary

Tell what you learned about surface area in this section.

ANSWERS TO SELECTED LESSON EXERCISES

2. (b) 40 square units
(c) Discuss whether she has counted all the surfaces.

3. (b) $18 \cdot 12 = 216 \text{ ft}^2$
(c) $18 \cdot 12 + 2 \cdot 12 \cdot 8 + 2 \cdot 18 \cdot 8 - 2 \cdot 3 \cdot 5 - 4 \cdot 7$
$= 638 \text{ ft}^2$
(d) 2

4. 660 m^2

5. (a) 3   (b) Circular   (c) Rectangular

6. (a) $2\pi r$ by h
(b) $A = 2\pi rh + 2\pi r^2$ square units

7. (b) 18π cm^2 (c) 48π cm^2 (d) 66π cm^2

8. (a) ph
(b) $A_{surface} = ph + 2B$

9. (a) 65 ft^2 (b) $s^2 + 2sl$ square units

10. (b) ns

11. (a) 3 cm   (b) 112 cm^2

13. 39π cm^2

10.5 HOMEWORK EXERCISES

Basic Exercises

1. A company is designing a juice can. Why would they be interested in surface area?

2. Explain in terms of measurement why a potato will bake more quickly if you cut it into smaller pieces and cook them separately.

3. (a) Draw a net for the rectangular prism below.
(b) What is the surface area of this rectangular prism?

4. (a) Draw a net for the rectangular prism below.
(b) What is the surface area of this rectangular prism?

5. (a) What are all the ways in which 20 cubes can be arranged into a rectangular prism for packing?
(b) If you want to use the smallest amount of material for a box that will hold cubes, which arrangement is best?

6. (a) What are all the ways in which 18 cubes could be arranged into a rectangular prism for packing?
(b) If you want to use the smallest amount of material for a box that will hold the cubes, which arrangement is best?

7. How much paper is needed to cover the box shown?

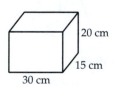

20 cm
15 cm
30 cm

8. A rectangular prism has dimensions l ft, w ft, and h ft. What is its total surface area?

9. A rectangular prism has dimensions of 3 m, 8 m, and W m. Its total surface is 246 m². What is W?

10. A rectangular prism has dimensions of 10 ft, 12 ft, and W ft. Its total surface area is 900 ft². What is W?

11. A box-shaped room is 16 ft long, 10 ft wide, and 8 ft high. There is one window that is 3 ft by 6 ft and a doorway that is 4 ft by 7 ft. If 1 gallon of paint covers 450 ft², how many gallon cans would you buy to paint the four walls and the ceiling?

12. Given the following floor/wall plan for an apartment, find the cost of redecorating it if you cover the ceiling with 1-ft² tiles costing $0.70 each, carpet the floor with carpet costing $8 per square yard, and paint the walls with paint costing $9 per gallon. Each gallon of paint covers 250 ft².

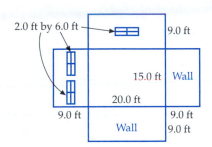

13. You want to carpet the steps as shown by the colored areas.

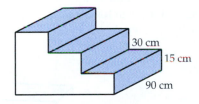

(a) How many square centimeters of carpet will you need?
(b) If the carpet costs $20 per square *meter,* how much will your carpet cost?

14. Carpet A sells for $6.99 per square yard. Carpet B is a remnant 9 ft by 12 ft, selling for $99. If you want a

carpet that is 9 ft by 12 ft, will it cost less to buy the remnant or to buy carpet A by the square yard?

15. Find the total surface area of the triangular prism.

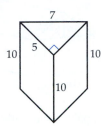

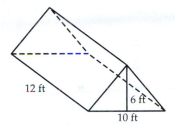

16. How much nylon (in square feet) would be needed to make the tent shown, including the bottom? (You may approximate square roots to one decimal place.)

17.

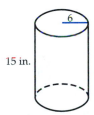

What is the surface area of the cylinder?

18. A cylinder has a circular base with a diameter of 12 ft and a height of 10 ft. What is its surface area?

19. Suppose a cylinder has a radius r and a height h. *Explain* how to find the formula for the total surface area of the cylinder.

20. Give two possible radii and heights for a cylinder with a surface area of 40π.

21. (a) Find the surface area of the sides, front, and back of the house. (Include the two triangles but not the roof.)

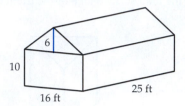

(b) If a gallon of paint covers 400 ft², how much paint will you need to paint the outside of the house?
(c) If paint comes in gallon cans, how many cans will you need?

22. Four basic roof designs used in the United States are flat, gable, hip, and pyramid hip. Find the total area of each roof shown. Exclude the shaded parts.

(a)
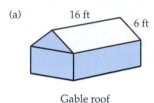

Gable roof

(b)
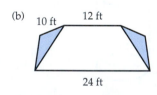

Hip roof

23. A rectangular pentagonal pyramid has a base with edges of length s and lateral surfaces with slant heights of l.
(a) What is the surface area if the base area is B?
(b) What is the perimeter of the base?

 24. A regular octagonal pyramid has a base with edges of length 6 ft and lateral surfaces with slant heights of 8 ft.
(a) What is the surface area if the base area is B?
(b) What is the perimeter of the base?

 25. A square pyramid has a base with edges of length 4 m. The edges connecting the base to the apex each have a length of 3 m.
(a) Sketch the pyramid.
(b) What is the slant height?
(c) What is the surface area?

 26. A square pyramid has a base with edges of length 26 ft. The edges connecting the base to the apex each have a length of 20 ft.
(a) Sketch the pyramid.
(b) What is the slant height?
(c) What is the surface area?

 27. An ice cream cone has a diameter of 3 in. and a slant height of 5 in. What is its lateral surface area?

28. A cone-shaped cup has a diameter of 8 cm and a slant height of 9 cm. What is its lateral surface area?

Extension Exercises

29.

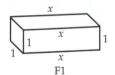 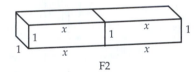

(a) Find the surface areas of F1 and F2.
(b) Draw F3 and find its surface area.
(c) What is the surface area of F10?
(d) What is a formula for the surface area of FN, in which N is a whole number?

30.

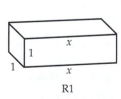 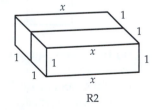

(a) Find the surface areas of R1 and R2.
(b) Draw R3 and find its surface area.
(c) What is the surface area of R10?
(d) What is a formula for the surface area of RN, in which N is a whole number?

31. If the length, width, and height of a rectangular prism are tripled, how does the surface area change? Give evidence to support your answer.

32. If the radius and height of a circular cylinder are doubled, how does the surface area change? Give evidence to support your answer.

33. The surface area A of a sphere of radius r is $A = 4\pi r^2$. (See Exercise 36 for a lab about this formula.) Suppose you are manufacturing two different-sized balls of the same material and thickness. The smaller ball has a radius of 2 in., and the larger one has a radius of 4 in.
(a) How do their surface areas compare?
(b) How do the costs of the material for the two balls compare?

34. The Earth has a diameter of about 7,926 miles. What is its approximate surface area?

Labs and Projects

35. (a) Design a floor plan for a one-story, two-bedroom apartment. Use a scale in which 1 cm represents 1 m. The apartment must fit on a 12-by-12-m square.
(b) Decide how high the walls will be and how many windows and doors the apartment will have. All the walls should be the same height.
(c) Estimate the cost of painting the interior of the apartment.

36. What is the surface area of a sphere? Try the following.
(a) Cut an orange in half. Trace the circumference of the orange three times on a sheet of waxed or regular paper.
(b) Peel half of the orange. Cut the peel into small pieces about 1 cm by 1 cm.
(c) How many circles can you fill with the pieces?
(d) Part (c) shows that the surface area of a hemisphere (half a sphere) is about _____.
(e) What is an approximate formula for the surface area of a sphere?

10.6 VOLUME

NCTM Standards

- develop strategies to determine the surface areas and volumes of rectangular solids (3–5)
- use two-dimensional representations of three-dimensional objects to visualize and solve problems such as those involving surface area and volume (6–8)
- develop strategies to determine the surface area and volume of selected prisms, pyramids, and cylinders (6–8)

 LE 1 Opener
Name an application in which it is useful to measure volume.

VOLUMES OF RIGHT RECTANGULAR PRISMS

In the late 1950s, college students enjoyed attempting to measure the volume of a phone booth in their own unique way (Figure 10–48 on page 582). **Volume** is the amount of space occupied by a three-dimensional figure. Volume is usually measured in cubic units. Whereas surface area is the total area of the faces of a solid, volume is the capacity of a solid.

Figure 10–48

Joe Munroe, Life Magazine. © 1959 Time, Inc.

Figure 10–49

LE 2 Concept

(a) What is the volume of the solid in Figure 10–49, in cubic units?

(b) What is the total surface area?

(c) Suppose the 3 cubes on the right-hand side of the first layer were moved up just to the right of the cubes on the second layer. The figure would then have 2 columns and 3 rows in each layer. How would this change the volume and the surface area?

In LE 2, one could compute the volume by counting how many cubes form the space figure. Children can first study examples like this using materials such as wooden cubes or multilink cubes. With rectangular prisms like the one in the following exercise, there is a shortcut (formula) for counting the cubes.

LE 3 Reasoning

You fill a box (Figure 10–50) with cubes. Each cube is 1 cm^3.

(a) How many cubes are there in a layer?

(b) How many layers are there?

(c) What is the volume of the rectangular prism?

(d) What are the length, width, and height of the rectangular prism in centimeters?

(e) How can you compute the volume using the length, width, and height?

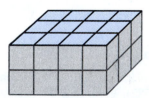

Figure 10–50

Children need experiences that help them see rectangular prisms in terms of layers of cubes. LE 3 shows how this understanding can be used to develop a formula for the volume of a rectangular prism. The formula lw (length times width) gives the number of cubes in a layer of a rectangular prism, and h (height) is the number of layers. So lwh gives the total number of cubes that fill the interior of the rectangular prism in Figure 10–50.

$$4 \times 3 = \text{number of cubes per layer}$$

$$2 \text{ layers}$$

$$4 \times 3 \times 2 = 24 \text{ cubes}$$

Volume of a Right Rectangular Prism

The volume V of a right rectangular prism that has dimensions l, w, and h is

$$V = lwh$$

Computing the volume of a right rectangular prism can help you decide which freezer is the best buy.

Classroom Connection

LE 4 Connection

(a) A store sells two types of freezers. Freezer A costs \$310 and measures 1.5 ft by 1.5 ft by 5.0 ft. Freezer B costs \$400 and measures 2.0 ft by 2.0 ft by 3.5 ft. Which freezer has the lower unit cost?

(b) A sixth grader asks whether you should compare dollars to feet or feet to dollars. How would you respond?

VOLUMES OF PRISMS AND CYLINDERS

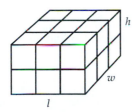

Figure 10–51

Are you ready for some good news? And no bad news? All right prisms and right circular cylinders have the same general volume formula!

As you have seen, you can compute the volume of a rectangular prism (Figure 10–51) by counting the number of cubes in each layer, which is the *area of the base* (lw), and multiplying by the number of layers, which is the *height* (h).

$$V = (l \cdot w) \cdot h$$

$$V = (\text{base area}) \cdot \text{height}$$

We can apply the same idea to all right prisms and cylinders. Each has two congruent bases. Imagine that the three solids in Figure 10–52 are glass containers.

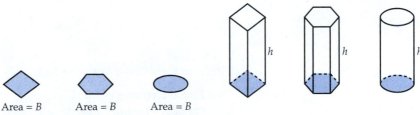

Area = B   Area = B   Area = B

Figure 10–52

Visualize filling them up with some water (or orange juice, if you prefer). The total amount of liquid can be measured by moving the base area from the bottom to the top of the container. In other words, the volume is the area of the base multiplied by the height.

Volume of Any Right Prism or Right Cylinder

The volume V of a right prism or a right cylinder that has a base of area B and height h is

$$V = Bh$$

Most geometry books use b to denote the *length* of a base (side) of a polygon and B to denote the *area* of a base that is the face of a polyhedron, cone, or cylinder. The general formula for prisms and cylinders can be used to derive specific volume formulas.

Discussion

LE 5   Reasoning

Consider a right circular cylinder with radius r and height h.

(a) What shape is the base?
(b) What is the area of the base?
(c) What is a formula for the volume obtained by substituting in $V = Bh$?

The following example illustrates how to find the volume of a triangular prism by using the general volume formula.

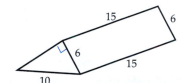

Figure 10–53

• EXAMPLE 1

Find the volume of the right triangular prism shown in Figure 10–53.

SOLUTION

$$V \text{ (volume)} = B \text{ (base area)} \times h \text{ (height of prism)}$$

The bases are right triangles. First, use the Pythagorean Theorem to find the length of the other leg.

$$6^2 + x^2 = 10^2$$
$$36 + x^2 = 100$$
$$x^2 = 64$$
$$x = 8$$

The base has area $B = \dfrac{1}{2} \cdot 6 \cdot 8 = 24$. The height of the prism is 15.

$$V_{\text{prism}} = 24 \cdot 15 = 360 \text{ cubic units}$$

LE 6   Skill

Refer to Figure 10–54 on page 585.

(a) Find the volume of the cylinder (it could be the design for a pipe).

(b) Find the volume of the prism (it could be the shape of a wedge of cheese).

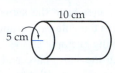

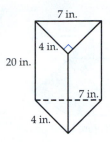

Figure 10–54

Think of the right prism in Figure 10–55 as a stack of extremely thin sheets of paper. The related oblique prism would be obtained by shifting the stack. The right cylinder and the oblique cylinder in Figure 10–55 are related in the same way.

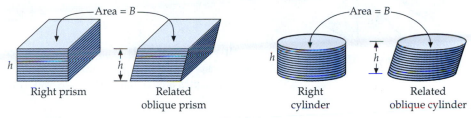

Right prism   Related oblique prism   Right cylinder   Related oblique cylinder

Figure 10–55

LE 7   Concept

(a) How do the volumes of the right prism and the oblique prism in Figure 10–55 compare?

(b) How do the volumes of the right cylinder and the oblique cylinder in Figure 10–55 compare?

Zu Chongzhi (A.D. 429–500) and his son Zu Geng of China were probably the first mathematicians to use the general relationship between a pair of figures like those in Figure 10–55. Bonaventuri Cavalieri (1598–1647), an Italian mathematician, later stated the general principle, and it is named after him.

Cavalieri's Principle

Two solids with bases in the same plane have equal volumes if every plane parallel to the two bases intersects the two solids in cross sections of equal area.

LE 7 and Cavalieri's Principle suggest why the volume formula for right prisms and right cylinders applies to all prisms and cylinders.

Volume of Any Prism or Cylinder

The volume V of a prism or cylinder that has a base of area B and height h is

$$V = Bh$$

VOLUMES OF PYRAMIDS AND CONES

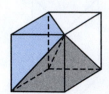

Figure 10–56

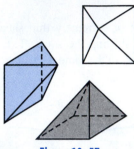

Figure 10–57

Pyramids and cones have the same general volume formula. Their volume formula is closely related to the volume formula for cylinders and prisms.

The following example shows how the volume of a pyramid is related to the volume of a prism with the same base and height. Take a cube, and draw the three diagonals from one vertex (Figure 10–56). The diagonals divide the cube into three congruent pyramids (Figure 10–57). Since the three pyramids are congruent, the volume of each pyramid is $\frac{1}{3}$ the volume of the cube. So

$$V_{\text{pyramid}} = \frac{1}{3} V_{\text{cube}}$$

This suggests the following general relationship between the volumes of pyramids and prisms that have the same base and height:

$$V_{\text{pyramid}} = \frac{1}{3} V_{\text{prism}} = \frac{1}{3} Bh$$

If you have models available, you can see that a prism holds about three times as much water or rice as a pyramid with the same base and height. The volumes of cylinders and cones are related in the same way (Figure 10–58).

$V = Bh$

$\frac{1}{3}$ volume $\frac{1}{3}$ volume

$V = \frac{1}{3} Bh$

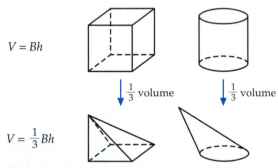

Figure 10–58

Volume of a Pyramid or Cone

The volume V of a pyramid or cone that has a base of area B and height h is

$$V = \frac{1}{3} Bh$$

The general volume formula $V = \frac{1}{3} Bh$ can be used to obtain the volumes of cones and pyramids.

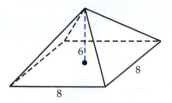

Figure 10–59

LE 8   Reasoning

What is the formula for the volume of a cone that has radius r and height h?

LE 9   Skill

Find the volume of the square pyramid in Figure 10–59. (It could be a building design.)

VOLUME OF A SPHERE

The Earth is approximately the shape of a sphere. A **sphere** is the set of points in space that are a fixed distance from a given point, the center. The Earth's diameter is about 7,926 miles. How would we find its volume?

First, find a formula for the volume of a sphere by comparing it to the volume of a cone and a cylinder with the same radius r and a height of r (Figure 10–60).

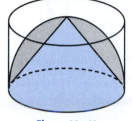

Figure 10–60

LE 10   Reasoning

(a) Write formulas for the volumes of the cone and the cylinder in Figure 10–60.
(b) The volume of the hemisphere (half sphere) in Figure 10–60 is exactly halfway between the volumes of the cone and the cylinder. Find a formula for the hemisphere.
(c) On the basis of part (b), what is a formula for the volume of a sphere?

As LE 10 suggests, the volume V of a sphere with radius r is $V = \frac{4}{3}\pi r^3$

LE 11   Skill

The Earth has a diameter of about 7,926 miles. What is its approximate volume?

LE 12   Summary

Tell what you learned about volume in this section.

ANSWERS TO SELECTED LESSON EXERCISES

2. (a) 12   (b) 38 square units
(c) V would be the same and A is 32 square units.

3. (a) 12   (b) 2   (c) 24 cm³
(d) 4 cm, 3 cm, 2 cm
(e) 4 cm × 3 cm × 2 cm = 24 cm³

4. (a) Freezer A ($27.56/ft³)
(b) You can compare them either way, but it changes whether you would prefer a larger or smaller number.

5. (a) Circular   (b) πr^2 (c) $V = \pi r^2 h$

6. (a) 250π cm³
(b) $20\left(\frac{1}{2} \cdot 4\sqrt{33}\right) = 40\sqrt{33}$ in.³

7. (a) They are equal.   (b) They are equal.

8. $V = \frac{1}{3}\pi r^2 h$

9. $\frac{1}{3}(8 \cdot 8 \cdot 6) = 128$ cubic units

10. (a) $V = \frac{1}{3}\pi r^2 h;\ V = \pi r^2 h$

(b) $V = \frac{2}{3}\pi r^3$ (c) $V = \frac{4}{3}\pi r^3$

11. 2.6×10^{11} mi³

10.6   HOMEWORK EXERCISES

Basic Exercises

1. What is the volume of the solid in cubic units?

2. What is the volume of the solid in cubic units?

3. (a) What is the volume of the rectangular prism?

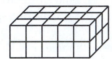

   (b) The most common incorrect answer to part (a) is 31. How would a child obtain this answer?

4. What is the volume of the swimming pool?

Reprinted by permission: Tribune Media Services.

 5. A store sells two types of freezers. Freezer A costs $350 and measures 2 ft by 2 ft by 4.5 ft. Freezer B costs $480 and measures 3 ft by 3 ft by 3.5 ft. Which freezer is the better buy?

6. A sketch of an apartment building follows. Find its volume.

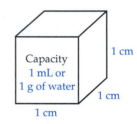

7. In the customary system, 1 pint = 1 lb of water, but these measures have no simple relationship to volume measures (such as 1 ft^3). The metric system relates volume (cm^3), liquid volume (mL), and mass (g) of water in a simple way.

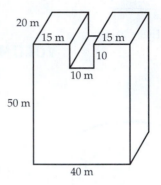

   (a) What is the volume of the container, in cubic centimeters?

   So 1 mL of water = 1 g of water = 1 cm^3 of water! Use this information to answer the following questions.

   (b) What is the mass of 4 L of water?
   (c) A fish tank is 70 cm by 50 cm by 40 cm. About how many liters of water are needed to fill it?
   (d) What is the mass of 1 m^3 of water?
   (e) What is the liquid volume of 1 m^3 of water?

8.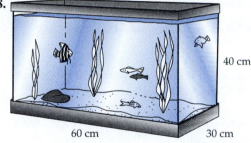

60 cm   40 cm   30 cm

(a) How much water can the fish tank hold?
(b) 1 L = 1000 cm^3. How many liters of water does the fish tank hold?
(c) Small tropical fish need about 1500 cm^3 of living space. How many fish can live in this tank?

9. Is each quantity related more to volume or to surface area?
(a) The amount of paper needed to make a bag
(b) The amount a bag will hold
(c) Your weight

10. Write a sentence that tells the difference between the surface area and volume of a prism.

11. Water and feed troughs for animals often have the shape of a prism. Find the volume of the watering trough.

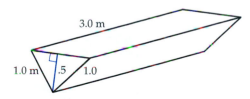

3.0 m
1.0 m .5 1.0

12. A 15-step staircase is made out of concrete. Three of the steps are shown. What is the volume of the staircase?

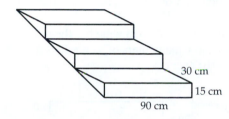

30 cm
15 cm
90 cm

13. A standard 46.00-oz can of juice has a radius of 5.3 cm and a height of 17.5 cm. What is its volume?

14. One can of juice is twice as tall as a second can but only half as wide. How do their volumes compare?

15. A cylindrical water tank has a radius of 6.0 m. About how high must it be filled to hold 400.0 m^3?

16. (a) Roll an $8\frac{1}{2}$-by-11-in. sheet of paper into a cylindrical tube. What is the diameter?
(b) Roll the sheet of paper into a cylindrical tube of a different size. What is the diameter?
(c) Which cylinder has the greater volume?

17. Find the volume of paper on a new roll of toilet paper.

18. (a) Find the volume and surface area of a box of tissues.
(b) Compute the ratio of the volume to the surface area.
(c) Would the manufacturer want the ratio in part (b) to be high or low?
(d) If possible, design a different tissue box with the same volume that has a smaller surface area.

19. Find the volume of a 12-oz (355-mL) soda can by
(a) measuring the radius and height and using a formula.
(b) using the conversion 1 mL = 1 cm^3.

20. A cylindrical hot water heater has a radius of 35 cm and a height of 170 cm. How many liters of water can it hold? (1 cm^3 = 1 mL.)

21. A cylindrical pipe is hollow inside. What is the volume of the pipe material?

3 in.   18 in.
30 ft

22. A cylindrical pipe has length L, inner radius I, and overall radius R. What is the volume of the pipe material?

23. Consider the following problem. "How heavy is a firefighter's hose? A large hose is 100 ft long and 5 in. in diameter when full. Empty, the hose weighs 200 lb. If 1 gallon (231 in.^{3}) of water weighs 8 lb, how much does a full hose weigh?" Devise a plan and solve the problem.

24. To calibrate a cylindrical beaker with an inside diameter of 2.2 cm to show cubic centimeters, how far apart should the calibration marks be?

25. Find the volume of the oblique prism shown. Its base area is 50 m².

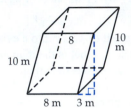

8   10 m

10 m

8 m   3 m

26. Find the volume of the oblique prism shown. Its base area is 30 ft².

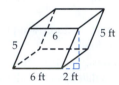

6   5 ft

5

6 ft   2 ft

27. (a) The formula $V = \frac{1}{3}Bh$ applies to what figures?

(b) The formula $V = Bh$ applies to what kind of figures?

28. A cylinder-shaped drinking cup holds _____ times more water than a cone-shaped cup that has the same radius and height.

 29. The largest pyramid in the world is in Cholula, Mexico. Its base area is 1,960,000 ft², and its height is 177 ft.
(a) What is its volume?
(b) About how many apartments, each 30 ft by 25 ft by 10 ft, would it take to make the same volume?

 30. The Great Pyramid of Egypt has a square base, with sides of 768 ft and a height of 482 ft.
(a) What is its volume?
(b) About how many apartments, each 30 ft by 25 ft by 10 ft, would it take to make the same volume?

31. *Explain* how to use the formula $V = Bh$ to find the volume formula for a cylinder.

32. *Explain* how to use the formula $V = \frac{1}{3}Bh$ to find the volume for a square pyramid whose base has edges of length *s*.

 33. An ice cream cone has a diameter of 2.0 in. and a height of 5.5 in. What is its volume?

34. A cone-shaped cup is filled to half its height. What fraction of the cup is filled?

35. (a) Complete the third example, repeating the error pattern from the completed examples.

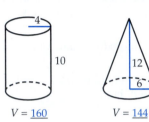   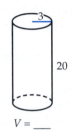

4   3

10   12   20

6

$V = \underline{160}$ $V = \underline{144}$ $V = \underline{}$

(b) Describe the error pattern.

 36. A conical tank has an inside diameter of 20 ft and a height of 12 ft. The tank is filled with liquid to a height of 9 ft. How much liquid is in the tank?

 37. A basketball has a diameter of about 10 in. What is its volume?

38. Suppose an igloo in the shape of a hemisphere has an inner diameter of 12 m. What is its volume?

39. A swimming pool 20 m wide has the cross section shown. How many liters of water does it hold? (1 L = 1000 cm³.)

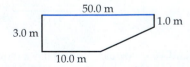

50.0 m

1.0 m

3.0 m

10.0 m

40. (a) How much land is needed for the house itself?
(b) What is the volume of the house?

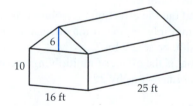

 41. A can of 3 tennis balls is 19.3 cm high with a radius of 3.8 cm, and each tennis ball has a radius of 3.2 cm. What percent of the can is occupied by the tennis balls? For a sphere, volume $V = \frac{4}{3}\pi r^3$. Use $\pi = 3.14$.

42. A balloon takes 3 seconds to inflate to a radius of 4 in. Assuming a constant flow of air, after 6 seconds, it has a radius of _____ in. (The answer is not 8.)

Extension Exercises

43.

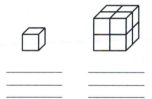

Length of edge	_____	_____
Volume	_____	_____
Surface area	_____	_____

(a) Fill in the blanks.
(b) As the cube grows in size, which increases faster, surface area or volume?

44. The lengths of the edges of a cube are increased by 20%.
(a) The surface area increases by _____%.
(b) The volume increases by _____%.

45. A city water inspector who is 6.0 ft tall inspected a spherical water tank. His head touched the tank when his feet were 18.0 ft away from the lowest point. How did he use this information to find the volume of the tank? (*Hint:* Use the Pythagorean Theorem.)

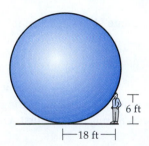

 46. A log has a diameter of 36 cm and a height of 60 cm. Find the volume of the largest rectangular solid with a square base that can be cut from it.

47. (a) Give the length, width, and height of three different rectangular prisms that have a volume of 8000 cm³.
(b) Find the surface area of each rectangular prism from part (a). What characteristic of the shape tends to increase the surface area?

48. In each graph, the region is revolved around the x-axis. Find the volume of the resulting figure.

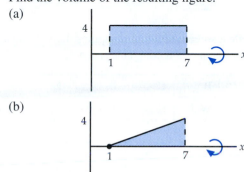

49. A cylindrical glass jar contains juice. Without measuring, how can you tell when the jar is half full?

 50. A plumber needs to deliver 150 steel pipes to an office building. Her truck can carry 800 kg. Each pipe is 120 cm long, with an outer diameter of 7 cm and an inner diameter of 5 cm. The steel weighs 7.7 g/cm³. How many trips will the plumber need to make?

Technology Exercise

51. You want to manufacture a cylindrical container with a volume of 84 cm³. To determine the radius and height that will give the minimum surface area, use a spreadsheet.
(a) Why would you want to find the minimum possible surface area of a container?
(b) A formula for the height h of a cylinder in terms of the volume V and radius r is $h =$ _____.
(c) A formula for the surface area A of a cylinder is $A =$ _____.

(Continued on page 592)

(d) Set up a spreadsheet as follows. Try different values of *r,* and use your formulas to compute *h* and *A.* Find the minimum value of *A.*

	A	B	C	D
1	Volume	*r*	*h*	*A*
2	84			
3	84			
4	84			

(e) Give the values of *r* and *h* that result in the minimum surface area.

52. (a) Create a spreadsheet that computes the volume and surface area of a rectangular prism.

	A	B	C	D	E
1	Length	Width	Height	Volume	Surface Area
2					
3					
4					
5					

(b) Write a formula for cells D2 and E2 in terms of cells A2, B2, and C2. Use these formulas in column D and E.

(c) Find out the effect on the volume and surface area of multiplying one of the dimensions (length, width, or height by 2.

(d) What is the effect of multiplying two or three of the dimensions by 2?

Labs and Projects

53. (a) Take an $8\frac{1}{2}$ in. by 11-in. piece of paper and cut off a 1-inch square from each corner.

(b) Fold and tape to make an open box (no top).

(c) Find the volume of the box.

(d) Repeat parts (a)–(c), but cut off a 2-inch square from each corner.

(e) Can you make a different open box with a larger volume?

54. (a) Go to the supermarket and collect data on the radii and heights of cylindrical cans. Compute the ratio of *h* to *r* for each can.

(b) Approximate the volume of a paper grocery bag and a plastic grocery bag from a local supermarket. How do they compare?

10.7   LENGTHS, AREAS, AND VOLUMES OF SIMILAR FIGURES

NCTM Standards

- develop, analyze, and explain methods for solving problems involving proportions, such as scaling and finding equivalent ratios (6–8)
- solve problems involving scale factors, using ratio and proportion (6–8)
- recognize and apply mathematics in contexts outside of mathematics (pre-K–12)

Do you want to invest in some imaginary real estate? It's safer than investing in real real estate. Consider the following situation.

LE 1   Opener

You are offered two right-triangular pieces of land that are the same shape. One has sides measuring 30 ft, 40 ft, and 50 ft. The other has sides measuring 60 ft, 80 ft, and 100 ft. The second piece of land costs 3 times as much as the first. Guess which piece of land is the better buy.

SIMILAR PLANE FIGURES

The preceding exercise involves two similar figures. The two triangles are the same shape, but all the sides of the second triangle are twice as long as the corresponding sides of the first. How are the areas of the triangles related?

LE 2   Skill

(a) If you have studied the Right-Triangle Test, show that both triangles in Figure 10–61, are right triangles.

(b) Find the area of each right triangle.

(c) The area of the larger triangle is _____ times the area of the smaller triangle.

(d) Find the perimeter of each triangle.

(e) The perimeter of the larger triangle is _____ times the perimeter of the smaller triangle.

(f) The second triangular piece of land costs 3 times as much as the first. Which is the better buy?

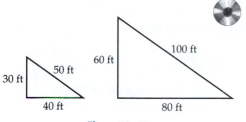

Figure 10–61

The area of the smaller triangle is $\frac{1}{2} \cdot 30 \cdot 40 = 600 \text{ ft}^2$, and the area of the larger triangle is $\frac{1}{2} \cdot (30 \cdot 2) \cdot (40 \cdot 2) = 2{,}400 \text{ ft}^2$. The triangle with twice the lengths had four times the area. What is the general relationship between all corresponding lengths and areas of similar figures? If the suspense is too much for you, do not hesitate to try the next exercise.

LE 3   Skill

The drawings in Figure 10–62 show two sizes of a photocopy. Enlarging a photocopy creates a *similar* rectangle.

(a) The ratio of corresponding sides is _____:_____.

(b) The areas of corresponding rectangles are _____ and _____.

(c) The ratio of their areas is _____.

(d) The perimeters of the two rectangles are _____ and _____.

(e) The ratio of their perimeters is _____.

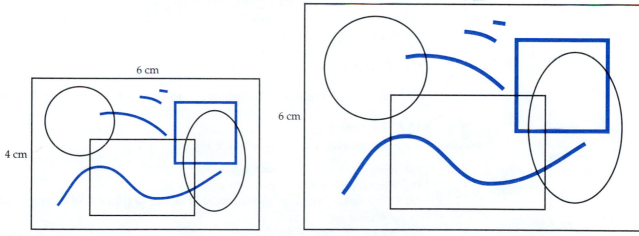

Figure 10–62

Try to generalize the results from LE 2 and LE 3 in the following exercise.

LE 4 Reasoning

Two similar figures have a ratio of corresponding sides $m:n$.

(a) What is the ratio of their areas?
(b) What is the ratio of their perimeters?

The results of the preceding exercises suggest the following rule.

Lengths and Areas of Similar Plane Figures

Figures A and B are similar plane figures with a ratio of corresponding length measurements $m:n$. The ratio of corresponding area measurements is $m^2:n^2$, and the ratio of their perimeters is $m:n$.

Use these properties in the following exercise.

Classroom Connection

LE 5 Reasoning

Two similar triangles have a ratio of corresponding sides of $4:1$. The area of the smaller triangle is 10 ft^2. A child says the area of the larger triangle is 40 ft^2. Is this correct? If not, what would you tell the child?

SIMILAR SPACE FIGURES

Relationships between corresponding areas and volumes in similar solids indicate the relative strengths, heat loss, and weights of animals with similar shapes. After examining relationships in similar geometric solids, you will use this knowledge to study similar shapes in nature.

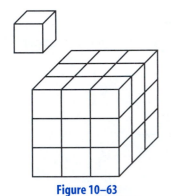

Figure 10–63

Discussion

LE 6 Skill

Suppose that you have a 1-cm cube and a 3-cm cube. (Use wooden blocks if they are available. See Figure 10–63.)

(a) What is the ratio of corresponding edge lengths?
(b) The surface areas of the two figures are _____ and _____.
(c) What is the ratio of their surface areas?
(d) Compute the volume of each figure.
(e) What is the ratio of their volumes?

Next, consider two similar rectangular prisms and see if you can generalize the results.

9 m

3 m

6 m

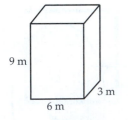

6 m

2 m

4 m

Figure 10–64

Discussion

LE 7 Reasoning

Consider the two rectangular prisms in Figure 10–64.

(a) What is the ratio of corresponding edge lengths?
(b) What is the ratio of their surface areas?
(c) Guess how the ratio of surface areas is related to the ratio of corresponding edges in similar solids.
(d) What is the ratio of their volumes?
(e) Guess how the ratio of volumes is related to the ratio of edges in similar solids.

As your results in LE 6 and LE 7 suggest, the ratio of volumes is the cube of the ratio of corresponding lengths, and the ratio of corresponding areas is the square of the ratio of corresponding lengths.

Lengths, Areas, and Volumes of Similar Solids

A and *B* are similar solids with a ratio of all corresponding length measurements of *m:n*. The ratio of all corresponding area measurements is $m^2:n^2$, and the ratio of volumes is $m^3:n^3$.

Use these properties to answer the following questions.

LE 8   Skill

Two similar solids have corresponding heights of 9 m and 12 m.

(a) What is the ratio of their total surface areas?
(b) What is the ratio of their volumes?
(c) The volume of the larger solid is 160 m^3. What is the volume of the smaller solid? (*Hint*: Write a proportion.)

SIMILARITY IN NATURE

Figure 10–65

Photo courtesy of Film Stills Archives.

Do you ever worry that a giant ape might visit your neighborhood (Figure 10–65)? Is this possible? Similar solids provide the answer! But first we'll consider some friendly dolphins (Figure 10–66).

Photo: Rob Mathewson. Courtesy of Department of Library Services, American Museum of Natural History.

Figure 10–66

The properties of similar solids can be applied to pairs of animals that are approximately the same shape. Young dolphins and adult dolphins are approximately the same shape.

Discussion

LE 9   Connection
Give some examples of measurements of dolphins that are length measurements.

In LE 10–LE 17, suppose that an adult dolphin is 2 times as long as a young dolphin of the same shape.

Discussion

LE 10   Opener
Guess how many times stronger the adult dolphin is.

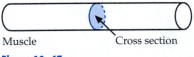

Muscle   Cross section

Figure 10–67

According to biologist D'Arcy Thompson, in *On Growth and Form,* "The strength of a muscle depends upon the size of a cross section (slice)" (Figure 10–67), just as the strength of a steel beam depends on the cross section of the beam.

The following exercise concerns the relationship between the strength of a cylindrical muscle and the area of its circular cross section.

Discussion

LE 11   Connection
(a) What is the formula for the area of a circle?
(b) If the cross section of the young dolphin's muscle is circular, with a radius of 1 in., the area of the cross section is _____.
(c) The radius of the adult's corresponding muscle is 2 in. Its area is _____.
(d) The adult has a muscle cross section that is _____ times larger in area than that of the young dolphin's, so the adult is _____ times stronger.

Discussion

LE 12   Concept
The ratio of corresponding lengths in the young dolphin and the adult dolphin is 1 : 2.

(a) The ratio of strengths is _____.
(b) The ratio of strengths is the same as the ratio of
    (1) corresponding lengths   (2) surface areas   (3) volumes

Another characteristic of interest is weight.

Discussion

LE 13   Connection

(a) Guess how many times heavier the adult dolphin in the preceding exercises is than the young dolphin.

(b) Would you guess that weight relationships are the same as length relationships, area relationships, or volume relationships?

(c) Would you like to revise your guess in part (a)? (Last chance.)

In fact, weight corresponds—is proportional—to volume. The following chart shows how a variety of characteristics relate to length, area, or volume.

> **Classification of Measurements**
>
> *Lengths*—sides, edges, perimeters, heights
>
> *Areas*—surface area, amount of skin, strength
>
> *Volumes*—weight

Now it is possible to deduce other facts about the two dolphins. The next two lesson exercises and the next example refer again to the young and adult dolphins.

LE 14   Connection

How many times more material would it take to make a rubber sweater (same thickness) for the adult dolphin than for the young dolphin?

• EXAMPLE 1

The adult's waist size is 27 guppy fins (dolphins don't use inches). What is the young dolphin's waist size?

SOLUTION

Understanding the Problem   We know the larger dolphin's waist size and want to find the smaller one's waist size based on how they are related.

Devising a Plan   First determine whether waist size corresponds to length, area, or volume. Then use proportional reasoning with the appropriate ratio.

Carrying Out the Plan   Waist size corresponds to length. The length ratio for the two dolphins is $2:1$. So the young dolphin's waist size is $\frac{1}{2}$ of 27, or 13.5 guppy fins.

Looking Back   If the problem had been about a ratio related to area or volume, we would have used a ratio of $4:1$ or $8:1$ instead.   ●

LE 15   Reasoning

The young dolphin can pull a weight of 8 shells with its dorsal fin. The adult can pull a weight of _____ of the same shells with its dorsal fin.

That's enough of the dolphins. Try the following more general question, and then we'll get back to the giant animals that may visit your neighborhood.

LE 16   Reasoning

In general, when an animal triples all of its length measurements, its weight becomes about _____ times as much.

STRENGTH VERSUS WEIGHT

Could a giant ape or giant insect from a horror film exist in real life? The following exercise will help you find out.

Discussion

LE 17   Connection

In a horror film, a cockroach eats radioactive brussels sprouts and grows 100 times longer in terms of all of its length measurements!

(a) How many times stronger are the cockroach's legs now?
(b) How many times heavier is the cockroach?
(c) Explain why such a giant cockroach could not even stand on its legs, let alone walk around your kitchen.

As animals get larger, supporting their weight with their own legs becomes more difficult. Consequently, heavy animals such as hippos tend to have legs that are thick in comparison to the rest of their bodies, and giant fictional apes such as King Kong would be unable to stand.

The tallest person on record was Robert Wadlow. He grew to a height of 8 ft 11 in. Unfortunately, his legs could not support his weight. He wore a leg brace to help support his weight, but his joints became diseased, and he died at the age of 22.

LE 18   Summary

Tell what you learned about similar solids in this section.

ANSWERS TO SELECTED LESSON EXERCISES

2. (a) $30^2 + 40^2 = 50^2$ and $60^2 + 80^2 = 100^2$
(b) 600 ft^2 and 2400 ft^2 (c) 4
(d) 120 ft and 240 ft   (e) 2
(f) The second

3. (a) 2:3   (b) 24 cm^2, 54 cm^2
(c) 4:9   (d) 20 cm, 30 cm
(e) 2:3

5. No. Discuss what the ratio of the areas would be. It's 16:1. Then how would you find the area of the larger triangle? (*Answer*: 160 ft^2.)

6. (a) 1:3   (b) 6 cm^2; 54 cm^2
(c) 1:9   (d) 1 cm^3, 27 cm^3 (e) 1:27

7. (a) 3:2   (b) 9:4   (d) 27:8

8. (a) 9:16   (b) 27:64   (c) 67.5 m^3

11. (b) π in.^{2} (c) 4π in.^{2} (d) 4; 4

12. (a) 1:4   (b) 2

14. 4

15. 32

16. 27

17. (a) 10,000   (b) 1,000,000
(c) The cockroach's weight has increased 100 times more than its strength.

10.7   HOMEWORK EXERCISES

Basic Exercises

1. The two rectangles shown are similar.

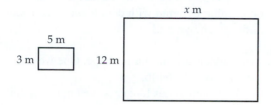

(a) Find x.
(b) The ratio of corresponding sides is _____.
(c) The ratio of the areas of the two rectangles is

_____.

(d) The ratio of the perimeters of the two rectangles is

_____.

2. The two triangles shown are similar.

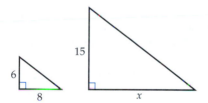

(a) Find x.
(b) The ratio of corresponding sides is _____.
(c) The ratio of the areas of the two triangles is

_____.

(d) The ratio of the perimeters of the two triangles is

_____.

3. At the Pizza Chalet, a small pizza has an 8-in. diameter, and a medium pizza has a 12-in. diameter.
(a) What is the ratio of their areas?
(b) What is the ratio of their perimeters?
(c) The medium pizza should cost about

_____ times more than the small pizza.

4. A scale drawing of a stop sign and the actual stop sign have a ratio of corresponding sides $m:n$.
(a) What is the ratio of their perimeters?
(b) What is the ratio of their areas?

 5. At Checkers, a pizza with a 9-in diameter costs $3.49, the 12-in. pizza costs $5.79, and the 16-in. pizza costs $7.99.
(a) Which is the best buy?
(b) Which is the worst buy?

6. A television screen with a 20-in. diagonal is a rectangle about 12 in. by 16 in. A television screen with a 27-in. diagonal is a rectangle about 16.2 in. by 21.6 in. Which would be a better buy, a 20-in. color television for $180 or a 27-in. color television for $320?

7. Suppose the length and width of a photograph are increased by 200%.
(a) The length and width of the enlarged photograph are _____ times as much as the original.
(b) The area of the enlarged photograph is _____ times as much as the original.

8. A triangle has a perimeter of 14 cm and an area of 12 cm^2. Find the perimeter and the area of a similar triangle with double the dimensions of the given triangle.

9. Consider the two rectangular prisms shown.

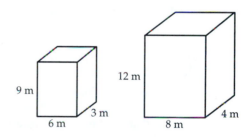

(a) What is the ratio of their corresponding edge lengths?
(b) The surface areas of the two figures are _____ and _____.
(c) What is the ratio of their surface areas?
(d) How is the ratio of their surface areas related to the ratio of their corresponding edges?
(e) Compute the volume of each figure.
(f) What is the ratio of their volumes?
(g) How is the ratio of volumes related to the ratio of edges?

10. Two similar solids have a ratio of corresponding heights $x:y$.
(a) What is the ratio of their surface areas?
(b) What is the ratio of their volumes?

11. The Earth has about 16 times the surface area of the moon.
(a) What is the ratio of their circumferences?
(b) What is the ratio of their volumes?

12. Two similar hexagons have corresponding heights of 3 m and 4 m. If the area of the smaller hexagon is 15 cm^2, what is the area of the larger hexagon?

13. Two similar cones have heights of 4 m and 7m.
(a) If the radius of the smaller cone is 2 m, what is the radius of the larger cone?
(b) The surface area of the larger cone is _____ times larger than the surface area of the smaller one.
(c) What is the ratio of the volume of the smaller cone to the volume of the larger cone?
(d) Do parts (a)–(c) involve induction or deduction?

14. Consider the following problem. "Two similar prisms have total surface areas of 100 cm^2 and 225 cm^2. If the smaller prism has a height of 10 cm, what is the height of the larger prism?" Devise a plan, and solve the problem.

15. The ratio of strengths in similar solids is the same as the ratio of corresponding
(a) lengths   (b) areas   (c) volumes

16. If a child grows so that all of her length measurements double, her new clothes will require _____ times as much of the same material as her old ones.

 17. A scale model of a car is built in the same shape and of the same materials as the actual car, but it is only $\frac{1}{8}$ as long. How many times heavier is the actual car?

 18. In a scale model of a building, 1 ft represents 40 ft. The volume of the scale model is 18 ft^3.
(a) What is the volume of the building?
(b) The roof of the scale model has a surface area of 7 ft^2. What is the surface area of the roof of the building?

 19. The surface area of a prism is 40 ft^2, and its volume is 24 ft^3. Find the surface area and volume of a similar prism with triple the dimensions of the given prism.

 20. The surface area of a pyramid is 82 cm^2, and its volume is 35 cm^3. Find the surface area and volume of a similar pyramid with double the dimensions of the given pyramid.

21. Two animals are similar in shape. One weighs 125 lb; the other weighs 64 lb.
(a) How many times longer is the larger animal?
(b) How many times stronger is the larger animal?
(c) How many times heavier is the larger animal?
(d) If the two animals walk on their legs, which one has an easier time walking?

22. Two animals are similar in shape. One is 6 ft long; the other is 4 ft long.
(a) How many times heavier is the larger animal?
(b) How many times more skin surface does the larger animal have?

23. Suppose a giant exists who is the same shape as you, except that all of the giant's length measurements are triple yours.
(a) The giant's waist size is _____ times yours.
(b) The giant is _____ times as strong as you are.
(c) The giant weighs _____ times as much as you do.

24. Explain why a giant cockroach could not exist.

 25. In *Gulliver's Travels,* Gulliver visits the tiny Lilliputians. The Lilliputian emperor finds out that all of Gulliver's length measurements are about 12 times the corresponding measurements of an average Lilliputian. The emperor says that Gulliver will need to eat as much food as 1,728 Lilliputians! Explain the emperor's reasoning.

26. In another scene in *Gulliver's Travels* (see the preceding exercise), the emperor wants to make a suit for Gulliver using the same fabric that the tiny Lilliputians wear. How many times more material will he need for Gulliver's suit than for an average Lilliputian's suit?

27. A child's bicycle might have a 20-in. wheel diameter, while an adult's bicycle might have a 26-in. wheel diameter.
(a) Is the cost of a bicycle proportional to length, area, or volume?
(b) If the bicycles are similar and the child's bicycle costs $60, what is a fair price for the adult's bicycle?

28. A regular can of Blando ravioli is 12 cm tall and sells for $1.25. A large can is the same shape and 16 cm tall, and it sells for $2.
(a) How many times more ravioli does the large can hold?
(b) Which can is the better buy?

Extension Exercises

29. Warm-blooded animals such as humans must maintain a reasonably constant body temperature. Animals must balance the heat gained from eating (calories) with the heat lost through their skin.
(a) Heat gained from eating is proportional to an animal's
(1) length (2) area (3) volume
(b) Heat lost through skin surfaces is proportional to an animal's
(1) length (2) area (3) volume

30.

Daddy   Baby

(a) Baby and Daddy are similar in shape. Complete the chart.

	Length	Surface Area (where heat is lost)	Weight (how heat is gained)
Baby	1 unit	1 square unit	1 cubic unit
Daddy	4 units		

(b) Who has an easier time gaining heat relative to the heat lost?
(c) Who has a harder time staying cool?
(d) As an animal grows larger, does it have more or less difficulty retaining heat?

31. A person eats about $\frac{1}{50}$ of his or her weight each day.
Would a smaller animal such as a mouse need to eat a larger or smaller fraction of its body weight each day to stay warm enough?

32. A large animal such as a hippopotamus or an elephant has more weight relative to surface area than a person. What problem does this create for these large animals?

Technology Exercise

33. How do the perimeter and area of two similar triangles compare?
(a) Construct two unequal segments $\overline{AB}$ and $\overline{CD}$. Make $\overline{CD}$ longer than $\overline{AB}$.
(b) Select $\overline{AB}$ and $\overline{CD}$, and mark CD/AB as a ratio with the Transform menu. Measure the ratio CD/AB. This will be the scale factor for a dilation.
(c) Construct a triangle and its interior.
(d) Construct a point outside the triangle. Double-click the point to designate it as the center of your dilation.
(e) Now select the sides, vertices, and the interior of your triangle. Dilate the triangle by the ratio CD/AB.
(f) In the Measure menu, use Calculate to find the perimeter and the area of each triangle. How do the results relate to the scale factor CD/AB? (If you are not sure, repeat the dilation with a different scale factor.)

CHAPTER 10 SUMMARY

Nearly every industrialized nation except the United States uses the metric system. In the United States, the metric system is used in most businesses other than companies dealing in building trades or consumer goods. Why? Because it is easier to work with. All metric conversions involve powers of 10, and the same set of prefixes is used for most metric measurements. In learning a new measurement system such as the metric system, it helps to begin by learning some benchmarks for the units.

Some carpentry, building, and consumer problems require finding length, area, or volume. Area is the measure of the surface of a solid or the space inside a polygon or circle. Volume is the measure of the capacity of a solid. Area and volume can be measured by counting squares and cubes, respectively, but formulas exist that simplify these computations for many common figures.

A number of area formulas are logically related. The parallelogram formula follows from the rectangle formula. The triangle and trapezoid formulas can be deduced from the parallelogram formula.

If you know the lengths of two sides of a right triangle, you can find the length of the third side using the Pythagorean Theorem. The Pythagorean Theorem is frequently applied to right triangles that appear in other figures.

There is one general volume formula and one general surface-area formula for all prisms and cylinders. There is also one general volume formula for all pyramids and cones.

All corresponding area measurements of two similar figures are in the same ratio; all corresponding volume measurements of two similar figures are also in a single ratio. These ratios are helpful in explaining the relationship between weight and muscle strength and in explaining how well warm-blooded animals of various sizes retain heat.

STUDY GUIDE

To review Chapter 10, see what you know about each of the following ideas or terms that you have studied. You can also use this list to generate your own questions about the chapter.

MEASUREMENT IN GRADES 1–8

The following chart shows at what grade levels selected measurement topics typically appear in elementary- and middle-school mathematics textbooks.

Topic	Typical Grade Level in Current Textbooks
Metric measure	1, 2, 3, 4, 5, 6, 7
Perimeter	2, 3, 4, 5, 6
Circumference	4, 5, 6, 7
Rectangle area	4, 5, 6
Parallelogram area	5, 6, 7, 8
Triangle area	5, 6, 7
Trapezoid area	7, 8
Circle area	6, 7, 8
Pythagorean Theorem	7, 8
Surface area	5, 6, 7, 8
Volume	3, 4, 5, 6, 7, 8
Similar solids	7, 8

REVIEW EXERCISES

1. Complete the following conversions.
   (a) 23.7 cm = _____ m
   (b) 36.45 L = _____ mL
   (c) 0.8 kg = _____ mg

2. The width of an adult woman's hand is about
   (a) 90 cm   (b) 9 mm
   (c) 90 mm   (d) 900 mm

3. On a hot summer day in Washington, D.C., the temperature is about
   (a) 5°C   (b) 35°C   (c) 65°C
   (d) 95°C   (e) 125°C

4. Give the GPE of the measurement 46.7 m.

5. Write a sentence that tells the difference between the perimeter and area of a convex polygon.

6. Find the area of the triangle. Tell how you did it.

7. The room shown has an area of 58 m². What is the length across the back of the room?

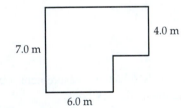

8. A store sells two kinds of wrapping paper. Package A costs \$5 and has 4 rolls, each 2 ft by 7 ft. Package B costs \$4 and has 3 rolls, each $3\frac{1}{2}$ ft by 4 ft. Which is the better buy? Explain how you made your decision.

9. (a) How many cm² in 1 m²?
   (b) Fill in the blank.
      8 cm² = _____ m².

10. Use the area formula for a parallelogram ($A = bh$) to explain why the area of the triangle shown is $A = \frac{1}{2} bh$.

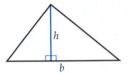

11. A triangular sail is measured as shown. What is the best approximation of its area?

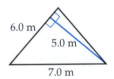

12. Find the shaded area.

13. The quarter circle has an area of 25π m². What is its perimeter?

14. The circumference of a circle is 50 cm. What is its exact area?

15. A 9-in. pizza costs \$5, and a 14-in. pizza costs \$11. Which is the better buy? Explain how you made your decision.

16. The figure is formed by putting together 8 congruent right triangles with sides of length a, b, and c.

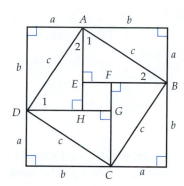

(a) Tell why $\angle DAB$ is a right angle.
(b) Part (a) confirms that $ABCD$ is a square. What is the area of square $ABCD$?
(c) The area of $ABCD$ is also equal to the sum of the areas of which smaller figures?
(d) How long is each side of $EFGH$?
(e) Write the area of $ABCD$ in terms of the five smaller figures.
(f) Set your expressions from parts (b) and (e) equal to each other, and derive the Pythagorean Theorem.

17. A rectangular prism is 8 m by 6 m by 7 m. How long is the diagonal shown?

 18. Find the shaded area. $\overline{AB}$ is a diameter.

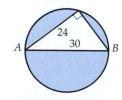

19. Tell whether each set of lengths could form a right triangle, an acute triangle, an obtuse triangle, or no triangle.
(a) 2, 3, 7   (b) 5, 6, 7   (c) 4, 5, $\sqrt{41}$

20. What is the surface area of the figure?

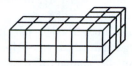

21. If the length, width, and height of a rectangular prism are each multiplied by 4, how does the surface area change? Give evidence to support your answer.

22. A rectangular prism has dimensions of 4 m, 5 m, and W m. Its total surface area is 166 m^2. What is W?

23. Imagine stacking centimeter cubes one on top of the other. What is the surface area of
(a) one cube?
(b) a stack of 2 cubes?
(c) a stack of 3 cubes?
(d) a stack of n cubes?
(e) What kind of reasoning is used to make the generalization in part (d)?

24. The formula $V = Bh$ applies to what kind of figures?

25. Find the volume and surface area of the triangular prism shown.

26. A cylinder has a radius of 6 ft and a height of 12 ft. Find its volume and surface area.

 27. A sphere has a volume of 288π cm^3. What is its radius?

28. Two similar spheres have radii of 4 ft and 7 ft. What is the ratio of their surface areas?

 29. Two similar solids have total surface areas of 400 m^2 and 900 m^2. The smaller solid has a volume of 200 m^3. What is the volume of the larger solid?

30. Two animals of the same type are the same shape. The larger animal has a waist size of 20 cm; the smaller animal has a waist size of 8 cm.
   (a) How many times heavier is the larger animal?
   (b) If the larger animal can lift 40 lb, how many pounds can the smaller animal lift?
   (c) Do parts (a) and (b) involve induction or deduction?

ALTERNATE ASSESSMENT

Do one of the following assessment activities: add to your portfolio, add to your journal, write another unit test, do another self-assessment, or give a presentation.

SUGGESTED READINGS

Beckmann, P. *A History of Pi.* 5th ed. New York: St. Martin's Press, 1976.

Blatner, David. *The Joy of* π. New York: Walker, 1997.

Erickson, T. *Get It Together.* Berkeley, CA: Equals, 1989.

Geddes, D., and others. *Measurement in the Middle Grades.* Reston, VA: NCTM, 1994.

Jacobs, H. *Geometry.* 3rd ed. New York: W. H. Freeman, 2003.

Mathematics Teaching in the Middle School, March–April, 1996 Focus Issue on Mathematical Connections. Reston, VA: NCTM, 1996.

National Council of Teachers of Mathematics. 2003 Yearbook. *Measurement.* Reston, VA: NCTM, 2003.

Navigating Through Geometry (4 vols.). Reston, VA: NCTM, 2001.

Navigating Through Measurement (4 vols.). Reston, VA: NCTM, 2003.

Schifter, D. et al. *Measuring Space in One, Two, and Three Dimensions.* Parsippany, NJ: Dale Seymour, 2002.

Serra, M. *Discovering Geometry,* 3rd ed. Emeryville, CA: Key, 2003.

Use Infotrac to find articles in *Teaching Children Mathematics* on teaching children about measurement.

11 | ALGEBRA AND COORDINATE GEOMETRY

Algebra is a mathematical language. You can represent some relationships between quantities with words, symbols, tables, and graphs. Translating from one representation to another is part of doing algebra.

Algebra is a language, but it is more than that. It is also a way of thinking. According to mathematics educator Mark Driscoll, algebraic thinking involves three processes: "1) doing and undoing, 2) building rules to represent functions, and 3) abstracting from computation."

Although the ancient Greeks had developed the major ideas of geometry by 300 B.C., algebra developed far more slowly after originating in Babylonia about 4000 years ago. Algebra was done first with words and later with a combination of words and symbols. Around A.D. 250, Diophantus was the first to use letters for variables and exponents. It was not until the late 1500s that mathematicians such as François Viète succeeded in solving fairly complicated algebraic equations.

The stage was then set for Descartes and Fermat to connect algebra and geometry using coordinate graphs. In coordinate geometry, algebraic equations are classified according to the shapes of their graphs, and geometric shapes can be represented by equations.

More recently, Emmy Noether (1882–1935), the greatest woman mathematician of her time (Figure 11–1), developed more advanced algebra that is now a fundamental part of graduate-school mathematics. Noether battled discrimination throughout her life. One of only a few female students in the entire University of Erlangen (Germany), she completed a doctorate in mathematics at the age of 24. Noether then did research and lectured at the university for no pay. In 1922, Noether was appointed an "extraordinary" professor at the University of Göttingen and received a modest salary. Eleven years later, in 1933, the Nazis came to power and dismissed her because she was Jewish. Noether emigrated to the United States and was

Courtesy of Library of Congress

Figure 11–1 *Amalie Emmy Noether*

appointed a visiting professor at Bryn Mawr College in Pennsylvania in the fall of 1933. She died suddenly in April 1935 after routine surgery.

In this chapter, you will use formulas, graphs, and tables to represent relations between variables and to solve problems. You will use models for solving equations. You will also see how graphs offer another way to study some geometric concepts.

11.1   THE LANGUAGE OF ALGEBRA

NCTM Standards
- represent and analyze patterns and functions, using words, tables, and graphs (3–5)
- represent the idea of a variable as an unknown quantity using a letter or symbol (3–5)
- represent, analyze, and generalize a variety of patterns with tables, graphs, words, and, when possible, symbolic rules (6–8)
- develop an initial conceptual understanding of different uses of variables (6–8)

In what ways is algebra like a language? What are the symbols, words, phrases, and sentences of algebra? First, consider variables and constants.

VARIABLES AND CONSTANTS

LE 1   Opener

(a) Give an example of a quantity that is variable and an example of a quantity that is constant.

(b) What are some different uses of variables in algebra?

Quantities whose value may change are called **variables.** A person's weight is variable. Quantities whose values cannot change are called **constants.** The number of eyes that a normal dog has is constant.

LE 2   Concept

Tell whether each quantity is a variable or a constant.

(a) The number of inches in a yard
(b) The time it takes a person to dress each morning

In algebra, we use expressions and equations to represent quantities. An **algebraic expression** such as $5N - 6$ is a mathematical phrase with an arithmetic operation or operations involving variables and numbers. The letters in algebraic expressions and equations are called variables or constants.

LE 3   Concept

In the formula $A = \pi r^2$, tell whether each of the following is a variable or a constant.

(a) A **(b)** π **(c)** r

Variables can be used in different ways.

LE 4   Concept

Tell how the variables are being used in each part.

(a) $y + 3 = 7$ **(b)** $x + y = y + x$
(c) $A = bh$ **(d)** $y = x^2 + 2$

LE 4 illustrates the main uses of a variable. Variables were first used to represent unknown numbers as in part (a). In the last 500 years, the uses of variables have expanded to represent a general property of numbers (part (b)), a formula that relates quantities (part (c)), or quantities that vary in relation to one another (the function in part (d)). More recently, variables have also been used in computer programs and spreadsheets.

A major objective in algebra is to learn to translate among words, tables, symbols, and graphs. These four modes of expression establish connections among arithmetic, algebra, and geometry.

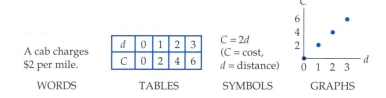

A cab charges $2 per mile.		$C = 2d$ ($C =$ cost, $d =$ distance)	
WORDS	TABLES	SYMBOLS	GRAPHS

In Section 2.3 or in algebra classes, you have generated tables of x- and y-values from equations and also found equations that fit the values in a table. This section focuses on translating words $\leftrightarrow$ symbols and words $\leftrightarrow$ a graph.

TRANSLATING WORDS INTO SYMBOLS

In applying mathematics to everyday situations, one may translate words into symbols. Algebra offers a precise way to communicate information about quantities. Some people enjoy the precision and logic of algebra. But not everyone has this reaction (Figure 11–2).

Figure 11–2

Mary Fairfax Sommerville (1780–1872), a self-educated Scotswoman, had a more positive reaction to algebra. Reading a fashion magazine at a tea party, she wondered about a puzzle filled with *x*'s and *y*'s. After learning that it was algebra, Sommerville decided to learn more about the subject. It was considered inappropriate for a young woman to study mathematics, so Sommerville had a man secretly obtain a copy of Euclid's *Elements* for her.

Later in life, Sommerville published some articles on mathematics. Her finest work was her 1831 translation of and commentary on Pierre LaPlace's *Méchanique Céleste,* a difficult text on the mathematics of astronomy. Sommerville continued doing mathematics until the day she died at age 92.

Algebra serves as a mathematical language. It is more concise, more abstract, and less ambiguous than English. Algebra is also limited; it can only describe relationships between quantities. I couldn't say $H = M + 5$, in which H = Horace and M = Mitzie. You'd say, "Now hold on, pal! Horace and Mitzie are more than just numbers." And you'd be right.

Children in the upper elementary grades learn to translate English phrases into algebraic expressions as illustrated in the fourth-grade textbook page from *Harcourt Math* (Figure 11–3). Some English phrases can be translated into algebraic expressions in two steps.

Step 1:   Select variables for unknown quantities.
Step 2:   Write an expression or equation using the variable(s).

• EXAMPLE 1

(a) Translate "five years older than Mitzie" into symbols.
(b) Find the value of the expression if Mitzie is 20 years old.

SOLUTION

(a) **1.** Mitzie's age is the unknown quantity. So let M = Mitzie's age in years.
  2. Then "five years older than Mitzie" is $M + 5$.

So the answer is: Let M = Mitzie's age

$$M + 5$$

(b) Algebraically, $M = 20$, so $M + 5$ is $20 + 5$, or 25 years.   ●

• EXAMPLE 2

Translate "funnier than a lead balloon" into symbols.

SOLUTION

This phrase does not involve relationships between *quantities,* so it cannot be translated into symbols.   ●

See whether you can translate each of the following English phrases into an algebraic expression.

Classroom Connection

LE 5   Skill

(a) Translate "25% of the recommended daily allowance of iron" into symbols.
(b) Find the value of the expression if the recommended daily allowance of iron is 20 mg.
(c) A child writes I = iron. What would you tell the child?

Classroom
Connection

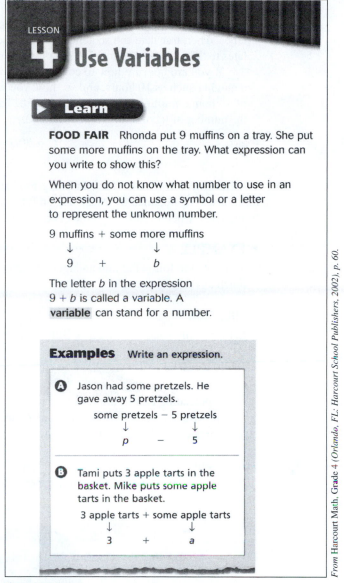

Figure 11–3

LE 6 Concept

Translate "a shower of commanded tears" into symbols.

You have seen how some English *phrases* translate into algebraic expressions (an arithmetic operation or operations involving variables and numbers). What does an English *sentence* look like after being translated into symbols?

● EXAMPLE 3

Translate the following into an algebraic sentence. "The total cost of the repair is $80 for parts plus $30 per hour for labor."

SOLUTION

The unknowns are total cost and hours of labor. Let C = total cost and H = hours of labor. Then translate the sentence. The phrase "the total cost of the repair is" translates to "$C =$," and this equals "$80 plus $30 per hour of labor."

If you are not sure how to compute the cost from the number of hours, try it with a number such as 10 hours, and see how you would find the cost. To compute the cost of 10 hours, multiply 30 by 10 and add 80. To compute the total cost, multiply 30 by the number of labor hours (H) and add 80.

$$C = 30H + 80$$ ●

One important reason for translating English into symbols is to obtain a formula. A formula can be obtained by translating an English-language description into an algebraic formula or by determining how the quantities in a problem are related mathematically.

● **EXAMPLE 4**

Translate the following into an algebraic sentence. "Sally's income is at least twice as much as Bill's income."

SOLUTION

Let S = Sally's income, and B = Bill's income. The phrase "at least" translates to "$\geq$."

$$S \geq 2B$$ ●

The preceding two examples illustrate the ways in which English sentences that describe quantities correspond to equations or inequalities. If possible, translate the English sentences in LE 7 and LE 8 into equations or inequalities.

Discussion

LE 7 Skill

Joe sells squirrel-powered vacuum cleaners. He earns $230 a week and also receives a commission of 14% of his sales.

(a) Write a formula for his weekly pay in terms of sales.
(b) How much must he sell in a week to earn $400?

LE 8   Skill

Translate the following statement into an equation or inequality. "The sum of the money spent by the U.S. government on transportation and the environment in the fiscal year 2004 was less than $\frac{1}{4}$ the amount spent on national defense."

An algebraic equation or inequality focuses upon some quantifiable aspect of a situation. In Example 4, the inequality does not tell us what Bill and Sally do, whether they enjoy their jobs, or whether their work is beneficial to society.

LE 9   Concept

Discussion

What are some questions that the inequality in LE 8 does not answer about the quantities involved?

Another important reason for translating English into symbols is to obtain concise descriptions of rules and properties. For example, instead of saying, "We can add two real numbers in either order and obtain the same sum," we could write "$x + y = y + x$ for real numbers x and y."

LE 10   Skill

Translate the following into an algebraic equation: "When two rational numbers are multiplied, the result is the product of the numerators divided by the product of the denominators."

TRANSLATING SYMBOLS INTO WORDS

You have practiced translating words into symbols. Now try translating symbols into words (Figure 11–4).

Figure 11–4

• EXAMPLE 5

(a) Describe $H = 0.2R$ in words.

(b) If H = cost of heat in dollars and R = rent in dollars, what does the equation mean?

SOLUTION

(a) An equals sign can be translated as "is" or "is equal to." Two possible answers are "H is the product of 0.2 and R" and "H is 0.2 times R."

(b) "The heat is 0.2 times the rent" or "Heating costs are 20% of the rent."   •

LE 11   Skill

(a) Describe $y = 4d + 16$ in words.

(b) If y = a dog's age in human years and d = a dog's age in years, what does the equation mean?

TRANSLATING WORDS INTO SYMBOLS AND TABLES

Suppose you are planning a concert at your school and want to see how much profit you might make in relation to ticket sales. The following exercise illustrates how a computer spreadsheet can be used to generate a table of possible outcomes after you translate from words into equations.

LE 12   Connection

The concert will cost $2,200 to set up and $2 in cleanup costs per person. Tickets will sell for $18 each.

(a) Write a formula for the total cost C of the concert in relation to the number of tickets sold, x.

(b) Write a formula for the total revenue R from the sale of x tickets.

(c) How is profit related to revenue and cost?

(d) Write a formula for the profit P in terms of x.

(e) If you have a computer spreadsheet, set it up as follows. Type the numbers 50, 100, 150, . . . 500 in column A. The top part of the spreadsheet follows.

	A	B	C	D
1	Concert			
2	Tickets sold	Cost	Revenue	Profit
3	50	$2300	$900	−$1400
4	100			

(f) Enter appropriate formulas in cells B3, C3, and D3. Then copy the formulas through the rest of your columns, B, C, and D. Compare your results for B3, C3, and D3 with those given in the spreadsheet.

(g) Comment on the results in each column.

TRANSLATING A GRAPH INTO WORDS

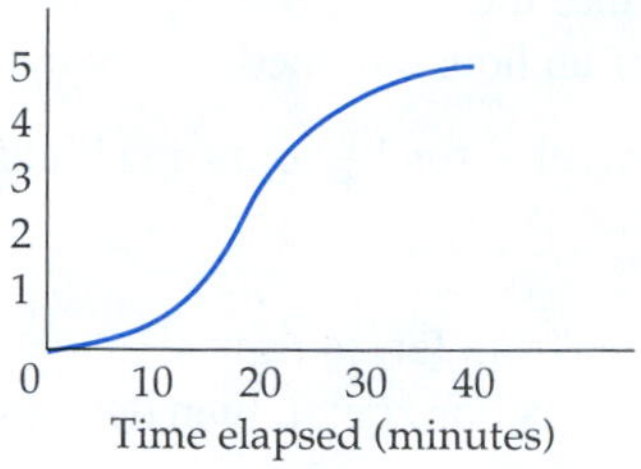

Total distance
traveled (miles)

Figure 11–5

Like equations and inequalities, a graph displays a relationship that can also be written in words.

LE 13 Communication

Figure 11–5 shows a graph of the progress of a runner. Which of the following describes the run?

(a) She started slowly and went faster and faster.
(b) She started fast, slowed down, and sped up again.
(c) She started slowly, sped up, and slowed down again.
(d) She started slowly and sped up at the end.

In the next exercise, see if you can interpret the graph in words.

LE 14 Communication

Figure 11–6 shows a graph of a journey by car.

(a) At what point might the driver have stopped to rest?
(b) When was the car traveling the fastest?
(c) Describe the progress of the car during the journey.
(d) What was the average speed of the car?

Total distance
traveled (miles)

90

60

30

0   1   2   3
Time elapsed (hours)

Figure 11–6

TRANSLATING WORDS INTO A GRAPH

If you have a description of how a variable changes in relation to another variable, you can represent the information with a graph.

LE 15 Connection

Sketch a graph that corresponds to each of the following descriptions.

(a) The more the apples weigh, the higher the price.
(b) The price of gasoline increased steadily from January to April; then it leveled off from April to October; there was a slight decrease from October to December.

LE 16 Summary

Tell what you learned about algebra in this section.

ANSWERS TO SELECTED LESSON EXERCISES

2. (a) Constant   (b) Variable

3. (a) Variable   (b) Constant   (c) Variable

5. (a) $0.25I$, in which I = RDA of iron
(b) 5 mg
(c) Iron is not a number. A variable must represent a quantity.

6. Cannot be translated

7. (a) Weekly pay $P = 230 + 0.14S$, where S is total sales
(b) \$1214.29

8. $T + E < \frac{1}{4}D$, in which T = trans. budget (\$),
E = env. budget (\$), D = def. budget (\$)

9. How the money is spent within each department; how much of the money goes to large campaign contributors

10. $\frac{a}{b} \times \frac{c}{d} = \frac{a \times c}{b \times d}$, in which $\frac{a}{b}$ and $\frac{c}{d}$ are rational.

11. (a) y is 16 more than 4 times d.
(b) A dog's age in human years is 16 more than 4 times the dog's age in years.

12. (a) $C = 2200 + 2x$ (b) $R = 18x$
(c) Profit $= R - C$ (d) $P = 16x - 2200$
(g) Profits made on 150, 200, . . ., 500 tickets sold.

13. (c)

14. (a) After 1 hour
(b) From 2 to 3 hours into the trip
(c) The car traveled for an hour, stopped for about 45 min, and then traveled for $1\frac{1}{4}$ hours at a higher speed.

15. (a) An increasing graph (from left to right)
(b) Graph increases, then is horizontal, then decreases slightly (from left to right).

11.1   HOMEWORK EXERCISES

Basic Exercises

1. Tell how the variables are being used in each part.
(a) $xy = yx$
(b) $y = 5x - 8$
(c) $d = rt$
(d) $3x - 4 = 11$

2. Tell how the variables are being used in each part.
(a) $V = \frac{1}{3}BH$
(b) $8y = 32$
(c) $y = 2^x$
(d) $x + y + z = z + y + x$

3. Tell whether each quantity is a variable or a constant.
(a) The number of faces on a rectangular prism
(b) The price of a dozen eggs

4. Tell whether each quantity is a variable or a constant.
(a) The time required to walk one mile
(b) The number of legs on a normal spider

5. In the general formula $C = 2\pi r$, tell whether each of the following is a variable or a constant.
(a) C (b) π (c) r

6. In the general formula $A = 2LW + 2LH + 2HW$, tell whether each of the following is a variable or a constant.
(a) A (b) L (c) 2

7. In the cartoon, why can't $X =$ Shobert Zangox?

8. Which of the following could be represented by an algebraic variable?
(a) Mary's age   (b) Transportation   (c) Mike

9. (a) Translate "twelve degrees warmer than the average for this day" into an algebraic expression.
(b) Find the value of the expression if the average for the day is 60 in °F.

10. (a) Translate "40% off the regular price" into an algebraic expression.
(b) Find the value of the expression if the regular price is $30.

If possible, translate the English sentences in Exercises 11–16 into algebraic equations or inequalities.

11. The total cost of movie tickets is $5 per adult ticket plus $3 per child's ticket.

12. The ratio of students to teachers is 12 to 1.

13. The perimeter of a special rectangle is three times its length.

14. "The fool thinks he is wise, but a wise man knows he is a fool." (Shakespeare, *As You Like It*)

15. Katie is at least twice as tall as she was last year.

16. There are, at most, 30 more girls than boys.

17. (a) Translate "Twenty dollars is 85% of the list price" into an equation.
(b) What doesn't this equation tell us about the quantities involved?

18. In each case, decide whether the quantity measured is (1) a significant or (2) a trivial feature of the person or object.
(a) What your score S on the last mathematics test tells about how much you know about the material covered
(b) What your score S on the last mathematics test tells about you
(c) What the time T it takes you to finish a mathematics test tells about your general mathematical ability

19. In middle school, children write their first algebra equations for situations such as the following. Write an equation for each situation, and tell what the variable represents.
(a) Venus has 8 dishes. She needs 13 for the party.
(b) CDs sell for $15 each. In one hour, a salesperson sold $90 worth of CDs. How many CDs is that?

20. Write an equation for each situation, and tell what the variable represents.
(a) Each page of a photo album holds 4 photos. How many pages will it take to display 36 photos?
(b) After Lleyton spent $12 for dinner, he had $20 left.

21. A man who is 5 ft 8 in. tall weighs 200 pounds.
(a) If he loses 5 pounds per week, what does he weigh after W weeks?
(b) Estimate a maximum reasonable value for W.

 22. An airplane at an altitude of 33,000 ft begins descending at a rate of 1,200 ft per minute.
(a) Write a formula relating altitude A (in ft) to M, the number of minutes after the descent begins.
(b) What is the domain for M?

23. Concert tickets are priced at $C each. There is a service charge of $5 per ticket. Jade buys 6 tickets.
(a) Write two equivalent expressions for the total cost of her tickets.
(b) What property confirms that the two expressions are equivalent?

24. Write two expressions, one in factored form and one in expanded form, for the area of the following figure. If you equate the two results, what property is illustrated?

25. A job pays $100 per week and a 5% commission of sales.
(a) Write a formula for total weekly salary in terms of sales.
(b) How much would a salesperson have to sell to earn $250 in a week.

26. (a) Under an installment plan, June makes a $400 down payment and also has a $50 monthly payment each month after that. Write a formula for the total amount paid in terms of the number of months.
(b) Suppose June has to pay a total of $1,600. Show how to use your formula to find the length of her installment plan.

 27. A rental van costs $120 per day plus $0.32 per mile. What is the range of distances you could drive the van for less than $280?
(a) Solve it with the guess-and-check strategy.
(b) Solve it with an equation.

28. The total bill in a restaurant, including 5% tax, was $16.17. How much was the meal without the tax? (*Hint:* The 5% tax was computed on what amount?)
(a) Solve it with the guess-and-check strategy.
(b) Solve it with an equation.

29. Women make up 36% of a college student body. There are 261 women. How many students are at the college?

30. How does your college bookstore price its textbooks? Suppose the bookstore takes the wholesale price and increases it by 30%. You buy a science textbook for $84. What was the wholesale price?

31. Translate the following sentence into algebra. "The product of a real number and the sum of two other real numbers is the sum of the products of the real number and each of the other two real numbers."

32. Translate the following sentence into algebra. "The sum of any real number and zero is that real number."

33. You ask some sixth graders to write "n is 4 times x" in symbols. A child writes $4n = x$. Is this correct? If not, what would you tell the child to help him understand the correct answer?

34. You ask some sixth graders to write "n is 4 more than x" in symbols. A child writes $n + 4 = x$. Is this correct? If not, what would you tell the child to help her understand the correct answer?

35. (a) Describe $R = 22D + 10$ in words.
(b) If R = rental rate in dollars and D = number of days, what does the equation mean?

36. (a) Describe $C = 0.025S$ in words.
(b) If C = commission in dollars and S = sales in dollars, what does the equation mean (use the word "percent" in your translation)?

37. (a) Describe $C > 2P$ in words.
(b) If C = current U.S. population and P = U.S. population in 1940, what does the inequality mean?

38. (a) Describe $0.1P$ in words.
(b) If P = Iowa's population, what does the expression mean?

39. Here is a graph of a journey by car.

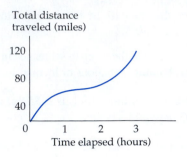

(a) Describe the progress of the car during the journey.
(b) What was the average speed of the car?

40. The graph shows the cost of producing a certain type of sneakers and the revenue from selling them.

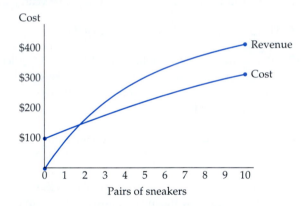

(a) What is the cost of producing 5 pairs of sneakers?
(b) What is the profit or loss on 5 pairs of sneakers?
(c) What is the profit or loss on 8 pairs of sneakers?
(d) At least how many pairs must be sold to make a profit?

41. Match each graph with its description.
(a) Outside temperature from 7:00 A.M. to 11:00 P.M.
(b) U.S. population from 1950 to 1990
(c) Speed of an airplane during a 2-hour flight

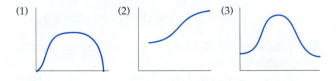

42. Match each graph with its description.
(a) Speed of a pitched baseball
(b) Falling object

(1)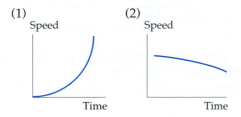
(2)

43. Make a reasonable sketch of each graph, showing a typical relationship between the two variables.

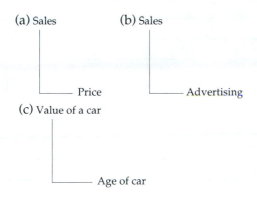

(a) Sales — Price
(b) Sales — Advertising
(c) Value of a car — Age of car

44. What does the graph suggest about the relationship between motivation and the difficulty of a lesson?

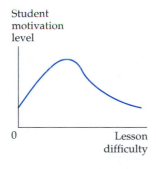

45. Sketch a graph that corresponds to each of the following stories.
(a) The more green peppers that are available, the lower the price
(b) Michael runs 6 mph for a half hour. Then he rests for 15 minutes. Then he runs 5 mph for 30 minutes.

46. Sketch a graph that corresponds to each of the following stories.
(a) The higher the price, the fewer bicycles you will sell.
(b) The temperature increased all morning until 2 P.M.; then it held steady for 2 hours; then it started to drop.

Extension Exercises

47. The graph shows the speed of a car on a highway.

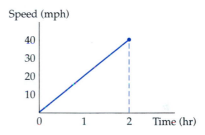

(a) How far did the car travel?
(b) What is the area of the rectangle on the graph?

48. The graph shows the speed of a car on a highway.

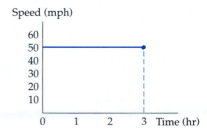

(a) How far did the car travel?
(b) What is the area of the triangle on the graph?
(c) On the basis of the preceding exercise and this one, it appears that the total distance traveled is equal to _____.

49. Bill made 50% more money this year than he did last year. Translate this sentence into algebra. (It cannot be translated word for word.)

50. (a) The length of a rectangle is 4 cm more than its width. Express this relationship with an equation.
(b) Write the area of this rectangle in terms of its width.

51. A school group already has $600 in a bank account. They need a total of $900 to finance a concert. If they sell concert tickets for D dollars, how many tickets must they sell to break even?

52. Do you know the day of the week on which you were born? The following formula gives the day of the week for any given date. D is the number of days in the latest year up to and including the date; Y is the year.

$$W = D + Y + \left[\frac{Y-1}{4}\right] - \left[\frac{Y-1}{100}\right] + \left[\frac{Y-1}{400}\right]$$

The notation [ ] means take the quotient without any remainder. For example,

$$\left[\frac{11}{4}\right] = 2$$

To obtain the day of the week, take the remainder when W is divided by 7 (1 = Sunday, 2 = Monday, . . . , 0 = Saturday).

On what day of the week was January 11, 1952?

$$W = 11 + 1952 + \left[\frac{1951}{4}\right] - \left[\frac{1951}{100}\right] + \left[\frac{1951}{400}\right]$$
$$= 11 + 1952 + 487 - 19 + 4 = 2435$$

$$7\overline{)2435} \quad 347R\,6 \rightarrow \text{sixth day} = \text{Friday}$$

It was a Friday.
(a) Use the formula to find out the day of the week on which you were born.
(b) On what day of the week will January 1, 2020, fall?

Technology Exercises

53. You plan to go on a 300-mile trip and stop for a 1-hour lunch. You want to find how the total time T for the trip is related to your average driving speed S.
(a) Find a formula for T in terms of S.
(b) Set up a computer spreadsheet as follows. Type the numbers 30, 35, 40, 45, 50, and 55 in column A. The top part of the spreadsheet follows.

	A	B
1	Trip	
2	Speed	Total time
3	30	

(c) Enter an appropriate formula in cell B3, and copy until the end of column B.
(d) Summarize the results.

54. In basketball, a player scores 1 point for a free throw, 2 points for a regular field goal, and 3 points for a long field goal. Find a formula for the total points T in terms of shots made, and use it to generate column E in the following spreadsheet.

	A	B	C	D	E
1	Name	1-Pt	2-Pt	3-Pt	Total
2	Nancy	5	7	2	
3	Tiwana	3	8	3	
4	Tim	6	11	0	
5	Shaq	9	10	0	
6	Kobe	4	6	4	

Magic Time

55. (a) Roll two dice. Compute the following four values: (1) the product of the two numbers on top of the dice, (2) the product of the two numbers on the bottom of the dice, (3) the product of the top number on one die and the bottom number on the other die, and (4) the product of the top number on the other die and the bottom number of the opposite die.
(b) Add the four numbers from part (a). I'll bet you got 49.
(c) See if you can prove why you always end up with 49. (*Hint:* Let x equal the top number on one die and y equal the top number on the other die.)

11.2   EQUATIONS AND LINEAR FUNCTIONS

NCTM Standards

- express mathematical relationships using equations (3–5)
- explore relationships between symbolic expressions and graphs of lines, paying particular attention to the meaning of intercept and slope (6–8)
- recognize and generate equivalent forms for simple algebraic expressions and solve linear equations (6–8)

Children first use equations to write basic facts such as $8 - 5 = 3$. Next, they may learn to check whether a given equation such as $3 + 2 = 7$ is true or false. The first algebra equations children solve are based on their knowledge of basic arithmetic facts. They can solve $3 + x = 7$ or $6n = 42$ with this basic knowledge. Children can also find solutions to equations such as $3t - 7 = 11$ by guessing and checking or by using the cover-up method (Exercises 3 and 4). Research suggests that the cover-up method is the best way to have children begin solving equations.

Next, children learn some properties that will enable them to solve many different equations with the same general method.

SOLVING EQUATIONS

Classroom Connection

LE 1   Opener

Consider the following two activities.

(a) A fifth grader has two piles with equal amounts of money. One pile has four nickels, and the other has two dimes. The child adds a dime to each pile, and the total amount in each pile is still equal.

(b) A fifth grader starts with the equation $3 \times 2 = 6$. Then the child multiplies each side by 5 and compares the results.

What properties are being developed in these two activities?

LE 1 illustrates two of the four properties of equality.

Addition, Subtraction, Multiplication, and Division Properties of Equality

For any real numbers a, b, and c, if $a = b$, then $a + c = b + c$.

For any real numbers a, b, and c, if $a = b$, then $a - c = b - c$.

For any real numbers a, b, and c, if $a = b$, then $ac = bc$.

For any real numbers a, b, and c, with $c \neq 0$, if $a = b$, then $a/c = b/c$.

These properties are used along with manipulatives or pictorial models to introduce equation solving in middle-school mathematics. Most middle-school textbooks use manipulatives or pictorial models to introduce equation solving. After solving some one-step equations, children might solve a problem like the following one.

● **EXAMPLE 1**

Suppose you have two puppies of equal weight and a hamster. The hamster weighs 2 kg, and the 3 animals weigh a total of 6 kg. How much do the puppies weigh?

(a) Write an equation that models this situation.

(b) Show how to solve the problem both algebraically and with a pictorial model for solving equations.

SOLUTION

(a) Let x = the weight of a puppy. Then $2x + 2 = 6$.

(b) The sixth-grade textbook of *Harcourt Math* (Figure 11–7) shows how to solve the equation with bars (rectangles) and squares. The solution is $x = 2$.   ●

Classroom Connection

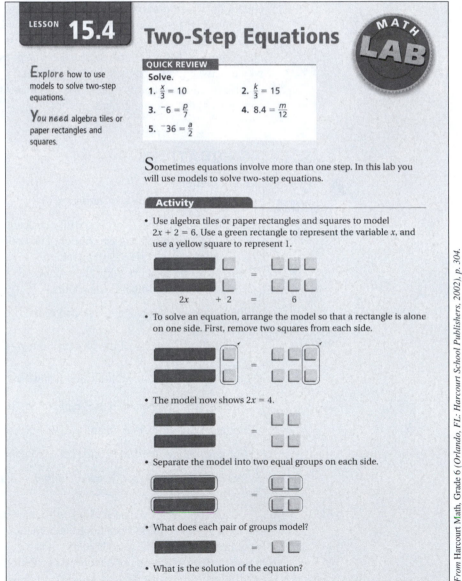

Figure 11–7

Some books use cups and pennies or envelopes and dots for the variable and the units, respectively. Now you try one.

LE 2 Connection

Consider the following problem. "A radio show has 14 minutes left. The producer plans to play 2 minutes of commercials and 3 songs of about the same length. How long should each song be?" Show how to solve the problem both algebraically and with a pictorial model for solving equations.

Some children have difficulty working with an equal sign. Consider the following.

Classroom Connection

LE 3 Reasoning

Two fifth graders try to solve the problem $2 + 4 = \underline{\hspace{1cm}} + 1$.

(a) One child says the answer is 6. Why might this error occur?
(b) A second child says the answer is 7. Why might this error occur?
(c) What don't the children in parts (a) and (b) understand?

Some children think an equation gives a set of computations to work out. They don't see each side of the equation as an expression for the same quantity. Research suggests that students who regularly make generalizations with equations are less likely to make the errors of LE 3.

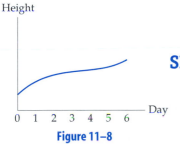

Figure 11–8

SLOPE

Slope measures the steepness of a line. Children may first describe the slope in graphs (Figure 11–8).

To be more precise, mathematicians assign numerical values to different slopes. What would be a good way to do that?

Classroom Connection

LE 4 Concept

(a) Which ramp is steeper, *A* or *B?*

(b) How could you measure the steepness of each ramp?
(c) The steepness can be measured with an angle or with two sides. Add to your answer to part (b).
(d) One way to measure the slope of the ramps is to use the two perpendicular sides. What relationship between the two perpendicular sides makes ramp *A* steeper than ramp *B?*
(e) How do the slopes of ramps *C* and *D* compare? What comparison of the lengths shows this?

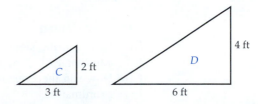

The slope compares the rise (vertical change) to the run (horizontal change). For lines with a positive slope, a steeper line has a vertical change that is larger relative to the horizontal change. For this reason, mathematicians define the slope as

$$\frac{\text{Rise}}{\text{Run}}$$

What makes the slope of a line negative? Imagine walking along a line from left to right (Figure 11–9). Lines rising from left to right (uphill) have a positive slope. Lines slanting downward from left to right (downhill) have a negative slope since the vertical change is negative and the horizontal change is positive. A horizontal line has a slope of 0.

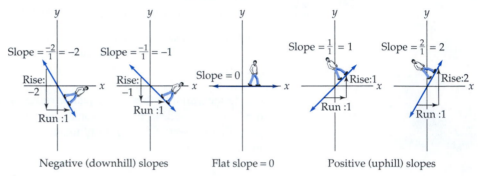

Negative (downhill) slopes   Flat slope = 0   Positive (uphill) slopes

Figure 11–9

LE 5 Concept
Classify the slope of each line as slope < 0, slope = 0, or slope > 0.

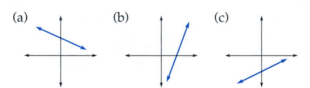

How can you compute the numerical value of the slope of a line using coordinates? Since the slope of a line is the same everywhere, you can compute $\frac{\text{rise}}{\text{run}}$ from any two points on the line! For example, (1, 2) and (4, 8) are on the line $y = 2x$, as shown in Figure 11–10.

Traveling from (1, 2) to (4, 8) involves moving vertically (up 6 in this case) and horizontally (right 3 in this case) a certain distance. In this example, the slope is $\frac{6}{3} = 2$.

You can compute the slope of a line joining two points by using the coordinates of the points.

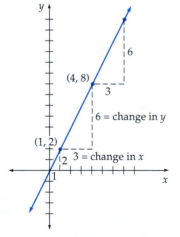

Figure 11–10

LE 6 Reasoning
(a) How can the slope of the line joining (1, 2) and (4, 8) be computed from the coordinates 1, 2, 4, and 8 (without making a drawing)?
(b) How can the slope of the line joining (x_1, y_1) and (x_2, y_2) be computed from the coordinates?

LE 6 concerns the following two-point slope formula.

> **The Two-Point Slope Formula**
>
> The slope of the line joining two points (x_1, y_1) and (x_2, y_2), in which $x_1 \neq x_2$, is
>
> $$m = \frac{\text{change in } y}{\text{change in } x} = \frac{y_2 - y_1}{x_2 - x_1}$$

Not all lines have a defined slope. Consider the following exercise.

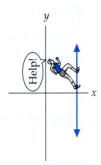

LE 7   Concept

(a) Why can't $x_1 = x_2$ in the two-point slope formula?
(b) Graph (2, 3) and (2, 4). What kind of line passes through these points?

When $x_1 = x_2$, the line passing through the two points is vertical. Imagine trying to walk up a vertical line (Figure 11–11)!

Vertical lines have an undefined (infinite) slope, corresponding to the fact that $x_1 = x_2$ makes the two-point slope formula come out undefined.

Figure 11–11

$y = mx$

Why are equations of the form $y = mx$ of interest? Many applications have formulas in which one variable (y) equals another variable (x) multiplied by a number ($y = mx$). Try the following exercise.

LE 8   Connection

(a) Suppose Juan walks 3 miles per hour. If y is the total distance he walks in x hours, what is a formula relating y to x?
(b) Find two points that satisfy your equation in part (a). Compute the slope from these two points.
(c) Valentina walks 4 miles per hour. If y is the total distance she walks in x hours, what is a formula relating y to x?
(d) Find two points that satisfy your equation in part (c). Compute the slope from these two points.
(e) In each case, how is the slope related to the equation?

As you may have guessed from the preceding exercise, the slope of $y = mx$ is m. The letter m is used because it is the first letter of the French word "monter," which means "to climb." What does the slope tell in an application?

LE 9   Connection

Waylon's walking formula is $y = 2x$, where y is the distance (miles) he walks in x hours.

(a) What is the slope of $y = 2x$?
(b) What does the slope tell about Waylon's walking?

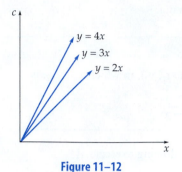

Figure 11–12

LE 10   Reasoning

Consider the graphs of $y = 3x$ (Juan), $y = 4x$ (Valentina), and $y = 2x$ (Waylon).

(a) Are the graphs of $y = 3x$, $y = 4x$, and $y = 2x$ (Figure 11–12) all the same shape?
(b) The graphs of $y = 3x$, $y = 4x$, and $y = 2x$ all pass through the point _____.
(c) In all three equations, $y = mx$ for a constant m. This equation can also be written

$$\frac{y}{x} = \underline{\hspace{2cm}}.$$

For each equation in LE 10, $\frac{y}{x}$ equals a constant. The two numbers in the ratio vary, but the rate stays the same. This means that distance walked and elapsed time are **directly proportional.** In symbols, when $\frac{y}{x} = k$, a constant, the equation $y = kx$ shows **direct variation.** Other examples of variables that are directly proportional include total pay and hours worked (when you earn the same pay for each hour), the number of dollars and the equivalent number of cents, and the length of a particular type of wire and its weight.

Equations of the Form $y = mx$

The graph of $y = mx$ for all real numbers x and constant m is a line with slope m through the origin $(0, 0)$, and y is directly proportional to x.

You have seen graphs relating distance walked and time. Consider the example of total pay and hours worked.

LE 11   Concept

Suppose you earn $9 per hour on a job. If you work a fraction of an hour, you earn that fraction of $9.

(a) Graph total pay (p) versus hours worked (t), with time on the x-axis.
(b) Is p proportional to t? Tell why or why not.
(c) Describe how p changes in relation to t.

$y = mx + b$

A rental car costs $89 per week plus $0.26 per mile. The cost C for a week's rental is $C = 0.26D + 89$, in which D is the distance driven (in miles). This equation has the form $y = mx + b$.

You have seen that an equation of the form $y = mx$ represents a line (or part of a line), with slope m passing through the origin, and that y is proportional to x. What happens when you add a constant b to the equation so that it has the form $y = mx + b$?

LE 12   Reasoning

(a) Graph $y = 2x$. (Use a graphing calculator if you have one.)
(b) On the same graph, plot three points that are solutions of $y = 2x + 2$, and connect them.
(c) On the same graph, plot three points that are solutions of $y = 2x - 2$, and connect them.
(d) What is the same about all three of the graphs?
(e) What is the relationship between the three equations you graphed?

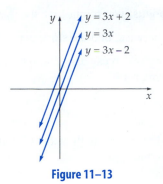

Figure 11–13

The graph of $y = mx + b$ is a line. What makes the line straight? It has a constant slope. For any line, the ratio of the change in y to the change in x is constant.

As LE 12 suggests, lines $y = mx + b$ with the same value of m have the same slope and are parallel. Consider the graphs of $y = 3x$, $y = 3x + 2$, and $y = 3x - 2$ shown in Figure 11–13. All three lines have the same slope, and they are parallel.

The graph of $y = 3x + 2$ is the same as the graph of $y = 3x$ translated up 2 units. The graph of $y = 3x - 2$ is the same as the graph of $y = 3x$ translated down 2 units. LE 12 and Figure 11–13 suggest the following two properties.

Slopes of Parallel Lines

Lines that have the same slope are parallel.

Relationship Between $y = mx$ and $y = mx + b$

The graph of $y = mx + b$ is the same as the graph of $y = mx$ translated b units vertically.

How can you tell where a line will cross the y-axis? The following exercise addresses this question.

LE 13   Connection

(a) A point on the y-axis has an x-coordinate of _____.
(b) Give the coordinates of the point where $y = mx + b$ crosses the y-axis.

LE 13 shows that b in $y = mx + b$ tells the y-coordinate of the point $(0, b)$ where the line intersects the y-axis. This point is called the **y-intercept.**

The Slope-Intercept Form of a Line

The graph of $y = mx + b$ is a line that has slope m and y-intercept $(0, b)$.

Some applications have a relationship of the form $y = mx + b$.

LE 14   Reasoning

A car has a ground clearance of 40 cm when it carries no weight. Every 10 kg of weight decreases the ground clearance by 1 cm.

(a) Complete the table.

Added weight, w (kg)	0	10	20	30
Ground clearance, g (cm)	40			

(b) A formula for g in terms of w is $g = 40 - $ _____. (*Hint:* Use the first two sentences in the exercise, or use the two-point slope formula.)
(c) Find the slope and g-intercept of your formula.
(d) What does the slope tell you about the relationship between ground clearance and added weight?
(e) Graph your equation from part (b).

To find a linear formula using two or more data points, you might first see if y is (directly) proportional to x. Then $\frac{y}{x} = m$, or $y = mx$. Otherwise, find the slope using the two-point slope formula. Then b is the y value when $x = 0$. You can also find b by substituting any point on the line into $y = mx + b$ and solving for b.

LINEAR FUNCTION TABLES

The application in LE 14 is a linear function. What kind of pattern in the table of points shows that it is linear? Try the following exercise to find out.

LE 15   Reasoning

(a) Tell whether each function is linear or nonlinear.

(1)

x	0	1	2	3
y	10	8	6	4

(2)

x	0	2	4	6
y	20	40	80	160

(3)

x	10	20	30	40
y	6	12	18	24

(b) Write a general rule for recognizing a table of points that represents a linear function.

You can recognize a linear pattern in data if y changes by a fixed amount when x changes by a fixed amount.

LE 16   Summary

Tell what you learned about linear functions in this lesson.

ANSWERS TO SELECTED LESSON EXERCISES

2.

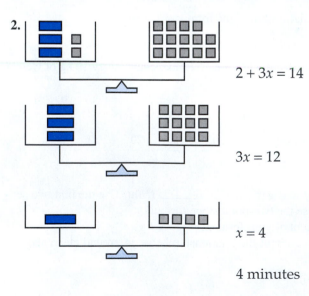

$2 + 3x = 14$

$3x = 12$

$x = 4$

4 minutes

3. (c) An equal sign represents a relation between numbers, not a signal to carry out a computation.

4. (e) They are equal. $\frac{2}{3} = \frac{4}{6}$

5. (a) Slope < 0 (b) Slope > 0 (c) Slope > 0

7. (a) It would make the fraction undefined.
   (b) Vertical

8. (a) $y = 3x$ (b) 3   (c) $y = 4x$ (d) 4
   (e) The slope $= m$.

9. (a) 2   (b) His average speed is 2 mph.

10. (a) Yes   (b) $(0, 0)$ (c) m

11. (b) Yes. $\frac{p}{t} = 9$.

(c) When t increases by 1 hour, p increases by \$9.

12. (d) Their slopes
(e) Their graphs are parallel lines.

13. (a) 0   (b) $(0, b)$

14. (a)

w	0	10	20	30
g	40	39	38	37

(b) $\frac{w}{10}$

(c) Slope $= \frac{-1}{10}$; g-intercept is $(0, 40)$

(d) The ground clearance decreases by 1 cm each time the weight increases by 10 kg.

(e)

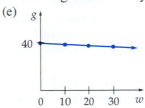

15. (a) (1) Yes, (2) no, (3) yes

11.2   HOMEWORK EXERCISES

1. Consider the following problem. "A family has 16 oz of cheese. The parents each have 5 oz of cheese. The 3 children each have the same amount. How much cheese does each child have?" Show how to solve the problem both algebraically and with a pictorial model for solving equations.

2. Show how to solve $2x + 3 = 11$ algebraically and with a pictorial model.

3. Solve $4x - 3 = 9$ using the cover-up method.
(a) Cover the term that contains the x. You have $\Box - 3 = 9$. What number goes in the box?
(b) If $\Box = 4x$, what is x?
(c) Try solving $-3x + 2 = 8$ using the cover-up method.

4. Solve the following equations using the cover-up method. (See the preceding exercise.)
(a) $-5x - 3 = 12$ (b) $8x + 4 = 9$

5. Tell how the following equations differ from one another.
(a) $x + 2x = 9$ (b) $x + y - x = y$ (c) $3x = y$

6. Tell how the following equations differ from one another.
(a) $xy = yx$ (b) $y = x + 7$ (c) $8x + 3 = 19$

7. A sixth grader solves $2x = 8$ by subtracting 2 from each side of the equation to obtain $x = 6$. What would you tell the child?

8. You ask a sixth grader for all the whole-number solutions to $x + y = 8$. The child does not include $x = 4$ and $y = 4$, because x and y cannot both have the same value. Is that right? If not, what would you tell the child?

9. Suppose $x < y$.
(a) On a standard number line, is x to the left of y or to the right of y?
(b) On a standard number line, is $x + 3$ to the left of $y + 3$ or to the right of $y + 3$? Tell why.
(c) On a standard number line, is $x - 3$ to the left of $y - 3$ or to the right of $y - 3$? Tell why.
(d) What properties of inequalities are suggested by parts (b) and (c)?

10. As you know, $4 < 5$.
(a) Multiply both sides by the same positive real number. How are the products related? Repeat this with two other positive numbers.
(b) Multiply both sides by the same negative real number. How are the products related? Repeat this with two other negative numbers.
(c) Write a multiplication property of inequalities.
(d) Develop a similar division property of inequalities.
(e) Inequalities can be solved the same way equations are solved except for what difference?

11. A swim club charges \$4 per visit or \$85 for a season pass. How many visits are necessary to make it cheaper to buy a season pass? Write an inequality for the situation, and solve it.

12. To qualify for an election, you must obtain at least 10,000 signatures. You now have 3,865 signatures. Write an inequality for the situation, and solve it.

13. How do builders describe the steepness of roofs? A roof with a pitch of "6 in 12" goes up 6 feet for every 12 feet of horizontal distance. Roof pitches usually run from "3 in 12" to "12 in 12."
    (a) What is the slope of a roof with a pitch of 6 in 12?
    (b) If a roof with slope 6 in 12 is 30 feet long, how much does it rise from the base of the roof to the top?

14. The Americans with Disabilities Act says that a wheelchair ramp cannot rise more than 1 inch for each 12 inches along the base of the ramp. What is the maximum slope of a legal ramp?

15. A building entrance is 3 ft above ground level. A wheelchair ramp with slope $\frac{1}{12}$ would have to be how long to go from the ground to the entrance?

16. A highway grade of 3% means that the highway rises 0.03 mile for every 1 mile of horizontal distance. How much does a highway with a 4% grade rise in 2 miles?

17. Classify the slope of each line as slope < 0, slope $= 0$, or slope > 0.

(a) 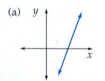 (b)   (c)

18. If a line goes down from left to right, what do you know about the slope?

19. What is the slope of each segment?

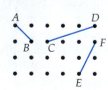

20. What is the slope of each segment?

21. Give the slope of each line.

(a) 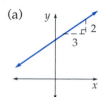 (b)   (c)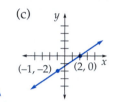

22. (a) In computing the slope, would $\frac{y_2 - y_1}{x_2 - x_1}$ and $\frac{y_1 - y_2}{x_1 - x_2}$ come out the same?

    (b) Explain the significance of your answer to part (a).

23. Water is being pumped into a tank. Readings are taken every 3 minutes.

Time (min)	0	3	6	9	12
Quarts of water	0	90	180	270	360

(a) Plot these 5 points.
(b) What is the slope of the line joining the 5 points?
(c) Estimate how much water is in the tank after 1 minute.
(d) At what rate is the water being pumped in?

24. The temperature in a room is measured every 30 minutes.

Time (min)	0	30	60	90	120
Temperature (°F)	54	56	58	60	62

(a) Plot these five points.
(b) What is the slope of the line joining these points?
(c) What is the average rate of change of the temperature during this time?

25. A marathon runner runs 12 km/hr.
 (a) Graph distance d (in kilometers) and time t (in hours) with t on the horizontal axis.
 (b) What is the slope of the graph?
 (c) What does the slope tell about the runner?
 (d) Why must $t \geq 0$?

26. A pump fills up an empty swimming pool at the rate of 4 m³/min.
 (a) A formula for the volume of water, V (in cubic meters), in the swimming pool, in terms of the elapsed time T (in minutes) is $V =$ _____.
 (b) Graph V vs. T.

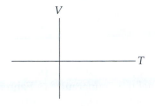

 (c) What is the slope of your graph?
 (d) What does the slope tell about the water?

27. A store estimates that the daily revenue (in dollars) $R = 4C$, in which C is the number of customers that day.
 (a) Describe in words what the formula says about customers and revenue.
 (b) Show why R is proportional to C.

28. A bug walks the shortest path from (0, 0) to (6, 12).
 (a) Draw the path on a graph.
 (b) Name two other points that the bug walks through. What pattern do you see in the coordinates?

29. What do the graphs of equations of the form $y = mx$ have in common?

30. Is a proportional relationship between x and y always a linear function?

31. Tell whether each table of values illustrates direct variation.

(a)
Number of apples	1	2	3	4
Cost ($)	$0.24	$0.48	$0.72	$0.96

(b)
Time (hr)	8	9	10	11
Temperature (°F)	54	58	60	62

32. Tell whether each table of values illustrates direct variation.

(a)
Age (yr)	4	5	6	7
Height (cm)	92	110	128	146

(b)
Time (min)	0	5	10	15
Depth (ft)	0	−20	−40	−60

33. You can time the interval between the lightning flash and the accompanying thunder to estimate how far it is from you. Every 5 seconds in the interval indicates an additional distance of about 1 mile.
 (a) A formula for distance D (in miles) in terms of time T (in seconds) is $D =$ _____.
 (b) Graph your formula.
 (c) What does the slope tell you about the relationship between distance and time?
 (d) The formula is based on the difference between the speed of light (lightning) and the speed of sound (thunder) in reaching us. The speed of light is 186,000 miles per second, and the speed of sound is 0.2 mile per second. Explain how the formula $D = \dfrac{T}{5}$ was obtained.

34. In scuba diving, the water pressure increases as a diver descends. For each additional 20 ft of depth, the pressure increases another 9 pounds per square inch (psi).
 (a) Assuming that water pressure P equals 0 at the water surface, a formula for P in terms of depth D is $P =$ _____.
 (b) Graph your formula.
 (c) What does the slope tell you about the relationship between water pressure and depth?
 (d) Suppose that you do not want to exceed 25 psi added pressure. Use your graph to estimate how deep you could go.

35. The graph of $y = 4x + 5$ is the same as the graph of $y = 4x$ translated _____.

36. What is the relationship between the graphs of $y = 5x + 2$ and $y = 5x + 6$?

37. How could you show that the two line segments on the square lattice are parallel?

38. Without graphing, tell whether $y = 3x - 2$ and $y = x - 2$ represent parallel lines.

39. (a) Name a point on the graph of $2x - y = 8$.
(b) Is $(2, -4)$ on the graph of $2x - y = 8$?
(c) How many different pairs of solutions does the equation $2x - y = 8$ have if x and y can be any real numbers?

40. Tell whether or not the graph of each equation passes through the origin. Explain how you know.
(a) $y = x + 4$ (b) $y = 3x$
(c) $y - 2x = 0$ (d) $y - x = 2$

 41. The cost of printing photos at Blurry's is $0.27 per picture plus $1.80 for development. So $C = 0.27P + 1.80$, in which C is cost and P is the number of pictures.
(a) Suppose that you got a bill for $3.69. How would you figure out, *without* using a formula, how many pictures were printed?
(b) Solve for P in the equation.
(c) Suppose that you bring in 12 negatives featuring your newborn nephew Drew Lalotte. Your total bill is $3.69 plus tax. Using one of the two formulas, compute how many pictures, P, of the baby they were able to print.

 42. Tickets for the Sonic Boom's rock concert sell for $12. The expenses for setting up the gig are $30,000. So the profit $P = 12 \cdot T - 30,000$, in which T is the number of tickets sold.
(a) Solve for T.
(b) Select one of the two formulas to solve the following problem. "The profit for the concert was $3,000. How many tickets were sold?"

43. Did you know that you can estimate the Fahrenheit temperature by counting cricket chirps?

(a) Let N equal the number of chirps per minute, and T equal the temperature in Fahrenheit. Derive a formula for T from the first frame of the cartoon.
(b) Solve for N.
(c) Use one of the formulas to figure out what the temperature is when crickets chirp 100 times per minute.
(d) It's 95°F outside. About how many times per minute are the crickets chirping?

44. A rectangle has a perimeter of 10 ft.
(a) Write a formula for the width W in terms of the length L.
(b) Is this formula exact or approximate?
(c) Use the formula to find the width when the length is 3 ft.
(d) What values can the length and width have?
(e) Graph W vs. L, with W on the vertical axis.

 45. Consider the following problem. "A job pays $6.10 an hour. If 23% is deducted for taxes, union dues, and benefits, how many hours must you work at this job in order to take home $100?" Devise a plan, and solve the problem.

 46. Your employer deducts 37% of your salary for taxes and benefits. You take home $311.98 for 40 hours of work. What does your job pay per hour?

47. What is the equation of the line?

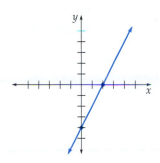

48. What is the equation of the line?

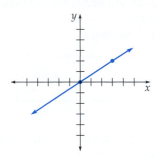

49.

Sales, x	0	100	200	300	400
Salary, y	100	110	120	130	140

A formula for salary in terms of sales is $y =$ _____.

50. Suppose that you want to cook a hot meal on a camping trip. The higher you go, the easier it is to boil water. At sea level, the boiling point of water is 212°F. For every additional 500 ft above sea level, the boiling point decreases by 1°.
(a) Complete the table.

Altitude, A (ft)	0	500	1000	1500
Boiling point, B (°F)	212	211		

(b) A formula for B in terms of A is
$B = 212 -$ _____.
(c) Graph your equation from part (b) with A on the x-axis.

51. The average weights of 6-year-old boys with selected heights are given in the table.

Height, H (in.)	42	44	46	48
Weight, W (lb)	43	49	55	61

A formula for W is $W =$ _____. (*Hint:* Find the slope.)

52.

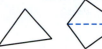

Any polygon can be divided into a *minimum* number of triangular regions by its diagonals. The following graph shows the relationship between S (the number of sides of a polygon) and T (the minimum number of triangular regions formed by diagonals).

(a) Give the coordinates of each point.
(b) Write a formula relating T to S.

53. The NCTM Algebra Working Group (1995) devised a problem like the following. Elaine makes square swimming pools of all different sizes with tile borders. Pools 1 and 2 are shown below.

Pool 1   Pool 2

(a) Draw Pool 3.
(b) Complete the following table.

Pool Number (x)	1	2	3	4
Number of Tiles (y)				

(c) Graph the 4 points. What pattern do you see?
(d) Find an equation relating x and y.
(e) Does part (d) involve induction or deduction?
(f) Use your equation from part (d) to find the number of tiles needed for Pool 12.

54. Elaine makes rectangular swimming pools of all different sizes with tile borders. Pools 1 and 2 are shown below.

Pool 1   Pool 2

(a) Draw Pool 3.
(b) Complete the following table.

Pool number (x)	1	2	3	4
Number of tiles (y)				

(c) Graph the 4 points. What pattern do you see.
(d) Find an equation relating x and y.
(e) Use your equation from part (d) to find the number of tiles needed for Pool 12.
(f) Does part (e) involve induction or deduction?

55. Without graphing, tell whether each set of points lies on a straight line.

(a)

x	1	2	3	4
y	5	7	9	11

(b)

x	5	10	15	20
y	1	2	4	7

(c)

x	0	3	6	9
y	11	8	5	2

56. Without graphing, tell whether each set of points lies on a straight line.

(a)

x	0	1	2	3
y	4	9	16	25

(b)

x	0	8	16	24
y	1	3	5	7

(c)

x	1	2	3	4
y	30	20	10	0

57. A seventh grader notices that the following function table has a straight line graph. She says that if the y-values form an arithmetic sequence, then the graph will be a straight line.

x	1	2	3	4
y	5	7	9	11

(a) Is this right?
(b) If this is right, explain why it works. If this is not right, tell the conditions under which it would not be true.

58. One eighth grader says a line with slope -1 has a greater slope than a line with slope -2. A second eighth grader says the opposite is true. What would you tell the children?

Extension Exercises

59.

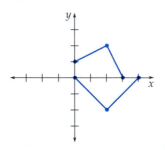

(a) Find the slopes of each pair of perpendicular line segments.
(b) Make a conjecture about the relationship between the slopes of perpendicular lines.

60. Use the properties of similar triangles to show that the slopes m_1 and m_2 of perpendicular lines L_1 and L_2, respectively, satisfy $m_1 m_2 = -1$.

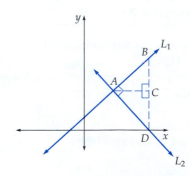

61. Motion geometry can be used to show why two perpendicular lines (which are not horizontal and vertical) have slopes with a product of -1. In the diagram that follows, the legs of $\triangle T$ show the slope of p. Answer parts (a) and (b) to see why the slopes of lines p and q have a product of -1.

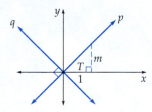

(a) What is the slope of line p?
(b) Rotate line x and $\triangle T$ counterclockwise 90°. Label the image of $\triangle T$ as $\triangle T'$. What is the slope of line q?

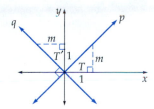

62. Use the result of the preceding exercise to explain why the two segments on the following square lattice form a right angle.

63. (a) Graph $y = 2x$.
(b) Draw the image of $y = 2x$ after a reflection through the y-axis.
(c) What is the equation of the image line?

64. (a) Graph $y = 3x - 2$.
(b) Draw the image of $y = 3x - 2$ after a reflection through the y-axis.
(c) What is the equation of the image line?

65. Use a graphing calculator (or graphing software) to investigate the relationship between the graphs of $y = mx + b$ and $y = -mx - b$ by graphing pairs of equations such as $y = 2x + 3$ and $y = -2x - 3$.

66. (a) Graph $y = 2x$.
(b) Draw the image of $y = 2x$ after a 90° clockwise rotation around $(0, 0)$.
(c) What is the equation of the image line?

Technology Exercises

67. (a) Use dynamic geometry software to construct $\overleftrightarrow{AB}$ and $\overleftrightarrow{CD}$ with an intersection point E.
(b) Now construct $\overleftrightarrow{CF}$ parallel to $\overleftrightarrow{AB}$.
(c) Measure the slopes of $\overleftrightarrow{AB}$ and $\overleftrightarrow{CF}$. How do they compare?

68. (a) Use dynamic geometry software to construct $\overleftrightarrow{AB}$. Now construct a point C above $\overleftrightarrow{AB}$.
(b) Construct a line through C that is perpendicular to $\overleftrightarrow{AB}$.
(c) Measure the slopes of the two lines. How do they compare?
(d) Rotate both lines, keeping them perpendicular, and record their slopes a second time. How do they compare? Make a generalization about slopes of perpendicular lines.

11.3 NONLINEAR FUNCTIONS

NCTM Standards

- identify and describe situations with constant or varying rates of change and compare them (3–5)
- investigate how a change in one variable relates to a change in a second variable (3–5)
- identify functions as linear or nonlinear and contrast their properties from tables, graphs, or equations (6–8)
- use graphs to analyze the nature of changes in quantities in linear relationships (6–8)

The idea of using graphs to represent functions is about 350 years old. In the seventeenth century, two Frenchmen, René Descartes (1596–1650) and Pierre de Fermat (1601–1665) (Figure 11–14), had the brilliant idea of using coordinates to relate algebraic equations (such as $y = 2x$) to geometric shapes (such as a line).

René Descartes   Pierre de Fermat

Figure 11–14

Descartes was a brilliant but sickly youngster who entered a prestigious private school at age 8. Because of his ill health, his teachers allowed him to lie in bed in the morning for rest and reflection. Descartes continued this practice throughout his life and came up with many of his philosophical and mathematical ideas during his morning meditations.

Although Fermat was one of the great mathematicians of his time, he was a lawyer who did mathematics as a hobby! He was not well known during his lifetime because he did not try to publish his work. Fermat's achievements include major discoveries in number theory, the invention (along with Descartes) of coordinate geometry, and helping to lay the groundwork for both probability and calculus.

The graph of a function $y = mx + b$ is a line. What about other functions, such as $y = x^2$ and $y = 2^x$?

LE 1   Opener

(a) Graph $y = x^2$ for $x = -2, -1, 0, 1,$ and 2. Is it a linear function?

(b) Graph $y = 2^x$ for $x = -1, 0, 1, 2,$ and 3. Is it a linear function?

(c) Describe how y changes in relation to x in each graph.

(d) Does either graph have a constant rate of change (slope)?

Linear functions have a constant rate of change (slope). The graphs in LE 1 are nonlinear functions with rates of change (slope) that are not constant.

QUADRATIC FUNCTIONS

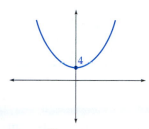

Figure 11–15

All equations of the form $y = mx + b$ have straight-line graphs. What happens if x is squared, as in the equation $y = x^2$? The graph of $y = x^2$ in LE 1 has the shape shown in Figure 11–15. Consider some other equations with x^2.

LE 2   Reasoning

(a) Graph $y = x^2 - 2$ after completing the following table, or use a graphing calculator.

x	−2	−1	0	1	2
y					

(b) How is this graph related to the graph of $y = x^2$?

(c) Figure 11–16 shows a graph of $y = x^2 + 4$. Use the graphs in this exercise to write a generalization about the relationship between the graph of $y = x^2$ and the graph of $y = x^2 + c$.

LE 2 illustrates how $y = x^2 + c$ is the curve $y = x^2$ translated c units vertically. The preceding graphs $y = x^2$, $y = x^2 - 2$, and $y = x^2 + 4$ are all the same shape: a parabola.

A **quadratic function** has the form $y = ax^2 + bx + c$, where $a \neq 0$. The graph of a quadratic function is a parabola that opens up or down. Parabolas model the path of a ball when it is thrown and the shape of the cable on some suspension bridges (Figure 11–17).

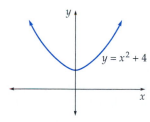

Figure 11–16

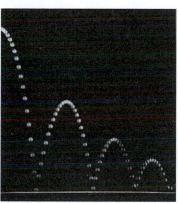 

Figure 11–17

Photo of bouncing ball from PSCC, Physics, 2d ed., 1965; © Education Development Center, Inc., and D.C. Heath & Co. Photo of Golden Gate Bridge Courtesy of Library of Congress.

Parabolas with the form $y = x^2 + c$ (such as $y = x^2$, $y = x^2 - 4$, and $y = x^2 + 2$) all open up (Figure 11–18). What happens with parabolas of the form $y = -x^2 + c$?

Figure 11–18

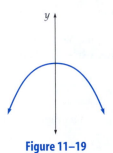

Figure 11–19

LE 3 Skill

(a) Graph $y = -x^2$ for $x = -3, -2, -1, 0, 1, 2, 3$. (You may use a graphing calculator.)
(b) How is $y = -x^2$ related to $y = x^2$?

Parabolas of the form $y = -x^2 + c$ open down (Figure 11–19). What kinds of applications are modeled by parabolas?

LE 4 Reasoning

(a) A formula for the area A of a square in terms of the length L of a side is
$A = $ _____.
(b) Which graph illustrates this relationship with length on the horizontal axis?

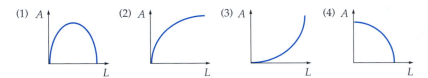

Classroom Connection

LE 5 Concept

An eighth grader is not sure which of $y = 2x^2$, $y = (2x)^2$, and $y = 2(x^2)$ are the same quadratic function. What would you tell the child?

EXPONENTIAL FUNCTIONS

Figure 11–20

Exponential functions are one of the most useful types of function other than linear functions. First, consider the simplest examples of these functions.

The graph of $y = 2^x$ in LE 1 has the shape shown in Figure 11–20. Consider another equation in which the exponent is a variable.

Discussion

LE 6 Reasoning

(a) Graph $y = 3^x$ after completing the following table, or use a graphing calculator.

x	−1	0	1	2	3
y					

(b) Describe the similarities between the tables for $y = 2^x$ and $y = 3^x$.
(c) Describe the similarities between the graphs of $y = 2^x$ and $y = 3^x$.
(d) Figure 11–21 shows a graph of $y = 4^x$. Use the graphs in this exercise to write general characteristics of the graph of $y = a^x$, where $a > 0$. Include a description of the change pattern (include the word "increasing").

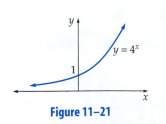

Figure 11–21

The function $y = a^x$, where $a > 0$ and $a \neq 1$, is called an **exponential function,** since the exponent is a variable. The graph of $y = a^x$, where $a > 1$, increases more rapidly as x increases, and it has a y-intercept at $(0, 1)$. In a table for $y = a^x$, if x increases by the same amount each time, y will multiply by the same amount each time.

What kind of applications are modeled by exponential functions? Exponential functions with $a > 1$ are used to analyze population growth and compound interest. Exponential functions with $0 < a < 1$ are decreasing functions used to model radioactive decay.

LE 7   Reasoning

(a) The population of the world was about 6 billion in 1999. At the current rate of growth, it will double about every 40 years. Assume this trend continues for the next 120 years, and complete the following table.

t (years after 1999)	0	40	80	120
P (population in billions)	6	12		

(b) Does the table have an exponential growth pattern?
(c) Graph the data from part (a).
(d) The equation for this function is $P = 6(2^{t/40})$. Estimate the world population in the year 2015.
(e) If you have a graphing calculator, graph this function and estimate when the world population will be 10 billion.

CHANGE PATTERNS IN LINEAR AND NONLINEAR FUNCTIONS

You have studied the change patterns in linear and exponential function tables. See if you can recognize the type of function represented by a table.

LE 8   Connection

Consider the population data for three different towns.

t (years after 1970)	0	10	20	30
P (population in thousands)	6	8	9	13
Q (population in thousands)	6	8	10	12
R (population in thousands)	6	12	24	48

(a) Describe the growth patterns of *P*, *Q*, and *R* in words.
(b) Tell whether *P*, *Q*, and *R* are linear, exponential, or neither.

LE 9   Summary

Tell what you learned about nonlinear functions in this section.

ANSWER TO SELECTED LESSON EXERCISES

5. Review order of operations. The first and third functions are the same.

7. (b) Yes.   (d) 7.9 billion   (e) In 2028

8. *P* is neither, *Q* is linear, and *R* is exponential.

11.3 HOMEWORK EXERCISES

Basic Exercises

1. (a) Graph $y = x^2 + 1$. Use $x = -3, -2, -1, 0, 1, 2,$ and 3.
(b) How is this graph related to $y = x^2$?
(c) For what values of x, $-3 \leq x \leq 3$, is it true that as x increases, y increases?

2. (a) Graph $y = -x^2 - 3$ for $x = -2, -1, 0, 1,$ and 2.
(b) How is this graph related to $y = -x^2$?

3. How is the graph of $y = x^2 + 5$ related to the graph of $y = x^2 + 1$?

4. How is the graph of $y = -x^2 - 1$ related to the graph of $y = x^2 + 1$?

5. When you drop a heavy object, its speed increases at every instant until it hits the ground. Which of the following graphs shows the correct relationship between h, the height above the ground, and t, the time elapsed since you dropped the object?

(a)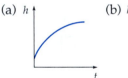
(b)
(c)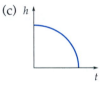

(d)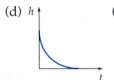
(e)
(f)

6. The graph of the price p of a television vs. the length of its screen diagonal d is approximately a parabola. Which graph is correct?

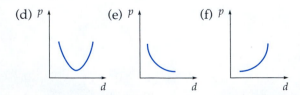
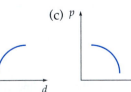

7. An object is tossed up in the air with a velocity of 80 ft/s. Its height h, in feet, after t seconds is $h = -16t^2 + 80t$.
(a) Find the height after 3 seconds.
(b) Graph h vs. t, with h on the vertical axis.
(c) Using your graph or a calculator, estimate when the object is at a height of 50 ft.

8. Stopping distance $D = 1.1S + 0.05S^2$, in which D is distance in feet and S is speed in miles per hour (mph). ($1.1S$ is the reaction distance, and $0.05S^2$ is the braking distance.)
(a) Find the stopping distance for a car traveling at 30 mph.
(b) Find the stopping distance for a car traveling at 50 mph.
(c) Graph D vs. S, with D on the vertical axis.
(d) Using your graph or a calculator, estimate the speed of a car that took 136 ft to stop.
(e) This formula is only an approximation. What factors in addition to speed should affect stopping distance?

9. Revenue R from ticket sales for a concert depends on the price charged, according to $R = -50P^2 + 600P$.
(a) Graph this equation.
(b) Approximately what price gives the maximum revenue?

10. A ball is dropped from a height of 144 ft. Its height H (in feet) at time T (in seconds) is $H = 144 - 16T^2$.
(a) Graph this equation.
(b) What shape is the graph?
(c) When does the ball hit the ground?
(d) How high does the ball go?

11. Write an equation for the graph.

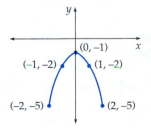

12. Use a trapezoid to estimate the shaded area.

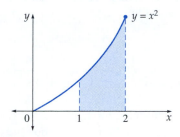

13. Sketch the graph of $y = 5^x$, and describe the characteristics it has in common with $y = 2^x$.

14. (a) Sketch the graph of $y = \left(\dfrac{1}{2}\right)^x$ for $x = -2, -1, 0, 1$ and 2.
   (b) How is this graph related to $y = 2^x$?

 15. An island has 50 moths, and the population is doubling each year.
   (a) Complete the table.

t (years)	0	1	2	3
m (moths)	50			

   (b) Which of the following equations fits the data?

   (1) $m = 50t$ (2) $m = 50^t$

   (3) $m = 50 \cdot (2^t)$ (4) $m = 2^t$

   (c) Use the equation from part (b) to predict the moth population after 1.5 years.

 16. (a) *E. coli* bacteria can double in population every 20 minutes. Suppose you begin with 100 bacteria. Complete the following table showing the bacteria population p after t minutes.

t (minutes)	0	20		
p (hundreds)	1			

   (b) Which of the following equations fits the data?

   (1) $p = 20^t$ (2) $p = 2^{t/20}$

   (3) $p = \dfrac{t}{20}$ (4) $p = \dfrac{(2^t)}{20}$

(Continued in the next column)

(c) Use the equation from part (b) to predict the bacteria population after 35 minutes.

 17. Norma deposits $3,000 in a bank account that earns 6% interest compounded annually. The exponential function $A = 3000(1.06)^t$ gives the balance A after t years.
   (a) How much money will be in the account after 20 years?
   (b) When will the balance be $4,000?

 18. All living things contain carbon. Some scientists use radioactive carbon-14 to estimate the age of fossils. Carbon-14 has a half-life of 5,600 years, meaning that *half its mass decays every 5,600 years.*
   (a) Complete the following table.

t (years)	0	5600	11,200
y (units of C-14)	1	$\dfrac{1}{2}$	

   (b) The data fit the equation $y = 2^{-\left(\frac{t}{5600}\right)}$. If you have a graphing calculator, graph this function and estimate the age of a fossil that has 0.35 units of C-14.

19. P, Q, and R are functions of x. Tell whether P, Q, and R are linear, exponential, or neither.

x	0	10	20	30
P	5	10	20	40
Q	2	3	5	8
R	-3	-1	1	3

20. P, Q, and R are functions of x. Tell whether P, Q, and R are linear, exponential, or neither.

x	0	10	20	30
P	14	11	8	5
Q	1	2	1	2
R	5	15	45	135

21. Another type of nonlinear function is useful in modeling postage rates and sales tax rates. The graph shows the first-class postage rates for different weight letters in 2003. This is an example of a **step function.**

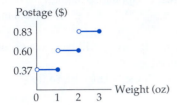

Postage ($)

Weight (oz)

(a) What is the rate for a letter that weighs 0.6 oz?
(b) Name two different weights that cost $0.60.
(c) A sales tax graph will also be a step function. Make a graph for a state that has a sales tax of 5%, where x is the amount in cents and y is the amount of tax in cents. Use $0 \le x \le 60$. Assume all taxes are rounded *up* to the nearest cent.

22. A parking garage charges $2 for parking up to 1 hour and $1 for each additional hour (or part of an hour). Make a graph of the parking fee y for x hours.

Extension Exercises

23. Suppose that you want to fence in a rectangular area of 30 m². Complete the following to investigate the length and width such a rectangular area could have.
(a) Fill in the table.

L	3	5	7	
W	10			

(b) Write a formula for W.
(c) Graph your formula.
(d) Describe the shape of the graph. (It is called a **hyperbola.**)
(e) What are the possible values of L and W?

 24. You plan to go on a 200-mile car trip. The time it takes is related to the rate at which you drive.
(a) Fill in the table.

Rate, R (mph)	20	30	40	50	60
Time, T (h)	10				

(b) Graph R (vertical axis) versus T.

25. The time needed to cook a piece of chicken in a microwave oven at a setting of 500 watts is about 6 minutes. Additional information is provided in the table.

W (watts)	200	300	400	500	600
Time, T (min)	15	10	7.5	6	

(a) Write a formula for T.
(b) How long would it take to cook the chicken at 600 watts?
(c) Graph the formula for T vs. W.

26. (a) In a vacuum, the product of the pressure P and the volume V is constant. For example, PV could always equal 4. Graph $PV = 4$, using positive numbers for P and V.
(b) The graph shows that, as P increases, V _____.

 27. Compare $x^2 - 9$ and $(x + 3)(x - 3)$ by doing the following.
(a) Plot $y = x^2 - 9$ and $y = (x + 3)(x - 3)$ with a graphing calculator.
(b) Compare the values of $x^2 - 9$ and $(x + 3)(x - 3)$ for $x = -2, -1, 0, 1,$ and 2.
(c) Multiply out $(x + 3)(x - 3)$.
(d) Which part or parts—(a), (b), or (c)—*prove* that $x^2 - 9 = (x + 3)(x - 3)$?

 28. Use a graphing calculator to estimate the solutions to
(a) $x^2 - 9x + 4 = 0$
(b) $x^2 - 9x + 4 = -8$

29. All lines have constant slope. Consider a line through (x_1, y_1) with slope m. Must its equation have the form $y = mx + b$?
(a) Let (x, y) represent any other point on the line. The two-point slope formula says that the slope from (x_1, y_1) to (x, y) is _____. Set this equal to m.
(b) Solve the last equation in part (a) for y.

11.4 SYSTEMS OF EQUATIONS AND MULTIPLE REPRESENTATIONS

NCTM Standards

- model problem situations with objects and use representations such as graphs, tables, and equations to draw conclusions (3–5)
- relate and compare different forms of representation for a relationship (6–8)
- model and solve contextualized problems using various representations such as graphs, tables, and equations (6–8)

SYSTEMS OF EQUATIONS

Some applications can be modeled with two equations. When two or more equations are considered together, they form a **system of equations.** One sometimes looks for a set of solutions that works with all of the equations.

Discussion

LE 1   Reasoning

Consider the following problem. "Suvi leaves her house and walks north on a trail at Noon. She walks at a speed of 3 miles per hour. Panu leaves at 2 P.M. and wants to catch up, so he runs at 6 miles per hour. At what time will he catch up?"

(a) If t is the number of hours after 12 noon, write a formula for the distance traveled y in terms of t, for each person.
(b) Make a table of values for each person for different values of t to find out when they will meet.
(c) Graph the two equations from part (a) to find out when they will meet. Use a graphing calculator if it is available.
(d) Solve the system of two equations algebraically to find the solution.

What method did you use to solve part (d) in LE 1? What problems can children solve to prepare them for the algebraic methods of solving a system of two equations?

Classroom Connection

LE 2   Connection

Consider the following problem. "Suppose two apples and one pear cost 59 cents and three apples cost 54 cents. Find the unit cost of an apple and the unit cost of a pear."

(a) Describe a method for solving this problem.
(b) Write a system of two equations that represents this problem, where x is the unit cost of an apple and y is the unit cost of a pear.
(c) Solve your system of equations from part (b) with the method you used in part (a).

LE 2 illustrates a situation that could help develop the substitution method of solving a system of equations. The next lesson exercise concerns the addition/subtraction method for solving a system of equations.

Classroom Connection

LE 3   Connection

Consider the following problem. "Suppose two grapefruits and one orange cost $0.92 and two grapefruits and four oranges cost $1.58. Find the unit cost of an orange and the unit cost of a grapefruit."

(a) Describe a method for solving this problem without writing equations.
(b) Write a system of two equations that represents this problem.
(c) Solve your system of equations from part (b) with the method you used in part (a).

Did you solve LE 3 by subtracting one equation from the other, or did you come up with a different method?

WHICH REPRESENTATION IS BEST?

In this chapter, you have solved problems involving words, symbols, tables, and graphs. What are the advantages and drawbacks of each form? The following problem uses all four representations together so you can compare them.

Discussion

LE 4   Connection

Four companies offer long-distance phone plans.
Frayed Fibers charges $6 per month plus $0.05 per minute.
The graph (Figure 11–22) shows the monthly cost of different numbers of minutes with Static Lines. The following table shows the monthly cost of different numbers of minutes with Hung Up.

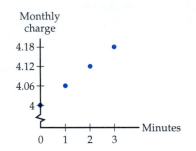

Monthly charge

Figure 11–22

Number of minutes	0	10	20	30	40	50	60
Monthly cost ($)	0	0.80	1.60	2.40	3.20	4	4.80

Loose Wires' monthly charge is $8 + 0.03n$ dollars for n minutes of calls.
In parts (a)–(d), for which company is it easiest to answer each of the following questions?

(a) What is the monthly cost for someone who makes 40 minutes worth of calls?
(b) What is the monthly cost for someone who makes 80 minutes worth of calls?
(c) What is the monthly cost for someone who makes 200 minutes worth of calls?
(d) Describe the monthly rate in words.
(e) Which two companies are easiest to compare?
(f) Describe the advantages and drawbacks of each of the four representations.

The verbal representation is useful for communicating everyday information about a problem and stating the final results. However, the verbal representation can be complicated and ambiguous.

The graphical representation gives the best visual representation and can make trends easy to recognize. However, the graphical representation lacks precision and makes it more difficult to read specific data values.

The tabular representation gives a set of concrete data that is easy to read and understand. However, it may be difficult to generalize or go much beyond this information.

The symbolic representation is straightforward, general, and precise. However, it is more abstract than the other representations, so its everyday meaning may be less clear. Research suggests that learning numerical and graphical representations will improve a student's understanding of equations and expressions.

LE 5   Reasoning

You want to compare the cost of two refrigerators. The Ice Queen costs $650 and has an average annual energy cost of $52. The Slush King costs $550 and has an average annual energy of cost of $64.

(a) For each model, write an equation for the total cost after T years.

(b) Assume each refrigerator could last up to 20 years. Graph both equations for this time interval. Use a graphing calculator if you have one.

(c) For what length of time is each model likely to cost less money?

A spreadsheet or graphing calculator can be used to generate a table that compares the costs of two or more alternatives.

LE 6   Reasoning

You decide to rent a copier from either Crisp Copy or Repro-Man. Crisp Copy charges $60 per month plus $0.02 per copy. Repro-Man charges $40 per month plus $0.025 per copy.

(a) Set up a computer spreadsheet (or use a calculator and make a table) as follows. Enter 1000, 2000, 3000, . . . , 10,000 in column A. The top part of the spreadsheet follows.

	A	B	C
1	Copies	Crisp Copy	Repro-Man
2	1000		
3	2000		
4	3000		
5	4000		

If you have a graphing calculator, enter equations for the two copier stores in y_1 and y_2. Then enter 1000 to 10000 for x. (Skip ahead to part (c).)

(b) Enter appropriate formulas in cells B2 and C2. Copy these formulas in rows 3 through 11.

(c) Tell how you would decide where to rent a copier.

LE 7   Reasoning

A store sells T-shirts for children and adults. The store orders adult T-shirts for $4 each and children's T-shirts for $3 each. They sell adult shirts for $15 and children's shirts for $9. It costs $190/day for salaries, rent, insurance, and utilities. Suppose they sell C children's shirts and A adult's shirts one day.

(a) Write an expression for the total cost of running the business for that day.

(b) Write an expression for the total revenue for that day.

(c) Write an expression for the profit for that day.

(d) Give a possible number of adult and children's shirts they could sell in a day to break even.

LE 8   Summary

Tell what you learned about multiple representations in this section.

ANSWERS TO SELECTED LESSON EXERCISES

1. (a) $y = 3t$, $y = 6(t - 2)$ (b) 4 P.M.

2. (b) $2x + y = 59$, $3x = 54$ (c) $x = 18¢$, $y = 23¢$

3. (b) $2x + y = \$0.92$, $2x + 4y = \$1.58$
(c) $x = \$0.35$, $y = \$0.22$

5. (a) $C_1 = 650 + 52T$, $C_2 = 550 + 64T$
(c) The Ice Queen would be cheaper for up to $8\frac{1}{3}$ years. Beyond $8\frac{1}{3}$ years, the Slush King would be cheaper.

6. (b)

Copies	Crisp Copy	Repro-Man
4000	140	140

(c) Repro-Man for less than 4,000 copies in a month; Crisp Copy for more than 4,000 copies in a month

7. (a) $3C + 4A + 190$ (b) $9C + 15A$
(c) $6C + 11A - 190$

11.4   HOMEWORK EXERCISES

Basic Exercises

1. Consider the following problem. "Compare two deals for T-shirts: either \$7 per shirt or \$50 plus \$4 per shirt."
(a) Write a formula for the total cost C for x shirts for each deal.
(b) Make a table of values for each deal for different values of x, and find out what circumstances make each deal better.
(c) Graph the two equations from part (a) to find out what circumstances make each deal better. Use a graphing calculator if it is available.
(d) Solve the system of two equations algebraically to find the solution.
(e) Which method did you prefer?

2. Consider the following problem. "You want to produce a 500-page book. Company A charges \$0.05 per page for typesetting and \$2 for binding each book. Company B charges \$0.03 per page for typesetting and \$2.50 for binding each book."
(a) Write a formula for the total cost C to produce x books for each company.
(b) Make a table of values for each company for different values of x and find out what circumstances make each deal better.
(c) Graph the two equations from part (a) to find out what circumstances make each deal better. Use a graphing calculator if one is available.

(Continued in the next column)

(d) Solve the system of two equations algebraically to find the solution.
(e) Which method did you prefer?

3. Consider the following problem. "Suppose two grapefruits and three oranges cost \$2.07, and two oranges cost \$0.78. Find the unit cost of an orange and the unit cost of a grapefruit." Describe a method for solving this problem.
(a) Describe a method for solving this problem without writing equations.
(b) Write a system of two equations that represents this problem.
(c) Solve your system of equations from part (b) with the method you used in part (a).

4. Consider the following problem. "Suppose two lemons and four peppers cost \$1.98, and three lemons cost \$0.75. Find the unit cost of a lemon and the unit cost of a pepper." Describe a method for solving this problem.
(a) Describe a method for solving this problem without writing equations.
(b) Write a system of two equations that represents this problem.
(c) Solve your system of equations from part (b) with the method you used in part (a).

5. Consider the following problem. "Suppose five apples and two pears cost $2.37, and two apples and two pears cost $1.32." Find the unit cost of an apple and the unit cost of a pear."
   (a) Describe a method for solving this problem without writing equations.
   (b) Write a system of two equations that represents this problem.
   (c) Solve your system of equations from part (b) with the method you used in part (a).

6. Consider the following problem. "Suppose four onions and three tomatoes cost $2.48, and four onions and five tomatoes cost $3.36." Find the unit cost of an onion and the unit cost of a tomato."
   (a) Describe a method for solving this problem without writing equations.
   (b) Write a system of two equations that represents this problem.
   (c) Solve your system of equations form part (b) with the method you used in part (a).

7. The three scales give information about the weight of three different-shaped blocks.

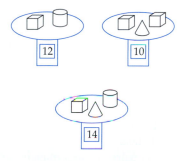

   (a) Find the weight of each shaped block.
   (b) Tell how you solved the problem.

8. The three scales give information about the weight of three different-shaped blocks.

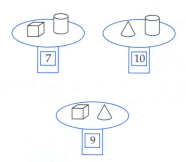

   (a) Find the weight of each shaped block.
   (b) Tell how you solved the problem.

9. You are choosing between two health insurance plans that cost the same amount. One pays 80% of all expenses *beyond* the first $500 (a $500 deductible). The second pays 60% of all expenses, with no deductible.
   (a) Write a formula for the amount A that each policy pays on a claim of C dollars.
   (b) Graph both equations on the same graph. (You may use a graphing calculator for parts (b) and (c).)
   (c) Find the intersection point.
   (d) Use the intersection point and state the conditions that make one plan more economical than the other.
   (e) What factors besides price might you consider in choosing a plan?

10. You are planning to buy a new air conditioner. Choice A sells for $400 and costs $30 per month to run. Choice B sells for $550 and is more energy-efficient, so it costs $20 per month to run.
   (a) Write a formula for the cost C of purchasing and operating each air conditioner for M months.
   (b) Graph both equations on the same graph. (You may use a graphing calculator for parts (b) and (c).)
   (c) Find the intersection point on the graph.
   (d) Find the intersection point algebraically.
   (e) Use the intersection point and state the conditions that make one air conditioner a better buy than the other.
   (f) What factors besides price might you consider in deciding which air conditioner to buy?

11. The Screeching Tires have been offered two different recording contracts. Contract A includes $20,000 for signing and $1 royalty per CD. Contract B includes $5000 for signing and $2 royalty per CD.
   (a) Write a formula for the total pay (y) for each contract in terms of the number of CDs sold (x).
   (b) Graph both equations on the same graph. (You may use a graphing calculator in parts (b) and (c).)
   (c) Find the intersection point.
   (d) Use the intersection point and state the conditions that make each contract the higher-paying one.

12. You are offered two sales jobs. Job A pays $800 per week plus 5% commission. Job B pays $300 per week plus 15% commission.
    (a) Write a formula for the weekly pay for each job.
    (b) Graph both equations on the same graph. (You may use a graphing calculator in parts (b) and (c).)
    (c) Find the intersection point.
    (d) Use the intersection point and state the conditions under which each job pays more than the other.
    (e) What factors besides pay might you consider in choosing a sales job?

13. Tickets for a college basketball game were $10 for adults and $4 for children. The paid attendance was 2460, and total ticket sales were $21,060. How many adult tickets and how many children's tickets were sold?

14. A sample of 1400 mg of automobile exhaust contains 36% carbon (from CO and CO_2). If CO is 43% carbon and CO_2 is 27% carbon, how many milligrams of exhaust are CO, and how many milligrams are CO_2?

15. You must choose between two different rental car companies for a 3-day trip. Drive-U-Crazy charges $39.95/day with unlimited mileage. Used-a-Bit charges $24.95/day and gives 100 free miles/day; mileage beyond that costs $0.10/mile. The sales tax rate at both places is 8%.
    (a) Set up a computer spreadsheet (or use a calculator and make a table) as follows. Enter 500, 700, 900, 1100, 1300, and 1500 in column A. The top part of the spreadsheet follows.

	A	B	C
1	Miles	Drive-U-Crazy	Used-a-Bit
2	500		
3	700		
4	900		

    (b) Enter appropriate formulas in cells B2 and C2. Copy these formulas in rows 3 through 7.
    (c) Tell how you would decide which car to rent.

16. You are offered two sales jobs. Fast Talkers pays $180/week plus 5% commission on sales. Wheeler Dealers pays $225/week plus 3% commission on sales.

    (a) Set up a computer spreadsheet (or use a calculator and make a table) as follows. Enter 1000, 2000, 3000, 4000, and 5000 in column A. The top part of the spreadsheet follows.

	A	B	C
1		Total	Pay
2	Weekly sales	Fast Talkers	Wheeler Dealers
3	1000		
4	2000		

    (b) Enter appropriate formulas in cells B3 and C3. Copy these formulas in rows 4 through 7.
    (c) Tell how you would decide which job to take.

17. A manufacturer of television sets has weekly expenses $E = 1500 + 100T$ dollars for manufacturing T televisions. The total revenue (money received) from selling T televisions is $180T$.
    (a) What is the *profit or loss* on manufacturing and selling 100 sets in a week?
    (b) What is the *profit or loss* on manufacturing and selling 20 sets in a week?
    (c) What number of sets is the break-even point?

18. A manufacturer of shoes has weekly expenses $E = 800 + 30x$ for manufacturing x pairs of shoes. The shoes sell for $50 a pair.
    (a) Write a formula for the revenue R from selling x pairs of shoes.
    (b) Write a formula for the profit P made from selling x pairs of shoes in a week.
    (c) How many pairs of shoes must be sold to make a profit? Tell how you figured this out.

Extension Exercises

19. Kari and Lisa are both north of the center of town after t hours. Kari's distance north (in miles) is $d_1 = 4t + 3$, and Lisa's is $d_2 = 3t + 9$.
    (a) What do these equations tell you about Kari and Lisa's locations?
    (b) When will they be together?

20. The equations $cx - ky = 13$ and $cx + ky = -31$ have a common solution of $(3, -2)$. What are the values of c and k?

21. In deciding between a National Motors Integer Maxima with a gas engine and one with a diesel engine, you collect the following information. First, the gas car costs $17,800 and the diesel car costs $18,500. Second, the gas car gets 32 miles per gallon (mpg) and the diesel car gets 48 mpg. Third, gas costs $1.60 per gallon and diesel fuel costs $1.68 per gallon. Under what conditions would each car be a better buy?

 22. Use a graphing calculator to find all intersection points of $y = 2^x$ and $y = x^2$.

Enrichment Topics

23. Read Chapters 1–7 of *The Education of T. C. Mits* by Lillian Lieber, and write a report that includes a summary of the main ideas and your reactions to them.

11.5   COORDINATE GEOMETRY AND NETWORKS

NCTM Standards

- use coordinate geometry to represent and examine the properties of geometric shapes (6–8)

- use coordinate geometry to examine special geometric shapes, such as regular polygons or those with pairs of parallel or perpendicular sides (6–8)

- use visual tools such as networks to represent and solve problems (6–8)

Coordinate geometry establishes a connection between algebra and geometry, enabling mathematicians to describe geometry problems algebraically.

THE DISTANCE FORMULA

How can we find the length of a line segment using coordinates?

 LE 1   Connection

(a) Plot $D(2, 1)$, $E(5, 3)$, and $F(5, 1)$ on a graph, and draw right triangle $\triangle DEF$.
(b) The legs of $\triangle DEF$ have lengths _____ and _____.
(c) Find DE using the Pythagorean Theorem.

 Classroom Connection   **LE 2   Skill**

A seventh grader computes DE in LE 1 as $\sqrt{3^2 + 2^2} = 3 + 2 = 5$. Is this right? If not, what would you tell the child?

The Pythagorean Theorem is the basis for computing the distance between any two points on a two-dimensional graph. Complete the following exercise to derive the distance formula.

 Discussion   **LE 3   Reasoning**

(a) For two points $A(x_1, y_1)$ and $B(x_2, y_2)$, construct a right triangle with hypotenuse $\overline{AB}$ as shown in Figure 11–23.
(b) What are the coordinates of C?
(c) Give the lengths AC and BC in terms of the coordinates.
(d) Using the results of part (c), $(AB)^2 = (AC)^2 + (BC)^2 =$ _____.
(e) Solve for AB, and write it in terms of the coordinates.
(f) Does part (e) involve induction or deduction?

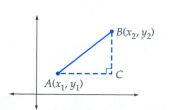

Figure 11–23

If the slope of $\overline{AB}$ is negative, the derivation is similar and the resulting formula is the same. The **distance formula** tells us how to find the length of a line segment using the coordinates of its endpoints.

The Distance Formula

The distance between two points $A(x_1, y_1)$ and $B(x_2, y_2)$ is

$$AB = \sqrt{(x_2 - x_1)^2 + (y_2 - y_1)^2}$$

LE 4 Connection

A quadrilateral $ABCD$ has coordinates $A(0, 0)$, $B(5, 0)$, $C(8, 4)$, and $D(3, 4)$. Use the distance formula to show that $ABCD$ is a rhombus.

THE MIDPOINT FORMULA

How are the coordinates of the midpoint of a line segment related to the coordinates of the endpoints?

LE 5 Reasoning

(a) Plot the points $A(3, 2)$, $B(-1, 4)$, and $C(5, 6)$, and draw $\triangle ABC$ on graph paper.
(b) Find the coordinates of the midpoint of $\overline{AC}$.
(c) Find the coordinates of the midpoint of $\overline{AB}$.
(d) Find the coordinates of the midpoint of $\overline{BC}$.
(e) What is the relationship between the coordinates of the endpoints of a line segment and the coordinates of the midpoint?
(f) Given two points $A(x_1, y_1)$ and $B(x_2, y_2)$, what are the coordinates of the midpoint of $\overline{AB}$?
(g) Using parts (b)–(d) to answer parts (e) and (f) is an example of what kind of reasoning?

The answer to LE 5(f) is the midpoint formula.

The Midpoint Formula

Given $A(x_1, y_1)$ and $B(x_2, y_2)$, then the midpoint M of $\overline{AB}$ is

$$\left(\frac{x_1 + x_2}{2}, \frac{y_1 + y_2}{2} \right)$$

The midpoint and distance formulas can be used to find properties of triangles and quadrilaterals.

Discussion

LE 6 Reasoning

$\triangle ABC$ is a right triangle with $A(0, 0)$, $B(8, 0)$, and $C(0, 6)$.

(a) Plot $\triangle ABC$ on a coordinate graph.
(b) Find the coordinates of M, the midpoint of the hypotenuse.
(c) Compare the lengths AM, BM, and CM.
(d) Repeat parts (a)–(c) with a new right triangle.
(e) Make a generalization based on your results.

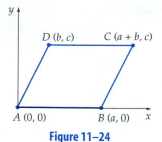

Figure 11–24

In Chapter 8, you studied properties of parallelograms. The following exercise shows how to use coordinate geometry to prove one such property: that the diagonals of a parallelogram bisect each other.

Discussion

LE 7   Reasoning

(a) What is the definition of a parallelogram?
(b) Prove that *ABCD* in Figure 11–24 is a parallelogram.
(c) Prove that the midpoint of $\overline{AC}$ is the midpoint of $\overline{BD}$.

NETWORKS

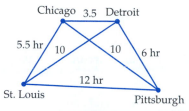

Figure 11–25

Suppose you live in Chicago, and you want to take a trip that includes Detroit, St. Louis, and Pittsburgh, in any order, and then return home. What is the fastest route? You can solve this problem by drawing a graph of the four cities (see Figure 11–25) called a network. A **network** is a graph whose edges are labeled with numbers such as lengths or times.

LE 8   Reasoning

Find the fastest route that starts in Chicago and goes through all 3 cities and then back to Chicago.

LE 9   Reasoning

A phone company wants to link up 4 cities (Figure 11–26). The network shows the cost (in millions of dollars) of different possible links. Find the cheapest possible network.

The points in a network are called **vertices,** and the curves are called **edges.** A network is **traversable** if it is possible to pass through each *edge* exactly one time. You do not have to end up at the vertex where you started. What makes a network traversable?

Figure 11–26

LE 10   Reasoning

(a) Which of the following networks are traversable? Mark where you started and where you finished.

(b) Determine whether the vertices of each network are odd or even. An even number of arcs meet at an **even vertex,** and an odd number of arcs meet at an **odd vertex.**
(c) How many even and odd vertices does each network have?
(d) Make a conjecture about a property of vertices that makes a network traversable.

To determine if a network is traversable, check whether each vertex is even or odd.

> **Traversable Network**
>
> A network is traversable if
>
> **1.** it has all even vertices or
>
> **2.** it has exactly two odd vertices. One of these will be where you start, and the other will be where you finish.

A business sometimes desires a traversable network where one can start and end in the same place.

LE 11   Reasoning

A street cleaner wants to find a route that travels along each street exactly once and starts and ends at the same place. Such a route is called an **Euler circuit.** What property of the vertices makes a network an Euler circuit?

Since an Euler circuit is traversable and has the same starting and ending point, it must have all even vertices.

LE 12   Reasoning

(a) Tell which of the following has an Euler circuit from point *A*.

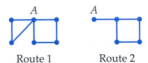

Route 1   Route 2

(b) Would it be possible to make an Euler circuit from a different starting point in either network?

LE 13   Reasoning

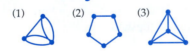

(1)   (2)   (3)

(a) Which of the networks are traversable?
(b) Which of the networks are Euler circuits?

> ### LE 14   Summary
> Tell what you learned about coordinate geometry and networks in this section.

ANSWERS TO SELECTED LESSON EXERCISES

1. (b) 2; 3   (c) $\sqrt{13}$

2. No. Do it the correct way, and compare the answers.

3. (b) (x_2, y_1) (c) $x_2 - x_1$ and $y_2 - y_1$
(d) $(x_2 - x_1)^2 + (y_2 - y_1)^2$
(e) $\sqrt{(x_2 - x_1)^2 + (y_2 - y_1)^2}$
(f) Deduction

4. $AB = BC = CD = DA = 5$, so $ABCD$ is a rhombus.

5. (b) (4, 4)   (c) (1, 3)   (d) (2, 5)
(e) The coordinates of the midpoint are the averages of the coordinates of the two endpoints.
(f) $\left(\dfrac{x_1 + x_2}{2}, \dfrac{y_1 + y_2}{2} \right)$
(g) Inductive

6. (b) $M(4, 3)$ (c) $AM = BM = CM = 5$
(e) The midpoint of the hypotenuse is equidistant from the three vertices.

7. (b) *Hint:* Compute slopes to show that the opposite sides are parallel.
(c) The midpoints of both diagonals are $\left(\dfrac{a + b}{2}, \dfrac{c}{2} \right)$.

8. 27 hours; possible route: Chicago, Detroit, Pittsburgh, St. Louis, Chicago

9. $7.4 million; A to C, B to C, D to C

10. (a) (1), (2), (4)

12. (a) Route 1
(b) Yes; you could start at any point on route 1.

13. (a) (1), (2)   (b) (2)

11.5   HOMEWORK EXERCISES

Basic Exercises

1. Explain why the distance between $A(x_1, y_1)$ and $B(x_2, y_2)$ is
$$AB = \sqrt{(x_2 - x_1)^2 + (y_2 - y_1)^2}$$

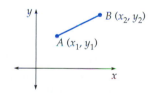

2. Explain how the distance formula is related to the Pythagorean Theorem.

3. Find the perimeter of $\triangle ABC$ with coordinates $A(3, 6)$, $B(3, 9)$, and $C(5, 7)$.

4. Determine whether or not (0, 4), (4, 5), and (6, −3) are the vertices of a right triangle.

5. Three vertices of a parallelogram are (1, −2), (0, 2), and (2, 0). Find three possible coordinates for the fourth vertex.

6. Three vertices of a rhombus are (1, 2), (4, 1), and (3, −2). What are the coordinates of the fourth vertex?

7. Point A has coordinates (4, −1), and point B has coordinates (2, 7). What are the coordinates of the midpoint of $\overline{AB}$?

8. Point M is the midpoint of $\overline{AB}$. Point A has coordinates (3, 7), and M has coordinates (−2, 1). Find the coordinates of B.

9. $ABCD$ is a rectangle.

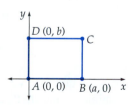

(a) What are the coordinates of C?
(b) What property do the diagonals have? Prove it.
(c) Name a second property that the diagonals have. Prove it.

10.

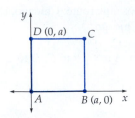

ABCD is a square.
(a) What must the coordinates of *C* be?
(b) What property do the diagonals have? Prove it.
(c) Name a second property that the diagonals have. Prove it.
(d) Name a third property that the diagonals have. Prove it.

11.

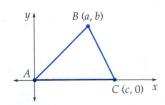

(a) Find the coordinates of the midpoints *D* of $\overline{AB}$ and *E* of $\overline{BC}$.
(b) Show that $\overline{DE} \parallel \overline{AC}$.
(c) How does *DE* compare to *AC*? Prove it.

12.

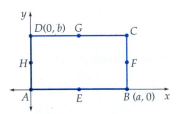

(a) Find the coordinates of the midpoints of the four sides, *E*, *F*, *G*, and *H*.
(b) What kind of shape is *EFGH?* Prove it.

13. Lyle is at home. He wants to go to the park, store, and library in any order and then return home. What is the fastest route he could take? How many different routes of the minimum time are there?

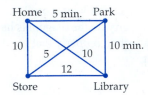

14. A phone company wants to link up 5 cities. The network shows the cost (in millions of dollars) of different possible links. Find the cheapest possible network.

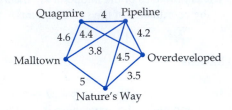

15. (a) What is a traversable circuit?
    (b) What is an Euler circuit?

16. Are all Euler circuits also traversable circuits? Explain why or why not.

17. (1)   (2)   (3)

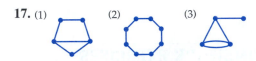

(a) Which of the networks are traversable?
(b) Which of the networks are Euler circuits?

18. (1)   (2)   (3)

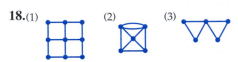

(a) Which of the networks are traversable?
(b) Which of the networks are Euler circuits?

19. A warehouse has 3 work areas. The diagram shows that there are 3 ways to go from area 1 to area 2 and 4 ways to go from area 2 to area 3. How many ways are there to go from area 1 to area 3?

20. A warehouse has 3 work areas. The diagram shows that there are 5 ways to go from area 1 to area 2 and 2 ways to go from area 2 to area 3. How many ways are there to go from area 1 to area 3?

21. A company wants to connect offices *A*, *B*, and *C* in three different cities with a telephone network. *A*, *B*, and *C* form the vertices of an equilateral triangle. The cities are about 400 miles from one another. Should they connect *A* to *B* and *B* to *C*, or should they install a phone at a fourth location (at the center of triangle *ABC*) and connect it to *A*, *B*, and *C?*

22. A company wants to connect offices *A*, *B*, *C*, and *D* in four different cities with a telephone network. *A*, *B*, *C*, and *D* form the vertices of a square. The cities are about 300 miles from one another. Should they connect *A* to *B*, *B* to *C*, and *C* to *D*, or should they install a phone at a fifth location (at the center of square *ABCD*) and connect it to *A*, *B*, *C*, and *D?*

Extension Exercises

23. Find the area of a triangle with vertices $(-1, 4)$, $(3, 1)$, and $(7, 2)$. (*Hint:* Enclose the triangle in a rectangle.)

24. If $(3, k)$ is equidistant from $A(5, -3)$ and $B(-1, -5)$, find the value of *k*.

25. A **parabola** is the set of points equidistant from a line and a point not on that line. Follow steps (a)–(b) to find the equation of the parabola that is equidistant from $y = -2$ and $(0, 4)$.

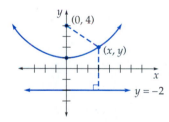

(a) Any point (x, y) on the parabola is equidistant from $(0, 4)$ and $y = -2$. What are the coordinates of the point where $y = -2$ intersects the perpendicular?
(b) Complete the following equation.

$(0, 4)$ to (x, y) dist. $= (x, -2)$ to (x, y) dist.
$$\sqrt{(\quad)^2 + (\quad)^2} = \sqrt{(\quad)^2 + (\quad)^2}$$

(c) Square both sides and combine like terms to obtain the equation of the parabola.
(d) Does part (c) involve induction or deduction?

26. Follow the instructions to create a curve by folding waxed paper.
(a) Crease a line segment $\overline{AB}$ and mark a point *C* on the waxed paper as shown.

$.C$

$A \rule{2cm}{0.4pt} B$

(b) Fold the point onto the segment and crease. Repeat this about 20 times, folding the point onto different places on the segment.
(c) What shape is suggested by the creases?
(d) How does this work? (*Hint:* See the preceding exercise.)

27. Find a point that is $\frac{3}{4}$ of the way from $(2, 5)$ to $(4, 1)$.

28. The **center of gravity** of a polygon on a two-dimensional graph is $(\bar{x}, \bar{y})$, in which $\bar{x}$ is the average (mean) of the *x*-coordinates of the vertices and $\bar{y}$ is the average (mean) of the *y*-coordinates of the vertices.
(a) Find the center of gravity of a quadrilateral with vertices at coordinates $A(-1, -3)$, $B(2, 4)$, $C(5, 4)$, and $D(6, -3)$.
(b) Cut out a piece of cardboard that has the shape of *ABCD*, and balance it on the tip of a pin. Is the center of gravity very close to the location you obtained in part (a)?

CHAPTER 11 SUMMARY

Algebra is a mathematical language used to express relationships between quantities. Algebra cannot be used to analyze nonquantifiable relationships or the merit of relationships that exist between quantities.

The most common applications of algebra involve formulas. Formulas can be derived from words, tables, or graphs. After they are derived, such formulas may be used to solve problems in which all but one of the quantities are known. Tables and graphs offer alternative ways to solve problems.

Children learn what operations will not disturb the balance between two sides of an equation. Equations can be grouped by their forms so that they correspond with graphs of a particular shape. All equations of the form $y = mx + b$ represent lines. Linear functions are a particularly useful class of functions with many applications. In linear functions of the form $y = mx$, y is proportional to x, and the graph passes through the origin and has a slope of m.

Nonlinear functions such as quadratic functions and exponential functions do not have a constant slope. Equations of the form $y = ax^2 + bx + c$ (in which $a \neq 0$) represent parabolas. Growth and decay are sometimes modeled by exponential functions of the form $y = a^x$ ($a > 0$ and $a \neq 1$). Exponential function tables have a distinctive change pattern.

Systems of linear equations offer a way of comparing certain alternatives. Comparing alternatives offers an excellent opportunity to solve the same problems with an equation, a table, and a graph.

Two thousand years after Euclid died, Fermat and Descartes devised coordinate geometry. Coordinate geometry unifies algebra and geometry. Coordinate geometry gives an algebraic perspective to many basic geometry concepts, including length, slope, and parallelism, and a new way of examining midpoints, right triangles, and quadrilaterals. Networks offer another graphical model for solving problems.

STUDY GUIDE

To review Chapter 11, see what you know about each of the following ideas or terms that you have studied. You can also use this list to generate your own questions about the chapter.

ALGEBRA AND COORDINATE GEOMETRY IN GRADES 1–8

The following chart shows at what grade levels selected algebra and graphing topics typically appear in elementary- and middle-school mathematics textbooks.

Topic	Typical Grade Level in Current Textbooks
Variables	5, 6, 7, 8
Translate words into symbols	4, 5, 6, 7, 8
Formulas and equations	4, 5, 6, 7, 8
Slope	7, 8
Linear functions	7, 8
Quadratic functions	8
Graphing in the coordinate plane	5, 6, 7, 8

REVIEW EXERCISES

1. Write an equation with variables x and y that is true for
(a) all real-number values of x and y.
(b) some real-number values of x and y.

2. Where possible, translate the following phrases or sentences into algebraic expressions, equations, or inequalities.
(a) Joe weighs more than twice as much as his sister.
(b) Sara Lee is the sweetest woman in Texas.

3. The graph shows distance traveled vs. time for a car trip.

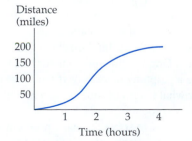

(a) What was the average speed for the trip?
(b) Did the car stop along the way?

4. Sketch a graph that corresponds to each of the following stories.
(a) A runner starts off slowly, and her speed increases steadily for a while. She runs at this maximum speed for some time, and then she tires out, and her speed gradually decreases.
(b) Your cup of very hot cocoa cools toward room temperature quickly at first. Then it decreases more slowly as it gets close to the room temperature.

5. Show how to solve $3x + 4 = 10$ algebraically and with a pictorial model.

6. Suppose Ben rides his bicycle at an average speed of 8 miles per hour.
(a) If y is the total distance he rides in x hours, what is a formula relating y to x?
(b) What is the slope of your equation in part (a)?
(c) Is y proportional to x? Tell why or why not.

7. (a) Graph $y = 2x + 1$.
   (b) What is the equation of the image of $y = 2x + 1$ after a reflection through the y-axis?

8. Give the equation of a line with a y-intercept of 5 that is parallel to $y = -4x + 3$.

9. The height of an automobile jack is a function of the number of times you crank it. Before cranking, a jack might be 10 in. high. Each crank raises the jack (and car) 0.5 in.
   (a) Write a formula for the height of the jack, h, in terms of the number of cranks, n.
   (b) Graph your formula.
   (c) What shape is your graph?

10. A waiter earns $1 per hour plus tips. Tips average $2 per table.
    (a) Write a formula for the total earnings E in terms of the number of hours H and number of tables T.
    (b) How many tables must he serve in an 8-hour shift to earn $40?

11. You have $50 to spend on a meal. You plan to leave 25% for the tax and tip. How much can you spend on the food?

 12. A salesperson earns a commission of 8% of her sales. Her boss then deducts 26% of the commission for taxes. How much must the salesperson sell to take home $500?

13. The typical weight W (in pounds) of a man of height H (in inches) can be estimated by $W = 5.5H - 220$.
    (a) What does the slope of the equation tell about the height and weight?
    (b) Solve for H.
    (c) Which formula would be easier to use to find the typical weight of a man 5 feet 8 inches tall?

14. A taxicab company charges $2.10 plus $0.80/mile.
    (a) A formula relating the total cost C (in $) to the distance d (in miles) is $C =$ _____.
    (b) Graph your equation from part (a), with d on the x-axis.
    (c) If a cab ride cost $11.70, how long was it?

15. How is the graph of $y = x^2 + 2$ related to the graph of $y = x^2$?

16. Without using a calculator, sketch a graph of $y = 3^x$ for $-2 \le x \le 2$. Label all x- and y-intercepts.

 17. (a) The population of a city doubles every 22 years. Complete the following table showing the city population p for t years after 1900.

t (years after 1900)	0			
p (in thousands)	20			

   (b) Which of the following equations fits the data?
   (1) $p = 22t$ (2) $p = 20 \cdot 2^t$
   (3) $p = 20 \cdot 2^{t/22}$ (4) $p = 22^t$
   (c) Use the equation from part (b) to predict the city population in 2000.

18. P, Q, and R are functions of x. Tell whether P, Q, and R are linear, exponential, or neither.

x	0	2	4	36
P	1	4	9	16
Q	5	25	125	625
R	70	60	50	40

19. Consider the following problem. "Suppose six peaches and two plums cost $2.40, and three peaches and two plums cost $1.44. Find the unit cost of a peach and the unit cost of a plum."
    (a) Describe a method for solving this problem without writing equations.
    (b) Write a system of two equations that represents this problem.
    (c) Solve your system of equations from part (b) with the method you used in part (a).

20. You want to compare the cost of two washing machines. The Clothes Horse costs $470 and has an average annual energy cost of $82. The Twister costs $398 and has an average annual energy of cost of $106.
    (a) Write an equation for the total cost of each washing machine after T years.
    (b) Assume each washing machine could last up to 12 years. Graph both equations for this time interval. Use a graphing calculator if you have one.
    (c) For what length of time is each model likely to cost less money?
    (d) Show how to solve the same problem with a table.
    (e) Show how to solve the same problem algebraically.

21. Explain why the distance between $A(x_1, y_1)$ and $B(x_2, y_2)$ is

$$AB = \sqrt{(x_2 - x_1)^2 + (y_2 - y_1)^2}$$

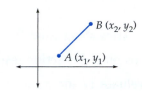

22.

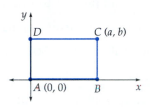

(a) *ABCD* is a rectangle. What are the coordinates of *B* and *D?*

(b) Show that the midpoint of $\overline{AC}$ is the midpoint of $\overline{BD}$.
(c) Show that the diagonals are equal in length.

23. After a party at your house (*H*), you want to drive three friends, Alberto (*A*), Blaine (*B*), and Chong (*C*), to their houses and then return back home. The network shows the driving times in minutes. What is the fastest route?

24. (a) Draw a network that is traversable but not an Euler circuit.
(b) Draw a network that is neither traversable nor an Euler circuit.

ALTERNATE ASSESSMENT

Do one of the following assessment activities: Add to your portfolio, add to your journal, write another unit test, do another self-assessment, or give a presentation.

SUGGESTED READINGS

Driscoll, M. *Fostering Algebraic Thinking.* Portsmouth, NH: Heinemann, 1999.

Edwards, R. *Algebra Magic Tricks* (2 vols.). Pacific Grove, CA: Critical Thinking Press, 1992.

Gardner, M. *The Colossal Book of Mathematics: Classic Puzzles, Paradoxes, and Problems.* New York: W. W. Norton, 2001.

Jacobs, H. *Algebra.* New York: W. H. Freeman, 1995.

Jenkins, R. *61 Cooperative Learning Activities for Algebra 1.* Portland, ME: Walch, 1999.

Lawrence, A., et al. *Lessons for Algebraic Thinking: Grades 6–8.* Sausalito, CA: Math Solutions, 2002.

Moses, B., Ed. *Algebraic Thinking, Grades K–12.* Reston, VA: NCTM, 1999.

Murdock, J., et al. *Discovering Algebra: An Investigative Approach.* Emeryville, CA; Key, 2002.

National Council of Teachers of Mathematics. *Communication in Mathematics, K–12 and Beyond.* 1996 Yearbook. Reston, VA: NCTM, 1996.

Navigating Through Algebra (4 vols.). Reston, VA: NCTM, 2001.

The Role of Representations in School Mathematics. 2001 Yearbook. Reston, VA: NCTM, 2001.

Use Infotrac to look for articles in *Teaching Children Mathematics* on teaching children about algebra. You might also look at recent issues of *Mathematics Teaching in the Middle School.*

12 | STATISTICS

Pick up a newspaper, and you'll read statistics about what the typical citizen eats, earns, and believes. Statistics help us predict election results and tomorrow's weather. People use statistics to evaluate TV shows, schools, and government programs. An educated citizen should be able to tell whether statistics are being used appropriately in all of these situations.

The first written records of statistics date back to 3000 B.C. when the ancient Egyptians compiled data on population and wealth. It wasn't until about 4,700 years later, in 1660, that Englishman John Graunt conducted the first formal statistical study. He tabulated the causes of death in different cities and compared them over many years. Graunt was able to document that many adults were dying due to the poor sanitation in London. When he wasn't doing statistics, Graunt made a living by selling buttons, needles, and other small items.

The term "statistics" refers to both a set of data and the methods used to analyze the data. Statistical methods enable us to organize and simplify a set of data to reveal essential characteristics and relationships.

First we select some quantifiable aspect to measure. Then data are collected. From there, statistical methods enable us to organize and interpret the data.

Throughout this process, certain information is discarded. For example, a graph or table removes the original order of the numbers. An average reduces a whole set of data to just one number. These methods can reveal underlying patterns, or they can obscure reality. That is the difference between good and bad statistics.

12.1   STATISTICAL GRAPHS AND TABLES

NCTM Standards

- represent data using tables and graphs such as line plots, bar graphs, and line graphs (3–5)

- compare different representations of the same data and evaluate how well each representation shows important aspects of the data (3–5)

- select, create, and use appropriate graphical representations of data, including histograms, box plots, and scatterplots (6–8)

LE 1   Opener

(a) Name all the types of graphs you know that are used to display data.

(b) Name several places where you regularly see graphs.

Statisticians often communicate information about data with a graph or a table (Figure 12–1). Reading and interpreting graphs and tables make up the most important part of statistics for most people. Graphs and tables can organize numerical data in a simple, clear way. Because graphs and tables simplify information, they also have the potential to mislead.

Figure 12–1

Graphs and tables are part of descriptive statistics. **Descriptive statistics** involves all of the techniques that are used to describe and summarize the characteristics of numerical data.

BAR GRAPHS, PICTOGRAPHS, AND LINE PLOTS

Statisticians generally use **bar graphs** to compare the values of several variables. Each bar shows the frequency of one of the variables. In most vertical bar graphs, a simple comparison of the heights of the bars reveals the relationships among the variables.

Three variations on bar graphs are pictographs, line plots, and stem-and-leaf plots. If the bars are replaced with pictures, the result is a pictograph. A **pictograph** employs pictures or symbols for quantities (Figure 12–2). A **line plot** is a number line with X's or dots placed above specific numbers to show their frequency. The data set {2, 3, 3, 3, 4} is shown as a line plot in Figure 12–2. Some teachers present line plots before having children work with bar graphs. Line plots are typically used to represent data sets with fewer than 40 values. The third variation on bar graphs, the stem-and-leaf plot, is introduced later in the section.

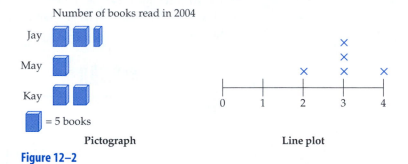

Figure 12–2

LE 2   Reasoning

A third-grade class conducts a survey asking, "How many people are in your family?" The results are 3, 4, 9, 8, 5, 3, 5, 6, 4, and 5. Below are two line plots of the data that children made.

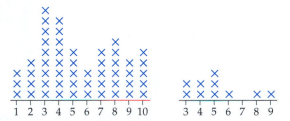

(a) Tell what each line plot shows about the data.
(b) Which display do you think is more useful?
(c) Which is more abstract?

At first, children tend to focus on individuals in data sets. An example is the first line plot in LE 2, which shows the relative size of each family. It is harder to consider the overall properties of a group, where the group is considered as a single entity. This is done in the second graph in LE 2, which shows the relative frequency of different family sizes.

CONSTRUCTING STEM AND LEAF PLOTS

A stem-and-leaf plot offers a quick way to display a small set of positive numbers. A **stem-and-leaf plot** uses the actual data numbers to make a type of bar graph. The fourth-grade textbook page of *Harcourt Math* (Figure 12–3) shows how to construct a stem-and-leaf plot.

C
Classroom Connection

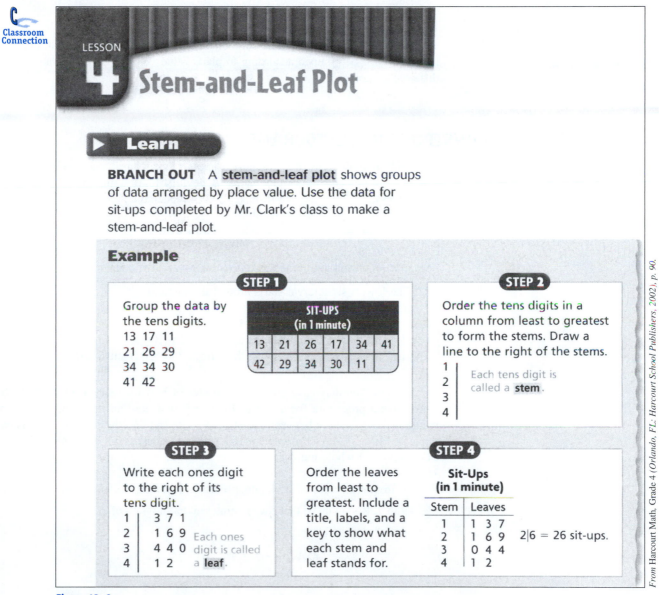

From Harcourt Math, Grade 4 (Orlando, FL: Harcourt School Publishers, 2002), p. 90.

Figure 12–3

A stem-and-leaf plot looks like a sideways bar graph. Note that each individual score can be read from the stem-and-leaf plot.

Now try one yourself.

LE 3   Skill

A researcher selects 20 U.S. families at random and records their annual family incomes (in thousands of dollars) as follows.

50	81	22	151	42	29	60	115	13	56
45	102	99	17	11	25	160	72	7	38

Make a stem-and-leaf plot of the family incomes.

Stem-and-leaf plots are more cumbersome for large data sets. In addition, stem-and-leaf data must be grouped according to place value. What if a statistician wants to use different groupings?

CONSTRUCTING HISTOGRAMS

To avoid the limitations of stem-and-leaf plots, statisticians usually select their own intervals to group data and display the frequency distribution in a histogram. To illustrate this process, let's use the data from LE 3. The first step is to decide how to put the data into 6 to 15 intervals of equal width.

LE 4   Concept

Suppose you want to put the data into 6 subintervals of equal size.

(a) The lowest subinterval starts around the number _____, and the highest subinterval ends around the number _____.
(b) About how big should each subinterval be?
(c) How many whole numbers are there from 1 to 3 (inclusive)?
(d) How many whole numbers are there from 7 to 160 (inclusive)?

In LE 4, you may have gotten a good idea of how to create 6 subintervals. Statisticians have a procedure for doing it. First they find how far apart the scores are. The **range** is the difference between the highest score and the lowest score. In this case, the range is $160 - 7 = 153$. Are there 153 whole numbers from 7 to 160 (inclusive)? No. The number of whole numbers from 7 to 160 (inclusive) is the range plus 1. There are $160 - 7 + 1 = 154$ scores.

To divide a width of 154 into 6 classes (subintervals) of equal width, compute $\frac{154}{6} = 25.7$. Round up so that each subinterval has a width of 26.

LE 5   Skill

Complete the grouped frequency distribution table for the income data.

Annual Income (thousands of dollars)	Frequency
7–32	
33–58	
59–84	
85–110	
111–136	
137–162	

Why is it reasonable to use 6 intervals? Ideally, the data values within each interval represent similar results and differ from the results in other intervals. A much larger or much smaller number of intervals would not give a meaningful display of the overall distribution. It would have many short bars or a few very big bars, respectively.

Statisticians usually select classes of equal width, making it possible to compare frequencies in different intervals by looking at the heights of the bars.

After tabulating the data in a table, statisticians usually display the frequency distribution of a *single* variable using a histogram. A **histogram** is a bar graph that shows the numbers of times data occur within a certain interval. It has the following properties: (1) the bars are always vertical, (2) the width of each bar is based on the size of the interval it represents, and (3) there are no gaps between adjacent bars (if bars of height 0 are included).

The histogram in Figure 12–4 shows the distribution of frequencies of family income. The histogram shows a pattern in which most families have incomes at the lower or middle level ($7,000 to $84,000) and a few are at the upper end.

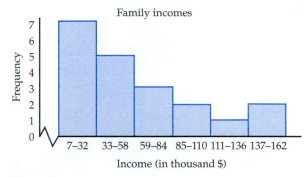

Figure 12–4

How would the graph change if the number of classes were changed?

LE 6   Skill

(a) Put the same income data into 9 classes of equal width, and construct a histogram.
(b) How does your histogram compare to the one with the 6 intervals?

Histograms have no gaps because their bases cover a continuous range of possible values of a variable, such as the days of the year. If all days of the year are included, there are no gaps between them. In a bar graph, the gaps between bars indicate that each bar describes a discrete item.

LE 7   Concept

Which of the following graphs are histograms?

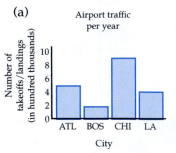

(a) Airport traffic per year

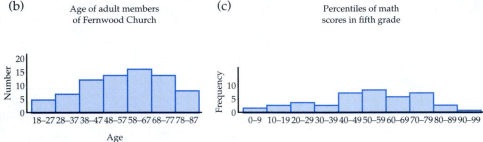

(b) Age of adult members of Fernwood Church

(c) Percentiles of math scores in fifth grade

The *area* of each bar is proportional to the number of items or the length of the corresponding interval. Usually, intervals are equal in size, so the reader can look at the *heights* of the bars to compare the frequencies of different intervals. When children study histograms in grades 5–8, the interval is usually written under each bar. In secondary school and college, many statistics texts label the two ends of each bar with boundary values. The boundary value is included in the interval to its right.

The next exercise uses live data from your class!

Discussion

LE 8   Skill

(a) If you asked each person in your class the month in which he or she was born, how do you think the results would come out?

(b) Check your prediction. Find out the month in which each person in your class has a birthday. Construct a frequency distribution table.

Month	Jan	Feb	Mar	Apr	May	Jun	Jul	Aug	Sep	Oct	Nov	Dec
Number of birthdays												

(c) Construct a bar graph of the frequency distribution. Assume that all intervals are the same size (although they differ slightly).

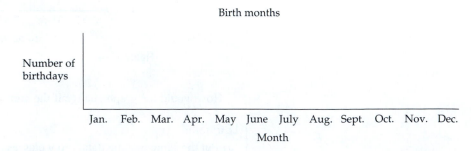

Birth months

Number of birthdays

(d) Do you see any pattern in the results? If so, why do you think the pattern occurs?

LE 8 illustrates a common process in statistics. Begin with a statistical question of interest and a hypothesis about the answer. Then collect data. Next, organize and display the data. Finally, analyze the data in relation to your hypothesis.

CONSTRUCTING CIRCLE GRAPHS

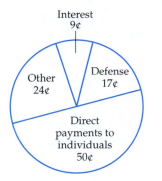

2005 federal budget dollar (estimated)

Figure 12–5

Circle graphs (or **pie charts**) are useful for showing what part of the whole falls into each of several categories. Governments often use circle graphs to show how they distribute their funds (Figure 12–5). The four sections of the graph are four **sectors** of the circle. If there are more than six or seven parts, a circle graph becomes difficult to subdivide and label.

LE 9 Skill

(a) Construct a circle graph from the data in LE 5. Calculate what percent each class is out of the total. In the circle graph, each class will be represented by a sector of the circle. The central angle for each class is obtained by computing the corresponding percent of 360°.
(b) Which display do you think is more useful, the histogram or the circle graph? Tell why.

BAR GRAPHS, LINE GRAPHS, AND CIRCLE GRAPHS

Three basic types of statistical graphs are bar graphs, line graphs, and circle graphs. Each type of graph is more suitable for presenting certain types of information.

LE 10 Opener

Which graph, (a) the line graph or (b) the bar graph, is more appropriate for displaying the data?

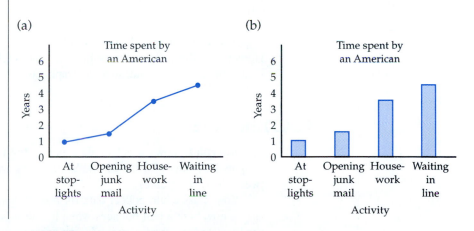

You have already reviewed bar graphs and circle graphs. Statisticians most often use **line graphs** to show changes over time (Figure 12–6).

Source: Kevin Phillips, Wealth and Democracy, New York: Broadway Books, 2002.

Figure 12–6

The following chart summarizes the types of graphs and their common uses.

Graph	Best for
Line	Showing changes over time
Bar	Comparing values of several categories or intervals
Circle	Showing parts of a whole

Line plots and stem-and-leaf plots work best with small data sets.

LE 11 Concept

For each of the following, would the best choice be a line graph, a bar graph, or a circle graph?

(a) Showing what percent of a family budget is devoted to each of the following: housing, clothing, food, taxes, and other
(b) Showing the change in the Consumer Price Index during the 12 months of 2004
(c) Showing the number of people who picked each of six sports as their favorite

A statistician would probably display the graph in LE 11(a) with a circle graph. However, if the budget were subdivided into eight or more categories, the circle graph would be difficult to read, so a bar graph would be better.

SCATTERPLOTS

A **scatterplot** shows a set of data points for two variables such as the heights and weights of adults (Figure 12–7). Scatterplots are used to display the relationship between the two variables.

Your class could collect data for a scatterplot by comparing arm span to height or seeing how high a ball bounces when it is dropped from 8 different heights. Otherwise, use the data from the following lesson exercise to create a scatterplot.

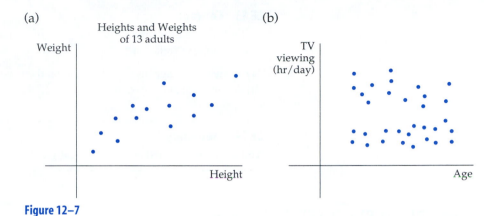

Figure 12–7

Discussion

LE 12 Connection

Do you think a student's attendance is related to his or her test scores?

(a) Graph the 6 data points to create a scatterplot for the following data.

Name	Jose	Jennifer	Jamaal	Jill	Jong	Jason
Classes Missed (x)	2	0	6	5	3	2
Test Score (y)	83	97	65	72	73	81

(b) Draw a line that goes close to most of the points.
(c) Does your line have a positive slope or a negative slope?
(d) As x increases, what does y tend to do?
(e) Estimate the slope of your line, and find its equation.
(f) What does the slope of your line tell you about the classes missed and the test scores?
(g) Use your equation to predict the test score for someone who has missed 4 classes.
(h) Why would it be risky to use your equation to predict the test score of someone who missed 10 classes?
(i) If you have a graphing calculator or a computer software program, use it to find the best-fitting (regression) line, and compare the result to part (e).

The line you drew in LE 12(b) is called a **trend line** or **line of best fit.** The trend line indicates that as the number of classes missed increases, the test score tends to decrease. This relationship shows a **negative correlation.** A graph of height versus weight (Figure 12–7(a)) shows a **positive correlation,** since weight tends to increase as height increases. However, suppose you made a scatterplot of age and hours spent watching TV (Figure 12–7(b)). You would probably find no significant positive or negative correlation.

LE 13   Concept

Tell whether each of the following pairs of variables would have a positive correlation or a negative correlation.

(a) A husband's age and a wife's age on their wedding day
(b) A car's age and its value

LE 14   Summary

Tell what you learned about statistical graphs in this section.

ANSWERS TO SELECTED LESSON EXERCISES

2. (c) The second graph

3.

0	7
1	137
2	259
3	8
4	25
5	06
6	0
7	2
8	1
9	9
10	2
11	5
12	
13	
14	
15	1
16	0

4. (a) 7; 160   (b) About 25 or 26
(c) $3 - 1 + 1 = 3$
(d) $160 - 7 + 1 = 154$

5.

Annual Income (thousands of dollars)	Frequency
7–32	7
33–58	5
59–84	3
85–110	2
111–136	1
137–162	2

7. (b) and (c)

9.

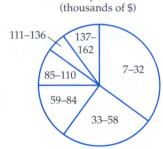

Family income
(thousands of $)

10. The bar graph

11. (a) A circle graph   (b) A line graph
(c) A bar graph

12. (a)

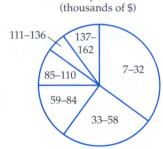

(c) Negative slope   (d) Decreases

(h) It is too far beyond the data range.

13. (a) Positive   (b) Negative

12.1 HOMEWORK EXERCISES

Basic Exercises

 1. The Perky Academy annual budget for 2003–2004 is as follows.

Revenues

Tuition and fees	$4,374,000
Alumni gifts	194,000
Endowment interest	109,000
Bookstore and vending machines	204,000
Summer day camp	90,000
Other	93,000
Total	**$5,064,000**

Expenditures

Teachers' salaries and benefits	$2,269,000
Administration salaries and benefits	769,000
Maintenance and renovation	654,000
Scholarships	504,000
Bookstore and vending machines	194,000
Summer day camp	63,000
Student activities	21,000
Utilities, phone, and postage	501,000
Insurance	114,000
Other	130,000
Total	**$5,219,000**

(a) What percent of the money spent goes for teachers' salaries and benefits?

(b) How much profit does the school make running the summer camp?

(c) What is the school's total deficit (loss) for the year?

2.

Pregnancies per 1000 15 to 19-Year-Olds (Alan Guttmacher Institute)	
United States	72/yr
Great Britain	47
Canada	39
Sweden	20
France	19

Percent of Teenagers Who Are Sexually Active		
	Age	
	15	**18**
United States	14%	63%
Great Britain	4%	62%
Canada	9%	53%
Sweden	12%	65%
France	7%	50%

(a) Which countries have the most sexually active teenagers?

(b) Which country has the most teenage pregnancies?

(c) To what extent do teenage pregnancies seem to be related to the degree of sexual activity?

(d) Why might two countries with equal rates of sexual activity among teenagers have different pregnancy rates?

(e) Speculate as to how these data were collected.

3.

Major and starting salary
for bachelor's degree in 2003

Major	Salary
Mechanical Engineering	$48,115
Computer Science	$44,678
Accounting	$42,005
Mathematics	$41,870
Business Adm.	$36,634
Marketing	$35,698
English	$35,538
Political Science	$34,594
Education	$29,890

(a) Which degree earned the lowest average starting salary?

(b) Which degree earned the highest average starting salary?

(c) Summarize the overall trend in starting salaries and the degrees shown.

(d) Is there any degree whose average salary surprises you?

(e) Do you think the data are accurate? Why or why not?

(f) What additional information would be useful?

4. The following pictograph shows how many books each child read in March.

Number of books read in March

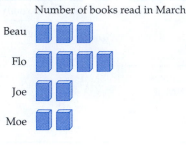

(a) How does a pictograph differ from a bar graph?
(b) What additional information would be useful?

5. Match each line plot with its description.
  (a) The number of questions each student got right on a fairly easy 10-question quiz
  (b) The number of heads each student obtained in 10 coin tosses
  (c) The number of days each student has been absent this month

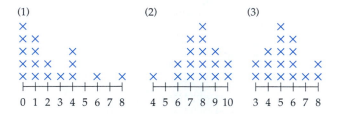

6. The heights (in inches) of the players on a professional basketball team are 70, 72, 75, 77, 78, 78, 80, 81, 81, 82, and 83. Make a line plot of the heights.

7. The following table lists estimated data on causes of premature deaths in the United States.

Cause of Death	Number of Premature Deaths
Tobacco use	400,000
Alcohol use	100,000
Preventable medical errors	70,000
Motor vehicle accident*	50,000
Suicide	29,000
Use of hard drugs	20,000
Murder	17,000
Air pollution	10,000
Food poisoning	5,000

*Includes 23,000 involving alcohol or drug use.

(a) Make a bar graph of the data for the top six categories.
(b) What surprises you about the data?
(c) What conclusions could be drawn from the data?
(d) What additional information would be useful?

8. Complete the double bar graph of fall-semester enrollments for the two years given in the table.

Student Enrollment	Math	English	Business	Psych.
Fall 2003	2421	2260	1572	874
Fall 2004	2580	2501	1610	710

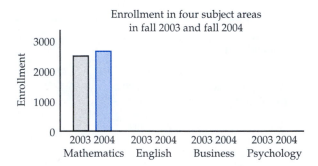

9. Test results for two mathematics classes are as follows.

Class A: 58, 62, 62, 70, 72, 75, 80, 81, 85, 92, 98
Class B: 42, 65, 65, 68, 75, 80, 82, 90, 91, 99

(a) Make a stem-and-leaf plot for class A.
(b) Make a stem-and-leaf plot for class B.
(c) Look at the stem-and-leaf plots from classes A and B, and describe the difference in the distribution of scores in the two classes.

10. The following list shows the retirement ages of the last 15 teachers who retired at Brain Power High.

64   52   68   72   65   47   59   43
56   48   49   51   58   66   58

(a) Make a stem-and-leaf plot of the data.
(b) Summarize in words what the stem-and-leaf plot indicates about the retirement ages.
(c) Make a bar graph of the same data.
(d) State one advantage of using the stem-and-leaf plot and one advantage of using the bar graph.

11. Which of the following graphs are histograms?

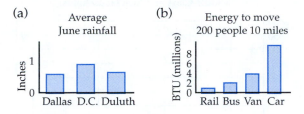

(a) Average June rainfall

(b) Energy to move 200 people 10 miles

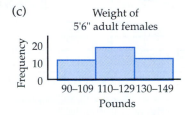

(c) Weight of 5'6" adult females

12. What is the difference between a histogram and a bar graph?

13. Following are the 2001 average annual salaries (in thousands of dollars) of U.S. elementary-school teachers, by state, from the *Statistical Abstract*.

AL	38.0	IL	45.9	MT	32.5	RI	48.5
AK	48.1	IN	43.4	NE	34.2	SC	37.9
AZ	36.3	IA	35.5	NV	40.4	SD	30.2
AR	35.2	KS	35.9	NH	38.3	TN	37.1
CA	52.1	KY	36.3	NJ	53.3	TX	38.0
CO	39.2	LA	33.6	NM	33.5	UT	36.4
CT	52.1	ME	36.2	NY	49.8	VT	38.1
DE	46.9	MD	45.1	NC	41.2	VA	40.2
DC	47.1	MA	47.8	ND	31.1	WA	42.2
FL	38.2	MI	50.7	OH	42.4	WV	35.7
GA	41.6	MN	42.7	OK	34.3	WI	41.9
HI	40.1	MS	31.5	OR	41.7	WY	34.7
ID	36.4	MO	35.9	PA	49.4		

(a) Select intervals, and make a table showing the frequency distribution.
(b) Construct a histogram of the data.

14. (a) Time the next 20 commercials you see on television.
(b) Select intervals and make a table for a frequency distribution of the commercial lengths.
(c) Construct a histogram showing the frequency distribution of commercial lengths.

Frequency

Length of commercial, in seconds

15. A class received the following test scores: 61, 63, 65, 68, 69, 75, 81, 82, 91, 92, 94, 97.
(a) Construct a table with score intervals that show that most of the scores were either high or low.
(b) Construct a shorter, deceptive table, using intervals that make the scores appear to be evenly distributed.

16. A poll is taken to see how people in different age groups feel about candidate Hope. One hundred people in each age group are asked whether they would vote for Hope rather than his opponent. The results are as follows.

Age	<21	21–30	31–40	41–50	51–60	>60
Votes for Hope	6	66	40	36	40	20

Make a table with different groupings that makes it appear that Hope receives similar support from young, middle-aged, and older voters.

 17. How big is the central angle for the region representing households that have no working computers?

USA TODAY Snapshots®

1 in 4 have at least 2 computers
The average number of working computers Americans have at home:

0 — 29%
1 — 47%
2 — 15%
3+ — 9%

18.

2001 world energy consumption

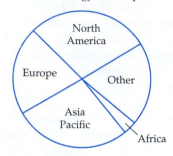

(a) Estimate the percent of energy consumption for each place.
(b) Measure each angle in the circle graph, and compute what percent of the 360° circle each place has.

 19. Juan Gomez's average monthly budget is broken down as follows:

Food	$150	Entertainment	$40
Clothing	50	Other	60
Rent	300		

(a) Find out what percent of the total monthly budget each category is.
(b) Use a protractor to construct a circle graph of Juan's average monthly budget.

 20. In 2000, the marital status of women ages 20–24 was as follows: 72.8% were single, 25.3% were married, and 1.9% were divorced or widowed.
(a) Use a protractor to draw a circle graph showing the marital status of women ages 20–24.
(b) What additional information would be useful?
(c) Guess what the data for men ages 20–24 would look like.

21. Construct a circle graph showing how you spend time during an average school week. Include the following categories: sleeping, eating, working, going to school, watching television, recreation, and other.

22. Construct a circle graph that shows major categories of spending in your monthly budget.

23. The multiple line graph from *Wealth and Democracy* by Kevin Phillips shows the change in income (in inflation adjusted dollars) of middle class and wealthy households from 1967 to 1997.

Inflation Adjusted Income of Middle-Class and Rich Households

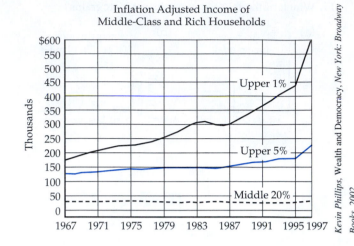

Kevin Phillips, Wealth and Democracy, New York: Broadway Books, 2002.

(a) Describe the overall trends.
(b) Can you explain why these trends occurred?

24. The following line graph shows the percent of persons living below the poverty level from 1984 to 2000.

Percent of people below poverty level in U.S.

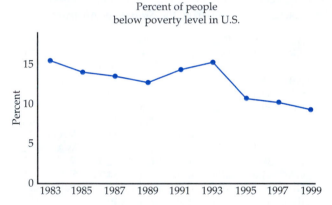

Source: U.S. Bureau of the Census, *2002 Statistical Abstract of the United States.*

(a) Describe the overall trends.
(b) Can you explain why these trends occurred?
(c) What additional information would be useful?

25. For each of the following, determine whether the best choice would be a line graph, a bar graph, or a circle graph.
(a) Showing the percent of people who sleep an average of 5 or fewer, 6, 7, 8, or 9 or more hours per night
(b) Showing the annual U.S. government budget for environmental protection from 1985 to 2004
(c) Showing the total sales of raffle tickets by each of grades 3, 4, and 5

26. For each of the following, determine whether the best choice would be a line graph, a bar graph, or a circle graph.
(a) Showing the percent of people who favor each of three candidates in an election
(b) Showing the favorite days of the week of students in your class
(c) Showing the change in the number of students taking high-school algebra over the last ten years

27. For each of the following data sets, list all types of graphs (bar, circle, line, stem-and-leaf plot) that could be made. Then tell which you think would be the best choice.
(a) The number of elementary schools in a county that use each of the four approved mathematics textbook series
(b) The annual salaries of teachers in your state
(c) Percent of electricity generated by solar power from 1990 to 2004

28. For each of the following data sets, list all types of graphs (bar, circle, line, stem-and-leaf plot) that could be made. Then tell which you think would be the best choice.
(a) The heights of everyone in your mathematics class
(b) The favorite flavor of ice cream for each person in your class
(c) The average annual salaries in your state from 1998 to 2004

29. A survey determines how many adults out of 400 would be willing to pay various prices for a hotel room. The results were as follows.

Price, x ($)	50	60	70	80	90
Number Willing, y	320	275	234	188	150

(a) Make a scatterplot of the five points.
(b) Draw a straight line that seems to fit the data best.
(c) Does your line have a positive slope or a negative slope?
(d) As x increases, what does y tend to do?
(e) Does this graph show a positive correlation or a negative correlation?
(f) Estimate the slope of your line, and find its equation.
(g) What does the slope of your line tell you about the price and the demand?
(h) Use your equation to predict the demand for a price of $55.

(Continued in the next column)

(i) Why would it be risky to use your equation to predict the demand for a price of $130?
(j) If it's available, use a calculator or computer to find the regression line, and compare it to your result in part (f).

30. On the basis of a survey, a student predicts the following ticket sales for a concert depending on the price charged per ticket.

Ticket Price, x ($)	5	10	15	20	25	30
Total Tickets Sold, y	700	500	350	250	190	100

(a) Make a scatterplot of the six points.
(b) Draw a straight line that seems to fit the data best.
(c) Does your line have a positive slope or a negative slope?
(d) As x increases, what does y tend to do?
(e) Does this graph show a positive correlation or a negative correlation?
(f) Estimate the slope of your line, and find its equation.
(g) What does the slope of your line tell you about the ticket price and the demand?
(h) Use your equation to predict the demand for a ticket price of $28.
(i) Why would it be risky to use your equation to predict the demand for a ticket price of $50?
(j) If it's available, use a calculator or a computer to find the regression line and compare it to your result in part (f).

31. The gas tank of a National Motors Titan holds 20 gallons of gas. The following data are collected during a week.

Fuel in tank (gal)	20	18	16	14	12	10
Dist. traveled (mi)	0	75	157	229	306	379

(a) Make a scatterplot of the six points.
(b) Draw a straight line that seems to fit the data best.
(c) Does this graph show a positive correlation or a negative correlation?
(d) Estimate the slope of your line, and find its equation.
(e) What does the slope of your line tell you about the fuel and distance?
(f) Use your equation to predict the distance traveled when 9 gallons are left.
(g) If it's available, use a calculator or computer to find the regression line, and compare it to your result in part (d).

32. A community tries different speed limits on a stretch of road and records the following numbers of accidents per month.

Speed limit, S (mph)	40	45	50	55	60	65
Accidents per month, A	3	7	13	18	21	25

(a) Make a scatterplot of the six points.
(b) Draw a straight line that seems to fit the data best.
(c) Does this graph show a positive correlation or a negative correlation?
(d) Find the slope and equation of your line.
(e) What does the slope of your line tell you about the speed limit and accidents?
(f) Use a calculator or computer to find the regression line, and compare it to your result in part (d).

33. Tell whether each of the following pairs of variables has a positive correlation or a negative correlation.
(a) The length of the side of a square and the area of the square
(b) The size of a car's gas tank and the number of stops needed on a long car trip

34. Tell whether each of the following pairs of variables has a positive correlation or a negative correlation.
(a) The number of people in a car pool and the cost per person of driving to work
(b) The speed of a car and the time it takes to travel 1 mile

Extension Exercises

35. The following table gives the percent frequencies of each letter in written English.

Letter	A	B	C	D	E	F	G
Frequency	8%	1%	3%	4%	13%	3%	2%

Letter	H	I	J	K	L	M	N
Frequency	5%	7%	0.2%	0.5%	4%	0.3%	4%

Letter	O	P	Q	R	S	T	U
Frequency	8%	2%	0.2%	6%	7%	9%	3y

Letter	V	W	X	Y	Z
Frequency	1%	2%	0.3%	2%	0.1%

(a) Name the four most frequently occurring letters.
(b) Plot the data for the 11 most frequently occurring letters as a bar graph.

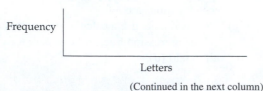

(Continued in the next column)

(c) Codes can sometimes be deciphered through analysis of the frequencies of the letters. Tabulate the frequency of each letter in the following coded message, and make a bar graph.

UAC   QCCUWPT   IWNN
XC   GU   UAC   ZWCS

(d) The most frequent letter in the coded message in part (c) represents E. Which letter is this?
(e) Try to decode the whole message. (*Hint:* W stands for I in the coded message.)

36. The frequencies of letters in Scrabble are as follows.

A—9	B—2	C—2	D—4	E—12	F—2
G—3	H—2	I—9	J—1	K— 1	L—4
M—2	N—6	O—8	P—2	Q— 1	R—6
S—4	T—6	U—4	V—2	W— 2	X—1
Y—2	Z—1				

(a) Construct a bar graph for the frequencies of the 11 most frequently occurring Scrabble letters.
(b) Compare this graph to the one in the preceding exercise. Which letters are significantly more frequent in Scrabble than in written English? Which letters are significantly less frequent in Scrabble?

Technology Exercises

37. Most computer spreadsheet programs can create bar charts (graphs) and pie charts (circle graphs) from spreadsheet data. A simple bar chart plots the values from one row or column of data in a spreadsheet. A pie chart tells what percent of the total in a row or column each entry represents. Enter the following data into a spreadsheet. Then select the appropriate type of chart (graph) from the choices offered by your spreadsheet program. Add a legend (title) to the graph.

	A	B	C
1	**Chandler Family Expenses in 2003**		
2	Rent/utilities		$13200
3	Food/clothes		$ 8950
4	Automobiles		$ 7280
5	Taxes		$11850
6	Other		$12710
7	Total		$53990

38. You can make a line graph with graphing calculators such as the TI-73 or 83. (You could generate data from Figure 12–6.) On the TI-83, press STAT and choose EDIT. Then enter your data in columns L1 and L2. Then choose a scale in WINDOW. Finally press STAT PLOT. Press ENTER to choose Plot 1. Then select On. Next, select, the line graph icon. Finally, press GRAPH.

39. (a) You can make a histogram on graphing calculators such as the TI-73 or 83. (You could use the data from LE 3.) On the TI-83, press STAT and choose EDIT. Then enter your data in columns L1. Then choose a scale in WINDOW. Finally, press STAT PLOT. Press ENTER to choose Plot 1. Then select On. Next, select the histogram. Select L1 for the Xlist, and 1 as the frequency. Finally, press GRAPH.

(b) Make a pictograph of the same data.

(c) Make a circle graph of the same data.

40. On the basis of a survey, a student predicts the following ticket sales for a concert depending upon the price charged per ticket.

Ticket Price, x ($)	5	10	15	20	25	30
Total Tickets Sold, t	700	500	350	250	190	100

(a) The concert will cost $800. Calculate the profit at each ticket price.

(b) Make a scatterplot of profit (y) versus price (x).

(c) On the TI-83, press STAT and select EDIT. Then enter your data in columns L1 and L2. Then press STAT, select CALC, and select QuadReg (quadratic regression). Press ENTER to fit a parabola to the data.

(d) Use your parabola to predict the profit for a price of $6.

(e) What ticket price would you recommend?

41. The population of the United States was 203 million in 1970, 227 million in 1980, 250 million in 1990, and 275 million in 2000.

(a) On the TI-83, press STAT and select EDIT. Then enter your data in columns L1 and L2. Then press STAT, select CALC, and select ExpReg (exponential regression). Press ENTER to fit an exponential function to the data.

(b) Use your formula to predict the U.S. population in 2010 and 2020.

12.2 MISLEADING GRAPHS AND STATISTICS

NCTM Standards

- analyze and evaluate the mathematical thinking and strategies of others (pre-K–12)
- recognize and apply mathematics in contexts outside of mathematics (pre-K–12)

Miracle Yellow Food Diet!
Eat only yellow foods . . . as much as you like.
LEMONS   POTATO CHIPS   SQUASH
GRAPEFRUITS   BANANAS   EGG YOLKS
Lose as much as 8 pounds in a week!

If you don't learn to recognize common statistical deceptions, some people will mislead you. By learning about the misleading graphs and statistics presented in this section, you will become a shrewder individual who can cut through the b.s. (butchered statistics).

MISLEADING GRAPHS

Figure 12–8

Look at how the average pay has increased at the Acme Electric Washcloth company (Figure 12–8)!

LE 1 Opener

(a) Is the graph in Figure 12–8 accurate or misleading? Explain why.
(b) Make a more accurate graph for the same data by changing the scale on the vertical axis.
(c) What additional information would be helpful in determining whether the average pay increase at Acme was significant?

Graphs convey visual information in a simple, dramatic way, but sometimes the display is distorted. People most commonly distort line and bar graphs by stretching and shrinking scales on the vertical axis without indicating what they have done.

The most common method of distorting graphs is to start from a number other than 0 on the vertical axis (Figure 12–8). Why might someone use a distorted graph? The person may want to mislead you. Many times, however, a distorted graph is used because it looks better and is easier to read. An undistorted graph such as the one in LE 1(b) may have a lot of wasted, empty space. If you use a distorted scale on a graph, alert the reader by distorting the *y*-axis with the $\gtrless$ symbol. The $\gtrless$ indicates a break in the numbering scale between 0 and the next number (Figure 12–9).

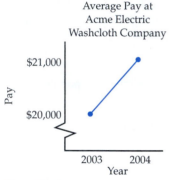

Figure 12–9

LE 2 Concept

Average Teachers' Salaries
(thousands of dollars)

Year	1997–98	1999–2000	2001–02	2003–04
Amount	39.3	41.8	44.7	47.5

(a) Construct an undistorted line graph of these data.
(b) Construct a misleading line graph that makes the salary increase from 1997–98 to 2003–04 look greater.

Someone who glances quickly at a distorted graph without reading the numbers on the axes may be deceived by the shape of the graph. If you don't want to be fooled by misleading graphs, read the numbers on the axes.

Classroom Connection

LE 3 Connection

A sixth grader asks, "If I start the vertical axis at 0, my graph will have a lot of empty space. Is it better to indicate a break on the vertical axis and use a distorted graph?" How would you respond?

DECEPTIONS INVOLVING PERCENTS

Some misleading statistics involve percents of numbers of different sizes.

Discussion

LE 4   Reasoning

I bought some pants last year. I liked them, so I went back to the store again this year to buy another pair. But the price had gone up 50%! I went to the manager to complain. She said, "Since you have been such a good customer, I'll take 50% off our new price for you."

(a) What do you think about the manager's offer?
(b) Are the pants back to last year's price? If not tell the percent increase or decrease from last year's price. (*Hint*: Assume that the pants cost $100 last year and see where the price ends up.)

Why don't the two 50%'s in LE 4 cancel each other out? The first 50% is of a smaller amount than the second one, so the 50% increase in price is a smaller amount than the 50% decrease in price. Be careful about comparing percents of different-sized amounts. Try this in the following exercises.

In LE 5 and LE 6, find the mistake in the italicized conclusion and correct it.

Discussion

LE 5   Skill

Suppose food prices went up 10% last year and 10% again this year. Over the 2 years, food prices went *up 20%*. (*Hint:* Assume that food prices started at 100, and see where they end up.)

Discussion

LE 6   Concept

Suppose the United States and Belgium both have GDP (gross domestic product) growth of 4% in 2004. *The two nations have increased their wealth (GDPs) by the same amount.*

Discussion

LE 7   Concept

Your landlord's heating bill went up 25% this year. Does this justify his raising your rent 25%? Why or why not?

LE 8   Reasoning

"SUDS 'N CRUD IS AMERICA'S FASTEST GROWING BEER!"

(a) Fill in the last column of the table.

Product	2002 Sales	2003 Sales	Percent Increase in Sales
Blubber Beer	$2,000,000	$2,200,000	
Lo-Cal Beer	1,500,000	2,100,000	
Slob's Beer	800,000	1,000,000	
Suds 'n Crud	300	600	

(b) What is misleading about the headline at the beginning of this problem?

LE 4–LE 8 concern deceptions involving percents. Remember that taking the same percent of a larger number will give a larger amount, so "all 50%'s are not equal." Beware of a percent standing alone. Every percent is a percent of *something*.

Percents of Numbers of Different Sizes

A, B, and C are positive numbers. If $A > B$, then $C\%$ of A is greater than $C\%$ of B.

DECEPTIONS INVOLVING MATHEMATICAL LANGUAGE

The following ad copy appeared in an issue of *Seventeen* magazine.

Beautiful Thick Hair

Your hair will be up to 136% thicker in 10 days. We guarantee it, or your money back.

Certain phrases, such as "up to" and "as much as," enable advertisers to include highly unlikely results in their slogans.

LE 9   Reasoning

"LOSE AS MUCH AS 8 POUNDS IN A WEEK!"

(a) If the claim is true, what is the maximum amount of weight you might lose in a week?
(b) If the claim is true, what is the minimum amount of weight you might lose in a week?

"Up To" and "As Much As"

The phrase "up to a number A" can mean any number $\leq A$. The phrase "as much as a number A" also can mean any number $\leq A$.

The statement "Nothing is better" makes it sound as if a product is the best.

LE 10   Reasoning

"NOTHING CLEANS SHOES BETTER THAN WASHOUT."

(a) Suppose that there are five different shoe-cleaning products on the market, all equally effective. Is the advertising headline true?

(b) Assume that there are five equally effective products. Would it also be true to say that "NO OTHER SHOE CLEANER IS WORSE AT CLEANING SHOES THAN WASHOUT"?

"Nothing Is More Than"

The statement "Nothing is more than a number A" can mean all numbers $\leq A$.

LE 9 and LE 10 concern deceptions involving mathematical language. As you have seen, the phrases "as much as a number" and "up to a number" mean that any amount is possible from 0 "up to" the number mentioned. "Nothing is better" means that other products may be the same as this one.

DECEPTIONS INVOLVING OMISSIONS OR TIME INTERVALS

In making a comparison based on statistics, consider unmentioned factors that are also important.

LE 11   Reasoning

"THE NATIONAL MOTORS PARAKEET GETS THE BEST MILEAGE OF ANY COMPACT CAR!"

What else should a prospective buyer find out about the Parakeet when comparing it to other compact cars?

People presenting historical statistics often select time intervals that are not representative. In examining data from a specific time period, consider whether there is anything special about that period that would affect the data.

LE 12   Reasoning

"SALARIES INCREASE MORE SLOWLY IN THESE HARD TIMES." A politician noted that U.S. salaries increased an average of 25% from 1980 to 1984, compared to an increase of 34% from 1976 to 1980. What additional information is needed?

LE 11 and LE 12 illustrate deceptions involving comparisons and time intervals. Advertisers often choose a particular factor that makes their products compare favorably by ignoring other important factors. In discussing trends over time, people sometimes select a time period that supports their viewpoint but is not truly representative.

LE 13   Summary

Tell what you learned about misleading statistics in this section.

ANSWERS TO LESSON EXERCISES

5. *21%* (100 → 110 → 121)

6. *Therefore, the United States has increased its wealth by a greater amount.*

7. No. The heating bill is only a small part of the total expenses.

8. (a) Last column has 10%, 40%, 25%, 100%
   (b) Suds 'n Crud has the highest percent increase but a much lower dollar increase in sales than the other three companies.

9. (a) 8 lb   (b) 0 lb

10. (a) Yes   (b) Yes

11. Price, repair record, safety record, trunk space

12. The rate of inflation during each time period; whether anything unusual happened between 1976 and 1984

12.2   HOMEWORK EXERCISES

Basic Exercises

1. The following graph appeared in the May–June 1989 issue of *Learning 89*. It compares the lengths of the school years in six countries.

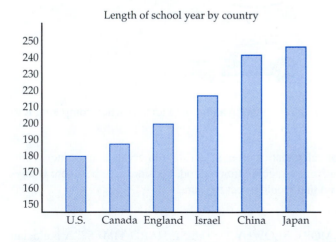

Length of school year by country

(a) Which country *appears* to have a school year twice as long as that in the U.S.? (Use a ruler.)

(b) Graph the same data on an undistorted graph.
(c) Which country is made to look the worst by the distortion in the original graph?
(d) What additional information would be helpful in comparing the school years in the different countries?

2. Governor Slick says that he has raised education spending during the past four years. Here are the figures.

Year	2000	2001	2002	2003
Spending (millions)	200	204	207	210

(a) He constructs a distorted graph to show that education spending has increased a lot. Draw a possible example of his graph.
(b) How many times higher does education spending in 2003 appear to be than education spending in 2002 on your graph in part (a)?
(c) Make an undistorted graph of the same data.

3. Why do people use distorted graphs in some instances?

4. Try stretching and shrinking the horizontal axis for your graph from Exercise 2(c). How much does it change the appearance of the graph?

5. Oil prices went up 20% one year and 30% the next. Over the two years, oil prices rose _____%.

6. My rent went down 10% last year and then rose 20% this year. Over the two years, my rent went up _____%.

7. An employee earning $500 per week was given a 20% pay cut. After a few weeks, she threatened to quit, and her boss offered her a 20% pay raise. What will her salary be after she receives the pay cut and the raise?

8. In 2003, Hospital A gave 50 people the wrong medication, and Hospital B gave 100 people the wrong medication. Was a patient twice as likely to receive the wrong medication at Hospital B? Explain why or why not.

9. Sidney's expenses for food, rent, and miscellaneous all increased 20% this year. Sidney figures that his total expenses went up 60%. Explain what is wrong with his reasoning.

10. The numbers c, x, and y are positive. If $x < y$, then c% of x _____ c% of y.
$$(<, >, =)$$

11. Last year, a company made a 10% profit. If its profit margin decreased by 60%, what percent profit did the company make this year?

12. Last year, the Social Security budget rose 12%. This year, the rate of increase is down 25%. This year's Social Security budget is _____%
_____ than last year's budget.
(higher, lower)

13. What important information is missing from the following statement? "After 100,000 miles of test driving, less than 1% of National Motors Cars needed any repairs."

14. What important information is missing from the following statement? "Eighty percent of all dentists surveyed recommend sugarless gum for their patients who chew gum."

15. "FALL SALE: SAVINGS UP TO 50%!"
 (a) If the claim is true, what is the most you will save on an item?
 (b) If the claim is true, what is the least you will save on an item?

16. Find any misleading statements in the following advertisement that once appeared in *Seventeen* magazine.

> LONDON UNIVERSITY CRASH-BURN WONDER DIET
> Of all medically safe and sound reducing programs
> **FASTEST WEIGHT-LOSS METHOD KNOWN TO MEDICAL SCIENCE!**
> (EXCEPT FOR TOTAL STARVATION)
> Compared to the Scarsdale diet, weight-watchers, Duke Univ., even Atkins or Pritikin, it burns away as much as:
> 4 TIMES FASTER THAN HIGH SPEED DIETS
> 11 TIMES FASTER THAN EXERCISE!"

17. The statement "A, B, and C are not more than X" means
 (a) $A = B = C = X$
 (b) A, B, and C are less than X
 (c) A, B, and C are greater than or equal to X
 (d) A, B, and C are less than or equal to X

18. "NO DETERGENT CLEANS CLOTHES BETTER THAN SLUDGE." According to the preceding headline, which of the following statements is true?
 (a) Sludge cleans clothes better than other detergents.
 (b) Sludge removes chocolate stains.
 (c) Sludge cleans clothes at least as well as other detergents.

19. "MORE PEOPLE DRINK JITTERS INSTANT COFFEE THAN ANY OTHER BRAND." If there are ten brands of instant coffee, the percent of all instant coffee drinkers drinking Jitters is between _____% and _____%.

20. "TANDY COMPUTERS ARE CLEARLY SUPERIOR." According to this slogan, which of the following is (are) true?
 (a) Tandy computers are the best computers.
 (b) Tandy computers are superior to Apple computers.
 (c) Tandy computers work well.

21. Explain the flaw in each claim or suggestion made about a product.
    (a) A store sells CDs with a list price of $16.99 for $13.99. You save 18%.
    (b) Alpo's beef-flavored dinner has "as much meat protein in a serving as 7 ounces of the finest steak."

22. An ad by the Beef Industry Council states that a 4-oz serving of beef contains the following percents of the recommended daily allowances: 76% of the protein, 35% of the niacin, 51% of the zinc, 19% of the iron, 100% of the vitamin B-12, and 13% of the calories. What important nutritional information about beef is omitted?

23. The following report describes an actual study involving Aim toothpaste.

> **Lever Brothers Company**
>
> Lever Brothers clinically tested Aim with fluoride .... Thousands of children were enrolled in two studies.... The dentists reported their findings as the reduction of dental decay of Aim with fluoride compared to Aim without fluoride in all of the teeth and their surfaces. After two years the dentists concluded that a significant reduction of dental decay resulted from the use of Aim with fluoride.

Summary of Clinical Trials of Aim with Fluoride

	Study I	Study II
Duration	2 years	2 years
Number of children	1107	1154

Reduction of Cavities of Aim with Fluoride versus Aim Without Fluoride

Whole mouth	25%	29%
Between teeth	38%	37%
Newly erupted teeth	21%	25%

(a) What two products are being compared in the study?
(b) Does the study show that Aim is better than any other fluoride toothpaste? If not, what does it show?

24. "More people die in hospitals than at home." From this statement you might conclude that you are better off staying at home when you are sick. Explain why this conclusion is unjustified.

25. In a speech on November 23, 1982, president Reagan used the following graph to illustrate the declining share of the federal budget spent on defense. Note that the defense budget was especially high from 1962 to 1967.

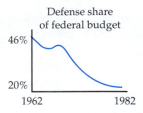

(a) Is there anything special about this time period that accounts for the higher defense spending?
(b) How is the graph distorted?
(c) Does the distortion favor or undermine Reagan's viewpoint?

26. On the same day, two New York newspapers published conflicting headlines. One said "ELECTRIC POWER USAGE INCREASES." The other said "ELECTRIC POWER USAGE DECLINES." Both were right! How is this possible?

27. Consider the following graph, which shows the average hourly earning for nonsupervisory workers from 1950 to 1998. (In 2003, average hourly earnings had increased to $15.20, or about $13.65 in 1997 dollars.)

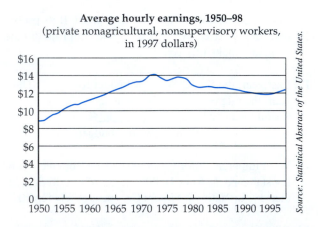

(a) If a politician wants to show that these earnings have declined, what time period might he select?
(b) If a politician wants to show that these wages have increased, what time period might he select?
(c) Give an overall description of what has happened to these wages from 1950 to 1998.

28. Consider the following graph, which shows the minimum wage from 1950 to 1998. (In 2003, the minimum wage was still $5.15, or about $4.45 in 1997 dollars.)

Minimum wage, 1950–98 (in 1997 dollars)

(a) If a politician wants to show that the minimum wage has declined, what time period might she select?
(b) If a politician wants to show that the minimum wage has increased, what time period might she select?
(c) Give an overall description of what has happened to the minimum wage from 1950 to 1998.

29. In 1970, roughly 331,000 people died from cancer in the United States. In 2003, roughly 540,000 people died from cancer in the United States. Explain why it is possible that deaths from cancer increased even though medical treatments for cancer were improving.

30. More students are taking mathematics courses than ever before. One could conclude that mathematics has become a well-liked subject. Why might the conclusion be incorrect?

Extension Exercises

31. A women's group claims that a university discriminates against women in its graduate-student admissions. The university admits 40% of the men who apply to graduate school and only 30% of the women who apply.

	Applied	Admitted	Percent Admitted
Men	150	60	40%
Women	100	45	30%

The university has two graduate departments. The university administration claims that departmental breakdowns of admissions data show that there is no discrimination.

		Applied	Admitted	Percent Admitted
Scuba Dept.	Men	50	10	20%
	Women	100	20	20%
Astrology Dept.	Men	100	50	50%
	Women	50	25	50%

In this situation, which argument is more valid, that of the women's group or that of the university?

32. A company hires men and women for jobs in three categories: executive, bureaucrat, and laborer.

	Men		Women	
Category	Number Applied	Number Hired	Number Applied	Number Hired
Executive	600	60	100	5
Bureaucrat	100	50	400	100
Laborer	100	20	200	30
Total	800	130	800	135

The company says it hires about the same overall proportion of men and women (130/800 compared to 135/800). A women's group claims that the company hires a higher % of men in each *category*. Verify the claim of the women's group, and then decide which argument is more valid.

33. What is misleading about the sizes of the regions in the following graph?

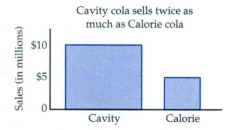

Cavity cola sells twice as much as Calorie cola

34. In the United States, women's average income is about 76% of men's average income. Speculate about what this statistic shows.

35. Consider the following problem. "A price P rises 20% and then rises 30%. Prove algebraically that the overall change is an increase of 56%. (*Hint:* Show that the final price is $1.56P$.)" Devise a plan and solve the problem.

36. A price P rises 10% and then declines 20%. Prove algebraically that the overall change is a decline of 12%.

Labs and Projects

37. Find an example of a misleading graph in a newspaper, magazine, or Web site. Write a paragraph telling what the graph is supposed to show and why it is misleading. Also, tell what additional information would be useful to have.

12.3 MEASURES OF THE CENTER

NCTM Standards

- use measures of the center, focusing on the median, and understand what each does and does not indicate about the data set (3–5)
- find, use, and interpret measures of center and spread, including mean and interquartile range (6–8)

It is possible to use a single number to describe a whole set of numbers. Sounds amazing, doesn't it? Whereas graphs and tables are useful for presenting statistical data, a set of numbers can be described more simply with an "average." However, information is also lost in this simple description of a data set.

THE MODE, MEAN, AND MEDIAN

What is the "average" amount of time it takes people in your class to go from home to class?

Discussion

LE 1 Opener

(a) Find out how long (in minutes) it takes each person in your class to go from home or from a dorm room to class. (If you have a large class, select 10 students.)
(b) Describe ways to find the "average" travel time for your data.

In my class, the commuting times in minutes were 30, 15, 12, 36, 12, 15, 15, 20, 45, 30, 15, and 20. How could we find the average? People may use "average" to refer to the most typical score, the mode. The **mode** is the score that occurs most frequently in a set. For my class, the mode is 15 minutes.

The mode is the first measure of the center that children study in elementary school. In statistics, the mode is used less often than the mean or the median and is most useful for categorical (nonnumerical) data such as peoples' favorite colors. Use the mode only for larger data sets, since slight changes in smaller data sets sometimes cause significant changes in the mode.

The most commonly used measure of the center is called the mean or average.

Definition: **Mean or Average**

The **mean** or **average** of a set of numbers is their sum divided by how many numbers there are. The mean $\bar{x}$ of the numbers $x_1, x_2, x_3, \ldots x_N$ is

$$\bar{x} = \frac{x_1 + x_2 + x_3 + \ldots + x_N}{N}$$

The mean (short for *arithmetic mean*) gives equal weight to all measurements. The word "average" usually refers to the mean. In my class, the commuting times, in minutes, were 30, 15, 12, 36, 12, 15, 15, 20, 45, 30, 15, and 20. The mean is the sum of the travel times, 265, divided by 12, or $\dfrac{265}{12} = 22\dfrac{1}{12}$ minutes.

Another useful measure of the center is the median.

> **Definition: Median**
>
> The **median** is the middle value of a set of numbers when the numbers are arranged in order. If there is an even number of values, the median is the mean of the two middle values.

To find the median in my class, arrange the numbers in order.

$$\{12, 12, 15, 15, 15, 15, 20, 20, 30, 30, 36, 45\}$$

$$\uparrow \quad \uparrow$$

The two middle numbers are 15 and 20, so the median is $\dfrac{15 + 20}{2} = 17\dfrac{1}{2}$ minutes.

LE 2   Skill

(a) Find the mode, mean, and median of your class data if you haven't already.

(b) Which of the mode, mean, and median do you feel is the best measure of your class's average travel time?

Classroom Connection

LE 3   Connection

(a) Suppose you asked a group of third graders to find the average number of pockets in the class using the following results.

$$0, 0, 0, 2, 3, 3, 3, 4, 4, 4, 4, 4, 5, 6, 8$$

What is the easiest type of average for them to use?

(b) Suppose you have a third-grade class line up according to their heights. Then you ask them what the average height is. Do you think the children will pick a height that approximates the mode, mean, or median?

Researchers have found that children want the average to be where most people are and around the middle. Which types of average do you think children would favor? The mean comes in third. When forced to choose, children will pick the mode over the median.

You can use counters or stacks of objects as models for the mode, mean, and median.

Discussion

LE 4   Connection

Suppose you have 5 stacks of math textbooks with 3, 5, 7, 2, and 3 books, respectively.

(a) How would you find the mode by looking at the stacks?

(b) How would you rearrange the stacks to show the median number of books in a stack?

(c) Suppose you want to give each of five groups the same number of books. You rearrange the books so that all the stacks are the same size. How many books will be in each stack?

(d) What type of average does part (c) illustrate?

MEAN OR MEDIAN?

How does a statistician decide whether to represent the center with the mean or the median? In my class, the median travel time $\left(17\frac{1}{2}\right)$ was slightly lower than the mean $\left(22\frac{1}{12}\right)$. Which is a better indicator of the "average" travel time? How do you know which one to use with a particular set of data?

Sometimes a set of data has a small subset of scores, called **outliers,** that are quite different from the rest of the scores. Would this group of scores be better represented by the mean or by the median? It depends on whether you want to include or minimize the effect of the outliers. In the travel-time example, the 36- and 45-minute travel times are particularly high. To give these scores equal weight, one would use the mean rather than the median.

Discussion

LE 5   Reasoning

My friend Bud is having trouble deciding which of two types of tires to buy. He has the results of a consumer group that tested 7 tires of each type. The table tells how many miles each tire lasted on the test machine.

Rocky Ride	56,000	48,000	48,000	56,000	56,000	44,000	56,000
Road Rubber	50,000	42,000	65,000	50,000	42,000	99,000	37,000

(a) Find the mean and median.
(b) Which tire is better? Give evidence to support your choice.

As LE 5 illustrates, the median is a better choice when you want to minimize the effect of an extreme value. Since the median involves only the middle value or values, it is not affected by a change in size of an extreme value. Economists usually compute "average" family income as a median so that a small number of very wealthy families do not have a large impact on the average. The median is also more appropriate than the mean for clothing sizes. A mean that comes out size 5.6 is confusing.

The mean, on the other hand, gives equal weight to all scores. The mean is a better choice if you want to include the effect of extreme values. When scores are scattered or there are very few scores, the mean may be a better indicator than the median. For example, the mean (72) of a set of test scores such as {30, 40, 90, 100, 100} is more representative than the median (90). When there are no extreme scores, the mean and median are usually about the same.

When asked to compare two groups (such as the heights of third- and fourth-graders), children's first inclination is to compare individuals rather than making comparisons of group measures such as means or medians.

LE 6   Reasoning

For each of the following sets, (1) without computing, tell whether the mean or median would be higher; and (2) tell which would be a better indicator of the "average." Justify your response to part (2).

(a) {23, 25, 28, 68, 71, 75, 78}
(b) {10, 12, 14, 16, 16, 23, 95} (Assume that you want to minimize the impact of the outlier, 95.)

People sometimes choose the "average" that promotes their viewpoint. They may even neglect to say whether they are using the mean or the median. LE 7 illustrates such a situation.

LE 7   Connection

The Washington Black-and-Blueskins have salaries (in thousands of dollars) as follows:

$$100 \quad 100 \quad 150 \quad 200 \quad 200 \quad 400 \quad 400 \quad 600 \quad 900$$

(a) The owner wants to demonstrate how high the players' "average" salary is. Would he prefer the mean or the median?
(b) The players' association wants to demonstrate how low the players' "average" salary is. Would they prefer the mean or the median?

AN INVESTIGATION: WORD LENGTHS

LE 8   Reasoning

Form groups of 2 to 4 people.

(a) Select two different books and guess which book has a higher reading level.
(b) Now decide a mathematical way of comparing a sample of words or sentences in the books to determine how the reading levels compare.
(c) Take your samples and tabulate the results.
(d) Write up the results, and use statistics and data displays to support your conclusion.

LE 9   Summary

Tell what you learned about measures of the center in this section.

ANSWERS TO SELECTED LESSON EXERCISES

3. (a) Mode   (b) Median

4. (a) Pick the most common size stack.
  (b) Arrange the stacks in size order and pick the middle (third) one.
  (c) 4   (d) Mean

5. (a) Rocky Ride: mean = 52,000, median = 56,000;
    Road Rubber: mean = 55,000, median = 50,000
  (b) Pick Road Rubber if you think that all the tires should be counted equally (mean). Pick Rocky Ride if you think the outlier (99,000) is suspicious, and you will use the median to lessen its impact on the comparison.

6. (a) (1) Median; (2) mean
    (b) (1) Mean; (2) median

7. (a) The mean   (b) The median

12.3   HOMEWORK EXERCISES

Basic Exercises

1. (a) Find the mode of the following set.

   {6, 8, 8, 9, 10, 10, 12, 12,15, 15, 15, 15, 18, 19, 20, 22}

   (b) How would you significantly lower the mode by changing just two scores?

2. Find the mode, mean and median of {4, 7, 10, 7, 5, 2, 7}.

3. Construct a set of numbers for which the mean and median are 5 and the mode is 4.

4. A family of five people has a median age of 9 years and a mean age of 17 years. Make up a set of possible ages of the family members.

5. (a) Use a shortcut to find the mean of {9, 10, 11}.
   (b) Use a shortcut to find the mean of {20, 21, 22, 23, 24, 25}.
   (c) Use a shortcut to find the mean of {36, 37, 38, . . . , 100}.

6. Find the mean for each set *without* adding up all the numbers and dividing.
   (a) {7, 7, 7, 7, 7}   (b) {56, 57, 58}
   (c) {5, 7, 9, 11, 13}   (d) {42, 42, 46, 46}

7. (a) A teacher wants to find the mean of test scores V, W, X, Y, and Z. What is the mean?
   (b) The mean of a set of five test scores is M. What is the *sum* of the five test scores?

8. If $\bar{x}$ is the mean of $x_1, x_2, \ldots, x_n$, and $\bar{y}$ is the mean of $y_1, y_2, \ldots, y_m$, what is the formula for the mean $\bar{z}$ of $x_1, x_2, \ldots, x_n, y_1, y_2, \ldots, y_m$ in terms of $\bar{x}$ and $\bar{y}$?

9. Suppose the mean of four test scores is 70. Is it possible that
   (a) none of the test scores is a 70?
   (b) all four test scores are the same?
   (c) all four scores are below 70?
   (d) three of the scores are above 70?
   (For each possible result, give an example.)

10. If possible, construct a set of six numbers in which the median is
   (a) greater than 50% of the numbers.
   (b) greater than $33\frac{1}{3}\%$ of the numbers.
   (c) greater than $66\frac{2}{3}\%$ of the numbers.

11. Match each of the four expressions in column A with one of the terms in column B.

	A	B
	Uses order of scores	Mean
	Most frequent	Median
	Equal weight to each score	Mode
	Middle	

12. (a) For a set of 7 numbers, which would usually be easiest for children to compute: the mean, the median, or the mode?
   (b) Which would usually be the hardest to compute?

13. A fourth grader says that if the mode number of books that kids have in their desks is 5, then most of the kids in the class have 5 books in their desk. Is that right? If not, how would you state correctly what the mode of 5 means?

14. Make up a set of numbers that has exactly two different modes.

15. A child says the median of {3, 14, 18, 20, 5} is 18.
   (a) What is the correct answer?
   (b) What doesn't the child understand about finding the median?

16. Where have you heard the word "median" used in everyday life? How does it relate to a median in mathematics?

17. A newspaper editorial says, "About half of our fifth graders scored below average on a standardized mathematics test. A major remediation effort must begin." Explain why these test results are not very alarming. (*Hint:* Look at the cartoon.)

David Pascal, 1978

18. Tell whether the mode, median, or mean best describes each of the following. Justify your answers.
  (a) The typical height of a fifth grader
  (b) The most typical color of sweaters in a store
  (c) The typical number of children is a U.S. family

19. You want to use counters to show a group of fourth graders how to find the mode, median, and mean of the set {2, 2, 2, 4, 5}.
  (a) Explain how you would show them the mode.
  (b) Explain how you would show them the median.
  (c) Explain how you would show them the mean.

20. A professor helped four students yesterday for 12, 18, 8, and 10 minutes, respectively.
  (a) Suppose the professor had spent the same total amount of time helping the four students but had spent an equal amount of time with each one. How much time would she have spent with each student?
  (b) What is another name for the number you found in part (a)?

 21. The Nielson ratings estimate the percent of all U.S. television sets that are tuned in to each show. The hourly prime-time Nielson ratings for two of the major networks from 3/31/03 to 4/4/03 were as follows.

NBC: 6.1, 6.6, 7.0, 5.4, 5.1, 5.7, 6.6, 8.4, 10.5, 11.4, 8.6, 12.4, 6.9, 6.9, 10.4
CBS: 8.9, 10.8, 11.4, 8.3, 7.8, 8.9, 6.3, 6.9, 7.3, 11.6, 16.6, 9.1, 5.2, 5.2, 5.1

How would you determine which network did the best in that week?

 22. How do prescription drug prices in Europe compare to those in the United States? A statistician collects average prices for seven different drugs.

Drug Name	U.S. price	Europe Price
Allegra	$23.33	$6.96
Cipro	$36.14	$15.20
Coumadin	$12.58	$2.74
Prilosec	$34.49	$17.56
Prozac	$23.98	$14.70
Vasotec	$9.38	$2.77
Zyrtec	$16.70	$5.91

Write a paragraph summarizing the difference in price.

 23. Monica has an average of 77 after 3 tests.
  (a) What score does she need on the fourth test to raise her average to 80?
  (b) If y represents Monica's average after 4 tests, write a formula for y in terms of t, her score on the fourth test.
  (c) Graph y versus t with t on the x-axis.
  (d) What conclusions can you draw from your graph?
  (e) How will an increase in t affect y? (*Hint:* The graph is a line.)

 24. Blair's bowling average after 3 games is 150.
  (a) Is a bowling average a mean or a median?
  (b) Blair then bowls 170. How much does his average go up?
  (c) If he bowls 170 again, how much does his average go up?
  (d) Why does the second 170 raise his average less than the first 170?
  (e) Blair has an average of A after N games. If he bowls 170, what is his new average?

25. In 1998, Senate incumbents running for reelection received the following campaign contributions over a two-year period.

Total Two-Year Donations for U.S. Senate Campaign (to nearest million dollars)	Frequency
1	4
2	3
3	6
4	7
5	2
6	3
7	2
12	1
16	1

(a) How many Senators are there in this study?
(b) Find the mode.
(c) Find the mean.
(d) How many Senators received donations above the mean amount?
(e) Find the median (*Hint:* You could imagine writing out all 29 numbers from lowest to highest.)

26. A class made the following scores on a test.

Score	Frequency
100	1
90	3
80	4
70	3
60	0
50	2

(a) How many students received a 70?
(b) How many students are in the class?
(c) Find the mode.
(d) Find the mean.
(e) Find the median.

 27. Following are Karen Svenson's grades for the fall semester. Compute her grade-point average (A = 4, B = 3, C = 2, D = 1, F = 0).

Course	Credits	Grade
Calculus	4	B
Biology	4	C
Literature	3	B
French II	3	A
Badminton	1	D

 28. (a) A class of 23 students had a mean of 78 on a mathematics test. The 10 boys in the class had a mean of 76.2. What was the mean of the girls' test scores?
(b) Does part (a) involve induction or deduction?

29. You have made test scores of 68, 79, 88, 74, and 82. The final test counts as $\frac{1}{4}$ of your grade. What is the minimum score you need on the final for an 80 (B) average?

30. In the math class you teach, the student's average quiz score counts for 10% of the grade, each unit test counts for 20%, and the final exam counts for 30%. Compute the mean for each student. (*Hint:* This is similar to computing the mean of a frequency distribution.)

Student	Quiz Average	Unit Test	Unit Test	Unit Test	Final Exam
Jose Cruz	80	70	60	90	90
Kathy Heid	75	80	92	93	70

31. In some large U.S. cities, the average commute is about 16 miles. How much would it help if people lived closer to where they work? Suppose that half the commuters reduced their average commute to 4 miles, and the other half still averaged 16 miles. How much would this reduce the overall commuting traffic?

 32. Ross wants to prepare a 20-quart orange-juice mixture that will sell for $3/qt. He will use Valencia OJ that sells for $3.75/qt and Parson OJ that sells for $2.50/qt. How many quarts of each should he use? (Make a table, and guess and check.)

33. Pat played in six basketball games, scoring the following numbers of points: 2, 22, 26, 28, 30, 30. She scored 2 points during a game in which she had the flu and played for only 3 minutes. Which would be a better indicator of her typical scoring game, the mean or the median score?

34. The salaries of the executives of a certain company follow. Find the mean and the median. Which statistic seems to describe the "average" salary best?

Position	Salary
President	$90,000
First vice-president	35,000
Second vice-president	35,000
Supervisor	30,000
Accountant	24,000
Personnel manager	24,000

35. A class has 10 students with brown hair, 6 with black hair, and 3 with blonde hair. Would you use the mean, median, or mode to find the "average" hair color?

36. A survey shows that green is a class's favorite color. Is this a mean, median, or mode?

 37. Each set shows the ages of a group of college students in a seminar. Which would be a better indicator of the "average" age, the mean or the median?
(a) {19, 20, 20, 20, 42, 47, 48}
(b) {18, 19, 19, 20, 20, 21, 62} (Assume that you want to minimize the effect of the outlier, 62.)
(c) Compute the mean and median in part (b) without the outlier. How much did each one change?

 38. A class has 8 students with heights, in inches, as follows:

$$\{60, 62, 64, 66, 68, 68, 70, 72\}$$

(a) Find the mean and the median.
(b) Suppose that a basketball recruit who is 86 inches tall joins the class. Find the new mean and median.
(c) Which "average" changed the most because of the new student?

39. The teachers' salaries (in thousands of dollars) at a school are

30 32 32 34 36 38 40 45 50

(a) The teachers' union wants to show how low the "average salary" is. Would they prefer the mean or the median?
(b) The school board wants to show that the "average salary" is high enough. Would they prefer the mean or the median?

40. A business has five employees. The owners say that the "average salary" is $30,000. The workers say that the "average salary" is $20,000. Make up a set of data that makes both groups right.

41. Without computing, tell which would be higher, the mean age or the median age in your mathematics class.

42. (a) A newspaper reports that "average family income in the United States in 2003 is $54,500." What important detail has been omitted from this report?
(b) Which would be higher, the mean family income or the median family income in the United States?

43. Which would be higher, the mean height or the median height of the three people in the following photograph?

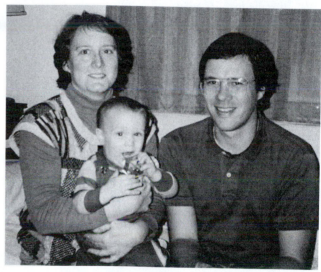

Photo by Tom Sonnabend.

44.

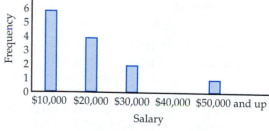

Salaries at Fatco (nearest $10,000)

(a) Which would be highest: the mean, median, or mode?
(b) Which would be lowest: the mean, median, or mode?

45. In a 1996 study of civil cases in 75 U.S. counties, the median award was $35,000, and the mean award was $300,000. Were most awards closer to $35,000 or $300,000?

46. An article says that the "average" number of persons in each car on the road is 1.3 persons. Is this figure a mean, median, or mode?

Extension Exercises

47. (a) During a 20-mile commute, Jill averages 50 mph for the first 10 miles on the expressway and 30 mph for the last 10 miles in town. What is her mean speed for the whole trip? (*Hint:* It's not 40 mph! Figure out the total time for the trip.)
(b) During a 20-mile commute, Jill averages M mph for the first 10 miles and N mph for the next 10 miles. What is her mean speed for the 20 miles?

48. In the first half (distance) of a trip, a truck travels at a speed of 60 km/hr. How fast must it go during the second half of the trip to average 80 km/hr for the entire trip?

49. Perry Peebles has taken 45 credits, and he has a GPA of 2.8 at a community college. Perry needs a 3.0 to transfer into his state's education program. This semester, he will take 15 credit hours. What GPA does he need this term to bring his average up to 3.0?

50. Two math classes take a test. The mean is 70 in class A and 80 in class B. What circumstance would guarantee that the mean for the total group is 75?

51. (Adapted from Walter Penney's problem in the 12/79 *Scientific American*)

A New Tax Plan

A society has four economic classes, each containing one quarter of the people. Class 1 is the richest, class 2 is the second richest, class 3 is the second poorest, and class 4 is the poorest.

Senator Trickledown proposes that each pair of consecutive classes, beginning with the richest (1 and 2, then 2 and 3, then 3 and 4), have their wealth averaged and redistributed evenly to everyone in those two classes. Senator Dragdown proposes the same thing but wants to begin the averaging with the poorest classes (3 and 4, then 2 and 3, then 1 and 2).

Suppose that the money is distributed as follows.

Class	Money, in Trillions of Quaggles
1	20
2	12
3	6
4	4

(a) How would the money be distributed after the completion of Senator Trickledown's proposed redistribution?
(b) How would the money be distributed after the completion of Senator Dragdown's proposed redistribution?
(c) Which classes will prefer which plan?
(d) Which class would lose under both plans?
(e) Which classes would gain under both plans?

52. Another way to compute the mean is to use the *assumed-mean method*. Suppose that you want to find the mean of {72, 86, 89, 90}. You might guess (assume) that the mean is 82. Then see how far each score is from the assumed mean.

Score	Assumed Mean	Difference
72	82	-10
86	82	4
89	82	7
90	82	8
		Total $+9$

The sum of the differences for the four scores is 9 above the assumed mean. If each score contributed equally to the $+9$, each score would be $\frac{9}{4}$ above the assumed mean. Thus, the actual mean is

$$82 + \frac{9}{4} = 84\frac{1}{4}.$$

(a) Find the mean of {65, 71, 75, 84} using the assumed-mean method.
(b) Find the mean of {25, 40, 42, 51, 58} using the assumed-mean method.
(c) Show why $\dfrac{72 + 86 + 89 + 90}{4}$ is the same as $82 + \dfrac{-10 + 4 + 7 + 8}{4}$.

$\left(\textit{Hint: } \text{The 82 results from } \dfrac{82 + 82 + 82 + 82}{4}\right).$

Technology Exercises

53. Use a graphing calculator to find the mean and the median for the data in Exercise 22.
    (a) On the TI-83, press STAT and press ENTER to Edit. Enter the values for the U.S. in L1 and the values for Europe in L2.
    (b) Press LIST (2nd STAT). Highlight MATH and choose mean(and press ENTER. Then press L1 (2nd 1) and press ENTER.
    (c) Return to the LIST menu and use it to compute the median of L1 and the mean and median of L2.

54. A teacher may keep students' grades in a spreadsheet. (Enter the following on a computer spreadsheet if you have one.)

	A	B	C	D	E	F
1	Student	Test1	Test2	Project	Total	Average
2	Nabila	86	90	70		
3	Dennis	73	84	65		
4	Candace	75	84	82		
5	Abbey	91	75	65		
6	Vince	91	68	65		

(a) What is the location of the 84?
(b) What does the number in cell B3 represent?
(c) What is the formula to determine the amount in cell E2?

(Continued in the next column)

(d) What is the formula to determine the amount in cell F2 if all three scores count equally?
(e) Use a spreadsheet to complete columns E and F.
(f) See if you can compute the mean, median, and mode with special spreadsheet functions called AVERAGE, MEDIAN, and MODE. For example enter =AVERAGE(B2:B6) into cell B7 to find the mean for test 1.
(g) See if you can find the median and mode for test 1 with spreadsheet functions.

Labs and Projects

55. Conduct a survey of your class or another group, and determine the mode, median, and mean. You could ask them, "How many brothers and sisters do you have?" or "How many hours a week do you work?" or devise your own question.

56. Find an article on the Web or in a newspaper that includes some type of average. Tell whether it is a mean, median, or mode. Do you think they chose the most appropriate measure of the center?

57. Use comparative data for ready-to-eat cereals, and develop a system for rating the nutritional value of these cereals.

58. Read "Milo and the Mathemagician" by Norton Juster in *The Mathematical Magpie*, edited by Clifton Fadiman, or in *The Phantom Tollbooth* by Norton Juster. Write a report that includes both a summary of the story and your reaction to it.

12.4   MEASURING SPREAD

NCTM Standards

- describe the shape and important features of a set of data and compare related data sets, with an emphasis on how the data are distributed (3–5)
- find, use, and interpret measures of center and spread, including mean and interquartile range (6–8)
- discuss and understand the correspondence between data sets and their graphical representations, especially histograms, stem-and-leaf plots, box plots, and scatterplots (6–8)

LE 1   Opener

A teacher wants to compare the work of two of her students named Milly and Billy, who each took eight math quizzes. Their scores are as follows.

Milly: 3, 5, 5, 6, 8, 9, 10, 10
Billy: 5, 6, 7, 7, 7, 7, 8, 9

(a) Find the mean and median for Milly and Billy.
(b) How do Milly's and Billy's scores differ?

As LE 1 illustrates, a mean or median is not always sufficient for comparing two sets of data. Milly and Billy have the same mean and median, yet their scores are different. What is a way to measure the difference?

Perhaps you computed the range in LE 1. The range is a simple way to measure the difference in spread between Milly's and Billy's scores. Milly's range is $10 - 3 = 7$, and Billy's range is $9 - 5 = 4$.

The range is not a very useful measure of spread, because it uses only the maximum and minimum values. Fortunately, statisticians have developed better ways to measure the spread.

PERCENTILES AND BOX-AND-WHISKER PLOTS

Statisticians usually describe a set of data using a measure of the center (mean or median) and a measure of the spread. In the case of Milly and Billy, their average scores are the same, but the spreads of their scores are different.

Statisticians sometimes describe the spread with percentiles. A score in the **pth percentile** of a distribution is greater than or equal to p percent of the scores. When we figure the percentile of someone's test score, there is usually more than one person with the same score. In such a case, half of the others with that score are assumed to have better scores, and half are assumed to have worse scores than the score being ranked.

LE 2   Concept

My friend Van Scott called the other day. She was upset because her child scored in the 60th percentile on a mathematics test. She said, "When I was in school, 60% right was a D or an F. I want my child to do better than that!"

(a) What would you tell her?
(b) Is her child's score above average or average?

As you may have known, Van Scott's child scored better than 60% of the test group. The child is above average on this mathematics test!

In describing the spread with percentiles, statisticians use a five-number summary that includes the median, the **upper extreme** (maximum) and **lower extreme** (minimum) scores, and the lower and upper quartiles. The median indicates the 50th percentile. The **lower quartile** indicates the 25th percentile, and the **upper quartile** indicates the 75th percentile.

> **The Five-Number Summary**
>
> **1.** Find the lower extreme (minimum), median, and upper extreme (maximum).
>
> **2.** The lower quartile is the median of the scores below the location of the median.
>
> **3.** The upper quartile is the median of scores above the location of the median.

Statisticians use these five numbers to construct a picture called a **box-and-whisker plot**. The sixth grade *Harcourt Math* textbook (Figure 12–10) introduces children to box-and-whisker plots (called "graphs"). How would these displays look for Milly and Billy?

Classroom Connection

Explore Box-and-Whisker Graphs

A **box-and-whisker graph** shows how far apart and how evenly data are distributed.

Activity

- Write each of the data in the table on a separate card.

NUMBER OF TROPHIES WON									
35	21	24	32	36	20	24	29	27	30

- Order the data from least to greatest. Draw a star on the card with the least value, or **lower extreme**, and the card with the greatest value, or **upper extreme**.

- Find the median of the data. If the median is not one of the numbers already written, write it on a card, put it in the middle of the data, and circle it.

- Find the median of the lower half of the data. This median is called the **lower quartile**. Circle it. Separate the data to the left of the lower quartile from the rest of the data.

- Find the median of the upper half of the data. This median is the **upper quartile**. Circle it. Separate the data to the right of the upper quartile from the rest of the data.

Think and Discuss

- Look at your cards. Into how many parts do the lower quartile, the median, and the upper quartile separate the data?

- What fraction of the data are to the left of the lower quartile? to the right of the upper quartile? What fraction of the data are between the lower quartile and the upper quartile?

You have found all you need to make this box-and-whisker graph.

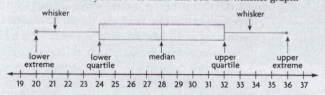

From Harcourt Math, Grade 6 (Orlando FL: Harcourt School Publishers, 2002), p. 129

Figure 12–10

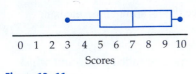

Figure 12–11

● **EXAMPLE 1**

Graph Milly's distribution in a box-and-whisker plot.

SOLUTION

Milly's lower extreme, median, and upper extreme are 3, 7, and 10, respectively. The lower quartile is the median of (3, 5, 5, 6) which is 5. The upper quartile is the median of (8, 9, 10, 10) which is 9.5. The five-number summary is 3, 5, 7, 9.5, and 10. The box-and-whisker plot uses these five numbers (Figure 12–11). The ends of the box are at the quartiles. The median determines the line segment inside the box. Line segments (called *whiskers*) extend outside the box to the extreme data values. ●

The preceding example used a very small data set to simplify computations. Statisticians draw box-and-whisker plots only for much larger data sets. Visually, the box-and-whisker plot shows the spread of the data. The five values of a box-and-whisker plot separate the graph into four parts that show the four quartiles. The thicker box section shows the center part of the distribution, and the skinny line segments extend to the high and low values.

Box-and-whisker plots provide an excellent way to compare two distributions.

LE 3 Skill

(**a**) Draw box-and-whisker plots for Milly's and Billy's scores (one above the other) on the same graph.

(**b**) Describe how the box-and-whisker plots for Milly and Billy differ.

Although box-and-whisker plots are useful for comparing distributions, a bar graph displays a single distribution with greater clarity and detail.

Classroom Connection

LE 4 Reasoning

A seventh grader computes Billy's five-number summary as 5, 6.25, 7, 7.75, and 9. What error is the child making?

THE STANDARD DEVIATION

The standard deviation is the most common measure statisticians use to describe the spread. The standard deviation complements the mean, and like the mean, it takes all scores into account.

The standard deviation measures how far scores tend to be from the mean. Consider two sets that both have means of 24: {22, 23, 24, 24, 24, 25, 25, 25} and {2, 13, 17, 24, 31, 33, 48}. The standard deviation is 1 for the first set and about 14 for the second set.

Visually, a frequency distribution that has a lower standard deviation has more scores closer to the mean, and a distribution that has a higher standard deviation has more scores farther from the mean. In Figure 12–12, compare a set of scores or a graph with a lower standard deviation to one with a higher standard deviation.

Determining the standard deviation usually requires some lengthy calculations with a calculator or computer. You can gain some understanding of the standard deviation by computing it for small sets of numbers.

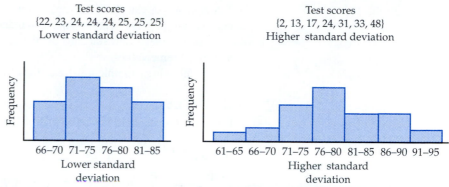

Figure 12–12

The standard deviation is based on the deviation of each score from the mean.

Definition: Deviation

The **deviation** of a score x from the mean $\bar{x}$ is $x - \bar{x}$.

For example, Milly's mean is 7. Her score of 3 has a deviation of $3 - 7 = -4$.

LE 5   Skill
Compute the deviation for each of Milly's scores.

Once you compute the deviations, is it possible simply to add the deviations and take the average (mean) of them?

LE 6   Skill
Compute the sum of the deviations for Milly's scores.

The sum of the deviations is always 0. The balance property of the mean suggests why positive and negative deviations cancel each other out.

How can we change the deviations so they won't cancel each other out? We can make them all positive by using absolute values or by squaring. Statisticians square the deviations. (Squaring the deviations gives greater weight to scores that have larger deviations.) Next, add the squared deviations and divide the sum by the total number of scores to obtain a mean of the squared deviations.

Computing the mean squared deviation squares the units of data. To obtain a number that has the same units as the original data, compute the **standard deviation**—the positive square root of the mean squared deviation.

Definition: Standard Deviation

The **standard deviation** $s = \sqrt{\text{mean of squared deviations}}$

A standard deviation of 0 indicates no variations from the mean. All other standard deviations are positive. The larger the standard deviation, the greater the spread of the data. Like the mean, the standard deviation includes all scores and is significantly affected by extreme values.

The following example illustrates how to compute the standard deviation.

• EXAMPLE 2

Calculate the standard deviation of Milly's scores.

SOLUTION

Step 1: The mean $\bar{x} = 7$.

Step 2: Find the deviation of each score from the mean.

Score	Deviation from the Mean
3	−4
5	−2
5	−2
6	−1
8	1
9	2
10	3
10	3

Step 3: Square each of the deviations.

Score	Deviation from the Mean	Squared Deviation
3	−4	16
5	−2	4
5	−2	4
6	−1	1
8	1	1
9	2	4
10	3	9
10	3	9

Step 4: The sum of the squared deviations is 48. The mean squared deviation is $\frac{48}{8} = 6$.

Step 5: The standard deviation $s = \sqrt{6} \approx 2.45$.

The five steps needed to compute the standard deviation are as follows.

Computing the Standard Deviation for *N* Scores

1. Calculate the mean.

2. Calculate the deviation of each score from the mean.

3. Square each of the deviations.

4. Add up the squared deviations and divide by *N*.

5. Take the square root to obtain the *standard deviation*.

The standard deviation measures how far scores are from the mean. It is something like an average deviation. (*Note:* In some situations, statisticians divide by $N - 1$ instead of N in step 4. We shall not concern ourselves with this distinction in this course.) Most scores will be somewhere around 1 standard deviation from the mean. Few scores will be more than 2 standard deviations from the mean.

LE 7   Skill

(a) Would you expect Billy's standard deviation to be higher than, the same as, or lower than Milly's?

(b) Follow the five-step procedure and compute Billy's standard deviation.

LE 8   Concept

The Teenytown Hamsters (basketball players) have heights of 60, 60, 64, 68, and 68 inches. The starting five for the Leantown Lizards have heights of 62, 62, 64, 65, and 67 inches.

(a) What is the mean height of each team?

(b) Without computing, tell which team's heights have a larger standard deviation.

Together, the mean and standard deviation tell a lot about a set of measurements that have an approximately normal distribution.

NORMAL DISTRIBUTION

Large random samples of measurements from homogeneous populations often have the same pattern! Consider the following exercise.

LE 9 Opener

Suppose that you took a random sample of 500 women at your college and measured their heights.

(a) Describe the pattern you would expect to find in the results.
(b) Sketch a histogram of your guess.

In the 1830s, the Belgian statistician Quetelet made a fascinating discovery. He had collected many different physical and mental measurements from a large sample of people. In nearly every case, the data sets had the same general pattern. Quetelet found that most of the measurements clustered around the mean and that there were fewer scores farther from the mean in either direction.

The data on 500 women's heights, in centimeters, might be as shown in the table and graphed in Figure 12–13.

Interval (cm)	Frequency	Relative Frequency
140–149	4	0.8%
150–159	90	18%
160–169	285	57%
170–179	112	22.4%
180–189	9	1.8%

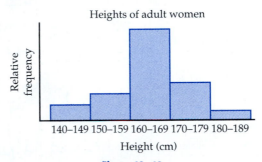

Figure 12–13

LE 10 Skill

Suppose you select one woman at random from the sample in Figure 12–13. What is the probability that her height is 170 cm or more?

If progressively larger samples were taken and the interval sizes decreased, the relative frequency histogram of women's heights would be expected to approach a certain shape, as shown in Figure 12–14.

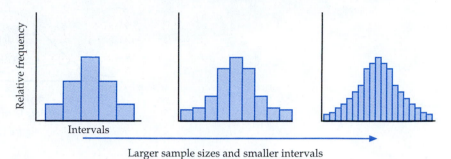

Figure 12–14

The histograms approach a theoretical graph, called a **normal distribution** or bell curve, that is continuous and perfectly symmetric on each side of the mean. Three examples of normal distributions are shown in Figure 12–15.

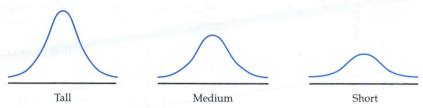

Figure 12–15

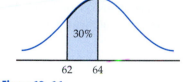

Figure 12–16

As with a histogram, the area under the curve for an interval is proportional to its frequency. Figure 12–16 shows the distribution of heights (in inches) of 16-year-old boys. Since 30%, or 0.30, of the heights fall between 62 and 64 inches, the area under the curve from 62 to 64 is 30% of the total area under the curve. The region looks like a bar with a curved top.

Statisticians use the standard deviation to describe the normal distribution in greater detail.

LE 11   Connection

College women have a mean height of about 5 ft 5 in. and a standard deviation of about 2.5 in.

(a) Estimate the percent of college women who are within 2.5 in. (one standard deviation) of 5 ft 5 in. That is, estimate the percent of women with heights between 5 ft 2.5 in. and 5 ft 7.5 in. (You could survey your class.)

(b) Estimate the percent of college women who are within 5 in. (two standard deviations) of 5 ft 5 in.

The theoretical normal distribution provides approximate answers to questions in LE 11. Those answers are given by the 68-95-99.7 Rule (Figure 12–17).

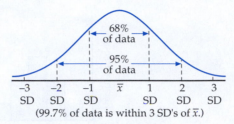

68%
of data
95%
of data
−3
SD
−2
SD
−1
SD
$\bar{x}$
1
SD
2
SD
3
SD
(99.7% of data is within 3 SD's of $\bar{x}$.)

Figure 12–17

The 68-95-99.7 Rule

In the normal distribution,

68% of the observations fall within one standard deviation of the mean.

95% of the observations fall within two standard deviations of the mean.

99.7% of the observations fall within three standard deviations of the mean.

Measurements, including heights, weights, IQs, and shoe sizes, have approximately normal distributions for large homogeneous populations. Just knowing the mean and standard deviation of a normal population enables one to estimate the percent of the population data that falls between two measurements.

Example 3 shows how to do this with the 68-95-99.7 rule.

• EXAMPLE 3

The mean height of U.S. adult men is 68.0 in., with a standard deviation of 2.5 in. Assume a normal distribution. What percent of adult men have the following heights?

(a) Between 65.5 and 70.5 in.
(b) Greater than 73 in.

SOLUTION

Sketch a graph of the distribution. It is a normal distribution, so it is bell-shaped, with the mean, 68, in the center (Figure 12–18). Since $s = 2.5$, count 2.5 units in each direction from the mean (Figure 12–19).

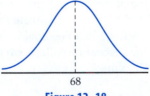

68

Figure 12–18

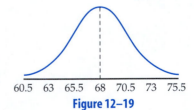

60.5 63 65.5 68 70.5 73 75.5

Figure 12–19

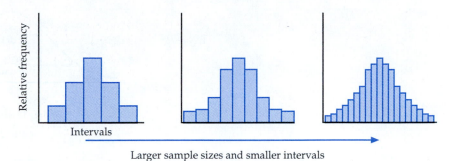

Figure 12–14

The histograms approach a theoretical graph, called a **normal distribution** or bell curve, that is continuous and perfectly symmetric on each side of the mean. Three examples of normal distributions are shown in Figure 12–15.

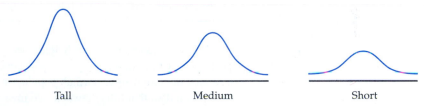

Figure 12–15

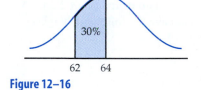

Figure 12–16

As with a histogram, the area under the curve for an interval is proportional to its frequency. Figure 12–16 shows the distribution of heights (in inches) of 16-year-old boys. Since 30%, or 0.30, of the heights fall between 62 and 64 inches, the area under the curve from 62 to 64 is 30% of the total area under the curve. The region looks like a bar with a curved top.

Statisticians use the standard deviation to describe the normal distribution in greater detail.

LE 11   Connection

College women have a mean height of about 5 ft 5 in. and a standard deviation of about 2.5 in.

(a) Estimate the percent of college women who are within 2.5 in. (one standard deviation) of 5 ft 5 in. That is, estimate the percent of women with heights between 5 ft 2.5 in. and 5 ft 7.5 in. (You could survey your class.)

(b) Estimate the percent of college women who are within 5 in. (two standard deviations) of 5 ft 5 in.

The theoretical normal distribution provides approximate answers to questions in LE 11. Those answers are given by the 68-95-99.7 Rule (Figure 12–17).

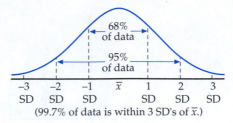

Figure 12–17

> ### The 68-95-99.7 Rule
>
> In the normal distribution,
>
> 68% of the observations fall within one standard deviation of the mean.
>
> 95% of the observations fall within two standard deviations of the mean.
>
> 99.7% of the observations fall within three standard deviations of the mean.

Measurements, including heights, weights, IQs, and shoe sizes, have approximately normal distributions for large homogeneous populations. Just knowing the mean and standard deviation of a normal population enables one to estimate the percent of the population data that falls between two measurements.

Example 3 shows how to do this with the 68-95-99.7 rule.

• EXAMPLE 3

The mean height of U.S. adult men is 68.0 in., with a standard deviation of 2.5 in. Assume a normal distribution. What percent of adult men have the following heights?

(a) Between 65.5 and 70.5 in.
(b) Greater than 73 in.

SOLUTION

Sketch a graph of the distribution. It is a normal distribution, so it is bell-shaped, with the mean, 68, in the center (Figure 12–18). Since $s = 2.5$, count 2.5 units in each direction from the mean (Figure 12–19).

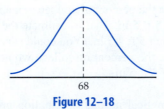

Figure 12–18

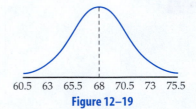

Figure 12–19

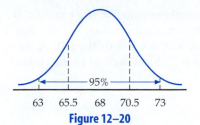

Figure 12–20

(a) 68% of the heights are between 65.5 and 70.5 in. (Figure 12–20).

(b) Since 73 in. is 2 standard deviations above the mean, we know that 95% of the heights are between 63 in. and 73 in. (Figure 12–20). This means that 100% − 95% = 5% of the heights are below 63 in. or above 73 in. Since the graph is symmetric, half of the 5%, or 2.5% of the heights, would be greater than 73 in.   ●

LE 12   Skill

You work in the summer as the mathematical consultant for a lightbulb factory.

(a) The boss wants to know what percent of the lightbulbs will last between 950 and 1,150 hours. You know that the mean life is 1,050 hours and the standard deviation is 50 hours. Make a sketch of the distribution, and make an estimate for the boss.

(b) The boss wants to offer a guarantee to replace bulbs that last less than 1,000 hours. What percent of the bulbs would this be?

(c) Do parts (a) and (b) involve induction or deduction?

You can also report measurements from normal distributions in percentiles.

LE 13   Skill

Using the data from Example 3, give the percentile for each man.

(a) N. B. Tween, 68 in. tall
(b) Hy Mann, 73 in. tall

Although the normal distribution was developed for continuous data (for example, all lengths between 0 and 10 cm), it can also be applied to large sets of discrete data (such as quiz scores that are all whole numbers between 0 and 10) (Figure 12–21).

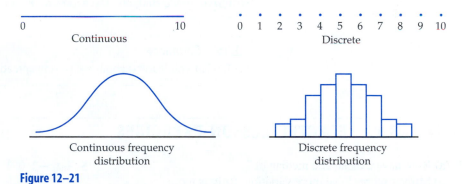

Figure 12–21

A data distribution can be classified as approximately **symmetric** (for example, the normal distribution) or **skewed** in one direction. When a nonsymmetric curve has a longer left-hand "tail," as shown in Figure 12–22(a), it is **skewed to the left.** A curve with a longer "tail" on the right, as shown in Figure 12–22(b), is **skewed to the right.**

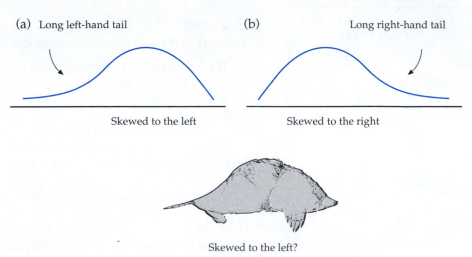

(a) Long left-hand tail   (b)   Long right-hand tail

Skewed to the left   Skewed to the right

Skewed to the left?

Figure 12–22

LE 14   Connection
Decide whether each set of discrete data is most likely to be approximately symmetric, skewed to the left, or skewed to the right.

(a) Second grade students' test scores on a fourth grade test
(b) Fourth grade students' test scores on the same test
(c) Sixth grade students' test scores on the same test

LE 15   Summary
Tell what you learned about measuring spread in this section.

ANSWERS TO SELECTED LESSON EXERCISES

1. (a) Both have a mean and median of 7.
   (b) Milly's scores have more variation. Billy is more consistent.

3. (a) The five-number summary for Billy is 5, 6.5, 7, 7.5, 9.
   (b) Milly's scores have a considerably larger spread than Billy's.

4. The child is using the mean instead of the median.

5. $-4, -2, 0, 1, 1, 4$

6. 0

7. (a) Lower   (b) $\sqrt{\frac{8}{6}} \approx 1.2$

8. (a) T. H. mean $= 64$, L. L. mean $= 64$
   (b) The Teenytown Hamsters

10. $\dfrac{121}{500}$

12. (a) 95%   (b) 16%   (c) Deduction

13. (a) 50th   (b) 97.5th

14. (a) Skewed to the right
(b) Symmetric
(c) Skewed to the left

12.4 HOMEWORK EXERCISES

Basic Exercises

1. A brochure for a new lakefront resort, Muddy Lake, says that the lake has a mean depth of 6 ft. Explain why it may not be possible to swim or canoe in the lake.

2. Make up two sets of 5 numbers that have the same mean and median, but have much different spreads.

3. Nick Durst scored in the 42nd percentile on a science test. This means that
(a) he got 42% of the questions right.
(b) he scored 42% better than the average student in his group.
(c) he did better than 42% of his group.
(d) his score is 42% of the average score.

4. Mary scored in the 80th percentile on a mathematics test. What does this mean?

5. Lyle scored in the 43rd percentile on a mathematics test. What does this mean?

6. Frederique scored in the 47th percentile in science. What does this mean?

7. How do the taxes in industrialized nations compare? The following table of 1999 data from OECD shows the percent of GDP (Gross Domestic Product) that goes for all taxes combined in a sample of 11 countries. (In 2001 and 2003, the United States had two more significant federal tax cuts.)

Country	Total Taxes as % of GDP	Country	Total Taxes as % of GDP
Austria	44%	Norway	42%
Canada	37%	Sweden	52%
Denmark	51%	Switzerland	35%
France	46%	United Kingdom	37%
Germany	38%	United States	29%
Japan	28%		

(a) Make a box-and-whisker plot of data.
(b) How does the United States compare to the other countries?
(c) What other factors should be considered in comparing taxes in these countries?

8. Following are the ages at inauguration of all U.S. presidents from Washington to G. W. Bush.

57 61 57 57 58 57 61 54 68
51 49 64 50 48 65 52 56 46
54 49 50 47 55 55 54 42 51
56 55 51 54 51 60 62 43 55
56 61 52 69 64 46 54

(a) Make a box-and-whisker plot of the data.
(b) Describe the overall pattern in the age data.

9. Two mathematics classes have the following test scores.

$A = \{56, 67, 73, 78, 81, 84, 90, 92, 97\}$
$B = \{68, 73, 75, 79, 80, 82, 84, 89, 91\}$

(a) Make box-and-whisker plots for the two classes on the same graph.
(b) Describe how their scores compare, based upon the box-and-whisker plots.

10. The hourly prime-time Nielsen ratings for NBC and CBS during the week of 3/31/03 were as follows.

NBC: 6.1, 6.6, 7.0, 5.4, 5.1, 5.7, 6.6, 8.4, 10.5, 11.4, 8.6, 12.4, 6.9, 6.9, 10.4

CBS: 8.9, 10.8, 11.4, 8.3, 7.8, 8.9, 6.3, 6.9, 7.3, 11.6, 16.6, 9.1, 5.2, 5.2, 5.1

  (a) Make box-and-whisker plots for the two networks on the same graph.

  (b) Describe how their ratings compare, based on the box-and-whisker plots.

11. (a) Give the five-number summary for the box-and-whisker plot shown.

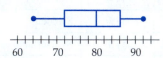

  (b) Draw a second box-and-whisker plot showing a 10% increase in all values in the set.

  (c) Compare the two box-and-whisker plots, and tell how they are related.

12. Give the five-number summary for the following box-and-whisker plot.

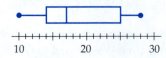

13. For each part, specify which type of graph you would use (bar graph, line graph, circle graph, box-and-whisker plot or scatterplot).

  (a) You want to study the relationship between the scores of a group of high school students on the mathematics and English sections of a standardized test.

  (b) You want to compare the test scores of two different sixth grade classes on a mathematics test.

14. For each part, specify which type of graph you would use (bar graph, line graph, circle graph, box-and-whisker plot, or scatterplot).

  (a) You want to study the monthly sales figures for a shoe store for the past 12 months.

  (b) You want to survey households about how many televisions sets they have.

15. Without computing the standard deviation, tell which histogram has a higher standard deviation.

  (a)

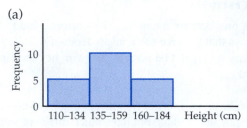

  (b)

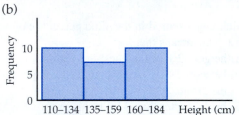

16. Without computing the standard deviation, tell which set of scores has a higher standard deviation.

  (a) {4, 5, 6, 7, 8}

  (b) {0, 3, 6, 9, 12}

17. Each of the following sets of numbers has a mean of 50. Without computing s, tell which has the largest standard deviation and the smallest standard deviation.

  (a) 0, 20, 40, 50, 60, 80, 100

  (b) 0, 48, 49, 50, 51, 52, 100

  (c) 0, 1, 2, 50, 98, 99, 100

18. Each of the following sets of numbers has a mean of 50. Without computing s, tell which has the largest standard deviation and the smallest standard deviation.

  (a) 47, 49, 50, 51, 53

  (b) 46, 48, 50, 52, 54

  (c) 46, 49, 50, 51, 54

19. Brand A lightbulbs have a mean life of 1,000 hours, with a standard deviation of 20 hours. Brand B lightbulbs have a mean life of 1,050 hours, with a standard deviation of 150 hours. Describe the advantages of each lightbulb.

20. My wife has two routes to work. She can drive on back roads that never have much traffic, or she can drive on an expressway that is faster when there is not much traffic but can take a lot longer when there is a traffic jam. Which would tend to have a higher standard deviation, a set of daily travel times on the back roads or a set of daily travel times on the expressway?

21. Match each class in column A to all appropriate descriptions in column B.

A: Class	B: Typical Test Scores
Class with wide ability range	High mean
Honors class	Low standard deviation
Students with very similar abilities	High standard deviation
Remedial class	Low median

22. A magazine rates 10 pizzas at a restaurant. Would you prefer a restaurant pizza with a
    (a) low mean and low standard deviation?
    (b) low mean and high standard deviation?
    (c) high mean and low standard deviation?
    (d) high mean and high standard deviation?

23. (a) Guess which has the larger standard deviation. Check your guesses by computing the standard deviation of both sets.
    (i) 1, 2, 4, 9   (ii) 2, 4, 5, 5
    (b) How many numbers in set (ii) are within one standard deviation of the mean?
    (c) What is the range (see Section 12.1) for each set?

 24. (a) Compute the mean and standard deviation of {3, 7, 8, 10}.
    (b) Use a calculator to compute the standard deviation of {3, 7, 8, 10}, and compare it to your answer.

25. (a) If the salary of each employee in a company is increased by $2,000, how does this change the mean and the standard deviation of the company's salaries?
    (b) If the salary of each employee in a company is increased by 10%, how does this change the mean and the standard deviation of the company's salaries?

26. (a) If you add 2 to each number in a set, how does this affect each of the following?
    (1) The mean
    (2) The standard deviation
    (b) If you add 10 to each number in a set, how does this affect each of the following?
    (1) The mean
    (2) The standard deviation
    (c) On the basis of parts (a) and (b), if you increase each score in a data set by a positive number k, how does this affect the mean and the standard deviation?
    (d) Does part (c) involve induction or deduction?

27. Take six coins.

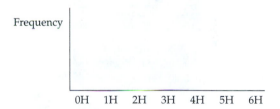

    (a) Toss them 40 times, and record the number of heads each time.
    (b) Graph a frequency distribution of the results.
    (c) Do your results suggest a normal curve?

28. An ERB standardized test report might include the following information for a sixth-grade class.

MEAN	S.D.	Q1	MED	Q3
426.7	6.2	422	428	431

    Tell what all the numbers represent.

29. Surveys indicate that U.S. adults watch television for a mean of 25 hours per week, with a standard deviation of 3 hours. Assume a normal distribution. Use the 68-95-99.7 rule to answer the following. What percent of adults watch television for the following lengths of time?
    (a) Between 22 and 28 hours per week
    (b) Less than 19 hours per week
    (c) More than 34 hours per week

30. A survey shows that the mean income in Boomtown is $21,000, with $s = \$2600$. Assume a normal distribution. Use the 68-95-99.7 rule to answer the following. What percent of people have the following incomes?
    (a) Over $21,000
    (b) Between $15,800 and $26,200
    (c) Under $18,400

31.

Sugar Fix cereal comes in a 16-oz package. The actual weights of the filled boxes are normally distributed, with a mean of 16.3 oz and $s = 0.3$ oz. What is the chance that a randomly selected box has each of the following weights?
    (a) Under 16.3 oz   (b) Under 16.0 oz
    (c) Above 16.9 oz

32. The length of a human pregnancy has a mean of 267 days and a standard deviation of 16 days. Assume a normal distribution. What percent of pregnancies
    (a) last over 283 days?
    (b) last over 235 days?

33. The weights of children in a town are normally distributed with a mean of 80 lb. and a standard deviation of 25 lb. Give the percentiles of the following scores.
    (a) 55 lb   (b) 130 lb   (c) 75 lb (Estimate it.)

34. A standardized test has normally distributed scores with $\bar{x} = 50$ and $s = 5$. Give the percentiles of the following scores.
    (a) 55   (b) 40   (c) 54 (Estimate it.)

35. A professor times the number of minutes it takes students to complete a test during a 50-minute period. The results were as follows:

    {28, 34, 39, 41, 41, 43, 45, 45, 45, 46, 47, 48, 49, 49, 50, 50, 50, 50, 50}

    (a) Summarize the data.
    (b) On the basis of this information, do you think the test was too long?

36. You grade a math test and find the following results.

    {46, 58, 62, 65, 73, 77, 77, 79, 82, 85, 85, 86, 89, 92, 94, 97}

    (a) Summarize the data.
    (b) On the basis of this information, do you think the test was too hard?

37. Describe the distribution of each of the following as approximately symmetric, skewed to the right, or skewed to the left. (*Hint:* Sketch a graph of each distribution.)
    (a) Typical scores on a classroom unit test graded 0 to 100%
    (b) Heights of adult males in the United States
    (c) Family incomes in the United States
    (d) Amounts of time students in this course will study mathematics in a given week

38. A group of gifted third-grade students takes a regular third-grade test. Would you expect their score distribution to be approximately symmetric, skewed to the right, or skewed to the left?

39. Suppose a health maintenance organization (HMO) keeps track of the numbers of visits made in one year by all of its members.

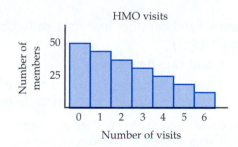

Is this distribution approximately symmetric, skewed to the right, or skewed to the left?

40. Select a page in any book.
(a) Tabulate the length of each word on the page, and display the results in a bar graph.
(b) Is your graph approximately symmetric, skewed to the right, or skewed to the left?

41. An advertisement boasts that the mean weight loss after using an exercise machine for two months was 10 pounds. You call and ask them for the median weight loss, which was 3 pounds. Does this suggest that the distribution of weight losses is symmetric, skewed to the left, or skewed to the right?

42. (a) What does it mean to say a distribution is "skewed to the left"?
(b) Would the distribution in part (a) have a higher mean or median?

Extension Exercises

43. Suppose that you read an article about a disease you have. The article says that a person with the disease lives an average (mean) of 4 years beyond diagnosis. You have already lived 3 years since being diagnosed.
(a) Explain why this article may be bad news.
(b) Explain why this article may be good news.

44. Prove that the sum of the deviations for a set of five numbers, A, B, C, D, and E, is 0.

45. There is another way to compute the standard deviation of a set. For four numbers, A, B, C, and D,

$$s = \sqrt{\frac{A^2 + B^2 + C^2 + D^2}{4} - (\bar{x})^2}$$

(a) Write a formula for $\bar{x}$ in terms of A, B, C, and D.
(b) Write s as the square root of a mean sum of squared deviations, using A, B, C, D, and $\bar{x}$.
(c) Square the binomials and combine terms to obtain the formula given at the beginning of the problem. (*Hint:* $A + B + C + D = 4\bar{x}$.)

46. Show that the standard deviation of $x_1 + k, x_2 + k, \ldots,$ $x_n + k$ is the same as the standard deviation of $x_1, x_2, \ldots, x_n$.

Technology Exercises

47. Use a graphing calculator to make a box-and-whisker plot of the data in Exercise 7. On the TI-83, press STAT and highlight Edit. Enter the data in L1. Then press STAT PLOT. Choose Plot 1. Highlight On and the box-and-whisker plot, L1 for the Xlist with 1 as the frequency. Set your WINDOW, and press GRAPH.

48. To use a TI-83 to find the standard deviation for the data from Exercise 47, press LIST, highlight MATH, and choose stdDev(. Then type L1 and a right parenthesis, and press ENTER. Use the TRACE key and the left and right arrow keys to see the five-number summary values.

Labs and Projects

49. (a) Select a book. Without looking inside, sketch a graph of the distribution of the lengths of the words. (Guess.)
(b) Now open the book somewhere around the middle. Tabulate the lengths of the first 50 words starting at the top of the page. Construct a histogram that shows the distribution of the lengths of the words.
(c) Construct a box-and-whisker plot of your data.

12.5 SAMPLES

NCTM Standards

- design investigations to address a question and consider how data-collection methods affect the nature of the data set (3–5)

- collect data using observations, surveys, and experiments (3–5)

- formulate questions, design studies, and collect data about a characteristic shared by two populations or different characteristics within one population (6–8)

Whom do people want as the next president? How do they feel about government spending on education? To answer questions such as these, would you need to ask every single adult what he or she thinks?

Fortunately, the answer is no. Taking a representative sample of about 1500 will usually give a good estimate of how 210 million U.S. adults stand on an issue! Sampling is a part of inferential statistics. **Inferential statistics** involves systematically drawing inferences (conclusions) about a larger group (population) based upon information about a subgroup (sample). The preceding lessons in this chapter covered descriptive statistics—techniques used to organize and summarize data. Descriptive statistics and inferential statistics are the two main areas of statistics.

SURVEYS

Newspapers frequently report survey results that give us information about our attitudes. A survey is one of the best ways to determine the political preferences, moral beliefs, and favorite television programs of a society.

LE 1 Opener

You want to conduct a five-question survey at your college. What would you do to make the results more reliable?

In this section, you will study some of the issues you probably considered in LE 1. The following experiment will help you understand how surveys work.

Discussion

LE 2 Connection

The class will need a paper bag with an identical small slip of paper for each class member.

(a) Select a question of interest to the class that has two possible responses, about which people in the class are likely to disagree. (For example, "Do you prefer plastic or paper bags at the supermarket?" "Would you prefer _____ or _____ in the upcoming election?")

(b) Have everyone write a response on a slip of paper and put it in the paper bag.

In a survey, the **population** is the total set of responses the researcher wants to study. In this example, the entire collection of responses from your class is the population.

LE 3   Skill

Appoint three poll takers. Each in turn will take five slips of paper from the bag, record the results, and return the slips to the bag.

A **sample** is a subset of the population, chosen to represent it. In this case, each poll taker collects a sample of five responses. Samples are used because they can provide simple, fairly accurate information about a much larger population. Samples are sometimes the only way to obtain information about a population. For example, would you expect a car company to crash-test every car it manufactures?

LE 4   Skill

(a) Ask each poll taker to give his or her sample results and to make a prediction about the population results based on *his or her sample*.
(b) Have the class predict the response totals for the class based on all three samples.
(c) Take all the papers out of the bag, and tabulate the actual results.

Discuss the following questions about the simulated survey.

LE 5   Concept

Was each poll taker's sample chosen randomly?

LE 6   Reasoning

(a) Was the sample large enough?
(b) How large a sample would be needed to predict with certainty the class's *majority* response?

Although a mathematical discussion of sample size is beyond the scope of this course, you probably have an intuitive feeling for it. If reliable sampling techniques are used, a larger sample tends to be more representative of a population than a smaller one. In nationwide surveys, pollsters usually select about 1,000 to 2,000 people to obtain results with an error of about 2% to 4%. If the sampling procedure is a biased one, such as voluntary responses, a larger sample will not improve it.

Daniel Yankelovich who ran a polling service for 40 years was asked what is the most important question to ask about the results of a survey. He replied, "Is the survey about an issue where people have made up their minds? If not, you can't rely on the results."

Figure 12–23 shows a survey activity from the fourth-grade *Harcourt Math* textbook.

Classroom Connection

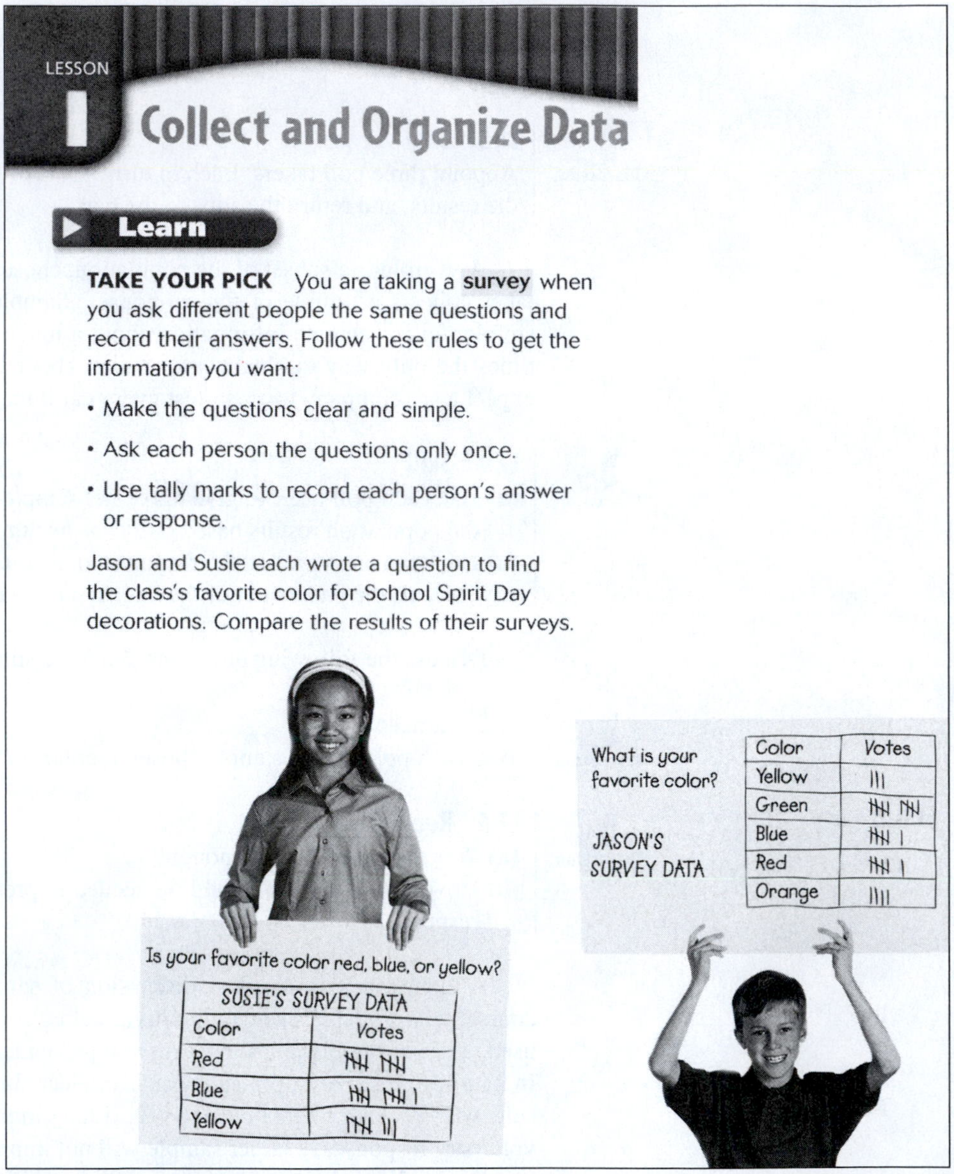

LESSON 1

Collect and Organize Data

Learn

TAKE YOUR PICK You are taking a **survey** when you ask different people the same questions and record their answers. Follow these rules to get the information you want:

- Make the questions clear and simple.
- Ask each person the questions only once.
- Use tally marks to record each person's answer or response.

Jason and Susie each wrote a question to find the class's favorite color for School Spirit Day decorations. Compare the results of their surveys.

Is your favorite color red, blue, or yellow?

SUSIE'S SURVEY DATA

Color	Votes
Red	卌 卌
Blue	卌 卌 I
Yellow	卌 III

What is your favorite color?

JASON'S SURVEY DATA

Color	Votes
Yellow	III
Green	卌 卌
Blue	卌 I
Red	卌 I
Orange	IIII

From Harcourt Math, Grade 4 *(Orlando, FL: Harcourt School Publishers, 2002), p. 82.*

Figure 12–23

SAMPLING METHODS

To be useful, a sample must be representative of the population. A large random sample has a good chance of being representative. In a **simple random sample of size *n*** from a population, each subgroup of size *n* has an equal chance of being selected. Also, each individual in a random sample has an equal chance of being selected. A random sample is the ideal. In practice, it is virtually impossible to obtain a perfectly random sample.

Discussion

LE 7   Opener

A college has 5 sections of Mathematics for Elementary Teachers, Course 1. You want to take a sample of 20 students and study their attitudes about the course. There are 20 students in each of the 5 sections. There are a total of 90 women and 10 men. Three of the classes meet during the day, and two meet in the evening.

(a) Describe some different ways to select 20 people for a sample.
(b) Which sampling methods are the best? Which are the most biased?

Which of the following sampling methods did you suggest in LE 7? To take a simple random sample, you would randomly select (with a computer) 20 names from a list of all 100 students. In a **systematic random sample,** you would randomly choose a student from the first 5 names as the starting point. Then, you would select every 5th name.

If you want a representative number of day and evening students, you could randomly select 12 students from the 60 day students and 8 students from the 40 evening students. Note that the number in each subgroup is proportional to the size of the subgroup. This is an example of a stratified random sample. In a **stratified random sample,** the pollster groups the population by a classification (stratum) such as region, ethnic group, or gender. Then the pollster takes an appropriately-sized random sample from each group.

Design a stratified random sample in the following exercise.

LE 8   Skill

A pollster wishes to select a stratified random sample totaling 200 children from three schools, in numbers proportional to the sizes of the schools. If the schools have 641, 410, and 549 students, respectively, how many children should be selected from each school?

The three preceding types of random sampling can work quite well. Two types of sampling that are likely to be biased are convenience sampling and voluntary response sampling. In LE 7, if you select 20 students whom you know or 20 students who are around at a particular time of day, that would be a convenience sample. In a **convenience sample,** the pollster selects the people who are easiest to reach. If you ask for 20 volunteers in LE 7, that is called a voluntary response sample. In a **voluntary response sample,** people volunteer to participate in response to a general appeal.

LE 9   Reasoning

(a) In a voluntary response sample, a TV station tells viewers to call one of two toll numbers to register their opinions about raising the sales tax. Why wouldn't we expect the results to be representative of what all adults think about raising the sales tax?

(b) In a convenience sample, a researcher questions some of the students coming out of the cafeteria one afternoon to determine their opinions about school parking facilities. Why wouldn't we expect the results to be representative of all the students in the college?

Classroom
Connection

LE 10   Concept

A sixth grader says that a random sample is one that has no particular pattern to it. Is that right? If not, what would you tell the child?

BIAS IN SURVEYS

In a biased sample, everyone does not have an equal chance of being selected. A small amount of bias is unavoidable in most surveys. You have already seen that convenience and voluntary response samples are both biased, yielding results that provide no useful information except about the sample itself.

What are some sources of bias when you try to select a random sample?

Discussion

LE 11   Reasoning

Suppose you want to conduct a survey of adults in your county by telephoning people from a randomly selected list of names. What problems would you have in obtaining a representative sample?

Three sources of bias are undercoverage, nonresponse, and response bias. **Undercoverage** refers to people who are left out. They cannot possibly be in the sample. In LE 10, this would include people who don't have telephones or who have unlisted numbers. Bias also results from **nonresponse.** Some people in the sample will not respond, or they may refuse to participate. **Response bias** results when people lie.

BIAS IN QUESTIONS

The wording or type of questions may result in misleading results.

LE 12   Concept

You are writing a survey question about whether or not people should be permitted to smoke cigarettes in restaurants.

(a) Write an unbiased question.
(b) Write a question that is biased against permitting smoking.
(c) Write a question that is biased in favor of permitting smoking.

It is virtually impossible to design a short survey on a complicated issue without significant bias of some sort. Opinions are usually superficial, rather than a reflection of deeply held beliefs. Polls don't usually measure the intensity of feelings very carefully. Polls rarely measure the depth of knowledge of various respondents.

LE 13   Summary

Tell what you learned about samples and surveys in this section.

ANSWERS TO SELECTED LESSON EXERCISES

5. Yes

8. 80, 51, and 69

9. (a) People who call tend to have stronger opinions about the sales tax or more positive feelings about this type of telephone survey. Also, TV station viewers do not accurately represent all adults.
(b) It is likely that the time of day and the location outside the cafeteria would eliminate certain groups of students from the sample.

10. No. Discuss the definition of a random sample. It may or may not have a pattern.

12. (a) Should people be permitted to smoke cigarettes in restaurants?
(b) Should people be permitted to smoke cigarettes in restaurants in spite of the threat to the health of others?
(c) Don't people have the right to smoke cigarettes in restaurants?

12.5   HOMEWORK EXERCISES

Basic Exercises

 1. Patti Englander is running for the U.S. Senate in Colorado. In a poll of 1,200 voters, 322 support her.
(a) What is the population for this poll?
(b) If P people vote in the election, what is a good prediction of the number that will vote for Patti?

 2. An aerial photograph of a forest is divided into 40 sections of equal area. Five sections are randomly selected, and they have 46, 38, 39, 52, and 47 trees, respectively. Estimate the total number of trees in the forest.

 3. A survey of shoppers shows that 20 prefer whole milk, 42 prefer low-fat milk, and 28 prefer skim milk.
(a) If 500 shoppers come to the store tomorrow, predict how many will buy each kind of milk.
(b) Explain how you made your prediction.

 4. Ms. Zhang wants to plan a sixth grade field trip. A random sample of sixth graders at William Gaddis Middle School finds that 15 want to go to a science museum, 12 want to go to an amusement park, and 8 want to go to a concert. There are 112 sixth graders at the school.
(a) Predict how many sixth graders want to go to the science museum, the amusement park, and the concert.
(b) Explain how you made your prediction.

5. You want to represent all U.S. households and their wealth with 10 people and 10 chairs. In the U.S. (2003), the richest 10% of all households have 70% of the wealth. How will you distribute the 10 chairs to the 10 people?

6. Suppose you took a representative sample of 100 people from the world. Guess the answers to the following.
(a) About how many can read?
(b) About how many have enough food to eat?
(c) How many have a college degree?
(d) The _____ richest people in the group have 50% of the wealth.

 7. A college has 250 freshmen, 220 sophomores, 200 juniors, and 180 seniors. If a stratified sample of 100 students is chosen, how many should be chosen from each class?

 8. A college faculty has 820 members. The college faculty wants to sample 40 faculty members using a stratified sample by rank. There are 205 full professors, 328 associate professors, 246 assistant professors, and 41 instructors. How many professors of each rank should be selected?

9. Classify each of the following as simple random sampling, systematic random sampling, stratified sampling, or convenience sampling.
(a) A student newspaper selects 10 names at random from the entire student body.
(b) A student newspaper interviews proportional numbers of freshmen, sophomores, juniors, and seniors.
(c) A dean interviews 10 students who visit her office.

10. Classify each of the following as simple random sampling, systematic random sampling, stratified sampling or convenience sampling.
   (a) A dean interviews every 10th student from a list of all the students.
   (b) A dean selects 5 men and 5 women from each of 4 English classes.
   (c) A researcher interviews all AIDS patients in the hospital closest to the researcher's home.

11. Classify each of the following as simple random sampling, systematic random sampling, stratified random sampling, or convenience sampling.
   (a) A researcher interviews a proportional number of mathematics students from the following age groups: 18–21, 22–30, and 31 and up.
   (b) A researcher interviews 20 students chosen at random from a list of all mathematics students.
   (c) A researcher interviews every 20th mathematics student on a list of all mathematics students.

12. Classify each of the following as simple random sampling, systematic random sampling, stratified random sampling, or convenience sampling.
   (a) A senator selects every 100th name from a list of voters in her district.
   (b) A student reporter interviews 20 students in the cafeteria.
   (c) A reporter selects 5 state senators to interview by writing the name of each senator on a card, shuffling the cards, and then picking 5 names.

13. The student newspaper wants to interview 100 education students at a four-year college. Give an example of
   (a) simple random sampling.
   (b) stratified sampling.

14. Describe how to obtain a sample of 30 fifth graders from 150 fifth graders, using
   (a) stratified sampling.
   (b) systemic sampling.

15. The children in a school conduct a random sample of fifth graders to determine their favorite lunch menu. The results could be used to draw valid conclusions about which of the following groups?
   (a) a fifth grade class in the school
   (b) the entire fifth grade in the school
   (c) their entire fourth grade in the school
   (d) the fifth-graders in that school district

16. We all make judgments based on sampling. For example, if you wanted to decide whether new Squeezie grapefruit juice is fresh and tasty, you might buy two cartons, try them, and then make your decision.
   Give an example of a sample you took and a conclusion you reached about
   (a) a restaurant.
   (b) a person.

17. A Harris Poll (4/2/03) asked 2,271 U.S. adults if various industries should be regulated more or less by the government. Each respondent could choose as many industries as he or she wanted from a list. Some of the results were as follows.

	Regulated More	Regulated Less
Health insurance co.	59%	7%
Drug co.	57%	7%
Oil co.	52%	7%
Tobacco co.	44%	11%
Hospitals	35%	11%
Banks	21%	10%
Supermarkets	10%	17%
None of these	20%	57%

Write a summary of these results.

18. A Harris Poll (3/16/03) asked 1,010 U.S. adults if the following groups have too much or too little power and influence in Washington, D.C. Some of the results were as follows.

	Too Much	Too Little	About Right	Not Sure
Big companies	80%	10%	4%	5%
Political lobbyists	72%	17%	6%	5%
Labor unions	45%	37%	8%	10%
Religious groups	27%	53%	10%	10%
Public opinion	19%	69%	6%	6%
Small business	4%	88%	3%	5%

Write a summary of these results.

19. A sample of 20 cars is obtained to determine the gas mileage for a 2004 Honda Civic. Tell whether each of the following is an example of undercoverage, nonresponse, or response bias.

   (a) A consumer group tests 20 cars from one dealership.

   (b) Suppose all the results came from 20 randomly selected dealerships that were likely to exaggerate the mileage of one of their cars.

20. A sample of 50 adults is obtained to rate Bubbly laundry detergent. Tell whether each of the following is an example of undercoverage, nonresponse, or response bias.

   (a) An interviewer visits a randomly selected set of homes and asks the first 50 adults who answer the door whether they use Bubbly.

   (b) An interviewer goes to the local Almost Fresh Market and observes what laundry detergent 50 shoppers select.

21. DOONESBURY   by Garry Trudeau

Doonesbury. Copyright 1980 G. B. Trudeau. Reprinted with permission of Universal Press Syndicate. All rights reserved.

   (a) In the cartoon, what does Doonesbury mean when he says polls "often become self-fulfilling prophecies"?

   (b) Do you agree with Doonesbury?

22. A telephone survey asking whether people favor or oppose the legalization of marijuana finds that 15% are in favor and 78% are opposed. Why are these results unreliable?

23. "IS OUR FACE RED!"

So read the headline of the November 14, 1936, issue of *The Literary Digest*. The magazine had conducted a poll just before the 1936 presidential election between Alfred Landon and Franklin Roosevelt. The sample came from a list of 10 million names, primarily from telephone directories, automobile registrations, and club membership lists. Each of these people was mailed a questionnaire. Of the 2.4 million people who sent in the questionnaires, 43% preferred Roosevelt, and 57% preferred Landon. This was one of the worst errors ever made in a major survey. Roosevelt won the election 62% to 38%!

   (a) Was the sample large enough?

   (b) Was the sample representative of the voting population?

   (c) Why do you think the prediction was so far off?

24. Ann Landers wrote an advice column. She asked her readers, "If you had it to do over again, would you have children?" About 70% of roughly 10,000 respondents said no. Then a national survey was taken, using a random sample of 1,400. About 90% of the respondents said yes. Which results should be more reliable? Why?

25. A school finds that 71% of the children in detention are boys even though only 50% of the children in the school are boys. Does this show that the school is biased against boys? Tell why or why not?

26. A police department checked the colors of clothing worn by pedestrians who were killed in traffic accidents at night. They found that $\frac{4}{5}$ were wearing dark-colored clothes and $\frac{1}{5}$ were wearing light-colored clothes. The police department concluded that pedestrians wearing white are less likely to be killed in accidents at night. Explain why this conclusion may not be correct.

27. Explain why the following question is biased: "Is tennis your favorite sport?"

28. Explain why the following question is biased: "Which pet would you prefer, an affectionate dog or an unresponsive cat?"

29. Give three versions of a question to determine if people think aluminum cans should be recycled.
(a) Write an unbiased question.
(b) Write a question that is biased in favor of recycling aluminum cans.
(c) Write a question that is biased against recycling aluminum cans.

30. A survey question reads as follows: "Would you favor or oppose building a peace shield that would protect us from nuclear attack?" Change the wording of the question so that more people will oppose it.

31. Changing the categories for survey responses can significantly affect the results. People are asked to respond to the following survey item. "Schoolteachers are more highly respected than stockbrokers."
(a) Agree   (b) Disagree
How would the results be affected if we added the following third choice?
(c) Don't know

32. Consider the following survey question, with no choices given for the response. "What should be the three main priorities of the federal government?" How would the results change if the following choices were given?
(a) Gun-control laws
(b) Balancing the budget
(c) Developing solar power
(d) Reducing taxes
(e) Aid to the poor and elderly
(f) Other

33. "How many languages do you speak?" What is unclear about this question?

34. "Do you favor or oppose recycling?" What is unclear about this question?

Extension Exercises

35. Following are the Nielsen ratings for the top ten regular network prime-time television shows during the week 3/31–4/6/03. The Nielsen ratings determine how around $50 billion is spent each year on television ads.

	Rating	Share
1. *CSI*	16.6	25
2. *Friends*	13.0	21
3. *American Idol* (Tu)	12.6	19
4. *ER*	12.4	20
5. *Anerical Idol* (W)	11.8	18
6. *Everyone Loves R.*	11.8	17
7. *Survivor*	11.6	18
8. *CSI: Miami*	11.4	19
9. *Law & Order*	10.5	18
10. *Law & Order: SVU*	10.4	18

The Nielsen ratings are based on the television-viewing habits of a sample of 5000 U.S. homes (out of 107 million). A box is attached to each television set in every home in order to record what show is watched each minute. The **rating** is the percentage of the 5000 homes in which a show is watched. The **share** considers only the homes with a television set on; it is the percent of these homes that are tuned to a particular show.
(a) Explain why the share is always higher than the rating.
(b) The Nielsen company attempts to obtain a random sample of 5000, but only about 65% of those they ask are willing to participate. How could they make a good case that the 5000 homes they end up with are representative?
(c) The Nielsen ratings record information for 15-minute time blocks. Any show that is watched for at least 5 minutes receives full credit for the 15-minute time block. What is the maximum number of shows that one person could be counted as watching during any 15-minute block?

 36. Suppose the 5000 homes in the Nielsen sample have 10,000 television sets. One Tuesday night, 3,300 households have sets in use, and 1,100 of these households tune in to *Friends*.
(a) What share does *Friends* receive?
(b) What percent rating does *Friends* receive?

37. *ER* had a 20 share on 4/2/03. Which of the following must be true?
    (a) 20% of all Americans watched *ER*.
    (b) *ER* is the third-best show on television.
    (c) Of all televisions in use, 20% were tuned to *ER*.
    (d) Of all the Nielsen households, 20% had sets tuned to *ER*.
    (e) Of the people living in Nielsen households, 20% were watching *ER*.

38. During the Monday night 8:00 P.M. slot, assume that 5,000 television sets are counted as being watched. Try to answer the following questions.
    (a) How many of the 5,000 homes have televisions on?
    (b) How many people are watching television?
    (c) How many people watched the whole program they were watching?
    (d) How many television sets were on?

39. In a 6-year study in Seattle, all firearm deaths were investigated. There were 743 firearm deaths, and investigators collected the following information.
    I. There were 469 suicides, 256 homicides, 11 accidents, and 7 deaths of unknown cause.
    II. Of the deaths, 473 occurred inside a house (343 involved a handgun).
    III. Of those 473 deaths, 398 happened in a house involving a firearm that was kept in the house.
    IV. The 398 deaths were divided as follows.

    333 suicides (68% from handguns)
    50 homicides (42 during a fight in the house)
    11 accidents
    4 self-inflicted deaths (not known whether they were suicide or accident)

    V. Of the 65 nonsuicides, 63 involved families or friends and 2 involved strangers (burglars).

    What conclusions can be drawn from the study?

40. The following data were collected from a variety of surveys.
    • About 45% of Americans do not read books at all.
    • The average number of books read by the reading population is 16 per year.
    • About 53% of Americans read fewer than 12 books a year.
    • About 25% of Americans read at least 20 books a year.

Assume that all of these results are accurate.
    (a) What percent of Americans read at least one book a year?
    (b) What percent of Americans read 12 to 19 books (inclusive) a year?

Labs and Projects

41. (a) Find an advertisement that makes a claim based on a survey or research study.
    (b) Write to the company for further details about the study.
    (c) If the company responds, analyze the study and decide whether the company has proved its claim and whether the claim justifies buying the product.

42. Find an example of a survey in a newspaper. Evaluate the survey questions, the sampling technique, the presentation of the data, and the interpretation of the data.

43. After the class has composed a list of items that are representative of what the class eats, have each person find the costs of these items at a neighborhood food store. Bring the results to class for a comparison.

44. Conduct a college or community survey on a topic of your choice. Write a summary of the results. Consider the following suggestions.
    What is your favorite radio station?
    What can be done to improve public schools in your county?
    How many hours do you work each week?

45. (a) Keep a log of your television watching, including hours spent watching, shows watched, and hours of commercials seen.
    (b) On the basis of your data from part (a), estimate the number of hours of television you have watched in your lifetime.
    (c) How does your answer in part (b) compare to the number of hours you have slept, eaten, or attended school?
    (d) Estimate how many hours of commercials you have seen in your lifetime.

12.6 STANDARDIZED TEST SCORES

NCTM Standards

- use the language of mathematics to express mathematical ideas precisely (pre-K–12)
- recognize and apply mathematics in contexts outside of mathematics (pre-K–12)

As a teacher, you will receive standardized test reports about your students similar to the one in the following table (which is based on the Stanford Achievement Test). Most standardized test scores are based on the concepts of the median, percentiles, and the normal distribution.

Test Score Report	Score Type	Math Comp.	Reading	Vocabulary	Math Appl.	Spell.	Lang.
Gr 5 Norms Gr 5.2	RS/NO POSS	35/44	42/60	26/36	38/40	33/40	32/53
Fran Tikley	NAT'L PR-S	68-6	59-5	64-6	97-9	73-6	48-5
Age 10-3	LOCAL PR-S	60-6	48-5	65-6	92-8	63-6	36-4
Test date 10/24/03	GRADE EQUIV	5.4	5.1	5.6	8.4	6.4	4.1

Discussion

LE 1 Opener

Try to guess the meanings of the numbers and symbols in the test score report just presented.

Fran Tikley's score report includes raw scores, percentiles, stanines, and grade-level equivalents. The **raw score** is the student's actual score on the exam. Fran received a raw score of 33/40 on spelling, which means she got 33 questions right out of 40.

LE 2 Concept

(a) What is Fran's raw score in math computation?
(b) What does this score mean?

The test report also gives percentiles. Fran's national percentile for spelling is 73. This means that she scored better in spelling than about 73% of the fifth grade second-month students in the United States on this exam.

LE 3 Concept

(a) What is Fran's local percentile score in math computation?
(b) What does this score mean?

LE 4 Reasoning

On the same test as Fran, Hy Skor had a national vocabulary percentile of 66. Does this show that Hy knows more of the vocabulary tested than Fran? Why or why not?

How accurate are these raw scores and percentiles? As you know from personal experience, a student's test score varies somewhat according to how the student feels on a particular day and which questions are chosen for the test. So a percent or percentile on an exam has some error in it, just as a percent in a survey does. Repeating a test or a survey on another day usually results in a slightly different outcome.

Percentile scores can be misleading because they appear more accurate (measuring to the nearest percent) than they really are. This problem led to the development of stanine scores. The term "stanine" is a contraction of "standard nine." **Stanines** divide normally distributed student scores into nine groups based on their percentiles, as shown in Figure 12–24.

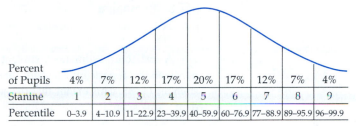

Percent of Pupils	4%	7%	12%	17%	20%	17%	12%	7%	4%
Stanine	1	2	3	4	5	6	7	8	9
Percentile	0–3.9	4–10.9	11–22.9	23–39.9	40–59.9	60–76.9	77–88.9	89–95.9	96–99.9

Figure 12–24

For example, a stanine of 6 means that the student is in the 60th to 76.9th percentile range. [*Note:* A 77th-percentile score on a test could represent a percentile such as 76.8 (6th stanine) or 77.3 (7th stanine) rounded to 77.]

In Fran Tikley's score report, the stanines are given to the right of the percentiles.

LE 5 Skill

What is Fran's national stanine score in math computation?

Stanine scores do not attempt to make "the fine distinctions which are frequently but improperly made" using percentile scores (NACOME report, 1975).

LE 6 Skill

One student has a stanine of 6, and another has a stanine of 7.

(a) What is the smallest possible difference between their percentile scores?
(b) What is the largest possible difference between their percentile scores?

Another type of score reports the student's standing in relation to grade levels. The **grade-level score** indicates the grade level, in years and months, at which the student's score would be average. Fran has a grade level of 6.4 in spelling. This indicates that she scored the same as the average student in the fourth month of sixth grade.

LE 7 Concept

(a) What is Fran's grade-level score in math computation?
(b) What does it mean?

Many educators question the value of grade-level scores because they are commonly misinterpreted in two ways. First, people expect most children to score at or above grade level even though about half the students should score below grade level.

Second, very high or low grade-level scores have little meaning. Fran Tikley scored so well on math applications that her grade-level score was 8.4. Just because she did extremely well on fifth-grade word problems, can we tell whether she can do eighth-grade word problems? A fifth-grade test would have so few eighth-grade-level questions that it could not measure Fran's ability at that level.

These drawbacks of grade-level scoring have led some educators to recommend that it be abandoned.

LE 8   Reasoning

A sixth-grade student receives a grade-level score of 3.2. Why is this score misleading?

A test score estimates a student's level. To account for possible error, some standardized test reports give a score such as 80 and a *confidence interval* or *percentile band* such as 77–83. This means that the student scored an 80 and it is very likely that the student's true score is between 77 and 83 (inclusive).

LE 9   Concept

A student has a test score of 420 and a confidence interval of 390-450. Tell what the confidence interval means.

CLASSROOM USE OF STANDARDIZED TESTS

Using standardized test scores in conjunction with other information, you can make a general evaluation of a new student's ability when he or she first comes to your school. The standardized test gives a general score that assesses student progress in subjects that are taught in most schools.

Classroom Connection

LE 10   Connection

The following student is transferring to your classroom. His score report (based on the Metropolitan Achievement Test) is given.

Name	Robert Mark	Grade	5
School	Fargo Elementary, N.D.	Date of Test	10/2/03

Score Summary Box

Test	Number Possible	Number Right	Grade Equiv.	Percentile	Stanine
Reading	60	56	6.1	85	
Mathematics	50	25	4.3	30	
Language	60	47	5.7	74	
Science	45	20	4.1	32	
Social Studies	45	28	5.3	62	

(a) Use the percentile information to fill in the stanine scores for each test.
(b) Give your general assessment of how this student will do at your school.

Some schools routinely use standardized tests to evaluate individual students. Since these tests do not measure student achievement in *specific* areas, and since they are not designed to test the particular curriculum the student has been studying, the test results have little value for measuring an individual student's progress in a school's curriculum.

IQ SCORES

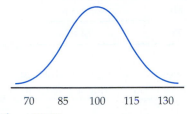

70   85   100   115   130

Figure 12–25

Psychologists disagree about what intelligence is and how to measure it. In spite of these disputes, some standardized tests attempt to measure intelligence.

The results of intelligence tests usually are reported as IQ scores. The IQ score compare a child's results with other children the same age. The average (mean) raw score is assigned a standardized IQ score of 100. A raw score one standard deviation above the mean is assigned a standardized IQ score of 115. **Standardized IQ** scores are normally distributed with a mean of 100 and a standard deviation of 15 (Figure 12–25).

LE 11   Skill
People who score more than two standard deviations below the mean are classified as mentally retarded. In the distribution in Figure 12–25, what IQ score is two standard deviations below the mean?

LE 12   Skill
What percent of children in an age group have IQs above 85?

LE 13   Summary
Tell what you learned about standardized tests in this section.

ANSWERS TO SELECTED LESSON EXERCISES

2. (a) 35/44   (b) 35 right out of 44

3. (a) 60
  (b) Fran scored higher in math computation than 60% of the students in her age group.

4. No. Their percentile scores are too close to be considered significantly different. Since the tests are multiple choice, guessing accounts for part of each child's score.

5. 6

6. (a) 0.1   (b) 28.9

7. (a) 5.4
  (b) He scored as well as the average student in the fourth month of fifth grade.

8. The student is being compared to third grade students based on a test with very few third-grade-level questions.

9. It is very likely that the student's true score is between 390 and 450 (inclusive).

10. (a) The stanines are 7, 4, 6, 4, and 6.
  (b) If your school is average, this student will be fairly average. He may be a little stronger than average in English and social studies and a little below average in science and math.

11. $100 - 2 \cdot 15 = 70$

12. 84%

12.6   HOMEWORK EXERCISES

Basic Exercises

Exercises 1–5 refer to the following test report.

Test Score Report	Score Type	Math Comp.	Reading
Gr 6 Norms Gr 6.2	RS/NO POSS	39/44	36/60
Susan Porter	NAT'L PR-S	75-6	41-5
Age 11-5	LOCAL PR-S	82-7	48-5
Test Date 10/22/03	GRADE EQUIV	7.4	5.8

1. (a) What is Susan's raw score in reading?
   (b) What does this score mean?

2. (a) What is Susan's local percentile score in reading?
   (b) What does this score mean?

3. The local percentiles are higher than the national percentiles. What does this indicate?

4. (a) What is Susan's national stanine score in math computation?
   (b) What does this score mean?

5. (a) What is Susan's grade-level score in math computation?
   (b) What does this score mean?

6. What are two common ways in which grade-level scores are misinterpreted?

7. If Alice scores in the 70th percentile in science, does this mean that she got 70% of the questions right?

8. Is it possible to miss 3 out of 60 questions on a standardized test and still score in the 99th percentile?

9. Complete the table. Refer to Figure 12–24 on page 723.

	Math	English	Science
Percentile	92	28	65
Stanine			

10. Why do percentiles tend to be more misleading than stanines?

11. Nancy's and Patty's mathematics scores follow. Is one score clearly better than the other? Explain why or why not.

	Math Score	Confidence Band
Nancy B.	80	76-84
Patty G.	78	74-82

12. A student has an IQ score of 102 with a confidence interval of 94-110. Tell what the confidence interval means.

13. Izzy Wright is transferring to your classroom. His score report (based on the Metropolitan Achievement Test) follows.
    (a) Use the percentile information to fill in the stanine score for each test.
    (b) Give your general assessment of how this student will do at your school.

Name   Izzy Wright _____   Grade _____ 4 _____

Score Summary Box

Test	Grade Equiv.	Percentile	Stanine
Reading	3.2	42	
Mathematics	3.4	43	
Language	2.9	31	
Science	2.7	24	
Social Studies	5.1	70	

14. Joe scored in the 75th percentile in math, and Moe scored in the 73rd percentile. Does this show that Joe knows more of the math that was tested than Moe? Why or why not?

15. People who score more than two standard deviations above the mean on intelligence tests are sometimes labeled "gifted." What is the minimum standardized IQ score for a "gifted" student?

16. (a) About what percent of children in an age group score over 115 on an IQ test?
    (b) What percentile is an IQ score of 115?

Extension Exercises

17. The percentile of an IQ score of 110 is about
(a) 30th   (b) 50th   (c) 60th   (d) 75th

18. Why isn't there a 100th percentile?

19. A standardized test has a mean of 65 and a standard deviation of 5. You want to place students in the upper 2.5% and the lower 2.5% into special programs. Which of the following scores qualify for special programs?
(a) 60   (b) 77   (c) 50   (d) 72

20. On a standardized test, the first quartile was 35, the median was 42, and the third quartile was 47.
(a) Sally scored in the 39th percentile. Estimate her score.
(b) Mike scored in the 9th percentile. Estimate his score.

CHAPTER 12 SUMMARY

Our society uses statistics to analyze trends and make decisions. Economists analyze financial trends, psychologists measure intelligence, pollsters ascertain our opinions, and educators assess achievement. Politicians and advertisers try to persuade us with statistics.

These situations make it vital for citizens to understand different ways of collecting, organizing, and interpreting data. People regularly collect data from samples and organize the results into graphs and tables. Data can be further analyzed by computing averages or measures of spread. People who understand these processes will not be fooled by those who collect, organize, or interpret data in a careless or misleading manner.

Even when they are reasonably accurate, statistics tell only part of the story. One also should know what other kinds of important information must be included to give a more complete picture.

Statistical methods are used to design and analyze surveys. After taking a sample that is as representative and random as possible, a well-made survey will still have an error of around $\pm 2\%$. In many large random surveys, the results approximate a special distribution called the normal distribution. In this case, the mean and the standard deviation can be used to approximate the percent of scores that fall within a particular interval.

Teachers regularly receive a collection of statistics about each student—a standardized test report. Since any test score has a certain amount of error, less precise stanine scores tend to be less misleading than percentiles. Grade-level scores that are much above or below the student's grade level are misleading, since most test items are related to the student's actual grade level.

Standardized test scores are useful for making general comparisons among schools or students. Standardized tests do not give any specific information about how well a student understands a particular topic in the student's school curriculum.

STUDY GUIDE

To review Chapter 12, see what you know about each of the following ideas or terms that you have studied. You can also use this list to generate your own questions about the chapter.

STATISTICS IN GRADES 1–8

The following chart shows at what grade levels selected statistics topics typically appear in elementary- and middle-school mathematics textbooks.

Topic	Typical Grade Level in Current Textbooks
Reading graphs	1, 2, 3, 4, 5, 6, 7, 8
Constructing graphs	2, 3, 4, <u>5</u>, <u>6</u>, 7, 8
Stem-and-leaf plot	4, 5, 6, 7
Misleading graphs	6, 7, 8
Mode	3, 4, 5, 6, 7, 8
Mean and median	4, 5, 6, 7, 8
Range	3, 4, 5, 6, 7, 8
Box-and-whisker plot	6, 7, 8
Surveys	2, 3, 4, 5, 6, 7, 8

REVIEW EXERCISES

1. If you were putting the following data into classes to make a bar graph, what intervals would you use? (List the classes, and do not make the bar graph.)

162	80	532	534	156	472	590	373	120
240	414	477	245	545	446	56	576	175
34	228	345	565	345	543	459	480	304
360	50	555	102	585	243	537	302	87

2. A school's monthly budget (in thousands of dollars) is broken down as follows.

Salaries	$420	Maintenance	$120
Insurance	$40	Utilities	$100
Supplies	$40	Other	$80

(a) Find what percent of the total monthly budget each category is.
(b) Use a protractor to construct a circle graph of the school's budget.

3. The third-biggest item in the federal budget is interest expense. The graph shows what percent of the U.S. government budget was spent on interest expense from 1980 to 2002. In 2002, the amount was about $171 billion.

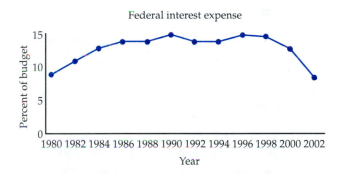

Federal interest expense

(a) Describe the overall trend and any significant exceptions to it.
(b) Why does the government have to pay interest?

4. For each of the following, determine whether the best choice would be a line graph, a bar graph, or a circle graph.
(a) Showing the number of cases of tuberculosis in the U.S. in 1950, 1960, 1970, 1980, 1990, and 2000.
(b) Showing the average life spans of ten different animals

5. A researcher studies the relationship between the number of days, x, that a worker is home sick and their productivity, y units.
(a) Would you expect x and y to have a positive correlation, a negative correlation, or no correlation? Tell why?
(b) Suppose the line of best fit is $y = -3.5x + 10$. What is the slope and what does it tell about productivity and sick days? Be specific.

6. The following table gives the median salary for employees at a restaurant in 1994, 1999, and 2004.

Median salary	$19,800	$21,500	$24,000
Year	1994	1999	2004

(a) Draw a graph that makes the salary increases look large.
(b) Draw a graph that makes the salary increases look small.

7. Sugar prices rose 40% one year and declined 20% the next year.
(a) Over the 2 years, prices rose _____%.
(b) Prove your answer, using P to represent the original price.

8. A Listerine ad says "98% of the people reading this ad have plaque. They can reduce it as much as 50% with Listerine." Explain what is deceptive about the ad.

 9. The heights of a group of children in centimeters are as follows:

128   124   132   118   126   125   140   115   128   130

(a) Make a stem-and-leaf plot of the data.
(b) Find the mode, mean, and median.

 10. The following table gives police response times to emergency calls.

Response time (minutes)	5	10	15	20	25	30
Frequency	3	8	7	2	1	2

Find the mean and median response times.

11. The mean of a set of 15 numbers is 9. One of the numbers is 10. What is the sum of the other 14 numbers?

12. Write a paragraph telling what the mean and median are and how you would decide which one to use.

13. Would you use the mode, mean, or median, to represent each of the following?
(a) "Average price" of a gallon of unleaded regular gas in Bismarck, North Dakota
(b) "Average year" of cars at a used-car dealer
(c) "Average weight" of a cereal box of Farmer Rick's Corn Cobs in a grocery store (it comes in 3 sizes)

14. A college has 2000 undergraduates who are ages 18 to 22, 500 undergraduates who are ages 23 to 40, and 100 undergraduates who are ages 41 to 60. Without computing them, explain which would be greater, the mean student age or the median student age.

15. Gar Guyton scored in the 59th percentile in reading. Explain what the 59th percentile means.

16. In a nutrition experiment, two groups want to lower their blood cholesterol level. Group C tries to do it by lowering the cholesterol in their diet. Group S tries to do it by lowering the saturated fat in their diet. The change in blood cholesterol level is shown for the ten people in each group.

$$C = \{-10, 8, 0, 2, 5, -2, -5, 4, 1, -2\}$$
$$S = \{-14, -7, 2, -7, -3, 0, -2, 1, -5, -4\}$$

(a) Draw box-and-whisker plots for the two groups on the same graph.
(b) Briefly describe how the two groups compare.

17. Show how to compute the mean and standard deviation of $\{5, 6, 10\}$.

18. Without computing the standard deviation, tell which set of scores has the highest standard deviation.
(a) $\{8, 9, 9, 10, 10, 11, 11, 12\}$
(b) $\{9, 9, 9, 10, 10, 11, 11, 11\}$
(c) $\{5, 8, 8, 10, 10, 12, 12, 15\}$

19. The governor of Maryland proposes to give all state employees a flat raise of $70 per month. What would this do to the mean and standard deviation of the monthly salary of state employees?

20. The diameter of a telephone cable is normally distributed with a mean of 1.20 cm and a standard deviation of 0.02 cm. Show how to use the 68-95-99.7 rule to find out what percent of the cables have diameters
(a) between 1.16 and 1.24 cm.
(b) less than 1.18 cm.

21. The mean weight of an adult woman is normally distributed with a mean of 146 lb and a standard deviation of 22 lb. Show how to use the 68-95-99.7 rule to find out the percentile for each of the following weights.
(a) 168 lb   (b) 190 lb

22. A county reports two "averages" for home prices in 2004: $97,920 and $113,422.
(a) Which is the mean and which is the median? Tell how you know.
(b) Would you expect the distribution to be symmetric, skewed to the left, or skewed to the right?

23. In a random sample of voters, a pollster selected proportional numbers of people from the following groups: men, ages 18–40; women, ages 18–40, men, ages 41 and up; and women, ages 41 and up. Is this an example of stratified sampling, simple random sampling, convenience sampling, or systematic sampling?

 24. A high school has 850 sophomores, 800 juniors, and 750 seniors. If a stratified random sample of 80 students is chosen, how many should be chosen from each class?

25. A city government wants to know how adult residents feel about the government. A researcher picks a simple random sample of 50 mailing addresses. A council member goes to each address and questions an adult living there. Tell why the results will have a
(a) nonresponse bias.
(b) response bias.

26. You are writing a survey question about whether or not people support an increase in funding for government regulation of food safety.
(a) Write an unbiased question.
(b) Write a question that is biased in favor of an increase in funding.
(c) Write a question that is biased against an increase in funding.

27. Consider the following test score report.

Test Score Report	Score Type	Spell.	Lang.
Gr 4 Norms Gr 4.2	RS/NO POSS	29/40	34/53
Kay O'Brien	NAT'L PR-S	39-4	53-5
Age 11-5	LOCAL PR-S	41-5	54-5
Test Date 10/22/03	GRADE EQUIV	3.7	4.5

(a) What is Kay's raw score in spelling?
(b) What does this score mean?
(c) What is Kay's local percentile score in language?
(d) What does this score mean?

28. Describe a situation in which grade-level scores are misleading, and explain why they are misleading.

29. According to an intelligence test, Mike Goldman has an IQ of 115. What is the percentile rank of his score?
(a) 34th   (b) 65th   (c) 84th   (d) 16th

ALTERNATE ASSESSMENT

Do one of the following assessment activities: add to your portfolio, add to your journal, write another unit test, do another self-assessment, or give a presentation.

SELECTED READINGS

Curcio, Frances, *Developing Data-Graph Comprehension in Grades K–8,* 2nd ed. Reston, VA: NCTM 2001.

Elementary Quantitative Literacy Project. *Exploring Statistics in the Elementary Grades,* 2 vols. White Plains, NY: Dale Seymour, 2003.

Freedman, D., R. Pisani, R. Purves, and A. Adhikari. *Statistics,* 3rd ed. New York: Norton, 1997.

Gnandeskian, M., and others. *Quantitative Literacy Series,* 5 vols. Palo Alto, CA: Dale Seymour, 1994.

Huff, D., and I. Geis. *How To Lie with Statistics,* New York: Norton, 1993.

Lindquist, M., and others. *Making Sense of Data,* Reston, VA: NCTM, 1992.

Moore, D. *The Basic Practice of Statistics,* 3rd ed. New York: W. H. Freeman, 2004.

Mosteller, F., and others. *Statistics: A Guide to the Unknown,* 3rd ed. New York: Holden Day, 1989.

National Council of Teachers of Mathematics, *Thinking and Reasoning with Data and Chance.* 2006 Yearbook (forthcoming). Reston, VA: NCTM, 2006.

Navigating Through Data Analysis, 2 vols. Reston, VA: NCTM, 2003.

Navigating Through Data Analysis and Probability, 2 vols. Reston, VA: NCTM, 2003.

Russell, S. J., and others. *Working with Data: Casebook.* Parsippany, NJ: Dale Seymour, 2002.

Teaching Children Mathematics. February, 1996 Focus Issue on Data Exploration. Reston, VA: NCTM, 1996.

Utts, J., and P. Heckard. *Mind on Statistics,* Pacific Grove, CA: Duxbury, 2002.

Look for articles in *Teaching Children Mathematics* and *Mathematics Teaching in the Middle School* about how to teach children about statistics. Infotrac also offers access to complete articles that discuss statistical studies in many science, social science, news, and business magazines.

13 | PROBABILITY

Our futures are uncertain. Will it rain today? Will you enjoy teaching for many years? Will the Cubs win the next World Series? These questions can be answered only with words such as "unlikely" or "probably" or with numerical probabilities.

People estimate probabilities using intuition and personal experience, but in some cases, it is possible to give a more precise numerical probability. If a future event is a repetition of (or very similar to) past events, a numerical estimate of its likelihood can be found. Probability is the study of random phenomena that are individually unpredictable but that have a pattern in the long run.

Although archaeologists have found dice that are 5,000 years old, the formal study of probability did not begin until the sixteenth and seventeenth centuries in Italy and France. In 1654, a French nobleman asked the mathematician Blaise Pascal (1623–1662) a question about a gambling game. Pascal wrote to another mathematician, Pierre de Fermat (1601?–1665), initiating a correspondence in which they solved gambling problems and developed probability theory (Figure 13–1 on page 734). In one letter, Pascal wrote [the French nobleman] "is very intelligent, but he is not a mathematician; this as you know is a great defect."

You already know Fermat as one of the inventors of coordinate geometry, but who was Blaise Pascal? Pascal's mathematical ability was evident at a young age. He excelled in geometry and wrote a major paper on the subject at the age of 25. But at 27, Pascal, already in poor health, decided to give up mathematics and devote his life to religion. During the rest of his life, he occasionally returned to mathematics. However, because he lived to be only 39, Pascal is known as the greatest "might-have-been" in the history of mathematics, and his best-known works concern religion.

Pascal and Fermat studied probability to understand gambling. People today use probability to estimate the chances of all kinds of repeated events that have patterns in the long run. Actuaries in insurance companies use

Photos courtesy of Library of Congress.

Blaise Pascal   *Pierre de Fermat*

Figure 13–1

probabilities of catastrophes to determine how much to charge for insurance. Testing services design and score standardized tests according to the probabilities of guessing correctly. Meteorologists utilize past frequencies of rain to predict the chance of rain tomorrow. Gambling casinos and some state governments use probabilities to design gambling games and lotteries.

13.1 EXPERIMENTAL AND THEORETICAL PROBABILITY

NCTM Standards

- collect data using observations, surveys, and experiments (3–5)
- describe events as likely or unlikely and discuss the degree of likelihood using such words as certain, equally likely, or impossible (3–5)
- predict the probability of outcomes of simple experiments and test the predictions (3–5)

Children first encounter probability when they learn how various events are "certain," "likely," "unlikely," or "impossible."

Discussion

LE 1 Concept

Suppose a bag contains 8 blue marbles and 2 black ones. You are going to pull 1 marble out of the bag. Describe an event that is

(a) likely.   (b) unlikely.   (c) impossible   (d) certain.

OUTCOMES AND SAMPLE SPACES

Next, children learn to list all the possible results of an experiment. For example, when you flip a coin, there are two possible results: heads (*H*) and tails (*T*). Each possible result, *H* or *T* in this case, is an outcome. The set of all possible outcomes, {*H*, *T*}, is called the **sample space** of an experiment.

LE 2   Skill

(a) Complete the sample space for flipping 2 coins. *HH* means a head on the first coin and a head on the second coin.

$$\{HH, \underline{\hspace{1cm}}, \underline{\hspace{1cm}}, \underline{\hspace{1cm}}\}$$

(b) Complete the diagram that shows the sample space.

Any subset of the sample space, such as {*HT*, *TH*}, is called an **event.** Probability questions deal with the probability of an event.

In a coin flip, a head or tail has the same chance of occurring. The outcomes "heads" and "tails" are **equally likely outcomes.** Probabilities for experiments that have equally likely outcomes are easier to compute. For this reason, it is important to determine whether or not outcomes are equally likely.

LE 3   Concept

(a) Are the outcomes 0, 1, and 2 equally likely to occur on the spinner in Figure 13–2?
(b) Sketch a spinner with regions numbered 0, 1, and 2 so that each outcome is equally likely.

Figure 13–2

THEORETICAL PROBABILITY

The terms "certain," "likely," "unlikely," and "impossible" describe probabilities in an informal way. Numerical probabilities are more precise measures of the likelihood of events. How do we find numerical probabilities of future events?

If someone asked you for the probability of heads on a coin flip, you wouldn't take a coin out and start flipping it, would you? You would state a theoretical probability. The **theoretical probability** assumes ideal conditions. In the following lesson exercise, find theoretical probabilities for each of the spinners from LE 3.

LE 4   Concept

What is the theoretical probability of spinning a 2 on each of the spinners in LE 3?

LE 4 illustrates **geometric probability** in which the probabilities are proportional to a geometric measurement of such an area. The second spinner has 3 equal areas so each outcome is equally likely. The probability of each outcome is $\frac{1}{3}$. Theoretical probabilities for an experiment involving equally likely outcomes can be computed as follows.

> **Definition:**   **Probability with Equally Likely Outcomes**
>
> If all outcomes in a sample space S of an experiment are equally likely, the **probability** of an event A is
>
> $$P(A) = \frac{\text{number of outcomes in } A}{\text{number of outcomes in } S}.$$

To use this definition, one must have a sample space that lists equally likely outcomes.

EXPERIMENTAL PROBABILITY

We can also approximate some probabilities by conducting experiments or surveys.

LE 5   Concept

(a) You are going to flip a coin 10 times. Predict how many heads you will get.
(b) Flip a coin 10 times and record the total number of heads.

Your results from LE 5(b) yield an experimental probability for getting heads on a coin toss. For example, if I tossed a coin 20 times and obtained heads seven times, my experimental probability for heads would be 7 out of 20, or $\frac{7}{20}$. The **experimental probability** of A, written $P(A)$, in an experiment of N trials is $\frac{\text{Number of times } A \text{ occurs}}{N}$. An experimental probability gives an estimate of the true (theoretical) probability.

LE 6   Concept

What is your experimental probability for getting heads on a coin toss, based on your results from LE 5(b)?

EXPERIMENTAL AND THEORETICAL PROBABILITY

How are experimental and theoretical probabilities related? In the following exercise, you will find an experimental probability for a coin flip. Will it come out the same as the theoretical probability?

Classroom Connection

LE 7   Connection

(a) You want to determine the experimental probability of getting 2 heads when you flip 2 coins. Flip 2 coins 20 times and obtain an experimental probability for 2 heads.
(b) You ask a sixth grade class to determine the theoretical probability of 2 heads. Bruce says, "The possible outcomes are {*HH, HT, TH, TT,*} so the probability of 2 heads is $\frac{1}{4}$." Angela says, "The possible outcomes are {0, heads, 1 head, 2 heads}, so the probability of 2 heads is $\frac{1}{3}$." Which child is right? Explain why.

(c) Which of the two spinners in LE 3 is a model for this coin-flipping experiment?

As with coins and spinners, the probabilities of outcomes with regular dice can be determined experimentally or theoretically. Consider, for example, the probability of rolling a sum of 7 on 2 dice (that is, hexahedral random digit generators).

Discussion

LE 8   Skill

(a) Estimate the probability of rolling a sum of 7 on 2 dice.
(b) Roll the dice 30 times, and make a line plot or bar graph showing the frequency of each sum from 2 to 12.
(c) On the basis of your results, what is the experimental probability for rolling a sum of 7?
(d) If possible, collect results from the whole class. Construct a line plot or bar graph for the class, and determine the class's experimental probability for a sum of 7.

To determine the theoretical probability of rolling a sum of 7 on 2 dice, complete the following lesson exercise.

Discussion

LE 9   Reasoning

(a) Complete the following list of all possible results for rolling 2 dice.

Sum of 2 Dice
First Die

	1	2	3	4	5	6
1	2	3	4	5	6	7
2						
3						
4						
5						
6						

Second Die

(b) How many (different) possible outcomes are there?
(c) Are they equally likely?
(d) The theoretical probability of rolling a sum of 7 is _____.
(e) How does your answer to part (d) compare to the experimental probability obtained by the class?

LE 10   Skill

(a) Use your table of possible dice outcomes to list the theoretical probability of each sum.

Sum	2	3	4	5	6	7	8	9	10	11	12
Probability											

(b) What is the pattern in your answers to part (a)?
(c) Predict how many sums of 11 you would get if you rolled 2 dice 50 times.

To predict how many times an event will occur, multiply the probability of the event times the total number of trials. In LE 10, $\frac{1}{18} \cdot 50 = 2.\overline{7} \approx 3$ times.

You have been finding both experimental and theoretical probabilities in a variety of situations. What is the relationship between these two types of probabilities?

Discussion

LE 11   Connection

Suppose you select 1 card at random from a regular deck of 52 cards. The theoretical probability of selecting the ace of spades is $\frac{1}{52}$. Which of the following conclusions does this probability lead you to make?

(a) If I repeat this experiment 52 times, I will pick an ace of spades 1 time.
(b) If I repeat this experiment 520 times, I will pick an ace of spades 10 times.
(c) If I repeat this experiment a large number of times, I *will* pick an ace of spades $\frac{1}{52}$ of that number of times.
(d) If I repeat this experiment a large number of times, I will *usually* pick an ace of spades $\frac{1}{52}$ of the time.
(e) If I repeat this experiment a large number of times, I will *usually* pick the ace of spades *about* $\frac{1}{52}$ of the time.

The correct statement in the preceding exercise contains three subtle but important parts: "repeat . . . a large number of times," "usually," and "about." Each is essential. First, the theoretical probability relates to a large number of trials. Second, no matter how many trials we perform, we can expect only certain kinds of results, since very unusual results are also possible. Third, we don't expect the experimental results to match the theoretical probability exactly. We must say "about" or "close to."

The following statement describes the relationship between the theoretical and experimental probabilities of an event.

Theoretical and Experimental Probability

If the theoretical probability of an event is $\frac{a}{b}$, then $\frac{a}{b}$ approximates the fraction of the time this event is expected to occur when the same experiment is repeated many times under uniform conditions.

It's amazing that the overall results of a *large number* of coin flips are fairly predictable, but an individual coin toss is completely unpredictable! That is why it is both fair and unpredictable to toss a coin at the beginning of a football game to decide who gets the ball.

In some situations, such as weather forecasting, there is no theoretical probability. A probability of rain tomorrow, such as 80%, is an experimental probability. The meteorologist looks at past records of similar weather conditions and sees that it rained the following day about 80% of the time.

LE 12   Reasoning

A principal is expecting a visitor 2 weeks from today who will evaluate the school. She wants to estimate the probability that all her teachers will be in school that day. How can she do it?

A survey is another common way to obtain experimental probabilities.

LE 13   Reasoning

Devise a plan, and solve the following problems. A survey of 1,000 elementary-school teachers regarding their preferred subject areas yielded the following results.

	Math/Science	Language/History
Male	14	22
Female	212	752

On the basis of this survey, estimate the probability of each of the following.

(a) An elementary-school teacher is female.
(b) An elementary-school teacher prefers the math/science area.
(c) A female elementary-school teacher prefers the math/science area.

A GAME: DICE PRODUCTS

LE 14   Reasoning

Dice Product is a game for two players that requires two dice.

Player 1 rolls both dice and computes the product. Player 2 rolls one die and squares the number. The player with the higher number wins.

(a) Guess whether this is a fair game. If it is not, which player do you think has the advantage?
(b) Play 15 games, and keep a record of who wins each time.
(c) Answer part (a) again. Construct sample spaces to support your answer.

LE 15   Summary

Tell what you learned about theoretical and experimental probability in this section.

ANSWERS TO SELECTED LESSON EXERCISES

1. (a) A blue marble   (b) A black marble
     (c) A red marble   (b) A marble

2. (a) *HT; TH; TT*
     (b)

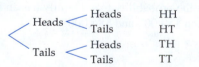

First toss	Second toss	Outcome
Heads	Heads	HH
	Tails	HT
Tails	Heads	TH
	Tails	TT

3. (a) No, 1 is more likely than 0 or 2.
     (b)

4. $\frac{1}{4}$ and $\frac{1}{3}$

6. The number of heads you got divided by 10.

7. (b) Bruce is right. His sample space shows equally likely events, while Angela's does not. The event "1 head" is more likely than 0 or 2 heads because there are two ways to do it.
(c) The spinner from part (a)

9. (b) 36   (c) Yes   (d) $\frac{6}{36} = \frac{1}{6}$

10. (a)

Sum	2	3	4	5	6	7	8	9	10	11	12
Probability	$\frac{1}{36}$	$\frac{2}{36}$	$\frac{3}{36}$	$\frac{4}{36}$	$\frac{5}{36}$	$\frac{6}{36}$	$\frac{5}{36}$	$\frac{4}{36}$	$\frac{3}{36}$	$\frac{2}{36}$	$\frac{1}{36}$

(c) About 3

11. (e)

12. She can look at attendance records for the past month of school days and see what percent of the days all her teachers were in school.

13. (a) $\frac{964}{1000} = 0.964$

(b) $\frac{226}{1000} = 0.226$

(c) $\frac{212}{964} = 0.22$

13.1 HOMEWORK EXERCISES

Basic Exercises

1. Suppose you roll a regular die. Describe an event that is
(a) likely.   (b) unlikely.
(c) impossible.   (d) certain.

2. Suppose you pick a card from a regular deck of cards. Describe an event that is
(a) likely.   (b) unlikely.
(c) impossible.   (d) certain.

3. You roll a fair die and read the result.
(a) What is the sample space?
(b) Is each outcome equally likely?

4. In an experiment, people rate 3 brands of orange juice, *A*, *B*, and *C*, from best to worst. What is the sample space?

5. Tell whether each experiment has equally likely outcomes:
(a) You shoot a basketball. You make or miss the shot.
(b) You get in your car. It starts or does not start.
(c) You guess on a true-false question. Your answer is right or wrong.

6. Tell whether each experiment has equally likely outcomes:
(a) A baby is born. It is a boy or a girl.
(b) You go to math class. You are on time or late.
(c) You roll a regular die. The result is 1, 2, 3, 4, 5, or 6.

7. A certain experiment consists of spinning a spinner like the one shown.

What is the probability of spinning each of the following?
(a) 1   (b) 2   (c) 5
(d) An even number

8. On the basis of the histogram in Figure 12–4 on page 665, what is the probability that a family has an income between $59,000 and $84,000?

9. A missing airplane crashed somewhere in the region shown. What is the probability that it is in the lake?

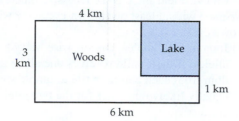

10. Suppose an object is dropped at random on the area shown. What is the probability it will land in the circle?

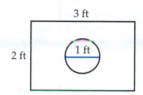

11. Draw a spinner that has four outcomes in which three are equally likely and the fourth has a probability that is three times that of each of the other three.

12. Draw a spinner that has four outcomes in which three are equally likely and the fourth has a probability that is two times that of each of the other three.

13. In In-Between, you are dealt 2 cards from a standard deck of 52 cards. Then you pick a third card from the deck. In order for you to win, its value must be between the values of the other 2 cards. What is the probability of winning if you are dealt each of the following?
(a) A 2 and a 9   (b) A 3 and a 7

14. In a poker game, you have 4 hearts and a spade and you do not know what anyone else has. You discard the spade and pick 1 card. What is the probability of picking a heart (making a "flush")?

15. A sixth grader says that when you roll 2 dice, the sum can be 2, 3, 4, 5, 6, 7, 8, 9, 10, 11, or 12. There are 11 possibilities. So $P(\text{sum is 7}) = \frac{1}{11}$. What is wrong with this reasoning?

16. A sixth grader says that when you flip three coins, the possible outcomes are 0 heads, 1 head, 2 heads, and 3 heads. Therefore, the probability of 0 heads is $\frac{1}{4}$. What is wrong with this reasoning.

17. You roll a regular die 5 times. Which result is more likely?
(a) 1 1 1 1 1   (b) 1 2 3 4 5
(c) Both results are equally likely.

18. A fair coin is tossed 5 times. Two possible results are *HHHHH* and *HTHTH*. Which is more likely to occur?
(a) *HHHHH*   (b) *HTHTH*
(c) Both are equally likely.

19. The following exercise requires flipping 3 coins.
(a) First, guess what the results will be. Then flip 3 coins 20 times and record the results.

Result	Number of Times
3 heads 0 tails	
2 heads 1 tail	
1 head 2 tails	
0 heads 3 tails	

(b) On the basis of part (a), what is your *experimental* probability for getting exactly 2 heads?
(c) List the different possible equally likely outcomes when flipping 3 coins all at once.

First Coin	Second Coin	Third Coin

(d) What is the *theoretical* probability of getting exactly 2 heads?
(e) How do your answers to parts (b) and (d) compare?

20. (a) Guess how many cards you would have to select from a regular deck of cards (without replacement) to obtain 2 cards of the same denomination (for example, two 7's).
(b) Try it 20 times with a deck of cards and obtain an experimental probability.

21. (a) Guess the probability that a thumbtack lands point up.
    (b) Devise an experiment and obtain an experimental probability.

22. Have you ever spun a penny? Stand a penny on its side holding the top with a finger of one hand. Now flick the side of the coin with a finger of your other hand so it spins.
    (a) If you spin it 20 times, guess how many heads and tails you will get.
    (b) Spin the penny 20 times, and give your experimental probability for heads.

23. Suppose you roll two regular dice and subtract the larger number minus the smaller or the difference of the two numbers if they are equal.
    (a) Construct a sample space of equally likely outcomes.
    (b) Roll the two dice 20 times, and graph your results with a line plot or bar graph.

24. Open this book to any page and find the first complete paragraph. Use that paragraph to estimate the probability that a word begins with a consonant.

25. Two dice are numbered 1, 2, 3, 4, 5, and 6 and 1, 1, 2, 2, 3, and 3, respectively.
    (a) Find the probability of rolling each sum from 2 through 9.
    (b) If you rolled these dice 100 times, predict how many times you would roll a sum of 3.

26. (a) What is the probability of rolling a *product* less than 9 on two regular dice?
    (b) If you rolled the dice 80 times, predict how many times you would roll a product less than 9?

27. Write a sample space for flipping a coin 3 times, and find the probability of getting *at least* 2 heads ("at least 2 heads" means 2 or more heads).

28. Assume that the probability of a pregnant woman's having a baby boy is 50%. If she has 3 children, what are the chances of having 2 boys and 1 girl in any order? (*Hint:* See the preceding exercise.)

29. Everyone has a combination of two alleles for eye color—BB, Bb, or bb, where B is the dominant brown-eye allele and b is the recessive blue-eye allele. A person with at least one brown-eye allele will have brown eyes.
    (a) John has Bb alleles, and his wife Mary has Bb alleles. If their child receives one allele at random from each parent, write a sample space of equally likely outcomes for the two alleles of the child.
    (b) What is the probability that John and Mary's child will have brown eyes as they do?
    (c) Craig has BB alleles, and his wife Lynda had Bb alleles. What is the probability that their child will have brown eyes?

30. In Chapter 8, you were introduced to regular polyhedra. Usually, dice are shaped like cubes, but any of the regular polyhedra could be used.

    Tetrahedron   Dodecahedron

    (a) Each of two tetrahedral dice has an equal chance of rolling a 1, 2, 3, or 4. What is the probability of rolling a sum of 6 on the 2 tetrahedral dice?
    (b) The faces of 2 dodecahedra are numbered 1 through 12. What is the probability of rolling a sum of 6 on 2 dodecahedral dice?

31. You have vulnerable backgammon men 4 and 6 spaces away from your opponent's pieces. What are your opponent's chances of knocking you off on the next roll of the dice? (Your opponent needs a sum of 4 or 6 on the two dice or a 4 or a 6 on either of the two dice.)

32. Mason wants to bet Juwan that he can roll a product of 8 on two regular dice more often than Juwan can roll a product of 12. Juwan asks your advice about whether to play. Write a note to Juwan explaining whether or not he should accept the bet and explain why.

33. (a) In flipping a coin, why does P (heads) $= \frac{1}{2}$?

(b) If I flip a coin 100 times, how many heads will I get? Tell what is wrong with each of the following children's answers.

Isaac says "I will get 50 heads."
Maya says "I will get close to 50 heads."

(c) Give a better answer to the question in part (b).

34. If I roll a regular die 120 times, how many 4's will I roll?

35. In rolling two regular dice, P (sum of 5) $= \frac{1}{9}$. If I roll two regular dice 90 times, how many times will I roll a sum of 5?

36. If I pick one card from a regular deck of cards, $P(\text{Ace}) = \frac{1}{13}$. If I repeat this event many times (always replacing the card and shuffling the deck), how many times will I pick an ace?

 37. In his 1989 bestseller *Innumeracy*, John Paulos compares the danger of being killed by terrorists on a trip abroad to the danger of being killed in a car accident at home. On the basis of the number of tourists killed abroad by terrorists in a year, Paulos estimated the probability of being killed by terrorists as 1 in 1.6 million. Paulos estimated that the probability of someone's being killed in a car crash in a year is 1 in 5,300.

(a) How many times more likely is death in a car crash than death on a trip abroad?

(b) How could you make a fairer comparison by considering the time intervals involved?

 38. The IRS audited 1 million out of 120 million taxpayers in 2001. Of those audited, 0.75 million had to make additional payments, and 120 were convicted of fraud.

(a) What is the estimated probability of having to make additional payments if you are audited?

(b) What is the estimated probability of being convicted of fraud if you are audited?

(c) The average IRS auditor collects $10 in back taxes for every $1 he or she is paid. On the basis of this fact, do you think it would pay to hire more IRS auditors?

39. A meteorologist reports that the chance of rain tomorrow is 60%. What does this mean?

40. How would you find the experimental probability of your mathematics professor continuing to teach beyond the end of the period?

41. The following table shows the results of a survey of 1,000 buyers of new or used cars of a certain model.

	Satisfied	Not Satisfied
New	300	200
Used	220	280

On the basis of this survey, what is the probability of each of the following?
(a) A new car buyer is satisfied.
(b) A used car buyer is satisfied.
(c) Someone who is not satisfied bought a used car.

 42. A survey of 400 community college students yielded the following results.

	Democrat	Republican	Other
Freshmen	95	70	40
Sophomores	86	84	25

On the basis of this survey, what is the probability of each of the following?
(a) A college student is a Republican.
(b) A freshman is a Democrat.

Extension Exercises

43. Pascal's triangle relates to a variety of mathematics problems.

$$
\begin{array}{ccccccccc}
 & & & & 1 & & 1 & & \\
 & & & 1 & & 2 & & 1 & \\
 & & 1 & & 3 & & 3 & & 1 \\
 & 1 & & 4 & & 6 & & 4 & & 1
\end{array}
$$

(a) Add a fifth row, continuing the same pattern.
(b) In flipping 1 coin, there is 1 way to get 0 heads and 1 way to get 1 head. In flipping 2 coins, how many ways are there to get 0 heads? 1 head? 2 heads?
(c) How does part (b) relate to Pascal's triangle?
(d) Explain how the third row of the triangle relates to flipping 3 coins.
(e) Using Pascal's triangle and the answer to part (c) to solve part (d) involves _____ reasoning.

44. (a) $(x + y)^1 = $ _____
(b) $(x + y)^2 = $ _____
(c) How are the results of parts (a) and (b) related to Pascal's triangle?
(d) Compute $(x + y)^4$ using Pascal's triangle.

45. Sicherman described the only nonstandard pair of cubical dice with counting numbers on each face that have the same probabilities for the sums 2 to 12 as regular dice. Can you figure out what numbers are on the faces? (*Hint:* The two dice have different numbers. One of them has only counting numbers from 1 to 4, inclusive.)

46. WIN, designed by Bradley Efron, is a game that uses unusual cubical dice. The faces are shown here.

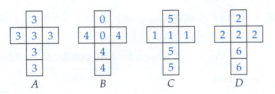

You pick 1 die, and your opponent picks 1 die. Whoever rolls the higher number wins.
(a) Suppose you pick A. Which die would a smart opponent then pick?
(b) Suppose your opponent picks B first. Which die would you then pick?
(c) Suppose your opponent picks C first. Which die would you then pick?

47. What is the probability of rolling a sum of 6 on three regular dice?

48. At the Rockville train station, trains run hourly in each direction. If you arrive at the train station at a random time, the next train is 3 times more likely to be northbound than southbound. Explain how this can be the case.

Labs and Projects

49. Read "Inflexible Logic" by Russell Maloney in *Fantasia Mathematica*, edited by Clifton Fadiman. Write a report that includes a summary of the story and your reaction to it.

13.2 PROBABILITY RULES AND SIMULATIONS

NCTM Standards

- understand the measure of likelihood of an event that can be represented by a number from 0 to 1 (3–5)
- understand and use appropriate terminology to describe complementary and mutually exclusive events (6–8)
- use proportionality and a basic understanding of probability to make and test conjectures about the results of experiments and simulations (6–8)

Probability has its own set of mathematical rules. For example, most numbers cannot represent probabilities. How large or small can a probability be? Next, consider arithmetic of probabilities. Under what circumstances can we add or subtract probabilities of events? (Multiplication of probabilities is covered later in the chapter.)

This section also presents another method for obtaining experimental probabilities, called simulation. Simulations are used because they are easier to perform than the actual experiments.

PROBABILITY VALUES

What numbers can represent probabilities? Read the cartoon in Figure 13–3, and then try LE 1.

Figure 13–3

By permission of Johnny Hart and NAS, Inc. © 1978 by Field Enterprises, Inc.

LE 1   Opener

What numbers can we have for probabilities?

(a) For example, what could the probability of rain next Tuesday be? Give all possible answers.
(b) Part (a) suggests that all probabilities are between _____ and _____ (inclusive).

LE 1 illustrates why probabilities can be as low as 0 or as high as 1. A probability tells what percent of the time an event is likely to happen, so it could be any number between 0% (or 0) and 100% (or 1). In the equally-likely-outcomes formula

$$P(A) = \frac{\text{number of outcomes in } A}{\text{number of outcomes in } S}$$

the number of outcomes in A can be as low as 0, making $P(A) = 0$, or as high as the number of outcomes in S, making $P(A) = 1$.

Probability Values of an Event

If A is any event, then $0 \leq P(A) \leq 1$.

Probabilities are normally expressed as fractions, decimals, or percents. (Odds and ratios are covered in Section 13.5.) For example, if I flip a coin, the probability of heads is given by

$$P(\text{heads}) = \frac{1}{2} = 0.50 = 50\%$$

LE 2   Concept

Which of the following could not be the probability of an event?

(a) $\frac{2}{3}$ **(b)** -3 **(c)** $\frac{7}{5}$ **(d)** 15%   **(e)** 0.7

In Section 13.1, you classified probabilities of events as "impossible," "unlikely," "likely," or "certain." How are these words related to numerical probabilities? Figure 13–4 shows possible probabilities of events and some corresponding verbal descriptions.

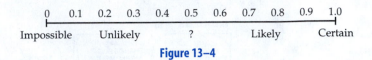

Figure 13–4

MUTUALLY EXCLUSIVE EVENTS

How are probabilities that involve more than one event computed? One of the simplest cases involves two events that do not intersect.

LE 3   Concept

Consider a college with 25% freshman, 25% sophomores, and 25% women.

(a) What is the probability that a student is a freshman *and* a sophomore?
(b) Name two nonintersecting groups at the college besides freshmen and sophomores.
(c) What is the probability of selecting a single student who has both of the characteristics you mentioned in part (b)?

LE 3(a) and (c) suggest the definition of nonoverlapping—that is, mutually exclusive—events. Mutually exclusive events cannot occur at the same time. Events A and B are **mutually exclusive** or **disjoint** if and only if $P(A \text{ and } B) = 0$ or $A \cap B = \phi$. The expression "$P(A \text{ and } B)$" means the probability that both events A and B occur.

$P(A \text{ or } B)$ is easier to compute for mutually exclusive events than for other events.

Classroom Connection

LE 4   Concept

A college has 25% freshman, 25% sophomores, and 25% women. A student is picked at random.

(a) Which of the following are mutually exclusive groups?
    (1) freshmen and sophomores (2) freshmen and women
(b) What is the probability that a randomly chosen student is a freshman or a sophomore? Draw a rectangular diagram to support your answer.
(c) A sixth grader says that the probability of choosing a freshman woman is 50%. Is this right? If not, what would you tell the child?

In part (b), you could add the probabilities (percentages) because the two groups, freshmen and sophomores, do not overlap. When two events A and B do overlap, as in part (c), you cannot simply add their probabilities to compute $P(A \text{ or } B)$.

The general rule for $P(A \text{ or } B)$ when A and B are mutually exclusive is as follows.

The Addition Rule for Mutually Exclusive Events

If A and B are mutually exclusive events, $P(A \text{ or } B) = P(A) + P(B)$.

The expression "*A* or *B*" combines all the outcomes from event *A* and event *B*. For this reason, *P(A* or *B)* can also be written as $P(A \cup B)$. However, current middle-school textbooks do not use set symbols in probability.

Probabilities have already been defined for equally likely outcomes. For a sample space with outcomes that are not equally likely, the addition rule can also be used to compute probabilities.

LE 5   Skill

Figure 13–5

Show how to determine the probability of spinning a 1 or a 2 on the spinner in Figure 13–5.

The preceding exercise illustrates the following.

Definition: The Probability of an Event
If *A* is any event, *P(A)* is the sum of the probabilities of all outcomes in set *A*.

A set of outcomes is mutually exclusive. This is why we can add the probabilities of the outcomes. The addition rule for mutually exclusive events can also be used to derive a formula for events that are complements.

COMPLEMENTARY EVENTS

The probability that an event will occur and the probability that it will not occur have a simple relationship.

LE 6   Opener

Suppose that the probability of rain tomorrow is $\frac{1}{4}$. What is the probability that it will *not* rain tomorrow?

The event "not rain" in the preceding exercise is the complement of the event "rain." The **complement** of any event *A* is the event that *A* does *not* occur, written "not *A*" or $\overline{A}$. (Some textbooks use the notation A'.) Complementary events have no outcomes in common (are mutually exclusive), and together they encompass all possible outcomes. In set notation, $A \cap \overline{A} = \phi$, and $A \cup \overline{A} = S$, the sample space.

LE 7   Reasoning

If the probability of an event is *n*, what is the probability of the complement of the event?

LE 7 generalizes the rain example. The probability of the complement of an event is 1 (or 100%) minus the probability of the event.

Probabilities of Complementary Events
$$P(\overline{A}) = 1 - P(A)$$

The preceding equation follows from the addition rule for mutually exclusive events.

LE 8 Reasoning

(a) A and $\overline{A}$ are mutually exclusive. By the addition rule, $P(A \text{ or } \overline{A}) = $ _____.

(b) What is the numerical value of $P(A \text{ or } \overline{A})$?

(c) Combine the equations in parts (a) and (b).

(d) How would you derive the probability rule for complementary events from your equation in part (c)?

(e) Do parts (a)–(d) involve induction or deduction?

SIMULATIONS

When the theoretical probability is difficult or impossible to compute, and it is impractical to find an experimental probability, a simulation can be designed. A **simulation** is a probability experiment that has the same kind of probabilities as the real-life event. Simulations are based on connections between mathematical models and everyday life.

A series of coin flips can be used to simulate whether a series of newborns will be boys or girls. The result of a coin flip has the same kind of probabilities as the sex of a baby. The probability of heads is $\frac{1}{2}$ and the probability of tails is $\frac{1}{2}$, just as the probability of a boy is about $\frac{1}{2}$ and the probability of a girl is about $\frac{1}{2}$. Furthermore, each coin flip has no effect on subsequent coin-flip probabilities. The same is usually true of the sexes of babies.

In the following simulation, each coin flip represents the birth of a new child. Furthermore, suppose that each head represents having a girl and each tail represents a boy.

Discussion

LE 9 Connection

My friends would like to have a baby girl. They are willing to have up to 3 children in an effort to have a girl.

(a) First, guess the probability that they will have a girl.
How can you calculate the probability for them? Use a coin-flipping simulation to approximate the answer. Form pairs to do this.

(b) Using H = girl and T = boy, simulate the event of the couple's having a family. (For example, if you flip heads on the first try, they have 1 girl and stop having children.) What family do you end up with?

(c) Repeat this experiment 19 more times so that you have created 20 families.

(d) Out of the 20 families, how many have a girl?

(e) Out of the 20 families, how many have 3 children?

(f) On the basis of your results, estimate the probability of the family's having a girl.

(g) On the basis of your results, estimate the probability of the family's having 3 children.

(h) If you have data from others in the class, use these data to revise your answers to parts (f) and (g).

A simulation is usually much easier to perform than the event it represents. The results of a simulation provide an experimental probability that is used to estimate the theoretical probability of the simulated event. A greater number of trials is likely to provide a better estimate.

A company is buying 8 automobiles from a manufacturer. From past experience, they know that about 1 out of every 6 automobiles needs some kind of adjustment. You can simulate determining whether or not automobiles need adjustment by rolling a die or by using a calculator or computer program.

Discussion

LE 10   Connection

Suppose you buy 8 automobiles, and you expect that about 1 out of every 6 will need an adjustment. What is the probability that 2 or more automobiles will need adjustments?

   Use a die to simulate 10 purchases of 8 automobiles like the one just described.

(a) What roll or rolls on the die will represent an automobile needing adjustment?

To perform the simulation, choose part (b) if you want to roll a die, part (c) if you want to use a random number table, parts (d)–(f) if you want to use a computer spreadsheet, or part (g) if you want to use a graphing calculator.

(b) Roll the die to simulate 10 orders of 8 cars each, and record your results. Proceed to part (h).

(c) Use the following random number table or generate your own random numbers.

74192	77567	88741	48409	41903	43909	99477	25330	64359
40085	16925	85117	36071	15689	14227	06565	14374	13352
49367	81982	87209	36579	58984	68288	22913	18638	54303
00795	08727	69051	64817	87174	09517	84534	06489	87201

Start at the left, and count each digit between 1 and 6 (inclusive) as representing a car. In part (a), you have decided which digit means that the car needs an adjustment. Continue through the random number table until you have finished buying 10 sets of 8 cars. Record your results, and proceed to part (h).

(d) You can generate random numbers with a spreadsheet. In EXCEL, use the command RAND( ) to select a number between 0 and 1. You can generate numbers from other sets with commands such as 6*RAND( ) or INT(6*RAND( )). The INT command rounds decimals down to the nearest integer. The command 6*RAND( ) selects numbers from what set?

(e) The command INT(6*RAND( )) selects numbers from what set?

(f) Write a command and select 10 sets of 8 random whole numbers from 1 to 6. Record your results, and proceed to part (h).

(g) On the TI-83, press MATH and then highlight PRB. Choose randInt(and press ENTER. Then type 1, 6) and press ENTER to choose a random integer between 1 and 6 inclusive. Choose 10 sets of 8 integers. Record your results.

(h) Make a frequency graph of the number of automobiles needing adjustment in each of your 10 shipments.

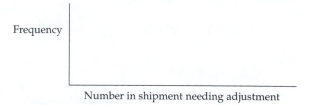

(i) What is your experimental probability that 2 or more cars in a shipment will need adjustments?

Statisticians often use simulations to approximate probabilities that are difficult to compute using theory or for which there is no theory. Computers offer the possibility of simulating a large number of trials in a short amount of time.

> **LE 11   Summary**
> Tell what you learned about probability rules and simulations in this section.

ANSWERS TO SELECTED LESSON EXERCISES

1. (a) 0% to 100% (inclusive)   (b) 0;1

2. (b) and (c)

3. (a) 0   (b) Men and women   (c) 0

4. (a) (1)   (b) 50%
(c) No. It cannot be determined, since some of the freshmen are also women.

5. $P(1 \text{ or } 2) = P(1) + P(2) = \frac{1}{4} + \frac{1}{2} = \frac{3}{4}$

6. $\frac{3}{4}$

7. $1 - n$

8. (a) $P(A) + P(\overline{A})$ (b) 1
(c) $P(A) + P(\overline{A}) = 1$
(d) Subtract $P(A)$ from both sides of the equation.
(e) Deduction

10. (e) $\{0, 1, 2, 3, 4, 5\}$

13.2   HOMEWORK EXERCISES

Basic Exercises

1. What is the probability that a resident of Iowa lives in the United States?

2. An event is very unlikely to happen. Its probability is about
(a) $\frac{1}{10}$ (b) $\frac{3}{10}$ (c) $\frac{1}{2}$
(d) $\frac{7}{10}$ (e) $\frac{9}{10}$

3. Which of the following numbers cannot be a probability?
(a) 0.6   (b) -1 (c) 3   (d) 0%
(e) $\frac{1}{1000}$

4. Estimate the probability that you will eat chicken in the next week.
(a) 0   (b) 0.3   (c) 0.5   (d) 0.9

5. Determine whether or not each pair of events is mutually exclusive.
(a) Getting an A in geometry and a B in English this semester
(b) Getting an A in geometry and a B in geometry this semester

6. Suppose that you flip a coin once. Name two events, A and B, that are mutually exclusive.

7. The probability that a person has black hair is 22%. The probability that a person has brown eyes is 72%. A child says that the probability that a person has either black hair or brown eyes is 94%. Is this right? If not, what would you tell the child?

8. A child says, "Since there are 7 continents, the probability of being born in North America is $\frac{1}{7}$. Is this right? If not, what would you tell the child?

9. Suppose Gordon Gregg and Debra Poese are the only two candidates for president. The probability that Gordon wins is 0.38. What is the probability that Debra wins?

10. If $P(N) = 0.03$, then $P(\text{not } N) = $ _____.

11. At a college, the probability that a randomly selected student is a freshman is 0.35, and the probability that the student is a sophomore is 0.2. What is the probability that a randomly selected student is neither a freshman nor a sophomore?

12. *Explain* why $P(\overline{A}) = 1 - P(A)$.

 13. A college has 1020 male and 1180 female students. What is the probability that a student selected at random will be female?

 14. A school has a raffle for the 130 students in the fifth and sixth grades. There are 62 fifth graders. What is the probability that the grand-prize winner selected at random from 130 students will be a sixth grader?

15. Use a coin flip or a computer program to simulate the birth of a boy or girl. Investigate four-child families by doing the following.
   (a) If there are 20 four-child families, guess how many will have exactly 1 girl.
   (b) Create 20 four-child families, using H = girl and T = boy. Write down the result for each family.
   (c) Use your results to complete the table.

Four-Child Families

Number of girls	0	1	2	3	4
Estimated probabilities					

   (d) Construct a histogram from your table.
   (e) Of the 20 families, how many have exactly 1 girl?

16. Suppose the chance of rain for each of the next three days is 50%. You are planning an outdoor activity and want to estimate the chance that it will not rain at least one of the days.
   (a) Guess the probability that it will not rain at least one of the three days.

(b) Use a coin or a random number generator to simulate the weather each day. Make 20 sets of three days and record the results.
(c) Make a histogram of the frequency of 0, 1, 2, 3 days with rain.
(d) What is the experimental probability that it will not rain at least one of the three days?

17. New Raisin Crumbles come with 1 of 6 different famous-mathematics-teacher cards in each box.
   (a) Guess the approximate number of boxes you would have to buy to get all 6 different teacher cards.
   (b) Using a die, simulate collecting all of the cards 8 times to obtain an experimental estimate.
   (c) Find the mean and the standard deviation of your data.
   (d) Make a box-and-whisker plot of your results.

18. Consider the following problem: "On a game show, a contestant gets to pick box 1, 2, or 3. One box contains $10,000. The other 2 boxes are empty. Use a die to simulate 20 groups of 5 shows (weekly), and estimate the probability that at least 4 people will win during a particular 5-show week." Devise a plan and solve the problem.

19. Two contestants on a game show must choose door 1, 2, 3, or 4. Three of the locations have prizes, and the other does not. A simulation is carried out with a spinner to find the probability that both contestants will win a prize. It is assumed that doors 1, 2, and 4 lead to prizes and door 3 is a dud. The results follow.

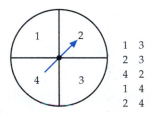

   (a) What does the first row of numbers represent?
   (b) On the basis of these results, what is the probability that both contestants will win a prize?

20. A particular forward on the Phoenix Suns makes 50% of his shots. Using a coin, someone simulates the player taking 6 shots. *H* (heads) means that he made the shot and *T* (tails) means that he missed the shot. On the basis of the following results, what is the experimental probability that the player makes exactly 3 out of 6 shots?

$$H\ H\ T\ T\ H\ T$$
$$H\ T\ T\ T\ T\ H$$
$$H\ T\ H\ T\ T\ T$$
$$H\ H\ T\ T\ H\ H$$
$$H\ H\ H\ H\ H\ T$$
$$H\ T\ T\ T\ H\ H$$
$$H\ T\ T\ H\ T\ T$$

21. Twenty people are at a party.
(a) Guess the probability that at least 2 people will have the same birthday.
(b) Use a random number table or technology to generate 20 random integers between 1 and 365 (inclusive), and see if at least 2 match. Repeat this 25 times to obtain an experimental probability.

22. You own two National Motors cars. Suppose that on a given morning, the probability that the Squealer starts is 0.4 and the probability that the Alley Cat starts is 0.2. Use a random number table or generator to estimate the probability that at least one of your cars will start.
(a) Simulate trying to start both cars 20 times.
(b) Estimate the probability that at least one car will start.

Extension Exercises

23. Draw a spinner, with sectors labeled 2, 3, 4, . . . , 12, that could be used to simulate rolling 2 dice and computing the sum.

24. Draw a spinner that has all the following characteristics.
(1) The probability of spinning a 2 is $\frac{1}{2}$.
(2) The probability of spinning an odd number is $\frac{1}{4}$.
(3) The probability of spinning a sum of 5 on 2 spins is $\frac{1}{8}$.

25. You are hired to determine how long a traffic signal should stay green in each direction at a particular intersection. First, consider the cars going from east to west (on a one-way street).
(a) Suppose you find out that 200 cars pass by during a 10-minute period. One car passes east to west every _____ seconds.
(b) Using a timer, you also compute that cars leave the intersection at a rate of 1 per second when the light is green. What if you make the light alternately red and green every 15 seconds (disregard yellow)? The east-west traffic pattern can be simulated with a die. This will give you an idea of how many cars will get backed up at the traffic light.

One car passes every 3 seconds. If each roll of the die represents 1 second, decide which die results represent "car" and which represent "no car," and write this information in the table.

Result on Die

	Result on Die
Car	
No car	

(c) Now for the simulation. Start with a red light, and assume that 1 car can leave the intersection each second. Roll 1 die 120 times (for 120 seconds), and fill in the results in a table like the one shown here. Compute how many cars are lined up each second.

Second	1	2	3	4	5	6	7	8	9	10	. . .
Result on die											
New car arrives?											
Car leaves? (when light is green)											
Total number of cars lined up											

(d) What was the largest number of cars you had backed up? Is this result satisfactory? If not, suggest an alternative timing for the signal.

26. The following supermarket-checkout simulation is adapted from NCTM's "Student Math Notes" of March 1986.

   You decide to open a supermarket. How many checkout lanes should you have? A die will be used to simulate the arrival of shoppers at the checkout counter. Assume that a new customer arrives at the checkout counter during 1 out of every 3 minutes. Assume that it takes the cashier 3 minutes to process each customer.

   (a) What die results would represent a customer arriving during a given minute?

   (b) What die results would represent a customer not arriving during a given minute?

   (c) Roll the die 30 times and simulate 30 minutes at a checkout counter (or use a computer program). See how long the line gets. You could use a chart like the following one to keep a record.

END OF MINUTE	1	2	3	4	5	6	...
Customer arrives	A	—	—	B	C		
Checking out	—	A	A	A	B	B	
Waiting	—	—	—	—	—	C	

   (d) Complete the following record of your results.

Customer	A	B	C	D	E	F	...
Minute of checkout arrival							
Minute checkout completed							
Minutes wait before checkout began							

   (e) Give the following estimates based on your 30 minutes of data.

   Number of customers arriving _____
   Total customer waiting time _____
   Average waiting time _____
   Total clerk idle time _____
   Total time to process everyone _____

   (f) Repeat parts (c)–(e) using two checkout counters.

   (g) Which number of checkout counters works out better? Consider the waiting time and the cost to the store.

27. Create a dice game that simulates baseball. Use probabilities something like the following.

   STRIKE OUT 17%   WALK 9%
   SINGLE 16.5%
   DOUBLE 4%   TRIPLE 0.5%
   HOME RUN 3%
   FLY OUT 22%
   OUT OR DOUBLE PLAY
   (IF RUNNER IS ON FIRST) 6%
   GROUNDOUT (RUNNERS ADVANCE
   1 BASE) 22%

Labs and Projects

28. Design a simulation game that uses dice or a spinner.

13.3 COUNTING

NCTM Standards

- use geometric models to solve problems in other areas of mathematics, such as number and measurement (3–5)
- recognize and use connections among mathematical ideas (pre-K–12)
- recognize and apply mathematics in contexts outside of mathematics (pre-K–12)

To compute theoretical probabilities, one often counts the number of equally likely results in a sample space. Sometimes the sample space is so large that shortcuts are needed to count all the possibilities.

LE 1   Opener

How do state officials know how many different license plates can be made using a certain number of letters and digits (Figure 13–6)?

Figure 13–6

To answer LE 1, people use methods to count all the possible arrangements of a sequence.

ORGANIZED LISTS AND TREE DIAGRAMS

To understand the shortcut for counting possible license plates, we'll begin by analyzing some simpler counting problems.

• EXAMPLE 1

You are taking 3 shirts (red, blue, yellow) and 2 pairs of pants (tan, gray) on a trip. How many different choices of outfits do you have (assuming everything matches)?

SOLUTION

There are 3 ways to solve this problem. If the shirts are s_1, s_2, and s_3 and the pants are p_1 and p_2, you can count the outfits in an organized list or a tree diagram (Figure 13–7). A tree diagram can be used when an event has two or more steps. Each step is shown with a set of branches. The third method uses the counting principle model from Section 3.3. You can multiply the number of ways to choose a shirt by the number of ways to choose a pair of pants.

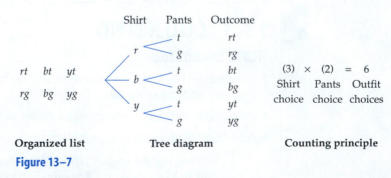

Organized list   Tree diagram   Counting principle

Figure 13–7

There are 6 different outfits.

LE 2   Connection

A cafe offers the following menu. Each customer chooses one of two appetizers and one main dish.

> **LUNCHEON MENU $4**
> **Appetizers**
> Canned fruit du jour
> Cream of bamboo soup
> -
> **Entrees**
> Blowfish thermidor
> Woodchuck pilaf
> Semi-boneless falcon
> Hippo in a blanket

(a) Use a tree diagram to show all possible selections a customer can make. (Abbreviate the item names.)
(b) Use an organized list to show all possible selections a customer can make.
(c) What is a shortcut for counting the total number of possible orders without using a tree diagram or an organized list?

THE FUNDAMENTAL COUNTING PRINCIPLE

Did you recognize the counting principle model of multiplication from Section 3.3? In Example 1 and LE 2, you could multiply the number of ways to make the first choice by the number of ways to make the second choice to obtain the total number of arrangements.

LE 3   Reasoning

Consider the following problem:

"Suppose I want to order a dessert at a restaurant. The menu offers 3 kinds of pie and 2 kinds of cake. How many dessert choices do I have?"

   A fifth grader says there are $2 \times 3 = 6$ choices. Is that right? If not, what would you tell the child?

   In Example 1 and LE 2, you can use the counting principle, also known as the *Fundamental Counting Principle.* It does not apply to the dessert choice in LE 3.

> **The Fundamental Counting Principle**
>
> If an event E can occur in e ways and, after it has occurred, an event F can occur in f ways, then event E followed by event F can occur in $e \cdot f$ ways.

   The Fundamental Counting Principle works not only for two events in sequence but also for any number of events in sequence. The Fundamental Counting Principle can be applied to a complex event if the event can be thought of as a *series of steps* with a specified order. Figure 13–8 shows how the sixth-grade *Harcourt Math* textbook uses a tree diagram and the Fundamental Counting Principle on a three-step menu.

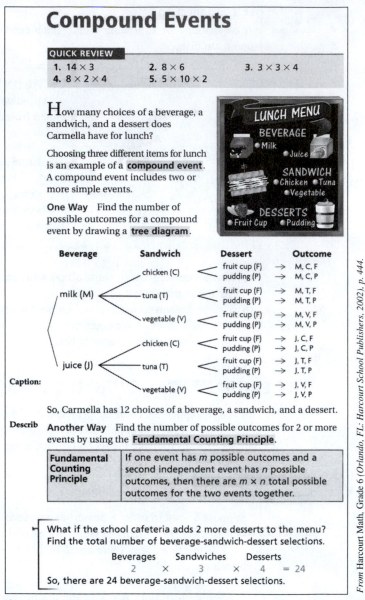

Classroom Connection

Compound Events

QUICK REVIEW

1. 14×3 **2.** 8×6 **3.** $3 \times 3 \times 4$
4. $8 \times 2 \times 4$ **5.** $5 \times 10 \times 2$

How many choices of a beverage, a sandwich, and a dessert does Carmella have for lunch?

Choosing three different items for lunch is an example of a **compound event**. A compound event includes two or more simple events.

One Way Find the number of possible outcomes for a compound event by drawing a **tree diagram**.

Beverage	Sandwich	Dessert		Outcome
	chicken (C)	fruit cup (F)	→	M, C, F
		pudding (P)	→	M, C, P
milk (M)	tuna (T)	fruit cup (F)	→	M, T, F
		pudding (P)	→	M, T, P
	vegetable (V)	fruit cup (F)	→	M, V, F
		pudding (P)	→	M, V, P
	chicken (C)	fruit cup (F)	→	J, C, F
		pudding (P)	→	J, C, P
juice (J)	tuna (T)	fruit cup (F)	→	J, T, F
		pudding (P)	→	J, T, P
	vegetable (V)	fruit cup (F)	→	J, V, F
		pudding (P)	→	J, V, P

So, Carmella has 12 choices of a beverage, a sandwich, and a dessert.

Another Way Find the number of possible outcomes for 2 or more events by using the **Fundamental Counting Principle**.

Fundamental Counting Principle	If one event has *m* possible outcomes and a second independent event has *n* possible outcomes, then there are *m* × *n* total possible outcomes for the two events together.

What if the school cafeteria adds 2 more desserts to the menu? Find the total number of beverage-sandwich-dessert selections.

Beverages		Sandwiches		Desserts		
2	×	3	×	4	=	24

So, there are 24 beverage-sandwich-dessert selections.

Caption:

Describ

From Harcourt Math, Grade 6 (Orlando, FL: Harcourt School Publishers, 2002), p. 444.

Figure 13–8

Example 2 returns to the question posed in LE 1.

• EXAMPLE 2

The state of Maryland has automobile license plates consisting of 3 letters followed by 3 digits. How many possible license plates are there?

SOLUTION

Understanding the Problem How many ways are there to choose 3 letters followed by 3 digits? A license plate can be created in 6 steps: picking each of the 3 letters and then selecting each of the 3 digits. For example, I might pick FUN 123. There is a certain number of choices for each step.

Devising a Plan   Using the Fundamental Counting Principle, one can compute the total number of choices for each step and then multiply these numbers together.

Carrying Out the Plan   There are 26 choices for each of the 3 letters and 10 choices for each of the 3 digits. The total number of possible license plates is

$$\underset{\text{Letter}}{(26)} \quad \underset{\text{Letter}}{(26)} \quad \underset{\text{Letter}}{(26)} \quad \underset{\text{Digit}}{(10)} \quad \underset{\text{Digit}}{(10)} \quad \underset{\text{Digit}}{(10)} = 17{,}576{,}000$$

Looking Back   The same technique would work for many other license plate designs.   ●

LE 4   Skill

A state has license plates with 2 letters followed by 4 digits. How many possible license plates can be made?

LE 5   Skill

How many possible telephone numbers can you form (in one area code) if you may choose any 10 digits, except that you may not select 0 or 1 as the first or fourth digit? (*Hint:* Write an example of a phone number, and see how many choices you have for each step.)

One special type of counting problem involves a series of steps in which each step uses up one of the choices.

● EXAMPLE 3

A club has 10 members. In how many ways can the club choose a president and vice-president if everyone is eligible?

SOLUTION

Understanding the Problem   Think of the elections as 2 steps. The first step is to elect a president. Then the club elects a vice-president from the remaining members.

Devising a Plan   Use the Fundamental Counting Principle.

Carrying Out the Plan   How many choices are there for president? 10. Once the president is selected, how many choices are there for vice-president? 9. Using the Fundamental Counting Principle, there are $10 \cdot 9 = 90$ ways to select a president followed by a vice-president.

Looking Back   This procedure works for any election in which each position is different.   ●

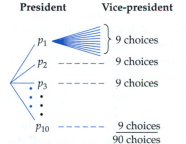

Figure 13–9

An ordered arrangement of people or objects (such as a choice of president and vice-president) is called a **permutation**. In the preceding example, the number of permutations is 90, as shown by the tree diagram in Figure 13–9. There are 10 branches for the first step, and each of these 10 choices has 9 branches for the second step ($10 \cdot 9 = 90$).

LE 6   Skill

A principal must select a head teacher, an assistant head teacher, and a workshop coordinator from her faculty of 30 people. In how many ways can she do this?

You can compute the number of permutations in Example 3 and LE 6 with a formula. In picking 2 people from 10, you compute $10 \cdot 9$. In picking 3 teachers from 30, you compute $30 \cdot 29 \cdot 28$. In each product, the first factor is the overall number of people. The number of factors is the number of people you are picking. More generally, the number of permutations or r object from n objects is $n \cdot (n - 1) \ldots \cdot (n - r + 1)$.

Permutation Formula

In n objects are chosen r at a time, the number of permutations (ordered arrangements) is

$$_nP_r = n \cdot (n - 1) \cdot (n - 2) \cdot \ldots \cdot (n - r + 1)$$

Try this formula out on the permutation problem on LE 6.

LE 7   Skill

(a) Solve LE 6 using the permutation formula.
(b) To solve the same problem with a calculator, first enter the value of n. Next, press the permutation key if your calculator has one. On the TI-83, press MATH and highlight PRB. Then choose nPr. Finally, on either type of calculator, enter the value of r.

FINDING PROBABILITIES USING THE FUNDAMENTAL COUNTING PRINCIPLE

People who design multiple-choice tests must figure out the probability that an examinee will guess a correct answer. A test is not very useful if someone who knows nothing about the material does well on it.

Suppose you give a 5-question multiple-choice quiz with 4 choices for each question. What is the probability that someone who randomly selects an answer from each set of choices will do well (that is, will get at least 4 out of 5 questions right)?

Example 4 shows how to solve part of this problem. You can complete the solution in the exercise that follows Example 4. You will need to use the Fundamental Counting Principle on various sets of test responses and then *add* together the number of ways of doing each set of responses.

● EXAMPLE 4

A quiz has 5 multiple-choice questions. Each question has 4 answer choices, of which 1 is the correct answer and the other 3 are incorrect. Suppose that you guess all the answers.

(a) How many ways are there to answer the 5 questions?
(b) What is the probability of getting all 5 questions right?
(c) What is the probability of getting exactly 4 questions right and 1 wrong?
(d) What is the probability of doing well (getting at least 4 right)?

SOLUTION

(a) In how many ways can you answer the 5 questions? Consider each question as a step in completing a test. There are 4 ways to do each step. Using the Fundamental Counting Principle,

$$(4)(4)(4)(4)(4) = 4^5 = 1024$$

(b) How many ways are there to get all 5 questions right? There is 1 way to answer each question correctly. Using the Fundamental Counting Principle, $(1)(1)(1)(1)(1) = 1$. There is 1 way to answer all 5 questions correctly out of 1,024 possibilities. So

$$P(\text{all 5 right}) = \frac{1}{1024}$$

(c) There is more than one way to get 4 right and 1 wrong, depending on *which question you get wrong*. In the following table, which lists all possible responses that involve at least 4 right answers, *R* stands for a right answer and *W* stands for a wrong answer.

Each type of response (for example, *WRRRR*) can occur in 3 ways based on the Fundamental Counting Principle. Since each of the 5 sets of responses is a complete test result, we add the five 3's together.

Five Responses	Number of Ways to Fill Out the Test
WRRRR	$(3)(1)(1)(1)(1) = 3$
RWRRR	$(1)(3)(1)(1)(1) = 3$
RRWRR	$(1)(1)(3)(1)(1) = 3$
RRRWR	$(1)(1)(1)(3)(1) = 3$
RRRRW	$(1)(1)(1)(1)(3) = \underline{\;3\;}$
	15 ways

So there are 15 ways out of the 1,024 possible ways that result in 4 right answers and 1 wrong answer.

$$P(4 \text{ right, 1 wrong}) = \frac{15}{1024} \approx 1.5\%$$

The chances of getting exactly 4 right by guessing are not very good. It makes more sense to study.

(d) "At least 4 right" means you can get either 4 right and 1 wrong or all 5 right. We add the probabilities (of mutually exclusive events).

$$P(\text{at least 4 right}) = P(4 \text{ right, 1 wrong}) + P(5 \text{ right})$$

$$= \frac{15}{1024} + \frac{1}{1024} = \frac{16}{1024} \approx 0.016 \qquad \bullet$$

People who construct multiple-choice tests know the probabilities that someone will get various numbers of questions right by guessing randomly. To decrease the probability of someone's doing well just by guessing, test constructors use a lot more than 5 questions.

LE 8   Reasoning

A true-false test has 6 questions.

(a) How many ways are there to answer the 6-question test?
(b) What is the probability of getting at least 5 right by guessing the answers at random?

The preceding example and lesson exercise both required multiplication *and* addition. Multiply numbers that represent the number of ways to do *each step*, such as answering each question on an exam. Add numbers that represent *complete results*, such as how many ways to answer 5 right on the whole test and how many ways to answer 6 right on the whole test.

PERMUTATIONS AND COMBINATIONS

In how many different ways can 8 horses finish in a race (assuming there are no ties)? Using the Fundamental Counting Principle, the number is

$$(8)(7)(6)(5)(4)(3)(2)(1) = 40,320$$

This is another example of an ordered arrangement called a permutation. Products such as $(8)(7)(6)(5)(4)(3)(2)(1)$ can be written in a shorthand notation called factorial. In this case, $8 \cdot 7 \cdot 6 \cdot 5 \cdot 4 \cdot 3 \cdot 2 \cdot 1 = 8!$ (read "8 factorial").

Factorial Notation

$n!$ is called **n factorial**, and $n! = n(n - 1)(n - 2) \cdot \ldots \cdot 3 \cdot 2 \cdot 1$, in which n is a positive integer. By definition, $0! = 1$.

The horse race example illustrates that the number of permutations of n objects is $n!$.

LE 9   Skill

(a) Compute 5! by hand.
(b) If your calculator has a factorial key, compute 5! on it.

Example 5 looks like Example 3, but it has a different answer.

• EXAMPLE 5

A club has 10 members. In how many ways can the club select 2 people for a committee?

SOLUTION

There won't be as many possibilities as in Example 3 because choosing members named Mr. Alonso and Ms. Brunett is the same as choosing Ms. Brunett and Mr. Alonso. These are *not ordered arrangements*. Now $10 \cdot 9 = 90$ counts each pair in both orders. Each two-person choice can be arranged in 2! or 2 ways. To obtain the correct result, compute

$$\frac{_{10}P_2}{_2P_2} = \frac{_{10}P_2}{2!} = \frac{10 \cdot 9}{2} = 45$$

●

Example 5 illustrates a combination. A **combination** is a group of items in which the order does not make a difference. In the example,

$$_{10}C_2 = \frac{_{10}P_2}{2!} = 45$$

The formula for combinations is as follows.

Combination

The number of ways of selecting a subset of r objects (order does not matter) from a set of n objects is

$$_{n}C_r = \frac{_{n}P_r}{r!}$$

Apply this formula to the following exercise, noting that the order of the arrangements does not matter.

LE 10 Skill

(a) A school has 30 teachers. In how many ways can the school choose 3 people to attend a national meeting?

(b) Find out the easiest way to solve this problem with your calculator.

As you can tell, permutations and combinations are quite similar. Both involve choosing r objects from n objects. See if you can distinguish between them. To do so, make up an arrangement you might choose and see if the order matters. If the order does matter, it's a permutation. If not, it's a combination.

LE 11 Skill

In each part, tell whether the problem concerns a permutation or a combination. Then find the answer using the appropriate formula.

(a) A college wants to hire a mathematics tutor, a calculus teacher, and a statistics teacher from a group of 10 applicants, each of whom is qualified for all of the jobs. In how many ways can the college fill the 3 positions?

(b) How many different 4-member committees can be formed from 100 U.S. senators?

(c) A meeting is to be addressed by 5 speakers: Al, Baraka, Clyde, Dana, and Elvira. In how many ways can the speakers be ordered? (*Hint:* $0! = 1$.)

LE 12 Summary

Tell what you learned about methods of counting in this section.

ANSWERS TO SELECTED LESSON EXERCISES

3. No. Make a list of possibilities.

4. $26^2 \cdot 10^4 = 6,760,000$

5. $8^2 \cdot 10^8 = 6,400,000,000$

6. $30 \cdot 29 \cdot 28 = 24,360$

8. (a) $2^6 = 64$ (b) $\dfrac{6}{64} + \dfrac{1}{64} = \dfrac{7}{64}$

9. (a) 120

10. (a) $_{30}C_3 = 4,060$

11. (a) Permutation; $_{10}P_3 = 720$
(b) Combination; $_{100}C_4 = 3,921,225$
(c) Permutation; $_5P_5 = 120$

13.3   HOMEWORK EXERCISES

Basic Exercises

1. Jane has 4 skirts and 2 shirts that match.
(a) Draw a tree diagram that shows all possible outfits.
(b) Make an organized list that shows all possible outfits.
(c) How many different outfits can she make?

2. A lottery allows you to select a two-digit number. Each digit may be either 1, 2, or 3. Use a tree diagram to show the sample space and tell how many different numbers can be selected.

3. The Grain Barn is known for its healthy 3-course dinner consisting of appetizer, entree, and vegetable. How many different dinners are possible if you choose 1 item for each course?

> **DINNER AT THE GRAIN BARN**
> **Appetizers**
> Sponge bread
> Yogurt gumbo
> Buckwheat balls
> -
> **Entrees**
> Twice-baked kelp
> Oat bran surprise
> Flax hash
> Marinated bulgur
> -
> **Vegetables**
> Scalloped kale
> Lima bean puree

4. A car dealer offers the National Motors Nag in 5 colors with a choice of 2 doors or 4 doors and a choice of 2 different engines. How many different possible models are there?

5.

PICK 4 NUMBERS GAME

DESCRIPTIONS OF POSSIBLE BETS.	EXAMPLE if you bet:	YOU ARE A WINNER if any of these combinations are selected (Example)
"STRAIGHT" bet is when your number matches the winning number selected in *exact order*. "BOX" bet is when your number matches the winning number selected in *any order*.		
STRAIGHT. To win, the four digit number you select must match in exact order the number drawn on TV. 1 way to win. (ODDS 1 IN 10,000)	1234	1234 ONLY EXACT MATCH WINS.
4 WAY BOX. Of the four digits you select, three are the same. To win, all four digits of your number must match the winning number in any order. 4 ways to win. (ODDS 1 IN 2,500)	1112	1112 1121 1211 2111

Explain how the probabilities (incorrectly called "odds") for the 2 types of payoffs are obtained.

6. In theory, a monkey randomly selecting keys on a typewriter would eventually type some English words. If a monkey is at a keyboard with 48 keys, what is the probability that 5 random keystrokes will produce the word "lucky"?

7. A fifth-grade class of 26 students wants to elect a president, vice-president, treasurer, and secretary. How many different ways are there to fill the 4 positions?

8. A club has 15 members. In how many different ways can the club select a president and a vice-president?

9. If 6 horses are entered in a race and there can be no ties, how many different orders of finish are there?

10. A photographer wants to arrange 3 children in a row for a family photograph. In how many ways can this be done?

11. (a) Substitute $r = n$ in the formula $_nP_r$ and simplify it to obtain the formula for $_nP_n$.
(b) Does part (a) involve induction or deduction?

12. Why can't you use the permutation formula in Exercise 2?

13. The boss asks you about the key design for the new SUV, the National Motors Guzzler. The car keys have 6 notch positions, and each notch has five different possible depths. How many different keys can be made?

 14. My wife's new bicycle lock is a combination lock with 5 dials, each numbered 1 to 9. If we forget the combination, how many possible combinations are there to try?

15. Until 1995, area codes could not have a first digit of 0 or 1, and the second digit had to be a 0 or a 1. The third digit could be any number.
(a) How many possible area codes were there? (In fact, some codes, such as 411 and 911, are used for other purposes.)
(b) In 1996, there were about 140 different area codes in the United States and Canada. Were the countries close to running out of possible area codes under the old system?
(c) Starting in 1995, the second digit was no longer restricted. How many possible area codes are there now?

16. A state license plate consists of 1 letter followed by 5 digits.
(a) How many different license plates can be made?
(b) If no digit could be repeated, how many possible license plates would there be?
(c) A witness to a crime saw the first letter and first three digits of a license plate correctly as A463 but could not see the last two digits. How many license plates could start with A463?

17. You want to design a format for license plates in a state with 4 million cars. How would you do it? Play it safe and make one that will cover 6 million cars.

18. You want to design a format for license plates in a state with 20 million cars. How would you do it? Play it safe and make one that will cover 25 million cars.

19. A true-false test has 5 questions. What is the probability of getting at least 4 right by guessing all the answers?

20. A true-false test has 4 questions. What is the probability of getting at least 3 right by guessing the answers randomly?

21. A quiz has 4 multiple-choice questions. Each question has 5 choices, 1 being correct and the other 4 being incorrect. Suppose you guess on all 4 questions.
(a) How many ways are there to answer the 4 questions?
(b) If someone got at least 3 right, exactly how many questions could he or she have gotten right?
(c) What is P(at least 3 right)?

22. (a) In the preceding exercise, how many ways are there to get no questions right?
(b) What is P(none right)?

 23. How many possible 5-card poker hands are there? (*Note:* A regular deck has 52 cards.)

24. Suppose a test allows a student to pick which questions to answer. In how many different ways can a student choose a set of 8 questions to answer from a group of 10 questions?

25. When do we use permutations rather than combinations in counting?

26. Which is usually greater, the number of combinations of a set of objects or the number of permutations?

27. A consumer group plans to select 2 televisions from a shipment of 8 to check the picture quality. In how many ways can they choose 2 televisions?

 28. A baseball manager wants to arrange 9 starting players into a batting order. How many different batting orders are possible?

 29. How many ways are there to rank (as first and second) the 2 greatest U.S. presidents in history from a list of 12 choices?

 30. In a state lottery game, you choose 6 different numbers from the set of counting numbers 1 to 50 (inclusive). The order does not matter. How many different choices can you make?

Extension Exercises

31. The first four rows of Pascal's triangle are

$$
\begin{array}{ccccccc}
 & & & 1 & & 1 & \\
 & & 1 & & 2 & & 1 \\
 & 1 & & 3 & & 3 & & 1 \\
1 & & 4 & & 6 & & 4 & & 1
\end{array}
$$

(a) Write the fifth row of Pascal's triangle.
(b) Use a calculator to find the values of $_5C_0$, $_5C_1$, $_5C_2$, $_5C_3$, $_5C_4$, and $_5C_5$. What do you notice about your results?
(c) Use Pascal's triangle to find $_4C_3$ and $_6C_2$. Check your results with a calculator.

32. A quiz has 5 multiple-choice questions. Each question has 4 choices; 1 is the correct answer and the other 3 are incorrect. Suppose that you guess all the answers.
(a) $P(\text{at least 1 right}) = 1 - P(\underline{\hspace{1.5cm}})$.
(b) Use the equation in part (a) to find the probability of getting at least 1 right.

33. A slot machine has the following symbols on its 3 dials. Each dial is spun independently at random and lands on 1 of 20 possibilities.

Dial 1	Dial 2	Dial 3
1 bar	1 bar	1 bar
1 bell	2 bells	2 bells
2 cherries	2 cherries	1 cherry
7 lemons	1 lemon	7 lemons
8 oranges	6 oranges	4 oranges
1 plum	8 plums	5 plums

The payoff for each nickel is as follows.

Payoffs

1 cherry	$.10	3 bells	$5
2 cherries	$.50	3 bars	$10
3 cherries	$1		

(a) How many possible outcomes are there for the 3 dials taken together?
(b) How many ways are there to get 3 cherries (1 on each dial)?
(c) How many ways are there to get 3 bars?
(d) How many ways are there to get 3 bells?
(e) Give the probabilities of the events in parts (b), (c), and (d).
(f) How many ways are there to spin exactly 1 cherry?
(g) How many ways are there to spin exactly 2 cherries?
(h) Give the probabilities of the events in parts (f) and (g).

34. A survey asks you to list your first 3 choices for the Democratic presidential nomination from a list of 6 contenders and your first 3 choices for the Republican presidental nomination from a list of 5 contenders. In how many ways can the survey be completed? (Assume that you choose all of your responses from the lists.)

35. Three couples are sitting together at a show. In how many ways can they sit without any of the couples being split up? Find out by answering the following.
(a) How many choices are there for the aisle seat?
(b) Once the aisle seat is filled, how many choices are there for the seat next to it?
(c) Now how many choices are there for the next seat in?
(d) Now how many choices for the fourth seat?
(e) Now how many choices are there for the fifth seat?
(f) Now how many are there for the sixth seat?
(g) How many ways are there, then, to seat all 6 people?

36. Follow the method of the preceding question to find how many ways there are to seat two couples (male-female) at a concert if a woman sits on the aisle and the remaining people are allowed to sit in any possible arrangement.

Labs and Projects

37. Photocopy the 3 characters shown here, and cut each one into 3 separate sections (head, midsection, legs).
(a) Construct some new characters using the pieces.
(b) How many possible characters could you construct?

13.4 INDEPENDENT AND DEPENDENT EVENTS

NCTM Standards

- compute probabilities for simple compound events, using such methods as organized lists, tree diagrams, and area models (6–8)

If you feel sick in the morning, it affects the probability that you will go to school or work. In the study of probability, it is important to know whether the outcome of one event affects the outcome of another.

LE 1 Opener

Consider the following events.

$$R = \text{rain tomorrow}$$

$$U = \text{you carry an umbrella tomorrow}$$

$$H = \text{coin flipped tomorrow lands on heads}$$

(a) Does the probability of R affect the probability of U?

(b) Does the outcome of the coin flip affect whether or not it will rain tomorrow?

In LE 1, R and H are independent events, whereas R and U are dependent events. Two events are **independent** if the probability of one remains the same regardless of how the other turns out. Events that are not independent are **dependent**.

LE 2 Concept

You roll a regular red die and a regular green die. Consider the following events.

$$A = \text{a 4 on the red die}$$

$$B = \text{a 3 on the green die}$$

$$C = \text{a sum of 9 on the two dice}$$

Tell whether each pair of events is independent or dependent.

(a) A and B (b) B and C

In LE 2, the outcome of the red die is not affected by the outcome of the green die. The events A and B are totally separate (independent). The probability of B is the same no matter how A turns out. (Similarly, the probability of A is the same no matter how B turns out.)

Classroom
Connection

LE 3 Reasoning

Suppose you flip a regular coin 6 times, and it comes up heads every time. Three sixth graders make the following claims. Sam says the next toss is more likely to be heads, because the coin has come up heads so much before. Mike says the next toss is more likely to be tails, because things have to even out. Janet says the chances of heads and tails on the next toss are each 50%.

(a) Who is right?

(b) How would you explain why to the other two children?

Figure 13–10

INDEPENDENT EVENTS

What is the probability of two independent events occurring in succession? Try the following exercise.

 LE 4   Reasoning

Suppose a spinner contains congruent regions numbered 1, 2, and 3 (Figure 13–10).

(a) In an experiment, you spin twice. Are the results of the two spins independent or dependent?

(b) Write a sample space of equally likely outcomes for the experiment. (Use an organized list or tree diagram.)

(c) Suppose A is the event that the first spin is a 1 and B is the event that the second spin is a 2. Compute $P(A)$, $P(B)$, and $P(A \text{ and } B)$.

(d) You can also find $P(A \text{ and } B)$ with an area model. Draw a rectangle in pencil. To show the result of the first spin, divide the rectangle into 3 equal parts. Label them 1, 2, and 3. To show the result of the second spin after a first spin of 1, divide the section of the rectangle labeled 1 into 3 parts and label or shade the part that would represent a second spin of 2. What fraction of the original rectangle represents a first spin of 1 and a second spin of 2?

(e) Suppose you get a 1 on the first spin $\frac{1}{3}$ of the time. Next, suppose that $\frac{1}{3}$ of those times that you get a 1 on the first spin, you also get a 2 on the second spin. The fraction of the time that you get a 1 on the first spin *and* a 2 on the second spin is

$$\frac{1}{3} \text{ of } \frac{1}{3} = \underline{\hspace{1cm}} \times \underline{\hspace{1cm}} = \underline{\hspace{1cm}}.$$

(f) What formula relating $P(A)$, $P(B)$, and $P(A \text{ and } B)$ is suggested by parts (c) (d), and (e)?

This spinner exercise is an example of the independent-events formula.

The Independent-Events Formula

If A and B are independent events, $P(A \text{ and } B) = P(A) \cdot P(B)$.

Apply this formula in the following exercise.

LE 5   Skill

A fair die is tossed twice.

(a) What is the probability of getting a 3 on the first toss followed by an odd number on the second toss?

(b) Draw a rectangular area model that shows the probability in part (a).

You can compute the probability of 3 or more independent events occurring in succession in the same way, by multiplying the probabilities of the individual events. Figure 13–11 shows how the sixth-grade *Harcourt Math* textbook introduces independent events.

LESSON 23.3

Independent and Dependent Events

Learn how to identify and find probabilities of dependent and independent events.

QUICK REVIEW

1. $\frac{1}{4} \times \frac{1}{2}$ 2. $\frac{2}{3} \times \frac{1}{6}$ 3. $\frac{1}{5} \times \frac{1}{4}$ 4. $\frac{1}{2} \times \frac{1}{2}$ 5. $\frac{3}{5} \times \frac{1}{6}$

Vocabulary

independent events

dependent events

The table below shows the contents of a bag of marbles. Juan randomly selects a marble from the bag, replaces it, and selects again.

RED	BLUE	GREEN	YELLOW
3	0	5	4

Because Juan replaces the marble after the first selection, the outcome of the second event does not depend on the outcome of the first event. These are **independent events**.

Marbles have been around for at least 4,000 years.

To find the probability of two independent events, multiply their probabilities.

> If A and B are independent events, then
> P(A, B) = P(A) × P(B).

EXAMPLE 1

Remember that to find the theoretical probability of an event with equally likely outcomes, you write the ratio of the number of favorable outcomes to the number of possible outcomes.

What is the probability that Juan first selects a blue marble, replaces it, and then selects a yellow marble?

Find P(blue, yellow).   *First blue, then yellow.*

first selection: P(blue) = $\frac{8}{20}$, or $\frac{2}{5}$ *Find the probability of blue.*

second selection: P(yellow) = $\frac{4}{20}$, or $\frac{1}{5}$ *Find the probability of yellow.*

P(blue, yellow) = $\frac{2}{5} \times \frac{1}{5} = \frac{2}{25}$ *Multiply the probabilities.*

So, the probability is $\frac{2}{25}$, 0.08, or 8%.

Figure 13–11

From Harcourt Math, Grade 6 (Orlando, FL: Harcourt School Publishers, 2002, p. 447.

DEPENDENT EVENTS

Now investigate the probability of two dependent events occurring in succession.

LE 6   Reasoning

A box contains cards numbered 1, 2, and 3. You pick 2 cards in succession *without replacement*. ("Without replacement" means you do not put back a card after you pick it.)

(a) In this experiment, are the first and second draws independent or dependent events?

(b) Write a sample space of equally likely outcomes for the experiment. (Use an organized list or a tree diagram.)

(c) Suppose A is the event that the first card is a 2 and B is the event that the second card is a 3. The $P(B$ given $A)$ means the probability of B, given that A has happened. Compute $P(A)$, $P(B$ given $A)$, and $P(A$ and $B)$.

(d) You can also find $P(A$ and $B)$ with an area model. To show the result of the first pick, divide a rectangle into 3 equal parts. Label them 1, 2, and 3. To show the result of the second pick after a first pick of 2, divide the section of the rectangle labeled 2 into 2 equal parts, and label or shade the part that would represent a second pick of 3. What fraction of the original rectangle represents a first pick of 2 and a second pick of 3?

(e) Suppose A occurs $\frac{1}{3}$ of the time. Next, suppose that $\frac{1}{2}$ of those times A occurs, B also occurs. The fraction of the time that A and B both occur is $\frac{1}{2}$ of $\frac{1}{3} =$ _____ $\times$ _____ $=$ _____.

(f) What formula relating $P(A)$, $P(B$ given $A)$, and $P(A$ and $B)$ is suggested by parts (c), (d), and (e)?

In LE 6(f), you may have guessed the formula for the Multiplication Rule for Probabilities.

Multiplication Rule for Probabilities

$P(A$ and $B) = P(A) \cdot P(B$ given $A)$

Use this formula in the next exercise.

LE 7   Reasoning

You have a drawer with 10 black socks and 6 white socks. If you select 2 socks at random, what is the probability of picking 2 white socks? (*Hint:* What is the probability that the first sock will be white?)

What is the connection between the independent-events formula and the Multiplication Rule for Probabilities?

LE 8   Reasoning

The independent-events formula is $P(A$ and $B) = P(A) \cdot P(B)$. The Multiplication Rule for Probabilities is $P(A$ and $B) = P(A) \cdot P(B$ given $A)$.

(a) What is the difference between the two formulas?

(b) If A and B are independent events, why would $P(B$ given $A) = P(B)$?

The preceding lesson exercise shows that the independent-events formula is a special case of the Multiplication Rule for Probabilities in which $P(B$ given $A)$ is simplified to $P(B)$.

Now see whether you can find an experimental and a theoretical probability for the following situation.

LE 9   Reasoning

You have 6 black socks and 4 white socks in a drawer. You pick 2 socks at random.

(a) Without computing probabilities, guess the probability of picking a matching pair.
(b) Simulate the experiment 30 times, using 2 colors of chips and a bag. What is your experimental probability for matching a pair of socks?
(c) Devise a plan and solve the following problem. Compute the theoretical probability of picking a matching pair.

LE 10   Summary

Tell what you learned about independent and dependent events in this section.

ANSWERS TO SELECTED LESSON EXERCISES

1. (a) Yes   (b) No

2. (a) Independent
(b) Dependent, since a 3 on the green die makes it more likely $\left(\dfrac{1}{6}\right)$ that you will get a sum of 9.

3. (a) Janet
(b) The coin has no memory of what happened on previous tosses (independent events). The probabilities for each toss are the same.

4. (a) Independent
(b)

First draw   Second draw   Outcome

```
         1           1        1,1
    1 <  2           2        1,2
         3           3        1,3
         1           1        2,1
    2 <  2           2        2,2
         3           3        2,3
         1           1        3,1
    3 <  2           2        3,2
         3           3        3,3
```

1,1 1,2 1,3
2,1 2,2 2,3
3,1 3,2 3,3

Organized list   **Tree diagram**

(c) $P(A) = \dfrac{1}{3}$, $P(B) = \dfrac{1}{3}$, $P(A \text{ and } B) = \dfrac{1}{9}$

(d) $\dfrac{1}{9}$

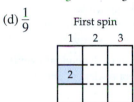

First spin

(e) $\dfrac{1}{3}$; $\dfrac{1}{3}$; $\dfrac{1}{9}$

5. (a) $\dfrac{1}{6} \cdot \dfrac{3}{6} = \dfrac{1}{12}$

(b) $\dfrac{1}{12}$

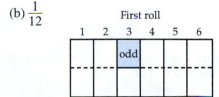

First roll

6. (a) Dependent

(b)

First draw   Second draw   Outcome

$$
\begin{array}{ccc}
1 & \begin{cases} 2 & 1, 2 \\ 3 & 1, 3 \end{cases} \\
2 & \begin{cases} 1 & 2, 1 \\ 3 & 2, 3 \end{cases} \\
3 & \begin{cases} 1 & 3, 1 \\ 2 & 3, 2 \end{cases}
\end{array}
$$

1, 2   1, 3
2, 1   2, 3
3, 1   3, 2

Organized list   Tree diagram

(c) $P(A) = \dfrac{1}{3}$, $P(B \text{ given } A) = \dfrac{1}{2}$, $P(A \text{ and } B) = \dfrac{1}{6}$

(d) $\dfrac{1}{6}$ First pick

1   2   3

	3	

(e) $\dfrac{1}{2}; \dfrac{1}{3}; \dfrac{1}{6}$

7. $\dfrac{6}{16} \cdot \dfrac{5}{15} = \dfrac{1}{8}$

9. (c) $\dfrac{4}{10} \cdot \dfrac{3}{9} + \dfrac{6}{10} \cdot \dfrac{5}{9} = \dfrac{7}{15}$

13.4 HOMEWORK EXERCISES

Basic Exercises

1. Consider the following events.

C = you eat cereal tomorrow
B = you wear blue shoes tomorrow
M = you drink milk tomorrow

(a) Are C and M independent or dependent events?
(b) Are M and B independent or dependent events?

2. A fair coin is tossed twice.

A = heads on the first toss
B = heads on the second toss
C = heads on both tosses

Tell whether each pair of events is independent or dependent.
(a) A and B (b) A and C (c) B and C

3. Suppose A = rolling a sum of 7 with two regular dice. Make up an event B so that
(a) A and B are independent.
(b) A and B are dependent.

4. Write a sentence that tells the difference between two independent events and two dependent events.

5. A box contains cards numbered 1 and 2. You pick 2 cards in succession with replacement. ("With replacement" means that the number selected first is replaced in the box and the cards are shuffled before the second draw is made.)

(a) Are the first and second draws independent or dependent events?
(b) Write a sample space of equally likely outcomes.
(c) Complete the tree diagram, showing all the possible outcomes.

First draw   **Second draw**   **Outcome**

$$
\begin{cases} 1 \\ 2 \end{cases}
$$

(d) Suppose A = 1 on the first pick and B = 2 on the second pick. $P(A \text{ and } B)$ = _____.
(e) Show how to determine the result of part (d) with an area model.
(f) Show how to solve part (d) with a formula.

6. A game requires that you spin two different spinners.

 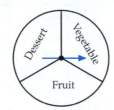

You win the number of food items that corresponds to what you spin.
(a) Write a sample space of equally likely outcomes for the experiment. (Use an organized list or a tree diagram.)
(b) What is the probability you will win 4 desserts?
(c) Show how to determine the result of part (b) with an area model.
(d) Show how to solve part (b) with a formula.

7. Two dice are rolled. What is the probability that both dice show a 5 or a 6? Draw an area model to support your answer.

8. Consider the following events.

A = college soccer team wins
B = college football team wins

Suppose the college soccer team wins $\frac{1}{2}$ of its games and the college football team wins $\frac{1}{3}$ of its games.

What fraction of the time would you expect both teams to win when they play on the same day?

Draw an area model to support your answer.

9. A man has 2 cars, a Recall and a Sea Bass Brougham. The probability that the Recall starts is 0.1. The probability that the Sea Bass Brougham starts is 0.7. (Assume that the cars operate independently of each other.)
(a) What is the probability that both cars start?
(b) What is the probability that neither car starts?
(c) What is the probability that exactly 1 car starts?

 10. Some drug tests are about 98% reliable. This means that there is a probability of 0.98 that the test will correctly identify a drug user or a nonuser. To be safe, each person is tested twice.
(a) What is the probability that a drug user will pass both tests?
(b) What is the probability that a drug user will fail at least 1 of the tests?

11. A sixth grader says that the probability of rolling two consecutive 1's on a die is $\frac{1}{6} + \frac{1}{6} = \frac{1}{3}$. Is this correct? If not, how would you help the child understand the correct solution?

12. At Risk High, the football coach also teaches probability. He knows that when the opposition calls the coin flip, they call heads 80% of the time. Since heads comes up only 50% of the time, he always lets the visiting team call the coin flip. Should he expect to come out ahead this way?

13. A box contains cards numbered 1, 2, 3, and 4. You pick 2 cards in succession without replacement.

(a) In this experiment, are the first and second draws independent or dependent?
(b) Write a sample space of equally likely outcomes. (Use an organized list or a tree diagram.)
(c) Suppose A = 3 on the first pick and B = 2 on the second pick. Find $P(A)$, $P(B$ given $A)$, and $P(A$ and $B)$.
(d) Show how to find $P(A$ and $B)$ with an area model.
(e) Show how to find $P(A$ and $B)$ using a formula.

14. You pick two cards in succession without replacement from a set of cards numbered 1 through 8 (inclusive).
(a) Are the first and second draws independent or dependent?
(b) Use an area model to determine the probability of drawing a 3 followed by an even number.
(c) Show how to solve part (b) with a formula.

15. You pick 2 cards from a regular deck *without* replacement. What is the probability of picking 2 aces?

16. A drawer contains a mixture of 10 black socks, 8 white socks, and 4 red socks. You randomly select 2 socks to wear. Determine the probability that
(a) both are black.
(b) both are white.
(c) you pick a matching pair.

17. (a) When drawing at random with replacement, are the draws independent or dependent?
(b) Without replacement, are the draws independent or dependent?

18. Two draws are made from a standard deck of cards without replacement. Are the draws independent or dependent?

19. Assume that for a nuclear power plant to have an accident, 4 systems must fail. The probabilities that the 4 individual systems fail are 0.01, 0.006, 0.002, and 0.002, respectively. What is the probability that the plant will have an accident?

20. What is the probability that a family with 8 children has 8 girls?

21. A group of mothers and their grown daughters were surveyed.

$$A = \text{mother attended college}$$
$$B = \text{daughter attended college}$$

$P(A) = 0.3$, $P(B) = 0.6$, and $P(B \text{ given } A) = 0.7$. What is the probability that a mother and her daughter both attended college?

22. On the average, it rains or snows about 114 days a year in Spokane, Washington.
(a) If you visit for a day, what is the probability that there will be precipitation (rain or snow)?
(b) If there is precipitation one day, the probability of precipitation the next day is 0.5. If you visit for 2 days, what is the probability that it will rain or snow both days?

23. A basketball team has a probability of 0.7 of making each free throw. If they shoot 5 free throws at the end of a game, determine the probability that they make
(a) all 5 shots.   (b) at least 4 shots.

24. A slot machine has 3 independent dials set up as follows.

Dial 1	Dial 2	Dial 3
1 bar	1 bar	1 bar
1 bell	2 bells	2 bells
2 cherries	2 cherries	1 cherry
7 lemons	1 lemon	7 lemons
8 oranges	6 oranges	4 oranges
1 plum	8 plums	5 plums

What is the probability of getting each of the following?
(a) 3 lemons   (b) 3 limes   (c) 0 cherries

25. A shipment of radios contains 97 that work and 3 that are defective. An inspector selects 5 at random. What is the probability that they all work?

26. How strange are coincidences? Suppose an event has a 1 in 500 chance of happening each day. Won't you be surprised if it occurs? But approximately what is the probability that this event will happen sometime in the next year? (*Hint:* Assume independence, and find the probability that it will not occur in the next year.)

27. The home team is behind by one point. There is one second on the clock as Ben steps up to the free-throw line to shoot "one on one." This means that he takes a one-point free throw (shot). He has to make the free throw to be allowed to take a second one. Ben makes about 60% of his free throws. Solve each of the following with a formula and a decimal square area model (Activity Card 3).
(a) What is the probability that Ben makes 0 free throws (and the team loses)?
(b) What is the probability that Ben makes 2 free throws (and the team wins)?
(c) What is the probability that Ben makes 1 free throw (leading to an overtime game)?

28. When it snows, the local schools are closed about 70% of the time. The forecast for tomorrow says there is a 40% chance of snow. Solve the following with a formula and a decimal square area model (Activity Card 3). What is the probability that the local schools will close?

Extension Exercises

29. Suppose tulip bulbs have a probability of 0.6 of flowering. How many bulbs should you buy to have a probability of 0.9 of obtaining at least 1 flower?

30. In a mathematics class that contains 50% boys and 50% girls, 60% of the children are 8 years old and 40% are 9 years old.
(a) What is the largest possible percentage of 9-year-old girls in the class?
(b) What is the smallest possible percentage of 9-year-old girls in the class?
(c) If age and sex are independent, then _____%
of the class is 9-year-old girls.

31. Two basketball teams play a best-2-out-of-3 series, meaning that the first team to win two games is the champion. Suppose that team A has a probability of 0.6 of winning each game. Find the probability that team A wins
(a) in 2 games.   (b) in 3 games.
(*Hint:* Draw a tree diagram.)

32. A World Series is a best-4-out-of-7 series. If the American League team has a probability of 0.5 of winning each game, determine the probability that they win the World Series in
(a) 4 games.   (b) 5 games.

33. Consider how $P(A \text{ and } B)$ is related to $P(A)$. Let A = Mike will take math next semester. If possible, make up an event B so that
(a) $P(A \text{ and } B) < P(A)$.
(b) $P(A \text{ and } B) = P(A)$.
(c) $P(A \text{ and } B) > P(A)$.

34. If possible, make up two events A and B that are
(a) independent and not mutually exclusive.
(b) independent and mutually exclusive.
(c) dependent and mutually exclusive.

13.5 EXPECTED VALUE AND ODDS

NCTM Standards

- propose and justify conclusions and predictions that are based on data and design studies to further investigate the conclusions or predictions (3–5)
- recognize and apply mathematics in contexts outside of mathematics (pre-K–12)

Have you ever played a lottery or gambled at a casino? Do you know how to compute how much you are going to *lose* in the long run? Do you know what the term "odds" means? This lesson introduces the mathematics of the expected value and odds.

In a casino, people risk their money and usually end up losing some of it. Although individual outcomes vary quite a bit, casinos consistently come out ahead at their games.

Many people are not interested in gambling, but the same mathematics applies to another area that does affect nearly everyone—insurance. In buying insurance, we also risk our money and usually end up losing some.

FAIR VERSUS UNFAIR GAMES

Would you play a gambling game in which you roll 2 dice and you win the amount you bet whenever the sum is greater than 10? This is an example of an unfair game.

In a **fair game** involving money, a player can expect to come out even in the long run. In a fair game between two players, neither player has an advantage due to the rules.

Discussion

LE 1   Opener

Consider the following game. Place four chips in a cup: two of one color and two of another color. Each player selects one chip from the cup without looking. Player 1 wins if both chips are the same color. Player 2 wins if the two chips are different colors.

(a) Do you think this is a fair game? If not, which player do you think has the advantage?
(b) If you have the game materials, play the game 8 times and tally the results. If not, skip to part (d).
(c) Answer part (a) again.
(d) Use a sample space, tree diagram, or area model to determine the theoretical probability of each player winning.

LE 2   Reasoning

Make up a fair game for 2 players that involves rolling 2 dice.

EXPECTED VALUE

Casino games are not fair games. If they were, casino owners would not make any money. Consider the simple model of a casino or lottery game in the following example.

● EXAMPLE 1

Suppose that you pick 1 card at random from the box shown and you win the amount of money printed on the card. The cards are replaced and reshuffled for each new game.

If you played this game often, how much would you expect to be paid on the average?

SOLUTION

In the long run, you would expect the number of times you picked each of the 10 cards to be about the same. To find the average, imagine playing the game 10 times and picking each of the different cards once. You would win

$$\$100 + \$10 + \$0 + \$0 + \$0 + \$0 + \$0 + \$0 + \$0 + \$0 = \$110$$

in 10 draws. The expected (average) winnings per draw would be $\frac{\$110}{10} = \$11$. ●

In Example 1, the expected average result in the long run was computed. This result is called the **expected value**.

LE 3   Reasoning

Consider a lottery game in which 7 out of 10 people lose, 1 out of 10 wins $50, and 2 out of 10 win $35. This can be modeled by selecting 1 card at random from the group shown.

(a) If you played 10 times, about how much would you expect to win?

(b) What is the expected (average) value of the payoff?

(c) Complete the following table, showing the outcomes.

Payoff	Frequency
$50	1
$35	
$ 0	

(d) Compute the average from the frequency table.

$$\text{Expected average} = \frac{50 \cdot 1 + 35 \cdot \underline{\ \ \ } + \underline{\ \ \ \ \ }}{10} = \underline{\ \ \ \ \ \ \ }$$

(e) If the fee were the same, would you prefer to play the game in Example 1 or the game in this exercise?

The first state lottery, held in New Hampshire in 1964, created a lot of controversy. Now about 41 states use lotteries to raise money for state programs, as an alternative to raising taxes. Is a gambling game a good way to raise money? Many people in the states without lotteries don't think so.

State officials in lottery states need to know how much money they can expect to raise in the long run—the expected value. People playing the lottery are also interested in the expected value of their lottery tickets (Figure 13–12).

Figure 13–12

The next exercise develops a more widely applicable method for computing expected value in gambling games. This method uses the probability of each possible financial outcome.

LE 4   Connection

Expected values are usually computed from probability tables.

(a) Complete the probability table for the lottery game described in the preceding exercise.

Payoff	Probability
$50	$\frac{1}{10}$
$35	
0	

(b) Compute the expected payoff from the probability table.

$$\text{Expected payoff} = 50 \cdot \frac{1}{10} + 35 \cdot \underline{\hspace{1cm}} + \underline{\hspace{1.5cm}} = \underline{\hspace{1cm}}$$

(c) If the game costs $15 to play, how much should you expect to lose, on the average, per play?

The formula used in part (b) of the preceding exercise is as follows.

Definition: Expected Value

If an experiment has the possible numerical outcomes $n_1, n_2, n_3, \ldots, n_r$ with corresponding probabilities $p_1, p_2, p_3, \ldots, p_r$, then the **expected value** E of the experiment is

$$E = n_1 p_1 + n_2 p_2 + n_3 p_3 + \ldots + n_r p_r$$

A simpler example makes it clear why we multiply each payoff times its probability to compute the expected value. Suppose a coin-flipping game pays $10 for a head and nothing for a tail. What is the average payoff? You would (expect to) be paid about $\frac{1}{2}$ the time.

$$\text{Average payoff} = (\$10)\left(\frac{1}{2}\right) = \$5$$

This is the payoff times its probability.

Use the expected-value formula in the following exercise.

LE 5   Reasoning

Suppose a lottery game allows you to select a 2-digit number. Each digit may be either 1, 2, 3, 4, or 5. If you pick the winning number, you win $10. Otherwise, you win nothing.

(a) What is the probability that you will pick the winning number?
(b) The notation $E(\text{payoff})$ means the expected (or average) payoff. What is $E(\text{payoff})$?
(c) If the lottery ticket costs $1, how much should you expect to lose on the average (per play)?

Insurance companies also use expected values. To determine their rates and payoffs, they estimate the probabilities of particular catastrophes.

• EXAMPLE 2

An insurance company will insure your dorm room against theft for a semester. Suppose the value of your possessions is $800. The probability of your being robbed of $400 worth of goods during a semester is $\frac{1}{100}$, and the probability of your being robbed of $800 worth of goods is $\frac{1}{400}$. Assume that these are the only possible kinds of robberies. How much should the insurance company charge people like you to cover the money they pay out and to make an additional $20 profit per person on the average?

SOLUTION

Understanding the Problem   The insurance company wants $20 per person to be left over after paying out all the claims.

Devising a Plan   Compute the expected payout for the insurance. Then, to make $20 profit, charge the expected payout plus an additional $20.

Carrying Out the Plan

$$E(\text{payout}) = (\$400) \cdot \frac{1}{100} + (\$800) \cdot \frac{1}{400} = \$4 + \$2 = \$6$$

They should charge $6 + $20 = $26 for the policy to make an average gain of $20 per policy.

Looking Back   The answer seems reasonable.   ●

Discussion

LE 6   Reasoning

Consider the following problem. "In a carnival game, you toss a single die. If you roll a 3, you win $3. If you roll a 6, you win $6. Otherwise, you win nothing. How much should the carnival operators charge you to have an expected gain of $0.50 per game?" Devise a plan, and solve the problem.

ODDS

In gambling games, probabilities are often stated using odds. Odds compare your chances of losing and winning on each play. For example, on the race card in the following table, the odds against Turf Tortoise winning are 12–1 (12 to 1). This expression means that Turf Tortoise is expected to lose about 12 times for every 1 time he wins. Whereas a probability compares your chance of losing or winning to the *total number of possibilities*, odds compare your possibilities of losing and winning to *each other*.

Post Time 1 P.M.

Race 1: 6 Furlongs 3YO; CLM $5000

Hot Air	6–5
Dog Lover	3–1
Wet Blanket	4–1
Lost Marbles	5–1
Lead Balloon	7–1
Turf Tortoise	12–1
Obstacle	20–1

Actually, the real chance of Turf Tortoise winning is slightly worse than 12 to 1. The track pays off bettors based upon the odds *after* taking out about 20% of the money for expenses. For this reason, the track adjusts the odds by a factor of about 20%.

Odds (against) can be defined as follows.

Definition: **Odds of Experiments with Equally Likely Outcomes**

Odds against = number of unfavorable outcomes
to number of favorable outcomes

LE 7 Concept

In the table just presented, the odds against Wet Blanket's winning are 4–1 (4 to 1). What does this statement mean?

Probabilities describe the frequency of a (favorable) result in relation to all possible outcomes. "Odds against" compare unfavorable and favorable results, such as losing and winning. Any probability can be converted to odds, and any odds can be converted to a probability.

• EXAMPLE 3

What is the probability that the director whose door is shown in Figure 13–13 is in his office at 1 P.M.?

SOLUTION

Odds of "25 to 1 against" mean that the director will not be there 25 times for each 1 time the director is there. The probability of the director's being there is 1 out of 26, or $\frac{1}{26}$. ●

The preceding example illustrates the relationship between the odds against E and the probability of E. You can compute the odds against E using the formula $\dfrac{P(\overline{E})}{P(E)}$.

© 1991 Carolina Biological Supply Company.

Figure 13–13

LE 8   Skill

The odds against Dead Weight winning the Crabgrass Derby today are 7 to 1. Express the probability of winning using a fraction, a decimal, and a percent.

LE 9   Skill

A roulette wheel has 38 numbers. If I bet on 1 number, what are the odds against my winning?

The odds are the basis for payoffs on bets. If you were paid off on the basis of the *true* odds, the game would be fair. You would tend to come out even in the long run. For example, in LE 9, suppose you won $37 for each $1 bet (from odds of 37 to 1) when you picked the winning number. The expected value could be computed from the following chart.

Gain	Probability
37	$\frac{1}{38}$
−$1	$\frac{37}{38}$

$$E(\text{Gain}) = \$37\left(\frac{1}{38}\right) + (-\$1)\left(\frac{37}{38}\right) = \$0$$

When the expected value is $0, it is a fair bet, since you would not expect to win or lose money in the long run.

Since casinos want to make money, they pay out the money at a slightly lower rate. If you made this bet in roulette, you would actually gain $35, rather than $37, on a $1 bet.

LE 10   Summary

Tell what you learned about fair games and expected value in this section.

ANSWERS TO SELECTED LESSON EXERCISES

1. (d) Player 2 has a probability of $\frac{2}{3}$ of winning each game.

3. (a) $120   (b) $12
   (c) Other frequencies are 2 and 7
   (d) $\dfrac{\$50 \cdot 1 + 35 \cdot 2 + \$0 \cdot 7}{10} = \$12$

4. (a) Other probabilities are $\frac{1}{5}$ and $\frac{7}{10}$.
   (b) $\$50 \cdot \dfrac{1}{10} + \$35 \cdot \dfrac{1}{5} + \$0 \cdot \dfrac{7}{10} = \$12$
   (c) $3

5. (a) $\frac{1}{25}$ (b) $.40   (c) $.60

6. $E(\text{Payoff}) = \dfrac{1}{6}(\$3) + \dfrac{1}{6}(\$6) = \$1.50$
   $0.50 + \$1.50 = \$2$

7. If the odds are accurate, Wet Blanket can be expected to lose 4 times for every 1 time he wins under these conditions.

8. $\frac{1}{8}$, 0.125, 12.5%

9. 37 to 1

13.5 HOMEWORK EXERCISES

Basic Exercises

1. In a dice game, you roll 2 dice. If the sum is divisible by 3 or 5, you win. Otherwise, your opponent wins. Whom does this game favor?

2. Make up a fair game for 2 players that involves flipping 2 coins.

3. Consider the following game. Place four chips in a cup: three of one color and one of another color. Each player selects one chip from the cup without looking. Player 1 wins if both chips are the same color. Player 2 wins if the two chips are different colors.
   (a) Do you think this is a fair game? If not, which player do you think has the advantage?
   (b) If you have the game materials, play the game 9 times and tally the results. If not, skip to part (d).
   (c) Answer part (a) again.
   (d) Use a sample space, tree diagram, or area model to determine the theoretical probability of each player winning.

4. Someone offers to play a dice game called "odd-even" with you. In this game, you roll two dice. If the product is odd, you win. If the product is even, your opponent wins. Is this a fair game? If not, whom does it favor?

5. A box contains the cards shown. You randomly select 1 card and win the amount written on that card. What is the expected (average) value of the payoff for a single draw?

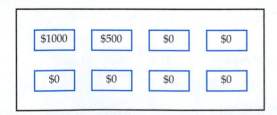

6. You have a job working for a mathematician. She pays you each day according to what card you select from a bag. Two of the cards say $200, five of them say $100, and three of them say $50. What is your expected (average) daily pay?

7. In a game, the number you spin is the number of dollars you win. Which of the following spinners would be the best one to spin?

(a) 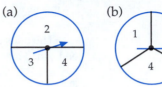 (b) 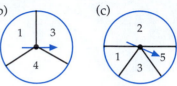 (c)

8. In a game, the number you spin is the number of dollars you win. Which of the following spinners would be the best one to spin?

(a) (b) (c)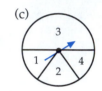

9. (a) The Unshakeable Insurance Company will insure your dorm room against theft. The insurance costs $25 per year. If the probability of getting robbed of an average of $500 worth of possessions is $\frac{1}{100}$, how much profit does the insurance company expect to make from your $25 insurance premium?
   (b) The company sells $100,000 worth of automobile liability insurance that costs $50 per year. In the past, about 1 out of every 1,000 people have collected on a claim. The average (mean) claim is $30,000. What is the expected payoff on your $50 investment?
   (c) Why might the insurance in part (b) be worth the investment, whereas the insurance in part (a) is probably not worth it?

10. A particular oil well costs $40,000 to drill. The probability of striking oil is $\frac{1}{20}$.
   (a) If an oil strike is worth $500,000, what is the expected gain from drilling?
   (b) Would you go ahead and drill?

11. An apartment complex has 20 air conditioners. Each summer, a certain number of them have to be replaced.

Number of Air Conditioners Replaced	Probability
0	0.21
1	0.32
2	0.18
3	0.11
4	0.11
5	0.07

What is the expected number of air conditioners that will be replaced in the summer?

12. The probability that a particular operation is successful is 0.32.
   (a) If the operation is performed 600 times a year at a hospital, about how many successful operations can the hospital expect?
   (b) If you decide to have this operation, give some reasons why the probability of its success for you might be somewhat higher or lower than 0.32.

13. Madame Wamma Jamma will predict the sex of a future baby for $5. If she is wrong, she even has a money-back guarantee! If Madame Wamma Jamma has no psychic power (Perish the thought!), what is her expected gain per customer? Assume that she cheerfully refunds the $5 whenever she is wrong.

14. (a) Suppose you buy $50,000 of air travel insurance against death. If the probability of your dying on the flight is 1 in 300,000, what is the mean payoff for your insurance policy?
   (b) If the insurance costs $1, what is the average gain or loss per policy for consumers?

15. State lotteries sell about $20 billion worth of tickets per year. In Vermont, you choose a 3-digit number. If it matches the state's number, the state pays you $500.
   (a) How many 3-digit numbers are there?
   (b) What is your chance of picking the winning number?
   (c) What is the average payoff you would receive in the long run?
   (d) Tickets cost $1. How much can you expect to lose per play?

16. A lottery game allows you to select a 3-digit number using the digits 0, 1, 2, and 3. You may use the same digit more than once. If you pick the winning number, you win $25 on a $1 bet.
   (a) What is the average payoff?
   (b) How much would you expect to lose, on the average, per $1 bet?

17. Recall that a roulette wheel has 38 slots. Eighteen are red, 18 are black, and 2 are green.
   (a) What is the probability of black?
   (b) If you bet $1 on black, you receive a $2 payoff if the result is black and $0 if the result is red or green. What is the average loss on a $1 bet on black?
   (c) You can also bet on 6 different numbers. If any of them comes up, you receive $6 back for each $1 bet. What is the expected gain or loss on a $1 bet?

18. A set of 3 coins is tossed 32 times. How many times would you expect the result to be 3 heads?

19. In gambling game, you receive a payoff of $10 if you roll a sum of 10 and $7 if you roll a sum of 7 on two dice. Otherwise, you receive no payoff. What is the average payoff per play?

20. In a carnival game, you roll 2 dice. If the sum is 5, you receive a $5 payoff. If the sum is 10, you receive a $10 payoff. What is the expected payoff?

21. In a gambling game, you pick 1 card from a standard deck. If you pick an ace, you win $10. If you pick a picture card (J, Q, or K), you win $5. Otherwise, you win nothing. How much should a carnival booth charge you to play this game if they want an average profit of $0.40 per game?

22. *T* tickets are sold for $1 each for a lottery drawing with a prize of $100. What is the expected payoff of each ticket in terms of *T*?

23. You are designing a spinner game for a carnival. You want to charge people $1 and estimate that 500 people will play. You would like to make about $100. Sketch a spinner, and give the rules and payoffs.

24. Read the Beetle Bailey cartoon, and answer the following questions.
  (a) If Beetle lives another 50 years, how much money will he win (after taxes)?
  (b) Assume that Beetle has a 1-in-5,000,000 chance of winning. What is the expected payoff for his mailing in the entry form?
  (c) Is this payoff worth the price of a stamp?

25. If you purchase a handgun to keep at home, the odds are 11 to 1 against using the gun on an intruder rather than on yourself, a family member, or someone you know. This means that about _____ of every _____ gun owners who use handguns will use it on an intruder.

26. Are you ready for a cheery problem? The following table gives the odds against dying in the next year.

Age	Odds Against Dying in the Next Year	
	Male	*Female*
15-24	576–1	1814–1
25-34	560–1	1489–1
35-44	384–1	743–1
45-54	140–1	269–1
55-64	57–1	111–1
65-74	25–1	49–1
75-84	11–1	17–1
85-up	6–1	7–1

What is the probability that you will live another year?

27. Data indicate that about 2 out of 5 marriages last at least 25 years. The odds against a marriage lasting at least 25 years are _____ .

28. A roulette wheel has 38 slots, of which 18 are red, 18 black, and 2 green. If I bet on red, what are the odds against my winning?

Extension Exercises

29. A company wants to test its employees for a drug. The test is 98% accurate. If someone is using the drug, that person tests positive 98% of the time. If someone is not using the drug, that person tests negative 98% of the time.
  (a) Suppose the company randomly tests all 10,000 of its employees. Also, assume that 50 people are actually using the drug. How many people who are not using the drug will test positive (false positive)?
  (b) In part (a), what is the probability that someone who tests positive is actually using the drug?
  (c) Suppose that, instead, the company tests only people who exhibit suspicious behavior. Suppose 100 people act suspiciously, and 30 of them are using the drug. How many people who are not using the drug will test positive (false positive)?
  (d) In part (c), what is the probability that someone who tests positive is actually using the drug?

30. A company wants to test its employees for a drug. The test is 95% accurate.
  (a) Suppose that the company randomly tests all 5,000 of its employees. Furthermore, assume that 100 people are actually using the drug. How many people who are not using the drug will test positive (false positive)?
  (b) In part (a), what is the probability that someone who tests positive is actually using the drug?
  (c) Suppose that, instead, the company tests only people who exhibit suspicious behavior. Suppose 150 people act suspiciously and 80 of them are using the drug. How many people who are not using the drug will test positive (false positive)?
  (d) In part (c), what is the probability that someone who tests positive is actually using the drug?

31. A patient has a stroke and must choose between brain surgery and drug therapy. Of 100 people who have the surgery, 20 die during surgery, 15 more die after 1 year, and 5 more die after 3 years. Of 100 people who receive drug therapy, 5 die almost immediately, 20 more die after 1 year, and 30 more die after 3 years. Which treatment would you advise someone to take? Why?

32. Refer to the slot machine in Homework Exercise 33 from Section 13.3. Use the probabilities you found to compute the expected payoff on a $0.05 bet.

33. Consider the following problem: "A football betting service picks 1 game each week. They charge $100 per pick, but only if they pick the game correctly. The service starts with 200 customers. They tell 100 customers to bet on one team and 100 customers to bet on the other team. Assume that this continues for 10 weeks, with the following results. Each week, only the customers who win continue the next week. Furthermore, the service attracts 40 new customers each week. How much can the service expect to make if none of the games are ties?" Devise a plan, and solve the problem.

34. Maria and Dionne have a basketball contest. The first one to make a shot wins. They each make their shots about half of the time. Since Maria brought the basketball, they decide she will shoot first. What is Maria's chance of winning? Show how to solve this with an area model.

Labs and Projects

35. Go to a Web site that gives information about state lotteries. Select two different state lotteries. See if you can find the expected payoff for each lottery and tell which would be better to play.

CHAPTER 13 SUMMARY

Uncertainty is part of our everyday lives. It has been said that nothing in life is certain but death and taxes. For this reason, probability theory is helpful for studying the likelihood of everyday events.

Probabilities tell us approximately what we can expect to happen when the same event is repeated many times under the same conditions. If a sample space of equally likely events can be written for an experiment, it may be possible to compute a theoretical probability.

Experimental probabilities are based on experimental results under identical or similar conditions. If it is impractical or costly to find an experimental probability of a random event, one can sometimes study a simulation of the event, using coins, dice, or a computer.

To decide how many letters and digits to put in license plates, phone numbers, and codes, people use counting techniques. These same techniques indicate how likely one is to guess correctly on a multiple-choice test. In probability, counting techniques are used to determine the sizes of sample spaces.

In computing the probability that events A and B will occur, one multiplies probabilities. If A and B are dependent, one must calculate how much one event affects the probability of the other:

$$P(A \text{ and } B) = P(B \text{ given } A) \cdot P(A)$$

If events A and B are independent, meaning they do not influence each other's probabilities, one can compute $P(A \text{ and } B)$ simply by multiplying $P(A) \cdot P(B)$.

How do casinos design gambling games, states design lotteries, and insurance companies set fees? They all use probabilities to estimate the expected average payoffs per person. They then use those payoffs to determine a fee that will cover the payoffs and other expenses.

STUDY GUIDE

To review Chapter 13, see what you know about each of the following ideas or terms that you have studied. You can also use this list to generate your own questions about the chapter.

PROBABILITY IN GRADES 1–8

The following chart shows at what grade levels selected probability topics typically appear in elementary- and middle-school mathematics textbooks.

Topic	Typical Grade Level in Current Texbooks
Basic probability	3, 4, 5, 6
Experimental and theoretical probability	5, 6, 7, 8
Simulations	6, 7, 8
Fundamental Counting Principle	5, 6, 7
Permutations and combinations	7, 8
Independent events	6, 7
Dependent events	7, 8

REVIEW EXERCISES

1. A judge rates the best and second-best orange juices out of 4 brands, *A*, *B*, *C*, and *D*. What is the sample space for his pair of choices?

2. (a) What is the probability of rolling a product less than 10 on 2 dice?
(b) Describe how you could determine the same probability experimentally.

3. What is the probability of getting 3 heads and 1 tail when you flip 4 coins?

4. If I toss 5 coins, the theoretical probability of getting 3 heads and 2 tails is $\frac{5}{16}$. If I flip 5 coins 80 times, how many times will I obtain 3 heads and 2 tails?

5. Write a paragraph that defines experimental and theoretical probability and tells how these two kinds of probabilities are related.

6. A pollster asked 300 students which of the following pizza toppings they prefer: mushrooms, pepperoni, or spinach.

	Mushroom	Pepperoni	Spinach
High school	50	80	25
College	50	60	35

On the basis of these results, what is the probability of each of the following?
(a) A college student prefers pepperoni.
(b) A student prefers spinach.

7. Suppose *A* = you pass the next math test. Make up an event *B* so that *A* and *B* are
(a) mutually exclusive.   (b) not mutually exclusive.

8. A caterer plans of offer people the choice of turkey, chicken, peanut butter, or vegetable sandwiches. On the basis of past orders, the probability someone will choose a chicken sandwich is $\frac{1}{4}$, and the probability someone will choose a turkey sandwich is $\frac{1}{5}$. What is the probability that someone chooses neither chicken nor turkey?

9. You have a spinner like the one shown. You want to simulate each *second* at a one-way traffic intersection. In preliminary work, you found that 150 cars passed in 10 minutes. How would you use the spinner in the simulation?

10. New Hay Checks come with one of 4 different collectible famous-thoroughbred cards in each box. Using a spinner with equal regions numbered 1 to 4, I simulated buying boxes until I had collected all 4 cards. The results follow.

$$1\ 3\ 1\ 2\ 4$$
$$2\ 4\ 1\ 1\ 2\ 2\ 3$$
$$4\ 3\ 1\ 2$$
$$4\ 3\ 4\ 1\ 2$$
$$1\ 3\ 3\ 1\ 4\ 1\ 3\ 2$$

On the basis of these results, how many boxes would you expect to buy to collect all 4 cards?

11. A state has 2 kinds of license plates: plates with 2 letters followed by 2 digits, and plates with 2 letters followed by 3 digits. How many different plates can the state make?

12. A true-false quiz has 3 questions. If I guess at random, what is the probability that I will get at least 2 right? (Assume that I answer each question either *true* or *false*.)

13. A jury of 12 is to be selected from 20 eligible jurors. How many different juries are possible?

14. A teacher wants to award a first prize, a second prize, and a third prize to a class of 25 children. How many ways are there to do this?

15. A fortune teller claims that the card she selects is related to your future. She would say the card and your future are _____ events. In fact, they are _____ events.

16. Suppose A = you pass the next math test. Make up an event B so that A and B are
(a) independent.   (b) dependent.

17. A princess walks through the following maze, choosing her path at random.

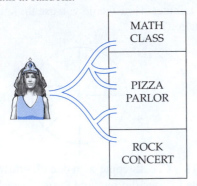

Use a rectangular area model to determine the probability that she ends up at
(a) the math class.   (b) the pizza parlor
(c) the rock concert.

18. Show how to solve exercise 17 without a rectangular area model.

19. A defective undergarment is inspected by inspector 12 and inspector 14. Each of them has a 0.9 chance of finding the defect. What is the probability that neither one of them will find the defect?

20. A motel chain has found that the probability that a customer pays by check is 0.2. The probability that the check is no good, given that a customer pays by check, is 0.1. What is the probability that
(a) a customer does not pay by check?
(b) a customer pays by check and it is no good?

21. Make up a fair game for 2 players that involves flipping 3 coins.

22. An insurance company will insure your home against theft. The value of your possessions that are insurable is $1,000. Suppose the probability of your being burglarized of $500 worth of goods is $\frac{1}{200}$, and the probability of your being burglarized of $1,000 worth of goods is $\frac{1}{1000}$. Assume that these are the only kinds of burglaries possible. How much should the insurance company charge people like you to make an average profit of $10 per policy?

23. A dice game pays $3 back on a $1 bet if you roll a sum of 7 on two dice and $10 back on a $1 bet if you roll a sum of 11. Otherwise, you lose the $1.
(a) What is E(payoff on $1 bet)?
(b) What is E(gain on $1 bet)?

24.

Courtesy of Library of Congress

Sure Thing	1–2
Hopeless	4–1
Bad Breath	6–1
Unsettled	7–1
Lethargic	10–1
Why Bother	50–1

(a) According to the odds, Unsettled (see photo) wins about 1 out of _____ races.
(b) The probability of Lethargic winning is about _____.

ALTERNATE ASSESSMENT

Do one of the following assessment activities: add to your portfolio, add to your journal, write another unit test, do another self-assessment, or give a presentation.

SELECTED READINGS

Hacking, I., *An Introduction to Probability and Inductive Logic.* New York: Cambridge, 2001.

Jacobs, H. *Mathematics: A Human Endeavor,* 3rd ed. New York: W. H. Freeman, 1994.

National Council of Teachers of Mathematics. *Thinking and Reasoning with Data and Chance.* 2006 Yearbook (forthcoming). Reston, VA: NCTM, 2006.

Navigating Through Probability, 2 vols. Reston, VA: NCTM, 2003.

Navigating Through Data Analysis and Probability, 2 vols. Reston, VA: NCTM, 2003.

Packel, E. *The Mathematics of Games and Gambling,* Washington, DC: MAA, 1981.

Paulos, J. *Innumeracy,* New York: Hill and Wang, 1988.

 Look for more articles in *Teaching Children Mathematics* and *Mathematics Teaching in the Middle School* about how to teach children about probability.

ANSWERS TO SELECTED EXERCISES

Chapter 1

1.1 Homework Exercises

1. (b) You end up with 6 more than the original number.

3. Yes

5. With a counterexample

7. False; Mike does not drink beer.

9. Reasonable

11. (a) 444,444,444   (b) 555,555,555
   (c) $12,345,679 \times 54 = 666,666,666$
   $12,345,679 \times 63 = 777,777,777$
   (d) Each expression can be rewritten as
   $(12,345,679 \times 9) \times C$, in which C is a one-digit
   number. Because $12,345,679 \times 9 = 111,111,111$,
   the final result is CCC,CCC,CCC in which each C
   represents a digit of the number.

13. (a) 16 units
   (b) The length of the pendulum is the square of the period of the swing.
   (c) $L = P^2$ (d) 2.2 sec

15. (a) Generalization 1 is based on the observation that
   all through recorded history there have been wars.
   Generalization 2 could be based on teaching a few
   classes in which the boys did better than the girls.
   Generalization 3 could be based on seeing a few
   movie sequels that were not as good as the
   originals.

17. (a) For the last 3 weeks, my teacher has given one pop
   quiz per week. I generalize that my teacher will always do this.

(b) A mathematician sees three examples in which the
   sum of the two odd numbers is an even number.
   The mathematician generalizes that this will always occur.

19. (a) True   (b) True   (c) True
   (d) The product of any N consecutive whole numbers
   is divisible by N.

21. No

1.2 Homework Exercises

1. All rectangles are quadrilaterals.

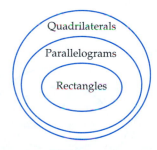

3. (a) $\overline{AB}$ is parallel to $\overline{CD}$, and $\overline{AD}$ is parallel to $\overline{BC}$.
   (b) *Hypotheses:* ABCD is a rectangle. The opposite
   sides of a rectangle are parallel.
   Conclusion: $\overline{AB}$ is parallel to $\overline{CD}$, and $\overline{AD}$ is parallel to $\overline{BC}$.

5. (a) If a person is a student, then the person wants the
   option to earn extra credit.
   (b) If a figure is a rectangle, then it has four sides.

7. True hypotheses

9. (a) −5 is less than 10.
(b) 12 is not a negative number.
(c)

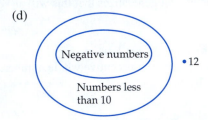

(d)

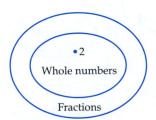

12 is not in the oval for numbers less than 10, so it cannot be in the oval for negative numbers.

11. (a) He gets at least a B in the course.
(b) No conclusion

13. Yes

15. No

17. Yes

19. 5, since one correct digit was lost by changing 5 to 4.

21. (a) 0, 1, 2, 6, 7, 9
(b) 8. The only digit that is dropped from the third guess to make the fourth guess is the 8.
(c) 3, 5   (e) 358

23. If you have your teeth cleaned twice a year, then you will lose fewer teeth.

25. Some adults don't have time to read.

27. Some students are intelligent.

31. (a) All mathematics teachers love mathematics.
(b) Time flies when you are having fun.

33. Sandy is a female dachshund that is not white.

35. Nancy, Ma, Igor, Migraine, Lurch

37. (a) None of your sons is fit to sit on a jury.
(b) None of my poultry is an officer.
(c) None of the birds in this aviary lives on mince pies.

39. (a) 1 honest, 99 crooked

41. Select one ball from "*W* and *Y*" and you know if it's really "*Y*" or "*W*" and then you can work out the rest.

43. (a)

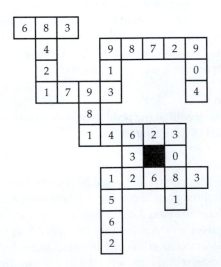

1.3 Homework Exercises

1. Inductive

3. Induction

5. Induction

7. Induction

9. (a) It will increase by 4.
(b) Because it may not work for some number that you did not try
(c) $N \to 3N \to 2N \to 2N + 8 \to N + 4$
(d) If you have $2N$ and then you add 8, you can deduce that the result is $2N + 8$.

11.

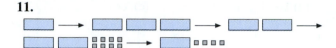

13. They make a conjecture from examples (induction), and then they prove it (deduction).

15. (a) Yes (in general)
(b) If you are sick, then you have a fever.
(c) No

17. (a) If you are smart, then you may be successful.
(b) If a triangle is equilateral, then it is isosceles.

19. If a triangle has two equal sides, then it has two equal angles. If a triangle has two equal angles, then it has two equal sides.

21. $2x = 4$ if and only if $x = 2$.

25. (a) Canaries like to sing. Blue jays like to sing. Cardinals like to sing. Therefore, all birds like to sing.
(b) Tweety is a canary. Canaries like to sing. Therefore, Tweety likes to sing.

27. (a) If $x + 3 \neq 5$ then $x \neq 2$.
(b) If I do not confess, then I am not guilty.
(c) Possible

29. (a) If $x \neq 2$ then $x + 3 \neq 5$.
(b) If I am not guilty, then I do not confess.
(c) Impossible

31. (a) If I drove to work, then it was raining.
(b) If I did not drive to work, then it was not raining.
(c) If it was not raining, then I did not drive to work.
(d) Part (b)

33. (c)

35. No

37. (a) If A, then B.
(b) If B, then A.
(c) If A, then B.
(d) If B, then A.
(e) If B, then A.

1.4 Homework Exercises

1. (a) No   (b) Yes, $d = -30$
(c) Multiply by 2

3. (a) 60, 100, 140, . . .   (b) Yes
(c) INITIAL = 60, CURRENT = PREVIOUS + 40

5. 9, 8, 7, 6, 5

7. (a) Yes
(b) INITIAL = 2, CURRENT = PREVIOUS + 7
(c) $2 + 7 \cdot 9 = 65$ (d) $2 + 7 \cdot 99 = 695$
(e) $2 + 7(n - 1)$ or $7n - 5$

9. (a) $586 - 6n$ (b) 406
(c) Inductive   (d) Deductive   (e) Yes

11. (a) 2, 4, 6, 8, . . .   (b) $2n$
(c) INITIAL = 2, CURRENT = PREVIOUS + 2

13. (a) $2(50) - 1 = 99$ (b) $50^2 = 2500$

15. (a) 14   (b) $8 + (n - 1) \cdot 2$ or $6 + 2n$

17. (a) nth term $= a + (n - 1) \cdot d$
(b) To reach the nth term, you start with a, and add the difference $n - 1$ times.

19. (a) (i) Arithmetic   (ii) Neither   (ii) Geometric
(b) (i) $a = 50, d = 11$ (iii) $a = 600, r = 0.2$
(c) (i) INITIAL = 50, CURRENT = PREVIOUS + 11
(iii) INITIAL = 600, CURRENT = PREVIOUS/5

21. (a)

Time (years)	0	5,600	11,200	16,800	22,400
Fraction of C-14 left	1	$\frac{1}{2}$	$\frac{1}{4}$	$\frac{1}{8}$	$\frac{1}{16}$

(b) About 16,000 years (since $\frac{1}{7}$ is between $\frac{1}{4}$ and $\frac{1}{8}$)

23. (a) 27   (b) 75

25. (a) Geometric   (b) $5 \cdot 7^{n-1}$ (c) $5 \cdot 7^{39}$

27. (a) Arithmetic   (b) $30 - 10(n-1)$
(c) -560

29. (a) The sum of the digits in each product is 9. The tens digit is one less than the first factor.
(b) Yes

31. (a) 27

33. (a) 8 or 7
(b) Multiplying by 2 or adding 1 more each time

35. (a) 1, 1, 2, 3, 5, 8, 13, 21, 34, 55
(b) The sum is 1 less than the term.
(c) The sum of the first n terms is 1 less than the $(n+2)$nd term.

37. (a) 625
(b) Induction to find the general rule of raising to the fourth power and deduction to apply the general rule to 5 and compute 5^4.

39. (a) There was another planet that exploded and formed the asteroids.
(b) $4 + (3 \times 2^6) = 196$ (c) Yes
(d) It is located where the "missing" planet should be.
(e) $4 + (3 \times 2^7) = 388$; not so close

41. (b)

43. (a)

(b) 3; 4   (c) 4; 5
(d) The sum of the first N even numbers is $\dfrac{N}{(N+1)}$.
(e) Inductive   (f) $31(32) = 992$

45. (a) $1 + 8 \cdot 10 = 9^2$; yes
(b) $(2n+1)^2$
(c) $1 + 8\left(\dfrac{n(n+1)}{2}\right) \overset{?}{=} (2n+1)^2$

$1 + 8\left(\dfrac{n^2+n}{2}\right) \overset{?}{=} 4n^2 + 4n + 1$

$1 + 4n^2 + 4n = 1 + 4n^2 + 4n$

47. (a) $5^2 - 3^2 = 16$; yes
(b) $C^2 - (C-2)^2 = 4(C-1)$
(c) $C^2 - (C-2)^2 \overset{?}{=} 4(C-1)$

$C^2 - (C^2 - 4C + 4) \overset{?}{=} 4C - 4$

$C^2 - C^2 + 4C - 4 \overset{?}{=} 4C - 4$

$4C - 4 \overset{?}{=} 4C - 4$

49. (a) 0   (b) 1   (c) $1 - 1 + 1 - 1 + \cdots$
(d) $1 - 1 + 1 - 1 + \cdots$ (e) Yes

51. (a) 1   (b) 3   (c) 12

53. No

55. (a) The height of each bounce decreases geometrically.

57. $50 \cdot 101 = 5{,}050$

59. (a) 21   (b) $\dfrac{N(N+1)}{2}$

61. $(15 + 29) \cdot 15/2 = 330$

1.5 Homework Exercises

1. Multistep translation

3. Puzzle

7. Briefly put, Polya's four steps for problem solving are: understand, plan, solve, and look back. First, read the problem and make sure you understand all the information. Next, devise a plan to solve the problem. Then carry out your plan. Finally, see if your result makes sense, and review what you have learned from working on the problem.

9. Use the area to find the length of one side. Then multiply by 4 to find the perimeter.

11. 5 socks

13. *Understand:* How are the cuts made? Vertically if the log is lying on the ground. *Plan:* Start making cuts and see what happens. *Solve:* 4 cuts. *Look Back:* The first cut creates two pieces, and each additional cut adds another piece.

15. $15 + 8 + 3 = 26$

17. (a) 5   (b) $T + N$

19. Move the two dots at the ends of the top row to the ends of the third row. Then move the bottom dot to the top.

21. "Find two ways to solve the problems." Or change it to "If the sides of the square are x feet and the rectangle has a width of 2 ft, what is the length of the rectangle?"

23. (a) 22   (b) $10 + (N - 1) \cdot 4$ or $4N + 6$

25. Divide the coins into 3 groups: 6, 5, and 5. Put 5 coins on each side of the balance and find which of the three groups has the lighter coin. In the worst case, you would have a group of 6 coins left containing the lighter coin. Divide them into 3 groups of 2, 2, and 2. Place two coins on each balance and find which of the three remaining groups has the lighter coin. Then place 1 coin on each side of the balance to find the lighter coin.

27. (a) 27   (b) 9

29. (a) Switch 2 with 7 and 4 with 9.   (b) 0   (c) 2
(d) For N pairs of checkers in which N is odd, you must move $N - 1$ checkers.
(e) No   (f) No
(g) For N pairs of checkers in which N is even, you must move N checkers.
(h) 24   (i) 50

1.6 Homework Exercises

1. In the guess-and-check strategy, you start by making a guess. Then check to see how accurate it is. Make a new guess using what you learned from your check. Continue guessing and checking until you find the answer.

3. 32 and 37

5. 4.6 or -9.6

7. 60

9. 33rd

11. (a) (possible answer) A worm goes up 4 ft and slides back 1 ft each day.

(b) (possible answer) A worm goes up 6 ft and slides back 3 ft each day.
(c) (possible answer) A worm goes up 9 ft and slides back 6 ft each day.

13. (possible answer) The assistant manager is off weeks 1 and 2; one secretary is off weeks 3 and 4; the other is off weeks 5 and 6. One agent is off weeks 1 and 2; the other is off weeks 3 and 4.

15. Guess and check

17. Draw a picture

19. 17 (make a table)

21. A is cheaper for 44 or fewer calls; B is cheaper for 45 or more calls.

23. (a) 40 yd (a square)   (b) $4\sqrt{N}$ yd

25. First, he takes the goose across. Then he takes the fox (or corn) across and picks up the goose when he drops off the fox (or corn). He brings the goose back to the starting side and picks up the corn (or fox). Finally, he goes back and brings the goose across again.

27. 2, 3, 4, 5, 6, 7, 8, 9, or 10

29. Minimum of $E - (C - G)$ and a maximum of G

31. $C = 4T + 2$

33. $C = 2T + 4$

Chapter 1   Review Exercises

2. If a teacher started class by checking attendance 10 days in a row, a student would generalize that the teacher always starts class this way.

3. Yes

4. (a) $142,857 \times 4 = 571,428$
(b) The answer always contains the digits 1, 2, 4, 5, 7, and 8.
(c) The pattern works for 1 to 6 as second factor.

5. The process of reaching a necessary conclusion from given hypotheses

6. (a) All dogs are animals.
    All poodles are dogs.
    Therefore, all poodles are animals.
  (b) All dogs are cats.
    All poodles are dogs.
    Therefore, all poodles are cats.

7. No

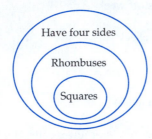

8. (a) You did not finish your fish.
  (b) Unknown

9. 483; from the first two guesses, the correct digits in the first guess are 4 and 3. One digit is in the correct position in the first guess, and one digit is in the correct position in the third. It must be 4 in the first guess, and 3 in third guess. The correct digit in the last guess must be in the second position: 8.

10. Induction

11. Deduction

12. Deduction

13. Inductive reasoning involves making a reasonable generalization from specific examples. Deductive reasoning is the process of drawing a necessary conclusion from given assumptions. To tell the difference, see if the conclusion requires a leap from specific to general (induction) or if the conclusion follows automatically from what is given to be true (deduction).

14. $N \to N + 5 \to 3N + 15 \to 3N + 6 \to 2N + 6 \to$
    $N + 3$

15. (a) If the ground gets wet, then it is raining.
  (b) No

16. (a) If it has four congruent sides, then a rectangle is a square.
  (b) If I call, then there is a problem.

17. (a) 54   (b) 150

18. (a) Arithmetic   (b) $20 - (n - 1)$ or $21 - n$
  (c) -19 (d) Inductive   (e) Deductive

19. (a) Geometric   (b) $a = 27{,}000$, $r = \dfrac{1}{3}$
  (c) INITIAL = 27,000, CURRENT = PREVIOUS/3
  (d) $27{,}000\left(\dfrac{1}{3}\right)^{n-1}$

20. (a) Neither   (d) 108

21. (a) 1, 2, 3   (b) $13 + 15 + 17 + 19 = 4^3$; yes

22. (a) $7^2 - 4^2 = 3 \cdot 11$
  (b) $N^2 - (N - 3)^2 = 3 \cdot (2N - 3)$
  (c) $N^2 - (N - 3)^2 \overset{?}{=} 3 \cdot (2N - 3)$
    $N^2 - (N^2 - 6N + 9) \overset{?}{=} 6N - 9$
    $N^2 - N^2 + 6N - 9 \overset{?}{=} 6N - 9$
    $6N - 9 = 6N - 9$

23. I had \$10. I spent \$3 and then found a \$5 bill. How much do I have now?

24. First, the mechanic would find out what was wrong with the car. Next, the mechanic would devise a plan to fix the problem. Then the mechanic would try out the repair plan. Finally, the mechanic would check to see if the problem no longer occurred.

25. 144

26. 18

27. (a) Guess and check
  (b) 3 tuna and 8 buckwheat

28. 6

29. 9 days

30. (a) (possible answer) A well is 40 ft deep. A snail goes up 7 ft and slides back 5 ft each day. On what day does it reach the top?
  (b) (possible answer) A well is 40 ft deep. A snail goes up 5 ft and slides back 3 ft each day. On what day does it reach the top?

31. Describe two ways to work out the answer.

32. Divide the coins into three groups of 3. Put 3 coins on each side of the balance, and determine which of the three groups has the lighter coin. Take the remaining group of 3. Place 1 coin on each side of the balance to find the lighter coin.

Chapter 2

2.1 Homework Exercises

1. Group, batch, bunch, pile, flock

3. (a) T   (b) F   (c) T   (d) F

5. (a) and (b)

7. $D = F$

9. (a) F   (b) {1, 2} and {3, 4}

11. (a) Infinite   (b) Finite

13. $\overline{A}$ = (science, Spanish); $\overline{B}$ = (history)

15.

17. (a), (c), and (d)

19. (a) T   (b) F   (c) T   (d) T

21. (a) ⊆   (b) ∈

23. {B, R, J}, {B, R, M}, {B, S, J}, {B, S, M}, {B, J, M}, {R, S, J}, {R, S, M}, {R, J, M}, {S, J, M}, {B, R, S, M}, {B, R, J, M}, {B, S, J, M}, {R, S, J, M}, {B, R, S, J, M}

25. (a) 2   (b) 6   (c) 24
(d) $N \cdot (N - 1) \cdot (N - 2) \cdot \ldots \cdot 2 \cdot 1$

27. (a) Match up each number with twice that number.
(b) Match up each number with 10 more than that number.

29. (a) 3   (b) 5   (c) 7   (d) $2N - 1$

31. $KN - K + 1$

2.2 Homework Exercises

1. (a) {3, 9}   (b) {1, 3, 5, 6, 7, 9, 11, 12, 15, 18}
(c) No

3. The two streets pass through the intersection, which is the region that is common to both streets.

5. (a) True   (b) False   (c) False

7. (a) Female education majors
(b) College students 22 years or older who are not education majors
(c) Set U (d) { }

9. Fish or spinach or both

11. $B \subseteq C$

13. (a) Yes; $(A \cup B) \cup C = A \cup (B \cup C)$
(b)

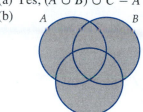
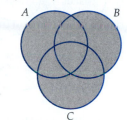

15. (a) {2, 3}   (b) {2, 3, 4, 6, 8}
(c) {2, 3}   (d) (a) and (c)
(e)

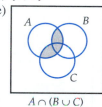
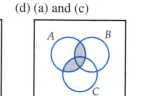

$A \cap (B \cup C)$ $(A \cap B) \cup (A \cap C)$

17. (a) 5   (b) 3

19. (a)

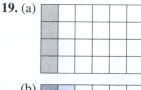

(b)

(c) 16

21. (a) 22   (b) 7

23. (a), (b), (c)

25. (a) $\in$ (b) $\cap$ (c) $\subseteq$

27. (a) Impossible   (b) $A = \{1\}, B = \{2\}$
   (c) $A = \{1\}, B = \{1\}$

29. (a)

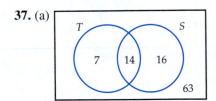

   (b)

   (c)

31. (a) All tomatoes are fruits   (b) $T \subseteq F$

33. (e)

35. (a) 125   (b) 350   (c) 10
   (d) $T \cap R; T \cap \overline{R}; \overline{T} \cap \overline{R}$

37. (a)

T	S
7	14	16
	63

   (b) 63

39. (b) 4, 5   (c) 2, 3   (d) 1, 2, 4, 5   (e) 6

41. (a) Gray blocks and triangle blocks
   (b) Gray triangles

43. I is small rectangles that are not blue; II is small blue
   shapes that are not rectangles; III is small, blue rectangles; and IV is blue rectangles that are not small.

45. (a) 2   (b) 7   (c) 8   (d) 3

47. (c); P contains polygons with the interior shaded

49. (a) 136%
   (b) Some people put down more than one response.

51. (a)

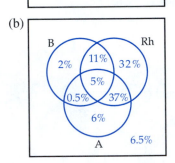

   (b)

   (c) 0

53. (a) 6   (b) (yellow, red), (yellow, blue),
               (green, red), (green, blue)

55. 70

57. (a) $\{(1, 0), (1, 2), (1, 4), (1, 6), (2, 0), (2, 2), (2, 4), (2, 6)\}$
   (b) 8

59. (b)

2.3 Homework Exercises

1. (a)

x	0	2	4	6	8
y	0	46	92	138	184

   (b) $y = 23x$ (c) Multiply x by 23
   (d)

3. Domain: $\{0, 2, 4, 6, 8\}$; range: $\{0, 46, 92, 138, 184\}$

5. (a) The set of feet measured
   (b) The measurements between 4 and 35 cm

7. (a) (1) and (3)
   (b) (1) Squaring   (2) "is a friend of"
      (3) "wants to be a"

9. (b) and (c)

11. (b) and (c)

13. (a) No   (b) Yes   (c) No

15. (a) $500 \le F \le 800$
   (b) $92{,}500 \le C \le 112{,}000$
   (c) Each F-value in the domain has exactly one
     C-value in the range.

17. (a) $P = 2N - 1$ (c) inductive
   (e) All the points lie on a line.
   (c) nth term $= 2n - 1$

19. (a) $V = L^3$ (b) 125   (c) n^3

21. (a) $N = 12\sqrt{S}$ (b) 96

23. (a) $y = 8$ (b) ▯▯▯▯ ; $y = 10$
   (c)

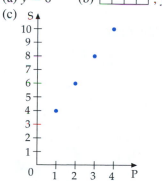

   (d) All the points lie on a line.
   (e) $y = 2N + 2$

25. $5; 8 - N$

27. "is taller than," "is heavier than"

29. (a) $\{(5, 9), (5, 18), (10, 18), (15, 18)\}$

Chapter 2 Review Exercises

1. (a) 8   (b) A set with 3 elements
   (d) A set with the two elements from U that are not in
     Set A.

2. $\{1, 2\}, \{a, b\}$

3. (a) 4   (b) 14

4. (a) $\cap$ (b) $\in$

5. All are true.

6.

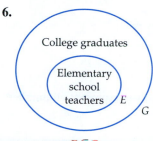

$E \subseteq G$

7. (a) 700   (b) 1000   (c) 300

8. (a) 5   (b) Deductive

9. (a) 8   (b) 6   (c) 1, 4

10. Not thick, gray, and not triangles

11. (a) (1) and (2)
   (b) (1): $y = 2x - 1$; (2): The person's birthplace

12. Yes

13. (a) $\{0, 1, 2, \ldots, 10\}$ (b) $\{-25, -15, -5, \ldots, 75\}$
   (c) Each x-value in the domain has exactly one
     P-value.

14. (a) 4   (b) 900   (c) $F = S^2/4$
   (d) Weight, direction   (e) $0 < S \le 100$ mph

15. $ST = 120$ or $T = 120/S$

16. (d)

Chapter 3
3.1 Homework Exercises

1. (a) Numeral   (b) Number

3. (a) 250   (b) 1,013

5. (a) Yes
   (b) Changing the position of different symbols in a Hindu-Arabic numeral changes the value of the numeral. Egyptian symbols can be written in any order.

7. (a) 𓍢𓍢𓍢∩∩IIII
   (b) 𓍢𓍢𓍢𓏤 —
   (c) 𒀭𒀭𒀭𒀭𒀭≪𒀭𒀭𒀭𒀭

9. (a) 9   (b) 120   (c) 247

11. A symbol for zero, ten digits to represent all numbers, place value

13. (a) 14   (b) 40   (c) 1613   (d) 1964

15. (a) 642   (b) 12   (c) 6

17. (a) 400 + 7
   (b) 3,000 + 100 + 20 + 5

19. $(A \times 1000) + (B \times 100) + (C \times 10) + D$

21. (a) 1324   (b) 207

23. (a) 3 greens, 7 blues, and 2 yellows
   (b) Green
   (c) Chip trading is more abstract because chips for 1, 10, 100, and 1000 are all the same size, whereas the value of a base-ten block is proportional to its size.

25. 1 hundred, 2 tens, 7 ones; 1 hundred, 1 tens, 17 ones; 1 hundred, 0 tens, 27 ones

27. (a) 3,007   (b) Writing multiples of 10 or 100
   (c) *Hint:* Go over the value of each place value and relate it to 20,010.

29. (a) Measure   (b) Count   (c) Measure

31. It's in between. Dollars are like a measure, but cents are like a count.

33. (a) $600   (b) $36,400

35. (a) Ten   (b) Hundred   (c) Ten thousand

37. (a) $120   (b) Down

39. (a) 7,500 and 8,499 (inclusive)

41. (a) 6,210,001,000

3.2 Homework Exercises

1. (a)

3. (a) Combine measures
   (b) The sale price, the current value, the size and condition of the house, the location of the house

5. (a) 2 + 3 = 5
   (b)

9. (a) 2 + ? = 7   (b) 37 + 72 = N
   (c) x = 4 + N

11. Compare measures

13. Missing part (sets)

15. 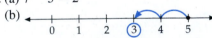   7 − 4 = 3

17. (a) 7 − 5 = 2
   (b)

19. (a) 8 − 5 = 3   (b) Take away sets

21. (a) ⊗⊗○○○○○○

   (b) ○○○○○○○○
      ⬆⬆
      ▢▢

   (c) ▢▢▢▢▢▢▢ → ▢▢▢▢▢▢
       ▢▢   ▢▢▢▢▢

27. (a) The child counts 8 as being the first number before 8.
   (b) Give the child 8 blocks. Then have the child count off as the child takes 3 away from the set.

29. (a) 10 − x   (b) x − 2   (c) x + 6

31. A − F − C + P

33. If $(a - b) \geq c$

35. (a) Subtraction; missing part (sets)

37. (a) $8 = 6 + 2$
(b) For whole numbers a and b, $a < b$ when $a = b - k$ for some counting number k.

39. (a) $3 = 1 + 2$, $5 = 2 + 3$, $6 = 1 + 2 + 3$, $7 = 3 + 4$, $9 = 2 + 3 + 4$, $10 = 1 + 2 + 3 + 4$
(b) Any counting number that is not a power of 2 can be written as the sum of two or more consecutive counting numbers.
(c) Inductive

41. The 2 should be subtracted from the 27, not added. The women first paid $30 and then got $3 back. So they ended up paying $27, $2 to the bellperson and $25 for the room.

43. (a)

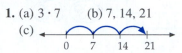

(b) They all lie on a straight line.
(c)

(d) They all lie on a straight line.
(e) If the ordered pairs of any fact family are plotted on a graph, the points will lie on a straight line.
(f) Induction

45. The set of inputs is the two addends, and the sum is the output.

47. (a)

4	3	8
9	5	1
2	7	6

(b) Same problem but change "1 to 9" to "5 to 13" and change "15" to "27"

3.3 Homework Exercises

1. (a) $3 \cdot 7$ (b) 7, 14, 21
(c)

3. (a) No, $2 \cdot 2 = 4$ (b) Yes   (c) Yes

5. Repeated measures

7. Repeated sets

9. Divide the interior into 1-ft squares as shown. It takes 6 squares to cover the interior, so the area is 6 ft².

11. (a) $3 \times 2 = 6$
(b)

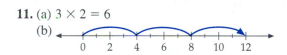

13. (a)   (b)

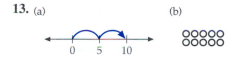

17. A school is charging $3 for lunch and $6 for a concert. How much will it cost to buy 7 tickets for both events?

19. (a) From adding 5 and 2
(b) The first factor tells you how many sets of the second factor so you need only sets of 2.

21. (a) 74 and $X^2 + 10$ (b) 38 and $4X - 2$
(c) Inductive

23. (a) 3^2 (b) 4^3

25. (a) $7 \times ? = 63$ (b) $N \times C = 0$

27. Partition measures

29. Repeated measures

31. (a) $8 \div 4 = 2$
(b)

33. (a)   (b)   (c)

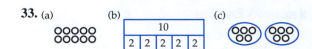

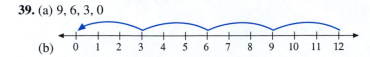

39. (a) 9, 6, 3, 0

(b)

41. (a) Pens come in packs of 4. How many full packs can be made with 19 pens?
(a) Each car takes 4 people. How many cars for 19 people?
(c) In part (a), how many pens are left over?

43. (a) If he puts a maximum of 4 on a page, how many pages will he need?
(b) If he puts 4 on each page, how many extra photographs will be left over?
(c) If 4 photographs fill a page, how many pages can he fill?

45. (a) 1   (b) 2   (c) Quotient   (d) Remainder

47. (a) 2   (b) Remainder

49. (a) Undefined
(b) The expression $7 \div 0 = ?$ is the same as $? \times 0 = 7$, which has no answer. So we make $7 \div 0$ undefined.

51. Some sort of error message; the problem is undefined.

53. No. Three is the divisor in the first example, and 6 is the divisor in the second example.

55. (a) How many 8s make 40?
$8 + 8 + 8 + 8 + 8 = 40$. Five. $40 \div 8 = 5$.
(b) How many 8s can you subtract from 40?
$40 - 8 = 32, 32 - 8 = 24, 24 - 8 = 16,$
$16 - 8 = 8, 8 - 8 = 0$. Five. $40 \div 8 = 5$.
(c) $? \times 8 = 40$. Since $? = 5$, then $40 \div 8 = 5$.

57. (a) 5 and 20   (b) Deductive

59. 36

61. Take out 1 chip. On successive turns, reduce the pot to 7 chips, 4 chips, and finally 1 chip.

63. (a) 2, 4; 2, 3; 6, 12

65. (a) 10   (b) 67   (c) 9

67. (a) 19   (b) 162

69. (a) $2t$ (b) $t - 5$ (c) $5t$

71. (a) 100   (b) 3   (c) 7

73. No

75. (a) Multiplication, array and subtraction, take away sets
(b) Multiplication, repeated sets and addition, combine sets

77. (a) Addition, combine sets and subtraction, take away sets $(10 - (2 + 3))$, or both steps are subtraction, take away sets $(10 - 2$ and $8 - 3)$
(b) Multiplication, array and subtraction, take away sets

79. $AB + CD$ dollars

81. The set of inputs is the two factors, and the output is the product.

83. (e) The first and last columns each contain 5, 4, 6, and 9. The second column contains the same four numbers each multiplied by 2. The third column contains the same 4 numbers each multiplied by 4. The product will always be $5 \cdot 4 \cdot 6 \cdot 9 \cdot 2 \cdot 4 = 8640$.
(f)

1	3	2
3	9	6
5	15	10

85. $n^2 - (n - 2)^2$ or $4n - 4$

87. Pour 3 cupfuls into the glass to get 6 oz. Then fill one cup. Pour from one cup into the other until both cups are at the same level. Then each has 1 oz. Pour one of them into the glass.

89. (a) Correct   (b) Incorrect

91. (a) Multiplication, repeated sets
(b) Division, partition sets

3.4 Homework Exercises

1. The commutative property says $x + y = y + x$ and the associative property says $(x + y) + z = x + (y + z)$

3. Associative property of addition

5. Addition, multiplication

7. (a) $(8 - 4) - 2 \neq 8 - (4 - 2)$
(b) The right side is greater because the third number is added instead of subtracted.

9. Each fact that you learn in one order has the same answer when the order of the factors is reversed.

11. (b)

13. Multiply $6 \times \$5$ first, and then multiply $\$30 \times 9 = \$270$.

15. $\$38 + \$2 = \$40$. Then $\$40 + \$57 = \$97$.

17. 0, identity, addition

19. (a) Associative property of multiplication
(b) Distributive property of multiplication over addition
(c) Commutative property of multiplication
(d) Associative property of addition

21. Order property = Commutative;
Zero property = Identity for addition;
Grouping property = Associative;
Property of one = Identity for multiplication

23. No. The child is trying to use the distributive property of multiplication over addition.

25. Nino understands the commutative property.

27. (a) and (b) $10(8 + 4) = 120$ and
$10 \cdot 8 + 10 \cdot 4 = 120$
(c) $10(8 + 4) = 10 \cdot 8 + 10 \cdot 4$ by the distributive property of multiplication over addition.

29. (a) True for $A = B$; false for $A \neq B$
(b) True if $B = 1$ or $C = 0$; false if $B \neq 1$ and $C \neq 0$
(c) True if $A = 1$ or $A = 0$ or $B = 0$ or $C = 0$; false if $A \neq 1$ and $A \neq 0$ and $B \neq 0$ and $C \neq 0$
(d) True if $A \geq B$; false if $A < B$
(e) True if A is divisible by B; false if A is not divisible by B
(f) True if $A = 1$ or $C = 0$; false if $A \neq 1$ and $C \neq 0$
(g) (possible answer) True if $B = 1$, $A = 2$, and C is any number; false for $A = B = C$ and many other cases

31. (a) It doubles.   (b) $2a + 2b = 2(a + b)$
(c) Inductive   (d) Deductive

33. (a) Yes   (b) Yes; 0   (c) Yes

3.5 Homework Exercises

1. (a) Correct; the child understands place value and breaking numbers apart
(b) Incorrect; use base-ten blocks to show regrouping
(c) Correct; the child understands place value and addition as counting on
(d) Correct; the child understands compensation with addends and adding a multiple of 10

3. Show 182 as 1 hundred, 8 tens, and 2 ones and 336 as 3 hundreds, 3 tens, and 6 ones. Add the ones. So 2 ones + 6 ones = 8 ones. Next add the tens. So 8 tens + 3 tens = 11 tens. Regroup 11 tens as 1 hundred 1 ten. Then, add the hundreds. So 1 hundred + 1 hundred + 3 hundreds = 5 hundreds. The sum is 5 hundreds, 1 ten, and 8 ones = 518. $182 + 336 = 518$

5. (a)
```
  357
 +529
   16
   70
  800
  886
```
(b)
```
 3 ¹5 7
 +5 2 9
  8 8 6
```
(c) Regrouping is easier to understand in the partial sums algorithm. The standard algorithm is faster.

7. Addition is associative; addition is commutative; addition is associate; addition is commutative; addition is associative.

9. (a) No. Only one number has to be greater than 300.
(b) Yes. The two largest numbers would.

11. (a) Subtraction as adding on
(b) Place value and breaking numbers apart
(c) Subtraction as adding on

13. 7; 12; 7

15. Show 336 as 3 hundreds, 3 tens, and 6 ones. Can you take away 2 ones? Yes. That leaves 4 ones. Next, can you take away 8 tens? No. Regroup 1 hundred as 10 tens. Subtract the tens. So 13 tens − 8 tens = 5 tens. Then subtract the hundreds. So 2 hundreds − 1 hundred = 1 hundred. The difference is 1 hundred, 5 tens, and 4 ones = 154. So $336 − 182 = 154$.

17. (a)
```
  814
 -391
  500
  -80
  + 3
  423
```
(b)
```
 ⁷8 ¹1 4
 -3 9 1
  4 2 3
```

(c) Regrouping is easier to understand in the partial differences algorithm. The standard algorithm is faster.

19. (a) $50 + 20$ (b) $50 - 20$

21. (a) 175
(b) Adding the number of ones in the second addend to the number of tens in the first addend
(c) Use base-ten blocks to show why we don't add the second number twice.

23. (a) 75
(b) Borrowing 1 ten without subtracting it from the tens column
(c) Discuss how regrouping involves an exchange of 1 ten for 10 ones.

25. (a)
$$
\begin{array}{rr}
29 & +\ 1 \\
30 & +30 \\
60 & +\ 2 \\
\hline
62 & +33 \\
\end{array}
$$
$$62 - 29 = 33$$

(b)
$$
\begin{array}{rr}
137 & +\ 3 \\
140 & +60 \\
200 & +12 \\
\hline
212 & +75 \\
\end{array}
$$
$$212 - 137 = 75$$

27. (a) Could use partial sums and counting on
(b) Could use partial differences and counting back

29. (a)
$$
\begin{array}{r}
{}^1\ \overset{2}{3}\ 8 \\
\overset{}{9}\,{}_4\overset{7}{}\,{}_5 \\
+2\ 4\ \overset{6}{6} \\
\hline
3\ 8\ 1 \\
\end{array}
$$
(b) It requires only the addition of one-digit numbers.

31. (a) 57   (b) Yes   (c) Yes

33. (a) 25   (b) 445

3.6 Homework Exercises

1. (a) Breaking apart numbers and the distributive property of multiplication over addition
(b) Breaking apart numbers and the distributive property of multiplication over addition
(c) Breaking apart numbers and the distributive property of multiplication over subtraction

3. Show 4 groups of 24

Combine the ones. You have 16 ones. Next, trade 10 ones for 1 ten, leaving 6 ones. Then combine the tens. You have 8 tens plus 1 ten for a total of 9 tens. The product is 9 tens 6 ones, or 96. So $24 \times 4 = 96$.

5. Multiplication over addition is distributive; multiplication over addition is distributive; addition is commutative; addition is commutative.

7. (b)

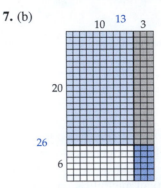

(c) 338

9. (a)
$$
\begin{array}{r}
49 \\
\times\ 62 \\
\hline
18 \\
80 \\
540 \\
2400 \\
\hline
3038 \\
\end{array}
$$

(b)
$$
\begin{array}{r}
49 \\
\times\ 62 \\
\hline
98 \\
294 \\
\hline
3038 \\
\end{array}
$$

(c) Easier to understand place value with partial products; standard algorithm is faster.

11. 52×4, 50×40

13. (a) Correct; repeated subtraction and breaking apart
(b) Incorrect; use base 10 blocks and go over place value.

15. You want to divide 246 into 2 equals groups. Show 246, and try to divide the hundreds into 2 equal groups. Put 1 hundred in each group. Next divide the 4 tens into 2 equal groups. Put 2 tens in each group. Then divide the 6 ones into 2 equal groups. Put 3 ones in each group. The quotient is 123. So $246 \div 2 = 123$.

17. (a)
$$
\begin{array}{r}
4)\overline{217} \\
\underline{200}\quad 50 \\
17 \\
\underline{16}\quad\ \ 4 \\
1\quad\ \ 54 \\
\end{array}
$$
54 R1

(b)
$$
\begin{array}{r}
54\ R1 \\
4)\overline{217} \\
\underline{20} \\
17 \\
\underline{16} \\
1 \\
\end{array}
$$

(c) Easier to understand place value in repeated subtraction; standard algorithm is faster

19. (a) 638

(b) Failing to carry the tens from the right-hand multiplication to the tens column

(c) Go over multiplying ones and regrouping.

21. (a) 46

(b) Placing the digits in the quotient from right to left

(c) Discuss place value and the placement of digits in the quotient.

23. (a) The child does only two of the four partial products.

(b)

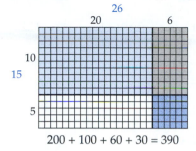

$$200 + 100 + 60 + 30 = 390$$

25. (a) 2812   (b) 4305

(c) It is harder to understand the underlying concepts of this algorithm.

27. (a) Could use lattice multiplication, Russian peasant multiplication, or multiply left to right

(b) Could use lattice multiplication, use Russian peasant multiplication, or list each partial product and add

(c) Could use counting by 20s or repeated subtraction algorithm

29. 372 R13 (Multiply the decimal remainder by 23.)

3.7 Homework Exercises

1. (a) $39 + 90 = 129$ and $129 + 7 = 136$ miles or $97 + 3 + 36 = 136$ miles or $39 + 97 = 36 + 100 = 136$ miles

(b) Commutative property of addition

3. $134 - 50 = 84$ and $84 - 8 = 76$; $134 - 58 = 136 - 60 = 76$; $130 - 50 = 80$ and $8 - 4 = 4$ and $80 - 4 = 76$

5. (a) $236 + 89 = 235 + 90 = 325$

(b) $82 - 58 = 84 - 60 = 24$

7. (a) $30 \times 6 + 9 \times 6 = 180 + 54 = 234¢$ or \$2.34

(b) $40 \times 6 - 6 = 240 - 6 = 234¢ = \$2.34$

9. $5^3 = 125$ and add 6 zeroes: 125,000,000

11. (a) You must estimate when you predict the future.

(b) You cannot tell exactly how much each person will weigh each time that people board the elevator. You also want to incorporate a safety factor.

13. (a) Wrong; the answer should be about $3 \cdot (600) = 1800$

(b) Reasonable, since $30 \times 50 = 1500$

15. Round it to $6000 + 8000 + 4000$. Add $6 + 8 + 4 = 18$ and attach three zeroes. 18,000.

17. (a) Change $437 \div 55$ to compatible numbers like $400 \div 50 = 8$ hours.

(b) Repeated measures

19. Too high

21. $469 \div 74 \approx 420 \div 70 = 6$

23. $59 + 42$ is about 100, plus 97 is about 200, plus 32 makes about 230.

25. (a) $8 \times 5 = 40$ and $22 \times 40 = 880$; commutative ($\times$)

(b) $2 + 78 = 80$ and $80 + 43 = 123$; associative ($+$)

27. Add $3 + 1 + 21 \approx \$25$ million (rounding)

29. (a) $32 \times 48 \approx 30 \times 40 = 1,200$ lightbulbs

(b) $32 \times 48 \approx 30 \times 50 = 1,500$ lightbulbs

(c) Rounding

31. (a) $4,872 - 3,194 \approx 5,000 - 3,000 = 2,000$

(b) $3,279 \div 65 \approx 3,000 \div 60 = 300 \div 6 = 50$

33. (a) Estimate   (b) Exact

35. (a) Calculator; too time-consuming to do another way

(b) Mental computation (add $750 + 250$ first)

(c) Paper-and-pencil or calculator, depending upon which is more convenient

37. (a) Between 1,500 and 2,400
(b) Between 10,000 and 12,000
(c) Between 30 and 40

39. Yes. Multiplying by 5 is the same as dividing by 2 and then multiplying by 10 (adding a zero).

41. (a) Taking ten 74s. Doubling that to get twenty 74s. Then adding two 74s to end up with twenty-two 74s.
(b) $31 \times 10 = 310$. Then $3 \times 310 = 930$. Then $930 + 93 = 1,023$

43. (a) $5 \cdot (800) = 4,000$ (b) $4 \cdot (40,000) = 160,000$

47. (a) 631×542 (guess and check)
(b) Same problem, but use 1, 2, 3, 7, 8, and 9 as the digits

49. (a) $n = 14$ (b) $x = 177$ (c) $y = 7$

51. (a) The distributive property of multiplication over subtraction
(b) 1,485   (c) $95 \times 100 - 95 = 9,405$

53. (a) 8   (b) 28   (c) 36   (d) 26   (e) 17

3.8 Homework Exercises

1. (a) 1 eight and 4 ones   (b) 2 fives and 2 ones

3. 1_{three}, 2_{three}, 10_{three}, 11_{three}, 12_{three}, 20_{three}, 21_{three}, 22_{three}, 100_{three}, 101_{three}, 102_{three}, 110_{three}

5. (a) 61   (b) 113   (c) 139

7. (a) 56_{eight} (b) 101_{five} (c) 2244_{five}

9. The last digit is 0.

11. 1314_{five}

13. (a)
$$
\begin{array}{r}
13_{five} \\
+\, 22_{five} \\
\hline
10 \\
30 \\
\hline
40_{five}
\end{array}
$$
(b) Show 13_{five} as 1 five and 3 ones and 22_{five} as 2 fives and 2 ones. Next add the ones. So 3 ones + 2 ones = 5 ones. Regroup 5 ones as 1 five. Then add the

fives. So 1 five + 1 five + 2 fives = 4 fives. The sum is 4 fives = 40_{five}. So $13_{five} + 22_{five} = 40_{five}$

15. (a)
$$
\begin{array}{r}
43_{five} \\
-\, 24_{five} \\
\hline
20 \\
-1 \\
\hline
14_{five}
\end{array}
$$
(b) Show 43_{five} as 4 fives and 3 ones. Can you take away 4 ones? No. Regroup 1 five as 5 ones. Next subtract the ones. So 8 ones − 4 ones = 4 ones. Then subtract the fives. So 2 fives − 1 five = 1 five. The difference is 1 five 4 ones = 14_{five}. So $43_{five} - 24_{five} = 14_{five}$.

17. (a) 112_{five} (b) 1011_{five} (c) 1220_{five}

19. (a)
$$
\begin{array}{r}
23_{five} \\
\times\, 14_{five} \\
\hline
22 \\
130 \\
30 \\
200 \\
\hline
432_{five}
\end{array}
$$
(b)
$$
\begin{array}{r}
23_{five} \\
\times\, 14_{five} \\
\hline
202 \\
23 \\
\hline
432_{five}
\end{array}
$$

21. The quotient is 22_{five}.

23. (a) 1442_{five} (b) $243,302_{five}$

25.

+	0	1	2	3	4	5	6	7
0	0	1	2	3	4	5	6	7
1	1	2	3	4	5	6	7	10
2	2	3	4	5	6	7	10	11
3	3	4	5	6	7	10	11	12
4	4	5	6	7	10	11	12	13
5	5	6	7	10	11	12	13	14
6	6	7	10	11	12	13	14	15
7	7	10	11	12	13	14	15	16

×	0	1	2	3	4	5	6	7
0	0	0	0	0	0	0	0	0
1	0	1	2	3	4	5	6	7
2	0	2	4	6	10	12	14	16
3	0	3	6	11	14	17	22	25
4	0	4	10	14	20	24	30	34
5	0	5	12	17	24	31	36	43
6	0	6	14	22	30	36	44	52
7	0	7	16	25	34	43	52	61

27. (a) 25_{eight} (b) 244_{eight} R2

29. (a) 13   (b) $10,001_{two}$ (c) $10,010_{two}$

31. (a) $101,101,101_{two}$
(b) Convert each base-eight digit to an equivalent three-digit base-two numeral, and place the resulting numerals in the same positions as the original base-eight digits.
(c) $110,100,010_{two}$

33. (a) 138   (b) 68_{twelve}

35. (a) 1220_{four} (b) 11_{two}

37. (a) 1, 2, 3, 4, . . . 15 pounds
(b) 1, 2, 4, and 8 pound weights
(c) Same problem with 1 to 100 pounds

41. b is any whole number greater than 7.

Chapter 3   Review Exercises

1. (a) 37   (b) 2   (c) 6

2. In a place-value system, the position of the digit determines its value. In a system that does not have place value, the value of a digit is the same regardless of its position.

3. Between 125,000 and 134,999 (inclusive)

4. Whole numbers

5.

6. (a) Sets   (b) Compare

7. (a) Sets   (b) Repeated sets

8. A repeated sets problem asks how many times you must repeat the amount of the divisor to obtain the total amount. A partition sets problem asks you to divide up the total into the number of equal-sized groups specified by the divisor.

9. (a) Sets
(b) Multiplication, repeated and subtraction, take away

10. (a) Measures
(b) Subtraction, take away and division, repeated

11. (a) Undefined
(b) $3 \div 0 = ?$ is the same as $? \times 0 = 3$, which has no solution. So we make $3 \div 0$ undefined.

12. See example in Section 3.2.

13. See example in Section 3.3.

14. See example in Section 3.2.

15. $HD - S$ dollars

16. $46

17. (a) Commutative property of addition
(b) Identity for multiplication
(c) Associative property of multiplication

18. (a) No   (b) $8 \div (4 \div 2) \neq (8 \div 4) \div 2$.

19. True for $A = 0$ and $B = C = 1$ and false for $A = B = C = 1$

20. (a) $8 \times 40 = 320$
(b) Distributive property of multiplication over addition

21. No. Have the child work it out by adding first and see if the answers are equal.

22. Show 326 as 3 hundred, 2 tens, and 6 ones and 293 as 2 hundreds, 9 tens, and 3 ones. Add the ones. So 6 ones + 3 ones = 9 ones. Next add the tens. So 2 tens + 9 tens = 11 tens. Regroup 11 tens as 1 hundred 1 ten. Then add the hundreds. So 1 hundred + 3 hundreds + 2 hundreds = 6 hundreds. The sum is 6 hundreds, 1 ten, and 9 ones = 619. So $326 + 293 = 619$.

23.
$$
\begin{array}{r}
127 \\
+246 \\
\hline
13 \\
60 \\
300 \\
\hline
373
\end{array}
$$

24. (a) Subtraction as adding on   (b) Compensation

25. $300 \times 5 = 1,500$; $20 \times 5 = 100$, and $4 \times 5 = 20$. Then $1,500 + 100 + 20 = 1,620$.

26. Show 3 groups of 26. Combine the ones. You have 18 ones. Next, trade 10 ones for 1 ten leaving 8 ones. Then combine the tens. You have 6 tens plus 1 ten for a total of 7 tens. The product is 7 tens 8 ones or 78. So $26 \times 3 = 78$.

27. (a) $14\overline{)312}$
$$\begin{array}{r} 280 \quad 20 \\ \hline 32 \\ 28 \quad 2 \\ \hline 4 \quad 22 \end{array}$$
   22 R4

   (b) $\quad$ 22 R4
$$14\overline{)312}$$
$$\begin{array}{r} 280 \\ \hline 32 \\ 28 \\ \hline 4 \end{array}$$

   (c) Easier to use place value in part (a); part (b) is faster.

28. (a) 1,513
   (b) Child does not regroup the ones as tens.
   (c) Show regrouping with base-ten blocks.

29. (a) 240
   (b) Child thinks a number minus zero is zero.
   (c) Go over one-digit subtraction problems with 0.

30. (a) 2,836
   (b) Child uses regrouping number from first partial product for both partial products.
   (c) Point out the error, and go over the correct method.

31. $167 + 3 = 170$. Then $170 + 30 = 200$. Then $200 + 39 = 239$. The answer is $3 + 30 + 39 = 72$. So $239 - 167 = 72$.

32.

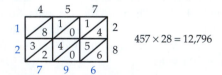

   $457 \times 28 = 12{,}796$

33. $80 - 40 = 40$ and $3 - 1 = 2$. Then $40 - 2 = 38$.

   $81 - 43 = 78 - 40 = 38$

   $81 - 40 = 41$ and $41 - 3 = 38$.

34. (a) Change to compatible $2400 \div 40 = 60$ gallons.
   (b) Measures
   (c) Repeated

35. In the rounding strategy, each number in the problem is rounded to the nearest number by place value. In the compatible numbers strategy, each number is rounded to a number close to it in size, and the resulting numbers must create a simple problem (usually division).

36. (a) $458 + 327 + 118 \approx 500 + 300 + 100 = 900$
   (b) $400 + 300 + 100 = 800$ and $58 + 27 + 18 \approx 100$. Then $800 + 100 = 900$.

37. 244_{six}

38. (a) $\quad 641_{\text{eight}}$
$$\begin{array}{r} 641_{\text{eight}} \\ -235_{\text{eight}} \\ \hline 400 \\ + \quad 10 \\ - \quad 4 \\ \hline 404_{\text{eight}} \end{array}$$

   (b) Show 641_{eight} as 6 sixty-fours, 4 eights, and 1 one. Can you take away 5 ones? No. Regroup 1 eight as 8 ones. Next, subtract the ones. So 9 ones − 5 ones = 4 ones. Then subtract the eights. So 3 eights − 3 eights = 0. Then subtract the sixty-fours. 6 sixty-fours − 2 sixty-fours = 4 sixty-fours. The difference is 4 sixty-fours 4 ones = 404_{eight}. So $641_{\text{eight}} - 235_{\text{eight}} = 404_{\text{eight}}$.

39. The quotient is 150_{seven} R4_{seven}

40. b is whole number ≥ 3; $a = 2b + 1$

Chapter 4

4.1 Homework Exercises

1. (a) 1, 2, 4, 7, 14, 28
   (b) There is no whole number c such that $6 \cdot c = 28$.

3. True

5. (a) 1 by 20, 2 by 10, and 4 by 5
   (b) Factoring 20

7. True; $A = 4$, $B = 8$, $K = 3$ is an example.

9. True; $A = 2$, $B = 10$, $C = 20$ is an example.

11. True; $A = 2$, $B = 4$, $C = 6$ is an example.

13. 3, B, A, B

15. Divisibility-of-a-Product Theorem

17. $8 \mid (C + 16)$

19. Since 12×71 and 48 are each divisible by 12, then $12 \times 71 + 48$ is divisible by 12.

21. (a) $A \mid 8C$ (b) Deduction

23. False; $B = 5$

25. True
  1. $A \mid B$ and $C \mid D$
  2. $AW = B$ and $CY = D$ in which W and Y are whole numbers.
  3. $ACWY = BD$ in which WY is a whole number.
  4. $AC \mid BD$

27. True
  1. $A \mid B$, $B \mid C$, and $A \neq 0$
  2. $A \cdot K = B$ and $B \cdot L = C$, where K and L are whole numbers
  3. $A \cdot K \cdot L = C$ and KL is a whole number
  4. $A \mid C$

29. True
  1. C and D are even numbers.
  2. $C = 2W$ and $D = 2X$, where W and X are whole numbers.
  3. $CD = 2W \cdot 2X$
  4. $CD = 2(2WX)$ where $2WX$ is a whole number.
  5. CD is an even number.

31. $x \mid x \rightarrow x \mid xy^2 \rightarrow x \mid (xy^2 \cdot x)$ or $x \mid x^2y^2$

33. (a) Abundant   (b) Perfect   (c) Deficient

35. $N^3 - N = N(N + 1)(N - 1)$. Either $N - 1$, N, or $N + 1$ must be divisible by 3. By the Divisibility-of-a-Product Theorem, $N^3 - N$ must be divisible by 3. So $3 \mid (N^3 - N)$.

4.2 Homework Exercises

1. 0. 7, 14, 21, . . .

3. There is no whole number C such that $6 \cdot C = 3$.

5. (a) Factor   (b) Factor   (c) Multiple
  (d) Factor

7. Say, "the number formed by the last two digits is divisible by 4."

9. (a) Divisible by 3 and 9
  (b) Divisible by 2, 3, 4, and 6
  (c) Divisible by 2, 5, and 10

11. No

13. 102, 108, 114, 120, 126

15. (a) 1 or 7   (b) 00, 20, 40, 60, or 80
  (c) 1, 4, or 7 followed by 0

17. (a) No. The total price of the stamps will end in a 0 or 5.
  (b) Any amount made with $0.25 and $0.15 stamps must be divisible by 5 cents.

19. 0 because it is not a counting number; 1 because it is not an even number; 4 because it is not a perfect cube; 8 because it is not a perfect square; and 64 because it is not a one-digit number

21. (a) False   (b) 15   (c)

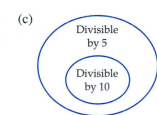

23. (a) False   (b) 24
  (c)

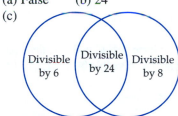

25. (a) True
  (b) A is divisible by 2, and 4 is divisible by 2. So A + 4 is divisible by 2.

27. 1, 3, and 5

29. 100A + 10B + C; 10B; C; 100A + 10B + C

31. It's a perfect cube. It's a perfect square. It's an odd number. It's divisible by 3.

33. $\underline{A}\ \underline{B}\ \underline{C}\ \underline{D} = 1000A + 100B + 10C + D$
$= 999A + 99B + 9C + A + B + C + D$
999A, 99B, 9C, and (A + B + C + D) are each divisible by 9. So $999A + 99B + 9C + A + B + C + D$ or $\underline{A}\ \underline{B}\ \underline{C}\ \underline{D}$ is divisible by 9.

35. 1. $\underline{A}\ \underline{B}\ \underline{C}\ \underline{D}$ is a four-digit number and $\underline{C}\ \underline{D}$ is divisible by 4.
2. $\underline{A}\ \underline{B}\ \underline{C}\ \underline{D} = 1000A + 100B + 10C + D$
3. 1000A, 100B, and 10C + D are each divisible by 4.
4. 1000A + 100B + 10C + D is divisible by 4.
5. $\underline{A}\ \underline{B}\ \underline{C}\ \underline{D}$ is divisible by 4.

37. (a) Last 2 digits are 0   (b) (1), (3), (4)

39. If a number is divisible by 3 and 5, then it is also divisible by 15.

41. 5, 17, 29, 41, . . .

43. (a) Yes   (b) Yes
(c) Let the larger number be $\underline{A}\ \underline{B}$. Then $\underline{A}\ \underline{B} - \underline{B}\ \underline{A} = (10A + B) - (10B + A) = 9A - 9B = 9(A - B)$, which must be divisible by 9.

45. (a) The last digit must be 0 or 4.
(b) For 2, the last digit must be 0, 2, 4, or 6; for 7, the sum of the digits must be divisible by 7.

47. (c) Yes   (d) Yes   (e) Yes   (g) 1001
(h) 382,382
(i) Any number of the form $\underline{A}\ \underline{B}\ \underline{C}\ \underline{A}\ \underline{B}\ \underline{C}$ is divisible by 1001, so it must also be divisible by 7, 11, and 13.

4.3 Homework Exercises

1. (a) Prime

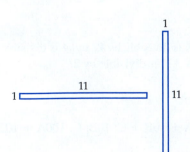

(b) Composite

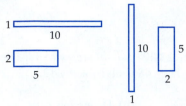

(c) Neither

3. All the results are prime.

5. (a) 257
(b) 4,294,967,297 and 4,294,967,297 ÷ 641 = 6,700,417

7. (a) 6   (b) 6
(c) Either at the beginning or second from the end of a row
(d) $(6n + 1)^2 + 17 = 36n^2 + 12n + 18 = 12(3n^2 + n + 1) + 6$, which will have remainder 6 when divided by 12. $(6n - 1)^2 + 17 = 36n^2 - 12n + 18 = 12(3n^2 - n + 1) + 6$, which will have remainder 6 when divided by 12.

9. (a) It is divisible by 3, 5, 7, 11, 13, etc.
(b) It is divisible by 2.   (c) It is divisible by 5.

11. 2, 3, 5, 7, 11, 13, 17, 19 (since $\sqrt{431} \approx 21$).

13. (a) Prime   (b) Composite (17 · 41)

15. (a) A whole number that has all 3s except for a last digit of 1 is a prime number.
(b) Inductive reasoning
(c) Composite, divisible by 17

17. Exactly one prime factorization

19. (a) $495 = 3^2 \cdot 5 \cdot 11$ (b) $320 = 2^6 \cdot 5$

21. (a) $6 \cdot 3 \cdot 2 = 36$ divisors
(b) $148 = 2^2 \cdot 37$. It has $3 \cdot 2 = 6$ divisors.

23. 101 and 103, 107 and 109, 137 and 139

25. (a) 5 + 7   (b) 13 + 17 or 11 + 19 or 23 + 7
(c) 5 + 103 is one answer.

27. 2 divisors: all prime, 3 divisors: 9, 25; 4 divisors; 8, 10, 14, 15, 21, 22; 5 divisors: 16; 6 divisors: 12, 18, 20; 8 divisors: 24. Perfect square numbers have an odd number of divisors. Prime numbers have two divisors. Squares of odd numbers have 3 divisors. (*Note:* This fails for 81.)

29. (a) $2 \mid 722$ (b) $3 \mid 723$ (c) $4 \mid 4$ so $4 \mid 724$
(d) $5 \mid 720$ and $5 \mid 5$ so $5 \mid 725$; $6 \mid 720$ and $6 \mid 6$ so $6 \mid 726$

31. (a) $x = 3$ (b) No solution   (c) $x = 7$
(d) The greatest prime number less than the divisor

33. (g) 2, 3, 5, 7, 11, 13, 17, 19, 23, 29, 31, 37, 41, 43, 47

4.4 Homework Exercises

1. (a)

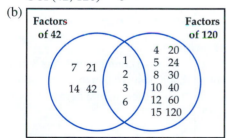

(b) 2

3. (a) Factors of 42: 1, 2, 3, 6, 7, 14, 21, 42
Factors of 120: 1, 2, 3, 4, 5, 6, 8, 10, 12, 15, etc.
GCF(42, 120) = 6
(b)

Factors of 42		Factors of 120
7 21	1	4 20
	2	5 24
14 42	3	8 30
	6	10 40
		12 60
		15 120

(c) $42 = 2 \cdot 3 \cdot 7$ and $120 = 2^3 \cdot 3 \cdot 5$.
GCF $= 2 \cdot 3 = 6$.

5. (a) 1, 2, 1, 4, 1, 2, 1, 4
(b) The sequence of GCF's will have the repeating pattern 1, 2, 1, 4, 1, 2, 1, 4, . . .
(c) Induction

7. The GCF must also include 3 and 5. It is $3 \cdot 5 \cdot 11 = 165$.

9. $2^2 \cdot 3 \cdot 5^6$

11. (a) and (c)

13. (a) $665 \div 627 = 1$ R38
$627 \div 38 = 16$ R19
$38 \div 19 = 2$
GCF(627, 665) = 19
(b) $2035 \div 851 = 2$ R333
$851 \div 333 = 2$ R185
$333 \div 185 = 1$ R148
$185 \div 148 = 1$ R37
$148 \div 37 = 4$
GCF(851, 2035) = 37
(c) $609 \div 551 = 1$ R58
$551 \div 58 = 9$ R29
$58 \div 29 = 2$
GCF(551, 609) = 29

15. (a) 60, 120, 180   (b) An infinite number
(c) 60

17. (a) Multiples of 20: 20, 40, 60, 80, 100, 120, 140, 160, . . .
Multiples of 32: 32, 64, 96, 128, 160, . . .
LCM(20, 32) = 160
(b) $20 = 2^2 \cdot 5$ and $32 = 2^5$.
So LCM $= 2^5 \cdot 5 = 160$.

19. (a) 120 cm   (b) 240, 360, 480 cm, . . .

21. 6 in. by 6 in. (GCF(30, 48) = 6)

23. (a) Yes   (b) No

25. (a) $2^3 \cdot 3^4 \cdot 5^7 \cdot 7 \cdot 11$ (b) $2 \cdot 3^2 \cdot 5 \cdot 11$

27. $2 \cdot 3^2 \cdot 5^4 \cdot 7$

29. (b)

31. $3y$

33. False; $A = 8$ and $B = 15$

35. (b) and (c)

37. (b) $ab = $ GCF$(a, b) \cdot$ LCM(a, b)

39. LCM (22, 40) = 440; GCF (242, 792) = 22

Chapter 4 Review Exercises

1. (a) is correct because of the Divisibility-of-a-Product Theorem.

2. Write the hypothesis as the first step and the conclusion as the last step. Then convert the hypothesis and conclusion into equations, and write these as the second and next-to-last steps, respectively. Finally, use properties of equations to work from the second step to the next-to-last step.

3. There is a whole number C such that $8 \cdot C = N$.

4. 0, 3, 6, or 9 followed by 0

5. (a) $\underline{A} \, \underline{B} \, \underline{5} = A \cdot 100 + B \cdot 10 + 5$ and $\underline{A} \, \underline{B} \, \underline{0} = A \cdot 100 + B \cdot 10$. Now $A \cdot 100$, $B \cdot 10$, and 5 are all divisible by 5. Therefore, $A \cdot 100 + B \cdot 10 + 5$ and $A \cdot 100 + B \cdot 10$ are divisible by 5.
(b) Deductive

6. 2, 3, 5, 7, 11, 13, 17, 19, 23 (since $\sqrt{577} \approx 24$)

7. (a) A-prime numbers, B-multiples of 10 or divisors of 20
(b) C-divisors of 20, D-prime numbers or odd numbers

8. $312 = 2^3 \cdot 3 \cdot 13$

9. True; $A = 2$, $B = 4$, $C = 5$ is an example

10. True; $A = 3$ and $B = 6$ is an example

11. True; GCF(3, 7) = 1 is an example

12. True, $XY \mid Z$ and $XY \mid XY$. So $XY \mid (XY + Z)$.

13. False; 2 and 4

14. (a) 35: 1, 5, 7, 35 and 56: 1, 2, 4, 7, 8, 14, 28, 56; GCF = 7
(b)

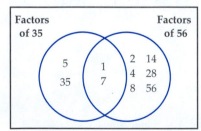

15. 3 and 7

16. (a) $2 \cdot 5^2 \cdot 11^2$ (b) $2^3 \cdot 3 \cdot 5^4 \cdot 11^3$

17. 14

18. 4, GCF

19. 840

Chapter 5

5.1 Homework Exercises

$-$**1.** (a), (b), and (d)

1. (a) I (b) { }

3. -97

5. (a) Loss of yardage
(b) More expenditures than receipts
(c) Below sea level

7. -6 means 6 less than the normal high; 7 means 7 more than the normal high.

9.

11. (a) 20 seconds before liftoff; -20
(b) A gain of \$3; 3   (c) 3 floors lower; -3

13. -5 and 5 are the same distance (5) from 0.

15. $(-4) + (-2) = -6$

17. Go to -5. Then move 3 to the right. You end up at -2. So $-5 + 3 = -2$.

(b) Combine measures
(c) Show -5 as 5 negative counters and 3 as 3 positive counters. Each positive counter cancels out a negative one. You are left with 2 negative counters. So $-5 + 3 = -2$.

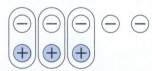

19. (c)

21. (a) $86 + (-30) + (-20) = \$36$
(b) Combine sets/measures

23. Draw a number line, and discuss how you can tell which of two numbers is greater.

25. (a) − 29   (b) 365   (c) −108

27. (b)

29. (a) Go to 4. Then move 6 to the left. You end up at −2. So $4 + (−6) = −2$.

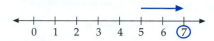

(b) Go to 5. Adding − 2 is a move of 2 to the left, so subtracting −2 is a move of 2 to the right. You end up at 7. So $5 − (−2) = 7$.

31. (a)

(b)

33. (a) $−5 − (−3) = −2$ (b) $2 − 5 = −3$
(c) $−3 − 2 = −5$

35. (a) Show −6 as 6 negative counters.

Now take away −2.

4 negative counters are left. So $−6 − (−2) = −4$.
(b) Show 2 as 2 positive counters. To be able to take away 6, we must change 2 to 6 positive counters and 4 negative counters.

Now take away 6 positive counters.

4 negative counters are left. So $2 − 6 = −4$.

37. (a) Number line   (b) Signed counters
(c) Number line   (d) Debatable

39. (a) 3, left   (b) 3, left
(c) Adding −3 is the same as subtracting 3.
(d) Adding −N is the same as subtracting N, where N is an integer.

41. (a) −73   (b) 38

43. (a) 6   (b) −6   (c) −4

45. (a) −14   (b) 8   (c) 10

47. (a) $−10 − 30 = −40$ (b) −40 ft
(c) Take away measures

49. (a) Loss of 7 yards
(b) Addition, combine measures or subtraction, take away measures

51. It is −6°, and the temperature drops 4°. What is the new temperature? −10°. So $−6 − 4 = −10$.

53. Missing numbers from top to bottom: −3, 4, −8, −2, −8

55. (b) deductive   (c) 5   (d) −3

57. (a) $6 = −2 + n; n = 8$
(b) $−3 = 2 + n; n = −5$

59. (a) All odd integers
(b) All positive and negative multiples of 3

61. (a) True, $x = 3, y = 2$
(b) False if $x < y$

63. (possible answer) First row: −2, −4, 6; second row: 8, 0, −8; third row: −6, 4, 2

65. $0 \leftrightarrow 0, 1 \leftrightarrow −1, 2 \leftrightarrow 1, 3 \leftrightarrow −2, 4 \leftrightarrow 2$, and so on.

5.2 Homework Exercises

1. (a) $4 \times (−3) = −3 + (−3) + (−3) + (−3) = −12$
(b) $4 \times (−3)$ means 4 sets of −3.

This makes − 12. So $4 \times (−3) = −12$.
(c)

3. (a) A negative times a positive is negative.
(b) Deductive

5. $-5 \cdot (-1) = 5$, $-5 \cdot (-2) = 10$

9. (a) -12 pounds   (b) $-3 \times 4 = -12$
(c) Repeated measures

11. (a) The temperature dropped 10°F during a 2-hour period. What was the average temperature change per hour? -5°F. So $-10 \div 2 = -5$.
(b) Show -10 with 10 negative counters. Divide the counters into two equal groups. How many are in each group? -5. So $-10 \div 2 = -5$.

13. $-10 \div 5 = -2$, $-10 \div (-2) = 5$

15. (a) $-6 \times$ _____ $= -54$; 9
(b) $-4 \times$ _____ $= 32$; -8

17. (a) $-\$40,000$ (b) $-480,000 \div 12 = -40,000$
(c) Partition sets/measures

19. (a) -24 (b) 240   (c) -3 (d) 1

21. (a) -7 (b) -37

23. (a) 0   (b) 2 or -2 (c) -5

25. (a) Positive   (b) Positive   (c) Negative

27. 2

29. Three rows: $1, -4, 2$ and $-2, -4, -2$ and $2, -2, -1$

31. For all x and y

33. $(-2 \div -2) + (-2 \div -2) = 2$;
$(-2 \times -2) - (-2 \div -2) = 3$;
$(-2 \times -2) + -2 - (-2) = 4$;
$(-2 \times -2) + (-2 \div -2) = 5$;
$-2 \times [-2 - (-2 \div -2)] = 6$;
$(-2 \times -2) + (-2 \times -2) = 8$

5.3 Homework Exercises

1. (a) Addition and multiplication
(b) Addition and multiplication

3. Compute $-5 \times (-8)$ first and then multiply by 7.

5. (a) -3 (b) 11
(c) Commutative and associative properties of addition

7. Commutative, multiplication

9. (a) Associative property of multiplication and distributive property of multiplication over addition
(b) Compute the answer on each side, and show that they are not equal. Discuss the number of different operations in the equation (one) and why the distributive property does not apply.

11. $(8 \div 4) \div 2 \neq 8 \div (4 \div 2)$

13. (a) Any counterexample showing that whole-number division is not associative would also show that integer division is not associative.
(b) Deductive

15. 1; identity element; multiplication

17. 0

19. Closure property of subtraction

21. (a) 5, 7   (b) The sum of two negatives is a positive.

23. Mixing up the addition and multiplication rules

25. (a) $a \div b = 1/(b \div a)$ (b) when $a = b$

27. (a) $0 = 2 \cdot 4 + (-2 \cdot 4)$
(b) $0 = 2 \cdot 4 + (-2 \cdot 4) = 8 + (-2 \cdot 4)$.
So $-2 \cdot 4 = -8$ because each integer has a unique additive inverse.

29. $(2m)(2n) = 4mn = 2(2mn)$, which is even.

Chapter 5 Review Exercises

1. The set of whole numbers is $\{0, 1, 2, 3, \ldots\}$. The set of whole numbers is a subset of the set of integers. The set of integers is the union of the set of whole numbers and their opposites: $\{\ldots, -3, -2, -1, 0, 1, 2, 3, \ldots\}$.

2. (a) Go to 2. To subtract 4, move 4 to the left.

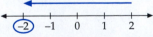

You end up at -2. So $2 - 4 = -2$.
(b) Show 2 as 2 positive counters. To be able to take away 4, we must change 2 to 4 positive counters and 2 negative counters.

Now take away 4.

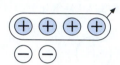

2 negative counters are left. So $2 - 4 = -2$.

3. (a) Show -4 as 4 negative counters. Now take away -2.

2 negative counters are left. So $-4 - (-2) = -2$.

(b) Go to -4. Subtracting -2 is the opposite of adding -2, so move 2 to the right. You end up at -2. So $-4 - (-2) = -2$.

4. (a) $-30 + 20 = -10$ ft
(b) Addition, combine measures

5. (a) 500 soldiers better
(b) $-300 - (-800) = 500$ (c) Compare sets

6. The temperature is now $-3°C$. If the temperature drops $6°C$, what will the new temperature be? $-9°C$. So $-3 - 6 = -9$.

7. (a) $3 \times (-2)$ means 3 sets of -2.

This makes -6. So $3 \times (-2) = -6$.
(b)

8. $-4 \times 2 \quad = -8$
$-4 \times 1 \quad = -4$
$-4 \times 0 \quad = 0$
$-4 \times (-1) = 4$
$-4 \times (-2) = 8$

10. $-18 \div (-3) = ?$ means $-3 \cdot ? = -18$. What number times -3 equals -18? 6. So $-18 \div (-3) = 6$.

11. (a) $-18 \div 9 = -2$ lb/month
(b) Partition measures

12. (a) -37 (b) -2 (c) 157   (c) -9

13. Addition and multiplication

14. $8 - (4 - 2) \neq (8 - 4) - 2$

15. $(-5 + 8)n$

16. Additive inverses

Chapter 6

6.1 Homework Exercises

1. (a) Shade 3 of 5 equal parts.

(b) Go $\frac{3}{5}$ of the way from 0 to 1.

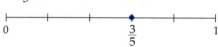

3. (a) One brownie; $2 \div 2 = 1$
(b) $\frac{2}{3}$ brownie; $2 \div 3 = \frac{2}{3}$
(c) $\frac{1}{2}$ brownie; $2 \div 4 = \frac{1}{2}$

5. The five parts are unequal.

7. $\frac{1}{3}$ represents:

Part of a whole   *Part of a set*

1 of 3 equal parts   1 of 3 equal groups

Division
$1 \div 3$

Point on a number line

Divide 1 into 3 equal parts.

Go $\frac{1}{3}$ of the way from 0 to 1.

9. (a)

0    $\frac{3}{8}$ 1

(b)

(c)

(d)

$3 \div 8$

11. Cut each brownie in half. Each person receives 3 halves. Or each person receives one whole plus one half of the third brownie.

13. (a) Cut each pizza in half. Each person receives half a pizza.
(b) Cut each pizza into 8 equal slices. Each person receives 3 slices.

15. 8 R25 means 8 measures of 50 and 25 out of a measure of 50 left over. And 25 out of 50 is $\frac{1}{2}$.

17. $\frac{0}{5}$ means $0 \div 5$, which is 0.

19. (b)

   or

(c) $\frac{1}{2}$

21. (a) All three
(b) Negative ÷ positive and positive ÷ negative equal a negative.

23.

25. Draw 6 cats (or circles). Show 3 of 5 equal parts. Each equal part has 2 cats.

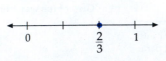

The whole has 5 equal parts with a total of 10 cats.

27.

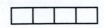

29.

0    $\frac{2}{3}$ 1

31. Eating 4 pieces of a 4-slice pizza is 4/4, and eating 8 slices of an 8-slice pizza is 8/8. 4/4 = 8/8.

33. (a)

(b) 3 wholes with 2 halves each make 3 × 2 halves. One more half makes 7 halves.

(c)

0   1   2   3 $3\frac{1}{2}$ 4

(d) $3\frac{1}{2}$

$2\overline{)7}$

35. (a) 3/1   (b) −3/1   (c) 9/2
(d) −56/10   (e) 25/100

37. (a) Yes   (b) W

39. (a)

 =

(b)

0    $\frac{3}{5}$ 1

0    $\frac{6}{10}$ 1

(c) $\frac{3}{5} = \frac{3 \times 2}{5 \times 2} = \frac{6}{10}$

41. 1600 say yes in the second group.

43. (a) $\frac{21}{58}$ (b) $\frac{1}{yz}$

45.

$\frac{1}{3}$

$\frac{2}{3}$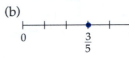

$\frac{1}{3} < \frac{2}{3}$

47.

$\frac{7}{8}$

$\frac{3}{4}$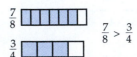

$\frac{7}{8} > \frac{3}{4}$

49. $\frac{4}{9} = \frac{20}{45}$ and $\frac{7}{15} = \frac{21}{45}$; $\frac{4}{9} < \frac{7}{15}$

51. $\frac{14}{23} = \frac{420}{690}$ and $\frac{17}{30} = \frac{391}{690}$; $\frac{14}{23} > \frac{17}{30}$

51. $\frac{14}{23} = \frac{420}{690}$ and $\frac{17}{30} = \frac{391}{690}$; $\frac{14}{23} > \frac{17}{30}$

53. $-\frac{1}{2}$

55. (a) $\frac{4}{9}$ is less; $\frac{7}{12}$ is greater.

   (b) See if the denominator is more or less than twice the numerator.

57. $\frac{3}{4} = \frac{24}{32}$ and $\frac{7}{8} = \frac{28}{32}$; $\frac{25}{32}, \frac{13}{16}, \frac{27}{32}$

59. No

61. (a) 6   (b) $\frac{1}{6}$ (c) $\frac{1}{2}$ (d) $\frac{2}{3}$

63. Yes

65.

$1\frac{1}{3}$	$\frac{2}{3}$	$\frac{2}{3}$	$\frac{1}{3}$
	$\frac{8}{4}$	$\frac{3}{9}$	$\frac{4}{10}$

6.2 Homework Exercises

1. (a) $\frac{4}{y}$ (b) $-\frac{x}{n}$

3.

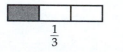

$\frac{1}{2}$ $\frac{1}{8}$ $\frac{2}{10}$

$\frac{1}{2} > \frac{2}{10}$ so $\frac{1}{2} + \frac{1}{8}$ does not equal $\frac{2}{10}$.

5. (a)

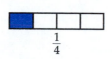

$\frac{1}{3}$ $\frac{1}{4}$

(b) $\frac{1}{3} + \frac{1}{4}$ would equal the shaded region shown, but we cannot name it unless we use a common denominator (twelfths).

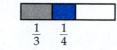

$\frac{1}{3}$ $\frac{1}{4}$

(c)

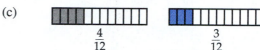

$\frac{4}{12}$ $\frac{3}{12}$

(d) $\frac{1}{3} + \frac{1}{4} = \frac{4}{12} + \frac{3}{12} = $ 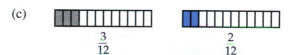 $= \frac{7}{12}$

7. Each represents a fraction of the *same* set (or whole). Draw one set of 3 circles. Shade $\frac{1}{3}$ of the set. Then shade another $\frac{1}{3}$ of the set. That makes $\frac{2}{3}$ of the set.

9. (a) $\frac{19}{24}$ (b) $\frac{b + 2a}{ab}$

11. (a)

$\frac{1}{4}$ $\frac{1}{6}$

(b) How much is left when we take $\frac{1}{6}$ away from $\frac{1}{4}$? We cannot tell. A common denominator is needed.

(c)

$\frac{3}{12}$ $\frac{2}{12}$

(d) $\frac{1}{4} - \frac{1}{6} = \frac{3}{12} - \frac{2}{12} = $ = $\frac{1}{12}$

13. (a) 140   (b) 280, 420   (c) $\frac{17}{70}$

15. (a) $2\frac{3}{4}$ hours

   (b) Subtraction, take away measures

17. $8\frac{3}{8}$

19. $4\frac{1}{2}$

21. You have a $5\frac{1}{4}$-hour job to do. You have been working for $2\frac{1}{2}$ hours. How much more time is needed? $2\frac{3}{4}$ hours.

23. (a) $\frac{1}{2}$ (b) $2\frac{6}{10} = 2\frac{3}{5}$

25. (a) 8   (b) $4\frac{2}{3}$

27. (a) $8\frac{2}{3} - 4 = 4\frac{2}{3}$ (b) $5\frac{1}{2} - 3 = 2\frac{1}{2}$

29. (a) $\frac{5}{12}$ (b) $\frac{3}{20}$

31. (a) $\frac{1}{4} = \frac{1}{5} + \frac{1}{20}$; yes   (b) $\frac{1}{N} = \frac{1}{N+1} + \frac{1}{N(N+1)}$

(c) Induction

(d) $\frac{1}{N+1} + \frac{1}{N(N+1)} = \frac{N+1}{N(N+1)} = \frac{1}{N}$

(e) Deduction

33. (a) $\frac{1}{2} + \frac{1}{4}$ (b) $\frac{1}{13} + \frac{1}{26}$ (c) $\frac{1}{2} + \frac{1}{8}$

(c) $\frac{1}{9} + \frac{1}{2} + \frac{1}{6}$

6.3 Homework Exercises

1. $4 \times \frac{1}{5} = \frac{1}{5} + \frac{1}{5} + \frac{1}{5} + \frac{1}{5} = \frac{4}{5}$

3. (a) The whole pizza   (b) $\frac{5}{8}$ of the pizza

5. $\frac{1}{5} \times \frac{1}{3}$ means $\frac{1}{5}$ of $\frac{1}{3}$. First show $\frac{1}{3}$.

Now darken $\frac{1}{5}$ of $\frac{1}{3}$.

$\frac{1}{15}$ of the figure is darkened. So $\frac{1}{5} \times \frac{1}{3} = \frac{1}{15}$.

7. $\frac{3}{4} \times \frac{4}{5}$ means $\frac{3}{4}$ of $\frac{4}{5}$. First show $\frac{4}{5}$.

Now darken $\frac{3}{4}$ of $\frac{4}{5}$.

$\frac{12}{20}$ of the figure is darkened. So $\frac{3}{4} \times \frac{4}{5} = \frac{12}{20} = \frac{3}{5}$.

9. $\frac{1}{4} \times \frac{4}{5} = \frac{1}{5}$

11. (a) $6 \times \frac{1}{3}$ means 6 measures of $\frac{1}{3}$

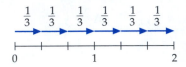

So $6 \times \frac{1}{3} = 2$

(b) $\frac{1}{3} \times 6$ is $\frac{1}{3}$ of 6.

So $\frac{1}{3}$ of 6 is 2.

13. $5\frac{1}{2} \times 2\frac{2}{3} = \frac{11}{2} \times \frac{8}{3} = \frac{11 \times 8}{2 \times 3}$. Now that it's all one fraction, you can divide the numerator and the denominator by a common factor, 2.

$\left(\text{Then } \frac{8}{2} \text{ becomes } \frac{4}{1}.\right)$

$= \dfrac{11 \times \overset{4}{8}}{\underset{1}{2} \times 3}$

15.

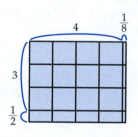

$$(3 \times 4) + \left(3 \times \frac{1}{8}\right) + \left(\frac{1}{2} \times 4\right) + \left(\frac{1}{2} \times \frac{1}{8}\right)$$

$$= 12 + \frac{3}{8} + 2 + \frac{1}{16} = 14\frac{7}{16}$$

17. (a) $3\frac{3}{4}$ (b) $8\frac{3}{4}$

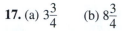

(c) $15\frac{3}{4}$

(d) The whole-number part is $b(b + 2)$, and the fraction part is $\frac{3}{4}$.

(e) $(18 \times 20) = 360 \rightarrow 360\frac{3}{4}$

19. (a) $\frac{1}{2} \times 8$ (b) $8 \div 2$

21. (a) $\frac{1}{2}$ (b) $1\frac{5}{9}$

23. (a) $2 \div \frac{1}{4}$ means how many $\frac{1}{4}$s does it take to make 2?

It takes 8 quarters to make 2. So $2 \div \frac{1}{4} = 8$.

(b) $2 \div \frac{1}{4} = \frac{8}{4} \div \frac{1}{4}$. How many $\frac{1}{4}$s make $\frac{8}{4}$s? 8.

So $2 \div \frac{1}{4} = 8$.

25. $\frac{2}{3} \div \frac{1}{6}$ means how many $\frac{1}{6}$s make $\frac{2}{3}$?

So $\frac{2}{3} \div \frac{1}{6} = 4$

27. $\frac{3}{4} \div 2$ means dividing $\frac{3}{4}$ into 2 equal parts.

Each of the two equal parts has $\frac{3}{8}$. So $\frac{3}{4} \div 2 = \frac{3}{8}$.

29. $1\frac{1}{4} \div \frac{1}{2}$ means how many $\frac{1}{2}$s make $1\frac{1}{4}$?

It takes $2\frac{1}{2}$ halves to make $1\frac{1}{4}$. So $1\frac{1}{4} \div \frac{1}{2} = 2\frac{1}{2}$.

31. (a) $\frac{3}{4}$ of the whole is shaded.

(b) The whole is $\frac{4}{3}$ of the shaded section.

(c) 1 of the 3 shaded parts

(d) How many shaded sections make the whole? $1\frac{1}{3}$

33. $\frac{10}{21} = \frac{2}{3} \times \frac{n}{m}; \frac{n}{m} = \frac{5}{7}$

35. (a) $1\frac{1}{4}$ (b) $-2\frac{4}{7}$ (c) $1\frac{2}{5}$

37. (a) 20   (b) 8

39. $\frac{1}{16}$

41. $\dfrac{1}{2} \div \dfrac{3}{4} = \dfrac{\frac{1}{2}}{\frac{3}{4}} = \dfrac{\frac{1}{2} \times \frac{4}{3}}{\frac{3}{4} \times \frac{4}{3}} = \dfrac{\frac{1}{2} \times \frac{4}{3}}{1} = \dfrac{1}{2} \times \dfrac{4}{3}$

43. $\dfrac{a}{b} \div \dfrac{c}{d} = \dfrac{\frac{a}{b}}{\frac{c}{d}} = \dfrac{\frac{a}{b} \times \frac{d}{c}}{\frac{c}{d} \times \frac{d}{c}} = \dfrac{\frac{ad}{bc}}{1} = \dfrac{ad}{bc}$

45. (a) $1\dfrac{1}{2}$ cups

(c) Multiplication, repeated measures

47. (a) 15   (b) Division, repeated measures

49. (a) $10 \div 4 = 2\dfrac{1}{2}$, $18 - 10 = 8$, and $8 \div 2\dfrac{1}{2} = 3\dfrac{1}{5}$.

You can stack 3 more.

(b) Division, partition measures; subtraction, compare measures; division, repeated measures.

51.

$\dfrac{1}{4} \times \dfrac{4}{5} = \dfrac{4}{20} = \dfrac{1}{5}$

53. $\dfrac{1}{6}$ cup

55. $154

57. $12

61. $\dfrac{2}{3} \div \dfrac{1}{3} = 2$

63. (a) $\left(\dfrac{1}{2} \text{ of } \dfrac{1}{4}\right) + \left(\dfrac{1}{4} \text{ of } \dfrac{1}{4}\right) + \left(\dfrac{1}{8} \text{ of } \dfrac{1}{4}\right) = \dfrac{7}{32}$

(b) Subdivide the entire region into the smallest right triangles $\left(\text{each } \dfrac{1}{32}\right)$. There are 7 blues ones. The answer is $\dfrac{7}{32}$.

65. (a) 8 for Wacky, 6 for Harpo, and 3 for Young Loopy II

(b) Because the fractions $\dfrac{4}{9} + \dfrac{1}{3} + \dfrac{1}{6} = \dfrac{17}{18}$ and $\dfrac{17}{18}$ of 18 is 17

(c) Debatable

67. (a) You have 46 cans of juice. How many complete six-packs can you make?

(b) You have 46 cans of juice that you want to pack in cartons of six. How many cartons are needed to pack all the cans?

(c) In part (a), how many cans are left over?

(d) How many cartons of juice do you have in part (b)?

69. (a) Both fractions have the same numerator. The denominator of the second fraction equals the sum of the numerator and denominator from the first fraction.

(b) $\dfrac{3}{5} - \dfrac{3}{8} = \dfrac{3}{5} \times \dfrac{3}{8}$ and $\dfrac{1}{2} - \dfrac{1}{3} = \dfrac{1}{2} \times \dfrac{1}{3}$

(c) $\dfrac{a}{b} - \dfrac{a}{a+b} = \dfrac{a(a+b) - ab}{b(a+b)}$

$= \dfrac{a^2 + ab - ab}{b(a+b)} = \dfrac{a^2}{b(a+b)} = \dfrac{a}{b} \times \dfrac{a}{a+b}$

71. (a) $10, 12, 30, \dfrac{2}{3} N$ (b) $35, 17\dfrac{1}{2}, 20, \dfrac{7}{2} N$

6.4 Homework Exercises

1. (a) Addition and multiplication

(b) Addition and multiplication

3. $\left(\dfrac{1}{2} \cdot 4\right) \cdot x$

5. $2\dfrac{1}{2}$

7. $\left(-\dfrac{4}{3} + \dfrac{2}{3}\right)n = -\dfrac{2}{3}n$

9. $50 \times 6 + 50 \times \dfrac{1}{2} = 300 + 25 = 325$ miles

11.

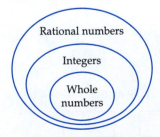

13. A counterexample for whole numbers (like $2 - 3 \neq 3 - 2$) would also be a counterexample for whole numbers, integers, and rational numbers.

15. $-\dfrac{2}{5}$

17. (a) Denseness   (b) Multiplicative inverses

19. $1\frac{3}{4} \times 8\frac{1}{4} \approx 2 \times 8 = 16$ cups

21. (d)

23. (a) $\frac{7}{8} + \frac{1}{16} \approx 1 + 0 = 1$ in.

(b) They are simply adding either the numbers in the numerators or the numbers in the denominators.

25. (a) Close to $\frac{1}{4}$ (b) Close to $\frac{3}{4}$

(c) Close to but less than $\frac{1}{2}$ (d) Close to 0

(e) Close to but greater than $\frac{1}{2}$

27. (a) The denominator is about 4 times the numerator.

(b) The numerator is slightly less than $\frac{1}{2}$ the denominator.

29. $\frac{41}{126} \approx \frac{1}{3}$

31. $\frac{1}{4}$ of $23 \approx 6$

33. $384\frac{1}{2} \div 17\frac{1}{4} \approx 400 \div 20 = 20$ panels

35. (d)

37. (c)

39. (a) $\frac{5}{12}; \frac{11}{20}$

(b) After obtaining the common denominator, the child adds the numerators and denominators.

41. (a) $1\frac{9}{3} = 4; 4\frac{9}{5} = 5\frac{4}{5}$

(b) In regrouping, the child adds 10 to the numerator of the fraction.

43. (a) $\frac{2}{3}$ and 3

(b) The child cancels common terms rather than common factors in the numerator and denominator.

45. The child is confusing fraction regrouping with whole-number regrouping.

47. (a) $\frac{2}{(1+1)} \neq \frac{2}{1} + \frac{2}{1}$

(b) $\frac{(10+10)}{2} \neq \frac{10}{2} + 10$

49. $\left(\frac{u}{v} \cdot \frac{w}{x}\right) \cdot \frac{y}{z} = \frac{uw}{vx} \cdot \frac{y}{z} = \frac{uwy}{vxz}$ and

$\frac{u}{v} \cdot \left(\frac{w}{x} \cdot \frac{y}{z}\right) = \frac{u}{v} \cdot \frac{wy}{xz} = \frac{uwy}{vxz}$

51. (a) False for $x = 1$ and $y = 1$

(b) False for $x = 1$ and $y = 1$

53. $>$

Chapter 6 Review Questions

1. $\frac{5}{6}$ represents:

Part of a whole   *Part of a set*

5 of 6 equal parts   5 of 6 equal groups

Division   *Number line location*
5 ÷ 6

Divide 5 into
6 equal parts.

Go $\frac{5}{6}$ of the way from 0 to 1.

2. (a)

$\frac{7}{4} = \quad = 1\frac{3}{4}$

(b)

$\frac{0}{4} \qquad \frac{4}{4} \qquad \frac{7}{4}$

(c) $1\frac{3}{4}$

$4\overline{)7}$

3. (a)   (b)

4. $\frac{5}{0}$ means $5 \div 0$. No number multiplied by 0 equals 5, so $5 \div 0$ and $\frac{5}{0}$ are both undefined.

5. $\frac{1}{2}$

6. The set of whole numbers is $\{0, 1, 2, 3, \ldots\}$. The set of whole numbers is a subset of the set of integers. The set of integers is the union of the set of whole numbers and their opposites: $\{\ldots, -3, -2, -1, 0, 1, 2, 3, \ldots\}$. The set of integers is a subset of the set of rational numbers. A rational number is a number that can be written in the form $\frac{p}{q}$ /q where p and q are integers and $q \neq 0$.

7. GCF(135, 162) = 27; $\frac{5}{6}$

8. (a) $\frac{3}{4}$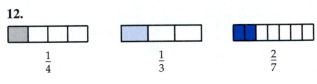
$\frac{3}{4} > \frac{2}{3}$

$\frac{2}{3}$

(b) $\frac{3}{4} = \frac{9}{12}$ and $\frac{2}{3} = \frac{8}{12}$; $\frac{3}{4} > \frac{2}{3}$

9. Science

10. (a) $b < d$ (b) Deduction

11. (a) $\frac{3}{4}, \frac{11}{14}, \frac{23}{28}$ (b) Denseness

12.

$\frac{1}{4}$ $\frac{1}{3}$ $\frac{2}{7}$

$\frac{1}{3} > \frac{2}{7}$ so $\frac{1}{3} + \frac{1}{4}$ does not equal $\frac{2}{7}$.

13. (a)

$\frac{2}{3}$ $\frac{1}{6}$

(b) $\frac{2}{3} + \frac{1}{6}$ would equal the shaded region shown, but we cannot name it unless we use a common denominator (sixths).

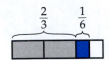

$\frac{2}{3}$ $\frac{1}{6}$

(c)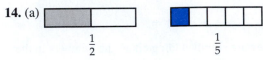

$\frac{4}{6}$ $\frac{1}{6}$

(d) $\frac{2}{3} + \frac{1}{6} = \frac{4}{6} + \frac{1}{6} =$  $= \frac{5}{6}$

14. (a)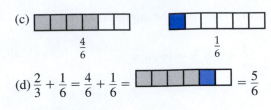

$\frac{1}{2}$ $\frac{1}{5}$

(b) How much is left when we take $\frac{1}{5}$ away from $\frac{1}{2}$? We cannot tell. A common denominator is needed.

(c)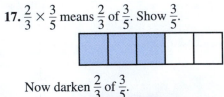

$\frac{5}{10}$ $\frac{2}{10}$

(d) $\frac{1}{2} - \frac{1}{5} = \frac{5}{10} - \frac{2}{10} =$ ⊠▯▯▯▯▯ $= \frac{3}{10}$

15. (a) $\frac{43}{72}$ (b) $17\frac{3}{4}$ (c) $-5\frac{3}{4}$

16. (a) $4 \div \frac{1}{3}$ means how many $\frac{1}{3}$s make 4?

▯▯▯ ▯▯▯ ▯▯▯ ▯▯▯ 12

(b) $4 \div \frac{1}{3} = \frac{12}{3} \div \frac{1}{3}$. How many $\frac{1}{3}$s make $\frac{12}{3}$? 12. So $4 \div \frac{1}{3} = 12$.

17. $\frac{2}{3} \times \frac{3}{5}$ means $\frac{2}{3}$ of $\frac{3}{5}$. Show $\frac{3}{5}$.

Now darken $\frac{2}{3}$ of $\frac{3}{5}$.

$\frac{6}{15}$ or $\frac{2}{5}$ is darkened. So $\frac{2}{3} \times \frac{3}{5} = \frac{2}{5}$.

18. $\frac{1}{4} \times \frac{8}{9} = \frac{1 \times 8}{4 \times 9}$. Now that it's all one fraction, you can divide the numerator and the denominator by a common factor, 4. $\left(\text{Then } \frac{8}{4} \text{ becomes } \frac{2}{1}.\right)$

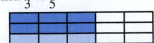

19.

$$(2 \times 3) + \left(2 \times \frac{1}{4}\right) + \left(\frac{1}{2} \times 3\right) + \left(\frac{1}{2} \times \frac{1}{4}\right)$$
$$= 6 + \frac{1}{2} + \frac{3}{2} + \frac{1}{8} = 8\frac{1}{8}$$

20. $\frac{2}{3} \div 4$ means divide $\frac{2}{3}$ into 4 equal parts.

Each of the four equal parts is $\frac{1}{6}$. So $\frac{2}{3} \div 4 = \frac{1}{6}$

21. $\frac{2}{5} \div \frac{7}{9} = \dfrac{\frac{2}{5}}{\frac{7}{9}} = \dfrac{\frac{2}{5} \times \frac{9}{7}}{\frac{7}{9} \times \frac{9}{7}} = \dfrac{\frac{2}{5} \times \frac{9}{7}}{1} = \frac{2}{5} \times \frac{9}{7}$

22. (a) $6\frac{2}{3}$ (b) Division, repeated measures

23. (a) $\frac{17}{30}$

(b) Addition, combine sets/measures and subtraction, take away measures/sets

24. (a) $\frac{7}{20}$ lb   (b) Division, partition measures

26. (a) $2\frac{1}{20}$ (b) 12   (c) $\frac{5}{6}$

27. $5 \times (3 \times 2) = (5 \times 3) \times 2$

28. The counterexample for whole-number division is also a counterexample for rational-number division.

29. Compute $24 \times 2 = 48$ and $24 \times \frac{1}{2} = 12$.
$48 + 12 = 60$.

30. Denseness or multiplicative inverses

31. (a)

32. (a) $1\frac{1}{3}$

(b) The child adds the denominators to obtain the common denominator. Then the child uses addition to obtain the new numerators.

(c) $\frac{1}{4} + \frac{2}{3} < \frac{1}{3} + \frac{2}{3}$ so the answer should be less than 1.

Chapter 7
7.1 Homework Exercises

1. $100; \frac{1}{100}$

3. (a) 41.16   (b) 7.005

5. Where is the decimal point in relation to the one's place?

7. (a) $0.89 lb   (b) Yes

9. (a) Because $11 > 2$

(b) To show 0.11, shade 11 of the 100 squares. To show 0.2, shade 2 of the 10 columns. The area shaded for 0.2 is larger, so $0.11 < 0.2$.

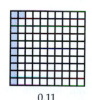 

0.11   <   0.2

(c) 0.2 is located to the right of 0.11 on the number line. So $0.11 < 0.2$.

(d)

	ones	tenths	hundredths
	0 $\cdot$	1	1
	0 $\cdot$	2	

Since 0.2 has a larger number in the tenths column, $0.11 < 0.2$.

11. To show 0.40, shade 40 out of 100 squares. To show 0.4, shade 4 out of 10 columns. The area shaded for 0.40 is equal to the area for 0.4 so 0.40 = 0.4.

13. (a) $0.90   (b) $0.80   (c) $0.90

15. (a) 700   (b) Rounding up

17. (a) $8847.6 - 7132.1 \approx 8800 - 7100 = 1700$ m
    (b) $8847.6 - 7132.1 \approx 8000 - 7000 = 1000$ m

19. $129 \div 46 \approx 120 \div 40 = 3¢$

21. (a)

23. $(\$1.79)(3)(4) \approx \$24$; (d)

25. 153.65; 153.75

27. $10^3 = 1000, 10^2 = 100, 10^1 = 10, 10^0 = 1$

29. (a) 5   (b) 1; $\frac{1}{5}$; $\frac{1}{25}$

31. (a) 2^2
    (b) Subtract the exponent of the denominator from the exponent of the numerator.
    (c) 10^3 (d) 5^{10} (e) x^4

33. (a) 36,200,000   (b) 5,600

35. $6,500,000,000,000

37. (a) 0.004268   (b) 0.00362

39. Move the decimal two places to the right.
    Answer: $329.

41. 0.0000054

43. 3.4×10^{12}

45. (a) 12,000,000   (b) 12 million

47. (a) 1.3×10^{-23} (b) It's shorter!
    (c) (possible answer) $\boxed{1.3 \quad -23}$

49. 6.4×10^{11} lb

51. 48 is greater than 10.

53. (a) Saturn   (b) 40   (c) 0.4

55. About 12 days

57. (a) About $13   (b) About $48

59. $5.75 \times 20 = (5 \times 20) + \left(\frac{3}{4} \times 20\right) = 100 + 15 =$
    $115

61. $12 \times \$10 + 12 \times \$1 - 12 \times \$0.02 =$
    $120 + 12 + 0.24 = \$131.76.$

63. (a) 46   (b) -278 (c) 4   (d) 5 or more in tenths place

7.2 Homework Exercises

1.

$0.2 + 0.3 = 0.5$

3. (a) Hundredths   (b) Tenths   (c) Yes

5. Use decimal squares.

7.

$0.23 - 0.08 = 0.15$

9. Subtraction, take away measures/sets

11. (a)

$2 \times 0.18 = 0.36$

(b) $2 \times 0.18 = \frac{2}{1} \times \frac{18}{100} = \frac{36}{100} = 0.36$

(c) 2×18 hundredths = 36 hundredths = 0.36

13. (a) 0.5×0.6 means 0.5 of 0.6
Dot 0.6.

Now darken 0.5 of 0.6.

0.3 is darkened. So $0.5 \times 0.6 = 0.3$

(b) $0.5 \times 0.6 = \frac{5}{10} \times \frac{6}{10} = \frac{30}{100} = 0.30 = 0.3$

(c) 5 tenths $\times$ 6 tenths = 30 hundredths = 0.30 = 0.3

15. (a) Yes
(b)

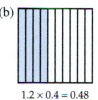

$1.2 \times 0.4 = 0.48$

17. (a) $34.56 \times 6.2 \approx 30 \times 6 = 180$ so the decimal point goes to the right of the 4.
(b) In 34.56×6.2, there are two decimal places in the first factor, and one decimal place in the second factor. The product will have 3 decimal places. The decimal goes between the 4 and the 2.

19. (a) 72 calories   (b) Repeated measures

21. It is right.

23.

$0.6 \div 3 = 0.2$

25.

$0.2 \div 0.04 = 5$

27. $6.4 \div 0.32 = \frac{6.4}{0.32} = \frac{6.4}{0.32} \times \frac{100}{100} = \frac{640}{32} = 640 \div 32$

29. (a), (c), and (d)

31. (a) The child put the remainder after the decimal point.
(b) That the remainder represents a fraction of the divisor

33. $44.237 \div 3.1 \approx 45 \div 3 = 15$. The decimal point goes to the right of the 4.

35. Division, repeated measures

37. Multiplication, repeated sets/measures and addition, combine sets/measures

39. 3.97   (b) 0.03   (c) 25

41. (a) $12.48   (b) $4.09
(c) One factor is less than 1.

43. (a) $xy < y$ (b) $\frac{y}{x} > y$ (c) $x^2 < x$

45. A serving of fish weighs 0.3 lb. How much would 5 servings weigh?

47. (a) 58.29, 1.17
(b) If one of the two numbers has no digit in a place-value column, the child copies the digit from the other number into the answer.

49. (a) 6.4, 4.6
(b) In each place-value column, the child subtracts the smaller digit from the larger.

51. (a) U.K.: 4.2 tons/person; U.S.: 8.0 tons/person; France: 4.9 tons/person

53. (a) 1.5, 2.5   (b) $3 \times 4 + 0.25 = (3.5)^2$
(c) Yes

55. (a) About 140 gallons more (b) About $220 more

57. (b) $12.00, $13.01, $14.02, $15.03, $16.04, $17.05, $18.06, or $19.07

61. (a) Rent check (b) F3 − D4
(c) $2476.10, $3551.50, $3266.39, $3204.14

63. 6; No

65. (a) 0.33333 . . . (b) 0.1_{three} (c) 0.2_{six}
(d) Nine; 0.3_{nine}

67. (a) $0.048 (b) $0.256 (c) $9.60
(d) $28 (e) $0.96

7.3 Homework Exercises

1. (a) 17 to 12 (b) 12 to 29

3. (a) 0 to 600 (b) The second number would be 0.

5. (b) $\dfrac{2}{5}$ (c) $1\dfrac{1}{2}$ (d) $\dfrac{2}{3}$

7. 1 to 4

9. (a) The mother is 20 years older than the daughter.
(b) The mother is twice as old as the daughter.

11. (a) 18.4; 14.7; 16.1; 14.7; 13.1
(b) Buffalo (c) San Francisco (d) Lower

13. No

15. (a) No (b) Yes

17. 3 platters for 10 people

19. (a) N.Y.: $407/\text{mile}^2$; Cal.: $229/\text{mile}^2$ (b) N.Y.

21. (a) $12.50/hr; $12.50/hr (b) They are the same.

(c) Number of Hours	1	10	20	30	40
Job 1 Pays ($)	12.50	125	250	375	500
Job 2 Pays ($)	12.50	125	250	375	500

23. $0.04/oz, the cost of juice per ounce; 24.8 oz/$, the amount of juice you get for $1

25. (a) $0.11/oz (b) $0.12/oz (c) 18-oz jar

27. (a) 9 to 12 (b) $\dfrac{3}{4} = \dfrac{9}{12}$

29. 28 comic books; writing a proportion, using a unit rate, using a multiplier

31. Work up to 35 in groups of 5. Use some manipulatives or pictures if necessary.

33. (a) 10 (b) 5 (c) 15

35. (a) $11\dfrac{2}{3}$ (b) $\pm\sqrt{48}$ (c) 12.6

37. $\dfrac{7}{5} = \dfrac{2}{N}, \dfrac{7}{2} = \dfrac{5}{N}, \dfrac{2}{7} = \dfrac{N}{5}$

39. (a) 1.2 (b) $\dfrac{8}{15}$ (c) $13\dfrac{1}{3}$

41. (b) $1\dfrac{5}{7}$ quarts of orange juice and $1\dfrac{1}{7}$ quarts of skim milk

(c) $\dfrac{3}{7} \times 4 = 1\dfrac{5}{7}$ qt juice and $\dfrac{2}{7} \times 4 = 1\dfrac{1}{7}$ qt milk

(d) $3 \times \dfrac{4}{7} = 1\dfrac{5}{7}$ qt juice and $2 \times \dfrac{4}{7} = 1\dfrac{1}{7}$ qt milk

43. (a) $\dfrac{9}{\frac{3}{4}} = \dfrac{x}{1}$; $x = \$12/\text{lb}$

(b) Draw a rectangle divided in fourths. Three of the fourths cost $9, so label each fourth as $3. The cost for the whole is $12 which is $12/lb.

45. 1195

47. About 5 feet 10 inches

49. $93\dfrac{1}{3}$ lb

51. $3529.41

53. 1.6 million cars

55. 33 fish

57. $0.94

59. The bus is more efficient than the car when it carries more than 4 times as many people. The train is more efficient than the bus when it carries more than 3 times as many people as the bus.

61. $\dfrac{MH}{G}$,

63. The function takes the weight as the input and gives the price as the output. The formula is price equals weight times 5.

7.4 Homework Exercises

1. Per 100

3. (a) 150   (b)

   (c) 30   (d) 3

5. (a) $\dfrac{34}{100}$, 0.34   (b) $1\dfrac{4}{5}$, 1.8   (c) $\dfrac{6}{10000}$, 0.0006

7. 1

9. (a) 23%   (b) 0.041%   (c) 2400%

11. (a) 4%   (b) 37.5%   (c) 175%

13. (a) 132; 200   (b) 66%   (c) $\dfrac{17}{50}$
   (d) 17 to 33   (e) 64

15. 99% watched television in the past week. In the survey, 490 more adults watched television than did not. In the survey, 99 times more adults watched television than did not. $\dfrac{99}{100}$ of the adults watched television.

17. (a) 150   (b) 62.5   (c) 70%

21. (a) 200   (b) 90%   (c) 14

23. (b) (0.12)($2400) = $288
   (c) $\dfrac{12}{100} = \dfrac{x}{\$2400}$; x = $288

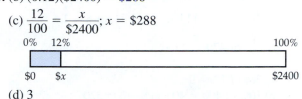

   (d) 3

25. (a) 82 ÷ 110 = 0.745 = 0.745 × 100% = 74.5%
   (b) $\dfrac{82}{110} = \dfrac{x}{100\%}$; x = 74.5%

27. (a) 0.08x = $798.40; x = $9980
   (b) $\dfrac{8}{100} = \dfrac{\$798.40}{x}$; x = $9980

29. Draw a row of 20 squares.

31. (a) U.S.: $4,630, Canada: $2,548
   (b) U.S.: 22%; Canada: 11%
   (c) $143 billion
   (d) Canadians tend to live a little longer than people in the U.S., and the U.S. has fewer people per physician.

33. $125,000

35. $150

37. (a) 1%   (b) 0.033%

39. (a) Interest: $480; balance: $8,480   (b) $8,988.80

41. (a) 2%   (b) Interest: $120; Balance: $6,120
   (c) $6,242.40

43. (a) Regressive   (b) Regressive   (c) Regressive

45. (a) 104%   (b) 1.04   (c) 2   (d) Multiply by 1.04^8.

47. (a) 0.5%   (b) 1.005   (c) 12   (d) $6,370.07

49. (a) 6.5 billion   (b) 7.5 billion   (c) 11.7 billion

51. Chicken salad

7.5 Homework Exercises

1. (a) $\frac{1}{2}$ of 222 = 111   (b) $\frac{1}{4}$ of 6 = 1.5

(c) $\frac{1}{10}$ of 470 = 47

3. (a) 3/4 of 12 = 9
(b) 10% of 400 = 40 and 8 · 40 = 320

5. 40% of 202 million ≈ 4 · (10% of 200 million) =
4 · 20 million = 80 million

7. (a) 44   (b) 320   (c) 550

9. (a) Two rectangles   (b) Four rectangles
(c) Three rectangles

11. 100; 2; left

13. (a) Remove two zeroes and multiply by 8: 32
(b) Remove two zeroes and multiply by 7: $63

15. $33\frac{1}{3}\%$

17. (a) 60% of 100,000 = 60,000
(b) 64% of 95,000 ≈ $\frac{2}{3}$ of 90,000 = 60,000

19. (a) Unreasonable, 15% of 724 would be less than 724
(b) Reasonable, $\frac{1}{5}$ of 60,000 = 12,000
(c) Unreasonable, 86% of 94 would be less than 94

21. $\frac{34}{48} \approx \frac{3}{4} = 75\%$

23. Double tax to $3 and add a little to get $3.25; 10% of
bill is $2.10 plus half ($1.05) makes $3.15;
$\frac{1}{6}$ of $21.00 ≈ $3.50

25. About 58%

27. (a) The new is $3,000 more than the old.
(b) The new is 10% higher than the old.

29. (a) $12.78
(b) $13.29

31. (a) 1.06   (b) 1.12

33. (a) Should be $42\frac{6}{7}$

(b) It is 30% of the unknown, larger number, not 30%
of 30.

35. $16

37. *Example:* Which has a higher percent discount, Hippo
or Bolt shoes?

39. (a) First row: $0, $100, $200, $300, $400;
second row: $0, $70, $140, $210, $280
(b)

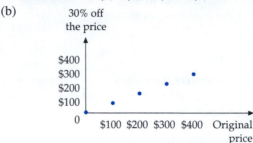

30% off
the price

(c) About $320

41. (a) Prices will rise in the next year.
(b) No, *b* is 17.6% higher than *a*.

43. 45% of 689 is close.

45. 4%

47. $48,500

49. $\dfrac{100(M - N)}{N}$

51. (a) The value of the REARGUARD fund on 1-1-04.
(b) D3 = C3 − B3
E3 = (C3 − B3) $* \dfrac{100}{B3}$

(c)

D	E
$337.49	4.0
−$600.65	−4.8

7.6 Homework Exercises

1. (a) $0.\overline{6}$ (b) $2.\overline{2}$

3. (a) $\dfrac{731}{1000}$ (b) $-13\dfrac{1}{25}$

5. $0.3\overline{4}$

7. (a) $\dfrac{4}{9}$ (b) $\dfrac{32}{99}$ (c) $\dfrac{267}{999} = \dfrac{89}{333}$

9. (a) $\dfrac{19}{90}$ (b) $\dfrac{338}{990} = \dfrac{169}{495}$

11. (a) $2.\overline{2}$ (b) $9N = 2$ (c) $\dfrac{2}{9}$ (d) $\dfrac{31}{99}$

13. A square has an area of 5. How long are its sides?

15. (b) $\sqrt{5}$ is an infinite decimal.

17. 24.2 in.

19. (a) and (c) are irrational; (b) and (d) are rational.

21. (a) Irrational   (b) Rational
     (c) Irrational   (d) Rational

23. You must convert a number to fraction form to determine if it is rational.

25. (a) 25 mph   (b) 36 mph
     (c) Irrational and rational

27. "Terminating" and "repeating" match with rational. "Infinite nonrepeating" matches with irrational.

29. (a) 3   (b) 5   (c) An infinite number

31. 1/3 or $\dfrac{3,333,333}{10,000,000}$

33. $\sqrt{13}$ is irrational and real; -6 is an integer, rational, real; $\sqrt{9}$ is whole, integer, rational, real; -0.317 is rational, real.

35.

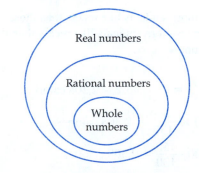

37. (a)

(b)

39. π or $\sqrt{12}$

41. Convert them both to decimals by dividing the numerator by the denominator. Then compare the size of the decimal values.

43. (b)

45. $\dfrac{1}{3} + \dfrac{1}{6}$

47. (a) $\dfrac{22}{7}$

49. π is an infinite nonrepeating decimal.

51. Commutative, addition

53. 0.41798

55. $3 - 2 \neq 2 - 3$

57. $6 \times 8 = 48$ plus $0.5 \times 8 = 4$ and $48 + 4 = 52$

59. (a) Some pairs   (b) All pairs
     (c) No real numbers   (d) Some pairs

61. (a) $\dfrac{1}{5^3} = 0.008,\ \dfrac{1}{5^4} = 0.0016$; both true

  (b) $\dfrac{1}{5^N} = 2^N \times 10^{-N}$

63. (a) (1), (2), and (4)   (b) (1), (2), and (4)
     (c) $5,\ 5^2,\ 3^2,\ 2^3 \cdot 5^2,\ 7$ (d) 2, 5

65. Only if $M = 0$ or $N = 0$

67. (a) $0.111111\ldots$ (b) Irrational, rational

69. Assume $\sqrt[3]{2}$ is rational. Then, $\dfrac{p}{q} = \sqrt[3]{2}$, in which p and q are counting numbers. Then $p = \sqrt[3]{2}q$ or $p^3 = 2q^3$. If p^3 is the cube of a counting number, then the number of prime factors it has is a multiple of 3. $2q^3$ has a number of prime factors equal to a multiple of 3 plus 1. So p^3 and $2q^3$ are equal numbers that have a different number of prime factors. This is impossible, so $\sqrt[3]{2}$ must be irrational.

Chapter 7   Review Exercises

1. (a) $(3 \times \dfrac{1}{1000}) + (7 \times \dfrac{1}{10,000})$
     (b) 3.7×10^{-3}

2. (a) Because $3 < 25$

(b) Represent 0.3 as 3 columns. Represent 0.25 as 25 small squares.

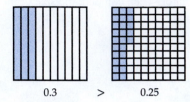

0.3   >   0.25

(c) 0.3 is located to the right of 0.25 on the number line. So $0.3 > 0.25$

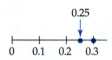

0.25

0   0.1   0.2   0.3

(d)

ones	tenths	hundredths
0 .	3	
0 .	2	5

Since 0.3 has a larger number in the tenths column, $0.3 > 0.25$.

3. By the addition rule, $4^2 \cdot 4^0 = 4^2$. Therefore, $4^0 = 1$. $4^3 = 64$, $4^2 = 16$, $4^1 = 4$. Each answer is 1/4 of the previous one. So $4^0 = 1$.

4. Move the decimal point two places to the left. Answer: \$0.0076.

5. 2.37×10^7

6. Work out the problem with a decimal square picture, and try to correct your mistake.

7.

$0.7 - 0.2 = 0.5$

8. (a) 0.2×0.4 means 0.2 of 0.4. Shade 0.4 of a decimal square.

Now darken 0.2 of the 0.4.

0.08 of the decimal square is darkened. So $0.2 \times 0.4 = 0.08$.

(b) $0.2 \times 0.4 = \dfrac{2}{10} \times \dfrac{4}{10} = \dfrac{8}{100} = 0.08$

(c) 2 tenths $\times$ 4 tenths = 8 hundredths = 0.08

9. $12.3 \div 0.2 \approx 12 \div \dfrac{1}{5} = 60$. The decimal goes to the right of the 1.

10. $42 \div 0.06 = \dfrac{42}{0.06} = \dfrac{42}{0.06} \times \dfrac{100}{100} = \dfrac{4200}{6} = 4{,}200 \div 6$

11. (a) 7.21   (b) 0.12   (c) 0.095

12. Division, partition measures

13. Subtraction, compare measures

14. 0.04

15. (a) Divide it into fifths, four for A and one for B.

(b) $\dfrac{4}{5}$ (c) $\dfrac{1}{4}$

16. (a) 12 more students like recess than those who do not.

(b) 3 times more students like recess than those who do not.

17. (a) $\dfrac{2000}{130} = \dfrac{x}{150}$ and $x = 2{,}308$ calories

(b) $150 \times \dfrac{2000}{130} = 2{,}308$ calories

(c) $2000 \cdot \dfrac{150}{130} = 2{,}308$ calories

18. Maria had 80 peanuts, and Debbie had 16.

19. $\dfrac{GC}{3}$ gallons

20. (a) 60%   (b) $\dfrac{2}{5}$ (c) 3 to 2

21. (a) 13.6   (b) 1.2%   (c) 215

22. Two columns of a decimal square represent 43. How much is the whole decimal square? $43 \cdot 5 = 215$.

23. 13.6   (b) 1.2%   (c) 215

24. Interest: $360; balance: $8,360

25. Draw $2\frac{1}{2}$ rectangles.

26. 40% of $7,000 = 4 \cdot (10\%$ of $7,000) = 4 \cdot \$700 = \$2,800$

27. (a) 27% of $489 \approx 30\%$ of $500 = 3 \cdot (10\%$ of $500) = 3 \cdot 50 = 150$

(b) 27% of $489 \approx \frac{1}{4}$ of $480 = 120$

28. (a) $30 = 80x$ so $x = 0.375 = 37.5\%$

(b) $\frac{110}{80} = \frac{x}{1}$ so $x = 1.375$, a 37.5% increase

29. 1.12

30. 6%

31. (a) $\frac{78}{99} = \frac{26}{33}$ (b) $\frac{924}{999} = \frac{308}{333}$

32. (a) Irrational   (b) Irrational
(c) Rational   (d) Rational

33. The irrational number π is the ratio of the circumference of a circle to its diameter.

34. (d) and (e)

35. (b), (c), (e)

36. The set of real numbers is the union of the set of rational numbers and the set of irrational numbers. The set of rational numbers contains all terminating and repeating decimals, while the set of irrational numbers contains all nonrepeating decimals.

37. (d)

38. $3(4 - 2) = (3 \cdot 4) - (3 \cdot 2)$

39. Additive inverses

Chapter 8

8.1 Homework Exercises

1. Earth measure

5. (a) Pentagon   (b) Triangular prism
(c) Line segment

7. (d)

9. A point has no length or width.

11. (a) False   (b)

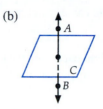

13. (a) 3   (b) 1   (c) 2

15. A line segment has two endpoints; a line has none.

17. (a) A
(b) Different endpoints, opposite directions

19. (a) $\angle BAC$ (b) $\overrightarrow{BD}$ (c) A

21. (a) 121°   (b) 32°

23.

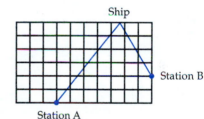

25. (a) $\angle A$ appears to have longer sides.
(b) Put the vertex of $\angle B$ on top of the vertex of $\angle A$ Compare the opening of each angle.

27. W, M, Z, N

29. (a) $\angle PIN$ and $\angle PIG$ or $\angle NIR$ and $\angle RIG$
(b) $\angle PIR$ and $\angle RIG$
(c) See the answer to part (a).

31. False

33. $\angle C$ and $\angle D$

35. $\overline{FG}$

37. (a) True
(b) Two parallel lines on the floor of a room that are perpendicular to a third line
(c) A carpenter could tell that two lines are parallel if they both form a right angle with a straight-edged tool.

39. Yes. See if the child can explain how the distance is measured.

41. A line running straight back from the front to the back of the parked car is perpendicular to the line of the curb.

43. (a) 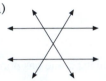   (b)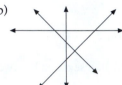

45. (a)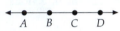

(b)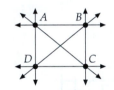

(c)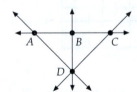

47. (a) 1   (b) 1   (c) 1 and 3

49. (a) 5 A.M. and 7 A.M.   (b) 2:30 A.M. and 9:30 A.M.

51.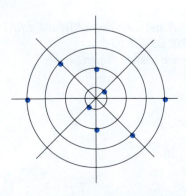

8.2 Homework Exercises

1.

3. (b) is a polygon. (a) is not a plane figure. (c) is not closed or simple. (d) is not formed by line segments.

5. (a) Octagon   (b) Trapezoid

7.

9. (b) *Hint:* Make a trapezoid.

11. (b) and (c)

13. (a)

(b) 9

15. (a) 6   (b) 6   (c) 8   (d) $2n$

17. (a) $15 + [15(12)/2] = 105$

8.3 Homework Exercises

1. Stability

3. (a) Isosceles and equilateral
(b) Isosceles and acute
(c) Scalene and obtuse

5. (a)   (b)

7.

9. Have the child rotate a plastic rectangle and discuss whether its shape changes.

11.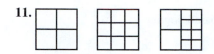

13. (a) 4 from Group 1, or 2 from Group 1 and 2 from Group 2
(b) 1 from Group 3 with 1 or 2 from Group 1 and 2 or 1 from Group 2
(c) 4 from Group 1

15. All of them

17. (b) They bisect each other.
(c) They are perpendicular.

19. (a) Can   (b) Must   (c) Can
(d) Must   (e) Can

21. Four sides; opposite sides congruent; opposite angles congruent; opposite sides parallel; angle sum is 360°.

23. Rectangle, parallelogram, pentagon, hexagon, triangle, kite, heptagon

25. (a) No, could have more than four sides.
(b) Yes   (c) Yes

27. Three

29. (a)   (b)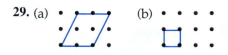

31. All of them

33. A square satisfies the definition of a rectangle. It is a parallelogram, and it has four right angles.

35. All squares are rectangles.

37. (b);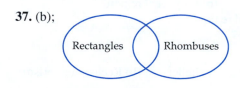

39. (a) S (b) Yes   (c) Q

41. Points; 8 in.; N

43. (a) $m\angle DCF = 90°$ (b) Isosceles   (c) $\overline{BG}$

45. Uniform curvature

47. (d) A parallelogram   (g) Induction

49. (a)

51. (a)   (b)
(c)   (d)
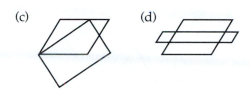

53. (e) The angle sum is 180°.

55. (c)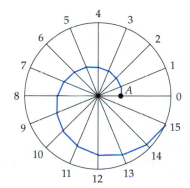

8.4 Homework Exercises

1. (c) The sum of the angles should be close to 180°, but it may not be exactly 180° because of drawing and measuring error.

3. A triangle with two right angles would have an angle sum greater than 180°

5. $m\angle PKA = 48°$, $m\angle PAK = 22°$, $m\angle AKR = m\angle R = m\angle KAR = 60°$

7. Their measures add up to 180°.

9. Divide the hexagon into 4 triangles as shown.

The sum of the angle measures of the 4 triangles is $4 \cdot 180° = 720°$. Therefore, the sum of the angle measures of the hexagon is also 720°.

11. The angles of the five triangles do not form the angles of the pentagon.

13. $38(180°) = 6840°$

15. (a) 120°   (b)

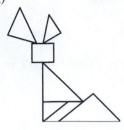

17. (a)

(b) The other six are: (1) 2 octagons, 1 square, (2) 1 square, 1 hexagon, 1 dodecagon, (3) 2 squares, 3 triangles, (4), 2 squares, 3 triangles (a second way), (5) 1 hexagon, 4 triangles, (6) 2 dodecagons, 1 triangle.

19. (a) Yes   (b) Yes

21. (a) 6   (b) 10   (c) 15

(d) $\dfrac{n(n+1)}{2}$ for n small rectangles

(e) Induction

(f) 36

23. (a) $\dfrac{(n-2)180°}{n}$

(b) Because $\dfrac{(n-2)}{n}$ is less than 1

25. 28

27. (b) Triangle and parallelogram   (c) $\dfrac{1}{8}$

(d)   (e)

(f)

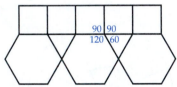

8.5 Homework Exercises

1. (a) False
(b) The three line segments that meet at the corner of a room

3. (a) An infinite number   (b) One   (c) One
(d) An infinite number

5. False

7. True

9. (a) Infinite number   (b) Infinite number
(c) One or infinite number

11. (a) $\overleftrightarrow{AB}$ and $\overleftrightarrow{EH}$ (b) $DCGH$
(c) ⊥   (d) $\overleftrightarrow{FG}$

13. (a) False
(b) Two parallel lines on the floor and a line from the ceiling to the floor that intersects one of them

15. True

17. (a) Polygon   (b) Polygonal region
(c) Polygonal region

19. (a) and (d)

21. (a) Right circular cylinder   (b) Rectangular prism

23. (a) Rectangle   (b) Isosceles

25. Two parallel faces

27. (a) 8s   (b) 18   (c) 12

29. 6 vertices, 6 faces

31. (a) $F = 9$, $V = 9$, $E = 16$
(b) $F = 7$, $V = 10$, $E = 15$
(c) $F = 18$, $V = 14$, $E = 30$

33. (a) Cone, cylinder   (b) Cylinders

35. Circle, can, two flat sides

37. Pentagonal pyramid

39. Uniform curvature

41. Not perfectly round or smooth

43. (a) Pentagonal prism   (b) Hexagonal pyramid
(c) Hexagonal prism

45. (a) 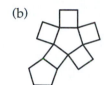   (b)

47. (a), (c), (d)

49. (a) Parallelogram
(b) The cylinder holds its shape more firmly.

51. (a) Plane   (b) Plane   (◀ Line

53. (a) $n + 2$ (b) $2n$ (c) $3n$ (d) Yes

55. (a) Octahedron   (b) Tetrahedron   (c) Cube
(d) y vertices, x faces, z edges

57. (c) The line on the regular loop is on one side. The line on the Möbius strip goes all around it on both "sides." It has only one side!

8.6 Homework Exercises

1. (b) CD is longer than AB.
(c) We tend to look for depth, a third dimension.

3. The room is not rectangular. The back of the room on the right side is closer to the camera.

5. (a)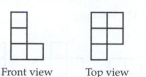
Front view   Top view   Right view

(b)
Front view   Top view   Right view

7. (a)   (b)

9.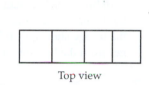
Front view   Top view

11. (a)
Top   Front   Right

(b)
Top   Front   Right

13.

15. Right view goes up 4, but the front view doesn't.

17. Hexagonal prism

19. (a) Triangle   (b) Rectangle

25. (b)

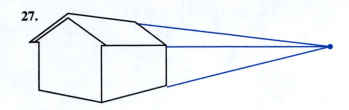

27.

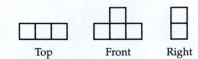

31. (a) Drawing 2   (b) Drawing 1

33. 4, 5, 7, 6, 2, 3, 1

35. The fisherman in the bottom right reaches the water with his pole. The man in the upper right on the hill lights the woman's lantern. The man on one side of the bridge shoots his rifle on the other side. The banner from the building hangs out among the trees.

37.

39.

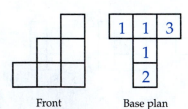

43. (a) 64   (b) 24   (c) 24   (d) 8
    (e) 0   (f) 8

45. (a) 96, 48, 8, 0, 64
    (b) $6(N-2)^2$, $12(N-2)$, 8, 0, $(N-2)^3$

Chapter 8 Review Exercises

1. To define space figures properly, one must first study the components of these figures.

2. It is finite, and it has thickness.

3. (a) $\angle ABC$ (b) { }

4. $m\angle A = m\angle C$ or $\angle A \cong \angle C$

5. (a) 26°   (b) Acute   (c) 64°

6.

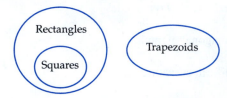

7. No

8. (a) Some   (b) All

9. (a)   (b) 5

10. (a) No, could be any parallelogram
    (b) Yes   (c) Yes
    (d) No, could be a trapezoid with one of the bases congruent to the two nonparallel sides

11.

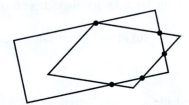

12. Could be a sphere (in three dimensions)

13. A pentagon can be divided into three triangles as shown. The angle sum of each triangle is 180°, so the angle sum of the pentagon is $3 \cdot (180°) = 540°$.

14. $\dfrac{7(180°)}{9} = 140°$

15.

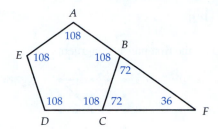

16. No, it may not have four congruent angles.

17. Yes;

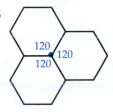

18. 2 hexagons (2 · 135°) + 1 square (90°) around each point

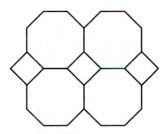

19. An infinite number

20. False

21. False

22. 7 faces, 7 vertices, and 12 edges

23. 9

24.

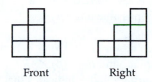

Front   Right

25.

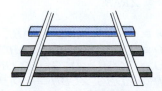

Chapter 9

9.1 Homework Exercises

1.

3. (a)   (b)

(c) $(a + 2, b - 4)$

5. (a)

(b) $A' = (-3, 3), B' = (-1, 0), C' = (-3, -1)$

(c)

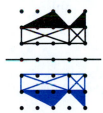

(d) Translation, left 4 units and up 1 unit

7.

9. (a)   (b)

 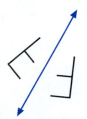

$(-1, -2)$ $(2, -3)$

(c) $(a, -b)$

11.

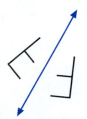

13. Draw a line segment from a point to its image point. Draw the ⊥ bisector of the line segment.

15. When it's reflected in someone's rearview mirror, it says AMBULANCE.

17. (d)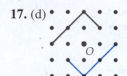

19. (a) 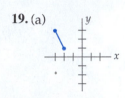   (b)

(c) $(-b, a)$

21. (a) C (b) F (c) E (d) E

23. (a) 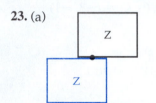   (b)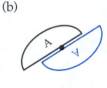

25. (a) 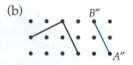   (b)

(c) 90° rotation counterclockwise around C which is 1 up and 1 right of B

27. (a) The same thing!   (b) TIM

29. (a) Rotation   (b) Reflection   (c) Translation

33. (a) Translation
(b) 180° rotation around a point midway between the bottom of each F.
(c) Not congruent

35. (a) 180° rotation around E (b) line m

37. $\overleftrightarrow{AB}$; $\overrightarrow{P'B}$

39. (b) Only the first pair is congruent.

41.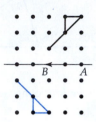

43. (a) 180° rotation   (b) Reflection

45. The boy is photographed in one mirror while holding another.

47. Translation

49. You input a shape, and the transformation produces a unique image.

53. (a) (1) and (2); Glide reflection
(c)

9.2 Homework Exercises

1. (a) ∠1, ∠3 or ∠2, ∠4 or ∠5, ∠7 or ∠6, ∠8
(b) ∠4, ∠6 or ∠3, ∠5
(c) ∠1, ∠5 or ∠2, ∠6 or ∠3, ∠7 or ∠4, ∠8
(d) ∠1, ∠2 (possible answer)
(e) $m\angle 1 = m\angle 5 = m\angle 7 = 110°$;
$m\angle 2 = m\angle 4 = m\angle 6 = m\angle 8 = 70°$
(f) ∠1, ∠2 (possible answer)

3. (a) 180° turn around E (b) ≅ in both blanks
(c) 180° turn around E (d) ∠ART
(e) A half turn around E indicates that
∠BTR ≅ ∠RAB.

5. (a) ∠ACB ≅ ∠AGF, ∠ADE ≅ ∠AHI,
∠ACD ≅ ∠AGH, ∠ADC ≅ ∠AHG
(b) ∠BCG ≅ ∠CGH, ∠DCG ≅ ∠CGF,
∠EDH ≅ ∠DHG, ∠CDH ≅ ∠DHI.

7. (a) ∠2, ∠8   (b) They are congruent
(c) ∠1 ≅ ∠3 and ∠3 ≅ ∠5 and ∠5 ≅ ∠7, so ∠1 ≅ ∠7.

9. (a) If they have one side of equal length
(b) All rays are congruent.

11. (a) $\overline{DP}$ (b) ∠ESU

13. (a) Rotate 120° clockwise around a point where their "noses" meet.
(b) Reflect through the body line that passes between the eyes.

19.

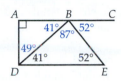

21. ∠1 ≅ ∠5 and ∠5 ≅ ∠3 so ∠1 ≅ ∠3.

9.3 Homework Exercises

1. (b) By the construction, $AN = CD$, $MA = BC$, and $MN = BD$. Then △MAN ≅ △BCD by SSS.

3. (a)

(b) It indicates that a quadrilateral is not rigid.

5. (b) *Hint:* Use SSS.   (c) Deduction

7. (5, −2), (5, 4), (7, −2), or (7, 4)

11. (a) No   (b) Yes   (c) No   (d) Yes (SAS)

13.

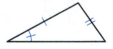

15. (b) *Hint:* Use SSS.

17. (a)

(b) Bisect $\overline{BC}$ to locate the midpoint. Then connect the midpoint to A.
(c) The three medians intersect at one point called the centroid.

19. *Hint:* Use the span of the compass.

21. (b) Fold one side of the angle onto the other.

25. Draw the angle bisector of ∠A, and label the point where it intersects side $\overline{BC}$ as point D. Since $AD = AD$, ∠B ≅ ∠C, and ∠BAD ≅ ∠CAD, △ABD ≅ △ACD by AAS. Since the triangles are congruent, corresponding sides $\overline{AB}$ and $\overline{AC}$ are congruent.

27. (a) The missing acute angles each measure 45°.

(b) 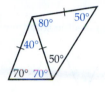   (c)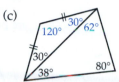

35.

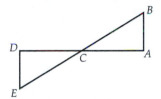

$AC = CD$, ∠A ≅ ∠D, ∠DCE ≅ ∠ACB. Therefore △ACB ≅ △DCE by AAS. Then $DE = AB$.

37. △ABC ≅ △EFG by SAS. So $AC = EG$ and ∠BAC ≅ ∠FEG. By subtraction, ∠DAC ≅ ∠GEH. Then △ADC ≅ △EHG by SAS. This means $DC=HG$ and ∠H ≅ ∠D. Since ∠ACD ≅ ∠EGH and ∠ACB ≅ ∠EGF, we can add to obtain ∠BCD ≅ ∠FGH. Therefore, $ABCD ≅ EFGH$.

39. *Hint:* Let ∠A = $x°$. Answer: 20°

41. (f) Yes

43. (e) Points on the perpendicular bisector are equidistant from A and B.

9.4 Homework Exercises

1. Yes; no

3. (a)

5. (d) The two lines of symmetry are perpendicular. The two lines of symmetry bisect the angles formed by the intersecting lines.
(e) Induction

7. (a) 4   (b) 2   (c) 2   (d) 0

9. Spells BIKE

11. If the rectangle is reflected through the diagonal, the image rectangle will not be in the same position as the original rectangle.

13. 0, 1, 2, 3, and 6 are possible. Possible solutions are:

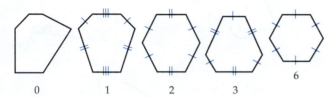

0   1   2   3   6

15. (a) 3   (b) 1   (c) 1   (d) 1

17. (a) and (c) have rotational and reflection; (b) has reflection.

19. (a) U, I, D (b) S, I

21. Usually: (a) all 2s, 4s, 10s, Js, Qs, Ks and the A, 3, 5, 6, 8, and 9 of diamonds
(b) None

23.

25. (Number of rotational symmetries) = (number of reflection symmetries) − 1

27. (a)   (b)

29. (a) They are congruent.
(b) They are congruent.

31. (a) 4   (b) 1

33. Right circular cone

35. 3 and 1

37. It may be possible to install it in two different ways.

39. (a) Usually rotational
(b) Can be laid down in a number of ways

41. (a) A tennis ball has rotational and reflection symmetry if one ignores the ridges; a tennis racket has reflection symmetry; a tennis court has rotational symmetry, and the surface has reflection symmetry.
(b) The tennis ball will bounce well and can be hit at any point without it making a difference; one can strike the ball with either face of the tennis racket; the court on either side of the net is the same, and the court is comparable on the left and right sides on each side of the net.

43. (a) Reflection   (b) Commutative   (c) Yes

45. Fold the circle in half from two different positions to create two different diameters. The intersection of the two diameters (folds) is the center.

47. (a)

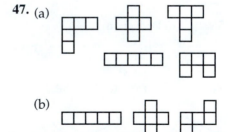

(b)

9.5 Homework Exercises

1. (a) Corresponding sides are not proportional.
$$\frac{3}{6} \neq \frac{4}{7} \neq \frac{5}{8}$$

3. Yes

5. (a) $\angle P$ (b) $NA; PE$

7. No. Corresponding angles may not be congruent.

9. No

11. $x = 17\frac{1}{7}$ $y = 22\frac{6}{7}$

13. 1.6

15. $x = 5$, $y = 11\frac{2}{3}$

17. About 297″ or 24′ 9″

19. (a) 6.8 cm by 10.2 cm   (b) Yes   (c) 46%

21. *Hint:* 8 m would be represented by 2 cm. 10 m would be represented by 2.5 cm.

23. 16.9 ft

25.

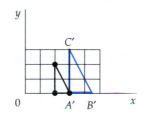

27.

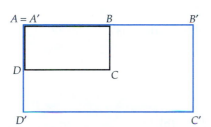

29.

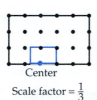

Center
Scale factor = $\frac{1}{3}$

31. (b) They are similar.
(c) Corresponding sides are proportional.
(d) Similar   (e) Induction

33. (c) Yes   (e) to be

35. 1.10

37. $13\frac{1}{2}$ in.

39. (b) Image is similar with scale factor = 3.
(c) Image is larger but not similar.

41. (g) They are similar.

Chapter 9   Review Exercises

1.

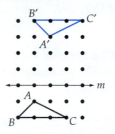

2.

3. (a) $A' = (-1, 2)$, $B' = (-6, 3)$, $C' = (-2, -3)$

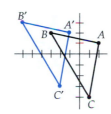

(b) Translation, left 3 units and up 1 unit

4. ⊥

5. Two figures are congruent if one can be mapped onto the other using a rotation, reflection, or translation.

6.

7. (a) ∠2, ∠4 or ∠5, ∠6 or ∠1, ∠3
(b) ∠4, ∠5
(c) ∠2, ∠5 or ∠4, ∠6
(d) $m\angle 4 = m\angle 5 = m\angle 6 = 64°$; $m\angle 1 = m\angle 3 = 116°$

8. (a) D

(b) Could say $\angle DAB \cong \angle BCD$, 180° turn around E

(c) $\angle AEB \cong \angle CED$, 180° turn around E

11. (b) From the construction, $AI = MI$, $TI = TI$, and $AT = MT$. So $\triangle AIT \cong \triangle MIT$ by SSS. Then, $\angle AIT \cong \angle MIT$, so $\overrightarrow{IT}$ is the angle bisector.

(c) Deduction

12. The angles of the hexagon are 120°; smaller angles of the triangle are 30°; the remaining two angles adjacent to the 30° angles are 90°.

13. Two, each passing through the midpoints of two opposite sides

14. A square

15.

16.

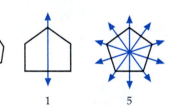

0   1   5

17. Yes, all their angles measure 108°, and corresponding sides are proportional.

18. 5

19. 26 ft 11 in.

20. $\dfrac{10L + 60}{L}$

21.

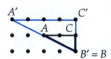

Chapter 10

10.1 Homework Exercises

7. (a) She may have counted 1, 2, 3, 4, 5, 6, 7.

(b) She counts unit marks instead of unit intervals.

9. So people can easily communicate measurements to one another

11. (a) 36   (b) 10,560   (c) $1\frac{2}{3}$

13. (a) The width of a finger

(b) The height of a doorknob

17. (a) Small paper clip   (b) An adult's foot

21. It will increase.

23. (a) 20   (b) 160   (c) 0.082

25. 20

27. 46 cm, 871 mm, 137 cm, 37 m, 3 km

29. 70 kg

31. (a) 5000   (b) Repeated measures

33. (a) 112   (b) 0.15   (c) 0.75

35. 3,240

37. (a) 9,400   (b) 0.037   (c) 82

39. 250 mL

41. 1.2 mL

43. (a) 12   (b) 2.5

45. (a) Slice of bread   (b) Carton of milk

47. (a) Meter stick   (b) 2 liters   (c) 21 by 28 cm

(d) 100 g

49. (a) g   (b) cm   (c) mm

51. (a)

53. (b)

55. (a) Approximate   (b) Exact

59. (a) 2   (b) 4   (c) 2

61. (a) 50 mm   (b) 49.5 and 50.5 mm

(c) 0.5 mm

63. (a) Nearest g; 61.5 and 62.5 g; 0.5 g
(b) Nearest hundredth of a cm; 5.115 and 5.125 cm; 0.005 cm
(c) Nearest tenth of a km; 4.55 and 4.65 km; 0.05 km

65. (a) 0.0081   (b) 0.00098   (c) 0.011

67. 0.17 km

69. (a) 3×10^{-6} sec   (b) 10^{-10} m
(c) 16 nanosec

10.2 Homework Exercises

1. (a) The second   (b) Height   (c) The first
(d) Area

3. 8.4 cm

5. (a) 12   (b) Largest: 16; smallest: 12

7. (a) The second has twice the perimeter.
(b) Same as (a)
(c) When the length of each side of a square is doubled, the perimeter also doubles.
(d) Induction

9. Because $2r = d$

11. 7,958 miles

13. (a) 5.8×10^8 miles   (b) 6.7×10^4 miles
(c) Gravity

15. (a) 325.7 m   (b) 6.3 m

17. $2x + 2y + 4$

19. 3

21. 4 square units

23. (a) 25 units^2 and 6 units^2 (b) The first

27. The space in its interior

29.
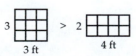

31. $50(10 + 20)$ or $50 \cdot 10 + 50 \cdot 20 = 1{,}500$ m^2

35. $0.2 \times 0.3 = 0.06$

37. (a) 9   (b) 9 ft^2 = 1 yd^2

39. (a) 100   (b) 0.0001

41. (a) 128   (b) About 162

43. Package B

45. 48 ft^2

47. (a) Draw two vertical line segments, or two horizontal ones, or one of each.
(b) $140,800

49. (a) 2.6
(b) Converting square units as if they were units

51. One answer is a 3-cm by 4-cm rectangle. Form other solutions by moving a square from one of the corners to various other positions where it shares one side with another border square.

53. Draw another square adjacent to the second highest square on its left.

55. Try to think of some more examples. Change both dimensions.

57. (a) 8 m   (b) 12 m   (c) $4\sqrt{N}$ m

59. (a) 4 by 4   (b) 3 by 6

61. (a) 4, 6, 8   (b) ⬚⬚⬚⬚ R4 ; 10
(c) $2N + 2$ (d) 22

63. Each rectangle could have $L = 42.5$ ft and $W = 60$ ft.

65. A line segment excluding its endpoints that connects (10, 0) and (0, 10).

67. (a) $a^2 + 4a + 4$ (b) $a^2 - 2ab + b^2$

69. The 500 miles is $\dfrac{7.5}{360}$ of the circle. The circumference would be $500 \cdot \dfrac{360}{7.5} = 24{,}000$ miles.

71. (a) The price/ft^2 of a 9-ft by 12-ft rug
   (b) C3 = A3*B3 and E3 = C3*D3
   (c) C3 = 108; E3 = $754.92; C4 = 200; E4 = $1798

10.3 Homework Exercises

1. (a) 6 square units   (b) 4.5 square units

3. (b)

5. (a) 50 in.^{2}, 32 in.   (b) 6 m^2, 12 m

7. See the answer to LE 4.

9. Put together two of the triangles as shown to form a parallelogram.

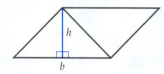

$A_\triangle = \dfrac{1}{2} A_{\text{parallelogram}} = \dfrac{1}{2} bh.$

11. (a) 360 in.^{2} (b) 12 square units   (c) $2.84

13. 2 square units

15. Put together two of the trapezoids as shown to form a parallelogram.

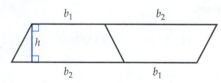

$A_{\text{trapezoid}} = \dfrac{1}{2} A_{\text{parallelogram}} = \dfrac{1}{2}(b_1 + b_2) \cdot h.$

17. About 250 mm^2

19. B ($28.57/m^2)

21. Squares and rectangles

23. (a) $\dfrac{1}{2} Cr = \dfrac{1}{2}(2\pi r) = \pi r^2$ (b) Deduction

25. (a) 20.25π square units   (b) 63.585 square units

   (c) $\left(\dfrac{8}{9} d\right)^2$ or $\left(\dfrac{16}{9} r\right)^2$ or $\dfrac{256}{81} r^2$

   (d) $\dfrac{256}{81} \approx 3.16$

27. (a) 1.96   (b) $11.76

29. (a) 2   (b) 4

33. No. Multiply $4 \cdot 10$ and take $\dfrac{1}{2}$ of the result. What do you get? Then discuss the distributive property.

35. (a) $\dfrac{8 + 2\pi}{3}$ (b) $\dfrac{40\pi}{3}$ square units

37. 100π ft^2

39. $7\pi \approx 22$ m^2

41. 2.25π square units

43. $1.5\pi + 1$ square units

45. $9\pi - 18$ square units

47. (c)

49. 56 square units

51. $A = \dfrac{C^2}{4\pi}$

53. (c) area $EFGH = \dfrac{1}{2}$ (area of ABCD)

10.4 Homework Exercises

1. (a) Right triangles
   (b) a and b are the lengths of the legs; c is the length of the hypotenuse.

3. (a) $\dfrac{1}{2} ab, \dfrac{1}{2} ab, \dfrac{1}{2} c^2$ (b) $\dfrac{1}{2}(a + b)(a + b)$

   (c) $\dfrac{1}{2} ab + \dfrac{1}{2} ab + \dfrac{1}{2} c^2 = \dfrac{1}{2}(a + b)(a + b)$

   $ab + \dfrac{1}{2} c^2 = \dfrac{1}{2} a^2 + ab + \dfrac{1}{2} b^2$

   $\dfrac{1}{2} c^2 = \dfrac{1}{2} a^2 + \dfrac{1}{2} b^2$

   $c^2 = a^2 + b^2$

   (d) Deduction

5. Draw a line segment that goes up 2 and over 3.

7. $3 + \sqrt{8} + 3 + \sqrt{8} = 6 + 2\sqrt{8}$ or $6 + 4\sqrt{2}$

9. 3.9 m

11. No

13. $4\sqrt{2}$ mph NE

15. 5 miles

17. $8\sqrt{20}$ or $16\sqrt{5}$ square units

19. 64π square units

21. 54m, 144m^2

23. (a) 10   (b) Finding the length of the hypotenuse

25. (a) Yes   (b) No

27. (a) Yes   (b) Yes
(c) If a, b, and c are the lengths of sides of a right triangle, then ka, kb, and kc ($k > 0$) also are the
lengths of sides of a right triangle.

29. Between 10 and 50 miles (inclusive)

31. (a) Obtuse   (b) Right   (c) Not   (d) Obtuse

33. (a) Yes   (b) 6, 8, 10; 5, 12, 13; 9, 12, 15
(c) If $a^2 + b^2 = c^2$ then $k(a^2 + b^2) = kc^2$ or
$ka^2 + kb^2 = kc^2$. So ka^2, kb^2, kc^2 is a Pythagorean
triple.

35. (a) $x = 3$, $y = 4$, $z = 5$; $x = 6$, $y = 8$, $z = 10$; $x = 9$,
$y = 12$, $z = 15$
(b) No

37. 50π square units

39. Triangles: 1 by 1 by $\sqrt{2}$, $\sqrt{2}$ by $\sqrt{2}$ by 2, and 2 by 2
by $2\sqrt{2}$; parallelogram: 1 by $\sqrt{2}$

41. $\sqrt{12}$ cm

43. (a) $a^2 + b^2 = c^2$
(b) They both equal $a^2 + b^2$.
(c) $\triangle GFE$; the SSS property
(d) m$\angle D$

45. (h) The two smaller areas add up to the largest area.

10.5 Homework Exercises

1. It affects the cost of the can.

3. (a)

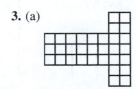

(b) 32 square units

5. (a) $1 \times 1 \times 20$; $1 \times 2 \times 10$; $2 \times 2 \times 5$
(b) $2 \times 2 \times 5$

7. 2700 cm^2

9. 9

11. 2 gallons

13. (a) 12,150 cm^2 (b) $24.30

15. $120 + 15\sqrt{24}$ or $120 + 30\sqrt{6}$ square units

17. 252π in.^{2}

19. A cylinder has 2 bases with area πr^2. The lateral surface is a rectangle with base $2\pi r$ and height h. The
area of the rectangle is $2\pi rh$. The total surface area is
$A = 2\pi r^2 + 2\pi rh$.

21. (a) 916 ft^2 (b) 2.29 gal   (c) 3

23. (a) $B + \dfrac{5}{2}sl$ (b) $5s$

25. (a)   (b) $\sqrt{5}$ m   (c) $16 + 8\sqrt{5}$ m^2

27. 9.75π in.^{2}

29. (a) $4x + 2$; $8x + 2$
(b)   ; $12x + 2$

F3

(c) $40x + 2$ (d) $4Nx + 2$

31. Multiplies by 9

33. (a) 16π and 64π in.^{2}; larger one has 4 times as much.
(b) The material for the larger ball costs about 4 times
as much.

10.6 Homework Exercises

1. 11 cubic units

3. (a) 30 cubic units
(b) The child counts the number of squares that are
visible.

5. B

7. (a) 1 cm^3 (b) 4000 g   (c) 140   (d) 1000 kg
(e) 1 kL

9. (a) Surface area   (b) Volume   (c) Volume

11. 1.30 m^3

13. 1544.3 cm^3

15. 3.54 m

19. (b) 355 cm^3

21. 16,200π in.^{3}

23. 1016 lb

25. 50$\sqrt{91}$ m^3

27. (a) Cones and pyramids
(b) Prisms and cylinders

29. (a) 115,640,000 ft^3 (b) 15,419

31. The base area of a cylinder is $B = \pi r^2$. So $V = \pi r^2 h$.

33. 5.8 in.^{3}

35. (a) 180   (b) Omitting π from the answer

37. 523.6 in.^{3}

39. 2,200,000 L

41. 47%

43. (a) 1, 2; 1, 8; 6, 24   (b) Volume

45. $r = 30$ ft; 113,097 ft^3

47. (b) Being closer to a cube

49. Lay it on its side.

51. (a) Least expensive   (b) $\dfrac{V}{\pi r^2}$
(c) $2\pi r^2 + 2\pi rh$ (d) $A = 106.18$ cm^2
(e) $r = 2.37$ cm, $h = 4.75$ cm

10.7 Homework Exercises

1. (a) 20   (b) 1:4   (c) 1:16   (d) 1:4

3. (a) 4:9   (b) 2:3   (c) 2.25

5. (a) 16 in.   (b) 9 in.

7. (a) 3   (b) 9

9. (a) 3:4   (b) 198 and 352 m^2 (c) 9:16
(d) It is the square of the ratio of the edges.
(e) 162 and 384 m^3 (f) 27:64
(g) It is the cube of the ratio of the edges.

11. (a) 4:1   (b) 64:1

13. (a) 3.5 m   (b) $3\frac{1}{16}$ (c) 64:343
(d) Deduction

15. (b)

17. 512

19. $A = 360$ ft^2, $V = 648$ ft^3

21. (a) $1\frac{1}{4}$ (b) $1\frac{9}{16}$ (c) $1\frac{61}{64}$
(d) The smaller one

23. (a) 3   (b) 9   (c) 27

25. Food consumption is proportional to volume. Cube the 12 from the ratio of lengths to obtain 1728.

27. (a) Volume   (b) $131.82

29. (a) 3   (b) 2

31. Larger

Chapter 10   Review Exercises

1. (a) 0.237   (b) 36,450   (c) 800,000

2. (c)

3. (b)

4. 0.05 m

5. The perimeter measures the sum of the lengths of all the sides (the border), and the area measures the size of the interior of the polygon.

6. 6.5 square units

7. 10.0 m

8. A

9. (a) 10,000   (b) 0.0008

10. Put together two of the triangles as shown to form a parallelogram.

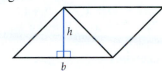

$$A_\triangle = \frac{1}{2} A_{parallelogram} = \frac{1}{2} bh.$$

11. 15.0 m^2

12. $100 - 25\pi$ m^2

13. $20 + 5\pi$ m

14. $625/\pi$ cm^2

15. 14 in. costs less ($0.07/in.^{2})

16. (a) In each triangle, $m\angle 1 + m\angle 2 = 90°$. So $\angle DAB$ is a right angle.
(b) c^2 (c) $EFGH$, $\triangle ADH$, $\triangle ABE$, $\triangle BCF$, $\triangle CDG$
(d) $b - a$ (e) $4\left(\frac{1}{2}ab\right) + (b-a)^2$

(f) $c^2 = 4\left(\frac{1}{2}ab\right) + (b-a)^2$
$c^2 = 2ab + b^2 - 2ab + a^2$
$c^2 = b^2 + a^2$

17. $\sqrt{149}$ m

18. $225\pi - 216$ square units

19. (a) No   (b) Acute   (c) Right

20. 72 units^2

21. Multiplied by 16

22. 7

23. (a) 6   (b) 10   (c) 14   (d) $4n + 2$
(e) Inductive

24. Prisms and cylinders

25. $V = 10\sqrt{20}$ or $20\sqrt{5}$ cubit units and
$A = 50 + 9\sqrt{20}$ or $50 + 18\sqrt{5}$ square units

26. 432π ft^3 and 216π ft^2

27. 6 cm

28. $16:49$

29. 675 m^3

30. (a) 15.625   (b) 6.4 lb   (c) Deduction

Chapter 11

11.1 Homework Exercises

1. (a) General property   (b) Function
(c) Formula   (d) Unknown value

3. (a) Constant   (b) Variable

5. (a) Variable   (b) Constant   (c) Variable

7. X must represent a quantity.

9. (a) $a + 12$, in which a is the average temperature for this day in degrees
(b) 72°F

11. Total cost $C = 5X + 3Y$, in which X is the number of adult tickets and Y is the number of children's tickets

13. $P = 3L$, in which L is the length and P is the perimeter.

15. $y \geq 2x$, in which x is last year's height and y is this year's height

17. (a) $20 = 0.85P$, in which P is the list price in dollars
(b) The wholesale price, the quality of the item, what the item is

19. (a) $8 + x = 13$, $x = $ # of additional dishes needed
(b) $\$90/\$15 = x$, $x = $ # of CDs sold

21. (a) $200 - 5W$ lb   (b) Possible answer: 10

23. (a) $6c + 6 \cdot 5 = 6(c + 5)$
(b) Distributive property of multiplication over addition

25. (a) $y = 100 + 0.05x$ (b) $\$3{,}000$

27. (a) Less than 500 miles
(b) $120 + 0.32x < 280$; $x < 500$ miles

29. 725

31. $a(b + c) = ab + ac$, in which a, b, and c are real.

33. No. Pick numbers for x and n, and test your equation.

35. (a) R is 10 more than 22 times D.
(b) The rental rate is $\$10$ plus $\$22$ per day.

37. (a) C is more than twice P.
(b) The current U.S. population is more than twice the 1940 U.S. population.

39. (a) The car went 120 miles in 3 hours, traveling more quickly at the beginning and the end of the trip.
(b) 40 mph

41. (a) (3)   (b) (2)   (c) (1)

43. (a) Negative slope   (b) Positive slope
(c) Negative slope

45. (a)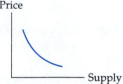

(b)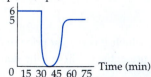

47. (a) 150 miles   (b) 150 square units

49. $S = 1.5x$, in which S is his new salary and x is his old one.

51. $\dfrac{300}{D}$

53. (a) $T = \dfrac{300}{S} + 1$ (c) $B3 = \dfrac{300}{A3} + 1$
(d) The total time decreases as the speed increases. The total time decreases more for a 5-mph increase at a lower speed than at a higher speed.

55. (c) $xy + (7 - x)(7 - y) + x(7 - y) + (7 - x)y = xy + 49 - 7x - 7y + xy + 7x - xy + 7y - xy = 49$

11.2 Homework Exercises

1.

3. (a) 12   (b) 3   (c) $x = -2$

5. (a) has one solution; (b) is true for all x and y; (c) is true for some x and y

7. Ask what $2x$ means. Then, what would you do to both sides of the equation to cancel out the 2?

9. (a) Left
(b) Left; move 3 to the right from x and y, and $x + 3$ is still to the left of $y + 3$.
(c) Left; move 3 to the left from x and y, and $x - 3$ is still to the left of $y - 3$.
(d) If $x < y$, then $x \pm c < y \pm c$ for all real numbers c.

11. $4x > 85$; $x > 21.25$; 22 visits

13. (a) $\dfrac{1}{2}$ (b) 15 ft

15. Ramp length is $\sqrt{36^2 + 3^2} \approx 36.1$ ft

17. (a) Slope > 0 (b) Slope < 0 (c) Slope $= 0$

19. $\overline{AB}$ has slope -1; $\overline{CD}$ has slope $\dfrac{1}{3}$; $\overline{EF}$ has slope 2.

21. (a) $\dfrac{2}{3}$ (b) $-\dfrac{4}{3}$ (c) $\dfrac{2}{3}$

23. (a)

Water

(b) 30   (c) About 30 quarts
(d) About 30 quarts/min

25. (a) A ray with endpoint at $(0, 0)$ going through $(1, 12)$
(b) 12   (c) Speed
(d) Time cannot be negative.

27. (a) The average revenue is \$4/customer.
(b) $\dfrac{R}{C} = 4$

29. They are all lines passing through the origin.

31. (a) Yes   (b) No

33. (a) $\dfrac{T}{5}$ (b) A line through $(0, 0)$ and $(5, 1)$

(c) Each change of 1 second is an additional $\dfrac{1}{5}$ mile of distance.
(d) The lightning (light) reaches one in virtually no time, so the formula simply estimates the distance the thunder (sound) travels at the rate of 0.2 mi/sec.

35. Up 5 units

37. Both have a slope of 1/2.

39. (a) $(4, 0)$ (b) Yes   (c) An infinite number

41. (a) Subtract \$1.80 to get \$1.89, and divide that by \$0.27 to get 7 photos
(b) $P = \dfrac{C - 1.80}{0.27}$
(c) Formula from part (b), $P = 7$

43. (a) $T = \dfrac{N}{4} + 40$ (b) $N = 4(T - 40)$
(c) $65°F$ (d) 220

45. 21.3 hours

47. $y = \dfrac{2}{3}x$

49. $y = 100 + \dfrac{x}{10}$

51. $W = 3H - 83$

53. (a)

Pool 3

(b)

Pool number (x)	1	2	3	4
Number of tiles (y)	8	12	16	20

(c) Linear   (d) $y = 4x + 4$ (e) Induction
(f) 52

55. (a) Yes   (b) No   (c) Yes

57. (a) No
(b) If the x-values also form an arithmetic sequence

59. (a) -1 and 1; $\dfrac{1}{2}$ and -2
(b) The product of the slopes is -1.

61. (a) m (b) $-\dfrac{1}{m}$

63. (a) A line through $(0, 0)$ and $(1, 2)$
(b) A line through $(0, 0)$ and $(-1, 2)$
(c) $y = -2x$

65. One is the image of the other for a reflection through the x-axis.

67. (c) They are equal.

11.3 Homework Exercises

1. (a) 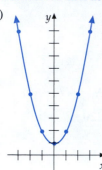   (b) Shift $y = x^2$ up 1 unit

(c) For $0 \le x \le 3$

3. Translate $y = x^2 + 1$ up 4 units to obtain $y = x^2 + 5$.

5. (c)

7. (a) 96 ft   (b)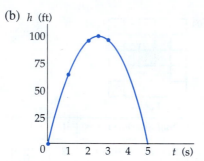

(c) $t = 0.7, 4.3$ s

9. (a)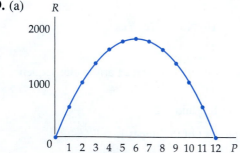

(b) $6

11. $y = -x^2 - 1$

13. y-intercept is 1; as x increases, y increases; $y > 0$

15. (a) $m = 50, 100, 200, 400$ (b) (3)   (c) 141

17. (a) $9621.41   (b) After 4.9 years

19. P is exponential; Q is neither; R is linear

21. (a) $0.37   (b) $1 < w \le 2$

(c)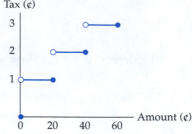

23. (a)

L	3	5	7	9	11
W	10	6	$\dfrac{30}{7}$	$\dfrac{10}{3}$	$\dfrac{30}{11}$

(b) $LW = 30$ or $W = \dfrac{30}{L}$

(c)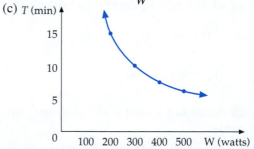

(d) It is all in the first quadrant, and y decreases less and less rapidly (from left to right). The curve gets very close to the x-axis.

(e) Any positive number

25. (a) $WT = 3000$ or $T = \dfrac{3000}{W}$ (b) 5 minutes

(c)

27. Parts (a), (b), and (c) come out the same for both expressions.

(d) Part (c)

29. (a) $\dfrac{y - y_1}{x - x_1}$ (b) $y = m(x - x_1) + y_1$

11.4 Homework Exercises

1. (a) $c_1 = 7x$, $c_2 = 50 + 4x$

(b) c_1 is lower when $x \le 16$; c_2 is lower when $x \ge 17$

3. (a) 1 orange is $0.39 → 3 oranges cost $1.17 →
   2 grapefruits cost $2.07 − $1.17 = $0.90 →
   1 grapefruit costs $0.45
   (b) $2x + 3y = 2.07$; $2y = 0.78$

5. (a) 3 apples cost $2.37 − $1.32 = $1.05 → 1 apple
   costs $0.35 → 1 pear cost $0.31
   (b) $5x + 2y = 2.37$; $2x + 2y = 1.32$

7. (a) Cube weighs 4; cylinder weighs 8; cone weighs 2

9. (a) A1 = $0.8(C − 500)$ and A2 = $0.6C$
   (b) Ray A1 starts at $(500, 0)$ and passes through
   $(1000, 400)$. Ray A2 starts at $(0, 0)$ and passes
   through $(1000, 600)$.
   (c) $(2000, 1200)$
   (d) The first plan is better for claims less than $2,000;
   the second is better for claims of more than $2,000.
   (e) Service, choice of doctor

11. (a) A: $y = 20,000 + x$ B: $y = 5,000 + 2x$
   (b)

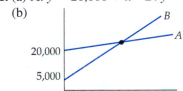

   (c) $(15000, 35000)$
   (d) Sign with A if you expect to sell fewer than 15,000
   CDs. Sign with B if you expect to sell more than
   15,000 CDs.

13. 1870 adults and 590 children

15. (b) B2 = $39.95 * 3$; C2 = $24.95 * 3 + .10 * (A − 100)$
   (c) Use Drive-U if I plan to drive 700 or more miles.
   Use Used-a-Bit if I plan to drive 500 or fewer
   miles, I could investigate more values between 500
   and 700 miles.

17. (a) $6,500 profit   (b) $100 profit   (c) $18.75 ≈ 19$

19. (a) Kari started 3 miles north and traveled 4 mph. Lisa
   started 9 miles north and traveled 3 mph.
   (b) After 6 hours

21. The diesel would be cheaper after driving about
   46,667 miles, so the diesel car is a better buy.

11.5 Homework Exercises

1. Form a right triangle with $C(x_2, y_1)$ as the third vertex.
   By the Pythagorean Theorem,
   $$AB = \sqrt{(AC)^2 + (CB)^2} = \sqrt{x_2 - x_1)^2 + (y_2 - y_1)^2}$$

3. $3 + \sqrt{8} + \sqrt{5}$ or $3 + 2\sqrt{2} + \sqrt{5}$

5. $(−1, 0), (1, 4), (3, −4)$

7. $(3, 3)$

9. (a) (a, b)
   (b)
   $$AC \overset{?}{=} BC$$
   $$\sqrt{(a − 0)^2 + (b − 0)^2} \overset{?}{=} \sqrt{(a − 0)^2 + (0 − b)^2}$$
   $$\sqrt{a^2 + b^2} = \sqrt{a^2 + b^2}$$
   (c) $\overline{AC}$ and $\overline{BD}$ bisect each other, since the midpoint
   of $\overline{AC}$ is $\left(\frac{a}{2}, \frac{b}{2}\right)$ and the midpoint of $\overline{BD}$ is $\left(\frac{a}{2}, \frac{b}{2}\right)$.

11. (a) $D(\frac{a}{2}, \frac{b}{2})$, $E(\frac{a + c}{2}, \frac{b}{2})$
   (b) Slope of $\overleftrightarrow{DE}$ = slope of $\overleftrightarrow{AC}$ = 0
   (c) $DE = \sqrt{\left[\frac{(a + c)}{2} - \frac{a}{2}\right]^2 + \left(\frac{b}{2} - \frac{b}{2}\right)^2}$
   $$= \sqrt{\frac{c^2}{4}} = \frac{c}{2} \text{ and } AC = c, \text{ so } DE = \frac{1}{2}AC.$$

13. 32 min; 2

15. (a) A network with a path that passes through each
   edge exactly one time
   (b) A traversible network with a path that starts and
   ends at the same place

17. (a) (1), (2)   (b) (2)

19. 12

21. Install a fourth locaiton

23. 8 square units

25. (a) $(x, −2)$
   (b) $\sqrt{x^2 + (y − 4)^2} = \sqrt{(0)^2 + (y + 2)^2}$
   (c) $x^2 = 12y − 12$ (d) Deduction

27. $(3.5, 2)$

Chapter 11 Review Exercises

1. (a) $x + y = y + x$ (b) $x + y = 3$

2. (a) $J > 2S$, in which J = Joe's weight and S = his
   sister's weight.
   (b) Untranslatable

3. (a) 50 mph   (b) No

4. (a)   (b)

5.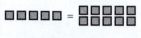

$3x = 4 + 10$

    $3x = 6$

$x = 2$

6. (a) $y = 8x$ (b) 8   (c) Yes, $\dfrac{y}{x} = 8$

7. (a) Line through $(0, 1)$ and $(1, 3)$
(b) $y = -2x + 1$

8. $y = -4x + 5$

9. (a) $h = 10 + 0.5n$
(b) A ray with endpoint $(0, 10)$ through $(2, 11)$
(c) A ray

10. (a) $E = H + 2T$ (b) 16

11. $40

12. $8445.95

13. (a) For each increase of 1 inch in height, there is a 5.5-lb increase in weight.
(b) $H = \dfrac{W + 220}{5.5}$ (c) The original

14. (a) $C = 2.10 + 0.80d$
(b)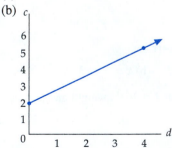
(c) 12 miles

15. It is $y = x^2$ translated up 2 units.

16.

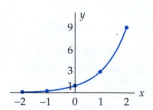

17. (a) $t = 0, 22, 44, 66; p = 20, 40, 80, 160$ (b) (3)
(c) 467,000

18. P is neither; Q is exponential; R is linear

19. (a) Three peaches cost $2.40 - $1.44 = $0.96. One peach costs $0.32. One plum costs $0.24.
(b) $6x + 2y = 2.40, 3x + 2y = 1.44$

20. (a) $C_1 = 470 + 82T, C_2 = 398 + 106T$
(c) The Twister costs less for up to 3 years. The Clothes Horse costs less for more than 3 years.
(e) $470 + 82T = 398 + 106T$
$72 = 24T$
$3 = T$

21. Form a right triangle with A, B, and C (x_2, y_1). The lengths of the legs are $(x_2 - x_1)$ and $(y_2 - y_1)$. By the Pythagorean Theorem, $AB = \sqrt{x_2 - x_1)^2 + (y_2 - y_1)^2}$.

22. (a) $(a, 0)$ and $(0, b)$
(b) The midpoint of $\overline{AC}$ is $\left(\dfrac{(a + 0)}{2}, \dfrac{(b + 0)}{2}\right)$ and the midpoint of $\overline{BD}$ is $\left(\dfrac{(a + 0)}{2}, \dfrac{(0 + b)}{2}\right)$. So they both have the same midpoint $\left(\dfrac{a}{2}, \dfrac{b}{2}\right)$.
(c) $\sqrt{(a - 0)^2 + (b - 0)^2} = \sqrt{(a - 0)^2 + (0 - b)^2}$
$\sqrt{a^2 + b^2} = \sqrt{a^2 + b^2}$

23. 39 minutes, H to B to A to C to H

24. (a)   (b)

Chapter 12

12.1 Homework Exercises

1. (a) 43.5%   (b) $27,000   (c) $155,000

3. (a) Education   (b) Mechanical engineering
(c) The salaries range from $29,890 to $48,115, the highest starting salary being about 60% higher than the lowest. Engineering and technological fields tend to pay higher salaries.
(f) Average salaries in each field

5. (a) (2)   (b) (3)   (c) (1)

7. (a)

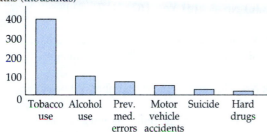

No. of premature deaths (thousands)

(c) Tobacco use is the major cause of premature deaths. Alcohol contributes to more deaths than illegal drug use. Suicide accounts for more deaths than murder.
(d) How numbers were derived, breakdown of types of murders.

9. (a) Class A

```
5 | 8
6 | 22
7 | 025
8 | 015
9 | 28
```
(b) Class B
```
4 | 2
5 |
6 | 558
7 | 5
8 | 02
9 | 019
```
(c) Class B has more extreme scores than Class A.

11. (c)

15. (a) Use 60–69, 70–79, 80–89, 90–99.
(b) Use 60–79 and 80–99.

17. 104.4°

19. (a) Food 25%, clothing 8.3%, rent 50%, entertainment 6.7%, other 10%
(b) Make the central angles: food 90°, clothing 30°, rent 180°, entertainment 24°, other 36°

23. (a) Wealthy households, especially the upper 1%, have had more income growth than middle-income households.
(b) Stock market gains, much greater salary increases for executives, limited salary increases for other workers

25. (a) Circle   (b) Line   (c) Bar

27. (a) Bar, circle, line   (b) Bar, stem-and-leaf
(c) Bar, line, stem-and-leaf

29. (a)

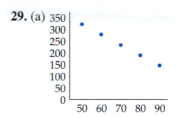

(c) Negative
(d) Decrease   (e) Negative   (f) About -4
(g) For each dollar increase in price, demand decreases by about 4.
(i) It's beyond the data range.
(j) $y = -4.27x + 532.3$

31. (d) About -38
(e) Distance traveled increases 38 miles for each gallon of gas used.
(f) After about 419 miles
(g) $y = -38x + 761$

33. (a) Positive   (b) Negative

35. (a) E, T, A, O
(b) *Hint:* The 11 most frequently occurring letters are A, D, E, H, I, L, N, O, R, S, and T.
(c) C—6, U—4, W—3, A—2, N—2, Q—1, P—1, T—1, I—1, X—1, G—1, Z—1, S—1
(d) C   (e) The meeting will be at the pier.

41. (a) $y = 204.0(1.01)^x$ (b) 304 million, 336 million

12.2 Homework Exercises

1. (a) Israel
   (b) Start the vertical axis at 0. The bars will be much closer in height.
   (c) U.S.   (d) Length of the school day

3. Carelessness, desire to deceive, trying to reduce the amount of empty space

5. 56%

7. $480

9. A percent is a ratio, so it would be more appropriate to average these numbers.

11. 4%

13. How far each car traveled

15. (a) 50%   (b) 0%

17. (d)

19. 10; 100

21. (a) No one sells them at list price.
   (b) The meat protein in steak has little to do with the appeal of steak.

23. (a) AIM with fluoride and AIM without fluoride
   (b) Fluoride is effective in reducing dental decay.

25. (a) The Vietnam War in the 1960s
   (b) The vertical axis does not start at 0.
   (c) Favors it

27. (a) 1950 to 1998   (b) 1978 to 1998
   (c) It increased steadily from 1950 to 1972, then leveled off until it began declining from 1976 to 1993. It increased slightly from 1995 to 1998.

29. Worsening external conditions such as water and air pollution, or few people can afford the new treatments, or medical improvements for heart disease are showing an even greater improvement

31. The university

33. The first region is 4 times bigger in area.

35. $P \rightarrow 1.2P \rightarrow 1.3(1.2)P = 1.56P$, which is 56% more than P.

12.3 Homework Exercises

1. (a) 15
   (b) Change two 15s into 6s or 8s.

3. $\{4, 4, 4, 5, 5, 6, 7\}$

5. (a) 10   (b) 22.5   (c) 68

7. (a) $\dfrac{(V + W + X + Y + Z)}{5}$ (b) $5M$

9. (a) Yes, $\{60, 60, 80, 80\}$ (b) Yes, $\{70, 70, 70, 70\}$
   (c) No   (d) Yes, $\{67, 71, 71, 71\}$

11. Mean: equal weight to each score; median: uses order of scores, middle; mode: most frequent

13. No. The most common number of books is 5.

15. (a) 14   (b) The numbers must be put in order.

17. In most large groups, close to 50% of the people score below average.

19. (a) Make stacks the size of each of the 5 numbers. The mode is the height (2) that occurs most frequently.
   (b) Put the stacks in size order. The median is the height (2) of the middle stack.
   (c) Even out the heights of the 5 stacks. The mean (3) is the height of all the stacks.

21. Compare the mean, median, and range of each set.

23. (a) 89   (b) $y = \dfrac{231 + t}{4}$
   (c) A line with slope $\dfrac{1}{4}$ and y-intercept $(0, 57.75)$
   (d) Her average could end up anywhere from 57.75 to 82.75.
   (e) Her average increases $\dfrac{1}{4}$ point for each additional point on the fourth test.

25. (a) 29   (b) $4 million   (c) $4.34 million
   (d) 9   (e) $4 million

27. 2.8

29. 85.4

31. Reduce it by 37.5%

33. Median

35. Mode

37. (a) Mean   (b) Median
(c) Mean decreases by 6.07; median decreases by 0.5

39. (a) Median   (b) Mean

41. Probably the mean

43. Median

45. $35,000

47. (a) 37.5 mph   (b) $\dfrac{10 + 10}{\dfrac{10}{M} + \dfrac{10}{N}}$ or $\dfrac{2MN}{M + N}$

49. 3.6

51. (a) Class 1: 16, Class 2: 11, Class 3: 7.5, Class 4: 7.5
(b) Class 1: 14.25, Class 2: 14.25, Class 3: 8.5, Class 4:5
(c) Class 4 prefers Trickledown's plan. Classes 2 and 3 prefer Dragdown's plan. Class 1 prefers to leave things as they are.
(d) 1   (e) 3 and 4

12.4 Homework Exercises

1. It may be 30 ft deep in a limited area and 2 ft deep everywhere else.

3. (c)

5. Lyle scored better in mathematics than 43% of those in his group.

7. (a) {28%, 35%, 38%, 46%, 52%}

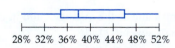

28% 32% 36% 40% 44% 48% 52%

(b) The U.S. is near the lower extreme.
(c) Relative size of the countries; military budgets

9. (a) *Hint:* The five-number summary for A is 56, 70, 81, 91, and 97, and the five-number summary for B is 68, 74, 80, 86.5, and 91.
(b) Class A's scores are more spread out.

11. (a) Lower extreme = 64, lower Q = 72, median = 80, upper Q = 86, upper extreme = 92
(b)

70 80 90 100 110

(c) The second box-and-whisker plot is shifted to the right and more spread out (10% more).

13. (a) Scatterplot   (b) Box-and-whisker plot

15. (b)

17. Largest; (c), smallest: (b)

19. *A* is more likely to last close to 1000 hours. *B* lasts longer than 1000 hours on the average.

21. Class with wide ability range → high standard deviation
Honors class → high mean, low standard deviation
Students with very similar abilities → low standard deviation
Remedial class → low standard deviation, low median

23. (a) 3.1, 1.2   (b) 3   (c) 8, 3

25. (a) The mean increases by $2000, and the standard deviation stays the same.
(b) The mean increases by 10% and the standard deviation increases by 10%.

29. (a) 68%   (b) 2.5%   (c) 16%

31. (a) 50%   (b) 16%   (c) 2.5%

33. (a) 16th   (b) 97.5th   (c) About 40th

35. (a) Five-number summary; {28, 41, 46, 50, 50}
(b) Probably; most of the class worked almost the entire period.

37. (a) Skewed to the left   (b) Normal
(c) Skewed to the right   (d) Skewed to the right

39. Skewed to the right

41. Skewed right

43. (a) If most people die after about 4 years, you may be close to death.
(b) If most people die before 4 years, you may have a good chance of surviving.

45. (a) $\dfrac{(A+B+C+D)}{4}$

(b) $s = \sqrt{\dfrac{(A-\bar{x})^2 + (B-\bar{x})^2 + (C-\bar{x})^2 + (D-\bar{x})^2}{4}}$

(c) $s = \sqrt{\dfrac{(A-\bar{x})^2 + (B-\bar{x})^2 + (C-\bar{x})^2 + (D-\bar{x})^2}{4}}$

$= \sqrt{\dfrac{A^2 + B^2 + C^2 + D^2 - 2\bar{x}(A+B+C+D) + 4\bar{x}^2}{4}}$

Use the hint.

$= \sqrt{\dfrac{A^2 + B^2 + C^2 + D^2}{4} - (\bar{x})^2}$

12.5 Homework Exercises

1. (a) Registered voters in Colorado
(b) $0.27P$

3. (a) Predict that 111 will buy whole milk, 233 will buy low-fat milk, and 156 will buy skim milk.

5. 1 person receives 7 chairs; the other 9 people share 3 chairs

7. 29 freshmen, 26 sophomores, 24 juniors, and 21 seniors

9. (a) Simple   (b) Stratified
(c) Convenience

11. (a) Stratified   (b) Simple   (c) Systematic

13. (b) *Hint:* Use gender or age.

15. (a), (b)

17. People want more government regulation in a number of areas and the same level of regulation in most other areas.

19. (a) Undercoverage   (b) Response bias

21. (a) The results of a poll can influence what people think, and, consequently, how they respond to subsequent polls.

23. (a) Yes   (b) No
(c) The respondents were not at all representative of voters.

25. No. Find out what % of the children who break school rules are boys.

27. Suggests tennis over other sports as a response

29. (a) Do you favor or oppose the recycling of aluminum cans?

31. Fewer people would agree or disagree.

33. It's not clear how well you speak each language.

35. (a) The rating is the percent of *all* TV sets; the share is out of a smaller sample, only those sets that are on.
(b) By showing that the 35% who do not respond are similar to the 65% who do.
(c) 3

37. None of these

39. The most likely kind of death that results from keeping a firearm at home is a suicide. The next most likely result is killing someone you know in a fight. These events are much more likely than using a firearm to defend your home against a burglar.

12.6 Homework Exercises

1. (a) 36 out of 60
(b) She got 36 questions right out of 60.

3. Students in this area score below the national average.

5. (a) 7.4
(b) She scored the same as the average child in the fourth month of seventh grade.

7. No

9. 8; 4; 6

11. No, their confidence bands overlap. Their scores are not significantly different.

13. (a) 5, 5, 4, 4, 6
(b) Above average in social studies and a little below average in everything else

15. 130 or above

17. (d)

19. (b) and (c)

Chapter 12   Review Exercises

1. Possible answer: 0–99, 100–199, 200–299, 300–399, 400–499, 500–599

2. (a)

Salaries	52.5%	Utilities	12.5%
Maintenance	15%	Supplies	5%
Insurance	5%	Other	10%

(b)

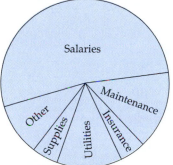

3. (a) The percent increased from 1980 to 1990, was steady from 1990 to 1998 and decreased from 1998 to 2002.
(b) The U.S. government spends more than it collects in taxes, so it borrows money and pays interest on it.

4. (a) Line   (b) Bar

5. (a) Negative; as sick days increase, productivity decreases.
(b) −3.5; productivity decreases about 3.5 units per sick day.

6. (a) Start the vertical axis at $19,000.
(b) Scale the vertical axis accurately.

7. (a) 12%
(b) $P \to 1.4P \to 1.12P$; overall change is 12%.

8. The phrase "as much as 50%" could mean anywhere from 0 to 50%.

9. (a)

11	58
12	45688
13	02
14	0

(b) Mode = 128, mean = 126.6, median = 127

10. Mean = 14.1 minutes, median = 15 minutes

11. 125

12. The mean is the sum of all the scores divided by the number of scores. The median is the middle score when all the scores are arranged in increasing order. I would use the mean if I want to count all the scores equally, and the median if I wanted to lessen the impact of a relatively small group of unusually high or low scores.

13. (a) Mean   (b) Median   (c) Mode

14. The mean age will be higher because the older students will pull it up, while the median age will be around 20.

15. Gar scored better in reading than 59% of the test group.

16. (a) The five number summaries are:
C: $\{-10, -2, 0.5, 4, 8\}$
S: $\{-14, -7, -3.5, 0, 2\}$
(b) The saturated fat group has more success lowering cholesterol. The other group shows little change on average.

17. Mean = 7, standard deviation = $\sqrt{\dfrac{14}{3}} \approx 2.2$

18. (c)

19. Mean increases $70; standard deviation does not change.

20. (a) 95%   (b) 16%

21. (a) 84th   (b) 97.5th

22. (a) $97,920 is the median and $113,422 is the mean. There are more house prices far above these numbers than below them (skewed to the right). These high prices pull up the mean more than the median.
(b) Skewed right

23. Stratified

24. 28 sophomores, 27 juniors, 25 seniors

25. (a) Some people will not be home or will refuse to participate.
(b) Some people will lie to a council member.

26. (a) Do you favor or oppose an increase in funding for government regulation of food safety?

27. (a) 29   (b) She got 29 questions right.   (c) 54
(d) In her region, she scored better than about 54% of those in her age group in language.

28. When the grade level is a few grades higher or lower than the general level of the test. Such a score would be based on test items that are mostly written for a significantly higher or lower grade level.

29. (c)

Chapter 13

13.1 Homework Exercises

1. (a) Roll a 1, 2, 3, 4, or 5   (b) Roll a 1
(c) Roll a 7   (d) Roll a counting number

3. (a) {1, 2, 3, 4, 5, 6}   (b) Yes

5. (a) No   (b) No   (c) Yes

7. (a) 1/4   (b) 1/6   (c) 1/8   (d) 11/24

9. $\dfrac{2}{9}$

11.

13. (a) $\dfrac{12}{25}$ (b) $\dfrac{6}{25}$

15. The 11 sums are not equally likely.

17. (c)

19. (c) *HHH, HHT, HTH, THH, TTH, THT, HTT, TTT*
(d) $\dfrac{3}{8}$

23. (a)

	1	2	3	4	5	6
1	0	1	2	3	4	5
2	1	0	1	2	3	4
3	2	1	0	1	2	3
4	3	2	1	0	1	2
5	4	3	2	1	0	1
6	5	4	3	2	1	0

25. (a) $P(2) = \dfrac{1}{18}$, $P(3) = \dfrac{1}{9}$, $P(4) = \dfrac{1}{6}$, $P(5) = \dfrac{1}{6}$,
$P(6) = \dfrac{1}{6}$, $P(7) = \dfrac{1}{6}$, $P(8) = \dfrac{1}{9}$, $P(9) = \dfrac{1}{18}$
(b) $\dfrac{1}{9}$ of $100 \approx 11$ times

27. 1/2

29. (a)

	B	**b**
B	BB	Bb
b	Bb	bb

(b) $\dfrac{3}{4}$ (c) 1

31. 13/18

33. (a) Heads is one of two equally likely outcomes.
(b) You won't usually get exactly 50 heads. You won't always get close to 50 beads.
(c) You will usually get close to 50 heads.

35. You will usually roll about 10 5s.

37. (a) About 300
(b) Also take into account how the average amount of time traveling abroad per person compares to the average time spent in a car per person.

39. It rained on 60% of the days that were like this in the past.

41. (a) $\dfrac{300}{500} = 0.60$ (b) $\dfrac{220}{500} = 0.44$ (c) $\dfrac{280}{480} \approx 0.58$

43. (a) 1 5 10 10 5 1   (b) 1, 2, 1
   (c) It's the second row.
   (d) If you flip a coin 3 times, add up the numbers in the third row to get the denominator (8), and the numerators of the probabilities are the numbers in the third row: 1 for 0 heads, 3 for 1 head, 3 for 2 heads, and 1 for 3 heads.
   (e) Deductive

45. One die has 1, 2, 2, 3, 3, and 4. The other has 1, 3, 4, 5, 6, and 8.

47. $\dfrac{10}{216} = \dfrac{5}{109}$

13.2 Homework Exercises

1. 1

3. (b) and (c)

5. (a) No   (b) Yes

7. No. If you add the percentages, you are assuming that no one has black hair and brown eyes, which is not true.

9. 0.62

11. 0.45

13. $\dfrac{1{,}180}{2{,}200} \approx 0.54$

19. (a) The contestants choose door 1 and door 3.
   (b) $\dfrac{3}{5}$

23. 2 and 12 are 10°, 3 and 11 are 20°, 4 and 10 are 30°, 5 and 9 are 40°, 6 and 8 are 50°, and 7 is 60°.

25. (a) 3

13.3 Homework Exercises

1. (a)

	Skirts	Shirts	Outfits
	1	1	1,1
		2	1,2
	2	1	2,1
		2	2,2
	3	1	3,1
		2	3,2
	4	1	4,1
		2	4,2

   (c) 8

3. 24

5. STRAIGHT, 10 choices for each of 4 digits
   $(10)(10)(10)(10) = 10{,}000$ choices, $\dfrac{1}{10{,}000}$ to pick the correct number
   4 WAY BOX, 4 ways to match out of 10,000 or $\dfrac{1}{2500}$

7. $_{26}P_4 = 26 \cdot 25 \cdot 24 \cdot 23 = 358{,}800$

9. 720

11. (a) $_nP_n = \dfrac{n!}{(n-n)!} = \dfrac{n!}{1} = n!$
   (b) Deduction

13. $5^6 = 15{,}625$

15. (a) 160   (b) Yes   (c) 800

19. $\dfrac{5}{32} + \dfrac{1}{32} = \dfrac{6}{32} = \dfrac{3}{16}$

21. (a) 625   (b) 3 or 4   (c) $\dfrac{17}{625}$

23. $_{52}C_5 = 2{,}598{,}960$

25. When the order matters

27. $_8C_2 = 28$

29. $_{12}P_2 = 132$

31. (a) 1, 5, 10, 10, 5, 1   (b) Same as part (a)
   (c) 4; 15

33. (a) 8000   (b) 4   (c) 1   (d) 4
   (e) $\dfrac{1}{2000}, \dfrac{1}{8000}, \dfrac{1}{2000}$ (f) 1692
   (g) 148   (h) $\dfrac{423}{2000}, \dfrac{37}{2000}$

35. (a) 6   (b) 1   (c) 4   (d) 1
   (e) 2   (f) 1   (g) 48

37. (b)

13.4 Homework Exercises

1. (a) Dependent (for most people)
   (b) Independent

3. (a) B = it rains tomorrow
   (b) B = the sum of two regular dice is odd

5. (a) Independent   (b) {1, 1), (1, 2), (2, 1), (2, 2)}

(c)

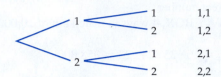

First draw   Second draw   Outcome

(d) $\dfrac{1}{4}$ (e)

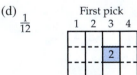

First pick
1   2

$\dfrac{1}{4}$

(f) $P(1 \text{ and } 2) = P(1) \cdot P(2) = \dfrac{1}{2} \cdot \dfrac{1}{2} = \dfrac{1}{4}$

7.

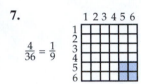

1 2 3 4 5 6

$\dfrac{4}{36} = \dfrac{1}{9}$

9. (a) 0.07   (b) 0.27   (c) 0.66

11. No. The two 1s are independent events. Their probabilities should be multiplied. Draw an area model.

13. (a) Dependent
(b) {(1, 2), (1, 3), (1, 4), (2, 1), (2, 3), (2, 4), (3, 1), (3, 2), (3, 4), (4, 1), (4, 2), (4, 3)}
(c) $P(A) = \dfrac{1}{4}$, $P(B \text{ given } A) = \dfrac{1}{3}$, $P(A \text{ and } B) = \dfrac{1}{12}$
(d) $\dfrac{1}{12}$

First pick
1   2   3   4

(e) $P(A \text{ and } B) = P(A) \cdot P(B \text{ given } A) = \dfrac{1}{4} \cdot \dfrac{1}{3} = \dfrac{1}{12}$

15. $\dfrac{1}{221}$

17. (a) Independent   (b) Dependent

19. $2.4 \cdot 10^{-10}$

21. $P(A \text{ and } B) = P(A) \cdot P(B \text{ given } A) = (0.3)(0.7) = 0.21$

23. (a) 0.17   (b) 0.53

25. 0.86

27. 40%   (b) 36%   (c) 24%

29. 3

31. (a) 0.36   (b) 0.288

33. (a) B = Mike will gain 5 pounds next semester.
(b) B = Mike will go to school next semester.
(c) Impossible

13.5 Homework Exercises

1. You

3. (d) $\dfrac{1}{2}$ for each player.

5. $187.50

7. (a)

9. (a) $20   (b) $30
(c) Part (b) protects you from a possible catastrophic loss, while part (a) does not.

11. 1.8

13. $2.50

15. (a) 1,000   (b) $\dfrac{1}{1,000}$ (c) $0.50   (d) $0.50

17. (a) 9/19   (b) $0.05 loss   (c) $0.05 loss

19. $2

21. $2.32

23. Answers will vary.

25. 1, 12

27. 3 : 2

29. (a) 199   (b) $\dfrac{49}{248}$ (c) About 1 (1.4)
(d) About $\dfrac{29.4}{30.8} \approx 0.95$

31. Discussion question; note that people choosing surgery have a better chance of surviving more than three years.

33. About $52,000

Chapter 13 Review Exercises

1. {AB, BA, AC, CA, AD, DA, BC, CB, BD, DB, CD, DC}

2. (a) $\dfrac{17}{36}$

(b) Roll two dice 30 or more times and see what fraction of the time a product less than 10 occurs.

3. $\dfrac{1}{4}$

4. You will usually get 3 heads and 2 tails about 25 times.

5. An experimental probability is the fraction of times an event occurs in an experiment. A theoretical probability is determined with a sample space or formula. For a large number of trials, the experimental probability is expected to be close to the theoretical probability.

6. (a) $\dfrac{60}{145} = \dfrac{12}{29}$ (b) $\dfrac{60}{300} = \dfrac{1}{5}$

7. (a) You fail the next test.
(b) You eat cereal for breakfast.

8. 0.55

9. 1 and 2 represent a car; 3, 4, 5, 6, 7, and 8 represent no car passing.

10. About 6 (expected value is 5.8)

11. 743,600

12. $\dfrac{1}{2}$

13. $_{20}C_{12} = 125,970$

14. $_{25}P_3 = 13,800$

15. Dependent; independent

16. (a) A coin flip lands on heads.
(b) You study hard for the test.

17. Path   (a) $\dfrac{1}{6}$ (b) $\dfrac{7}{12}$ (c) $\dfrac{1}{4}$

18. (a) $\dfrac{1}{2}\cdot\dfrac{1}{3} = \dfrac{1}{6}$ (b) $\dfrac{1}{2}\cdot\dfrac{2}{3} + \dfrac{1}{2}\cdot\dfrac{1}{2} = \dfrac{7}{12}$
(c) $\dfrac{1}{2}\cdot\dfrac{1}{2} = \dfrac{1}{4}$

19. 0.01

20. (a) $1 - 0.2 = 0.8$ (b) $(0.2)(0.1) = 0.02$

22. $13.50

23. (a) $1.06   (b) $0.06

24. (a) 8   (b) $\dfrac{1}{11}$

INDEX

Activity Card 1: Base-Ten Blocks

Activity Card 2: Base-Five Blocks

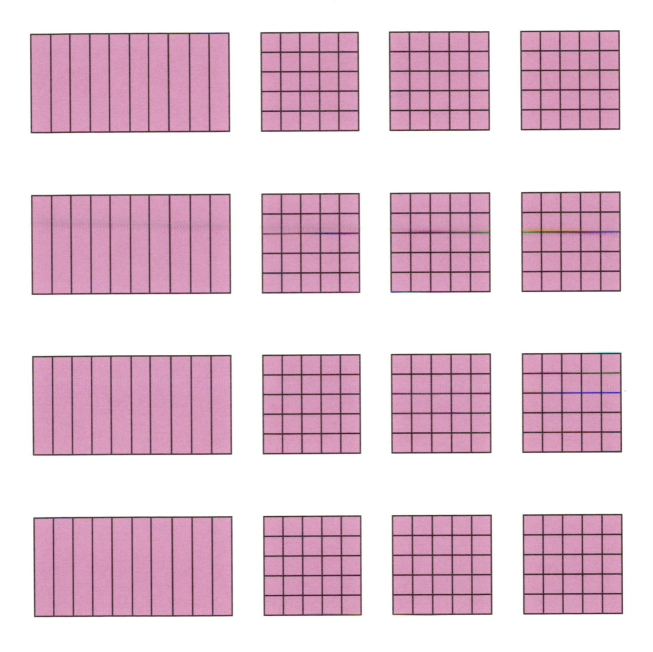

Activity Card 3: Decimal Squares

Unit Squares	Tenths Squares	Hundredths Squares

Activity Card 4: Grids

Geoboard Square Dot Grid

Isometric Grid

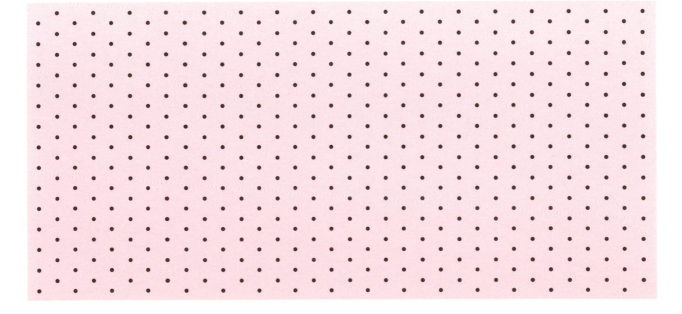

Activity Card 5: Regular Polygons

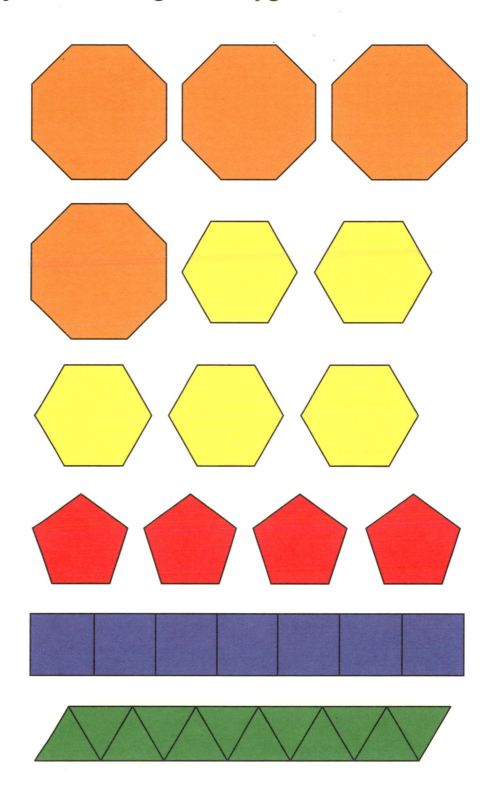

Activity Card 6: Pattern Blocks

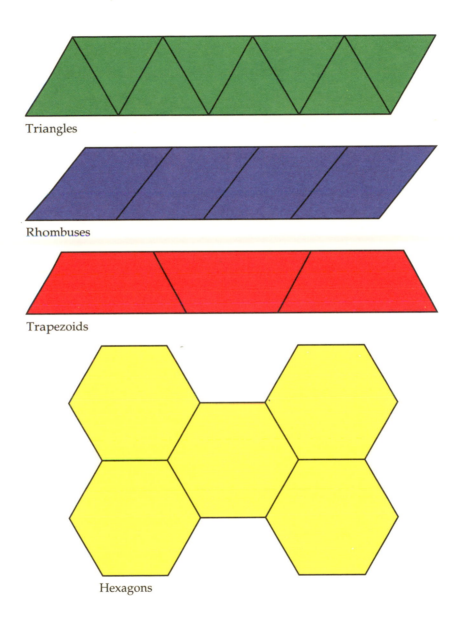

Triangles

Rhombuses

Trapezoids

Hexagons

Activity Card 7: Tangram Pieces

Activity Card 8: Nets

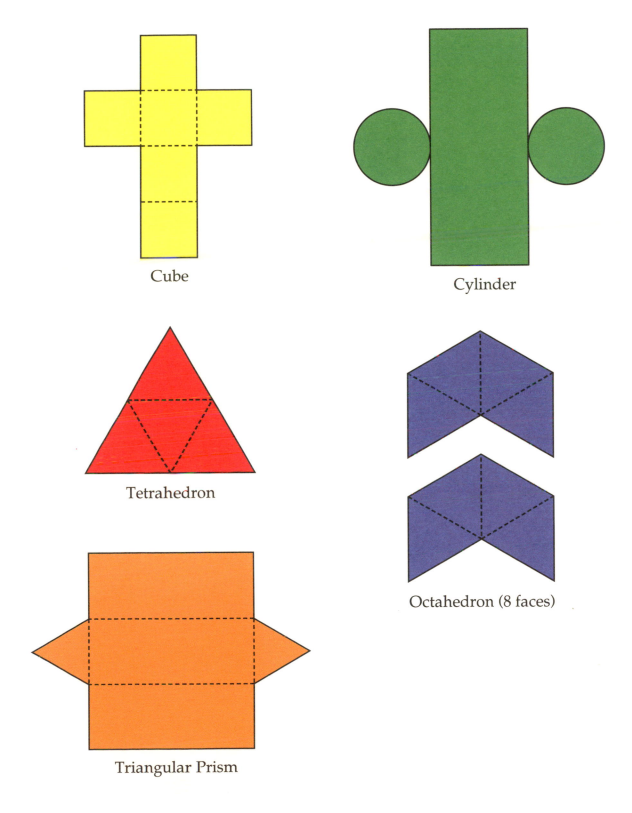

Cube

Cylinder

Tetrahedron

Octahedron (8 faces)

Triangular Prism